Statistical Mechanics of Lattice Systems

A Concrete Mathematical Introduction

This motivating textbook gives a friendly, rigorous introduction to fundamental concepts in equilibrium statistical mechanics, covering a selection of specific models, including the Curie–Weiss and Ising models, the Gaussian Free Field, O(N) models, and models with Kać interactions. Using classical concepts such as Gibbs measures, pressure, free energy and entropy, the book describes the main features of the classical description of large systems in equilibrium, in particular the central problem of phase transitions. It treats such important topics as Peierls' argument, the Dobrushin Uniqueness, Mermin–Wagner and Lee–Yang theorems, and develops from scratch such workhorses as correlation inequalities, the cluster expansion, Pirogov–Sinai theory and reflection positivity. Written as a self-contained course for advanced undergraduate or beginning graduate students, the detailed explanations, large collection of exercises (with solutions), and appendix of mathematical results and concepts also make it a handy reference for researchers in related areas.

Sacha Friedli is Associate Professor of Mathematics at the Federal University of Minas Gerais in Brazil. His current research interests are in statistical mechanics, mathematical physics and Markov processes.

Yvan Velenik is Professor of Mathematics at the University of Geneva. His current work focuses on applications of probability theory to the study of classical statistical mechanics, especially lattice random fields and random walks.

Statistical Mechanics of Lattice Systems

A Concrete Mathematical Introduction

S. FRIEDLI
Universidade Federal de Minas Gerais, Brazil

Y. VELENIK
Université de Genève

CAMBRIDGE
UNIVERSITY PRESS

CAMBRIDGE
UNIVERSITY PRESS

University Printing House, Cambridge CB2 8BS, United Kingdom

One Liberty Plaza, 20th Floor, New York, NY 10006, USA

477 Williamstown Road, Port Melbourne, VIC 3207, Australia

314–321, 3rd Floor, Plot 3, Splendor Forum, Jasola District Centre, New Delhi - 110025, India

79 Anson Road, #06-04/06, Singapore 079906

Cambridge University Press is part of the University of Cambridge.

It furthers the University's mission by disseminating knowledge in the pursuit of education, learning, and research at the highest international levels of excellence.

www.cambridge.org
Information on this title: www.cambridge.org/9781107184824
DOI: 10.1017/9781316882603

First published 2018

A catalog record for this publication is available from the British Library.

Library of Congress Cataloging-in-Publication Data
Names: Friedli, Sacha, 1974– author. | Velenik, Yvan, 1970– author.
Title: Statistical mechanics of lattice systems : a concrete mathematical introduction / Sacha Friedli (Universidade Federal de Minas Gerais, Brazil), Yvan Velenik (Université de Genáeve).
Description: Cambridge, United Kingdom ; New York, NY : Cambridge University Press, 2017.
Identifiers: LCCN 2017031217| ISBN 9781107184824 (hardback) | ISBN 1107184827 (hardback)
Subjects: LCSH: Statistical mechanics.
Classification: LCC QC174.8 .F65 2017 | DDC 530.13–dc23
LC record available at https://lccn.loc.gov/2017031217

ISBN 978-1-107-18482-4 Hardback

For the sunshine and smiles: Janet, Jean-Pierre, Mimi and Kathryn.

Aos amigos e colegas do Departamento de Matemática.

À Agnese, Laure et Alexandre, ainsi qu'à mes parents.

Contents

Preface

Equilibrium statistical mechanics is a field that has existed for more than a century. Its origins lie in the search for a microscopic justification of equilibrium thermodynamics, and it developed into a well-established branch of mathematics in the second half of the twentieth century. The ideas and methods that it introduced to treat systems with many components have now permeated many areas of science and engineering, and have had an important impact on several branches of mathematics.

There exist many good introductions to this theory designed for physics undergraduates. It might, however, come as a surprise that textbooks addressing it from a *mathematically rigorous* standpoint have remained rather scarce. A reader looking for an introduction to its more advanced mathematical aspects must often either consult highly specialized monographs or search through numerous research articles available in peer-reviewed journals. It might even appear as if the mastery of certain techniques has survived from one generation of researchers to the next only by means of oral communication, through the use of chalk and blackboard...

It seems a general opinion that pedagogical introductory mathematically rigorous textbooks simply do not exist. This book aims at starting to bridge this gap. Both authors graduated in physics before turning to mathematical physics. As such, we have witnessed this lack from the student's point of view, before experiencing it, a few years later, from the teacher's point of view. Above all, this text aims to provide the material we would have liked to have had at our disposal when entering this field.

Although we hope that it will also be of interest to students in theoretical physics, this is in fact a book on *mathematical physics*. There is no general consensus on what this term actually refers to. In rough terms, what it means for us is: the analysis of problems originating in physics, at the level of rigor associated with mathematics. This includes the introduction of concepts and the development of tools enabling such an analysis. It is unfortunate that mathematical physics is often held in rather low esteem by physicists, many of whom see it as useless nitpicking and as dealing mainly with problems that they consider to be already fully understood. There are, however, very good reasons for these investigations. First, such an approach allows a very clear separation between the assumptions (the basic principles of the underlying theory, as well as the particulars of the model analyzed) and the actual derivation: once the proper framework is set, the entire analysis is done without further assumptions or approximations. This is essential in order to ensure that the phenomenon that has been derived is indeed a consequence of the starting hypotheses and not an artifact of the approximations made along the way. Second, to provide a complete mathematical analysis requires us to understand the phenomenon of interest in a much deeper and detailed way. In particular, it forces one to provide precise definitions and statements. This is highly useful in clarifying issues that are sometimes puzzling for students and, occasionally, researchers.

Let us emphasize two central features of this work.

- The first has to do with content. Equilibrium statistical mechanics has become such a rich and diverse subject that it is impossible to cover more than a fraction of it in a single book. Since our driving motivation is to provide an easily accessible introduction in a form suitable for self-study, our first decision was to focus on some of the most important and relevant examples rather than to present the theory from a broad point of view. We hope that this will help the reader build the necessary intuition, in concrete situations, as well as provide background and motivation for the general theory. We also refrained from introducing abstractions for their own sake and have done our best to keep the technical level as low as possible.
- The second central feature of this book is related to our belief that the main value of the proof of a theorem is measured by the extent to which it enhances understanding of the phenomena under consideration. As a matter of fact, the concepts and methods introduced in the course of a proof are often at least as important as the claim of the theorem itself. The most useful proof, for a beginner, is thus not necessarily the shortest or the most elegant one. For these reasons, we have strived to provide, throughout the book, the arguments we personally consider the most enlightening in the most simple manner possible.

These two features have shaped the book from its very first versions. (They have also contributed, admittedly, to the lengthiness of some chapters.) Together with the numerous illustrations and exercises, we hope that they will help the beginner to become familiar with some of the central concepts and methods that lie at the core of statistical mechanics.

As underlined by many authors, one of the main purposes of writing a book should be one's own pleasure. Indeed, leading this project to its conclusion was by and large a very enjoyable albeit long journey! But, beyond that, the positive feedback we have already received from students, and from colleagues who have used early drafts in their lectures, indicates that it may yet reach its goal, which is to help beginners enter this beautiful field.

Acknowledgements. This book benefited both directly and indirectly from the help and support of many colleagues. First and foremost, we would like to thank Charles Pfister who, as a PhD advisor, introduced both authors to this field of research many years ago. We have also learned much of what we know from our various co-authors during the past two decades. In particular, YV would like to express his thanks to Dima Ioffe, for a long, fruitful and very enjoyable ongoing collaboration.

Our warmest thanks also go to Aernout van Enter, who has been a constant source of support and feedback and whose enthusiasm for this project has always been highly appreciated!

We are very grateful to all the people who called to our attention various errors they found in preliminary versions of the book - in particular, Costanza Benassi, Quentin Berthet, Tecla Cardilli, Loren Coquille, Margherita Disertori, Hugo Duminil-Copin, Mauro Mariani, Philippe Moreillon, Sébastien Ott, Ron Peled, Sylvie Roelly, Costanza Rojas-Molina and Daniel Ueltschi.

We also thank Claudio Landim, Vladas Sidoravicius and Augusto Teixeira for their support and comments. Our warm thanks to Maria Eulalia Vares for her constant encouragement since the earliest drafts of this work.

SF thanks the Departamento de Matemática of the Federal University of Minas Gerais for its long-term support, Hans-Jörg Ruppen (CMS, EPFL), as well as the Section de Mathématiques of the University of Geneva for hospitality and financial support on countless visits during which large parts of this work were written. Both authors are also grateful to the Swiss National Science Foundation for its support, in particular through the NCCR SwissMAP.

Finally, writing this book would have been considerably less enjoyable without the following fantastic pieces of open source software: bash, GNU/Linux (openSUSE and Ubuntu flavors), GCC, GIMP, git, GNOME, gnuplot, Inkscape, KDE, Kile, LaTeX 2_ε, PGF/Tikz, POV-Ray, Processing, Python, Sketch (for LaTeX), TeXstudio, Vim and Xfig.

Sacha Friedli
Yvan Velenik

Conventions

$a \stackrel{\text{def}}{=} b$	a is defined as being b
$\mathbb{R}^d$	d-dimensional Euclidean space
$\mathbb{Z}^d$	d-dimensional cubic lattice
$\mathbb{R}_{\geq 0}$	nonnegative real numbers
$\mathbb{R}_{>0}$	positive real numbers
$\mathbb{Z}_{\geq 0}$	nonnegative integers: $0, 1, 2, 3, \ldots$
$\mathbb{N}, \mathbb{Z}_{>0}$	positive integers: $1, 2, 3, \ldots$
i	$\sqrt{-1}$
$\mathfrak{Re}\, z$, $\mathfrak{Im}\, z$	real and imaginary parts of $z \in \mathbb{C}$
$a \wedge b$	minimum of a and b
$a \vee b$	maximum of a and b
$\log$	natural logarithm, that is, in base $e = 2.718\ldots$
$A \subset B$	A is a (not necessarily proper) subset of B
$A \subsetneq B$	A is a proper subset of B
$A \triangle B$	symmetric difference
$\#A$, $\lvert A \rvert$	number of elements in the set A (if A is finite). At several places, also used to denote the Lebesgue measure.
$\delta_{m,n}$	Kronecker symbol: $\delta_{m,n} = 1$ if $m = n$, 0 otherwise
δ_x	Dirac measure at x: $\delta_x(A) = 1$ if $x \in A$, 0 otherwise
$\lfloor x \rfloor$	largest integer smaller than or equal to x
$\lceil x \rceil$	smallest integer larger than or equal to x

Asymptotic equivalence of functions will follow the standard conventions. For functions f, g, defined in the neighborhood of x_0 (possibly $x_0 = \infty$),

$f(x) \sim g(x)$	means $\lim_{x \to x_0} \frac{\log f(x)}{\log g(x)} = 1$,
$f(x) \simeq g(x)$	means $\lim_{x \to x_0} \frac{f(x)}{g(x)} = 1$,
$f(x) \approx g(x)$	means $0 < \liminf_{x \to x_0} \frac{f(x)}{g(x)} \leq \limsup_{x \to x_0} \frac{f(x)}{g(x)} < \infty$,
$f(x) = O(g(x))$	means $\limsup_{x \to x_0} \left\lvert \frac{f(x)}{g(x)} \right\rvert < \infty$,
$f(x) = o(g(x))$	means $\lim_{x \to x_0} \left\lvert \frac{f(x)}{g(x)} \right\rvert = 0$.

As usual, A^B is identified with the set of all maps $f \colon B \to A$. A sequence of elements $a_n \in E$ will usually be denoted as $(a_n)_{n \geq 1} \subset E$. In several places, we will set

$0 \log 0 \stackrel{\text{def}}{=} 0$. Sums or products over empty families are defined as follows:

$$\sum_{i \in \varnothing} a_i \stackrel{\text{def}}{=} 0, \qquad \prod_{i \in \varnothing} a_i \stackrel{\text{def}}{=} 1.$$

Several important notations involving geometrical notions on $\mathbb{Z}^d$ will be defined at the end of the introduction and at the beginning of Chapter 3.

1 Introduction

Statistical mechanics is the branch of physics that aims at bridging the gap between the *microscopic* and *macroscopic* descriptions of large systems of particles in interaction, by combining the information provided by the microscopic description with a probabilistic approach. Its goal is to understand the relations existing between the salient macroscopic features observed in these systems and the properties of their microscopic constituents. Equilibrium statistical mechanics is the part of this theory that deals with macroscopic systems *at equilibrium* and is, by far, the best understood.

This book is an introduction to some classical mathematical aspects of equilibrium statistical mechanics, based essentially on some important examples. It does not constitute an exhaustive introduction: many important aspects, discussed in a myriad of books (see those listed in Section 1.6.2 below), will not be discussed. Inputs from physics will be restricted to the terminology used (especially in this introduction), to the nature of the problems addressed, and to the central probability distribution used throughout, namely the **Gibbs distribution**. The latter provides the probability of observing a particular microscopic state ω of the system under investigation, when the system is at equilibrium *at a fixed temperature* T. It takes the form

$$\mu_\beta(\omega) = \frac{e^{-\beta \mathscr{H}(\omega)}}{\mathbf{Z}_\beta},$$

where $\beta = 1/T$, $\mathscr{H}(\omega)$ is the *energy* of the microscopic state ω and $\mathbf{Z}_\beta$ is a normalization factor called the **partition function**.

Saying that the Gibbs distribution is well suited to understanding the phenomenology of large systems of particles is an understatement. This book provides, to some extent, a proof of this fact by diving into an in-depth study of this distribution when applied to some of the most important models studied by mathematical physicists since the beginning of the twentieth century. The many facets and the rich variety of behavior that will be described in the following chapters should, by themselves, constitute a firm justification for the use of the Gibbs distribution for the description of large systems at equilibrium.

An *impatient reader* with some basic notions of thermodynamics and statistical mechanics, or who is willing to consider the Gibbs distribution as a postulate and is not interested in additional motivations and background, can jump directly to the following chapters and learn about the models presented throughout the book. A quick glance at Section 1.6 might be useful since it contains a reading guide.

The rest of this introduction is written for a reader interested in obtaining more information on the origin of the Gibbs distribution and the associated terminology, as well as an informal discussion of thermodynamics and its relations to equilibrium statistical mechanics.

One of the main themes to which this book is devoted, phase transitions, is illustrated on gases and magnets. However, we emphasize that, *in this book, the main focus is on the mathematical structure of equilibrium statistical mechanics,* and the many possible interpretations of the models we study will most of the time not play a very important role; they can, nevertheless, sometimes provide intuition.

Although this book is undeniably written for a mathematically inclined reader, in this introduction we will avoid delving into too many technicalities; as a consequence, it will not be as rigorous as the rest of the book. Its purpose is to provide intuition and motivation behind several key concepts, relying mainly on physical arguments. The content of this introduction is not necessary for the understanding of the rest of the book, but we hope that it will provide the reader with some useful background information.

We will start with a brief discussion of the first physical theory describing macroscopic systems at equilibrium, equilibrium thermodynamics, and also present examples of one of the most interesting features of thermodynamic systems: phase transitions. After that, starting from Section 1.2, we will turn our attention to equilibrium statistical mechanics.

1.1 Equilibrium Thermodynamics

Equilibrium thermodynamics is a phenomenological theory, developed mainly during the nineteenth century. Its main early contributors include Carnot, Clausius, Kelvin, Joule and Gibbs. It is based on a few empirical principles and does not make any assumptions regarding the microscopic structure of the systems it considers (in fact, the atomic hypothesis was still hotly debated when the theory was developed).

This section will briefly present some of its basic concepts, mostly on some simple examples; it is obviously not meant as a complete description of thermodynamics, and the interested reader is advised to consult the detailed and readable account in the books of Callen [58] and Thess [329], in Wightman's introduction in [176] or in Lieb and Yngvason's paper [224].

1.1.1 On the Description of Macroscopic Systems

Gases, liquids and solids are the most familiar examples of large physical systems of particles encountered in nature. In the first part of this introduction, for the sake of simplicity and concreteness, we will mainly consider the system of a gas contained in a vessel.

Let us thus consider a specific, homogeneous gas in a vessel (this could be, say, one liter of helium at standard temperature and pressure). We will use Σ to denote such a specific system. The **microscopic state**, or **microstate**, of the gas is the complete microscopic description of the state of the gas. For example, from the point of view of Newtonian mechanics, the microstate of a gas of monoatomic molecules is specified by the position and the momentum of each molecule (called hereafter *particle*). Since a gas contains a huge number of particles (of the order of 10^{22} for our liter of helium), the overwhelming quantity of information contained in its microstate makes a complete analysis not only challenging, but in general impossible. Fortunately, we are not interested in the full information contained in the microstate: the precise behavior of each single particle is irrelevant at our scale of observation.

It is indeed an empirical fact that, in order to describe the state of the gas at the macroscopic scale, only a much smaller set of variables, of a different nature, is needed. This is particularly true when the system is in a particular kind of state called *equilibrium*. Assume, for instance, that the gas is **isolated**, that is, it does not exchange matter or energy with the outside world, and that it has been left undisturbed for a long period of time. Experience shows that such a system reaches a state of **thermodynamic equilibrium**, in which macroscopic properties of the gas do not change anymore and there are no macroscopic flows of matter or energy (even though molecular activity never ceases).

In fact, by definition, isolated systems possess a number of **conserved** quantities, that is, quantities that do not change through time evolution. These are particularly convenient to describe the macroscopic state of the system. For our example of a gas, these conserved quantities are the **volume** V of the vessel,[1] the **number of particles** N and the **internal energy** (or simply: **energy**) U.

We therefore assume, from now on, that the **macroscopic state** (or **macrostate**) of the gas Σ is determined by a triple:

$$\mathbf{X} = (U, V, N).$$

The variables (U, V, N) can be thought of as the quantities one can control in order to alter the state of the gas. For example, U can be changed by cooling or heating the system, V by squeezing or expanding the container and N by injecting or extracting particles with a pump or similar device. These variables determine the state of the gas in the sense that setting these variables to some specific values always yields, once equilibrium is reached, systems that are macroscopically indistinguishable. They can thus be used to put the gas in a chosen macroscopic state reproducibly. Of course, what is reproduced is the macrostate, *not the microstate*: there are usually infinitely many different microstates corresponding to the same macrostate.

Now, suppose that we split our system Σ into two subsystems Σ_1 and Σ_2, by adding a wall partitioning the vessel. Each of the subsystems is of the same type

[1] We assume that the vessel is large enough and that its shape is simple enough (for example: a cube), so as not to influence the macroscopic behavior of the gas and boundary effects may be neglected.

as the original system and only differs from it by the values of the corresponding variables (U^m, V^m, N^m), $m = 1, 2$. Observe that the total energy U, total volume V and total number of particles N in the system satisfy: [1]

$$U = U^1 + U^2, \quad V = V^1 + V^2, \quad N = N^1 + N^2. \tag{1.1}$$

Variables having this property are said to be **extensive.**[2]

Note that the description of a system Σ composed of two subsystems Σ_1 and Σ_2 now requires six variables: $(U^1, V^1, N^1, U^2, V^2, N^2)$. Below, a system will always come with the set of variables used to characterize it. Note that this set of variables is not unique: we can always split a system into two pieces in our imagination, without doing anything to the system itself, only to our description of it.

One central property of equilibrium is that, when a system is at equilibrium, each of its (macroscopic) subsystems is at equilibrium too, and all subsystems are in equilibrium with each other. Namely, if we imagine that our system Σ is partitioned by an imaginary wall into two subsystems Σ_1, Σ_2 of volume V^1 and V^2, then the thermodynamic properties of each of these subsystems do not change through time: their energy and the number of particles they contain remain constant.[3]

Assume now that Σ_1, Σ_2 are originally separated (far apart, with no exchanges whatsoever), each isolated and at equilibrium. The original state of the union of these systems is represented by $(\mathbf{X}^1, \mathbf{X}^2)$, where $\mathbf{X}^m = (U^m, V^m, N^m)$ is the macrostate of Σ_m, $m = 1, 2$. Suppose then that these systems are put in contact, allowing them to exchange energy and/or particles, while keeping them, as a whole, isolated from the rest of the universe (in particular, the total energy U, total volume V and total number of particles N are fixed). Once they are in contact, the whole system goes through a phase in which energy and particles are redistributed between the subsystems and a fundamental problem is to determine which new equilibrium macrostate $(\overline{\mathbf{X}}^1, \overline{\mathbf{X}}^2)$ is realized and how it relates to the initial pair $(\mathbf{X}^1, \mathbf{X}^2)$.

The core postulate of equilibrium thermodynamics is to assume the existence of a function, assigned to any system Σ, which describes how the new equilibrium state is selected among the a-priori infinite number of possibilities. This function is called *entropy*.

1.1.2 Thermodynamic Entropy

Let us assume that Σ is the union of two subsystems Σ_1, Σ_2 and that some *constraints* are imposed on these subsystems. We model these constraints by the set

2 The identity is not completely true for the energy: part of this comes, in general, from the interaction between the two subsystems. However, this interaction energy is generally negligible compared to the overall energy (exceptions only occur in the presence of very long-range interactions, such as gravitational forces). This will be quantified once we study similar problems in statistical mechanics.

3 Again, this is not true from a microscopic perspective: the number of particles in each subsystem does fluctuate, since particles constantly pass from one subsystem to the other. However, these fluctuations are of an extremely small relative size and are neglected in thermodynamics (if there are of order N particles in each subsystem, then statistical mechanics will show that these fluctuations are of order $\sqrt{N}$).

$\mathbb{X}_c$ of allowed pairs $(\mathbf{X}^1, \mathbf{X}^2)$. We expect that the system selects some particular pair in $\mathbb{X}_c$ to realize equilibrium, in some optimal way. The main postulate of thermostatics is that this is done by choosing the pair that maximizes the entropy:

Postulate (Thermostatics). *To each system Σ, described by a set of variables $\mathbf{X}$, is assigned a differentiable function S^Σ of $\mathbf{X}$, called the* **(thermodynamic) entropy***; it is specific to each system. The entropy of a system Σ composed of two subsystems Σ_1 and Σ_2 is* **additive:**

$$S^\Sigma(\mathbf{X}^1, \mathbf{X}^2) = S^{\Sigma_1}(\mathbf{X}^1) + S^{\Sigma_2}(\mathbf{X}^2). \tag{1.2}$$

Once the systems are put in contact, the pair $(\overline{\mathbf{X}}^1, \overline{\mathbf{X}}^2)$ realizing equilibrium is the one that maximizes $S^\Sigma(\mathbf{X}^1, \mathbf{X}^2)$, among all pairs $(\mathbf{X}^1, \mathbf{X}^2) \in \mathbb{X}_c$.

The principle by which a system under constraint realizes equilibrium by maximizing its entropy will be called the **extremum principle**.

Remark 1.1. The entropy function is characteristic of the system considered (in our example, it is the one associated with helium; if we were considering a piece of lead or a mixture of gases, such as the air, the entropy function would be different). It also obviously depends on the set of variables used for its description (mentally splitting the system into two and using six variables instead of three yields a different entropy function, even though the underlying physical system is unchanged). However, if one considers two systems Σ_1 and Σ_2 of the same type and described by the same set of variables (the latter possibly taking different values), then $S^{\Sigma_1} = S^{\Sigma_2}$. ◇

Remark 1.2. Let us emphasize that although the thermodynamic properties of a system are entirely contained in its entropy function (or in any of the equations of state or thermodynamic potential derived later), *thermodynamics does not provide tools to determine what this function should be for a specific system* (it can, of course, be measured empirically in a laboratory). As we will see, the determination of these functions for a particular system from first principles is a task that will be devolved to equilibrium statistical mechanics. ◇

Let us illustrate with some examples how the postulate is used.

Example 1.3. In this first example, we suppose that the system Σ is divided into two subsystems Σ_1 and Σ_2 of volume V_1, respectively V_2, by inserting a hard, impermeable, fixed wall. These subsystems have, initially, energy U_1, respectively U_2, and contain N_1, respectively N_2, particles. We assume that the wall allows the two subsystems to exchange energy, but not particles. So, by assumption, the following quantities are kept fixed: the volumes V_1 and V_2 of the two subsystems, the number N_1 and N_2 of particles they contain and the total energy $U = U_1 + U_2$; these form the constraints. The problem is thus to determine the values $\overline{U}^1, \overline{U}^2$ of the energy in each of the subsystems once the system has reached equilibrium.

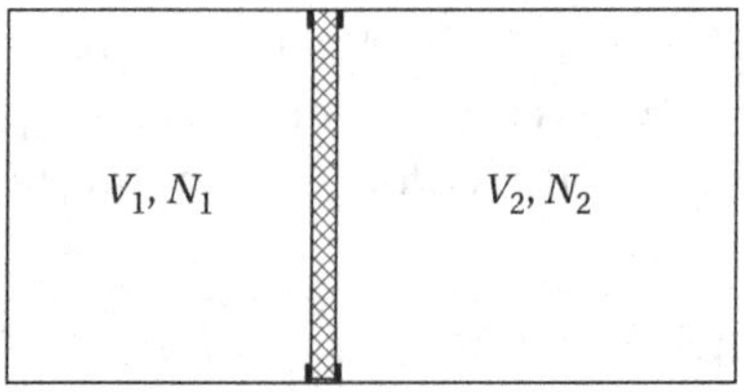

Since V_1, N_1, V_2, N_2 are fixed, the postulate states that the equilibrium values $\overline{U}_1$ and $\overline{U}_2$ are found by maximizing

$$(\widetilde{U}_1, \widetilde{U}_2) \mapsto S(\widetilde{U}_1, V_1, N_1) + S(\widetilde{U}_2, V_2, N_2) = S(\widetilde{U}_1, V_1, N_1) + S(U - \widetilde{U}_1, V_2, N_2).$$

We thus see that equilibrium is realized when $\overline{U}_1$ satisfies

$$\left\{\frac{\partial S}{\partial \widetilde{U}_1}(\widetilde{U}_1, V_1, N_1) + \frac{\partial S}{\partial \widetilde{U}_1}(U - \widetilde{U}_1, V_2, N_2)\right\}\bigg|_{\widetilde{U}=\overline{U}_1} = 0.$$

Therefore, equilibrium is realized when $\overline{U}_1, \overline{U}_2$ satisfy

$$\frac{\partial S}{\partial U}(\overline{U}_1, V_1, N_1) = \frac{\partial S}{\partial U}(\overline{U}_2, V_2, N_2).$$

The quantity[4]

$$\beta \stackrel{\text{def}}{=} \left(\frac{\partial S}{\partial U}\right)_{V,N} \tag{1.3}$$

is called the **inverse (absolute) temperature**. The **(absolute) temperature** is then defined by $T \stackrel{\text{def}}{=} 1/\beta$. It is an empirical fact that the temperature thus defined is positive, that is, the entropy is an increasing function of the energy.[5] We will therefore assume from now on that *$S(U, V, N)$ is increasing in U*:

$$\left(\frac{\partial S}{\partial U}\right)_{V,N} > 0. \tag{1.4}$$

We conclude that, *if two systems that are isolated from the rest of the universe are allowed to exchange energy, then, once they reach equilibrium, their temperatures (as defined above) will have equalized.*

Note that this agrees with the familiar observation that there will be a heat flow between the two subsystems, until both reach the same temperature. ◇

Example 1.4. Let us now consider a slightly different situation, in which the wall partitioning our system Σ is allowed to slide:

[4] In this introduction, we will follow the custom in thermodynamics and keep the same notation for quantities such as the entropy or temperature, even when seen as functions of different sets of variables. It is thus important, when writing down partial derivatives, to specify which are the variables kept fixed.

[5] Actually, there are very special circumstances in which negative temperatures are possible, but we will not discuss them in this book. In any case, the adaptation of what we discuss to negative temperatures is straightforward.

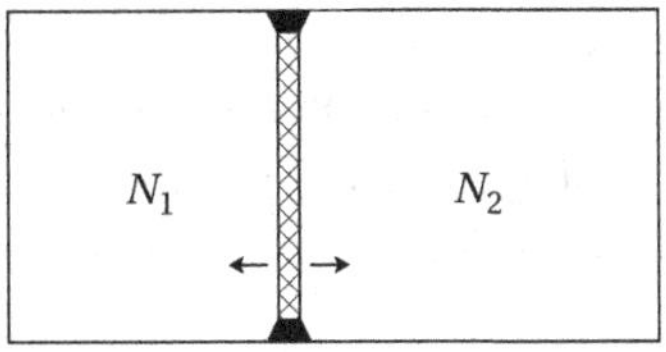

In this case, the number of particles on each side of the wall is still fixed to N_1 and N_2, but the subsystems can exchange both energy and volume. The constraint is thus that $U = U_1 + U_2$, $V = V_1 + V_2$, N_1 and N_2 are kept fixed. Proceeding as before, we have to find the values $\overline{U}_1, \overline{U}_2, \overline{V}_1, \overline{V}_2$ maximizing

$$(\widetilde{U}_1, \widetilde{U}_2, \widetilde{V}_1, \widetilde{V}_2) \mapsto S(\widetilde{U}_1, \widetilde{V}_1, N_1) + S(\widetilde{U}_2, \widetilde{V}_2, N_2).$$

We deduce this time that, once equilibrium is realized, $\overline{U}_1, \overline{U}_2, \overline{V}_1$ and $\overline{V}_2$ satisfy

$$\begin{cases} \frac{\partial S}{\partial U}(\overline{U}_1, \overline{V}_1, N_1) = \frac{\partial S}{\partial U}(\overline{U}_2, \overline{V}_2, N_2), \\ \frac{\partial S}{\partial V}(\overline{U}_1, \overline{V}_1, N_1) = \frac{\partial S}{\partial V}(\overline{U}_2, \overline{V}_2, N_2). \end{cases}$$

Again, the first identity implies that the temperatures of the subsystems must be equal. The quantity

$$p \stackrel{\text{def}}{=} T \cdot \left(\frac{\partial S}{\partial V}\right)_{U,N} \tag{1.5}$$

is known as the **pressure** (T, in this definition, is introduced as a convention). We conclude that, *once two systems that can exchange both energy and volume reach equilibrium, their temperatures and pressures will have equalized.* ◇

Example 1.5. For the third and final example, we suppose that the system is partitioned into two subsystems by a fixed permeable wall, which allows exchange of both particles and energy. The constraints in this case are that $U = U_1 + U_2$, $N = N_1 + N_2$, V_1 and V_2 are kept fixed. This time, we thus obtain that, at equilibrium, $\overline{U}_1, \overline{U}_2, \overline{N}_1$ and $\overline{N}_2$ satisfy

$$\begin{cases} \frac{\partial S}{\partial U}(\overline{U}_1, V_1, \overline{N}_1) = \frac{\partial S}{\partial U}(\overline{U}_2, V_2, \overline{N}_2), \\ \frac{\partial S}{\partial N}(\overline{U}_1, V_1, \overline{N}_1) = \frac{\partial S}{\partial N}(\overline{U}_2, V_2, \overline{N}_2). \end{cases}$$

Once more, the first identity implies that the temperatures of the subsystems must be equal. The quantity

$$\mu \stackrel{\text{def}}{=} -T \cdot \left(\frac{\partial S}{\partial N}\right)_{U,V} \tag{1.6}$$

is known as the **chemical potential** (the sign, as well as the introduction of T, is a convention). We conclude that, *when they reach equilibrium, two systems that can exchange both energy and particles have the same temperature and chemical potential.* ◇

We have stated the postulate for a very particular case (a gas in a vessel, considered as made up of two subsystems of the same type), but the postulate extends to *any* thermodynamic system. For instance, it can be used to determine how equilibrium is realized when an arbitrary large number of systems are put in contact:

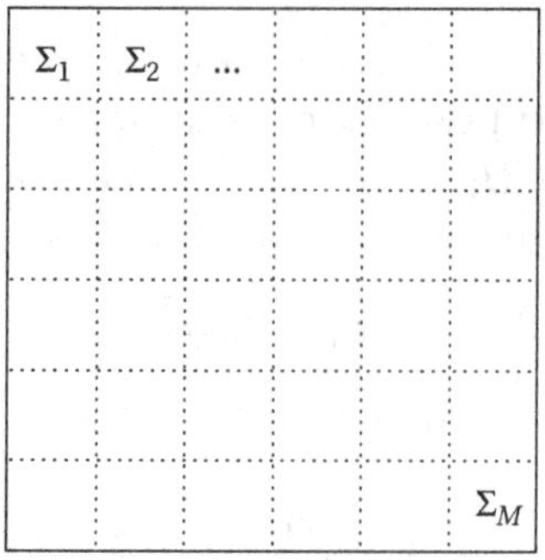

We have discussed a particular case, but (1.3), (1.5) and (1.6) provide the definition of the temperature, pressure and chemical potential for any system characterized by the variables U, V, N (and possibly others) whose entropy function is known.

Several further fundamental properties of the entropy can be readily deduced from the postulate.

Exercise 1.1. Show that the entropy is **positively homogeneous of degree 1**, that is,

$$S(\lambda U, \lambda V, \lambda N) = \lambda S(U, V, N), \quad \forall \lambda > 0. \tag{1.7}$$

Hint: Consider first $\lambda \in \mathbb{Q}$.

Exercise 1.2. Show that the entropy is **concave**, that is, for all $\alpha \in [0,1]$ and any $U_1, U_2, V_1, V_2, N_1, N_2$,

$$S\big(\alpha U_1 + (1-\alpha)U_2, \alpha V_1 + (1-\alpha)V_2, \alpha N_1 + (1-\alpha)N_2\big) \\ \geq \alpha S(U_1, V_1, N_1) + (1-\alpha)S(U_2, V_2, N_2). \tag{1.8}$$

1.1.3 Conjugate Intensive Quantities and Equations of State

The temperature, pressure and chemical potential defined above were all defined via a partial differentiation of the entropy: $\frac{\partial S}{\partial U}, \frac{\partial S}{\partial V}, \frac{\partial S}{\partial N}$. Generally, if X_i is any extensive variable appearing in S,

$$f_i \stackrel{\text{def}}{=} \frac{\partial S}{\partial X_i}$$

is called the **variable conjugate to** X_i. It is a straightforward consequence of the definitions that, in contrast to U, V, N, the conjugate variables are not extensive, but **intensive**: they remain unchanged under a global scaling of the system: for all $\lambda > 0$,

$$\begin{aligned} T(\lambda U, \lambda V, \lambda N) &= T(U, V, N), \\ p(\lambda U, \lambda V, \lambda N) &= p(U, V, N), \\ \mu(\lambda U, \lambda V, \lambda N) &= \mu(U, V, N). \end{aligned}$$

In other words, T, p and μ are **positively homogeneous of degree 0**.

Differentiating both sides of the identity $S(\lambda U, \lambda V, \lambda N) = \lambda S(U, V, N)$ with respect to λ, at $\lambda = 1$, we obtain

$$S(U, V, N) = \frac{1}{T}U + \frac{p}{T}V - \frac{\mu}{T}N. \tag{1.9}$$

The latter identity is known as the **Euler relation**. It allows one to reconstruct the entropy function from a knowledge of the functional dependence of T, p, μ on U, V, N:

$$T = T(U, V, N), \quad p = p(U, V, N), \quad \mu = \mu(U, V, N). \tag{1.10}$$

These relations are known as the **equations of state**.

1.1.4 Densities

Using homogeneity, we can write

$$S(U, V, N) = VS\big(\tfrac{U}{V}, 1, \tfrac{N}{V}\big) \quad \text{or} \quad S(U, V, N) = NS\big(\tfrac{U}{N}, \tfrac{V}{N}, 1\big). \tag{1.11}$$

This shows that, when using densities, the entropy can actually be considered as a density as well and seen as a function of two variables rather than three. For example, one can introduce the **energy density** $u \stackrel{\text{def}}{=} \frac{U}{V}$ and the **particle density** $\rho \stackrel{\text{def}}{=} \frac{N}{V}$, and consider the **entropy density**:

$$s(u, \rho) \stackrel{\text{def}}{=} \frac{1}{V}S(uV, V, \rho V). \tag{1.12}$$

Alternatively, using the **energy per particle** $e \stackrel{\text{def}}{=} \frac{U}{N}$ and the **specific volume** (or **volume per particle**) $v \stackrel{\text{def}}{=} \frac{V}{N}$, one can consider the **entropy per particle**:

$$s(e, v) \stackrel{\text{def}}{=} \frac{1}{N}S(eN, vN, N).$$

In particular, its differential satisfies

$$\mathrm{d}s = \frac{\partial s}{\partial e}\,\mathrm{d}e + \frac{\partial s}{\partial v}\,\mathrm{d}v = \frac{1}{T}\mathrm{d}e + \frac{p}{T}\mathrm{d}v. \tag{1.13}$$

The entropy can thus be recovered (up to an irrelevant additive constant) from the knowledge of two of the three equations of state. This shows that the equations of state are not independent.

Example 1.6 (The Ideal Gas). Consider a gas of N particles in a container of volume V, at temperature T. An **ideal gas** is a gas which, at equilibrium, is described by the following two equations of state:

$$pV = RNT\,, \qquad U = cRNT\,,$$

where the constant c, the **specific heat capacity**, depends on the gas and R is some universal constant known as the **gas constant**.[6] Although no such gas exists, it turns out that most gases approximately satisfy such relations when the temperature is not too low and the density of particles is small enough.

When expressed as a function of $v = V/N$, the first equation becomes

$$pv = RT\,. \tag{1.14}$$

An **isotherm** is obtained by *fixing* the temperature T and studying p as a function of v:

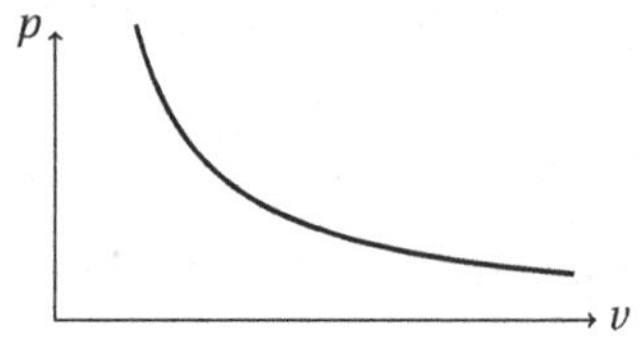

Figure 1.1. An isotherm of the equation of state of the ideal gas: at fixed temperature, the pressure p is proportional to $\frac{1}{v}$.

Let us explain how the two equations of state can be used to determine the entropy for the system. Notice first that the equations can be rewritten as

$$\frac{1}{T} = cR\frac{N}{U} = \frac{cR}{e}\,, \qquad \frac{p}{T} = R\frac{N}{V} = \frac{R}{v}\,.$$

Therefore, (1.13) becomes

$$\mathrm{d}s = \frac{cR}{e}\mathrm{d}e - \frac{R}{v}\mathrm{d}v\,.$$

Integrating the latter equation, we obtain

$$s(u, v) - s_0 = cR\log(e/e_0) - R\log(v/v_0)\,,$$

where e_0, v_0 is some reference point and s_0 is an undetermined constant of integration. We have thus obtained the desired fundamental relation:

$$S(U, V, N) = Ns_0 + NR\log\bigl[(U/U_0)^c(V/V_0)(N/N_0)^{-(c+1)}\bigr]\,,$$

where we have set $U_0 \stackrel{\text{def}}{=} N_0e_0$, $V_0 \stackrel{\text{def}}{=} N_0v_0$ for some reference N_0.

Later in this introduction, the equation of state (1.14) will be derived from the microscopic point of view, using the formalism of statistical mechanics. ◇

[6] Gas constant: $R = 8.3144621\,\mathrm{J\,m^{-1}K^{-1}}$.

It is often convenient to describe a system using certain thermodynamic variables rather than others. For example, in the case of a gas, it might be easier to control pressure and temperature rather than volume and internal energy, so these variables may be better suited for the description of the system. Not only is this possible, but there is a systematic way of determining which **thermodynamic potential** should replace the entropy in this setting and finding the corresponding extremum principle.

1.1.5 Alternative Representations, Thermodynamic Potentials

We now describe how alternative representations corresponding to other sets of variables, both extensive and intensive, can be derived. We treat explicitly the two cases that will be relevant for our analysis based on statistical mechanics later.

The Variables (β, V, N). We will first obtain a description of systems characterized by the set of variables (β, V, N), replacing U by its conjugate quantity β. Note that, if we want to have a fixed temperature, then the system must be allowed to exchange energy with the environment and is thus *not isolated anymore*. One can see such a system as being in permanent contact with an infinite *thermal reservoir* at fixed temperature $1/\beta$, with which it can exchange energy, but not particles or volume.

We start with some heuristic considerations.

We suppose that our system Σ is put in contact with a much larger system Σ_R, representing the thermal reservoir, with which it can only exchange energy:

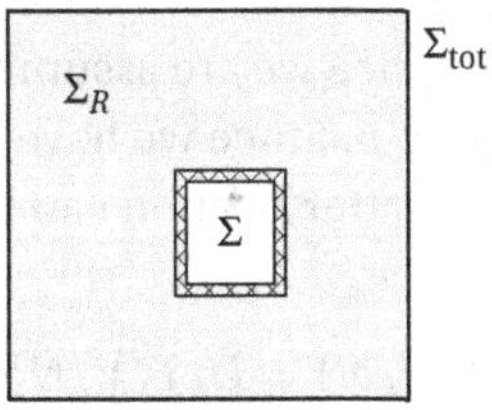

Figure 1.2. A system Σ, in contact with a reservoir.

We know from Example 1.3 that, under such conditions, both systems must have the same inverse temperature, denoted by β. The total system Σ_{tot}, of energy U_{tot}, is considered to be isolated. We denote by U the energy of Σ; the energy of the reservoir is then $U_{tot} - U$.

We assume that the reservoir is so much larger than Σ that we can ignore the effect of the actual state of Σ on the values of the intensive parameters associated with the reservoir: $\beta_R(= \beta)$, p_R and μ_R remain constant. In particular, by the Euler relation (1.9), *the entropy of the reservoir satisfies*

$$S^R(U_{tot} - U) = \beta_R U_R + \beta_R p_R V_R - \beta_R \mu_R N_R = \beta(U_{tot} - U) + \beta p_R V_R - \beta \mu_R N_R .$$

Observe that, under our assumptions, the only term in the last expression that depends on the state of Σ *is* $-\beta U$.

To determine the equilibrium value $\overline{U}$ *of the energy of* Σ, *we must maximize the total entropy (we only indicate the dependence on* U, *since the volumes and numbers of particles are fixed):* $\overline{U}$ *is the value realizing the supremum in*

$$S^{\text{tot}}(U_{\text{tot}}, \overline{U}) = \sup_U \{S^\Sigma(U) + S^R(U_{\text{tot}} - U)\},$$

which, from what was said before, is equivalent to finding the value of U *that realizes the infimum in*

$$\hat{F}^\Sigma(\beta) \stackrel{\text{def}}{=} \inf_U \{\beta U - S^\Sigma(U)\}.$$

◇

In view of the preceding considerations, we may expect the function

$$\hat{F}(\beta, V, N) \stackrel{\text{def}}{=} \inf_U \{\beta U - S(U, V, N)\} \tag{1.15}$$

to play a role analogous to that of the entropy, when the temperature, rather than the energy, is kept fixed. The thermodynamic potential $F(T, V, N) \stackrel{\text{def}}{=} T\hat{F}(1/T, V, N)$ is called the **Helmholtz free energy** (once more, the presence of a factor T is due to conventions).

In mathematical terms, $\hat{F}$ is, up to minor differences,[7] the **Legendre transform** of $S(U, V, N)$ with respect to U; see Appendix B.2 for the basic definition and properties of this transform. The Legendre transform enjoys several interesting properties. For instance, it has convenient convexity properties (see the exercise below), and it is an involution (on convex function, Theorem B.19). In other words, $\hat{F}$ and S contain the same information.

Observe now that, since we are assuming differentiability, the infimum in (1.15) is attained when $\frac{\partial S}{\partial U} = \beta$. Since we have assumed that the temperature is positive (remember (1.4)), the latter relation can be inverted to obtain $U = U(T, V, N)$. We get

$$F(T, V, N) = U(T, V, N) - TS(U(T, V, N), V, N). \tag{1.16}$$

As a shorthand, this relation is often written simply

$$F = U - TS. \tag{1.17}$$

The thermodynamic potential $\hat{F}$ inherits analogues of the fundamental properties of S.

[7] Since S is concave, $-S$ is convex. Therefore, indicating only the dependence of S on U,

$$\inf_U \{\beta U - S(U)\} = -\sup_U \{(-\beta)U - (-S(U))\},$$

which is minus the Legendre transform (defined in (B.11)) of $-S$, at $-\beta$.

Exercise 1.3. Show that $\hat{F}$ is convex in V and N (the extensive variables), and concave in β (the intensive variable).

Notice that, since $\hat{F}$ is convex in V, we have $\frac{\partial^2 \hat{F}}{\partial V^2} \geq 0$. But, by (1.16),

$$\left(\frac{\partial \hat{F}}{\partial V}\right)_{T,N} = \beta\left(\frac{\partial U}{\partial V}\right)_{T,N} - \underbrace{\left(\frac{\partial S}{\partial U}\right)_{V,N}}_{=\beta}\left(\frac{\partial U}{\partial V}\right)_{T,N} - \left(\frac{\partial S}{\partial V}\right)_{U,N} = -\frac{p}{T}.$$

Therefore, differentiating again with respect to V yields

$$\left(\frac{\partial p}{\partial V}\right)_{T,N} \leq 0, \tag{1.18}$$

a property known as **thermodynamic stability**.

A system for which $\left(\frac{\partial p}{\partial V}\right)_{T,N} > 0$ *would be* unstable *in the following intuitive sense: any small increase in V would imply an increase in pressure, which in turn would imply an increase in V, etc.* ◇

To state the analogue of the extremum principle in the present case, let us consider a system Σ, kept at temperature T, and composed of two subsystems Σ_1, with parameters T, V^1, N^1, and Σ_2, with parameters T, V^2, N^2. Similarly as before, we assume that there are constraints on the admissible values of V^1, N^1, V^2, N^2 and are interested in determining the corresponding equilibrium values $\overline{V}^1$, $\overline{V}^2$, $\overline{N}^1, \overline{N}^2$.

Exercise 1.4. Show that $\hat{F}$ satisfies the following extremum principle: the equilibrium values $\overline{V}^1, \overline{V}^2, \overline{N}^1, \overline{N}^2$ are those minimizing

$$\hat{F}(T, \widetilde{V}^1, \widetilde{N}^1) + \hat{F}(T, \widetilde{V}^2, \widetilde{N}^2) \tag{1.19}$$

among all $\widetilde{V}^1, \widetilde{N}^1, \widetilde{V}^2, \widetilde{N}^2$ compatible with the constraints.

The Variables (β, V, μ). We can proceed in the same way for a system characterized by the variables (β, V, μ). Such a system must be able to exchange both energy and particles with a reservoir.

This time, the thermodynamic potential associated with the variables $\beta, V, \hat{\mu} \stackrel{\text{def}}{=} -\mu/T$ is defined by

$$\hat{\Phi}_{\text{G}}(\beta, V, \hat{\mu}) \stackrel{\text{def}}{=} \inf_{U,N}\{\beta U + \hat{\mu} N - S(U, V, N)\}. \tag{1.20}$$

The function $\Phi_G(T, V, \mu) \stackrel{\text{def}}{=} T\hat{\Phi}_G(1/T, V, -\mu/T)$ is called the **grand potential**. As for the Helmholtz free energy, it can be shown that $\hat{\Phi}_G$ is concave in β and $\hat{\mu}$, and convex in V. The extremum principle extends also naturally to this case. Proceeding as before, we can write

$$\Phi_G = U - \mu N - TS \tag{1.21}$$

(with an interpretation analogous to the one done in (1.17)). Since, by the Euler relation (1.9), $TS = U + pV - \mu N$, we deduce that

$$\Phi_G = -pV, \tag{1.22}$$

so that $-\Phi_G/V$ coincides with the pressure of the system (expressed, of course, as a function of (T, V, μ)).

Generically,[8] the descriptions of a given system, in terms of various sets of thermodynamic variables, yield the same result at equilibrium. For example, if we start with a microstate (U, V, N) and compute the value of β at equilibrium, then starting with the macrostate (β, V, N) (with that particular value of β) and computing the equilibrium value of the energy yields U again.

In the following section, we leave aside the general theory and consider an example, in which an equation of state is progressively obtained from a combination of experimental observations and theoretical considerations.

1.1.6 Condensation and the Van der Waals–Maxwell Theory

Although rather accurate in various situations, the predictions made by the equation of state of the ideal gas are no longer valid at low temperature or at high density. In particular, the behavior observed for a *real* gas at low temperature is of the following type (compare Figure 1.3 with Figure 1.1): When v (the volume per particle) is large, the density of the gas is low, it is homogeneous (the same density is observed in all subsystems) and the pressure is well approximated by the ideal gas behavior. However, decreasing v, one reaches a value $v = v_g$, called the **condensation point**, at which the following phenomenon is observed: macroscopic droplets of liquid start to appear throughout the system. As v is further decreased, the fraction of the volume occupied by the gas decreases while that of the liquid increases. Nevertheless, the pressures inside the gas and inside the droplets are equal and constant. This goes on until another value $v = v_l < v_g$ is reached, at which all the gas has been transformed into liquid. When decreasing the volume to values $v < v_l$, the pressure starts to increase again, but at a much higher rate due to the fact that

[8] The word "generically" is used to exclude first-order phase transitions, since, when these occur, the assumptions of smoothness and invertibility that we use to invert the relations between all these variables fail in general. Such issues will be discussed in detail in the framework of equilibrium statistical mechanics in later chapters.

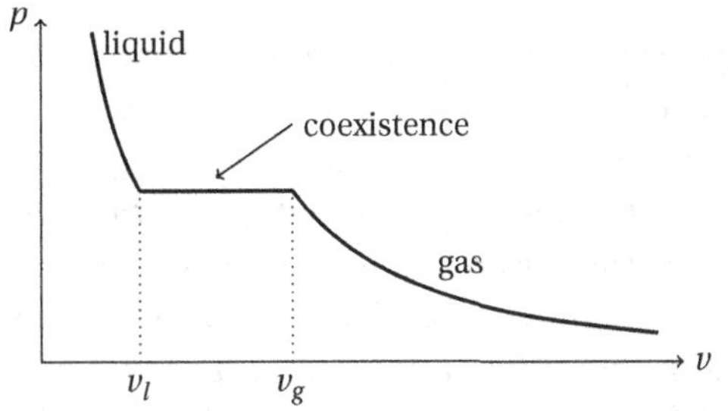

Figure 1.3. An isotherm of a real gas at low enough temperature. See Figure 1.6 for a plot of realistic values measured in the laboratory.

the system now contains only liquid and the latter is almost incompressible.[9] The range $[v_l, v_g]$ is called the **coexistence plateau.**

The condensation phenomenon shows that, from a mathematical point of view, the equations of state of a system are not always smooth in their variables (the pressure at the points v_l and v_g, for instance). The appearance of such singularities is the signature of *phase transitions*, one of the main themes studied in this book.

For the time being, we will only describe the way by which the equation of state of the ideal gas can be modified to account for the behavior observed in real gases.

Van der Waals Theory of Condensation

The first theory of condensation originated with van der Waals' thesis in 1873. Van der Waals succeeded in establishing an equation of state that described significant deviations from the equation of the ideal gas. His analysis is based on the following two fundamental hypotheses on the *microscopic* structure of the system:[10]

1. The gas has microscopic constituents, the *particles*. Particles are extended in space. They interact *repulsively* at short distances: particles do not overlap.
2. At larger distances, the particles also interact in an *attractive* way. This part of the interaction is characterized by a constant $a > 0$ called *specific attraction.*

The short-distance repulsion indicates that, in the equation of state, the volume v available to each particle should be replaced by a smaller quantity $v - b$ taking into account the volume of space each particle occupies. In order to deal with the attractive part of the interaction, van der Waals assumed that *the system is homogeneous*, a drastic simplification to which we will return later. These two hypotheses led van der Waals to his famous equation of state:

$$\left(p + \frac{a}{v^2}\right)(v - b) = RT\,. \tag{1.23}$$

[9] At even smaller specific volumes, the system usually goes through another phase transition at which the liquid transforms into a solid, but we will not discuss this issue here.

[10] The interaction between two particles at distance r is often modeled by a **Lennard-Jones potential**, that is, an interaction potential of the form $Ar^{-12} - Br^{-6}$, with $A, B > 0$. The first term models the short-range repulsion between the particles, while the second one models the long-range attraction.

The term $\frac{a}{v^2}$ can be understood intuitively as follows. Let $\rho \stackrel{\text{def}}{=} \frac{N}{V} = \frac{1}{v}$ denote the density of the gas (number of particles per unit volume). We assume that each particle only interacts with particles in its neighborhood, up to some large, finite distance. By the homogeneity assumption, the total force exerted by the other particles on a given particle deep inside the vessel averages to zero. However, for a particle in a small layer along the boundary of the vessel, the force resulting from its interaction with other particles has a positive component away from the boundary, since there are more particles in this direction. This inward force reduces the pressure exerted on the boundary of the vessel. Now, the number of particles in this layer along a portion of the boundary of unit area is proportional to ρ. The total force on each particle in this layer is proportional to the number of particles it interacts with, which is also proportional to ρ. We conclude that the reduction in the pressure, compared to an ideal gas, is proportional to $\rho^2 = 1/v^2$. A rigorous derivation will be given in Chapter 4. ◇

Of course, the ideal gas is recovered by setting $a = b = 0$. The analysis of (1.23) (see Exercise 1.5 below) reveals that, in contrast to those of the ideal gas, the isotherms present different behaviors depending on the temperature being above or below some **critical temperature** T_c; see Figure 1.4.

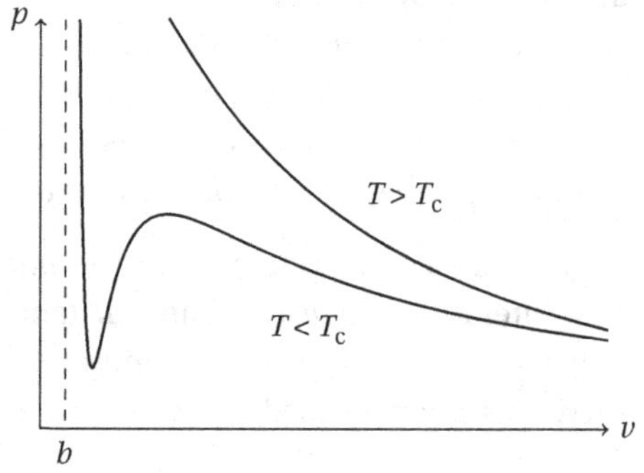

Figure 1.4. Isotherms of the van der Waals equation of state (1.23) at low and high temperature.

For **supercritical** temperatures, $T > T_c$, the behavior is qualitatively the same as for the ideal gas; in particular, p is strictly decreasing in v. However, for **subcritical** temperatures, $T < T_c$, there is an interval over which $\left(\frac{\partial p}{\partial v}\right)_T > 0$, thereby violating (1.18); this shows that this model has unphysical consequences. Moreover, in real gases (remember Figure 1.3), there is a plateau in the graph of the pressure, corresponding to values of v at which the gas and liquid phases coexist. This plateau is absent from van der Waals' isotherms.

Exercise 1.5. Study the equation of state (1.23) for $v > b$ and show that there exists a critical temperature,

$$T_c = T_c(a,b) \stackrel{\text{def}}{=} \frac{8a}{27Rb},$$

such that the following occurs:

- When $T > T_c$, $v \mapsto p(v,T)$ is decreasing everywhere.
- When $T < T_c$, $v \mapsto p(v,T)$ is increasing on some interval.

Maxwell's Construction

Van der Waals' main simplifying hypothesis was the assumption that the system remains *homogeneous*, which by itself makes the theory inadequate to describe the inhomogeneities that must appear at condensation.

In order to include the condensation phenomenon in van der Waals' theory, Maxwell [235] proposed a natural, albeit ad hoc, procedure to modify the low-temperature isotherms given by (1.23). Since the goal is to allow the system to split into regions that contain either gas or liquid, the latter should be at equal temperature and pressure; he thus replaced $p(v)$, on a well-chosen interval $[v_l, v_g]$, by a constant p_s, called **saturation pressure**.

From a physical point of view, Maxwell's determination of p_s, and hence of v_g and v_l, can be understood as follows. The integral $\int_{v_l}^{v_g} p(v)\,\mathrm{d}v$ represents the area under the graph of the isotherm, between v_l and v_g, but it also represents the amount of *work* necessary to compress the gas from v_g down to v_l. Therefore, if one is to replace $p(v)$ by a constant value between v_l and v_g, this value should be chosen such that the work required for that compression be the *same* as the original one. That is, v_l, v_g and p_s should satisfy

$$\int_{v_g}^{v_l} p(v)\,\mathrm{d}v = \int_{v_g}^{v_l} p_s\,\mathrm{d}v,$$

which gives

$$\int_{v_g}^{v_l} p(v)\,\mathrm{d}v = p_s \cdot (v_g - v_l). \tag{1.24}$$

This determination of p_s can also be given a geometrical meaning: it is the unique height at which a coexistence interval can be chosen in such a way that the two areas delimited by the van der Waals isotherm and the segment are equal. For that reason, the procedure proposed by Maxwell is usually called **Maxwell's equal-area rule**, or simply **Maxwell's construction**. We denote the resulting isotherm by $v \mapsto \mathrm{MC}\,p(v,T)$; see Figure 1.5.

Although it relies on a mixture of two conflicting hypotheses (first assume homogeneity, build an equation of state, and then modify it by plugging in the condensation phenomenon, by hand, using the equal-area rule), the van der

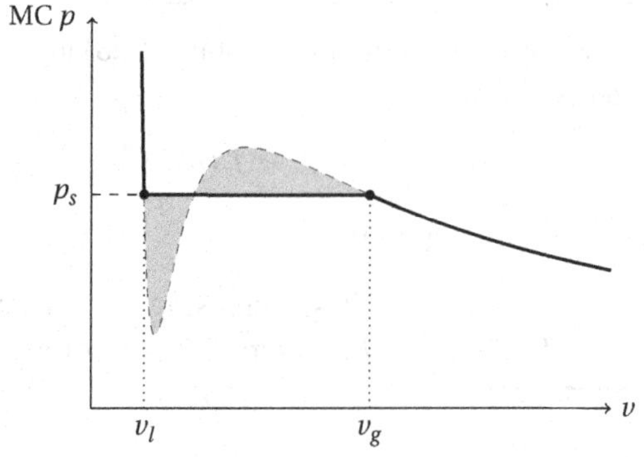

Figure 1.5. Maxwell's construction. The height of the segment, p_s, is chosen in such a way that the two connected regions delimited by the graph of p and the segment (shaded in the picture) have equal areas.

Waals–Maxwell theory often yields satisfactory quantitative results. It is a landmark in the understanding of the thermodynamics of the coexistence of liquids and gases and remains widely taught in classrooms today.

There are many sources in the literature where the interested reader can find additional information about the van der Waals–Maxwell theory; see, for instance, [89, 130]. We will return to the liquid–vapor equilibrium in a more systematic (and rigorous) way in Chapter 4.

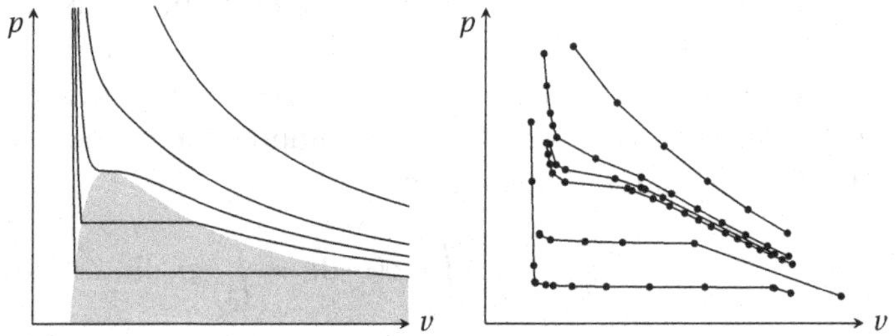

Figure 1.6. Left: some of the isotherms resulting from the van der Waals–Maxwell theory, which will be discussed in detail in Chapter 4. The shaded region indicates the pairs (v, p) for which coexistence occurs. Right: the isotherms of carbonic acid, measured by Thomas Andrews in 1869 [12].

1.2 From Micro to Macro: Statistical Mechanics

Just as was the case for thermodynamics, the object of statistical mechanics is the description of macroscopic systems. In sharp contrast to the latter, however, statistical mechanics relies on a reductionist approach, whose goal is to derive the macroscopic properties of a system from the microscopic description provided by the fundamental laws of physics. This ambitious program was initiated in the second half of the nineteenth century, with Maxwell, Boltzmann and Gibbs as its main contributors. It was essentially complete, as a general framework, when Gibbs published his famous treatise [137] in 1902. The theory has known many important developments since then, in particular regarding the fundamental problem of phase transitions.

As we already discussed, the enormous number of microscopic variables involved renders impossible the problem of deriving the macroscopic properties of a system directly from an application of the underlying fundamental theory describing its microscopic constituents. However, rather than giving up, one might try to use this to our advantage: indeed, the huge number of objects involved makes it conceivable that a *probabilistic* approach might be very efficient (after all, in how many areas does one have samples of size of order 10^{23}?). That is, the first step in statistical mechanics is to abandon the idea of providing a complete deterministic description of the system and to search instead for a *probability distribution* over the set of all microstates, which yields predictions compatible with the observed macrostate. Such a distribution should provide the probability of observing a given microstate and should make it possible to compute the probability of events of interest or the averages of relevant physical quantities.

This approach can also be formulated as follows: suppose that the only information we have on a given macroscopic system (at equilibrium) is its microscopic description (namely, we know the set of microstates and how to compute their energy) and the values of a few macroscopic variables (the same fixed in thermodynamics). Equipped with this information, and only this information, what can be said about a typical microstate?

The main relevant questions are therefore the following:

- What probability distributions should one use to describe large systems at equilibrium?
- With a suitable probability distribution at hand, what can be said about the other macroscopic observables? Is it possible to show that their distribution *concentrates* on their most probable value, when the system becomes large, thus yielding deterministic results for such quantities?
- How can this description be related to the one provided by equilibrium thermodynamics?
- Can phase transitions be described within this framework?

Note that the goal of equilibrium statistical mechanics is not limited to the computation of the fundamental quantities appearing in equilibrium thermodynamics. The formalism of statistical mechanics allows one to investigate numerous problems outside the scope of the latter theory. It allows for example to analyze *fluctuations* of macroscopic quantities in large finite systems and thus obtain a finer description than the one provided by thermodynamics.

In this section, we will introduce the central concepts of equilibrium statistical mechanics, valid for general systems, not necessarily gases and liquids, although such systems (as well as magnets) will be used in our illustrative examples.

As mentioned above, we assume that we are given the following basic inputs from a more fundamental theory describing the system's microscopic constituents:

1. *The set Ω of microstates.* To simplify the exposition and since this is enough to explain the general ideas, we assume that the set Ω of microstates describing the system is finite:
$$|\Omega| < \infty. \tag{1.25}$$
2. *The interactions between the microscopic constituents,* in the form of the energy $\mathscr{H}(\omega)$ associated with each microstate $\omega \in \Omega$:
$$\mathscr{H} : \Omega \to \mathbb{R},$$
called the **Hamiltonian** of the system. We will use $\mathscr{U} \overset{\text{def}}{=} \{U = \mathscr{H}(\omega) \ : \ \omega \in \Omega\}$.

Assumption (1.25) *will require space to be discretized. The latter simplification, which will be made throughout the book, may seem rather extreme. It turns out, however, that many phenomena of interest can still be investigated in this setting, while the mathematical analysis becomes much more tractable.* ◇

We denote the set of probability distributions on Ω by $\mathscr{M}_1(\Omega)$. Since we assume that Ω is finite, a distribution[11] $\mu \in \mathscr{M}_1(\Omega)$ is entirely characterized by the collection $(\mu(\{\omega\}))_{\omega \in \Omega}$ of the probabilities associated with each microstate $\omega \in \Omega$; we will usually abbreviate $\mu(\omega) \equiv \mu(\{\omega\})$. By definition, $\mu(\omega) \geq 0$ for all $\omega \in \Omega$ and $\sum_{\omega \in \Omega} \mu(\omega) = 1$. We call **observable** the result of a measurement on the system. Mathematically, it corresponds to a random variable $f\colon \Omega \to \mathbb{R}$. We will often denote the **expected value** of an observable f under $\mu \in \mathscr{M}_1(\Omega)$ by
$$\langle f \rangle_\mu \overset{\text{def}}{=} \sum_{\omega \in \Omega} f(\omega)\mu(\omega),$$
although some alternative notations will occasionally be used in later chapters.

1.2.1 Microcanonical Ensemble

The term "ensemble" was originally introduced by Gibbs.[2] *In more modern terms, it could simply be considered as a synonym of "probability space". In statistical mechanics, working in a specific* ensemble *usually means adopting either of the three descriptions described below: microcanonical, canonical, grand canonical.* ◇

In Section 1.1.1, we saw that it was convenient, when describing an isolated system at equilibrium, to use extensive conserved quantities as macroscopic variables, such as U, V, N in our gas example.

[11] Although we already use it to denote the chemical potential, we also use the letter μ to denote a generic element of $\mathscr{M}_1(\Omega)$, as done very often in the litterature.

We start our probabilistic description of an isolated system in the same way. The relevant conserved quantities depend on the system under consideration, but always contain the energy. Quantities such as the number of particles and the volume are assumed to be encoded into the set of microstates. Namely, we denote by $\Omega_{\Lambda;N}$ the set of all microstates describing a system of N particles located inside a domain Λ of volume $|\Lambda| = V$ (see the definition of the lattice gas in Section 1.2.4 for a specific example). For our current discussion, we assume that the only additional conserved quantity is the energy.

So, let us assume that the energy of the system is fixed to some value U. We are looking for a probability distribution on $\Omega_{\Lambda;N}$ that is concentrated on the set of all microstates compatible with this constraint, that is, on the **energy shell**

$$\Omega_{\Lambda;U,N} \stackrel{\text{def}}{=} \{\omega \in \Omega_{\Lambda;N} \, : \, \mathscr{H}(\omega) = U\} \, .$$

If this is all the information we have on the system, then the simplest and most natural assumption is that all configurations $\omega \in \Omega_{\Lambda;U,N}$ are *equiprobable*. Indeed, if we consider the distribution μ as a description of the *knowledge* we have of the system, then the uniform distribution faithfully represents the totality of our information. This is really just an application of Laplace's Principle of Insufficient Reason.[3] It leads naturally to the following definition:

Definition 1.7. Let $U \in \mathscr{U}$. The **microcanonical distribution (at energy** U**)**, $\nu^{\text{Mic}}_{\Lambda;U,N}$, associated with a system composed of N particles located in a domain Λ, is the uniform probability distribution concentrated on $\Omega_{\Lambda;U,N}$:

$$\nu^{\text{Mic}}_{\Lambda;U,N}(\omega) \stackrel{\text{def}}{=} \begin{cases} \frac{1}{|\Omega_{\Lambda;U,N}|} & \text{if } \omega \in \Omega_{\Lambda;U,N}\,, \\ 0 & \text{otherwise.} \end{cases} \tag{1.26}$$

Although the microcanonical distribution has the advantage of being natural and easy to define, it can be difficult to work with, since counting the configurations on the energy shell can represent a challenging combinatorial problem, even in simple cases. Moreover, one is often more interested in the description of a system at a fixed temperature T, rather than at fixed energy U.

1.2.2 Canonical Ensemble

Our goal now is to determine the relevant probability distribution to describe a macroscopic system at equilibrium *at a fixed temperature*. As discussed earlier, such a system is not isolated anymore, but in contact with a thermal reservoir with which it can exchange energy.

Once we have the microcanonical description, the problem of constructing the relevant probability distribution describing a system at equilibrium with an infinite *reservoir with which it can exchange energy and/or particles becomes conceptually straightforward from a probabilistic point of view. Indeed, similarly to what we did in Section 1.1.5, we can consider a system Σ in contact with a reservoir Σ_R, the union of the two forming a system Σ_{tot} isolated from the rest of the universe, as in Figure 1.2. Since Σ_{tot} is isolated, it is described by the microcanonical distribution, and the probability distribution of the subsystem Σ can be deduced from this microcanonical distribution by integrating over all the variables pertaining to the reservoir. That is, the measure describing Σ is the* marginal *of the microcanonical distribution corresponding to the subsystem Σ. This is a well-posed problem, but also a difficult one. It turns out that implementing such an approach rigorously is possible, once the statement is properly formulated, but a detailed discussion is beyond the scope of this book. In Section 4.7.1, we consider a simplified version of this problem; see also the discussion (and the references) in Section 6.14.1.*

Instead of following this path, we are going to use a generalization of the argument that led us to the microcanonical distribution to derive the distribution describing a system interacting with a reservoir. ◇

When we discussed the microcanonical distribution above, we argued that the uniform measure was the proper one to describe our knowledge of the system when the only information available to us is the total energy. We would like to proceed similarly here. The problem is that the temperature is not a mechanical quantity as was the energy (that is, there is no observable $\omega \mapsto T(\omega)$), so one cannot restrict the set $\Omega_{\Lambda;N}$ to microstates with fixed temperatures. We thus need to take a slightly different point of view.

Even though the energy of our system is not fixed anymore, one might still measure its *average*, which we also denote by U. In this case, extending the approach used in the microcanonical case corresponds to looking for the probability distribution $\mu \in \mathcal{M}_1(\Omega_{\Lambda;N})$ that best encapsulates the fact that our only information about the system is that $\langle \mathscr{H} \rangle_\mu = U$.

Maximum Entropy Principle. In terms of randomness, the outcomes of a random experiment whose probability distribution is uniform are the *least predictable.* Thus, what we did in the microcanonical case was to choose the most unpredictable distribution on configurations with fixed energy.

Let Ω be an arbitrary finite set of microstates. One convenient (and essentially unique, see below) way of quantifying the unpredictability of the outcomes of a probability distribution μ on Ω is to use the notion of *entropy* introduced in information theory by Shannon [301]:

Definition 1.8. The **(Shannon) entropy** [4] of $\mu \in \mathcal{M}_1(\Omega)$ is defined by

$$S_{Sh}(\mu) \stackrel{\text{def}}{=} -\sum_{\omega \in \Omega} \mu(\omega) \log \mu(\omega). \tag{1.27}$$

Exercise 1.6. $\mathrm{S}_{\mathrm{Sh}} : \mathscr{M}_1(\Omega) \to \mathbb{R}$ is concave: for all $\mu, \nu \in \mathscr{M}_1(\Omega)$ and all $\alpha \in [0,1]$,

$$\mathrm{S}_{\mathrm{Sh}}\big(\alpha\mu + (1-\alpha)\nu\big) \geq \alpha \mathrm{S}_{\mathrm{Sh}}(\mu) + (1-\alpha)\mathrm{S}_{\mathrm{Sh}}(\nu) .$$

The Shannon entropy provides a characterization of the uniform distribution through a variational principle.

Lemma 1.9. *The uniform distribution on* Ω, $\nu^{\mathrm{Unif}}(\omega) \stackrel{\mathrm{def}}{=} \frac{1}{|\Omega|}$, *is the unique probability distribution at which* S_{Sh} *attains its maximum:*

$$\mathrm{S}_{\mathrm{Sh}}(\nu^{\mathrm{Unif}}) = \sup_{\mu \in \mathscr{M}_1(\Omega)} \mathrm{S}_{\mathrm{Sh}}(\mu) . \tag{1.28}$$

Proof Consider $\psi(x) \stackrel{\mathrm{def}}{=} -x \log x$, which is concave. Using Jensen's inequality (see Appendix B.8.1) gives

$$\begin{aligned}\mathrm{S}_{\mathrm{Sh}}(\mu) &= |\Omega| \sum_{\omega \in \Omega} \frac{1}{|\Omega|} \psi(\mu(\omega)) \\ &\leq |\Omega| \psi\Big(\sum_{\omega \in \Omega} \frac{1}{|\Omega|}\mu(\omega)\Big) = |\Omega|\psi\Big(\frac{1}{|\Omega|}\Big) = \log|\Omega| = \mathrm{S}_{\mathrm{Sh}}(\nu^{\mathrm{Unif}}) ,\end{aligned}$$

with equality if and only if $\mu(\cdot)$ is constant, that is, if $\mu = \nu^{\mathrm{Unif}}$. □

Since it is concave, with a unique maximum at ν^{Unif}, the Shannon entropy provides a way of measuring how far a distribution is from being uniform. As shown in Appendix B.11, the Shannon entropy is the unique (up to a multiplicative constant) such function (under suitable assumptions). We can thus use it to select, among all probability distributions in some set, the one that is "the most uniform".

Namely, assume that we have a set of probability distributions $\mathscr{M}_1'(\Omega) \subset \mathscr{M}_1(\Omega)$ representing the set of all distributions compatible with the information at our disposal. Then, the one that best describes our state of knowledge is the one maximizing the Shannon entropy; this way of selecting a distribution is called the **maximum entropy principle**. Its application to statistical mechanics, as an extension of the Principle of Insufficient Reason, was pioneered by Jaynes [181].

For example, the microcanonical distribution $\nu^{\mathrm{Mic}}_{\Lambda;U,N}$ has maximal entropy among all distributions concentrated on the energy shell $\Omega(U)$:

$$\mathrm{S}_{\mathrm{Sh}}(\nu^{\mathrm{Mic}}_{\Lambda;U,N}) = \sup_{\substack{\mu \in \mathscr{M}_1(\Omega_{\Lambda;N}): \\ \mu(\Omega_{\Lambda;U,N})=1}} \mathrm{S}_{\mathrm{Sh}}(\mu) . \tag{1.29}$$

Canonical Gibbs Distribution. We apply the maximum entropy principle to find the probability distribution $\mu \in \mathscr{M}_1(\Omega_{\Lambda;N})$ that maximizes S_{Sh}, under the constraint that $\langle \mathscr{H} \rangle_\mu = U$. From an analytic point of view, this amounts to searching for the collection $(\mu(\omega))_{\omega \in \Omega_{\Lambda;N}}$ of nonnegative real numbers that solves the following optimization problem:

$$\text{Minimize} \sum_{\omega\in\Omega_{\Lambda;N}} \mu(\omega)\log\mu(\omega) \text{ when } \begin{cases} \sum_{\omega\in\Omega_{\Lambda;N}} \mu(\omega) = 1\,, \\ \sum_{\omega\in\Omega_{\Lambda;N}} \mu(\omega)\mathscr{H}(\omega) = U\,. \end{cases} \tag{1.30}$$

For this problem to have a solution, we require that $U \in [U_{\min}, U_{\max}]$, where $U_{\min} \stackrel{\text{def}}{=} \inf_\omega \mathscr{H}(\omega)$, $U_{\max} = \sup_\omega \mathscr{H}(\omega)$. Such problems with constraints can be solved by using the method of Lagrange multipliers. Since there are two constraints, let us introduce two Lagrange multipliers, λ and β, and define the following Lagrange function:

$$L(\mu) \stackrel{\text{def}}{=} \sum_{\omega\in\Omega_{\Lambda;N}} \mu(\omega)\log\mu(\omega) + \lambda \sum_{\omega\in\Omega_{\Lambda;N}} \mu(\omega) + \beta \sum_{\omega\in\Omega_{\Lambda;N}} \mu(\omega)\mathscr{H}(\omega)\,.$$

The optimization problem then turns into the analytic study of a system of $|\Omega_{\Lambda;N}|+2$ unknowns:

$$\begin{cases} \nabla L = 0\,, \\ \sum_{\omega\in\Omega_{\Lambda;N}} \mu(\omega) = 1\,, \\ \sum_{\omega\in\Omega_{\Lambda;N}} \mu(\omega)\mathscr{H}(\omega) = U\,, \end{cases}$$

where ∇ is the gradient involving the derivatives with respect to each $\mu(\omega)$, $\omega \in \Omega_{\Lambda;N}$. The condition $\nabla L = 0$ thus corresponds to

$$\frac{\partial L}{\partial\mu(\omega)} = \log\mu(\omega) + 1 + \lambda + \beta\mathscr{H}(\omega) = 0\,, \qquad \forall\omega \in \Omega_{\Lambda;N}\,.$$

The solution is of the form $\mu(\omega) = e^{-\beta\mathscr{H}(\omega)-1-\lambda}$. The first constraint $\sum\mu(\omega) = 1$ implies that $e^{1+\lambda} = \sum_{\omega\in\Omega_{\Lambda;N}} e^{-\beta\mathscr{H}(\omega)}$. In conclusion, we see that the distribution we are after is

$$\mu_\beta(\omega) \stackrel{\text{def}}{=} \frac{e^{-\beta\mathscr{H}(\omega)}}{\sum_{\omega'\in\Omega_{\Lambda;N}} e^{-\beta\mathscr{H}(\omega')}}\,,$$

where the Lagrange multiplier β must be chosen such that

$$\sum_{\omega\in\Omega_{\Lambda;N}} \mu_\beta(\omega)\mathscr{H}(\omega) = U\,. \tag{1.31}$$

Note that this equation always possesses exactly one solution $\beta = \beta(U)$ when $U \in (U_{\min}, U_{\max})$; this is an immediate consequence of the following:

Exercise 1.7. Show that $\beta \mapsto \langle\mathscr{H}\rangle_{\mu_\beta}$ is continuously differentiable, decreasing and

$$\lim_{\beta\to-\infty} \langle\mathscr{H}\rangle_{\mu_\beta} = U_{\max}\,, \qquad \lim_{\beta\to+\infty} \langle\mathscr{H}\rangle_{\mu_\beta} = U_{\min}\,.$$

Since β can always be chosen in such a way that the average energy takes a given value, it will be used from now on as the natural parameter for the canonical distribution. To summarize, the probability distribution describing a system at equilibrium that can exchange energy with the environment and possesses an average energy is assumed to have the following form:

Definition 1.10. The **canonical Gibbs distribution** at parameter β associated with a system of N particles located in a domain Λ is the probability distribution on $\Omega_{\Lambda;N}$ defined by

$$\mu_{\Lambda;\beta,N}(\omega) \stackrel{\text{def}}{=} \frac{e^{-\beta\mathscr{H}(\omega)}}{\mathbf{Z}_{\Lambda;\beta,N}}.$$

The exponential $e^{-\beta\mathscr{H}}$ is called the **Boltzmann weight**, and the normalizing sum

$$\mathbf{Z}_{\Lambda;\beta,N} \stackrel{\text{def}}{=} \sum_{\omega\in\Omega_{\Lambda;N}} e^{-\beta\mathscr{H}(\omega)}$$

is called the **canonical partition function.**

We still need to provide an interpretation for the parameter β. As will be argued below, in Section 1.3 (see also Exercise 1.12), β *can in fact be identified with the inverse temperature.*

Exercise 1.8. Using the maximum entropy principle, determine the probability distribution of maximal entropy, $\mu = (\mu(1), \dots, \mu(6))$, for the outcomes of a dice whose expected value is 4.

1.2.3 Grand Canonical Ensemble

Let us generalize the preceding discussion to the case of a system at equilibrium that can exchange both energy and particles with the environment. From the thermodynamic point of view, such a system is characterized by its temperature and its chemical potential.

One can then proceed exactly as in the previous section and apply the maximum entropy principle to the set of all probability distributions with prescribed average energy and average number of particles. Let us denote by $\Omega_\Lambda \stackrel{\text{def}}{=} \bigcup_{N\geq 0} \Omega_{\Lambda;N}$ the set of all microstates with an arbitrary number of particles all located inside the region Λ. A straightforward adaptation of the computations done above (with two Lagrange multipliers β and $\hat{\mu}$) shows that the relevant distribution in this case should take the following form (writing $\hat{\mu} = -\beta\mu$):

Definition 1.11. The **grand canonical Gibbs distribution** at parameters β and μ associated with a system of particles located in a region Λ is the probability distribution on Ω_Λ defined by

$$\nu_{\Lambda;\beta,\mu}(\omega) \stackrel{\text{def}}{=} \frac{e^{-\beta(\mathscr{H}(\omega)-\mu N)}}{\mathbf{Z}_{\Lambda;\beta,\mu}}, \quad \text{if } \omega \in \Omega_{\Lambda;N}.$$

The normalizing sum

$$\mathbf{Z}_{\Lambda;\beta,\mu} \stackrel{\text{def}}{=} \sum_N e^{\beta\mu N} \sum_{\omega\in\Omega_{\Lambda;N}} e^{-\beta\mathscr{H}(\omega)}$$

is called the **grand canonical partition function.**

Similarly as before, the parameters β and μ have to be chosen in such a way that the expected value of the energy and number of particles match the desired values. In Section 1.3, we will argue that β *and* μ *can be identified with the inverse temperature and chemical potential.*

1.2.4 Examples: Two Models of a Gas

We now present two examples of statistical mechanical models of a gas. The first one, although outside the main theme of this book, will be a model in the continuum, based on the description provided by Hamiltonian mechanics. The second one will be a lattice gas model, which can be seen as a simplification of the previous model and will be the main topic of Chapter 4.

Continuum Gas

We model a gas composed of N particles, contained in a vessel represented by a bounded subset $\Lambda \subset \mathbb{R}^d$. As a reader familiar with Hamiltonian mechanics might know, the state of such a system consists in the collection $(p_k, q_k)_{k=1,\dots,N}$ of the momentum $p_k \in \mathbb{R}^d$ and the position $q_k \in \Lambda$ of each particle. In particular, the set of microstates is

$$\Omega_{\Lambda;N} = (\mathbb{R}^d \times \Lambda)^N.$$

The Hamiltonian takes the usual form of a sum of a kinetic energy and a potential energy:

$$\mathscr{H}(p_1, q_1, \dots, p_N, q_N) \stackrel{\text{def}}{=} \sum_{k=1}^N \frac{\|p_k\|_2^2}{2m} + \sum_{1\le i<j\le N} \phi(\|q_j - q_i\|_2),$$

where m is the mass of each particle and the potential ϕ encodes the contribution to the total energy due to the interaction between the ith and jth particles, assumed to depend only on the distance $\|q_j - q_i\|_2$ between the particles.

Let us consider the canonical distribution associated with such a system. Of course, our discussion above does not apply verbatim, since we assumed that $\Omega_{\Lambda;N}$

was finite, while here it is a continuum. Nevertheless, the conclusion in this case is the natural generalization of what we saw earlier. Namely, the probability of an event B under the canonical distribution at inverse temperature β is also defined using the Boltzmann weight:

$$\mu_{\Lambda;\beta,N}(B) \stackrel{\text{def}}{=} \frac{1}{\mathbf{Z}_{\Lambda;\beta,N}} \int_{\Omega_{\Lambda;N}} \mathbf{1}_B \, e^{-\beta \mathscr{H}(p_1,q_1,\dots,p_N,q_N)} \mathrm{d}p_1 \mathrm{d}q_1 \cdots \mathrm{d}p_N \mathrm{d}q_N \,,$$

where $\mathbf{1}_B = \mathbf{1}_B(p_1, q_1, \dots, q_N, p_N)$ is the indicator of B and

$$\mathbf{Z}_{\Lambda;\beta,N} \stackrel{\text{def}}{=} \int_{\Omega_{\Lambda;N}} e^{-\beta \mathscr{H}(p_1,q_1,\dots,p_N,q_N)} \mathrm{d}p_1 \mathrm{d}q_1 \cdots \mathrm{d}p_N \mathrm{d}q_N \,.$$

Note that we cannot simply give the probability of each individual microstate, since they all have zero probability. Thanks to the form of the Hamiltonian, the integration over the momenta can be done explicitly:

$$\mathbf{Z}_{\Lambda;\beta,N} = \Big\{ \int_{\mathbb{R}} e^{-\frac{\beta}{2m}p^2} \,\mathrm{d}p \Big\}^{dN} \int_{\Lambda^N} e^{-\beta \mathscr{H}^{\text{conf}}(q_1,\dots,q_N)} \mathrm{d}q_1 \cdots \mathrm{d}q_N = \Big(\frac{2\pi m}{\beta}\Big)^{dN/2} \mathbf{Z}^{\text{conf}}_{\Lambda;\beta,N},$$

where we have introduced the **configuration integral**

$$\mathbf{Z}^{\text{conf}}_{\Lambda;\beta,N} \stackrel{\text{def}}{=} \int_{\Lambda^N} e^{-\beta \mathscr{H}^{\text{conf}}(q_1,\dots,q_N)} \mathrm{d}q_1 \cdots \mathrm{d}q_N \,,$$

and $\mathscr{H}^{\text{conf}}(q_1, \dots, q_N) \stackrel{\text{def}}{=} \sum_{1 \le i < j \le N} \phi(\|q_j - q_i\|_2)$.

In fact, when an event B only depends on the positions of the particles, not on their momenta, the factors originating from the integration over the momenta cancel in the numerator and in the denominator, giving

$$\mu_{\Lambda;\beta,N}(B) = \frac{1}{\mathbf{Z}^{\text{conf}}_{\Lambda;\beta,N}} \int_{\Lambda^N} \mathbf{1}_B e^{-\beta \mathscr{H}^{\text{conf}}(q_1,\dots,q_N)} \mathrm{d}q_1 \cdots \mathrm{d}q_N \,.$$

We thus see that the difficulties in analyzing this gas from the point of view of the canonical ensemble come from the positions: the momenta have no effect on the probability of events depending only on the positions, while they only contribute an explicit prefactor to the partition function. The position-dependent part, however, is difficult to handle as soon as the interaction potential ϕ is not trivial.

Usually, ϕ is assumed to contain two terms, corresponding to the short- and long-range parts of the interaction:

$$\phi(x) = \phi_{\text{short}}(x) + \phi_{\text{long}}(x) \,.$$

If we assume that the particles are identified with small spheres of fixed radius $r_0 > 0$, a simple choice for ϕ_{short} is the following *hard-core* interaction:

$$\phi_{\text{short}}(x) \stackrel{\text{def}}{=} \begin{cases} +\infty & \text{if } |x| \le 2r_0 \,, \\ 0 & \text{otherwise.} \end{cases}$$

The long-range part of the interaction can be of any type, but it should at least vanish at long distance:

$$\phi_{\text{long}}(x) \to 0 \quad \text{when } |x| \to \infty \,.$$

The decay at $+\infty$ should be fast enough to guarantee the existence of $\mathbf{Z}^{\mathrm{conf}}_{\Lambda;\beta,N}$, but we will not describe this in any further detail.

Unfortunately, even under further strong simplifying assumptions on ϕ, the mathematical analysis of such systems, in particular the computation of $\mathbf{Z}^{\mathrm{conf}}_{\Lambda;\beta,N}$, remains as yet intractable in most cases. This is the reason for which we consider *discretized* versions of these models. The model we will now introduce, although representing a mere caricature of the corresponding continuum model, is based on an interaction embodying van der Waals' two main assumptions: *short-range repulsion* and *long-range attraction.*

Lattice Gas

The lattice gas is obtained by first ignoring the momenta (for the reasons explained above) and assuming that the particles' positions are restricted to a discrete subset of $\mathbb{R}^d$. In general, this subset is taken to be the **d-dimensional cubic lattice**

$$\mathbb{Z}^d \stackrel{\mathrm{def}}{=} \big\{ i = (i_1, \dots, i_d) \in \mathbb{R}^d \ : \ i_k \in \mathbb{Z} \text{ for each } k \in \{1, \dots, d\} \big\} .$$

In other words, one imagines that $\mathbb{R}^d$ is composed of small cells and that each cell can accommodate at most one particle. To describe the microstates of the model, we consider a finite region $\Lambda \subset \mathbb{Z}^d$ representing the vessel and associate an **occupation number** η_i taking values in $\{0, 1\}$ with each cell $i \in \Lambda$: the value 0 means that the cell is empty, while the value 1 means that it contains a particle. The set of microstates is thus simply

$$\Omega_\Lambda \stackrel{\mathrm{def}}{=} \{0,1\}^\Lambda .$$

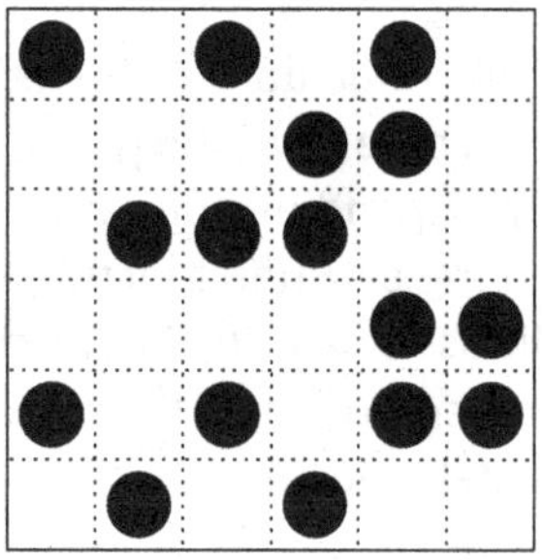

Note that this model automatically includes a short-range repulsion between the particles, since no two particles are allowed to share the same cell. The attractive part of the interaction can then be included in the Hamiltonian: for any $\eta = (\eta_i)_{i \in \Lambda} \in \Omega_\Lambda$,

$$\mathscr{H}(\eta) \stackrel{\mathrm{def}}{=} \sum_{\{i,j\} \subset \Lambda} J(j-i)\, \eta_i \eta_j ,$$

which is completely similar to the Hamiltonian $\mathscr{H}^{\mathrm{conf}}(q_1, \dots, q_N)$ above, the function $J : \mathbb{Z}^d \to \mathbb{R}$ playing the role of ϕ_{long} (one may assume that $J(j-i)$ depends only on the distance between the cells i and j, but this is not necessary). Note that the contribution of a pair of cells $\{i,j\}$ is zero if they do not both contain a particle.

The number of particles in Λ is given by

$$N_\Lambda(\eta) \stackrel{\text{def}}{=} \sum_{i\in\Lambda} \eta_i .$$

It will be useful to distinguish the partition functions in the various ensembles. The canonical partition function will be denoted

$$\mathbf{Q}_{\Lambda;\beta,N} = \sum_{\eta\in\Omega_{\Lambda;N}} e^{-\beta\mathscr{H}(\eta)} ,$$

where $\Omega_{\Lambda;N} \stackrel{\text{def}}{=} \{\eta \in \Omega_\Lambda : N_\Lambda(\eta) = N\}$, and the grand canonical one will be denoted

$$\Theta_{\Lambda;\beta,\mu} = \sum_N e^{\beta\mu N} \sum_{\eta\in\Omega_{\Lambda;N}} e^{-\beta\mathscr{H}(\eta)} .$$

Example 1.12. The simplest instance of the lattice gas is obtained by setting $J \equiv 0$, in which case the Hamiltonian is identically 0. Since its particles only interact through short-range repulsion, this model is called the **hard-core lattice gas.** ◇

1.3 Linking Statistical Mechanics and Thermodynamics

So far, we have introduced the central probability distributions of statistical mechanics. With these definitions, the analysis of specific systems reduces to an application of probability theory. Nevertheless, if one wishes to establish a link with thermodynamics, then one must identify the objects in statistical mechanics that correspond to the quantities in thermodynamics that are not observables, that is, not functions of the microstate, such as the entropy or the temperature.

This will be done by making certain identifications, making one assumption (Boltzmann's principle) and using certain analogies with thermodynamics. The real justification that these identifications are meaningful lies in the fact that the properties derived for and from these quantities in the rest of the book parallel precisely their analogues in thermodynamics. A reader unconvinced by these analogies can simply take them as motivations for the terminology used in statistical mechanics.

Since our discussion of thermodynamics mostly dealt with the example of a gas, it will be more convenient to discuss the identifications below in a statistical mechanical model of a lattice gas as well. But everything we explain can be extended to general systems.

1.3.1 Boltzmann's Principle and the Thermodynamic Limit

Consider the lattice gas in a region $\Lambda \subset \mathbb{Z}^d$ with $|\Lambda| = V$, composed of N particles and of total energy U. How should the entropy, a function of the macrostate

(U, V, N), be defined? We are looking for an additive function, as in (1.2), associated with an extremum principle that determines equilibrium.

Let us first generalize the energy shell and consider, for each macrostate (U, V, N), the set of all **microstates compatible with** (U, V, N):

$$\Omega_{\Lambda;U,N} \stackrel{\text{def}}{=} \{\eta \in \Omega_\Lambda : \mathscr{H}(\eta) = U,\ N_\Lambda(\eta) = N\} .$$

Definition 1.13. The **Boltzmann entropy** associated with a system of N particles in Λ with total energy U is defined by [12]

$$S_{\text{Boltz}}(\Lambda; U, N) \stackrel{\text{def}}{=} \log |\Omega_{\Lambda;U,N}| .$$

We motivate this definition with the following discussion.

Consider two lattice gases at equilibrium, in two separate vessels with equal volumes $|\Lambda_1| = |\Lambda_2| = V$, containing N_1^0 and N_2^0 particles, respectively. For simplicity, assume that the particles interact only through hard-core repulsion ($\mathscr{H} \equiv 0$) and that $N \stackrel{\text{def}}{=} N_1^0 + N_2^0$ is even.

Let us now put the two vessels in contact so that they can exchange particles. To reach a new equilibrium state, the N particles are redistributed among the two vessels, using the total volume at their disposal:

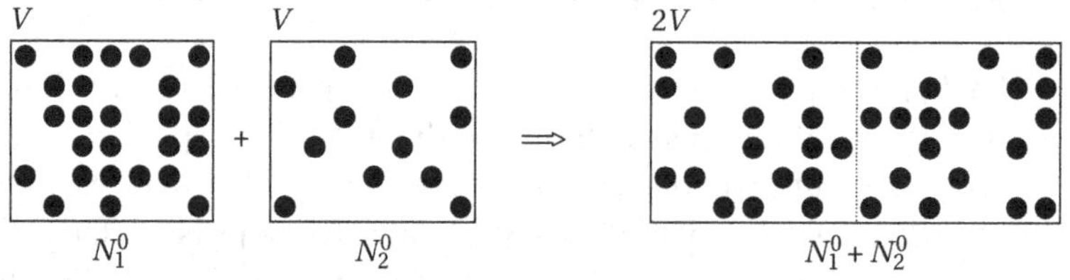

According to the postulate of thermostatics, equilibrium is realized once the pair giving the number of particles in each vessel, (N_1, N_2), maximizes the sum of the entropies of the vessels, under the constraint that $N_1^0 + N_2^0 = N$. Let us see why Boltzmann's definition of entropy is the most natural candidate for the function describing this extremum principle.

From the point of view of statistical mechanics, once the vessels have been put in contact, the whole system is described by $\nu^{\text{Mic}}_{\Lambda;N_1+N_2}$ (we set $\Lambda \stackrel{\text{def}}{=} \Lambda_1 \cup \Lambda_2$), under which the probability of observing N_1 particles in Λ_1, and thus $N_2 = N - N_1$ in Λ_2, is given by

$$\frac{|\Omega_{\Lambda_1;N_1}| \cdot |\Omega_{\Lambda_2;N_2}|}{|\Omega_{\Lambda;N}|} . \tag{1.32}$$

We are interested in the pairs (N_1, N_2) that maximize this probability, under the constraint $N_1 + N_2 = N$. In (1.32), only the numerator depends on N_1, N_2, and $|\Omega_{\Lambda;N_k}| = \binom{V}{N_k}$. As can be easily verified (see Exercise 1.9 below),

[12] Physicists usually write this condition as $S_{\text{Boltz}}(\Lambda; U, N) \stackrel{\text{def}}{=} k_B \log |\Omega(\Lambda; U, N)|$, where k_B is **Boltzmann's constant**. In this book, we will always assume that the units are chosen so that $k_B = 1$.

$$\max_{\substack{N_1,N_2:\\ N_1+N_2=N}} \binom{V}{N_1}\binom{V}{N_2} = \binom{V}{\frac{N}{2}}\binom{V}{\frac{N}{2}}, \tag{1.33}$$

meaning that the most probable configuration is the one in which the two vessels have the same number of particles. In terms of the Boltzmann entropy, (1.33) takes the form

$$\max_{\substack{N_1,N_2:\\ N_1+N_2=N}} \Big\{ \mathrm{S}_{\mathrm{Boltz}}(\Lambda_1; N_1) + \mathrm{S}_{\mathrm{Boltz}}(\Lambda_2; N_2) \Big\} = \mathrm{S}_{\mathrm{Boltz}}\big(\Lambda_1; \tfrac{N}{2}\big) + \mathrm{S}_{\mathrm{Boltz}}\big(\Lambda_2; \tfrac{N}{2}\big), \tag{1.34}$$

which, since this is a discrete version of the postulate of thermodynamics, makes $\mathrm{S}_{\mathrm{Boltz}}$ a natural candidate for the entropy of the system.

Unfortunately, this definition still suffers from one important defect. Namely, if the system is large, although having the same number of particles in each half, $(N_1, N_2) = (\frac{N}{2}, \frac{N}{2})$, is more likely than any other repartition (N_1', N_2'), it is nevertheless an event with *small* probability! (Of order $\frac{1}{\sqrt{N}}$, see exercise below.) Moreover, any other pair (N_1', N_2') such that $N_1' + N_2' = N$, $|N_1' - \frac{1}{2}N| \ll N^{1/2}$ and $|N_2' - \frac{1}{2}N| \ll N^{1/2}$ has essentially the same probability.

Exercise 1.9. Prove (1.33). Then, show that the probability of having the same number of particles, $\frac{N}{2}$, in each vessel, is of the order $\frac{1}{\sqrt{N}}$.

What must be considered, in order to have a deterministic behavior, is not the *number* of particles in each vessel but their *densities*. Let therefore

$$\rho_{\Lambda_1} \stackrel{\mathrm{def}}{=} \frac{N_{\Lambda_1}}{V}, \qquad \rho_{\Lambda_2} \stackrel{\mathrm{def}}{=} \frac{N_{\Lambda_2}}{V}$$

denote the random variables giving the densities of particles in each of the two halves. The constraint $N_1 + N_2 = N$ translates into

$$\frac{\rho_{\Lambda_1} + \rho_{\Lambda_2}}{2} = \frac{N}{2V} \stackrel{\mathrm{def}}{=} \overline{\rho},$$

which is the overall density of the system.

For a large system, ρ_{Λ_1} and ρ_{Λ_2} are both close to $\overline{\rho}$, but they always undergo microscopic fluctuations around $\overline{\rho}$: the probability of observing a fluctuation of size at least $\epsilon > 0$, $\nu^{\mathrm{Mic}}_{\Lambda;N}\big(|\rho_{\Lambda_1} - \overline{\rho}| \geq \epsilon\big)$, is always positive, even though it might be very small.

If we are after a more *macroscopic* statement, of the type "at equilibrium, the densities in the two halves are (exactly) equal to $\overline{\rho}$", then some limiting procedure is necessary, similar to the one used in the Law of Large Numbers.

The natural setting is that of a large system with a fixed density. Let us thus fix $\overline{\rho}$, which is the overall density of the system. We let the size of the system $|\Lambda| = V$ increase indefinitely, $V \to \infty$, and also let the total number of particles $N \to \infty$, in such a way that

$$\frac{N}{2V} \to \overline{\rho}.$$

This procedure is called the **thermodynamic limit**. One might then expect, in this limit, that the densities in the two subsystems concentrate on $\overline{\rho}$, in the sense that

$$\nu^{\mathrm{Mic}}_{\Lambda;N}\big(|\rho_{\Lambda_1}-\overline{\rho}|\geq\epsilon\big)=\nu^{\mathrm{Mic}}_{\Lambda;N}\big(|\rho_{\Lambda_2}-\overline{\rho}|\geq\epsilon\big)\to 0\,,\qquad\text{for all }\epsilon>0\,.$$

Let us see how this concentration can be obtained and how it relates to Boltzmann's definition of entropy. Keeping in mind that $N_2=N-N_1$,

$$\begin{aligned}\nu^{\mathrm{Mic}}_{\Lambda;N}\big(|\rho_{\Lambda_1}-\overline{\rho}|\geq\epsilon\big)&=\sum_{\substack{N_1:\\ |\frac{N_1}{V}-\overline{\rho}|\geq\epsilon}}\frac{|\Omega_{\Lambda_1;N_1}|\cdot|\Omega_{\Lambda_2;N_2}|}{|\Omega_{\Lambda;N}|}\\&=\sum_{\substack{N_1:\\ |\frac{N_1}{V}-\overline{\rho}|\geq\epsilon}}\exp\big(S_{\mathrm{Boltz}}(\Lambda_1;N_1)+S_{\mathrm{Boltz}}(\Lambda_2;N_2)-S_{\mathrm{Boltz}}(\Lambda;N)\big)\,.\end{aligned}\tag{1.35}$$

It turns out that, in the case of the hard-core gas we are considering, the Boltzmann entropy has a well-defined *density* in the thermodynamic limit. Namely, since $|\Omega_{\Lambda;N}|=\binom{2V}{N}$, by a simple use of Stirling's formula (Lemma B.3),

$$\lim\frac{1}{V}S_{\mathrm{Boltz}}(\Lambda;N)=\mathrm{s}^{\mathrm{hard}}_{\mathrm{Boltz}}(\rho)\stackrel{\mathrm{def}}{=}-\rho\log\rho-(1-\rho)\log(1-\rho)\,.$$

We can therefore use the **entropy density** $\mathrm{s}^{\mathrm{hard}}_{\mathrm{Boltz}}(\cdot)$ in each of the terms appearing in the exponential of (1.35). Letting $\rho_k\stackrel{\mathrm{def}}{=}\frac{N_k}{V}$ and remembering that $\frac{\rho_1+\rho_2}{2}=\overline{\rho}$,

$$\nu^{\mathrm{Mic}}_{\Lambda;N}\big(|\rho_{\Lambda_1}-\overline{\rho}|\geq\epsilon\big)=e^{o(1)V}\sum_{\substack{N_1:\\ |\rho_1-\overline{\rho}|\geq\epsilon}}\exp\big\{\big(\mathrm{s}^{\mathrm{hard}}_{\mathrm{Boltz}}(\rho_1)+\mathrm{s}^{\mathrm{hard}}_{\mathrm{Boltz}}(\rho_2)-2\mathrm{s}^{\mathrm{hard}}_{\mathrm{Boltz}}(\overline{\rho})\big)V\big\}\,,$$

where $o(1)$ tends to 0 in the thermodynamic limit.[13] Now, $\mathrm{s}^{\mathrm{hard}}_{\mathrm{Boltz}}$ is concave and so

$$\frac{\mathrm{s}^{\mathrm{hard}}_{\mathrm{Boltz}}(\rho_1)+\mathrm{s}^{\mathrm{hard}}_{\mathrm{Boltz}}(\rho_2)}{2}\leq\mathrm{s}^{\mathrm{hard}}_{\mathrm{Boltz}}(\overline{\rho})\,.$$

In fact, it is *strictly* concave, which implies that there exists $c(\epsilon)>0$ such that

$$\inf\Big\{\mathrm{s}^{\mathrm{hard}}_{\mathrm{Boltz}}(\overline{\rho})-\frac{\mathrm{s}^{\mathrm{hard}}_{\mathrm{Boltz}}(\rho_1)+\mathrm{s}^{\mathrm{hard}}_{\mathrm{Boltz}}(\rho_2)}{2}\;:\;\frac{\rho_1+\rho_2}{2}=\overline{\rho},\,|\rho_1-\overline{\rho}|\geq\epsilon\Big\}=c(\epsilon)\,.$$

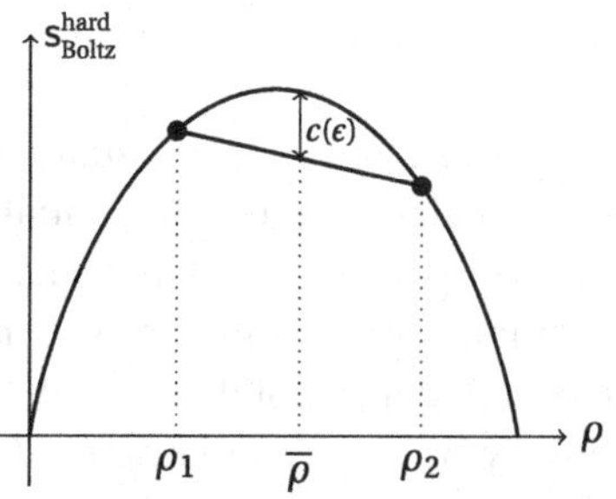

[13] Strictly speaking, one should treat values of ρ close to 0 and 1 separately. To keep the exposition short, we ignore this minor issue here. It will be addressed in Chapter 4.

Therefore, since the number of terms in the sum is bounded above by V, we conclude that

$$\nu^{\mathrm{Mic}}_{\Lambda;U,N}(|\rho_1 - \overline{\rho}| \geq \epsilon) \leq V e^{-(2c(\epsilon)-o(1))V} .$$

The latter quantity tends to 0 in the limit $V \to \infty$, for any $\epsilon > 0$. We conclude that the densities in each of the two subsystems indeed concentrate on $\overline{\rho}$ as $V \to \infty$.

In other words, to make the parallel with the discrete version (1.34), we have proven that in the thermodynamic limit the densities in the two boxes become both equal to $\overline{\rho}$; these are the *unique* densities that realize the supremum in

$$\sup_{\substack{\rho_1,\rho_2:\\ \frac{\rho_1+\rho_2}{2}=\overline{\rho}}} \left\{ \mathrm{s}^{\mathrm{hard}}_{\mathrm{Boltz}}(\rho_1) + \mathrm{s}^{\mathrm{hard}}_{\mathrm{Boltz}}(\rho_2) \right\} = \mathrm{s}^{\mathrm{hard}}_{\mathrm{Boltz}}(\overline{\rho}) + \mathrm{s}^{\mathrm{hard}}_{\mathrm{Boltz}}(\overline{\rho}) .$$

The above discussion was restricted to the hard-core lattice gas, but it shows that, while the Boltzmann entropy $\mathrm{S}_{\mathrm{Boltz}}$ does not fully satisfy our desiderata, considering its density in the thermodynamic limit yields a function that correctly describes the equilibrium values of the thermodynamic parameters, as the unique solution of an extremum principle.

Let us then consider a generic situation of a system with macrostate (U, V, N). To treat models beyond the hard-core lattice gas, the definition of the thermodynamic limit must include a limit $U \to \infty$ with $U/V \to u$.

Definition 1.14. Fix u and ρ. Consider the **thermodynamic limit**, $U \to \infty$, $V \to \infty$, $N \to \infty$, in such a way that $\frac{U}{V} \to u$ and $\frac{N}{V} \to \rho$, and let Λ be increasing, such that $|\Lambda| = V$. The **Boltzmann entropy density** at energy density u and particle density ρ is defined by the following limit, when it exists:

$$\mathrm{s}_{\mathrm{Boltz}}(u, \rho) \stackrel{\mathrm{def}}{=} \lim \frac{1}{V} \mathrm{S}_{\mathrm{Boltz}}(\Lambda, U, N) .$$

Of course, two nontrivial claims are hidden in the definition of the Boltzmann entropy density: the existence of the limit and the fact that it does not depend on the chosen sequence of sets Λ. This can be proved for a very large class of models, at least for sufficiently regular sequences (cubes would be fine, but much more general shapes can be accommodated). Several statements of this type will be proved in later chapters.

In view of the above discussion, it seems natural to consider the Boltzmann entropy density as the statistical mechanical analogue of the thermodynamic entropy density $s(u, \rho) = \frac{1}{V} S(U, V, N)$.

Boltzmann's Principle. *The thermodynamic entropy density associated with the macrostate (U, V, N), corresponding to densities $u = \frac{U}{V}$, $\rho = \frac{N}{V}$, will be identified with the Boltzmann entropy density:*

$$s(u, \rho) \leftrightarrow \mathrm{s}_{\mathrm{Boltz}}(u, \rho) .$$

It is possible, at least for reasonably well-behaved systems, to prove that s_{Boltz} possesses all the properties we established for its thermodynamic counterpart. For this introduction, we will only give some plausibility arguments to show that s_{Boltz} is concave in general (similar arguments will be made rigorous in Chapter 4).

Consider again a gas Σ contained in a cubic domain Λ, with parameters $V = |\Lambda|, U, N$. Let $u = \frac{U}{V}$, $\rho = \frac{N}{V}$. Fix $\alpha \in (0,1)$ and consider u_1, u_2 and ρ_1, ρ_2 such that

$$u = \alpha u_1 + (1-\alpha)u_2 , \quad \rho = \alpha\rho_1 + (1-\alpha)\rho_2 .$$

We think of Σ as being composed of a large number M of subsystems, $\Sigma_1, \dots, \Sigma_M$, each contained in a sub-cube of volume $V' = V/M$. If one can neglect the energy due to the interaction between the subsystems (which is possible if we assume that the latter are still very large), then one way of having energy and particle densities u and ρ in Σ is to have energy and particle densities u_1 and ρ_1 in a fraction α of the subsystems, and energy and particle densities u_2 and ρ_2 in the remaining ones. We then have

$$|\Omega_{\Lambda;U,N}| \geq |\Omega_{\Lambda';u_1 V',\rho_1 V'}|^{\alpha M} |\Omega_{\Lambda';u_2 V',\rho_2 V'}|^{(1-\alpha)M} ,$$

where Λ' denotes a cube of volume V'. Therefore,

$$\frac{1}{V}\log|\Omega_{\Lambda;U,N}| \geq \alpha\frac{1}{V'}\log|\Omega_{\Lambda';u_1 V',\rho_1 V'}| + (1-\alpha)\frac{1}{V'}\log|\Omega_{\Lambda';u_2 V',\rho_2 V'}| .$$

Letting first $M \to \infty$ and then taking the thermodynamic limit $V' \to \infty$ yields

$$s_{\text{Boltz}}\big(\alpha u_1 + (1-\alpha)u_2, \alpha\rho_1 + (1-\alpha)\rho_2\big) \geq \alpha s_{\text{Boltz}}(u_1,\rho_1) + (1-\alpha)s_{\text{Boltz}}(u_2,\rho_2) ,$$

as desired.

Assuming that $s_{\text{Boltz}}(u,\rho)$ exists and satisfies the relations we have seen in thermodynamics, and using Boltzmann's principle, we will now motivate the definition of the thermodynamic potentials studied in the canonical and grand canonical ensembles of statistical mechanics, namely the *free energy* and *pressure*.

Canonical Ensemble

Observe first that the canonical partition function at parameter β of a lattice gas with N particles in a vessel of size $|\Lambda| = V$ can be rewritten as

$$\mathbf{Q}_{\Lambda;\beta,N} = \sum_{U\in\mathscr{U}} e^{-\beta U}|\Omega_{\Lambda;U,N}| = \sum_{U\in\mathscr{U}} e^{-\beta U + S_{\text{Boltz}}(\Lambda;U,N)} = e^{o(1)V}\sum_{U\in\mathscr{U}} e^{-(\beta u - s_{\text{Boltz}}(u,\rho))V} ,$$

where we introduced $u \stackrel{\text{def}}{=} U/V$, $\rho \stackrel{\text{def}}{=} N/V$ and used the definition of the Boltzmann entropy density:

$$S_{\text{Boltz}}(\Lambda;U,N) = (s_{\text{Boltz}}(u,\rho) + o(1))V .$$

One can then bound the sum from above and from below by keeping only its largest term

$$e^{-V\inf_u\{\beta u - s_{\text{Boltz}}(u,\rho)\}} \leq \sum_{U\in\mathscr{U}} e^{-(\beta u - s_{\text{Boltz}}(u,\rho))V} \leq |\mathscr{U}|e^{-V\inf_u\{\beta u - s_{\text{Boltz}}(u,\rho)\}} . \quad (1.36)$$

To be more specific, let us assume that the lattice gas has **finite-range** interactions, meaning that $J(j-i) = 0$ as soon as $\|j-i\|_2$ is larger than some fixed constant.

In this case, $|\mathscr{U}|$ is bounded by a constant times V, which gives $\frac{1}{V}\log|\mathscr{U}| \to 0$. Therefore, in the thermodynamic limit ($N, V \to \infty, N/V \to \rho$), we obtain

$$\lim \frac{1}{V} \log \mathbf{Q}_{\Lambda;\beta,N} = -\inf_u \{\beta u - \mathrm{s}_{\mathrm{Boltz}}(u,\rho)\} \stackrel{\text{def}}{=} -\hat{f}(\beta,\rho). \tag{1.37}$$

In order to make the desired identifications with thermodynamics, we first argue that, *under the canonical Gibbs distribution, the energy density of the system concentrates, in the thermodynamic limit, on the value* $\overline{u} = \overline{u}(\beta,\rho)$ *minimizing* $\beta u - \mathrm{s}_{\mathrm{Boltz}}(u,\rho)$. (Note the similarity between the argument below and the discussion in Section 1.1.5.) Indeed, for $\epsilon > 0$, let

$$\mathscr{U}_\epsilon \stackrel{\text{def}}{=} \{U \in \mathscr{U} : \big|\beta u - \mathrm{s}_{\mathrm{Boltz}}(u,\rho) - \beta\{\overline{u} - \mathrm{s}_{\mathrm{Boltz}}(\overline{u},\rho)\}\big| \le \epsilon\}$$

denote the set of values of the energy for which $\beta u - \mathrm{s}_{\mathrm{Boltz}}(u,\rho)$ differs from its minimum value by at most ϵ. Then, repeating (1.36) for the sum over $U \in \mathscr{U} \setminus \mathscr{U}_\epsilon$,

$$\begin{aligned}\mu_{\Lambda;\beta,N}\big(\tfrac{\mathscr{H}}{V} \notin \mathscr{U}_\epsilon\big) &= \frac{\sum_{U\in\mathscr{U}\setminus\mathscr{U}_\epsilon} e^{-\beta U + S_{\mathrm{Boltz}}(\Lambda;U,N)}}{\mathbf{Q}_{\Lambda;\beta,N}} \\ &\le \frac{|\mathscr{U}|e^{-V\{\beta\overline{u} - \mathrm{s}_{\mathrm{Boltz}}(\overline{u},\rho)+\epsilon\}}}{e^{-V\{\beta\overline{u} - \mathrm{s}_{\mathrm{Boltz}}(\overline{u},\rho)\}}} e^{o(1)V} \le e^{-(\epsilon - o(1))V},\end{aligned}$$

which tends to 0 as $V \to \infty$, for any $\epsilon > 0$.

Up to now, the arguments were purely probabilistic. We are now going to use Boltzmann's principle in order to relate relevant quantities to their thermodynamic counterparts.

First, since the energy density concentrates on the value $\overline{u}$, it is natural to identify the latter with the thermodynamic equilibrium energy density. Now, note that $\overline{u}$ is also the value such that (assuming differentiability)

$$\beta = \frac{\partial \mathrm{s}_{\mathrm{Boltz}}}{\partial u}(\overline{u},\rho). \tag{1.38}$$

Using Boltzmann's Principle to identify $\mathrm{s}_{\mathrm{Boltz}}$ with the thermodynamic entropy density s and comparing (1.38) with the right-hand side of (1.3), we see that the parameter β of the canonical distribution should indeed be interpreted as the **inverse temperature**.

In turn, comparing (1.37) and (1.15), we see that $f(T,\rho) = T\hat{f}(1/T,\rho)$ can be identified with the Helmholtz free energy density. We conclude that, when it exists, the limit

$$-\lim \frac{1}{\beta V} \log \mathbf{Q}_{\Lambda;\beta,N} \tag{1.39}$$

is the relevant thermodynamic potential for the description of the canonical lattice gas. It will simply be called the **free energy**.

Exercise 1.10. Show that $\hat{f}(\beta,\rho)$ defined in (1.37) is concave in β, in agreement with the result of Exercise 1.3 obtained in the thermodynamic context.

Exercise 1.11. Let $\nu^{\text{Mic}}_{\Lambda;U,N}$ denote the microcanonical distribution in Λ associated with the parameters U, N. Show that its Shannon entropy coincides with the Boltzmann entropy:

$$\mathrm{S}_{\text{Sh}}(\nu^{\text{Mic}}_{\Lambda;U,N}) = \mathrm{S}_{\text{Boltz}}(\Lambda; U, N).$$

Therefore, Boltzmann's principle actually identifies the Shannon entropy density associated with the microcanonical distribution with the thermodynamic entropy density.

Exercise 1.12. Let $\mu_{\Lambda;\beta_U,N}$ be the canonical Gibbs distribution associated with the parameter β_U for which (1.31) holds. Show that

$$\frac{\partial \mathrm{S}_{\text{Sh}}(\mu_{\Lambda;\beta_U,N})}{\partial U} = \beta_U.$$

Therefore, identifying, in analogy with what is done in Exercise 1.11, the Shannon entropy of the canonical distribution with the thermodynamic entropy yields an alternative motivation to identify the parameter β with the inverse temperature.

Compute also $U - T(U)\mathrm{S}_{\text{Sh}}(\mu_{\beta_U})$ and verify that it coincides with the definition of free energy in the canonical ensemble.

Grand Canonical Ensemble

We can do the same type of argument for the grand canonical partition function at parameter β, μ of a gas in a region of volume $|\Lambda| = V$:

$$\begin{aligned}\Theta_{\Lambda;\beta,\mu} &= \sum_{N,U} e^{\beta\mu N - \beta U} |\Omega_{\Lambda;U,N}| \\ &= \sum_{N,U} e^{\beta\mu N - \beta U + \mathrm{S}_{\text{Boltz}}(\Lambda;U,N)} = e^{o(1)V} \sum_{\rho,u} e^{-\{\beta u - \beta\mu\rho - s(u,\rho)\}V}.\end{aligned}$$

Arguing as before, we conclude that, in the thermodynamic limit $V \to \infty$,

$$\lim \frac{1}{V} \log \Theta_{\Lambda;\beta,\mu} = -\inf_{u,\rho}\{\beta u - \beta\mu\rho - s(u,\rho)\} \stackrel{\text{def}}{=} -\hat{\phi}_{\text{G}}(\beta, -\mu/T). \tag{1.40}$$

Again, the particle and energy densities concentrate on the values $\overline{u} = \overline{u}(\beta, \mu)$ and $\overline{\rho} = \overline{\rho}(\beta, \mu)$ such that

$$\beta = \frac{\partial s}{\partial u}(\overline{u}, \overline{\rho}), \qquad \beta\mu = \frac{\partial s}{\partial \rho}(\overline{u}, \overline{\rho}).$$

In view of (1.3) and (1.6), this allows us to interpret the parameters β and μ of the grand canonical distribution as the **inverse temperature** and the **chemical potential**, respectively. Moreover, comparing (1.40) with (1.20), we see that $\phi_{\text{G}}(T, \mu) \stackrel{\text{def}}{=} T\hat{\phi}_{\text{G}}(1/T, -\mu/T)$ can be identified with the density of the grand potential, which, by (1.22), corresponds to minus the pressure $p(T, \mu)$ of the model.

We thus see that, when the limit exists,

$$\lim \frac{1}{\beta V} \log \Theta_{\Lambda;\beta,\mu} \tag{1.41}$$

is the relevant thermodynamic potential for the description of the grand canonical ensemble; it will be called simply the **pressure**.

In later chapters, we will see precise (and rigorous) versions of the kind of argument used above.

1.3.2 Deriving the Equation of State of the Ideal Gas

Computing the free energy or the pressure of a given model is not trivial, in general, and will be done for several interesting cases in later chapters. Nevertheless, if we consider the simplest possible case, the hard-core lattice gas, then some explicit computations can be done.

Fix $\beta > 0$ and $\mu \in \mathbb{R}$. Since $\mathscr{H} \equiv 0$, the grand canonical partition function is easily computed:

$$\Theta_{\Lambda;\beta,\mu} = \sum_{N=0}^{V} \binom{V}{N} e^{\beta\mu N} = \left(1 + e^{\beta\mu}\right)^V .$$

It follows from (1.41) that, in the thermodynamic limit $V \to \infty$, the pressure is given by

$$p(T,\mu) = T \log(1 + e^{\beta\mu}) . \tag{1.42}$$

The average number of particles is given by

$$\left\langle N_\Lambda \right\rangle_{\nu_{\Lambda;\beta,\mu}} = \frac{1}{\beta} \frac{\partial \log \Theta_{\Lambda;\beta,\mu}}{\partial \mu} = \frac{e^{\beta\mu}}{1 + e^{\beta\mu}} V .$$

In particular, in the thermodynamic limit $V \to \infty$,

$$\rho(\beta,\mu) \stackrel{\text{def}}{=} \lim_{V\to\infty} \left\langle \tfrac{N_\Lambda}{V} \right\rangle_{\nu_{\Lambda;\beta,\mu}} = \frac{e^{\beta\mu}}{1 + e^{\beta\mu}} .$$

Using this in (1.42), we obtain the equation of the isotherms:

$$p = -T \log(1 - \rho) . \tag{1.43}$$

For a diluted gas, $\rho \ll 1$, a Taylor expansion gives $p = \rho T + O(\rho^2)$, which in terms of the specific volume $v = 1/\rho$ becomes

$$pv = T + O(\tfrac{1}{v}) ,$$

and we recover the equation of state for an ideal gas, see (1.14).

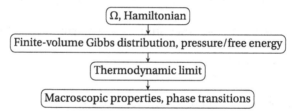

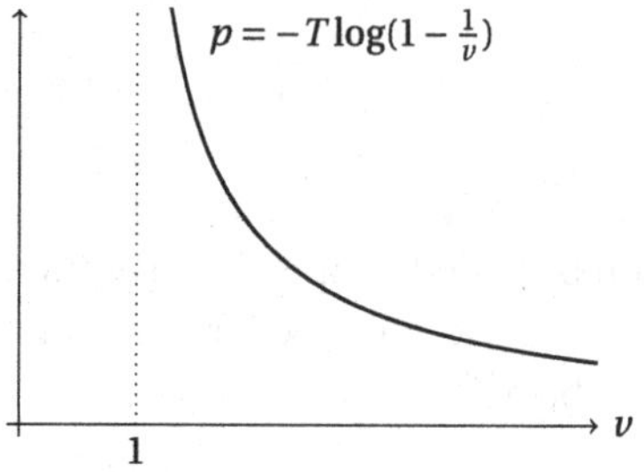

The reason we observe deviations from the ideal gas law at higher densities (small v) is due to the repulsive interaction between the particles, which comes from the fact that there can be at most one of them in each cell. Note, also, that we do not find a coexistence plateau in this model, see (1.43). This is due to the absence of attractive interaction between the particles. More general situations will be considered in Chapter 4.

1.3.3 The Basic Structure

The structure provided by the formalism presented so far, applied to the lattice gas, can be summarized as follows:

Ω, Hamiltonian
↓
Finite-volume Gibbs distribution, pressure/free energy
↓
Thermodynamic limit
↓
Macroscopic properties, phase transitions

Many models will be introduced and analyzed in the rest of the book, based on this structure. These models will be used to study various aspects of equilibrium statistical mechanics: macroscopic features, first-order phase transitions, fluctuations, equivalence of the different ensembles, etc. Since it is mathematically simpler and physically often more relevant, we will mostly work at fixed temperature rather than fixed energy; the basic object for us will thus be the canonical and grand canonical Gibbs distributions.

Let us now move on to another type of phenomenon which can be studied in this formalism and which will be one of the main concerns of later chapters.

1.4 Magnetic Systems

In this section, we describe another important class of macroscopic systems encountered in this book: *magnets*. We will discuss the two main types of behavior magnets can present, *paramagnetism* and *ferromagnetism*, and introduce one of the main models used for their description.

1.4.1 Phenomenology: Paramagnets vs. Ferromagnets

Consider a sample of some material whose atoms are arranged in a regular crystalline structure.[5] We suppose that each of these atoms carries a magnetic moment (picture a small magnet attached to each atom) called its *spin*. We assume that each spin has the tendency of aligning with its neighbors and with an external magnetic field.

If the magnetic field points in a fixed direction, the spins are globally ordered: they tend to align with the field and thus all point roughly in the same direction. If we then slowly decrease the intensity of the external field to zero, two behaviors are possible.

Paramagnetic Behavior. In the first scenario, the global order is progressively lost as the field decreases and, when the latter reaches zero, the spins' global order is lost. Such behavior is called *paramagnetism*:

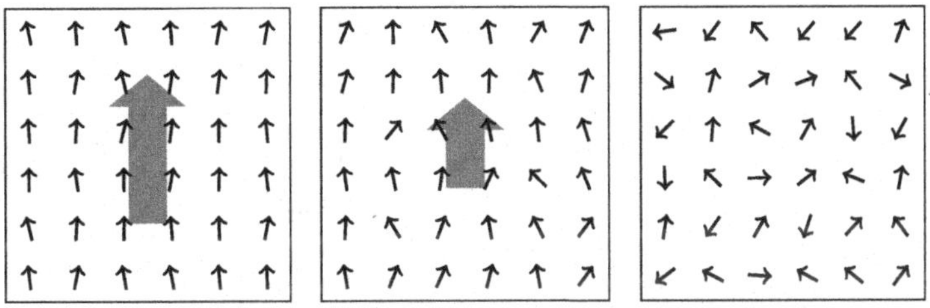

This phenomenon can be measured quantitatively by introducing the **magnetization**, which is the average of the spins, projected along the direction of the magnetic field. For a paramagnet, when the field decreases from a positive value to zero (maintaining the direction fixed), or similarly if it increases from a negative value to zero, the magnetization tends to zero:

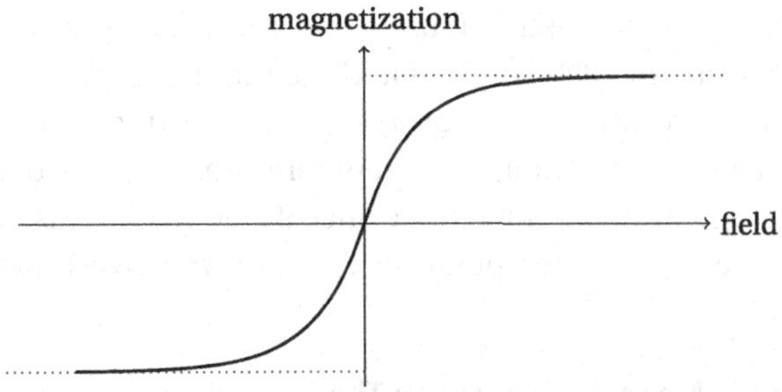

Ferromagnetic Behavior. But another scenario is possible: as the external field decreases, the global order decreases, but the local interactions among the spins are strong enough for the material to maintain a globally magnetized state even after the external field has vanished. Such behavior is called *ferromagnetism*:

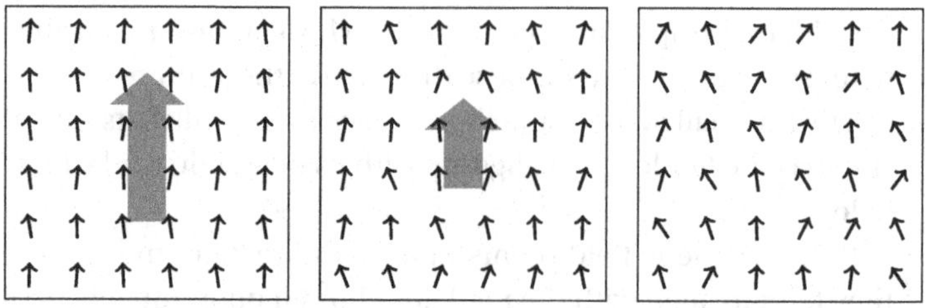

A ferromagnet thus exhibits *spontaneous magnetization*, that is, global ordering of the spins even in the absence of an external magnetic field. The value of the spontaneous magnetization, $\pm m^*$, depends on whether the external field approached zero from positive or negative values:

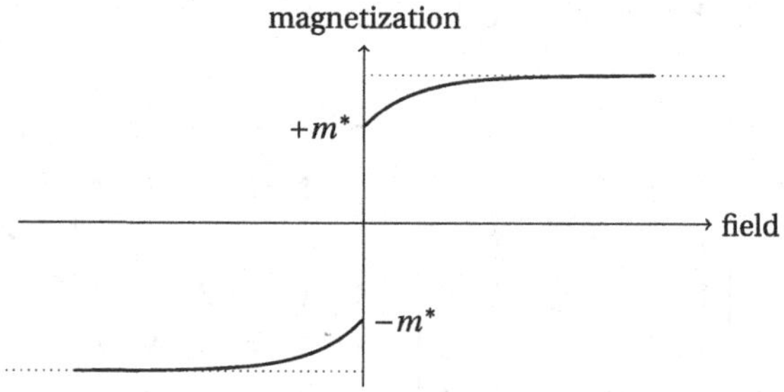

Observe that, as the field goes through zero, the magnetization suffers a discontinuity: it jumps from a strictly positive to a strictly negative value. This corresponds to a *first-order phase transition.*

Using the process described above, one can in principle prepare a ferromagnetic material with a spontaneous magnetization pointing in an arbitrary direction, by simply applying a magnetic field in that direction and slowly decreasing its intensity to zero.

The distinction between these two types of magnetic behavior was first made by Pierre Curie in 1895 and initiated the modern theory of magnetism. Among other important results, Curie observed that the same material can present both types of behavior, depending on the temperature: its behavior can suddenly change from ferromagnetic to paramagnetic once its temperature is raised above a well-defined, substance-specific temperature, now known as the *Curie temperature.*

1.4.2 A Simple Model for a Magnet: The Ising Model

The Ising model was introduced by Wilhelm Lenz in 1920 [221], in view of obtaining a theoretical understanding of the phase transition from ferromagnetic to paramagnetic behavior described above. The name "Ising model" (sometimes, but much less frequently, more aptly called the Lenz–Ising model, as suggested by Ising himself) was coined in a famous paper by Rudolph Peierls [266] in reference to

Ernst Ising's 1925 PhD thesis [175], which was undertaken under Lenz's supervision and devoted to the one-dimensional version of the model.

A major concern, and a much debated issue, in the theoretical physics community at the beginning of the twentieth century, was to determine whether phase transitions could be described within the framework of statistical mechanics, still a young theory at that time.[6]

This question was settled using the Ising model. This is indeed *the first system of locally interacting units for which it was possible to prove the existence of a phase transition*. This proof was given in the above-mentioned paper by Peierls in 1936, using an argument that would later become a central tool in statistical mechanics.[7]

Its simplicity and the richness of its behavior have turned the Ising model into a preferred laboratory to test new ideas and methods in statistical mechanics. It is nowadays, undoubtedly, the most famous model in this field, and has been the subject of thousands of research papers. Moreover, through its numerous interpretations in physics as well as in many other fields, it has been used to describe qualitatively, and sometimes quantitatively, a great variety of situations.[8]

We model the regular crystalline structure corresponding to the positions of the atoms of our magnet by a finite, nonoriented graph $G = (\Lambda, \mathscr{E})$, whose set of vertices Λ is a subset of $\mathbb{Z}^d$. A typical example, often used in this book, is the **box of radius *n***:

$$\mathrm{B}(n) \stackrel{\text{def}}{=} \{-n, \dots, n\}^d.$$

For example, B(4) is represented on Figure 1.7. The edges of the graph will most often be between **nearest neighbors**, that is, pairs of vertices i, j with $\|j - i\|_1 = 1$, where the norm is defined by $\|i\|_1 \stackrel{\text{def}}{=} \sum_{k=1}^{d} |i_k|$. We write $i \sim j$ to indicate that i and j are nearest neighbors. So, the set of edges in the box $\mathrm{B}(n)$ is $\{\{i,j\} \subset \mathrm{B}(n) : i \sim j\}$, as depicted in Figure 1.7.

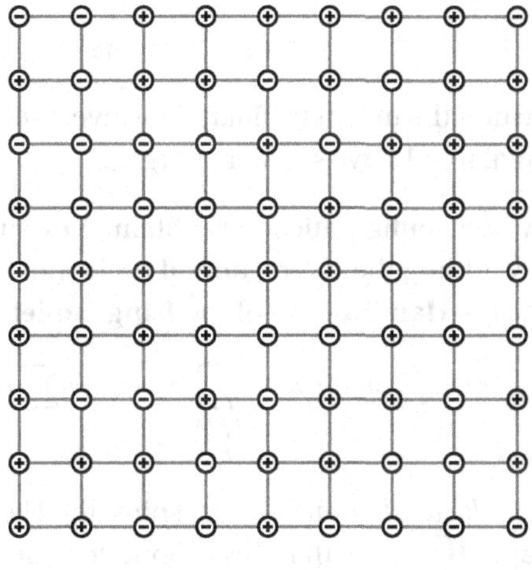

Figure 1.7. A spin configuration $\omega \in \Omega_{\mathrm{B}(4)}$.

The Ising model is defined by first assuming that a spin is located at each vertex of the graph $G = (\Lambda, \mathscr{E})$. One major simplification is the assumption that, unlike the pictures of Section 1.4.1, the spins are restricted to one particular direction, pointing either "up" or "down"; the corresponding two states are traditionally denoted by $+1$ ("up") and -1 ("down"). It follows that, to describe a microstate, a variable ω_i taking two possible values ± 1 is assigned to each vertex $i \in \Lambda$; this variable will be called the **spin** at i.

A microstate of the system, usually called a **configuration**, is thus an element $\omega \in \Omega_\Lambda$, where

$$\Omega_\Lambda \stackrel{\text{def}}{=} \{-1, 1\}^\Lambda .$$

The microscopic interactions among the spins are defined in such a way that:

1. *There is only interaction between pairs of spins located at neighboring vertices.* That is, it is assumed that the spins at two distinct vertices $i, j \in \Lambda$ interact if and only if the pair $\{i, j\}$ is an edge of the graph.
2. *The interaction favors agreement of spin values.* In the most common instance of the model, to which we restrict ourselves here, this is done in the simplest possible way: a pair of spins at the endpoints i and j of an edge *decreases* the overall energy of the configuration if they agree ($\omega_i = \omega_j$) and *increases* it if they differ; more precisely, the spins at the endpoints of the edge $\{i, j\}$ contribute to the total energy by an amount

$$-\omega_i \omega_j .$$

   Therefore, configurations in which most pairs of neighbors are aligned have smaller energy.
3. *Spins align with the external magnetic field.* Assume that a constant external magnetic field of intensity $h \in \mathbb{R}$ (oriented along the same direction as the spins) acts on the system. Its interaction with the spin at i contributes to the total energy by an amount

$$-h\omega_i .$$

   That is, when the magnetic field is positive, the configurations with most of their spins equal to $+1$ have smaller energy.

The **energy** of a configuration ω is obtained by summing the interactions over all pairs and by adding the interaction of each spin with the external magnetic field. This leads to the **Hamiltonian of the Ising model**:

$$\mathscr{H}_{\Lambda;h}(\omega) \stackrel{\text{def}}{=} -\sum_{\substack{i,j \in \Lambda \\ i \sim j}} \omega_i \omega_j - h \sum_{i \in \Lambda} \omega_i , \qquad \omega \in \Omega_\Lambda . \tag{1.44}$$

Since it favors *local alignment* of the spins, the Hamiltonian of the model is said to be **ferromagnetic**. (Note that this terminology does not necessarily imply that the model *behaves* like a ferromagnet.)

The Gibbs distribution is denoted by

$$\mu_{\Lambda;\beta,h}(\omega) = \frac{e^{-\beta \mathscr{H}_{\Lambda;h}(\omega)}}{\mathbf{Z}_{\Lambda;\beta,h}},$$

where $\mathbf{Z}_{\Lambda;\beta,h}$ is the associated partition function. The expectation of an observable $f\colon \Omega_\Lambda \to \mathbb{R}$ under $\mu_{\Lambda;\beta,h}$ is denoted $\langle f \rangle_{\Lambda;\beta,h}$.

An important observation is that, in the absence of a magnetic field (that is, when $h = 0$), even though local spin alignment is favored by the Hamiltonian, neither of the orientations ($+1$ or -1) is favored globally. Namely, if $-\omega$ denotes the spin-flipped configuration in which $(-\omega)_i \stackrel{\text{def}}{=} -\omega_i$, then $\mathscr{H}_{\Lambda;0}(-\omega) = \mathscr{H}_{\Lambda;0}(\omega)$; this implies that

$$\mu_{\Lambda;\beta,0}(-\omega) = \mu_{\Lambda;\beta,0}(\omega).$$

The model is then said to be **invariant under global spin flip**. When $h \neq 0$, this symmetry no longer holds.

1.4.3 Thermodynamic Behavior

Our goal is to study the Ising model in a large region Λ and eventually to take the thermodynamic limit, for instance taking $\Lambda = \mathrm{B}(n)$ and letting $n \to \infty$.

To simplify the discussion, we will first consider the model in the absence of a magnetic field: $h = 0$. A natural question, which will be a central theme in this book, is: under which circumstances does the ferromagnetic nature of the model, whose tendency is to align the spins locally, induce order also at the global/macroscopic scale? To make this question more precise, we need suitable ways to quantify global order. One natural such quantity is the **total magnetization**

$$M_\Lambda(\omega) \stackrel{\text{def}}{=} \sum_{i \in \Lambda} \omega_i .$$

Then, the **magnetization density**

$$\frac{M_\Lambda(\omega)}{|\Lambda|} \in [-1, 1]$$

equals the difference between the fractions of spins that take the values $+1$ and -1 respectively; it therefore provides some information on the balance between the two spin values in the whole system.

As we already pointed out, the Gibbs distribution is invariant under a global spin flip when $h = 0$. As a consequence, the *average* magnetization is zero at all temperatures:

Exercise 1.13. Show that

$$\langle M_\Lambda \rangle_{\Lambda;\beta,0} = 0 . \qquad (1.45)$$

The interpretation of (1.45) is that, *on average*, the densities of + and − spins are equal. However, as we will see below, this does not necessarily mean that the densities of two species of spins are equal in typical configurations of the model. As a first natural step, let us study the fluctuations of M_Λ around this average value. Since the spins are dependent, this is a subtle question.

To approach the problem of understanding the dependence on the temperature, we will first study the fluctuations of M_Λ in two limiting situations, namely those of **infinite temperature** ($\beta = 1/T \downarrow 0$) and **zero temperature** ($\beta = 1/T \uparrow \infty$). Although these two cases are essentially trivial from a mathematical point of view, they will already provide some hints as to what might happen at other values of the temperature, in the infinite-volume Ising model. For the sake of concreteness, we take $\Lambda = \mathrm{B}(n)$.

Infinite Temperature. Consider the model on $\mathrm{B}(n)$ (with a fixed n). In the limit $\beta \downarrow 0$, the Gibbs distribution converges to the uniform distribution on $\Omega_{\mathrm{B}(n)}$: for each $\omega \in \Omega_{\mathrm{B}(n)}$,

$$\lim_{\beta \downarrow 0} \mu_{\mathrm{B}(n);\beta;0}(\omega) = \mu_{\mathrm{B}(n);0,0}(\omega) = \frac{1}{|\Omega_{\mathrm{B}(n)}|}\,. \tag{1.46}$$

Therefore, after $\beta \downarrow 0$, $M_{\mathrm{B}(n)}$ is a sum of independent and identically distributed random variables. Its behavior in regions of increasing sizes can thus be described using the classical limit theorems of probability theory. For instance, the Law of Large Numbers implies that, for all $\epsilon > 0$,

$$\mu_{\mathrm{B}(n);0,0}\Big(\frac{M_{\mathrm{B}(n)}}{|\mathrm{B}(n)|} \notin [-\epsilon, \epsilon]\Big) \longrightarrow 0 \quad \text{as } n \to \infty\,. \tag{1.47}$$

Looking at a finer scale, the Central Limit Theorem states that, for all $a < b$,

$$\mu_{\mathrm{B}(n);0,0}\left(\frac{a}{\sqrt{|\mathrm{B}(n)|}} \leq \frac{M_{\mathrm{B}(n)}}{|\mathrm{B}(n)|} \leq \frac{b}{\sqrt{|\mathrm{B}(n)|}}\right) \longrightarrow \frac{1}{\sqrt{2\pi}} \int_a^b e^{-x^2/2}\,\mathrm{d}x\,. \tag{1.48}$$

Zero Temperature. In the opposite regime, in which $\beta \uparrow \infty$ in a fixed box $\mathrm{B}(n)$, the distribution $\mu_{\mathrm{B}(n);\beta,0}$ concentrates on those configurations that minimize the Hamiltonian, the so-called **ground states**. It is easy to check that the Ising model in $\mathrm{B}(n)$ has exactly two ground states: the constant configurations $\eta^+, \eta^- \in \Omega_{\mathrm{B}(n)}$, defined by

$$\eta_i^+ \stackrel{\text{def}}{=} +1 \quad \forall i \in \mathrm{B}(n)\,, \qquad \eta_i^- \stackrel{\text{def}}{=} -1 \quad \forall i \in \mathrm{B}(n)\,.$$

For any configuration ω different from η^+ and η^-, there exists at least one pair $\{i,j\}$ of nearest neighbors in $\mathrm{B}(n)$ such that $\omega_i \neq \omega_j$. Therefore,

$$\mathscr{H}_{\mathrm{B}(n)}(\omega) - \mathscr{H}_{\mathrm{B}(n)}(\eta^\pm) = \sum_{\substack{i,j \in \mathrm{B}(n) \\ i \sim j}} (1 - \omega_i\omega_j) \geq 2\,. \tag{1.49}$$

Consequently,

$$\frac{\mu_{B(n);\beta,0}(\omega)}{\mu_{B(n);\beta,0}(\eta^{\pm})} = \frac{e^{-\beta \mathscr{H}_{B(n)}(\omega)}}{e^{-\beta \mathscr{H}_{B(n)}(\eta^{\pm})}} \leq e^{-2\beta} \to 0 \quad \text{as } \beta \uparrow \infty .$$

Since $\mu_{B(n);\beta,0}(\eta^-) = \mu_{B(n);\beta,0}(\eta^+)$, we thus get

$$\lim_{\beta \uparrow \infty} \mu_{B(n);\beta,0}(\omega) = \begin{cases} \frac{1}{2} & \text{if } \omega \in \{\eta^+, \eta^-\}, \\ 0 & \text{otherwise,} \end{cases} \tag{1.50}$$

which means that, in the limit of very low temperatures, the Gibbs distribution "freezes" the system in either of the ground states.

The two very different behaviors observed above in the limits $\beta \uparrow \infty$ and $\beta \downarrow 0$ suggest two possible scenarios for the high- and low-temperature behavior of the Ising model in a large box $B(n)$:

1. When β is small (high temperature), the global magnetization density is close to zero: with high probability,
$$\frac{M_{B(n)}}{|B(n)|} \cong 0 .$$
In this scenario, in a typical configuration, the fractions of + and − spins are essentially equal.
2. When β is large (low temperature), $\mu_{B(n);\beta,0}$ concentrates on configurations that mostly coincide with the ground states η^+, η^-. In particular, with high probability,
$$\text{either} \quad \frac{M_{B(n)}}{|B(n)|} \cong +1 , \quad \text{or} \quad \frac{M_{B(n)}}{|B(n)|} \cong -1 .$$
In this scenario, **spontaneous magnetization/global order** is observed, since a majority of spins have the same sign. Observe that the Law of Large Numbers would not hold in such a regime. Namely, each spin has an average value equal to zero and, nevertheless, the observation of the system as a whole shows that $|B(n)|^{-1} \sum_{i \in B(n)} \omega_i$ is not close to 0. The symmetry under a global spin flip is **spontaneously broken**, in the sense that typical configurations favor one of the two types of spins, even though the Gibbs distribution is completely neutral with respect to both species of spins.

The main problem is to determine which of these behaviors (if any) gives the correct description of the system for intermediate values $0 < \beta < \infty$.

From the physical point of view, the question we will be most interested in is to determine whether the global alignment of the spins observed at $\beta = \infty$ survives, in arbitrarily large systems, for large but finite values of β. This is a delicate question, since the argument given above actually consisted of fixing n and observing that the ground states were dominant when $\beta \uparrow \infty$. But the true limiting procedure we are interested in is to take the thermodynamic limit at fixed temperature, that is, to fix β (possibly very large) and *then* let $n \to \infty$. It turns out that, in this limit, the ground states are in fact very *unlikely* to be observed. Indeed, let us denote by

$\Omega^k_{\mathrm{B}(n)}$ the set of configurations coinciding everywhere with either η^+ or η^-, except at exactly k vertices where the spins disagree with the ground state. Such local deformations away from a ground state are often called **excitations**. Then, for each $\omega \in \Omega^k_{\mathrm{B}(n)}$,

$$\mathscr{H}_{\mathrm{B}(n);0}(\omega) - \mathscr{H}_{\mathrm{B}(n);0}(\eta^\pm) \le 4dk\,.$$

(This bound is saturated when none of these k misaligned spins are located at neighboring vertices and none are located along the boundary of $\mathrm{B}(n)$.) Observe that $|\Omega^k_{\mathrm{B}(n)}| = \binom{|\mathrm{B}(n)|}{k}$ and that k is always at most equal to $|\mathrm{B}(n)|/2$. This means that, for any $k \ge 1$,

$$\begin{aligned}\frac{\mu_{\mathrm{B}(n);\beta,0}(\Omega^k_{\mathrm{B}(n)})}{\mu_{\mathrm{B}(n);\beta,0}(\eta^\pm)} &= \sum_{\omega\in\Omega^k_{\mathrm{B}(n)}} e^{-\beta(\mathscr{H}_{\mathrm{B}(n);0}(\omega)-\mathscr{H}_{\mathrm{B}(n);0}(\eta^\pm))}\\ &\ge \binom{|\mathrm{B}(n)|}{k} e^{-4d\beta k} \ge \frac{1}{k!}\left(\tfrac{1}{2}|\mathrm{B}(n)|e^{-4d\beta}\right)^k \gg 1\,,\end{aligned}$$

for all n large enough (at fixed β). In other words, even at very low temperature, it is always much more likely to have misaligned spins in large regions. This discussion shows that there are two competing aspects when analyzing typical configurations under a Gibbs distribution at low temperature. On the one hand, configurations with low energy are favored, since the latter have a larger individual probability; this is the *energy* part. On the other hand, the number of configurations with a given number of excitations grows fast with the size of the system and rapidly outnumbers the small number of ground states; this is the *entropy* part. This *competition between energy and entropy* is at the heart of many phenomena described by equilibrium statistical mechanics (note that it can already be witnessed in (1.37)), in particular in methods to prove the existence of a phase transition.

These questions will be investigated in detail in Chapter 3. As we will see, the *dimension* of the underlying graph $\mathbb{Z}^d$ will play a central role in the analysis. In the next section, we discuss this dependence on d on the basis of numerical simulations.

Behavior on $\mathbb{Z}^d$

One-Dimensional Model. The following figure shows simulations of typical configurations of the one-dimensional Ising model on $\mathrm{B}(50)$, for increasing values of the inverse temperature β (at $h = 0$). For the sake of clarity, $+$ and $-$ spins are represented by black white dots respectively:

As the value of β increases, we see that spins tend to agree over ever larger regions; locally, a configuration looks like either of the ground states η^+, η^-. Increasing β even more would yield, with high probability, a configuration with all spins equal.

Nevertheless, for any value of β, one observes that, taking the system's size sufficiently large, regions of $+$ and $-$ spins even out and the global magnetization is always zero on the macroscopic scale.

As seen before, global order can be conveniently quantified using $\frac{M_{\mathrm{B}(n)}}{|\mathrm{B}(n)|}$. Since the latter has an expectation value equal to zero (Exercise 1.13), we consider the expectation of its absolute value.

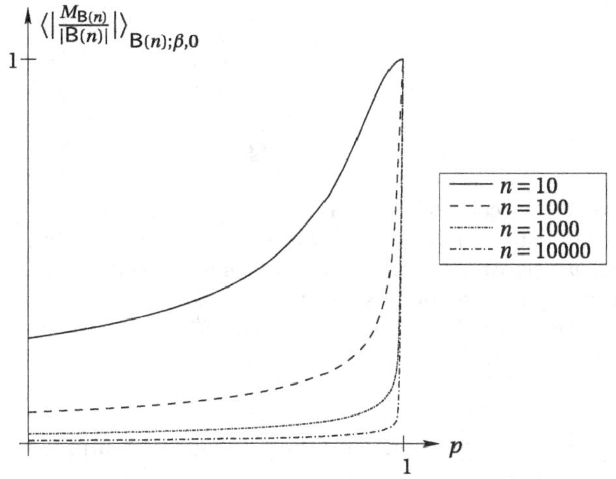

Figure 1.8. The expected value of the absolute value of the magnetization density, as a function of $p = 1 - e^{-2\beta}$, for the one-dimensional Ising model, tending to zero as $n \to \infty$ for all $p \in [0, 1)$.

This reflects the fact that in $d = 1$ the Ising model exhibits paramagnetic behavior at all positive temperatures. See the discussion below.

Model in Dimensions $d \geq 2$. In contrast to its one-dimensional version, the Ising model in higher dimensions exhibits ferromagnetic and paramagnetic behaviors, as the temperature crosses a critical value, similarly to what Curie observed in real magnets.

The phase transition is characterized by two distinct regimes (low and high temperatures), in which the large-scale behavior of the system presents important differences which become sharper as the size of the system increases. A few simulations will reveal these behaviors in $d = 2$. Consider first the model without a magnetic field ($h = 0$), in a square box $\mathrm{B}(n)$. A few typical configurations for $n = 100$ are shown in Figure 1.9, for various values of the inverse temperature $\beta \geq 0$. For convenience, instead of varying β, we vary $p = p(\beta) \stackrel{\text{def}}{=} 1 - e^{-2\beta}$, which has the advantage of taking values in $[0, 1)$. Values of p near 0 thus correspond to high temperatures, while values of p near 1 correspond to low temperatures.

Figure 1.9 shows that, in contrast to what we observed in the one-dimensional case, the large-scale behavior of the system in two dimensions depends strongly on the temperature. When p is small, the symmetry under global spin flip is preserved

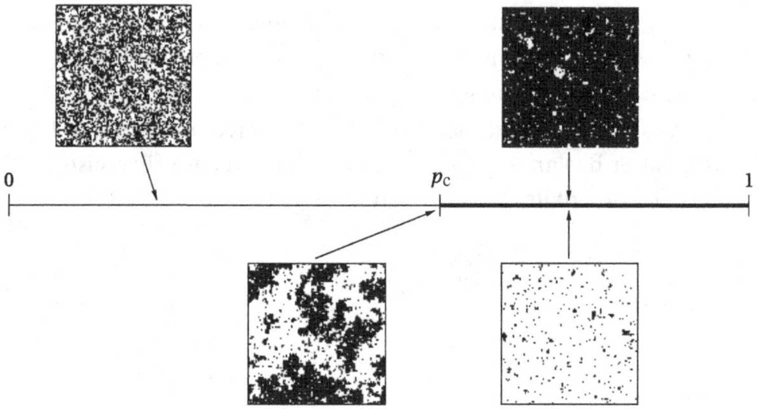

Figure 1.9. Typical configurations of the two-dimensional Ising model in the box B(100), for different values of $p = 1 - e^{-2\beta}$. Black dots represent $+$ spins, white dots represent $-$ spins. When p is close to 0 (β small), the spins behave roughly as if they were independent and thus appear in equal proportions. When p is close to 1 (β large), a typical configuration is a small perturbation of either of the ground states η^+, η^-; in particular, it has a magnetization near either $+1$ or -1.

in typical configurations and the fractions of $+$ and $-$ spins are essentially equal. When p is close to 1, this symmetry is spontaneously broken: one of the two spin types dominates the other. The simulations suggest that this change of behavior occurs when p is near 0.58, that is, when β is near 0.43. Therefore, the high- and low-temperature behaviors conjectured on the basis of the limiting cases $\beta = 0$ and $\beta \uparrow \infty$ are indeed observed, at least for a system in B(100). In Figure 1.10, $\langle|\frac{M_{B(n)}}{B(n)}|\rangle_{B(n);\beta,0}$ is represented as a function of p, for different values of n.

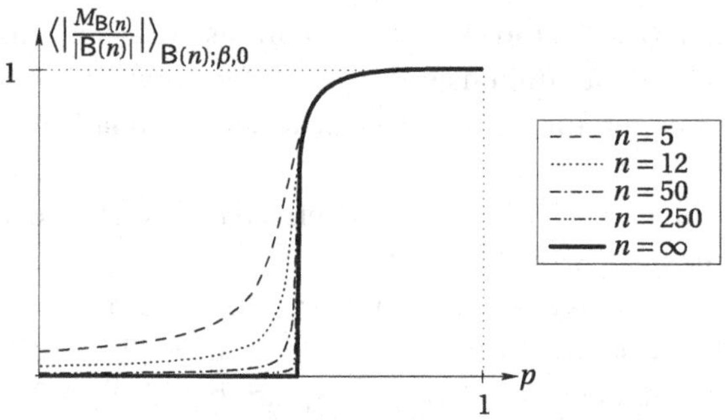

Figure 1.10. The expected value of the absolute value of the magnetization density, as a function of p, for the two-dimensional Ising model.

The simulations suggest that, as n increases, the sequence of functions $p \mapsto \langle|\frac{M_{B(n)}}{B(n)}|\rangle_{B(n);p}$ converges to some limiting curve. This is indeed the case and the limiting function can be computed explicitly:[9]

$$p \mapsto m_p^* \stackrel{\text{def}}{=} \begin{cases} 0 & \text{if } p < p_c\,, \\ \left[1 - \left(\frac{2(1-p)}{p(2-p)}\right)^4\right]^{1/8} & \text{if } p \geq p_c\,, \end{cases} \tag{1.51}$$

where

$$p_c \stackrel{\text{def}}{=} \frac{\sqrt{2}}{1+\sqrt{2}} \cong 0.586$$

is the critical point, to which corresponds the critical inverse temperature

$$\beta_c \stackrel{\text{def}}{=} -\tfrac{1}{2}\log(1-p_c) \cong 0.441\,.$$

The above explicit expression implies, in particular, that the limiting magnetization density is continuous (but not differentiable) at p_c.

Exercise 1.14. Using (1.51), show that the behavior of $m^*_{p(\beta)}$ as $\beta \downarrow \beta_c$ is

$$m^*_{p(\beta)} \sim (\beta - \beta_c)^{1/8}\,,$$

in the sense that $\lim_{\beta\downarrow\beta_c} \frac{\log m^*_{p(\beta)}}{\log(\beta-\beta_c)} = \frac{1}{8}$.

Concerning the dependence of the Ising model on the magnetic field when $h \neq 0$, two main quantities of interest will be considered: the average magnetization density $\left\langle \frac{M_{B(n)}}{|B(n)|} \right\rangle_{B(n);\beta,h}$ and the **pressure**

$$\psi_{B(n)}(\beta, h) \stackrel{\text{def}}{=} \frac{1}{\beta|B(n)|} \log \mathbf{Z}_{B(n);\beta,h}\,.$$

Remark 1.15. The reader might wonder why the term *pressure* is used for the magnet. In fact, one can establish a one-to-one correspondence between the microstates of the lattice gas and those of the Ising model, by

$$\omega_i \leftrightarrow 2\eta_i - 1\,.$$

In the Ising model, the number of spins is of course fixed and equal to the size of the region on which it is defined. But the number of $+$ (or $-$) spins is not fixed and can vary. Therefore, the number of particles in the lattice gas, under the above correspondence, can also vary; it thus corresponds to a grand canonical description, in which the natural thermodynamic potential is the pressure.

The relation between the lattice gas and the Ising model will be fully described, and exploited, in Chapter 4. ◇

The following **infinite-volume limits** will be considered:

$$m(\beta, h) \stackrel{\text{def}}{=} \lim_{n\to\infty} \left\langle \frac{M_{B(n)}}{|B(n)|} \right\rangle_{B(n);\beta,h}, \qquad \psi(\beta, h) \stackrel{\text{def}}{=} \lim_{n\to\infty} \psi_{B(n)}(\beta, h)\,.$$

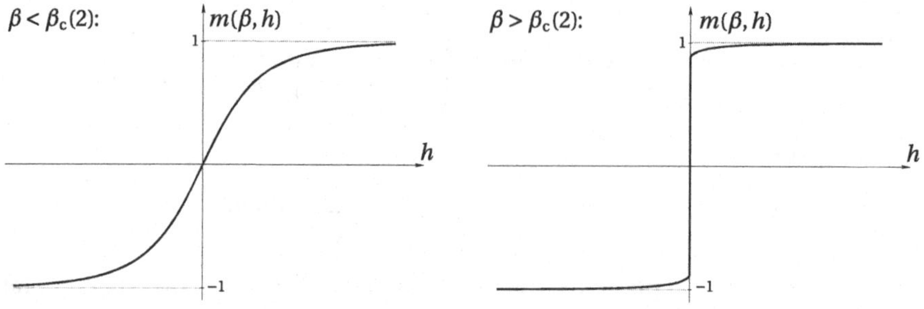

Figure 1.11. Dependence of the magnetization density of the two-dimensional Ising model on the magnetic field (obtained from numerical simulations). Left: paramagnetic behavior at high temperature; Right: ferromagnetic behavior at low temperature.

Besides showing that the above limits exist, we will show that, for all $h \neq 0$,

$$\frac{\partial \psi(\beta, h)}{\partial h} = m(\beta, h).$$

The map $h \mapsto m(\beta, h)$ is plotted in Figure 1.11 for sub- and supercritical temperatures.

The presence of two different typical behaviors when $h = 0$ and $\beta > \beta_c$ shows how sensitive the system becomes to perturbation by an external field. On the one hand, when $\beta < \beta_c$, a small magnetic field $h > 0$ induces a positive magnetization density which is approximately proportional to h: the response of the system is linear for small h and vanishes when $h \to 0$. On the other hand, when $\beta > \beta_c$, the introduction of an infinitesimal magnetic field $h > 0$ (resp. $h < 0$) induces a magnetization density close to $+1$ (resp. -1)! This implies that, in contrast to the one-dimensional case, the pressure is not differentiable at $h = 0$: the phase transition is of first order in the magnetic field. Informally, one can say that, in the absence of magnetic field, the system "hesitates" between two different behaviors and the introduction of a nonzero, arbitrarily small magnetic field is enough to tip the balance in the corresponding direction.

1.5 Some General Remarks

1.5.1 Role of the Thermodynamic Limit

In Section 1.3.1, the thermodynamic limit has been introduced as a way of establishing a precise link between statistical mechanics and thermodynamics.

Approximating a large system by an infinite one might seem a rather radical step, since real systems are always finite (albeit quite large: a cube of iron with a sidelength of 1 cm contains roughly 10^{23} iron atoms).

It turns out that taking a limit of infinite volume is important for other reasons as well.

Deterministic Macroscopic Behavior. As we have seen, one of the main assumptions in thermodynamics is that, once a small set of thermodynamic quantities has been fixed (say, the pressure and the temperature for an ideal gas), the values of all other macroscopic quantities are in general completely determined. In statistical mechanics, macroscopic observables associated with large finite systems are random variables which are only approximately determined: they still undergo fluctuations, although the latter decrease with the system's size. As we will see in Chapter 6, it is only in the thermodynamic limit that all macroscopic observables take on deterministic values (actually, we already saw concentration of some observables: the density of particles in a subsystem for the hard-core lattice gas in the microcanonical ensemble, the energy density in the canonical ensemble, etc.).

The emergence of deterministic behavior should be reminiscent of certain central results in mathematics, such as the Law of Large Numbers or the Ergodic Theorem.

Equivalence of Ensembles. On the one hand, we have seen that, in thermodynamics, many choices are possible for the thermodynamic parameters used to describe a system. In the case of a gas, for example, one might use (U, V, N), (T, V, N), etc. If the values are suitably chosen, all these approaches lead to the same predictions for the equilibrium properties (except, possibly, at phase transitions).

On the other hand, we have seen that, in statistical mechanics, to each particular set of thermodynamic parameters corresponds an ensemble (microcanonical, canonical, grand canonical), that is, a particular probability distribution on the set of microstates. Obviously, the latter do not coincide for finite systems. It turns out that they indeed become equivalent, in general, but *only* once the thermodynamic limit has been taken: in this limit, the local behavior of the system in different ensembles generally coincides, provided that the thermodynamic parameters are chosen appropriately. In this limit, one says that there is *equivalence of ensembles.* Although equivalence of ensembles will not be described in full generality, we will come back to it in Sections 4.4, 4.7.1 and 6.14.1.

Phase Transitions. One additional major reason to consider infinite-volume limits is that it is the only way the formalism of equilibrium statistical mechanics can lead to the singular behaviors thermodynamics associates with phase transitions, such as the coexistence plateau in the liquid–vapor equilibrium, or the discontinuity of the magnetization in a ferromagnet.

Notice that the dependence of a finite system on its parameters is *always* smooth. Consider, for example, the Ising model in $\mathrm{B}(n)$. From an algebraic point of view, its partition function can be written (up to an irrelevant smooth prefactor) as a polynomial in the variables $e^{-2\beta}$ and e^{-2h}, with nonnegative (real) coefficients. It follows that the pressure $\psi_{\mathrm{B}(n)}(\beta, h)$ is real-analytic for all values of β and h. Of course, the same is true of the magnetization in $\mathrm{B}(n)$. An analytic singularity, such as a discontinuity of the magnetization when going from $h > 0$ to $h < 0$, can *only* occur if the thermodynamic limit is taken.

In view of the above, one might wonder how this can be compatible with our everyday experience of various types of phase transitions. The crucial point is that, although finite-volume thermodynamic quantities are always smooth, in very large systems their behavior will be closely approximated by the singular behavior of the corresponding infinite-volume quantities. This was already witnessed in Figure 1.10, in which the finite-volume magnetization of a system in a box as small as B(250) already displays a near-singular behavior. For real macroscopic systems, the behavior will be experimentally indistinguishable from a genuine singularity.

Genuine Long-Range Order vs. Apparent Long-Range Order. In our discussion of the one-dimensional Ising model, we mentioned that, for a box of arbitrary size, with probability close to 1, configurations of this model will be perfectly ordered (all spins being equal) as soon as the temperature is low enough. Nevertheless, we will prove in Chapter 3 that the infinite one-dimensional Ising model is disordered at all positive temperatures. This shows that looking at finite systems might lead us to "wrong" conclusions. Of course, real systems are finite, so a "real" one-dimensional Ising model would typically display order. But this ordering would be a finite-size effect only. Being able to distinguish between such effects and genuine ordering is essential to obtain a conceptual understanding of these issues (for example, the role of the dimension). An important example will be discussed in Chapter 9.

This short discussion shows that, from the point of view of statistical mechanics, thermodynamics is only an approximate theory dealing (very effectively!) with idealized systems of infinite size. In order to recover predictions from the latter in the framework of the former, it is thus necessary to take the limit of infinite systems. Of course, once a system is well understood in the thermodynamic limit, it can be of great interest to go beyond thermodynamics, by analyzing the finite-volume corrections provided by equilibrium statistical mechanics.

1.5.2 On the Role of Simple Models

The lattice gas and the Ising model share an obvious feature: they are extremely crude models of the systems they are supposed to describe. In the case of the lattice gas, the restriction of the particles to discrete positions is a dramatic simplification and the interaction only keeps very superficial resemblance with the interactions between particles of a real gas. Similarly, in a real magnet, the mechanism responsible for the alignment of two spins is of a purely quantum mechanical nature, which the Ising model simply ignores; moreover, the restriction of the spin to one direction is also not satisfied in most real ferromagnets.

One may thus wonder about the purpose of studying such rough approximations of real systems. This was indeed a major preoccupation of physicists in the early twentieth century, who believed that such models might be of interest to mathematicians, but are certainly irrelevant to physics.[10]

Nevertheless, the point of view on the role of models and on the actual goal of theoretical physics changed substantially at that time. In the realm of statistical mechanics, the mathematical analysis of realistic models of physical systems is in general of such a degree of complexity as to be essentially hopeless. As a consequence, one must renounce obtaining, in general, a complete *quantitatively precise* description of most phenomena (for example, computing precisely the critical temperature of a real magnet). However, it is still possible and just as important to try to understand complex phenomena *at a qualitative level*: What are the mechanisms underlying some particular phenomenon? What are the relevant features of the real system that are responsible for its occurrence? For this, simple models are invaluable. [11] We will see all through this book that many subtle phenomena can be reproduced qualitatively in such models, without making any further uncontrolled approximation.

One additional ingredient that played a key role in this change of perspective is the realization that, in the vicinity of a critical point, the behavior of a system becomes essentially independent of its microscopic details, a phenomenon called *universality*. Therefore, in such a regime, choosing a simple model as the representative of the very large class of systems (including the real ones) that share the same behavior allows one to obtain even a *quantitative* understanding of these real systems near the critical point.

Finally, one nice side-effect of considering very simple descriptions is that they often admit many different interpretations. Already in the 1930s, the Ising model was used as a model of a ferromagnet, of a fluid and of a binary alloy, and to model an adsorbed monolayer at a surface. The fact that the same model qualitatively describes a wide variety of different systems clarifies the observations made at the time that these very different physical systems exhibit very similar behavior.

1.6 About This Book

We wrote this book because we believe that there does not yet exist, in the literature, a book that is self-contained, starts at an elementary level, and yet provides a detailed analysis of some of the main ideas, techniques and models of the field.

The target reader we have in mind is an advanced undergraduate or graduate student in mathematics or physics, or anybody with an interest in learning more about some central concepts and results in rigorous statistical mechanics.

Let us list some of the main characteristic features of this book.

- *It is mostly self-contained.* It is only assumed that the reader has basic notions of analysis and probability (only Chapter 6 requires notions from measure theory, and these are summarized in Appendix B).
- *It discusses only the equilibrium statistical mechanics of classical lattice systems.* Other aspects of statistical mechanics, not treated here, can be found in the books listed in Section 1.6.2 below.

- *It favors the discussion of specific enlightening examples over generality.* In each chapter, the focus is on a small class of models that we consider to be the best representatives of the topic discussed. These are listed right below in Section 1.6.1.
- *It aims at conveying understanding and not only proofs.* In particular, the proofs given are not always the shortest, most elegant ones, but those we think best help to understand the underlying mechanisms. Moreover, the methods, ideas and concepts introduced in the course of the proof of a statement are often as important as the statement itself.

1.6.1 Contents, Chapter by Chapter

The first chapters are devoted mainly to the study of models whose spin variables are discrete and take values in a finite set:

- **Chapter 2: The Curie–Weiss Model.** Mean-field models play a useful role, both from the physical and mathematical points of view, as first approximation to more realistic ones. This chapter gives a detailed account of the Curie–Weiss model, which can be seen as the mean-field version of the Ising model. The advantage is that this model exhibits a phase transition between paramagnetic and ferromagnetic behaviors that can be described with elementary tools.
- **Chapter 3: The Ising Model.** As we already said, the Ising model is the simplest "realistic" model which exhibits a nontrivial collective behavior. As such, it has played, and continues to play, a central role in statistical mechanics. This chapter uses it to introduce several very important notions for the first time, such as the notion of infinite-volume state or precise definitions of phase transitions. Then, the complete phase diagram of the model is constructed, in all dimensions, using simple mathematical tools developed from scratch.
- **Chapter 4: Liquid–Vapor Equilibrium.** Historically, the liquid–vapor equilibrium played a central role in the first theoretical studies of phase transitions. In this chapter, the mathematical description of the lattice gas is given in detail, as well as its mean-field and nearest-neighbor (Ising) versions. The mean-field (Kać) limit is also studied in a simple case, providing a rigorous justification of the van der Waals–Maxwell theory of condensation.
- **Chapter 5: Cluster Expansion.** The cluster expansion remains the most important perturbative technique in mathematical statistical mechanics. It is presented in a simple fashion and several applications to the Ising model and the lattice gas are presented. It is also used several times later in the book and plays, in particular, a central role in the implementation of the Pirogov–Sinai theory of Chapter 7.
- **Chapter 6: Infinite-Volume Gibbs Measures.** In this chapter, we present a probabilistic description of infinite systems of particles at equilibrium, nowadays known as the theory of Gibbs measures or the DLR (Dobrushin–Lanford–Ruelle) formalism. This theory is developed from scratch, using the Ising model as a guiding example. Several important aspects, such as Dobrushin's Uniqueness

Theorem, spontaneous symmetry breaking, extremal measures and the extremal decomposition, are also described in detail. At the end of the chapter, the variational principle is introduced; this is closely linked with the basic concepts of equilibrium thermodynamics.
- **Chapter 7: Pirogov–Sinai Theory.** The Pirogov–Sinai theory is one of the very few general approaches to the rigorous study of first-order phase transitions. It yields, under weak assumptions, a sharp description of such phase transitions in perturbative regimes. This theory is first introduced in a rather general setting and then implemented in detail on one specific three-phase model: the Blume–Capel model.

The last three chapters are devoted to models whose variables are of a continuous nature:

- **Chapter 8: The Gaussian Free Field.** In this chapter, the lattice version of the Gaussian Free Field is analyzed. Several features related to the noncompactness of its single-spin space are discussed, exploiting the Gaussian nature of the model. The model has a random walk representation, whose recurrence properties are crucial in the study of the behavior of the model in the thermodynamic limit.
- **Chapter 9: Models with Continuous Symmetry.** An important class of models with a continuous symmetry, including the *XY* and Heisenberg models, is studied in this chapter. The emphasis is on the implications of the presence of the continuous symmetry on long-range order in these models in low dimensions. In particular, a strong form of the celebrated Mermin–Wagner Theorem is proved in a simple way.
- **Chapter 10: Reflection Positivity.** Reflection positivity is another tool that plays a central role in the rigorous study of phase transitions. We first explain it in detail, proving its two central estimates: the infrared bound and the chessboard estimate. We then apply the latter to obtain several results of importance. In particular, we prove the existence of a phase transition in the anisotropic *XY* model in dimensions $d \geq 2$, as well as in the (isotropic) $O(N)$ model in dimensions $d \geq 3$. Combined with the results of Chapter 9, this provides a detailed description of this type of system in the thermodynamic limit.

In order to facilitate the reading of the content of each chapter, which can sometimes be pretty technical, the bibliographical references have been placed at the end of the chapter, in a section called **Bibliographical References**. Some chapters also contain a section **Complements and Further Reading**, in which the interested reader can find further results (usually without proofs) and suggestions for further reading. The goal of these complements is to provide information about some more advanced themes that cannot be treated in detail in the book.

The book ends with three appendices:

- **Appendix A: Notes.** This appendix regroups short **Notes** that are referred to in the text.

- **Appendix B: Mathematical Appendices.** Since we want the book to be mostly self-contained, we introduce various mathematical topics used throughout the book, which might not be part of all undergraduate curricula. For example: elementary properties of convex functions, some aspects of complex analysis, measure theory, conditional expectation, random walks, etc., are briefly introduced, not always in a self-contained manner, often without proofs, but with references to the literature.
- **Appendix C: Solutions to Exercises.** Exercises appear in each of the chapters, with various levels of difficulty. Hints or solutions for most of them can be found in this appendix.

We would like to emphasize that Chapter 3 plays a central role, since it introduces several important concepts that are then used constantly in the rest of the book; it should be considered as a priority for a novice reader. The only other true constraint is that Chapter 5 should be read before Chapter 7. Besides that, the chapters can mostly be read independently of each other, and any path following the arrows in the picture below represents a possible way through the book:

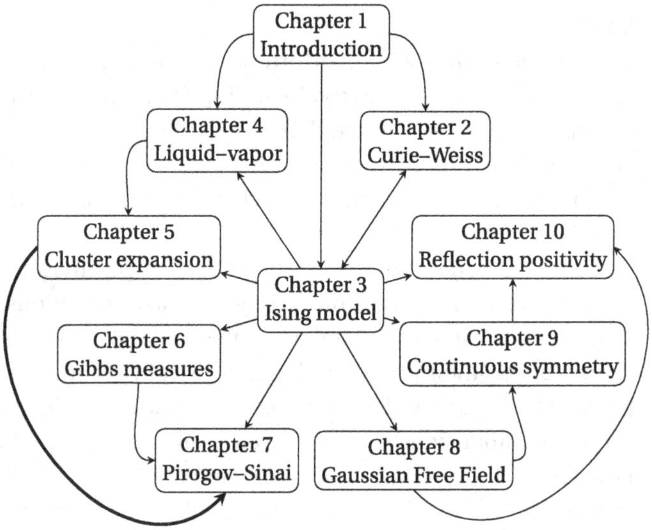

Warning: As stated previously, we have strived to make the book as self-contained as possible, and accessible to readers with little to no prior knowledge. Moreover, we have tried to make the chapters independent from each other, respecting whenever possible the conventional notations used in the field. This has had the following consequences on the final form of the book.

- Together with the fact that we have avoided developing too general a theory, writing essentially independent chapters has had the inevitable consequence of introducing various repetitions: the partition function of a model, for instance, or its Gibbs distribution in finite volume, is always defined in a way suited for the particular analysis used for that model. The same holds for the pressure and other recurring quantities. We therefore warn the reader that corresponding notions might be written slightly differently from one chapter to another.

- As in many areas, the notational conventions in statistical mechanics are different in the mathematical and physical communities. For example, probabilists define the free energy as $\frac{1}{V}\log \mathbf{Z}$ whereas in physics it is written as $-\frac{1}{\beta V}\log \mathbf{Z}$, respecting the structure that appeared in the analogies with thermodynamics.

  In this book, we have adopted one convention or the other, depending on the physical relevance of the theory developed in the chapter. Chapter 4, for example, was a natural place in which to use the physicists' conventions, since it describes the liquid–vapor equilibrium.

  The choices made are always indicated at the beginning of the chapters and we hope that this will not generate too much confusion when jumping from one chapter to another.

1.6.2 The Existing Literature

This book does not aim to present the most recent developments in statistical mechanics. Rather, it presents a set of models and methods, most of which were already known in the 1980s. However, these classical topics form the backbone of this subject and should still be learned by new researchers entering this field. The absence of an introductory text aimed at beginners was deplored by many colleagues and prompted us to write this book.

Statistical mechanics is now such a wide field that it has become impossible to cover more than a fraction of it in one book. In this section, we provide some references to other works covering the various aspects that are discussed either not at all in the present text or only very superficially. Note that we mostly restrict ourselves to books aimed at mathematicians and mathematical physicists.

Books Covering Similar Areas. There exist several books covering some of the areas discussed in the present text. Although the distinction is a bit subjective, we split the list into two, according to what we consider to be the intended audience.

The first set of books is aimed at mathematical physicists. Ruelle wrote the first book [289] on rigorous equilibrium statistical mechanics in 1969. Discussing both classical and quantum systems, in the continuum and on the lattice, this book played a major role in the development of this field. Israel's book [176] provides an in-depth discussion of the variational principle, Gibbs measures as tangent functionals, and the role of convexity in equilibrium statistical mechanics. It contains many abstract results found nowhere else in book form. Sinai's book [312] discusses the general theory of Gibbs measures on a lattice with an emphasis on phase transitions and includes perturbative expansions, the Pirogov–Sinai theory, as well as a short introduction to the renormalization group (mostly in the context of hierarchical lattices). Minlos' short book [247] covers similar ground. The book [227] by Malyshev and Minlos deals with more or less the same topics, but with an approach based systematically on the cluster expansion. Simon's book [308] provides an extensive discussion of the pressure, Gibbs states and their basic properties, and perturbative expansions, for both classical and quantum lattice systems. Presutti's book [279] proposes an alternative approach to several of the topics covered in the present book, but with a strong emphasis on models with Kać interactions. Lavis' book [207] provides coverage of a wide class of models and techniques.

The second set of books is aimed at probabilists. For this audience, Georgii's remarkable book [134] has become the standard reference for the theory of Gibbs measures; although less accessible than the present text, it is highly recommended to more advanced readers interested in very general results, in particular on the topics covered in our Chapter 6. The shorter book [282] by Prum covers similar ground, but in less generality. Preston's book [278] contains an interesting early account of Gibbs measures, aimed at professional probabilists and limited to rather abstract general results. Kindermann and Snell's very pedagogical monograph [192] includes a clear and intuitive exposition of phase transition in the two-dimensional Ising model.

Disordered Systems. One of the important topics in equilibrium statistical mechanics which is not even touched upon in the present book is disordered systems, in which the Gibbs measures considered depend on additional randomness (such as random interactions). In spite of the activity in this domain, there are only a limited number of books available for mathematically inclined readers. The book [254] by Newman discusses short-range models for spin glasses; see also [321]. Talagrand's books [325, 326, 327] provide a comprehensive account of mean-field models for spin glasses. Bovier's book [37] starts with an introduction to equilibrium statistical mechanics (including a discussion of the DLR formalism and cluster expansion) and then moves on to discuss both mean-field and lattice models of disordered systems.

Large Deviations. Large deviations theory plays an important role in equilibrium statistical mechanics, both at a technical level and at a conceptual level, providing the natural framework to relate thermodynamics and statistical mechanics. This theme will be recurrent in the present text. Nevertheless, we do not develop the general framework here. There are now many books on large deviation theory, with various levels of emphasis on the applications to statistical mechanics, such as the books by Deuschel and Stroock [77], Dembo and Zeitouni [74], den Hollander [75], Ellis [100], Rassoul-Agha and Seppäläinen [283] and Olivieri and Vares [258], as well as the lecture notes by Lanford [205], Föllmer [108] and Pfister [274]. Georgii's book also has a section on this topic [134, Section 15.5]. Let us also mention the more elementary introduction by Touchette [334].

Quantum Systems. In this book, we consider only classical lattice spin systems. A discussion of quantum lattice spin systems can be found, for example, in the books by Sewell [300], Simon [308] and Bratteli and Robinson [43, 44].

Historical Aspects. Except in some remarks, we do not discuss historical aspects in this book. Good references on the general history of statistical mechanics are the books by Brush [54] and Cercignani [63]; Gallavotti's treatise [130] also provides interesting information on this subject. More specific references to historical aspects of lattice spin systems are given in the articles by Brush [55], Domb [89] and Niss [255, 256, 257].

Percolation. Bernoulli percolation is a central model in probability theory, with strong links to equilibrium statistical mechanics. These links (which we only superficially address in Section 3.10.6) lead to an alternative approach to the analysis of some lattice spin systems (such as the Ising and Potts models), reinterpreting the phase transition as a percolation transition. The percolation model is discussed in detail in the books by Kesten [189], Grimmett [149, 151] and by Bollobás and Riordan [31]. The link with the Ising and Potts models is explained in the books by Grimmett [150] and Werner [350], in the review paper [132] by Georgii, Häggström and Maes and in the lecture notes by Duminil-Copin [91].

Thermodynamic Formalism. Some core ideas from equilibrium statistical mechanics have been successfully imported into the theory of dynamical systems, where it is usually known as the thermodynamic formalism. An excellent early reference is Bowen's book [40]. Other references are the books by Ruelle [291] and Keller [187], or the lecture notes by Sarig [293].

Stochastic Dynamics. An area that is closely related to several problems studied in this book is the analysis of the stochastic dynamics of lattice spin systems. In that, one considers Markov chains on set Ω of microscopic configurations, under which the Gibbs distributions are invariant. The book [225] by Liggett and the lecture notes [232] by Martinelli provide good introductions to this topic.

Critical Phenomena. This topic is one of the major omissions in this book. See the short discussion and the bibliographical references given in Section 3.10.11.

Exactly Solvable Models. A discussion of the exact (but not always necessarily rigorous) solutions of various models of statistical mechanics can be found in the books by McCoy and Wu [239], Baxter [17] or Palmer [261].

Foundations of Equilibrium Statistical Mechanics. There are several books on the foundations of statistical mechanics and its relations to thermodynamics. We refer the reader, for example, to those by Gallavotti [130], Martin-Löf [230], Khinchin [190] and Sklar [314].

2 The Curie–Weiss Model

In statistical mechanics, a *mean-field approximation* is often used to approximate a model by a simpler one, whose global behavior can be studied with the help of *explicit* computations. The information thus extracted can then be used as an indication of the kind of properties that can be expected from the original model. In addition, this approximation turns out to provide quantitatively correct results in sufficiently high dimensions.

The Ising model, which will guide us throughout the book, is a classical example of a model with rich behavior but with no explicit solution in general (the exceptions being the one-dimensional model, see Section 3.3, and the two-dimensional model when $h = 0$). In this chapter, we consider its mean-field approximation, in the form of the *Curie–Weiss model.* Although it is an over-simplification of the Ising model, the Curie–Weiss model still displays a phase transition, with distinct behavior at high and low temperature. It will also serve as an illustration of various techniques and show how the probabilistic behavior is intimately related to the analytic properties of the thermodynamic potentials (free energy and pressure) of the model.

2.1 Mean-Field Approximation

Consider a system of Ising spins living on $\mathbb{Z}^d$, described by the Ising Hamiltonian defined in (1.44). In that model, the spin ω_i located at vertex i interacts with the rest of the system via its neighbors. The contribution to the total energy coming from the interaction of ω_i with its $2d$ neighbors can be written as

$$-\beta \sum_{j:\, j\sim i} \omega_i \omega_j = -2d\beta\, \omega_i \cdot \frac{1}{2d} \sum_{j:\, j\sim i} \omega_j \,. \tag{2.1}$$

Written this way, one can interpret the contribution of ω_i to the total energy as an interaction of ω_i with a local magnetization density, produced by the average of its $2d$ nearest neighbors:

$$\frac{1}{2d} \sum_{j:\, j\sim i} \omega_j \,.$$

Of course, this magnetization density is *local* and varies from one point to another. The **mean-field approximation** assumes that each local magnetization density can be approximated by the *global* magnetization density,

$$\frac{1}{N}\sum_{j=1}^{N}\omega_j\,,$$

where N is the number of spins in the system. The mean-field approximation of the Ising model thus amounts to doing the following transformation on the Hamiltonian (1.44) (up to a multiplicative constant that will be absorbed in β):

$$\text{Replace}\quad -\beta\sum_{i\sim j}\omega_i\omega_j \quad\text{by}\quad -\frac{d\beta}{N}\sum_{i,j}\omega_i\omega_j\,.$$

The term involving the magnetic field, on the other hand, remains unchanged. This leads to the following definition.

Definition 2.1. The **Curie-Weiss Hamiltonian** for a collection of spins $\omega = (\omega_1,\dots,\omega_N)$ at inverse temperature β and with an external magnetic field h is defined by

$$\mathscr{H}^{\mathrm{CW}}_{N;\beta,h}(\omega) \stackrel{\mathrm{def}}{=} -\frac{d\beta}{N}\sum_{i,j=1}^{N}\omega_i\omega_j - h\sum_{i=1}^{N}\omega_i\,. \tag{2.2}$$

In contrast to those of the Ising model, the interactions of the Curie-Weiss model are global: each spin interacts with all other spins in the same way, and the relative positions of the spins can therefore be ignored. Actually, due to this lack of geometry, one may think of this model as defined on the **complete graph** with N vertices, which has an edge between any pair of distinct vertices:

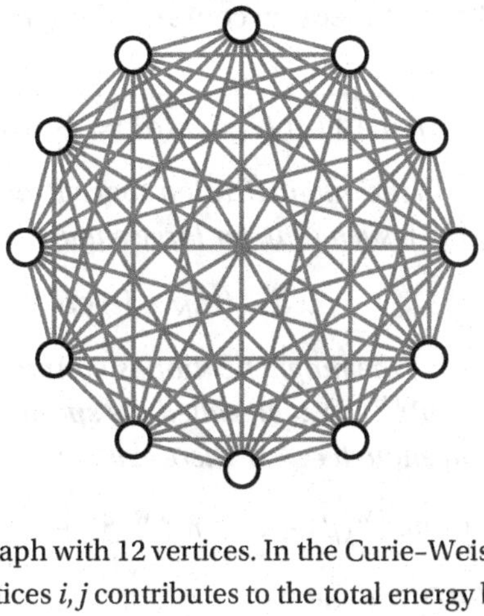

Figure 2.1. The complete graph with 12 vertices. In the Curie-Weiss model, all spins interact: a pair of spins living at vertices i, j contributes to the total energy by an amount $-\frac{d\beta}{N}\omega_i\omega_j$.

We denote by $\Omega_N \stackrel{\mathrm{def}}{=} \{\pm 1\}^N$ the set of all possible configurations of the Curie-Weiss model. The Gibbs distribution on Ω_N is written

$$\mu^{\mathrm{CW}}_{N;\beta,h}(\omega) \overset{\mathrm{def}}{=} \frac{e^{-\mathscr{H}^{\mathrm{CW}}_{N;\beta,h}(\omega)}}{Z^{\mathrm{CW}}_{N;\beta,h}}, \quad \text{where} \quad Z^{\mathrm{CW}}_{N;\beta,h} \overset{\mathrm{def}}{=} \sum_{\omega\in\Omega_N} e^{-\mathscr{H}^{\mathrm{CW}}_{N;\beta,h}(\omega)}.$$

As mentioned in the introduction, the expectation (or average) of an observable $f\colon \Omega_N \to \mathbb{R}$ under $\mu^{\mathrm{CW}}_{N;\beta,h}$ will be denoted by $\langle f\rangle^{\mathrm{CW}}_{N;\beta,h}$.

Our aim, in the rest of the chapter, is to show that the Curie–Weiss model exhibits paramagnetic behavior at high temperature and ferromagnetic behavior at low temperature.

2.2 Behavior for Large N When $h = 0$

We will first study the model in the absence of a magnetic field. The same heuristic arguments given in Section 1.4.3 for the Ising model also apply here. For instance, when $h = 0$, the Hamiltonian is invariant under the global spin flip $\omega \mapsto -\omega$ (which changes each ω_i into $-\omega_i$), which implies that the magnetization density

$$m_N \overset{\mathrm{def}}{=} \frac{M_N}{N}, \quad \text{where} \quad M_N \overset{\mathrm{def}}{=} \sum_{i=1}^{N} \omega_i,$$

has a symmetric distribution: $\mu^{\mathrm{CW}}_{N;\beta,0}(m_N = -m) = \mu^{\mathrm{CW}}_{N;\beta,0}(m_N = +m)$. In particular,

$$\langle m_N\rangle^{\mathrm{CW}}_{N;\beta,0} = 0. \tag{2.3}$$

As discussed in Section 1.4.3, we expect that the spins should essentially be independent when β is small, but that, when β is large, the most probable configurations should have most spins equal and thus be close to one of the two ground states. The following theorem confirms these predictions.

Theorem 2.2. ($h = 0$) *Let $\beta_c = \beta_c(d) \overset{\mathrm{def}}{=} \frac{1}{2d}$. Then, the following holds.*

1. *When $\beta \le \beta_c$, the magnetization concentrates at zero: for all $\epsilon > 0$, there exists $c = c(\beta,\epsilon) > 0$ such that, for large enough N,*

$$\mu^{\mathrm{CW}}_{N;\beta,0}\big(m_N \in (-\epsilon,\epsilon)\big) \ge 1 - 2e^{-cN}.$$

2. *When $\beta > \beta_c$, the magnetization is bounded away from zero. More precisely, there exists $m^{*,\mathrm{CW}}(\beta) > 0$, called the* ***spontaneous magnetization****, such that, for all small enough $\epsilon > 0$, there exists $b = b(\beta,\epsilon) > 0$ such that if*

$$J_*(\epsilon) \overset{\mathrm{def}}{=} \big(-m^{*,\mathrm{CW}}(\beta) - \epsilon, -m^{*,\mathrm{CW}}(\beta) + \epsilon\big) \cup \big(m^{*,\mathrm{CW}}(\beta) - \epsilon, m^{*,\mathrm{CW}}(\beta) + \epsilon\big),$$

then, for large enough N,

$$\mu^{\mathrm{CW}}_{N;\beta,0}\big(m_N \in J_*(\epsilon)\big) \ge 1 - 2e^{-bN}.$$

β_c is called the ***inverse critical temperature*** *or* ***inverse Curie temperature****.*

In other words, when N is large,

$$\forall \beta \le \beta_c\,, \quad m_N \simeq 0 \quad \text{with high probability,}$$

whereas

$$\forall \beta > \beta_c\,, \quad m_N \simeq \begin{cases} +m^{*,\mathrm{CW}}(\beta) & \text{with probability close to } \frac{1}{2}\,, \\ -m^{*,\mathrm{CW}}(\beta) & \text{with probability close to } \frac{1}{2}\,. \end{cases}$$

This behavior is understood easily by simply plotting the distribution of m_N:

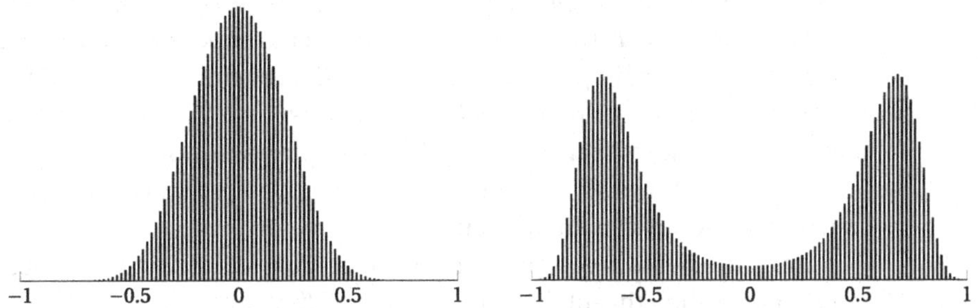

Figure 2.2. The distribution of the magnetization of the Curie–Weiss model, $\mu^{\mathrm{CW}}_{N;\beta,0}(m_N = \cdot)$, with $N = 100$ spins, when $h = 0$, plotted using (2.9) below.

At high temperature (on the left, $2d\beta = 0.8$), m_N concentrates around zero. At low temperature (on the right, $2d\beta = 1.2$), the distribution of m_N becomes bimodal, with two peaks near $\pm m^{*,\mathrm{CW}}(\beta)$. In both cases, $\langle m_N \rangle^{\mathrm{CW}}_{N;\beta,0} = 0$. The width of the peaks in the above pictures tends to 0 when $N \to \infty$, which means that

$$\lim_{N\to\infty} \mu^{\mathrm{CW}}_{N;\beta,0}(m_N \in \cdot) = \begin{cases} \delta_0(\cdot) & \text{if } \beta \le \beta_c\,, \\ \frac{1}{2}\big(\delta_{+m^{*,\mathrm{CW}}(\beta)}(\cdot) + \delta_{-m^{*,\mathrm{CW}}(\beta)}(\cdot)\big) & \text{if } \beta > \beta_c\,, \end{cases}$$

where δ_m is the Dirac mass at m (that is, the probability measure on $[-1, 1]$ such that $\delta_m(A) = 1$ or 0, depending on whether A contains m or not).

Remark 2.3. We emphasize that, when $\beta > \beta_c$ and N is large, the above results say that the typical values of the magnetization observed when sampling a configuration are close to either $+m^{*,\mathrm{CW}}(\beta)$ or $-m^{*,\mathrm{CW}}(\beta)$. Of course, this does not contradict the fact that it is always zero *on average*: $\langle m_N \rangle^{\mathrm{CW}}_{N;\beta,0} = 0$. The proper interpretation of the latter average comes from the Law of Large Numbers. Namely, let us fix N and sample an infinite sequence of independent realizations of the magnetization density: $m_N^{(1)}, m_N^{(2)}, \dots$, each distributed according to $\mu^{\mathrm{CW}}_{N;\beta,0}$. Then, by the Strong Law of Large Numbers, the empirical average over the first n samples converges almost surely to zero as $n \to \infty$:

$$\frac{m_N^{(1)} + \cdots + m_N^{(n)}}{n} \longrightarrow \langle m_N \rangle^{\mathrm{CW}}_{N;\beta,0} = 0\,.$$

There is another natural Law of Large Numbers that one might be interested in in this context. When $\beta = 0$, the random variables

$$\sigma_i(\omega) \stackrel{\text{def}}{=} \omega_i$$

are independent Bernoulli random variables of mean 0, which also satisfy a Law of Large Numbers: their empirical average $\frac{1}{N}\sum_{i=1}^{N}\sigma_i$ converges to $\langle\sigma_1\rangle_{N;0,0}^{\text{CW}} = 0$ in probability. One might thus wonder whether this property survives the introduction of an interaction between the spins: $\beta > 0$. Since $\frac{1}{N}\sum_{i=1}^{N}\sigma_i = m_N$, Theorem 2.2 shows that this is the case if and only if $\beta \le \beta_c$. ◇

To the inverse critical temperature β_c corresponds the critical temperature $T_c = \frac{1}{\beta_c}$. The range $T > T_c$ (that is, $\beta < \beta_c$) is called the **supercritical regime**, while the range $T < T_c$ (that is, $\beta > \beta_c$) is the **subcritical regime**, also called the regime of **phase coexistence** . The value $T = T_c$ corresponds to the **critical regime**. The result above shows that the Curie–Weiss model is not ordered in this regime; it can be shown, however, that the magnetization possesses peculiar properties at T_c, such as non-Gaussian fluctuations.

The Curie-Weiss model possesses a remarkable feature, which makes its analysis much easier than that of the Ising model on $\mathbb{Z}^d$: since

$$\sum_{i,j=1}^{N}\omega_i\omega_j = \Big(\sum_{i=1}^{N}\omega_i\Big)^2 \equiv M_N^2\,,$$

the Hamiltonian $\mathscr{H}_{N;\beta,0}^{\text{CW}}$ is entirely determined by the magnetization density:

$$\mathscr{H}_{N;\beta,0}^{\text{CW}} = -d\beta m_N^2 N\,. \tag{2.4}$$

This property will make it possible to compute explicitly the thermodynamic potentials (and other quantities) associated with the Curie–Weiss model.

The thermodynamic potential that plays a central role in the study of the Curie-Weiss model is the free energy:

Definition 2.4. Let $e(m) \stackrel{\text{def}}{=} -dm^2$ and

$$s(m) \stackrel{\text{def}}{=} -\frac{1-m}{2}\log\frac{1-m}{2} - \frac{1+m}{2}\log\frac{1+m}{2}\,.$$

Then

$$f_\beta^{\text{CW}}(m) \stackrel{\text{def}}{=} \beta e(m) - s(m) \tag{2.5}$$

is called the **free energy** of the Curie-Weiss model.

The claims of Theorem 2.2 will be a direct consequence of the following proposition, which shows the role played by the free energy in the asymptotic distribution of the magnetization.

Proposition 2.5. *For any β,*

$$\lim_{N\to\infty} \frac{1}{N} \log Z^{\mathrm{CW}}_{N;\beta,0} = -\min_{m\in[-1,1]} f^{\mathrm{CW}}_\beta(m)\,. \tag{2.6}$$

Moreover, for any interval $J \subset [-1,1]$,

$$\lim_{N\to\infty} \frac{1}{N} \log \mu^{\mathrm{CW}}_{N;\beta,0}(m_N \in J) = -\min_{m\in J} I^{\mathrm{CW}}_\beta(m)\,, \tag{2.7}$$

where

$$I^{\mathrm{CW}}_\beta(m) \stackrel{\mathrm{def}}{=} f^{\mathrm{CW}}_\beta(m) - \min_{\tilde m\in[-1,1]} f^{\mathrm{CW}}_\beta(\tilde m)\,. \tag{2.8}$$

One can write (2.7) roughly as follows:

$$\textit{For large } N, \quad \mu^{\mathrm{CW}}_{N;\beta,0}(m_N \in J) \simeq \exp\Bigl(-\bigl\{\min_{m\in J} I^{\mathrm{CW}}_\beta(m)\bigr\}N\Bigr)\,.$$

(In the language of large deviations theory, I^{CW}_β is called a ***rate function****.) Notice that $I^{\mathrm{CW}}_\beta \geq 0$ and*

$$\min_{m\in[-1,1]} I^{\mathrm{CW}}_\beta(m) = 0\,.$$

Thus, if $J \subset [-1,1]$ is such that I^{CW}_β is uniformly strictly positive on J,

$$\min_{m\in J} I^{\mathrm{CW}}_\beta(m) > 0\,,$$

then $\mu^{\mathrm{CW}}_{N;\beta,0}(m_N \in J)$ converges to zero exponentially fast when $N \to \infty$, meaning that the magnetization is very likely to take values outside *J. This shows that the typical values of the magnetization correspond to the regions where I^{CW}_β vanishes.* ◇

Proof of Proposition 2.5. Observe that, for a fixed N, m_N is a random variable taking values in the set

$$\mathscr{A}_N \stackrel{\mathrm{def}}{=} \Bigl\{-1 + \tfrac{2k}{N} \,:\, k = 0,\dots,N\Bigr\} \subset [-1,1]\,.$$

Let $J \subset [-1,1]$ be an interval. Then,

$$\mu^{\mathrm{CW}}_{N;\beta,0}(m_N \in J) = \sum_{m\in J\cap\mathscr{A}_N} \mu^{\mathrm{CW}}_{N;\beta,0}(m_N = m)\,.$$

Since there are exactly $\binom{N}{\frac{1+m}{2}N}$ configurations $\omega \in \Omega_N$ that have $m_N(\omega) = m$, one can express explicitly the distribution of m_N using (2.4):

$$\mu^{\mathrm{CW}}_{N;\beta,0}(m_N = m) = \sum_{\substack{\omega\in\Omega_N:\\ m_N(\omega)=m}} \frac{e^{-\mathscr{H}^{\mathrm{CW}}_{N;\beta,0}(\omega)}}{Z^{\mathrm{CW}}_{N;\beta,0}} = \frac{1}{Z^{\mathrm{CW}}_{N;\beta,0}} \binom{N}{\frac{1+m}{2}N} e^{d\beta m^2 N}\,. \tag{2.9}$$

In the same way,

$$Z^{\mathrm{CW}}_{N;\beta,0} = \sum_{m\in\mathscr{A}_N} \binom{N}{\frac{1+m}{2}N} e^{d\beta m^2 N} . \tag{2.10}$$

Since we are interested in its behavior on the exponential scale and since it is a sum of only $|\mathscr{A}_N| = N+1$ positive terms, $Z^{\mathrm{CW}}_{N;\beta,0}$ can be estimated by keeping only its *dominant term*:

$$\max_{m\in\mathscr{A}_N} \binom{N}{\frac{1+m}{2}N} e^{d\beta m^2 N} \le Z^{\mathrm{CW}}_{N;\beta,0} \le (N+1) \max_{m\in\mathscr{A}_N} \binom{N}{\frac{1+m}{2}N} e^{d\beta m^2 N} .$$

To study the large N behavior of the binomial factors, we use Stirling's formula. This implies the existence of two constants $c_-, c_+ > 0$ such that, for all $m \in \mathscr{A}_N$,

$$c_- N^{-1/2} e^{Ns(m)} \le \binom{N}{\frac{1+m}{2}N} \le c_+ N^{1/2} e^{Ns(m)} . \tag{2.11}$$

Exercise 2.1. *Verify* (2.11).

We can thus compute an upper bound as follows:

$$\begin{aligned} Z^{\mathrm{CW}}_{N;\beta,0} &\le c_+ (N+1) N^{1/2} \exp\bigl(N \max_{m\in\mathscr{A}_N} \{d\beta m^2 + s(m)\}\bigr) \\ &\le c_+ (N+1) N^{1/2} \exp\bigl(-N \min_{m\in[-1,1]} f^{\mathrm{CW}}_\beta(m)\bigr) , \end{aligned}$$

which yields

$$\limsup_{N\to\infty} \frac{1}{N} \log Z^{\mathrm{CW}}_{N;\beta,0} \le - \min_{m\in[-1,1]} f^{\mathrm{CW}}_\beta(m) .$$

For the lower bound, we first use the continuity of $m \mapsto \{d\beta m^2 + s(m)\}$ on $[-1,1]$ and consider some $m' \in [-1,1]$ for which $f^{\mathrm{CW}}_\beta(m') = \min_m f^{\mathrm{CW}}_\beta(m)$. Fix $\epsilon > 0$, and choose some $m \in \mathscr{A}_N$ such that $|f^{\mathrm{CW}}_\beta(m) - f^{\mathrm{CW}}_\beta(m')| \le \epsilon$, which is always possible once N is large enough. We then have

$$Z^{\mathrm{CW}}_{N;\beta,0} \ge \frac{c_-}{\sqrt{N}} \exp\bigl\{-N(f^{\mathrm{CW}}_\beta(m') + \epsilon)\bigr\} .$$

This yields

$$\liminf_{N\to\infty} \frac{1}{N} \log Z^{\mathrm{CW}}_{N;\beta,0} \ge - \min_{m\in[-1,1]} f^{\mathrm{CW}}_\beta(m) - \epsilon .$$

Since ϵ was arbitrary, (2.6) follows.

A similar computation can be done for the sum over $m \in J \cap \mathscr{A}_N$,

$$\lim_{N\to\infty} \frac{1}{N} \log \sum_{m\in J\cap\mathscr{A}_N} \binom{N}{\frac{1+m}{2}N} e^{d\beta m^2 N} = - \min_{m\in J} f^{\mathrm{CW}}_\beta(m) ,$$

and we get (2.7). □

Proof of Theorem 2.2. As discussed after the statement of Proposition 2.5, we must locate the zeros of I_β^{CW}. Since it is smooth and $I_\beta^{\mathrm{CW}} \geq 0$, the zeros correspond to the solutions of $\frac{\partial I_\beta^{\mathrm{CW}}}{\partial m} = 0$. After a straightforward computation, we easily see that this condition is equivalent to the **mean-field equation**:

$$\tanh(2d\beta m) = m\,. \tag{2.12}$$

Since $\lim_{m\to\pm\infty} \tanh(\beta m) = \pm 1$, there always exists at least one solution and, as an analysis of the graph of $m \mapsto \tanh(\beta m)$ shows (see below), the number of solutions of (2.12) depends on whether $2d\beta$ is larger or smaller than 1, that is, whether β is larger or smaller than β_c.

On the one hand, when $\beta \leq \beta_c$, (2.12) has a unique solution, given by $m = 0$. On the other hand, when $\beta > \beta_c$, there are two additional nontrivial solutions, $+m^{*,\mathrm{CW}}(\beta)$ and $-m^{*,\mathrm{CW}}(\beta)$ (which depend on β):

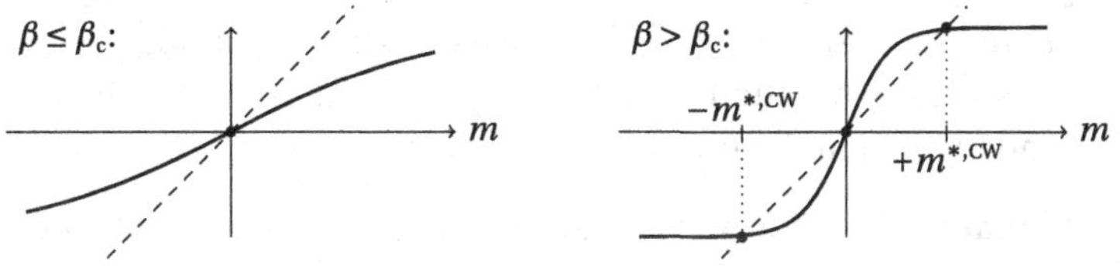

The trivial solution $m = 0$ is a local maximum of I_β^{CW}, whereas $+m^{*,\mathrm{CW}}(\beta)$ and $-m_\beta^{*,\mathrm{CW}}$ are global minima (see Figure 2.3).

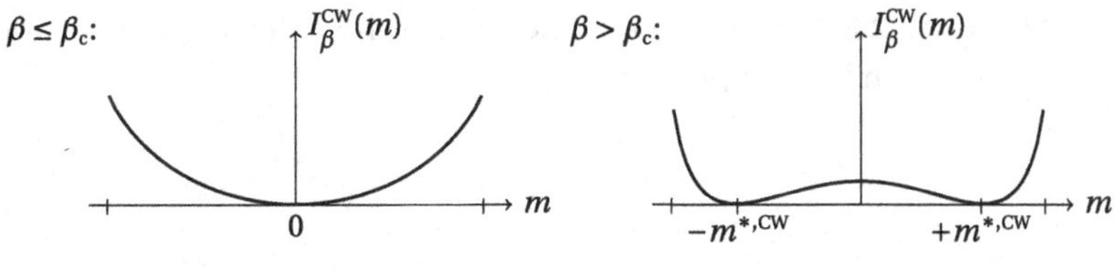

Figure 2.3. The rate function of the Curie–Weiss model. The values taken by the magnetization density of a very large system lie in the neighborhood of the points m at which I_β^{CW} vanishes, with probability very close to 1, as seen in Figure 2.2. In the supercritical and critical phases ($\beta \leq \beta_c$), there exists a unique global minimum $m = 0$. In the subcritical phase ($\beta > \beta_c$), there exist two nonzero typical values $\pm m^{*,\mathrm{CW}}(\beta)$: there is a *phase transition* at β_c.

Combined with (2.7), this analysis proves the theorem. □

Remark 2.6. One clearly sees from the graphical characterization of $m^{*,\mathrm{CW}}(\beta)$ that

$$m^{*,\mathrm{CW}}(\beta) \downarrow 0 \quad \text{as} \quad \beta \downarrow \beta_c^{\mathrm{CW}}\,. \tag{2.13}$$

A more quantitative analysis is provided in Section 2.5.3. ◇

The above analysis reveals that the typical values of the magnetization of the model are those near which the function I_β^{CW} vanishes. Since I_β^{CW} differs from f_β^{CW} only by a constant, this means that *the typical values of the magnetization are those that minimize the free energy*, a property typical of the thermodynamic behavior studied in Section 1.1.5 (when letting a system exchange energy with a heat reservoir), or as was already derived nonrigorously in Section 1.3.1.

The bifurcation of the typical values taken by the magnetization in the Curie–Weiss model at low temperature originated in the appearance of two global minima in the free energy. On the one hand, $e(m) = -dm^2$ is the **energy density** associated with configurations of magnetization density m; it is minimal when $m = +1$ or -1 (all spins equal). On the other hand, $s(m)$ is the **entropy density**, which measures the number of configurations with a magnetization density m; it is maximal at $m = 0$ (equal proportions of $+$ and $-$ spins). Since $\beta e(m)$ and $s(m)$ are both concave, the convexity/concavity properties of their difference depend on the temperature. When β is small, entropy dominates and $f_\beta^{\mathrm{CW}}(m)$ is strictly convex. When β is large, energy starts to play a major role by favoring configurations with small energy: $f_\beta^{\mathrm{CW}}(m)$ is not convex and has two global minima.

As already mentioned in Chapter 1, this interplay between energy and entropy is fundamental in the mechanism leading to phase transition.

Remark 2.7. The nonconvex free energy observed at low temperature in the Curie–Weiss model is a consequence of the lack of geometry in the model. As will be seen later, the free energy of more realistic systems (such as the Ising model on $\mathbb{Z}^d$, or the lattice gas of Chapter 4) is always convex. ◇

Exercise 2.2. Let $\zeta : \mathbb{N} \to \mathbb{R}^+$. Consider the following modification of the Curie–Weiss Hamiltonian (with $h = 0$):

$$\widetilde{\mathscr{H}}_{N;\beta,0}(\omega) \stackrel{\text{def}}{=} -\frac{\beta}{\zeta(N)} \sum_{i,j=1}^{N} \omega_i \omega_j \,.$$

Denote by $\tilde{\mu}_{N;\beta}$ the corresponding Gibbs distribution. Show that the following hold.

1. If $\lim_{N\to\infty} \frac{\zeta(N)}{N} = \infty$, then m_N tends to 0 in probability for all $\beta \geq 0$.
2. If $\lim_{N\to\infty} \frac{\zeta(N)}{N} = 0$, then $|m_N|$ tends to 1 in probability for all $\beta > 0$.

This shows that the only scaling leading to a nontrivial dependence in β is when $\zeta(N)$ is of the order of N.

2.3 Behavior for Large N When $h \neq 0$

In the presence of an external magnetic field h, the analysis is similar. The relevant thermodynamic potential associated with the magnetic field is the *pressure*.

Theorem 2.8. *The* ***pressure***

$$\psi_\beta^{\mathrm{CW}}(h) \stackrel{\text{def}}{=} \lim_{N\to\infty} \frac{1}{N} \log Z_{N;\beta,h}^{\mathrm{CW}}$$

exists and is convex in h. Moreover, it equals the ***Legendre transform of the free energy****:*

$$\psi_\beta^{\mathrm{CW}}(h) = \max_{m\in[-1,1]} \{hm - f_\beta^{\mathrm{CW}}(m)\}. \tag{2.14}$$

Proof We start by decomposing the partition function as in (2.10):

$$Z_{N;\beta,h}^{\mathrm{CW}} = \sum_{m\in\mathscr{A}_N} \sum_{\substack{\omega\in\Omega_N:\\ m_N(\omega)=m}} e^{-\mathscr{H}_{N;\beta,h}^{\mathrm{CW}}(\omega)} = \sum_{m\in\mathscr{A}_N} \binom{N}{\frac{1+m}{2}N} e^{(hm+d\beta m^2)N}.$$

We can then proceed as in the proof of Theorem 2.2. For example,

$$Z_{N;\beta,h}^{\mathrm{CW}} \le \frac{c_+(N+1)}{\sqrt{N}} \exp\left\{N \max_{m\in[-1,1]} \{hm - f_\beta^{\mathrm{CW}}(m)\}\right\}.$$

A lower bound of the same type is not difficult to establish, yielding (2.14) in the limit $N \to \infty$. As shown in Appendix B.2.3, a Legendre transform is always convex. □

We first investigate the behavior of the pressure as a function of the magnetic field and later apply it to the study of typical values of the magnetization density.

Again, since $hm - f_\beta^{\mathrm{CW}}(m)$ is smooth (analytic, in fact) in m, we can find the maximum in (2.14) by explicit differentiation. Before that, let us plot the graph of $m \mapsto hm - f_\beta^{\mathrm{CW}}(m)$ for different values of h (here, at low temperature):

h < 0:
h = 0:
h > 0:
$m_\beta^{\mathrm{CW}}(h)$
$m_\beta^{\mathrm{CW}}(h)$

When $h \neq 0$, the supremum of $hm - f_\beta^{\mathrm{CW}}(m)$ is attained at a unique point which we denote by $m_\beta^{\mathrm{CW}}(h)$. This point can be computed by solving $\frac{\partial}{\partial m}\{hm - f_\beta^{\mathrm{CW}}(m)\} = 0$, which is equivalent to $\frac{\partial f_\beta^{\mathrm{CW}}}{\partial m} = h$, and can be written as the modified mean-field equation:

$$\tanh(2d\beta m + h) = m. \tag{2.15}$$

Again, this equation always has at least one solution. Let $\beta_c (= 1/2d)$ denote the inverse critical temperature introduced before. When $\beta < \beta_c$, the solution to (2.15)

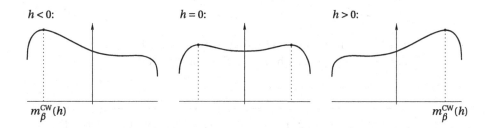

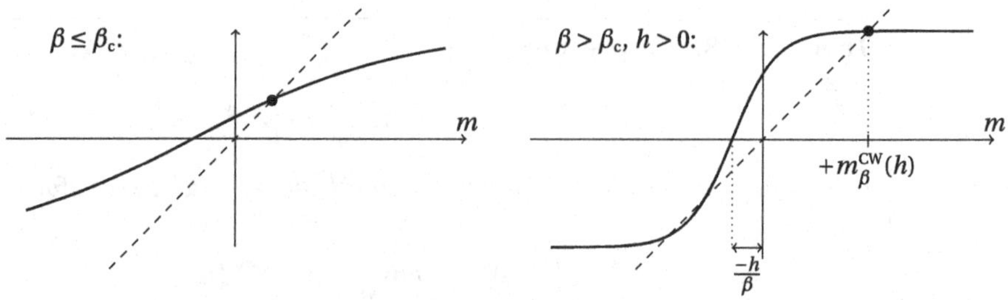

Figure 2.4. Equation (2.15) has a unique solution when $\beta \le \beta_c$ (left), but up to three different solutions when $\beta > \beta_c$ (right) and one must choose the largest (resp. smallest) one when $h > 0$ (resp. $h < 0$).

is unique. When $\beta > \beta_c$, there can be more than one solution, depending on h; in every case, $m_\beta^{CW}(h)$ is the largest (resp. smallest) one if $h > 0$ (resp. $h < 0$).

On the one hand, a glance at the above graph shows that, when $\beta \le \beta_c$,

$$\lim_{h\uparrow 0} m_\beta^{CW}(h) = \lim_{h\downarrow 0} m_\beta^{CW}(h) = 0 \,. \tag{2.16}$$

On the other hand, when $\beta > \beta_c$,

$$\lim_{h\uparrow 0} m_\beta^{CW}(h) = -m^{*,CW}(\beta) < +m^{*,CW}(\beta) = \lim_{h\downarrow 0} m_\beta^{CW}(h) \,. \tag{2.17}$$

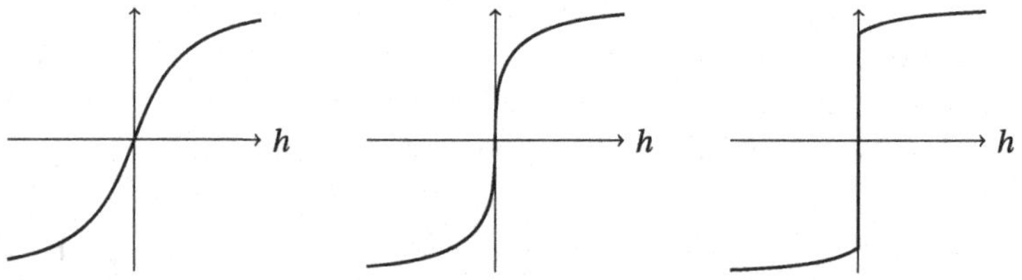

Figure 2.5. The magnetization $h \mapsto m_\beta^{CW}(h)$ for $\beta < \beta_c$ (left), $\beta = \beta_c$ (center), $\beta > \beta_c$ (right). These pictures were made by a numerical study of (2.15).

Exercise 2.3. Show that $h \mapsto m_\beta^{CW}(h)$ is analytic on $(-\infty, 0)$ and on $(0, \infty)$.

Exercise 2.4. Using (2.14), show that the pressure can be written explicitly as

$$\psi_\beta^{CW}(h) = -d\beta m_\beta^{CW}(h)^2 + \log\cosh\bigl(2d\beta m_\beta^{CW}(h) + h\bigr) + \log 2 \,.$$

Conclude, in particular, that it is analytic on $(-\infty, 0)$ and $(0, +\infty)$.

Using the terminology of Section 1.4.1, we thus see that the Curie–Weiss model provides a case in which is observed *paramagnetism* at high temperature and *ferromagnetism* at low temperature.

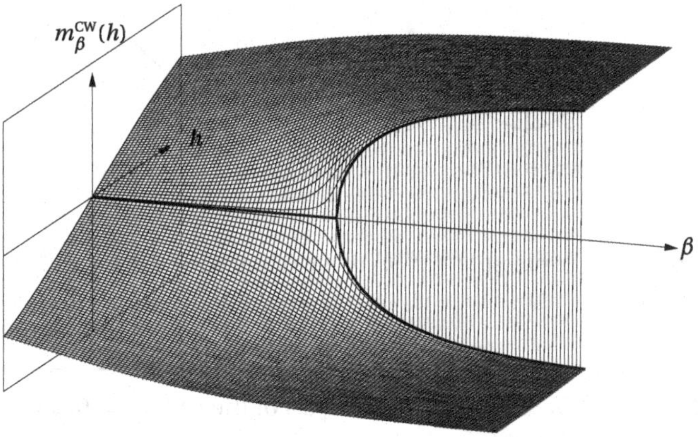

Figure 2.6. The graph of $(\beta, h) \mapsto m_\beta^{\mathrm{CW}}(h)$. At fixed $\beta > 0$, one observes the curves $h \mapsto m_\beta^{\mathrm{CW}}(h)$ of Figure 2.5; in particular, these are discontinuous when $\beta > \beta_c$.

We then move on to the study of the pressure, by first considering nonzero magnetic fields: $h \neq 0$. In this case, we can express $\psi_\beta^{\mathrm{CW}}(h)$ using the Legendre transform:

$$\psi_\beta^{\mathrm{CW}}(h) = h \cdot m_\beta^{\mathrm{CW}}(h) - f_\beta^{\mathrm{CW}}(m_\beta^{\mathrm{CW}}(h)),$$

from which we deduce, using Exercise 2.3 and the analyticity of $m \mapsto f_\beta^{\mathrm{CW}}(m)$ on $(-1, 1)$, that $h \mapsto \psi_\beta^{\mathrm{CW}}(h)$ is analytic on $(-\infty, 0) \cup (0, \infty)$. Differentiating with respect to h yields, when $h \neq 0$,

$$\frac{\partial \psi_\beta^{\mathrm{CW}}}{\partial h}(h) = m_\beta^{\mathrm{CW}}(h). \tag{2.18}$$

To study the behavior at $h = 0$, we first notice that, since ψ_β^{CW} is convex, Theorem B.28 guarantees that its one-sided derivative, $\frac{\partial \psi_\beta}{\partial h^+}\big|_{h=0}$ (resp. $\frac{\partial \psi_\beta}{\partial h^-}\big|_{h=0}$), exists and is right-continuous (resp. left-continuous). If $\beta \leq \beta_c$, (2.16) gives

$$\begin{aligned}\frac{\partial \psi_\beta^{\mathrm{CW}}}{\partial h^-}\Big|_{h=0} = \lim_{h \to 0^-} \frac{\partial \psi_\beta^{\mathrm{CW}}}{\partial h} &= \lim_{h \to 0^-} m_\beta^{\mathrm{CW}}(h) \\ &= 0 \\ &= \lim_{h \to 0^+} m_\beta^{\mathrm{CW}}(h) = \lim_{h \to 0^+} \frac{\partial \psi_\beta^{\mathrm{CW}}}{\partial h} = \frac{\partial \psi_\beta^{\mathrm{CW}}}{\partial h^+}\Big|_{h=0}.\end{aligned}$$

As a consequence, ψ_β^{CW} is differentiable at $h = 0$. Assume then that $\beta > \beta_c$. By (2.18) and (2.17), the same argument yields

$$\frac{\partial \psi_\beta^{\mathrm{CW}}}{\partial h^-}\Big|_{h=0} = -m^{*,\mathrm{CW}}(\beta) < 0 < m^{*,\mathrm{CW}}(\beta) = \frac{\partial \psi_\beta^{\mathrm{CW}}}{\partial h^+}\Big|_{h=0},$$

and so ψ_β is not differentiable at $h = 0$.

Finally, we let the reader verify that when $h \neq 0$ the magnetization density m_N concentrates exponentially fast on $m_\beta^{\mathrm{CW}}(h)$.

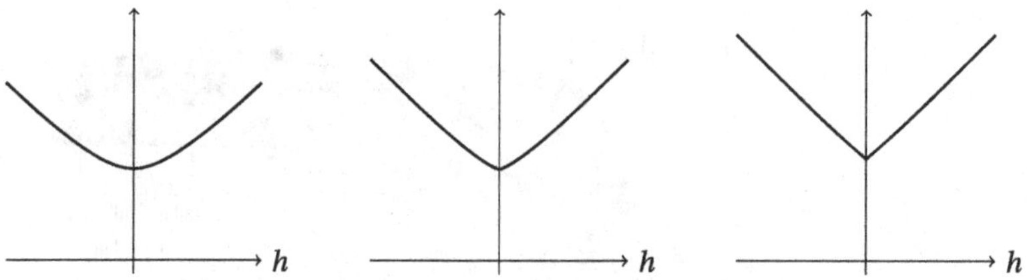

Figure 2.7. The pressure $\psi_\beta^{\mathrm{CW}}(h)$ of the Curie–Weiss model, with the same values of β as in Figure 2.5.

Exercise 2.5. Adapting the analysis of the case $h = 0$, show that an expression of the type (2.7) holds:

$$\lim_{N\to\infty} \frac{1}{N} \log \mu^{\mathrm{CW}}_{N;\beta,h}(m_N \in J) = -\min_{m\in J} I^{\mathrm{CW}}_{\beta,h}(m), \tag{2.19}$$

with

$$I^{\mathrm{CW}}_{\beta,h}(m) \stackrel{\mathrm{def}}{=} f^{\mathrm{CW}}_\beta(m) - hm - \min_{\tilde m\in[-1,1]} \big(f^{\mathrm{CW}}_\beta(\tilde m) - h\tilde m\big).$$

Show that, when $h \neq 0$, the rate function $I^{\mathrm{CW}}_{\beta,h}(m)$ has a unique global minimum at $m^{\mathrm{CW}}_\beta(h)$, for all $\beta > 0$ (see the figure below).

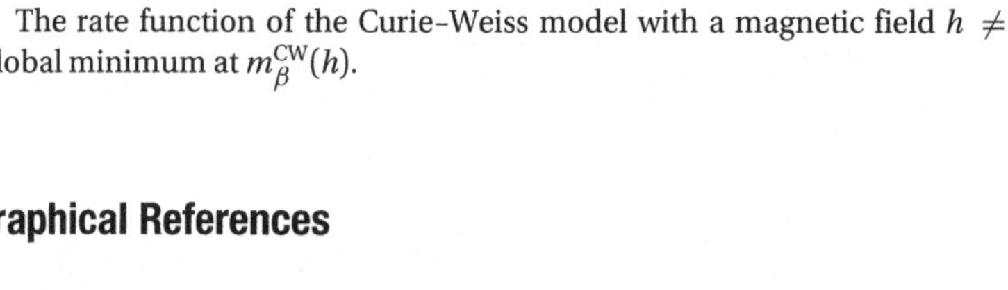

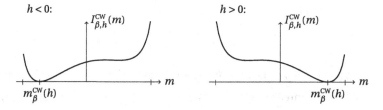

Figure 2.8. The rate function of the Curie–Weiss model with a magnetic field $h \neq 0$ has a unique global minimum at $m^{\mathrm{CW}}_\beta(h)$.

2.4 Bibliographical References

The Curie–Weiss model, as it is described in this chapter, has been introduced independently by many people, including Temperley [328], Husimi [167] and Kać [183]. There exist numerous mathematical treatments where the interested reader can get much more information, such as Ellis' book [100].

In our study of the van der Waals model of a gas in Section 4.9, we will reinterpret the Curie–Weiss model as a model of a lattice gas. There, we will derive, in a slightly different language, additional information on the free energy, the pressure and the relationship between these two quantities.

2.5 Complements and Further Reading

2.5.1 The "Naive" Mean-Field Approximation

This approximation made its first appearance in the early twentieth century work of Pierre-Ernest Weiss, based on earlier ideas of Pierre Curie, in which the method now known as *mean-field theory* was developed. This is somewhat different from what is done in this chapter, but leads to the same results.

Namely, consider the nearest-neighbor Ising model on $\Lambda \Subset \mathbb{Z}^d$. The distribution of the spin at the origin, conditional on the values taken by its neighbors, is given by

$$\mu_{\Lambda;\beta,h}(\sigma_0 = \pm 1 \mid \sigma_j = \omega_j, j \neq 0) = \frac{1}{Z} \exp\{\pm(\beta \sum_{j \sim 0} \omega_j + h)\},$$

where $Z \stackrel{\text{def}}{=} 2\cosh(\beta \sum_{j\sim 0} \omega_j + h)$ is a normalization factor. The naive mean-field approximation corresponds to assuming that each of the neighboring spins ω_j can be replaced by its mean value m. This yields the following distribution:

$$\nu(\sigma_0 = \pm 1) \stackrel{\text{def}}{=} \frac{1}{Z'} \exp\{\pm(2d\beta m + h)\},$$

with the normalization $Z' \stackrel{\text{def}}{=} 2\cosh(2d\beta m + h)$. The expected value of σ_0 under ν is equal to $\tanh(2d\beta m + h)$. However, for this approximation to be self-consistent, this expected value should also be equal to m. This yields the following consistency condition:

$$m = \tanh(2d\beta m + h),$$

which is precisely (2.15). Notice that the approximation made above, replacing each ω_j by m, seems reasonable in large dimensions, where the average of the $2d$ nearest-neighbor spins is expected to already have a value close to the expected magnetization m.

2.5.2 Alternative Approaches to Analyze the Curie–Weiss Model

Our analysis of the Curie–Weiss model was essentially combinatorial. We briefly describe two other alternative approaches, whose advantage is to be more readily generalizable to more complex models.

Hubbard–Stratonovich Transformation

The first alternative approach has a more analytic flavor; it relies on the Hubbard-Stratonovich transformation [166, 322].

Observe first that, for any $\alpha > 0$, a simple integration yields

$$\exp\{\alpha x^2\} = \frac{1}{\sqrt{\pi\alpha}} \int_{-\infty}^{\infty} \exp\Bigl\{-\frac{y^2}{\alpha} + 2yx\Bigr\}\, dy. \tag{2.20}$$

This can be used to express the interactions among spins in the Boltzmann weight as

$$\exp\Big\{\frac{d\beta}{N}\Big(\sum_{i=1}^{N}\omega_i\Big)^2\Big\} = \sqrt{\frac{N}{\pi d\beta}}\int_{-\infty}^{\infty}\exp\Big\{-\frac{N}{d\beta}y^2 + 2y\sum_{i=1}^{N}\omega_i\Big\}\,\mathrm{d}y\,.$$

The advantage of this reformulation is that the quadratic term in the spin variables has been replaced by a linear one. As a consequence, the sum over configurations, in the partition function, can now be performed as in the previous subsection:

$$\begin{aligned} Z^{\mathrm{CW}}_{N;\beta,h} = \sum_{\omega\in\Omega_N} e^{-\mathscr{H}^{\mathrm{CW}}_{N;\beta,h}(\omega)} &= \sqrt{\frac{N}{\pi d\beta}}\int_{-\infty}^{\infty} e^{-Ny^2/d\beta}\prod_{i=1}^{N}\sum_{\omega_i=\pm1}\exp\{(2y+h)\omega_i\}\mathrm{d}y \\ &= \sqrt{\frac{N}{\pi d\beta}}\int_{-\infty}^{\infty} e^{-N\varphi_{\beta,h}(y)}\,\mathrm{d}y\,, \end{aligned}$$

where

$$\varphi_{\beta,h}(y) \stackrel{\mathrm{def}}{=} y^2/d\beta - \log(2\cosh(2y+h))\,.$$

Exercise 2.6. Show that

$$\lim_{N\to\infty}\sqrt{N}\int_{-\infty}^{\infty} e^{-N(\varphi_{\beta,h}(y)-\min_y\varphi_{\beta,h}(y))}\,\mathrm{d}y > 0\,.$$

Hint: Use second-order Taylor expansions of $\varphi_{\beta,h}$ around its minima.

We then obtain

$$\psi^{\mathrm{CW}}_{\beta}(h) = \lim_{N\to\infty}\frac{1}{N}\log Z^{\mathrm{CW}}_{N;\beta,h} = -\min_{y}\varphi_{\beta,h}(y)\,,$$

and leave it as an exercise to check that this expression coincides with the one given in Exercise 2.4 (it might help to minimize over $m \stackrel{\mathrm{def}}{=} y/d\beta$).

Stein's Methods for Exchangeable Pairs

The second alternative approach we mention, which is more probabilistic, relies on *Stein's method for exchangeable pairs*. We describe only how it applies to the Curie–Weiss model and refer the reader to Chatterjee's paper [66] for more information.

We start by defining a probability measure P on $\Omega_N \times \Omega_N$ by sampling (ω, ω') as follows: (i) ω is sampled according to the Gibbs distribution $\mu^{\mathrm{CW}}_{N;\beta,h}$; (ii) an index $I \in \{1, \dots, N\}$ is sampled uniformly (with probability $\frac{1}{N}$); (iii) we set $\omega'_j = \omega_j$, for all $j \neq I$, and then let ω'_I be distributed according to $\mu^{\mathrm{CW}}_{N;\beta,h}$, conditionally on the other spins $\omega_j, j \neq I$. That is, $\omega'_I = +1$ with probability

$$\frac{\exp(2d\beta\check{m}_I + h)}{\exp(2d\beta\check{m}_I + h) + \exp(-2d\beta\check{m}_I - h)}\,,$$

where

$$\check{m}_i = \check{m}_i(\omega) \stackrel{\text{def}}{=} \frac{1}{N}\sum_{j\neq i}\omega_j\,.$$

The reader can easily check that the pair (ω, ω') is **exchangeable**:

$$P\big((\omega,\omega')\big) = P\big((\omega',\omega)\big) \qquad \text{for all } (\omega,\omega') \in \Omega_N \times \Omega_N\,.$$

Let $F(\omega,\omega') \stackrel{\text{def}}{=} \sum_{i=1}^N(\omega_i - \omega_i')$. The pairs (ω,ω') with $P((\omega,\omega')) > 0$ differ on at most one vertex, and so $|F(\omega,\omega')| \leq 2$. Denoting by E the expectation with respect to P, let

$$f(\omega) \stackrel{\text{def}}{=} E\big[F(\omega,\omega') \mid \omega\big] = \frac{1}{N}\sum_{i=1}^N\big\{\omega_i - \tanh\big(2d\beta\check{m}_i(\omega)+h\big)\big\}\,.$$

Again, for a pair (ω,ω') with nonzero probability,

$$|f(\omega) - f(\omega')| \leq \frac{2+4d\beta}{N}\,.$$

(We used $|\tanh x - \tanh y| \leq |x-y|$.) The next crucial observation is the following: for any function g on Ω_N,

$$E\big(f(\omega)g(\omega)\big) = E\big(F(\omega,\omega')g(\omega)\big) = E\big(F(\omega',\omega)g(\omega')\big) = -E\big(F(\omega,\omega')g(\omega')\big)\,.$$

We used the tower property of conditional expectation in the first identity and exchangeability in the second. Combining the first and last identities,

$$E\big[f(\omega)g(\omega)\big] = \tfrac{1}{2}E\big[F(\omega,\omega')(g(\omega)-g(\omega'))\big]\,.$$

In particular, since each element of the pair (ω,ω') has distribution $\mu^{\mathrm{CW}}_{N;\beta,h}$, and since $E[f] = 0$,

$$\mathrm{Var}^{\mathrm{CW}}_{N;\beta,h}(f) = E[f(\omega)^2] = \tfrac{1}{2}E\big[F(\omega,\omega')(f(\omega)-f(\omega'))\big] \leq \frac{2+4d\beta}{N}\,.$$

Therefore, by Chebyshev's inequality (B.18), for all $\epsilon > 0$,

$$\mu^{\mathrm{CW}}_{N;\beta,h}(|f| > \epsilon) \leq \frac{2+4d\beta}{N\epsilon^2}\,.$$

Finally, since $|\check{m}_i(\omega) - m_N(\omega)| \leq 2/N$ for all i,

$$\big|f(\omega) - \big\{m_N(\omega) - \tanh\big(2d\beta m_N(\omega)+h\big)\big\}\big| \leq \frac{4d\beta}{N}\,,$$

from which we conclude that, for all N large enough,

$$\mu^{\mathrm{CW}}_{N;\beta,h}\big(\big|m_N(\omega) - \tanh\big(2d\beta m_N(\omega)+h\big)\big| > 2\epsilon\big) \leq \mu^{\mathrm{CW}}_{N;\beta,h}(|f| > \epsilon) \leq \frac{2+4d\beta}{N\epsilon^2}\,.$$

This implies that the magnetization density m_N concentrates, as $N \to \infty$, on the solution of (2.15). Further refinements can be found in [66], such as much stronger concentration bounds and the computation of the distribution of the fluctuations of the magnetization density in the limit $N \to \infty$.

2.5.3 Critical Exponents

As we have seen in this chapter, the Curie–Weiss model exhibits two types of phase transitions:

- When $\beta > \beta_c = 1/2d$, the magnetization, as a function of h, is discontinuous at $h = 0$: there is a *first-order phase transition.*
- When $h = 0$, the magnetization, as a function of β is continuous, but not analytic at β_c: there is a *continuous phase transition.*

It turns out that the behavior of statistical mechanical systems at continuous phase transitions displays remarkable properties, which will be briefly described in Section 3.10.11. In particular, the different models of statistical mechanics fall into broad *universality classes,* in which all models share the same type of *critical behavior,* characterized by their *critical exponents.*

In this section, we will take a closer look at the critical behavior of the Curie–Weiss model in the neighborhood of the point $\beta = \beta_c, h = 0$ at which a continuous phase transition takes place. This will be done by defining certain critical exponents associated with the model. Having the graph of Figure 2.6 in mind might help the reader understand the definitions of these exponents.

To start, let us approach the transition point by varying the temperature from high to low. We know that $h \mapsto m_\beta^{CW}(h)$ is continuous at $h = 0$ when $\beta \le \beta_c$, but discontinuous when $\beta > \beta_c$. We can consider this phenomenon from different points of view, each associated with a way of fixing one variable and varying the other. First, one can see how the derivative of $m_\beta^{CW}(h)$ with respect to h at $h = 0$ diverges as $\beta \uparrow \beta_c$. Let us thus consider the **magnetic susceptibility**,

$$\chi(\beta) \stackrel{\text{def}}{=} \frac{\partial m_\beta^{CW}(h)}{\partial h}\Big|_{h=0},$$

which is well defined for all $\beta < \beta_c$. Since $\chi(\beta)$ must diverge when $\beta \uparrow \beta_c$, one might expect a singular behavior of the form

$$\chi(\beta) \sim \frac{1}{(\beta_c - \beta)^\gamma}, \qquad \text{as } \beta \uparrow \beta_c, \tag{2.21}$$

for some constant $\gamma > 0$. More precisely, the last display should be understood in terms of the following limit:

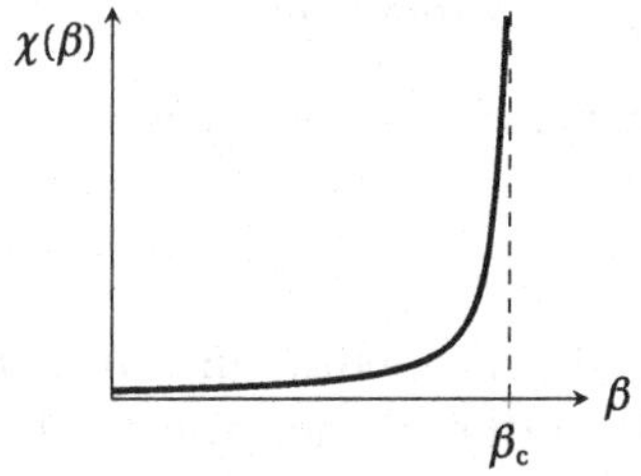

Figure 2.9. Magnetic susceptibility of the Curie–Weiss model.

$$\gamma \overset{\text{def}}{=} -\lim_{\beta\uparrow\beta_c} \frac{\log \chi(\beta)}{\log(\beta_c - \beta)}.$$

On the other hand, one can fix $\beta = \beta_c$ and consider the fast variation of the magnetization at $h = 0$:

$$m_{\beta_c}^{CW}(h) \sim h^{1/\delta}, \qquad \text{as } h \downarrow 0, \tag{2.22}$$

with δ defined by

$$\delta^{-1} \overset{\text{def}}{=} \lim_{h\downarrow 0} \frac{\log m_{\beta_c}^{CW}(h)}{\log h}.$$

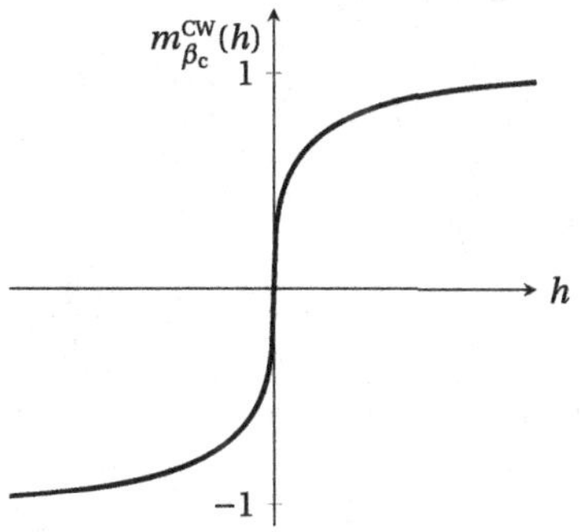

Figure 2.10. Magnetization of the Curie–Weiss model as a function of h at β_c.

But one can also approach the transition by varying the temperature from low to high. So, for $\beta > \beta_c$, consider the magnetization, $m^{*,CW}(\beta)$, in the vicinity of β_c. We have already seen in Remark 2.6 that $m^{*,CW}(\beta)$ vanishes as $\beta \downarrow \beta_c$, and one is naturally led to expect some behavior of the type:

$$m^{*,CW}(\beta) \sim (\beta - \beta_c)^b, \tag{2.23}$$

with b defined by

$$b \overset{\text{def}}{=} \lim_{\beta\downarrow\beta_c} \frac{\log m^{*,CW}(\beta)}{\log(\beta - \beta_c)}.$$

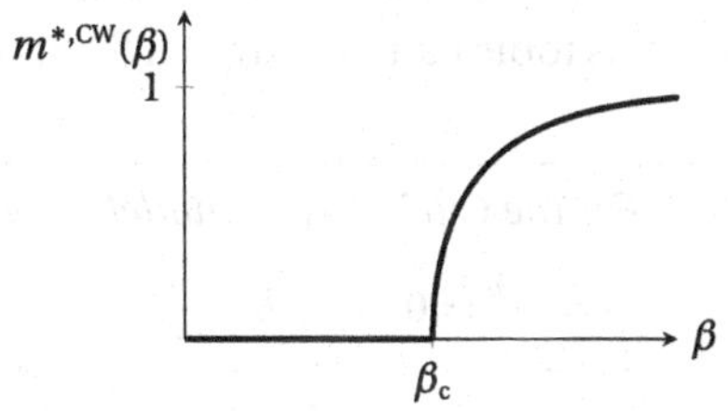

Figure 2.11. Spontaneous magnetization of the Curie–Weiss model at $h = 0$, as a function of β.

(Usually, the letter used for b is β, but we prefer to use b for obvious reasons.)

Let us introduce one last pair of exponents. This time, in order for our definition to match the standard one in physics, we consider the dependence on the temperature $T = \beta^{-1}$ and on H, defined by $h \equiv \beta H$. Let us define the **internal average energy density** by

$$u(T,H) \stackrel{\text{def}}{=} \lim_{N\to\infty} \frac{1}{\beta N} \langle \mathscr{H}^{\text{CW}}_{N;\beta,\beta H} \rangle^{\text{CW}}_{N;\beta,\beta H},$$

and define the **heat capacity**

$$c_H(\beta) \stackrel{\text{def}}{=} \frac{\partial u}{\partial T}.$$

The exponents α and α' are defined through

$$c_{H=0}(\beta) \sim \begin{cases} (\beta_c - \beta)^{-\alpha} & \text{as } \beta \uparrow \beta_c, \\ (\beta - \beta_c)^{-\alpha'} & \text{as } \beta \downarrow \beta_c, \end{cases}$$

or, more precisely, by

$$\alpha \stackrel{\text{def}}{=} -\lim_{\beta\uparrow\beta_c} \frac{\log c_{H=0}(\beta)}{\log(\beta_c - \beta)},$$

and similarly for α'.

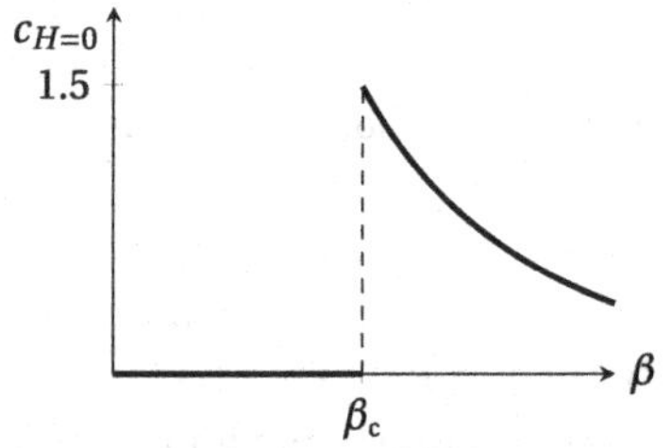

Figure 2.12. Heat capacity of the Curie-Weiss model at $H = 0$.

The numbers $\alpha, \alpha', b, \gamma, \delta$ are examples of **critical exponents.** Similar exponents can be defined for any model at a continuous phase transition, but are usually difficult to compute. These exponents can vary from one model to the other, but coincide for models belonging to the same universality class. We have seen, for instance, in Exercise 1.14, that $b = 1/8$ for the two-dimensional Ising model (more information on this topic can be found in Section 3.10.11).

Theorem 2.9. *For the Curie-Weiss model,*

$$\alpha = \alpha' = 0, \qquad b = \tfrac{1}{2}, \qquad \gamma = 1, \qquad \delta = 3.$$

Proof We start with b. Since $m^{*,\text{CW}}(\beta) > 0$ is the largest solution of (2.12) and since $\beta m^{*,\text{CW}}(\beta)$ is small when β is sufficiently close to β_c, we can use a Taylor expansion for $\tanh(\cdot)$:

$$
\begin{aligned}
m^{*,\mathrm{CW}}(\beta) &= \tanh(2d\beta m^{*,\mathrm{CW}}(\beta)) \\
&= 2d\beta m^{*,\mathrm{CW}}(\beta) - \tfrac{1}{3}(2d\beta m^{*,\mathrm{CW}}(\beta))^3 + O\big((\beta m^{*,\mathrm{CW}}(\beta))^5\big) \\
&= 2d\beta m^{*,\mathrm{CW}}(\beta) - (1+o(1))\tfrac{(2d\beta)^3}{3}(m^{*,\mathrm{CW}}(\beta))^3 ,
\end{aligned}
$$

where $o(1)$ tends to zero when $\beta \downarrow \beta_c$. We thus get

$$
m^{*,\mathrm{CW}}(\beta) = (1+o(1))\Big(\frac{3(\beta-\beta_c)}{4d^2\beta^3}\Big)^{1/2} \tag{2.24}
$$

(using the fact that $\beta_c = 1/2d$), which shows that $b = 1/2$.

To study $\chi(\beta)$ with $\beta < \beta_c$, we start with the definition of $m_\beta^{\mathrm{CW}} = m_\beta^{\mathrm{CW}}(h)$, as the unique solution to the mean-field equation (2.15), which we differentiate implicitly with respect to h, to obtain

$$
\chi(\beta) = \frac{\partial m_\beta^{\mathrm{CW}}}{\partial h}\Big|_{h=0} = \frac{1-\tanh^2(2d\beta m_\beta^{\mathrm{CW}})}{1-2d\beta(1-\tanh^2(2d\beta m_\beta^{\mathrm{CW}}))}\Big|_{h=0} = \frac{\beta_c}{\beta_c-\beta} ,
$$

which shows that $\gamma = 1$.

Let us now turn to the exponents α, α'. The internal energy density of the Curie-Weiss model at $H = 0$ is given by

$$
u = -d \lim_{N\to\infty} \langle m_N^2 \rangle_{N;\beta,0}^{\mathrm{CW}} .
$$

Now, by Theorem 2.2,

$$
\lim_{N\to\infty} \langle m_N^2 \rangle_{N;\beta,0}^{\mathrm{CW}} = \begin{cases} 0 & \text{if } \beta < \beta_c , \\ m^{*,\mathrm{CW}}(\beta)^2 & \text{if } \beta > \beta_c . \end{cases}
$$

We immediately deduce that $u = 0$ when $\beta < \beta_c$ and, thus, $\alpha = 0$. When $\beta > \beta_c$,

$$
\frac{\partial u}{\partial T} = 2d\beta^2 m^{*,\mathrm{CW}}(\beta)\frac{\partial m^{*,\mathrm{CW}}(\beta)}{\partial \beta} .
$$

Differentiating (2.12) with respect to β, we get

$$
\frac{\partial m^{*,\mathrm{CW}}(\beta)}{\partial \beta} = \big(1-\tanh^2(2d\beta m^{*,\mathrm{CW}}(\beta))\big)\Big(2dm^{*,\mathrm{CW}}(\beta) + 2d\beta\frac{\partial m^{*,\mathrm{CW}}(\beta)}{\partial \beta}\Big) .
$$

Rearranging, expanding the hyperbolic tangents to leading order, and using (2.24), we obtain that, as $\beta \downarrow \beta_c$,

$$
\frac{\partial m^{*,\mathrm{CW}}(\beta)}{\partial \beta} = \frac{1-\tanh^2(2d\beta m^{*,\mathrm{CW}}(\beta))}{1-2d\beta+2d\beta\tanh^2(2d\beta m^{*,\mathrm{CW}}(\beta))} 2dm^{*,\mathrm{CW}}(\beta) = \frac{(1+o(1))}{2(\beta-\beta_c)} m^{*,\mathrm{CW}}(\beta) .
$$

Using once more (2.24), we conclude that, as $\beta \downarrow \beta_c$,

$$
c_{H=0}(\beta) = \frac{d\beta^2}{\beta-\beta_c}(m^{*,\mathrm{CW}}(\beta))^2(1+o(1)) = \frac{3}{2}\frac{\beta_c}{\beta}(1+o(1)) ,
$$

so that $\lim_{\beta\downarrow\beta_c} c_{H=0}(\beta) = 3/2$ and $\alpha' = 0$.

We leave the proof that $\delta = 3$ to the reader. □

2.5.4 Links with Other Models on $\mathbb{Z}^d$

One of the main reasons for the interest in mean-field models is that the results obtained often shed light on the type of behavior that might be expected in more realistic lattice models on $\mathbb{Z}^d$, of the type discussed in the rest of this book. In view of the approximation involved, one might expect the agreement between a lattice spin system on $\mathbb{Z}^d$ and its mean-field version to improve as the number of spins with which one spin interacts increases, which happens when either the range of the interaction becomes large, or the dimension of the lattice increases. It turns out that, in many cases, this can in fact be quantified rather precisely. We will not discuss these issues in much detail, but will rather provide some references. Much more information can found in Sections II.13–II.15 and V.3–V.5 of Simon's book [308] and in Section 4 of Biskup's review [22].

Rigorous Bounds. A first type of comparison between models on $\mathbb{Z}^d$ and their mean-field counterpart is provided by various bounds on some quantities associated with the former in terms of the corresponding quantities associated with the latter.

First, the mean-field pressure is known to provide a rigorous lower bound on the pressure in very general settings. For example, we will show in Theorem 3.53 that

$$\psi_\beta^{\text{Ising on } \mathbb{Z}^d}(h) \geq \psi_\beta^{\text{CW}}(h).$$

(Remember that the dimensional parameter d also appears in the definition of the Curie–Weiss model in the right-hand side.) See also Exercise 6.28 for a closely related result.

Second, the mean-field critical temperature is known to provide a rigorous upper bound on the critical temperature of models on $\mathbb{Z}^d$. Again, this is done for the Ising model, via a comparison of the magnetizations of these two models, in Theorem 3.53. More information and references on this topic can be found in [308, Sections V.3 and V.5].

Convergence. The bounds mentioned above enable a general comparison between models in arbitrary dimensions and their mean-field approximation, but do not yield quantitative information about the discrepancy. Here, we consider various limiting procedures, in which actual convergence to the mean-field limits can be established.

First, one can consider spin systems on $\mathbb{Z}^d$ with spread-out interactions (for example, such that any pair of spins located at a distance less than some large value interact). A prototypical example is models with *Kać interactions.* An example of these is discussed in detail in Section 4.10, and a proof that the corresponding pressure converges to the corresponding mean-field pressure when the range of the interactions diverges is provided in Theorem 4.31. Additional information and references on this topic can be found at the end of Chapter 4.

Another approach is to consider models on $\mathbb{Z}^d$ and prove convergence of the pressure or the magnetization as $d \to \infty$. A general result can be found in [308, Theorem II.14.1] with the relevant bibliography. Alternatively, one might try to provide quantitative bounds for the difference between the magnetization of a model on $\mathbb{Z}^d$ and the magnetization of its mean-field counterpart. This is the approach developed in [23, 24, 68]; see also the lecture notes by Biskup [22]. This approach is particularly interesting for models in which the mean-field magnetization is discontinuous. Indeed, once the dimension is large enough (or the interaction is sufficiently spread out), the error term becomes small enough that the magnetization of the corresponding model on $\mathbb{Z}^d$ must necessarily also be discontinuous. This provides a powerful technique to prove the existence of first-order phase transitions in some models.

Critical Exponents. Finally, as mentioned in Section 2.5.3 and as will be discussed in more detail in Section 3.10.11, when a continuous phase transition occurs, describing quantities exhibiting singular behavior qualitatively is of great interest. In particular, a challenging problem is to determine the corresponding critical exponents, as we did for the Curie–Weiss model in Section 2.5.3. It is expected that the critical exponents of models on $\mathbb{Z}^d$ *coincide* with those of their mean-field counterpart *for all large enough dimensions* (and not only in the limit!). Namely, there exists a critical dimension d_u, known as the **upper critical dimension**, such that the critical exponents take their mean-field values for all $d > d_u$. This has been proved in several cases, such as the Ising model, for which $d_u = 4$. A thorough discussion can be found in the book by Fernández, Fröhlich and Sokal [102].

3 The Ising Model

In this chapter, we study the Ising model on $\mathbb{Z}^d$, which was introduced informally in Section 1.4.2. We provide both precise definitions of the concepts involved and a detailed analysis of the conditions ensuring the existence or absence of a phase transition in this model, therefore providing full rigorous justification to the discussion in Section 1.4.3. Namely,

- In Section 3.1, the Ising model on $\mathbb{Z}^d$ is defined, together with various types of boundary conditions.
- In Section 3.2, several concepts of fundamental importance are introduced, including: the thermodynamic limit, the pressure and the magnetization. The latter two quantities are then computed explicitly in the case of the one-dimensional model (Section 3.3).
- The notion of infinite-volume Gibbs state is given a precise meaning in Section 3.4. In Section 3.6, we discuss correlation inequalities, which play a central role in the analysis of ferromagnetic systems like the Ising model.
- In Section 3.7, the phase diagram of the model is analyzed in detail. In particular, several criteria for the presence of first-order phase transitions, based on the magnetization and the pressure of the model, are introduced in Section 3.7.1. These are used to prove the existence of a phase transition when $h = 0$ (Sections 3.7.2 and 3.7.3) and the absence of a phase transition when $h \neq 0$ (Section 3.7.4). A summary with a link to the discussion in the Introduction is given in Section 3.7.5.
- Finally, in Section 3.10, the reader can find a series of complements to this chapter, in which a number of interesting topics, related to the core of the chapter but usually more advanced or specific, are discussed in a somewhat less precise manner.

We emphasize that some of the ideas and concepts introduced in this chapter are not only useful for the Ising model, but are also of central importance for statistical mechanics in general. They are thus fundamental for the understanding of other parts of the book.

3.1 Finite-Volume Gibbs Distributions

In this section, the Ising model on $\mathbb{Z}^d$ is defined precisely and some of its basic properties are established. As a careful reader might notice, some of the definitions in this chapter differ slightly from those of Chapter 1. This is done for later convenience.

▶ **Finite Volume with Free Boundary Condition.** The configurations of the Ising model in a finite volume $\Lambda \Subset \mathbb{Z}^d$ with the free boundary condition are the elements of the set

$$\Omega_\Lambda \stackrel{\text{def}}{=} \{-1,1\}^\Lambda .$$

A configuration $\omega \in \Omega_\Lambda$ is thus of the form $\omega = (\omega_i)_{i\in\Lambda}$. The basic random variable associated with the model is the **spin** at a vertex $i \in \mathbb{Z}^d$, which is the random variable $\sigma_i \colon \Omega_\Lambda \to \{-1,1\}$ defined by $\sigma_i(\omega) \stackrel{\text{def}}{=} \omega_i$.

We will often identify a finite set Λ with the graph that contains all edges formed by nearest-neighbor pairs of vertices of Λ. We denote the latter set of edges by

$$\mathscr{E}_\Lambda \stackrel{\text{def}}{=} \big\{\{i,j\} \subset \Lambda : i \sim j\big\} .$$

To each configuration $\omega \in \Omega_\Lambda$, we associate its **energy**, given by the Hamiltonian

$$\mathscr{H}^{\varnothing}_{\Lambda;\beta,h}(\omega) \stackrel{\text{def}}{=} -\beta \sum_{\{i,j\}\in\mathscr{E}_\Lambda} \sigma_i(\omega)\sigma_j(\omega) - h \sum_{i\in\Lambda} \sigma_i(\omega) ,$$

where $\beta \in \mathbb{R}_{\geq 0}$ is the inverse temperature and $h \in \mathbb{R}$ is the magnetic field. The superscript $\varnothing$ indicates that this model has the **free boundary condition**: spins in Λ do not interact with other spins located outside of Λ.

Definition 3.1. The Gibbs distribution of the Ising model in Λ with **free boundary condition**, at parameters β and h, is the distribution on Ω_Λ defined by

$$\mu^{\varnothing}_{\Lambda;\beta,h}(\omega) \stackrel{\text{def}}{=} \frac{1}{\mathbf{Z}^{\varnothing}_{\Lambda;\beta,h}} \exp\big(-\mathscr{H}^{\varnothing}_{\Lambda;\beta,h}(\omega)\big) .$$

The normalization constant

$$\mathbf{Z}^{\varnothing}_{\Lambda;\beta,h} \stackrel{\text{def}}{=} \sum_{\omega\in\Omega_\Lambda} \exp\big(-\mathscr{H}^{\varnothing}_{\Lambda;\beta,h}(\omega)\big)$$

is called the **partition function** in Λ with free boundary condition.

▶ **Finite Volume with Periodic Boundary Condition.** We now consider the Ising model on the torus $\mathbb{T}_n$, defined as follows. Its set of vertices is given by

$$V_n \stackrel{\text{def}}{=} \{0,\dots,n-1\}^d ,$$

and there is an edge between each pair of vertices $i = (i_1,\dots,i_d), j = (j_1,\dots,j_d)$ such that $\sum_{r=1}^d |(i_r - j_r) \mod n| = 1$; see Figure 3.1 for illustrations in dimensions 1 and 2. We denote by $\mathscr{E}^{\text{per}}_{V_n}$ the set of edges of $\mathbb{T}_n$.

Configurations of the model are now the elements of $\{-1,1\}^{V_n}$ and have an energy given by

$$\mathscr{H}^{\mathrm{per}}_{V_n;\beta,h}(\omega) \stackrel{\mathrm{def}}{=} -\beta \sum_{\{i,j\}\in\mathscr{E}^{\mathrm{per}}_{V_n}} \sigma_i(\omega)\sigma_j(\omega) - h\sum_{i\in V_n}\sigma_i(\omega)\,.$$

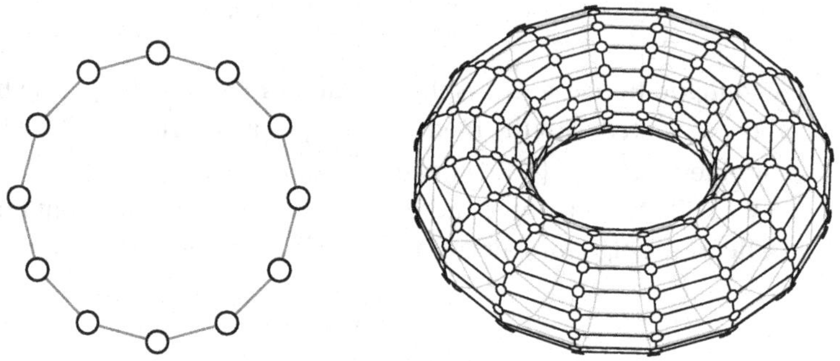

Figure 3.1. Left: the one-dimensional torus $\mathbb{T}_{12}$. Right: the two-dimensional torus $\mathbb{T}_{16}$.

Definition 3.2. The Gibbs distribution of the Ising model in V_n with **periodic boundary condition**, at parameters β and h, is the probability distribution on $\{-1,1\}^{V_n}$ defined by

$$\mu^{\mathrm{per}}_{V_n;\beta,h}(\omega) \stackrel{\mathrm{def}}{=} \frac{1}{\mathbf{Z}^{\mathrm{per}}_{V_n;\beta,h}} \exp\bigl(-\mathscr{H}^{\mathrm{per}}_{V_n;\beta,h}(\omega)\bigr)\,.$$

The normalization constant

$$\mathbf{Z}^{\mathrm{per}}_{V_n;\beta,h} \stackrel{\mathrm{def}}{=} \sum_{\omega\in\Omega_{V_n}} \exp\bigl(-\mathscr{H}^{\mathrm{per}}_{V_n;\beta,h}(\omega)\bigr)$$

is called the **partition function** in V_n with periodic boundary condition.

▶ **Finite Volumes with Configurations as Boundary Condition.** It will turn out to be useful to consider the Ising model on the full lattice $\mathbb{Z}^d$, but with configurations that are frozen outside a finite set.

Let us thus consider configurations of the Ising model on the infinite lattice $\mathbb{Z}^d$, that is, elements of

$$\Omega \stackrel{\mathrm{def}}{=} \{-1,1\}^{\mathbb{Z}^d}\,.$$

Fixing a finite set $\Lambda \Subset \mathbb{Z}^d$ and a configuration $\eta \in \Omega$, we define a **configuration of the Ising model in Λ with boundary condition η** as an element of the finite set

$$\Omega^{\eta}_{\Lambda} \stackrel{\mathrm{def}}{=} \{\omega\in\Omega : \omega_i = \eta_i,\ \forall i\notin\Lambda\}\,.$$

The **energy** of a configuration $\omega \in \Omega_\Lambda^\eta$ is defined by

$$\mathscr{H}_{\Lambda;\beta,h}(\omega) \stackrel{\text{def}}{=} -\beta \sum_{\{i,j\}\in\mathscr{E}_\Lambda^{\mathrm{b}}} \sigma_i(\omega)\sigma_j(\omega) - h\sum_{i\in\Lambda}\sigma_i(\omega), \tag{3.1}$$

where we have introduced

$$\mathscr{E}_\Lambda^{\mathrm{b}} \stackrel{\text{def}}{=} \big\{\{i,j\} \subset \mathbb{Z}^d : \{i,j\} \cap \Lambda \neq \varnothing,\, i \sim j\big\}. \tag{3.2}$$

Note that $\mathscr{E}_\Lambda^{\mathrm{b}}$ differs from $\mathscr{E}_\Lambda$ by the addition of all the edges connecting vertices inside Λ to their neighbors outside Λ.

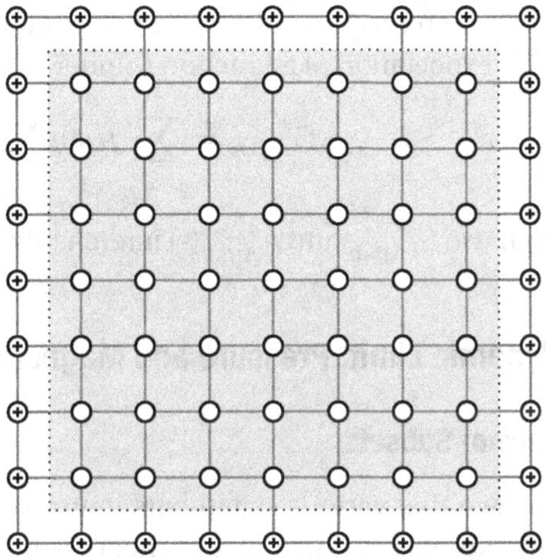

Figure 3.2. The model in a box Λ (shaded) with $+$ boundary condition.

Definition 3.3. The Gibbs distribution of the Ising model in Λ **with boundary condition** η, at parameters β and h, is the probability distribution on Ω_Λ^η defined by

$$\mu_{\Lambda;\beta,h}^\eta(\omega) \stackrel{\text{def}}{=} \frac{1}{\mathbf{Z}_{\Lambda;\beta,h}^\eta} \exp\left(-\mathscr{H}_{\Lambda;\beta,h}(\omega)\right).$$

The normalization constant

$$\mathbf{Z}_{\Lambda;\beta,h}^\eta \stackrel{\text{def}}{=} \sum_{\omega\in\Omega_\Lambda^\eta} \exp(-\mathscr{H}_{\Lambda;\beta,h}(\omega))$$

is called the **partition function** with η-boundary condition.

It will be seen later (in particular in Chapter 6) why defining $\mu_{\Lambda;\beta,h}^\eta$ on configurations in infinite volume is convenient (here, we could as well have defined it on Ω_Λ and included the effect of the boundary condition in the Hamiltonian).

Two boundary conditions play a particularly important role in the analysis of the Ising model: the **+ boundary condition** η^+, for which $\eta_i^+ \stackrel{\text{def}}{=} +1$ for all i (see Figure 3.2), and the **− boundary condition** η^-, similarly defined by $\eta_i^- \stackrel{\text{def}}{=} -1$ for all i. The corresponding Gibbs distributions will be denoted simply by $\mu^+_{\Lambda;\beta,h}$ and $\mu^-_{\Lambda;\beta,h}$; similarly, we will write Ω^+_Λ, Ω^-_Λ for the corresponding sets of configurations.

On the Notations Used Below. In the following, we will use the symbol # to denote a generic type of boundary condition. For instance, $\mathbf{Z}^{\#}_{\Lambda;\beta,h}$ can denote $\mathbf{Z}^{\varnothing}_{\Lambda;\beta,h}$, $\mathbf{Z}^{\text{per}}_{\Lambda;\beta,h}$ or $\mathbf{Z}^{\eta}_{\Lambda;\beta,h}$. In the case of the periodic boundary condition, Λ will always implicitly be assumed to be a cube (see below).

Following the custom in statistical physics, expectation of a function f with respect to a probability distribution μ will be denoted by a bracket: $\langle f\rangle_\mu$. When the distribution is identified by indices, we will apply the same indices to the bracket. For example, expectation of a function f under $\mu^{\#}_{\Lambda;\beta,h}$ will be denoted by

$$\langle f\rangle^{\#}_{\Lambda;\beta,h} \stackrel{\text{def}}{=} \sum_{\omega\in\Omega^{\#}_\Lambda} f(\omega)\mu^{\#}_{\Lambda;\beta,h}(\omega)\,.$$

We will often use $\langle\cdot\rangle^{\#}_{\Lambda;\beta,h}$ and $\mu^{\#}_{\Lambda;\beta,h}(\cdot)$ interchangeably.

3.2 Thermodynamic Limit, Pressure and Magnetization

3.2.1 Convergence of Subsets

It is well known that various statements in probability theory, such as the Strong Law of Large Numbers or the Ergodic Theorem, take on a much cleaner form when considering infinite samples. For the same reason, it is convenient to have some notion of the Gibbs distribution for the Ising model on the whole of $\mathbb{Z}^d$. The theory describing *Gibbs measures* of infinite lattice systems will be discussed in detail in Chapter 6.

In this chapter, we adopt a more elementary point of view, using a procedure that approaches an infinite system by a sequence of growing sets. This procedure, crucial for a proper description of thermodynamics and phase transitions, is called the *thermodynamic limit.*

To define the Ising model on the whole lattice $\mathbb{Z}^d$ (one often says "in infinite volume"), the thermodynamic limit will be considered along sequences of finite subsets $\Lambda_n \Subset \mathbb{Z}^d$ which **converge to** $\mathbb{Z}^d$, denoted by $\Lambda_n \uparrow \mathbb{Z}^d$, in the sense that

1. Λ_n is *increasing*: $\Lambda_n \subset \Lambda_{n+1}$,
2. Λ_n *invades* $\mathbb{Z}^d$: $\bigcup_{n\geq 1}\Lambda_n = \mathbb{Z}^d$.

Sometimes, in order to control the influence of the boundary condition and of the shape of the box on thermodynamic quantities, it will be necessary to impose a further regularity property on the sequence Λ_n. We say that a sequence $\Lambda_n \uparrow \mathbb{Z}^d$ **converges to $\mathbb{Z}^d$ in the sense of van Hove**, which we denote by $\Lambda_n \Uparrow \mathbb{Z}^d$, if and only if

$$\lim_{n\to\infty} \frac{|\partial^{\text{in}}\Lambda_n|}{|\Lambda_n|} = 0, \tag{3.3}$$

where $\partial^{\text{in}}\Lambda \stackrel{\text{def}}{=} \{i \in \Lambda : \exists j \notin \Lambda, j \sim i\}$. The simplest sequence to satisfy this condition is the sequence

$$\mathrm{B}(n) \stackrel{\text{def}}{=} \{-n, \dots, n\}^d .$$

Exercise 3.1. Show that $\mathrm{B}(n) \Uparrow \mathbb{Z}^d$. Give an example of a sequence Λ_n that converges to $\mathbb{Z}^d$, but not in the sense of van Hove.

3.2.2 Pressure

The partition functions introduced above play a very important role in the theory, in particular because they give rise to the pressure of the model.

Definition 3.4. The **pressure in** $\Lambda \Subset \mathbb{Z}^d$, with boundary condition of the type $\#$, is defined by

$$\psi_\Lambda^\#(\beta, h) \stackrel{\text{def}}{=} \frac{1}{|\Lambda|} \log \mathbf{Z}^\#_{\Lambda;\beta,h} .$$

Exercise 3.2. Show that, for all $\Lambda \Subset \mathbb{Z}^d$, all $\beta \geq 0$ and all $h \in \mathbb{R}$,

$$\psi_\Lambda^\varnothing(\beta, h) = \psi_\Lambda^\varnothing(\beta, -h), \quad \psi_\Lambda^{\text{per}}(\beta, h) = \psi_\Lambda^{\text{per}}(\beta, -h), \quad \psi_\Lambda^+(\beta, h) = \psi_\Lambda^-(\beta, -h).$$

The following simple observation will play an important role in the sequel.

Lemma 3.5. *For each type of boundary condition #, $(\beta, h) \mapsto \psi_\Lambda^\#(\beta, h)$ is convex.*

Proof We consider $\psi_\Lambda^\eta(\beta, h)$, but the other cases are similar. Let $\alpha \in [0,1]$. Since $\mathscr{H}_{\Lambda;\beta,h}$ is an affine function of the pair (β, h), Hölder's inequality (see Appendix B.1.1) yields

$$\begin{aligned}\mathbf{Z}^\eta_{\Lambda;\alpha\beta_1+(1-\alpha)\beta_2,\alpha h_1+(1-\alpha)h_2} &= \sum_{\omega\in\Omega^\eta_\Lambda} e^{-\alpha\mathscr{H}_{\Lambda;\beta_1,h_1}(\omega)-(1-\alpha)\mathscr{H}_{\Lambda;\beta_2,h_2}(\omega)} \\ &\leq \Big(\sum_{\omega\in\Omega^\eta_\Lambda} e^{-\mathscr{H}_{\Lambda;\beta_1,h_1}(\omega)}\Big)^\alpha \Big(\sum_{\omega\in\Omega^\eta_\Lambda} e^{-\mathscr{H}_{\Lambda;\beta_2,h_2}(\omega)}\Big)^{(1-\alpha)} .\end{aligned}$$

Therefore, ψ_Λ^η is convex:

$$\psi_\Lambda^\eta\big(\alpha\beta_1 + (1-\alpha)\beta_2, \alpha h_1 + (1-\alpha)h_2\big) \leq \alpha\psi_\Lambda^\eta(\beta_1, h_1) + (1-\alpha)\psi_\Lambda^\eta(\beta_2, h_2). \quad \square$$

Of course, the finite-volume pressure $\psi_\Lambda^\#$ depends on Λ and on the boundary condition used. However, as the following theorem shows, when Λ is so large that

$|\Lambda| \gg |\partial\Lambda|$, the boundary condition and the shape of Λ only provide negligible corrections: there exists a function $\psi(\beta, h)$ such that

$$\psi_\Lambda^\#(\beta, h) = \psi(\beta, h) + O(|\partial\Lambda|/|\Lambda|)\,.$$

$\psi(\beta, h)$ then provides a better candidate for the corresponding thermodynamic potential, since the latter does not depend on the "details" of the observed system, such as its shape.

Theorem 3.6. *In the thermodynamic limit, the* ***pressure***

$$\psi(\beta, h) \stackrel{\text{def}}{=} \lim_{\Lambda \Uparrow \mathbb{Z}^d} \psi_\Lambda^\#(\beta, h)$$

is well defined and independent of the sequence $\Lambda \Uparrow \mathbb{Z}^d$ and of the type of boundary condition. Moreover, ψ is convex (as a function on $\mathbb{R}_{\geq 0} \times \mathbb{R}$) and is even as a function of h.

Proof ▶ *Existence of the limit.* We start by proving convergence in the case of the free boundary condition. The proof is done in two steps. We first show existence of the limit

$$\lim_{n\to\infty} \psi_{D_n}^{\varnothing}(\beta, h)\,,$$

where $D_n \stackrel{\text{def}}{=} \{1, 2, \ldots, 2^n\}^d$. After that, we extend the convergence to any sequence $\Lambda_n \Uparrow \mathbb{Z}^d$. Since the pair (β, h) is fixed, we will omit it from the notations most of the time, until the end of the proof.

The pressure associated with the box D_{n+1} will be shown to be close to the one associated with the box D_n. Indeed, let us decompose D_{n+1} into 2^d disjoint translates of D_n, denoted by $D_n^{(1)}, \ldots, D_n^{(2^d)}$:

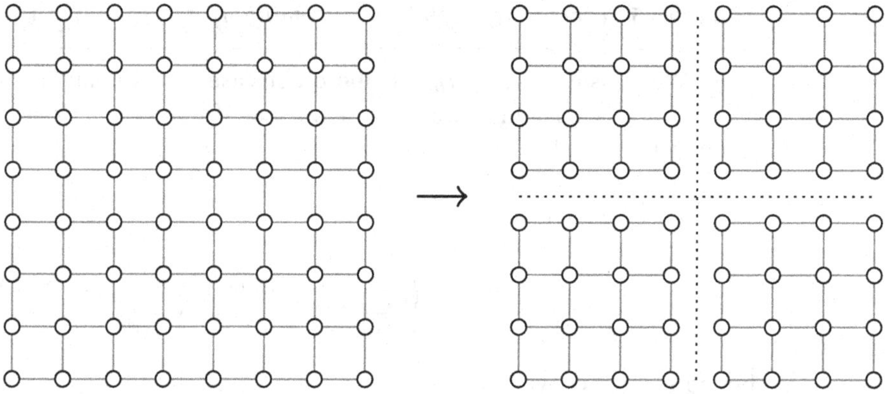

Figure 3.3. A cube D_{n+1} and its partition into 2^d translates of D_n. The interaction between different sub-boxes is denoted by $R_n(\omega)$.

The energy of ω in D_{n+1} can be written as

$$\mathscr{H}^{\varnothing}_{D_{n+1}} = \sum_{i=1}^{2^d} \mathscr{H}^{\varnothing}_{D_n^{(i)}} + R_n \,,$$

where R_n represents the energy of interaction between pairs of spins that belong to different sub-boxes. Since each face of D_{n+1} contains $(2^{n+1})^{d-1}$ points, we have $|R_n(\omega)| \leq \beta d (2^{n+1})^{d-1}$. To obtain an upper bound on the partition function, we can write $\mathscr{H}^{\varnothing}_{D_{n+1}} \geq -\beta d (2^{n+1})^{d-1} + \sum_{i=1}^{2^d} \mathscr{H}^{\varnothing}_{D_n^{(i)}}$, which yields

$$\mathbf{Z}^{\varnothing}_{D_{n+1}} \leq e^{\beta d 2^{(n+1)(d-1)}} \sum_{\omega \in \Omega_{D_{n+1}}} \prod_{i=1}^{2^d} \exp\big(-\mathscr{H}^{\varnothing}_{D_n^{(i)}}(\omega)\big) \,.$$

Splitting the sum over $\omega \in D_{n+1}$ into 2^d sums over $\omega^{(i)} \in D_n^{(i)}$,

$$\sum_{\omega \in \Omega_{D_{n+1}}} \prod_{i=1}^{2^d} \exp\big(-\mathscr{H}^{\varnothing}_{D_n^{(i)}}(\omega)\big) = \prod_{i=1}^{2^d} \sum_{\omega^{(i)} \in \Omega_{D_n^{(i)}}} \exp\big(-\mathscr{H}^{\varnothing}_{D_n^{(i)}}(\omega^{(i)})\big) = \big(\mathbf{Z}^{\varnothing}_{D_n}\big)^{2^d} \,,$$

where we have used the fact that $\mathbf{Z}^{\varnothing}_{D_n^{(i)}} = \mathbf{Z}^{\varnothing}_{D_n}$ for all i. A lower bound can be obtained in a similar fashion, leading to

$$e^{-\beta d 2^{(n+1)(d-1)}} \big(\mathbf{Z}^{\varnothing}_{D_n}\big)^{2^d} \leq \mathbf{Z}^{\varnothing}_{D_{n+1}} \leq e^{\beta d 2^{(n+1)(d-1)}} \big(\mathbf{Z}^{\varnothing}_{D_n}\big)^{2^d} \,.$$

After taking the logarithm, dividing by $|D_{n+1}| = 2^{d(n+1)}$ and taking n large enough,

$$|\psi^{\varnothing}_{D_{n+1}} - \psi^{\varnothing}_{D_n}| \leq \beta \, d \, 2^{-(n+1)} \,.$$

This implies that ψ_{D_n} is a Cauchy sequence: for all $n \leq m$,

$$|\psi^{\varnothing}_{D_m} - \psi^{\varnothing}_{D_n}| \leq \beta \, d \sum_{k=n+1}^{m} 2^{-k} = \beta d (2^{-n} - 2^{-m}) \,.$$

Therefore, $\lim_{n\to\infty} \psi^{\varnothing}_{D_n}$ exists; we denote it by ψ.

Let us now consider an arbitrary sequence $\Lambda_n \Uparrow \mathbb{Z}^d$. We fix some integer k and consider a partition of $\mathbb{Z}^d$ into adjacent disjoint translates of D_k. For each n, consider a minimal covering of Λ_n by elements $D_k^{(j)}$ of the partition, and let $[\Lambda_n] \stackrel{\text{def}}{=} \bigcup_j D_k^{(j)}$:

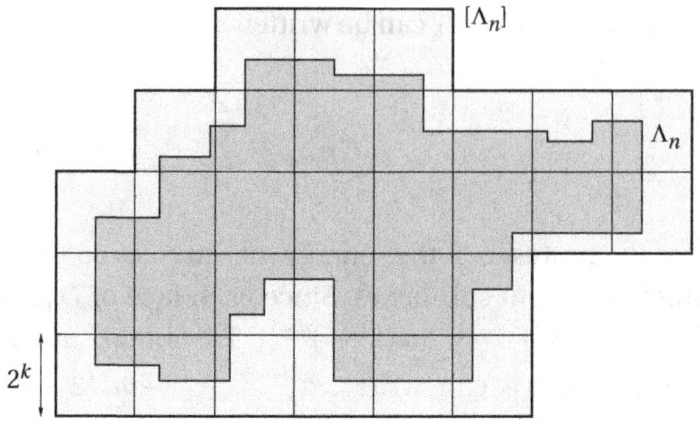

We use the estimate

$$|\psi^{\varnothing}_{\Lambda_n} - \psi| \le |\psi^{\varnothing}_{\Lambda_n} - \psi^{\varnothing}_{[\Lambda_n]}| + |\psi^{\varnothing}_{[\Lambda_n]} - \psi^{\varnothing}_{D_k}| + |\psi^{\varnothing}_{D_k} - \psi| . \tag{3.4}$$

Fix $\epsilon > 0$. Since $\psi^{\varnothing}_{D_k} \to \psi$ when $k \to \infty$, there exists k_0, depending on β, h and ϵ, such that $|\psi^{\varnothing}_{D_k} - \psi| \le \epsilon/3$ for all $k \ge k_0$. We then compute $\psi^{\varnothing}_{[\Lambda_n]}$ by writing

$$\mathscr{H}^{\varnothing}_{[\Lambda_n]} = \sum_j \mathscr{H}^{\varnothing}_{D_k^{(j)}} + W_n ,$$

where $|W_n| \le \beta \frac{|[\Lambda_n]|}{|D_k|} d(2^k)^{d-1} = \beta\, d\, 2^{-k} |[\Lambda_n]|$. Therefore, there exists k_1 (also depending on β and ϵ) such that

$$|\psi^{\varnothing}_{[\Lambda_n]} - \psi^{\varnothing}_{D_k}| \le \beta d 2^{-k} < \epsilon/3 ,$$

for all $k \ge k_1$. Let us then fix $k \ge \max\{k_0, k_1\}$. Let us write $\Delta_n \overset{\text{def}}{=} [\Lambda_n] \setminus \Lambda_n$. We observe that

$$\left| \mathscr{H}^{\varnothing}_{\Lambda_n} - \mathscr{H}^{\varnothing}_{[\Lambda_n]} \right| \le (2d\beta + |h|)\, |\Delta_n| .$$

Therefore,

$$\begin{aligned}\mathbf{Z}^{\varnothing}_{[\Lambda_n]} = \sum_{\omega \in \Omega_{[\Lambda_n]}} e^{-\mathscr{H}^{\varnothing}_{[\Lambda_n]}(\omega)} &\le \sum_{\omega \in \Omega_{\Lambda_n}} e^{-\mathscr{H}^{\varnothing}_{\Lambda_n}(\omega)} \sum_{\omega' \in \Omega_{\Delta_n}} e^{(2d\beta + |h|)\, |\Delta_n|} \\ &= e^{(2d\beta + |h| + \log 2)\, |\Delta_n|}\, \mathbf{Z}^{\varnothing}_{\Lambda_n} .\end{aligned}$$

Proceeding similarly to get a lower bound and observing that Δ_n contains at most $|\partial^{\text{in}} \Lambda_n||D_k|$ vertices, this yields

$$\left| \log \mathbf{Z}^{\varnothing}_{\Lambda_n} - \log \mathbf{Z}^{\varnothing}_{[\Lambda_n]} \right| \le |\partial^{\text{in}} \Lambda_n||D_k| \big(2d\beta + |h| + \log 2\big) . \tag{3.5}$$

Since

$$1 \le \frac{|[\Lambda_n]|}{|\Lambda_n|} \le 1 + \frac{|\partial^{\text{in}} \Lambda_n||D_k|}{|\Lambda_n|}$$

and since $\psi^{\varnothing}_{\Lambda}$ is uniformly bounded (for example, by $2d\beta + |h| + \log 2$), it follows from (3.3) and (3.5) that

$$|\psi^{\varnothing}_{\Lambda_n} - \psi^{\varnothing}_{[\Lambda_n]}| \le \epsilon/3 ,$$

for all n large enough. Combining all these estimates, we conclude from (3.4) that, when n is sufficiently large,

$$|\psi^{\varnothing}_{\Lambda_n} - \psi| \le \epsilon\,.$$

(An alternative proof of convergence, using a subadditivity argument, is proposed in Exercise 3.3.)

▶ *Independence of boundary condition.* The fact that all boundary conditions lead to the same limit follows from the observation that, for any $\Lambda \Subset \mathbb{Z}^d$ and any η, $|\mathscr{H}^{\eta}_{\Lambda} - \mathscr{H}^{\varnothing}_{\Lambda}| \le 2d\beta|\partial^{\text{in}}\Lambda|$. Indeed, the latter implies that

$$e^{-\beta 2d|\partial^{\text{in}}\Lambda|}\,\mathbf{Z}^{\varnothing}_{\Lambda} \le \mathbf{Z}^{\eta}_{\Lambda} \le e^{\beta 2d|\partial^{\text{in}}\Lambda|}\,\mathbf{Z}^{\varnothing}_{\Lambda}\,.$$

Applying this to each Λ_n and using (3.3) shows that $\lim_{\Lambda_n \Uparrow \mathbb{Z}^d} \psi^{\eta}_{\Lambda_n}$ exists and coincides with ψ. A completely similar argument, comparing $\mathbf{Z}^{\varnothing}_{V_n}$ and $\mathbf{Z}^{\text{per}}_{V_n}$, shows that $\lim_{n\to\infty} \psi^{\text{per}}_{V_n} = \psi$.

▶ *Convexity.* Since $(\beta, h) \mapsto \psi^{\#}_{\Lambda}(\beta, h)$ is convex (Lemma 3.5), its limit $\Lambda \Uparrow \mathbb{Z}^d$ is also convex (Exercise B.3).

▶ *Symmetry.* The fact that $h \mapsto \psi(\beta, h)$ is even is a direct consequence of the above and Exercise 3.2. □

The following exercise provides an alternative proof for the existence of the pressure (along a specific sequence of boxes), using a subadditivity argument. [1]

Exercise 3.3. Let $\mathscr{R}$ be the set of all parallelepipeds of $\mathbb{Z}^d$, that is, sets of the form $\Lambda = [a_1, b_1] \times [a_2, b_2] \times \cdots \times [a_d, b_d] \cap \mathbb{Z}^d$.

1. By writing $\sigma_i\sigma_j = (\sigma_i\sigma_j - 1) + 1$, express the Hamiltonian as $\mathscr{H}^{\varnothing}_{\Lambda} = \widetilde{\mathscr{H}}^{\varnothing}_{\Lambda} - \beta|\mathscr{E}_{\Lambda}|$, and observe that, for any disjoint sets $\Lambda_1, \Lambda_2 \Subset \mathbb{Z}^d$,

$$\widetilde{\mathscr{H}}^{\varnothing}_{\Lambda_1\cup\Lambda_2} \ge \widetilde{\mathscr{H}}^{\varnothing}_{\Lambda_1} + \widetilde{\mathscr{H}}^{\varnothing}_{\Lambda_2}\,.$$

   Conclude that

$$\widetilde{\mathbf{Z}}^{\varnothing}_{\Lambda_1\cup\Lambda_2} \le \widetilde{\mathbf{Z}}^{\varnothing}_{\Lambda_1}\widetilde{\mathbf{Z}}^{\varnothing}_{\Lambda_2}\,. \tag{3.6}$$

2. Use (3.6) and Lemma B.6 to show the existence of $\lim_{n\to\infty} \frac{1}{|\Lambda_n|} \log \widetilde{\mathbf{Z}}^{\varnothing}_{\Lambda_n}$ along any sequence $\Lambda_n \uparrow \mathbb{Z}^d$ with $\Lambda_n \in \mathscr{R}$ for all n.

3.2.3 Magnetization

As we already emphasized in the previous chapters, another quantity of central importance is the **magnetization density** in $\Lambda \Subset \mathbb{Z}^d$, which is the random variable

$$m_{\Lambda} \stackrel{\text{def}}{=} \frac{1}{|\Lambda|} M_{\Lambda}\,,$$

where $M_\Lambda \stackrel{\text{def}}{=} \sum_{i\in\Lambda} \sigma_i$ is the **total magnetization**. We also define, for any $\Lambda \subset \mathbb{Z}^d$,

$$m^\#_\Lambda(\beta, h) \stackrel{\text{def}}{=} \langle m_\Lambda \rangle^\#_{\Lambda;\beta,h} .$$

As can be easily checked,

$$m^\#_\Lambda(\beta, h) = \frac{\partial \psi^\#_\Lambda}{\partial h}(\beta, h) . \tag{3.7}$$

Exercise 3.4. Check that, more generally, the **cumulant generating function** associated with M_Λ (see Appendix B.8.3) can be expressed as

$$\log\langle e^{tM_\Lambda}\rangle^\#_{\Lambda;\beta,h} = |\Lambda| \left(\psi^\#_\Lambda(\beta, h+t) - \psi^\#_\Lambda(\beta, h)\right).$$

Deduce that the rth cumulant of M_Λ is given by

$$c_r(M_\Lambda) = |\Lambda| \frac{\partial^r \psi^\#_\Lambda}{\partial h^r}(\beta, h) .$$

The observation made in the previous exercise explains the important role played by the pressure, a fact that might surprise a reader with little familiarity with physics; after all, the partition function is just a normalizing factor. Indeed, we explain in Appendix B.8.3 that the cumulant generating function of a random variable encodes all the information about its distribution. In view of the central importance of the magnetization in characterizing the phase transition, as explained in Chapters 1 and 2, the pressure should hold precious information about the occurrence of a phase transition in the model. ◇

It will turn out to be important to determine whether (3.7) still holds in the thermodynamic limit. There are really two issues here: on the one hand, one has to address the existence of $\lim_{\Lambda\Uparrow\mathbb{Z}^d} \frac{\partial \psi^\#_\Lambda}{\partial h}(\beta, h)$ and whether the limit depends on the chosen boundary condition; on the other hand, there is also the problem of interchanging the thermodynamic limit and the differentiation with respect to h, that is, to verify whether it is true that

$$\lim_{\Lambda\Uparrow\mathbb{Z}^d} \frac{\partial \psi^\#_\Lambda}{\partial h} \stackrel{?}{=} \frac{\partial}{\partial h} \lim_{\Lambda\Uparrow\mathbb{Z}^d} \psi^\#_\Lambda = \frac{\partial \psi}{\partial h} .$$

These issues are intimately related to the differentiability of the pressure as a function of h. This is a delicate matter, which will be investigated in Section 3.7. Nevertheless, partial answers can already be deduced from the convexity properties of the pressure.

For instance, the one-sided derivatives of $h \mapsto \psi(\beta, h)$,

$$\frac{\partial \psi}{\partial h^-}(\beta, h) \stackrel{\text{def}}{=} \lim_{h'\uparrow h} \frac{\psi(\beta, h') - \psi(\beta, h)}{h' - h}, \qquad \frac{\partial \psi}{\partial h^+}(\beta, h) \stackrel{\text{def}}{=} \lim_{h'\downarrow h} \frac{\psi(\beta, h') - \psi(\beta, h)}{h' - h},$$

exist everywhere (by item 1 of Theorem B.12) and are, respectively, left- and right-continuous (by item 5). Of course, the pressure will be differentiable with respect to h if and only if these two one-sided derivatives coincide. It is thus natural to introduce, for each β, the set

$$\begin{aligned}\mathfrak{B}_\beta &\stackrel{\text{def}}{=} \{h \in \mathbb{R} : \psi(\beta, \cdot) \text{ is not differentiable at } h\} \\ &= \{h \in \mathbb{R} : \tfrac{\partial\psi}{\partial h^-}(\beta, h) \neq \tfrac{\partial\psi}{\partial h^+}(\beta, h)\}.\end{aligned}$$

It follows from item 6 of Theorem B.12 that, for each β, the set $\mathfrak{B}_\beta$ is at most countable. On the complement of this set, one can answer the question raised above.

Corollary 3.7. *For all $h \notin \mathfrak{B}_\beta$, the **average magnetization density***

$$m(\beta, h) \stackrel{\text{def}}{=} \lim_{\Lambda \Uparrow \mathbb{Z}^d} m^\#_\Lambda(\beta, h)$$

is well defined, independent of the sequence $\Lambda \Uparrow \mathbb{Z}^d$ and of the boundary condition, and satisfies

$$m(\beta, h) = \frac{\partial\psi}{\partial h}(\beta, h). \tag{3.8}$$

Moreover, the function $h \mapsto m(\beta, h)$ is nondecreasing on $\mathbb{R} \setminus \mathfrak{B}_\beta$ and is continuous at every $h \notin \mathfrak{B}_\beta$. It is however discontinuous at each $h \in \mathfrak{B}_\beta$: for any $h \in \mathfrak{B}_\beta$,

$$\lim_{h' \downarrow h} m(\beta, h') = \tfrac{\partial\psi}{\partial h^+}(\beta, h), \quad \lim_{h' \uparrow h} m(\beta, h') = \tfrac{\partial\psi}{\partial h^-}(\beta, h). \tag{3.9}$$

*In particular, the **spontaneous magnetization***

$$m^*(\beta) \stackrel{\text{def}}{=} \lim_{h \downarrow 0} m(\beta, h)$$

is always well defined.

Proof When $h \notin \mathfrak{B}_\beta$,

$$\frac{\partial\psi}{\partial h}(\beta, h) = \frac{\partial}{\partial h} \lim_{\Lambda \Uparrow \mathbb{Z}^d} \psi^\#_\Lambda(\beta, h) = \lim_{\Lambda \Uparrow \mathbb{Z}^d} \frac{\partial}{\partial h} \psi^\#_\Lambda(\beta, h) = \lim_{\Lambda \Uparrow \mathbb{Z}^d} m^\#_\Lambda(\beta, h),$$

which proves (3.8), the existence of the thermodynamic limit of the magnetization and the fact that it depends neither on the boundary condition nor on the sequence of volumes. Above, the second equality follows from item 7 of Theorem B.12 and the third one from (3.7).

The monotonicity and continuity of $h \mapsto m(\beta, h)$ on $\mathbb{R} \setminus \mathfrak{B}_\beta$ follow from (3.8) and items 4 and 5 of Theorem B.12.

Suppose now that $h \in \mathfrak{B}_\beta$ and let $(h_k)_{k \geq 1}$ be an arbitrary sequence in $\mathbb{R} \setminus \mathfrak{B}_\beta$ such that $h_k \downarrow h$ (there are always such sequences, since $\mathfrak{B}_\beta$ is at most countable). By (3.8), $\frac{\partial\psi}{\partial h^+}(\beta, h_k) = m(\beta, h_k)$ for all k. The claim (3.9) thus follows from (3.8) and item 5 of Theorem B.12. □

3.2.4 A First Definition of Phase Transition

The above discussion shows that the average magnetization density is discontinuous precisely when the pressure is not differentiable in h. This leads to the following:

> **Definition 3.8.** The pressure ψ exhibits a **first-order phase transition** at (β, h) if $h \mapsto \psi(\beta, h)$ fails to be differentiable at that point.

Later, we will introduce another notion of first-order phase transition, of a more probabilistic nature. Determining whether phase transitions occur or not, and at which values of the parameters, is one of the main objectives of this chapter.

3.3 One-Dimensional Ising Model

Before pursuing the general case, we briefly discuss the one-dimensional Ising model, for which explicit computations are possible.

> **Theorem 3.9.** $(d = 1)$ *For all $\beta \geq 0$ and all $h \in \mathbb{R}$, the pressure $\psi(\beta, h)$ of the one-dimensional Ising model is given by*
>
> $$\psi(\beta, h) = \log\Big\{e^{\beta}\cosh(h) + \sqrt{e^{2\beta}\cosh^2(h) - 2\sinh(2\beta)}\Big\}. \qquad (3.10)$$

The explicit expression (3.10) shows that $h \mapsto \psi(\beta, h)$ is differentiable (real-analytic in fact) everywhere, for all $\beta \geq 0$, thus showing that $\mathfrak{B}_\beta = \varnothing$ when $d = 1$.

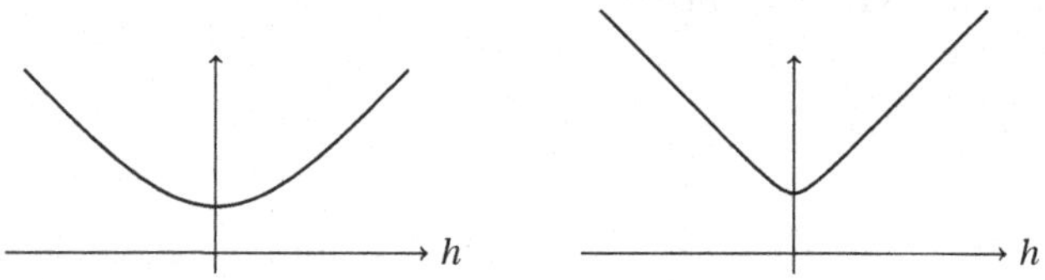

Figure 3.4. The pressure $h \mapsto \psi(\beta, h)$ of the one-dimensional Ising model, analytic in h at all temperatures ($\beta = 0.8$ on the left, $\beta = 2$ on the right).

Consequently, as seen in Corollary 3.7, the average magnetization density $m(\beta, h)$ is given by

$$m(\beta, h) = \frac{\partial \psi}{\partial h}(\beta, h), \qquad \forall h \in \mathbb{R}.$$

Since $h \mapsto \psi(\beta, h)$ is analytic, its derivative $h \mapsto m(\beta, h)$ is also analytic, in particular continuous. Therefore, $m^*(\beta) = \lim_{h \downarrow 0} m(\beta, h) = m(\beta, 0)$. But, since (see Exercise 3.2) $\psi(\beta, h) = \psi(\beta, -h)$, we get $\frac{\partial \psi}{\partial h}(\beta, 0) = 0$. This shows that

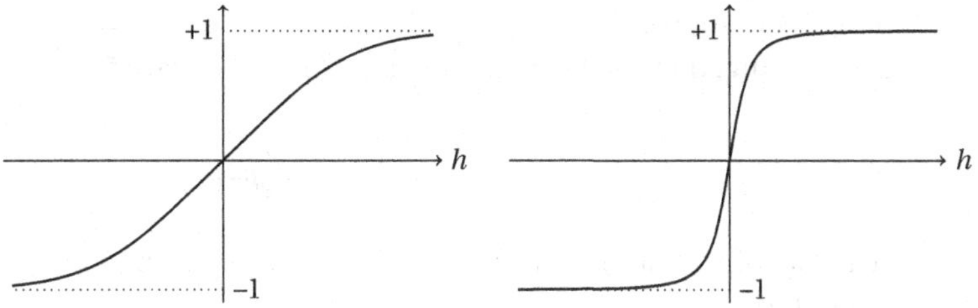

Figure 3.5. The average magnetization density $m(\beta, h)$ of the one-dimensional Ising model (for the same values of β as in Figure 3.4).

the spontaneous magnetization of the one-dimensional Ising model is zero at all temperatures:

$$m^*(\beta) = 0\,, \qquad \forall \beta > 0\,.$$

In particular, the model exhibits *paramagnetic* behavior at all nonzero temperatures (remember the discussion in Section 1.4.3). We will provide an alternative proof of this fact in Section 3.7.3.

Only in the limit $\beta \to \infty$ does $\psi(\beta, h)$ become nondifferentiable at $h = 0$, as seen in the following exercise.

Exercise 3.5. Using (3.10), compute $m(\beta, h)$. Check that

$$\lim_{h\to\pm\infty} m(\beta, h) = \pm 1\,, \quad \forall \beta \geq 0\,,$$

$$\lim_{\beta\to\infty} m(\beta, h) = \begin{cases} +1 & \text{if } h > 0\,,\\ 0 & \text{if } h = 0\,,\\ -1 & \text{if } h < 0\,. \end{cases}$$

Proof of Theorem 3.9. As seen in Theorem 3.6, the pressure is independent of the choice of boundary condition and of the sequence of volumes $\Lambda \Uparrow \mathbb{Z}$. The most convenient choice is to work on the torus $\mathbb{T}_n$, that is, to use $V_n = \{0, \dots, n-1\}$ with periodic boundary conditions; see Figure 3.1 (left). The advantage of this particular choice is that $\mathbf{Z}^{\text{per}}_{V_n;\beta,h}$ can be written as the trace of a 2×2 matrix. Indeed, writing $\omega_n \equiv \omega_0$,

$$\begin{aligned}
\mathbf{Z}^{\text{per}}_{V_n;\beta,h} &= \sum_{\omega\in\Omega_{V_n}} e^{-\mathscr{H}^{\text{per}}_{V_n;\beta,h}(\omega)}\\
&= \sum_{\omega_0=\pm1} \cdots \sum_{\omega_{n-1}=\pm1} \prod_{i=0}^{n-1} e^{\beta\omega_i\omega_{i+1}+h\omega_i}\\
&= \sum_{\omega_0=\pm1} \cdots \sum_{\omega_{n-1}=\pm1} \prod_{i=0}^{n-1} A_{\omega_i,\omega_{i+1}}\,,
\end{aligned}$$

where the numbers $A_{+,+} = e^{\beta+h}$, $A_{+,-} = e^{-\beta+h}$, $A_{-,+} = e^{-\beta-h}$ and $A_{-,-} = e^{\beta-h}$ can be arranged in the form of a matrix, called the **transfer matrix**:

$$A \stackrel{\text{def}}{=} \begin{pmatrix} e^{\beta+h} & e^{-\beta+h} \\ e^{-\beta-h} & e^{\beta-h} \end{pmatrix}. \tag{3.11}$$

The useful observation is that $\mathbf{Z}^{\text{per}}_{V_n;\beta,h}$ can now be interpreted as the trace of the nth power of A:

$$\mathbf{Z}^{\text{per}}_{V_n;\beta,h} = \sum_{\omega_0=\pm 1} (A^n)_{\omega_0,\omega_0} = \text{Tr}\,(A^n).$$

A straightforward computation shows that the eigenvalues λ_+ and λ_- of A are given by

$$\lambda_\pm = e^\beta \cosh(h) \pm \sqrt{e^{2\beta}\cosh^2(h) - 2\sinh(2\beta)}.$$

Writing $A = BDB^{-1}$, with $D = \left(\begin{smallmatrix} \lambda_+ & 0 \\ 0 & \lambda_- \end{smallmatrix}\right)$, and using the fact that $\text{Tr}\,(GH) = \text{Tr}\,(HG)$, we get

$$\mathbf{Z}^{\text{per}}_{V_n;\beta,h} = \text{Tr}\,(A^n) = \text{Tr}\,(BD^nB^{-1}) = \text{Tr}\,(D^n) = \lambda_+^n + \lambda_-^n.$$

Since $\lambda_+ > \lambda_-$, this gives $\psi(\beta,h) = \log\lambda_+$ and (3.10) is proved. (An interested reader with some familiarity with discrete-time, finite-state Markov chains can find some additional information on this topic in Section 3.10.4.) □

When $h = 0$, there exist several simple ways of computing the pressure of the one-dimensional Ising model: two are proposed in the following exercise and another one will be proposed in Exercise 3.26.

Exercise 3.6. (Assuming $h = 0$.)

1. Configurations can be characterized by the collection of edges $\{i, i+1\}$ such that $\omega_i \neq \omega_{i+1}$. What is the contribution of a configuration with k such edges? Use that to compute the pressure.
2. Express the partition function in terms of the variables $(\omega_1, \tau_1, \ldots, \tau_{n-1})$, where $\tau_i = \omega_{i-1}\omega_i$. Use this to compute the pressure.

Hint: Since this does not affect the end result, one should choose a boundary condition that simplifies the analysis. We recommend using the free boundary condition.

With an explicit analytic expression for the pressure, we can extract information on the typical values of the magnetization density in large finite boxes. We will only consider the case $h = 0$; the extension to an arbitrary magnetic field is left as an exercise.

A consequence of the next theorem is that m_{Λ_n} concentrates on 0 under $\mu^{\#}_{\Lambda_n;\beta,0}$, as $n \to \infty$, for any type of boundary condition.

Theorem 3.10. *($d=1$) Let $0<\beta<\infty$ and consider any sequence $\Lambda_n \Uparrow \mathbb{Z}$, with an arbitrary boundary condition #. For all $\epsilon>0$, there exists $c=c(\beta,\epsilon)>0$ such that, for large enough n,*

$$\mu^{\#}_{\Lambda_n;\beta,0}(m_{\Lambda_n} \notin (-\epsilon,\epsilon)) \le e^{-c|\Lambda_n|}. \tag{3.12}$$

Proof of Theorem 3.10. We start by writing

$$\mu^{\#}_{\Lambda_n;\beta,0}(m_{\Lambda_n} \notin (-\epsilon,\epsilon)) = \mu^{\#}_{\Lambda_n;\beta,0}(m_{\Lambda_n} \ge \epsilon) + \mu^{\#}_{\Lambda_n;\beta,0}(m_{\Lambda_n} \le -\epsilon).$$

These two terms can be studied in the same way. The starting point is to use Chernov's inequality (B.19): for all $h \ge 0$,

$$\mu^{\#}_{\Lambda_n;\beta,0}(m_{\Lambda_n} \ge \epsilon) \le e^{-h\epsilon|\Lambda_n|}\langle e^{hm_{\Lambda_n}|\Lambda_n|}\rangle^{\#}_{\Lambda_n;\beta,0}.$$

Since $\langle e^{hm_{\Lambda_n}|\Lambda_n|}\rangle^{\#}_{\Lambda_n;\beta,0} = \mathbf{Z}^{\#}_{\Lambda_n;\beta,h}/\mathbf{Z}^{\#}_{\Lambda_n;\beta,0}$, we have

$$\begin{aligned}\limsup_{n\to\infty} \frac{1}{|\Lambda_n|}\log \mu^{\#}_{\Lambda_n;\beta,0}(m_{\Lambda_n} \ge \epsilon) &\le \lim_{n\to\infty}\left(\psi^{\#}_{\Lambda_n}(\beta,h) - \psi^{\#}_{\Lambda_n}(\beta,0)\right) - h\epsilon \\ &= I_\beta(h) - h\epsilon,\end{aligned}$$

where $I_\beta(h) \stackrel{\text{def}}{=} \psi(\beta,h) - \psi(\beta,0)$. Since $h \ge 0$ was arbitrary, we can minimize over the latter:

$$\limsup_{n\to\infty} \frac{1}{|\Lambda_n|}\log \mu^{\#}_{\Lambda_n;\beta,0}(m_{\Lambda_n} \ge \epsilon) \le -\sup_{h\ge 0}\{h\epsilon - I_\beta(h)\}. \tag{3.13}$$

In order to prove that $\mu^{\#}_{\Lambda_n;\beta}(m_{\Lambda_n} \ge \epsilon)$ decays exponentially fast in n, one must establish that $\sup_{h\ge 0}\{h\epsilon - I_\beta(h)\} > 0$. Remember that the explicit expression for ψ provided by Theorem 3.9 is real-analytic in h. Moreover, $I_\beta(0)=0$ and, if $I'_\beta = \frac{\partial}{\partial h}I_\beta$, then $I'_\beta(0)=0$ and $I'_\beta(h) \to 1$ as $h\to\infty$, as was seen in Exercise 3.5. Therefore, for each $0<\epsilon<1$, there exists some $h_*>0$, depending on ϵ and β, such that $\sup_{h\ge 0}\{h\epsilon - I_\beta(h)\} = h_*\epsilon - I_\beta(h_*) > 0$ (see Figure 3.6). This proves (3.12). □

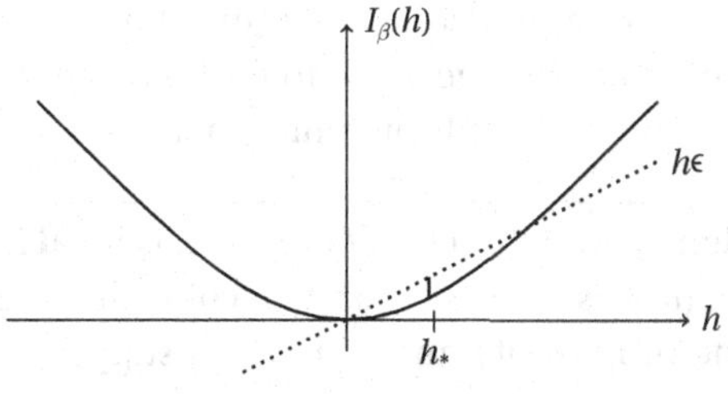

Figure 3.6. A picture showing the graphs of $h \mapsto I_\beta(h) = \psi(\beta,h) - \psi(\beta,0)$ and $h \mapsto h\epsilon$, on which it is clear that $\sup_{h\ge 0}\{h\epsilon - I_\beta(h)\} > 0$ as soon as $\epsilon > 0$.

Exercise 3.7. Proceeding as above, show that, under $\mu^{\#}_{\Lambda_n;\beta,h}$ with $h \ne 0$, m_{Λ_n} converges to $m(\beta,h)$ as $n\to\infty$ (in the same sense as in (3.12)), for any boundary condition.

As explained above, the pressure contains a lot of information on the magnetization density. We will see in the following sections that smoothness of the pressure also guarantees *uniqueness of the infinite-volume Gibbs state.*

As we have seen in this section, explicitly computing the pressure yields useful information on the system. Unfortunately, computing the pressure becomes much more difficult, if at all possible, in higher dimensions. In fact, in spite of much effort, the only known results are for the two-dimensional Ising model with $h = 0$. In that case, Onsager determined, in a celebrated work, the explicit expression for the pressure:

$$\psi(\beta,0) = \ln 2 + \frac{1}{8\pi^2}\int_0^{2\pi}\int_0^{2\pi} \ln\bigl\{(\cosh(2\beta))^2 - \sinh(2\beta)(\cos\theta_1 + \cos\theta_2)\bigr\}\, d\theta_1 d\theta_2 . \tag{3.14}$$

If we want to gain some understanding of the behavior of the Ising model on $\mathbb{Z}^d$, $d \geq 2$, other approaches are therefore required. This will be our main focus in the remainder of this chapter.

3.4 Infinite-Volume Gibbs States

The pressure only provides information about the thermodynamic behavior of the system in large volumes. If one is interested in the statistical properties of general observables, such as the fluctuations of the magnetization density in a finite region or the correlations between far-apart spins, one needs to understand the behavior of the Gibbs distribution $\mu^{\#}_{\Lambda;\beta,h}$ in large volumes.

One way of doing this is to define infinite-volume Gibbs measures by taking some sequence $\Lambda_n \uparrow \mathbb{Z}^d$ and by considering the accumulation points (if any) of sequences of the type $(\mu^{\eta_n}_{\Lambda_n;\beta,h})_{n\geq 1}$. This is possible and will be done in detail in Chapter 6, by introducing a suitable notion of convergence for sequences of probability measures. Such an approach necessitates, however, rather abstract topological and measure-theoretic notions. In the present chapter, we avoid this by following a more hands-on approach: a *state* (in infinite volume) will be identified with an assignment of an average value to each *local function*, that is, to each observable whose value depends only on finitely many spins.

> **Definition 3.11.** A function $f\colon \Omega \to \mathbb{R}$ is **local** if there exists $\Delta \Subset \mathbb{Z}^d$ such that $f(\omega) = f(\omega')$ as soon as ω and ω' coincide on Δ. The smallest[1] such set Δ is called the **support** of f and denoted by $\mathrm{supp}(f)$.

For example, the value taken by the spin at the origin, σ_0, or the magnetization density in a set $\Lambda \Subset \mathbb{Z}^d$, $m_\Lambda = \frac{1}{|\Lambda|}\sum_{i\in\Lambda}\sigma_i$, are local functions with supports given respectively by $\{0\}$ and Λ.

[1] The reason one can speak about the *smallest* such set is the following observation: if a function f is characterized by $(\omega_i)_{i\in\Delta_1}$ and is also characterized by $(\omega_i)_{i\in\Delta_2}$, then it is characterized by $(\omega_i)_{i\in\Delta_1\cap\Delta_2}$.

Remark 3.12. Subsequently, we will occasionally make the following mild abuse of notation: if $f : \Omega \to \mathbb{R}$ is a local function and $\Delta \supset \operatorname{supp}(f)$, then, for any $\omega' \in \Omega_\Delta$, $f(\omega')$ is defined as the value of f evaluated at any configuration $\omega \in \Omega$ such that $\omega_i = \omega'_i$ for all $i \in \Delta$. (Clearly, that value does not depend on the choice of ω.) ⋄

Definition 3.13. An **infinite-volume state** (or simply a **state**) is a mapping associating to each local function f a real number $\langle f \rangle$ and satisfying:

Normalization: $\langle 1 \rangle = 1$.

Positivity: If $f \geq 0$, then $\langle f \rangle \geq 0$.

Linearity: For any $\lambda \in \mathbb{R}$, $\langle f + \lambda g \rangle = \langle f \rangle + \lambda \langle g \rangle$.

The number $\langle f \rangle$ is called the **average of f in the state $\langle \cdot \rangle$**.

Definition 3.14. Let $\Lambda_n \uparrow \mathbb{Z}^d$ and $(\#_n)_{n\geq 1}$ be a sequence of boundary conditions. The sequence of Gibbs distributions $(\mu^{\#_n}_{\Lambda_n;\beta,h})_{n\geq 1}$ is said to **converge to the state $\langle \cdot \rangle$** if and only if

$$\lim_{n\to\infty} \langle f \rangle^{\#_n}_{\Lambda_n;\beta,h} = \langle f \rangle ,$$

for every local function f. The state $\langle \cdot \rangle$ is then called a **Gibbs state (at (β, h))**.

We simply write, as a shorthand,

$$\langle \cdot \rangle = \lim_{n\to\infty} \langle \cdot \rangle^{\#_n}_{\Lambda_n;\beta,h}$$

to indicate that $\langle \cdot \rangle^{\#_n}_{\Lambda_n;\beta,h}$ converges to $\langle \cdot \rangle$.

The above notion of convergence is natural. Indeed, from a thermodynamic perspective, it is expected that the properties of large systems at equilibrium should be well approximated by those of the corresponding infinite systems. In particular, finite-size effects, such as those resulting from the macroscopic shape of the system, should not affect local observations made far from the boundary of the system. The notion of convergence stated above corresponds precisely to a formalization of this principle, by saying that the measurement of a local quantity in a large system, corresponding to $\langle f \rangle^{\#_n}_{\Lambda_n;\beta,h}$, is well approximated by the corresponding measurement $\langle f \rangle$ in the infinite system. This is discussed in a more precise manner in Section 3.10.8. ⋄

Remark 3.15. The reader familiar with functional analysis will probably have noticed that, using the *Riesz–Markov–Kakutani Representation Theorem*, the average $\langle f \rangle$ of a local function f in a state $\langle \cdot \rangle$ can always be seen as the expectation of f under some probability measure μ (on $\{+1, -1\}^{\mathbb{Z}^d}$):

$$\langle f \rangle = \int f \, d\mu .$$

We are mostly interested in states $\langle\cdot\rangle$ that can be constructed as limits of finite-volume Gibbs distributions: $\langle\cdot\rangle = \lim_{n\to\infty}\langle\cdot\rangle^{\eta_n}_{\Lambda_n;\beta,h}$. We will see that, in this case, the corresponding measure μ coincides with the weak limit of the probability measures $\mu^{\eta_n}_{\Lambda_n;\beta,h}$:

$$\mu^{\eta_n}_{\Lambda_n;\beta,h} \Rightarrow \mu\,.$$

This will be explained in Chapter 6, where the necessary framework for weak convergence of probability measures on $\{+1,-1\}^{\mathbb{Z}^d}$ will be introduced. ◇

Since states are defined on the infinite lattice, it is natural to distinguish those that are *translation invariant*. The **translation** by $j \in \mathbb{Z}^d$, $\theta_j\colon \mathbb{Z}^d \to \mathbb{Z}^d$ is defined by

$$\theta_j i \stackrel{\text{def}}{=} i + j\,.$$

Translations can naturally be made to act on configurations: if $\omega \in \Omega$, then $\theta_j\omega$ is defined by

$$(\theta_j\omega)_i \stackrel{\text{def}}{=} \omega_{i-j}\,. \tag{3.15}$$

Definition 3.16. A state $\langle\cdot\rangle$ is **translation invariant** if $\langle f\circ\theta_j\rangle = \langle f\rangle$ for every local function f and for all $j \in \mathbb{Z}^d$.

The first important question is: *can Gibbs states be constructed for the Ising model with parameters* (β, h)*?* The following theorem shows that the constant-spin boundary conditions η^+ and η^- can be used to construct two states that will play a central role subsequently.

Theorem 3.17. *Let $\beta \geq 0$ and $h \in \mathbb{R}$. Along any sequence $\Lambda_n \uparrow \mathbb{Z}^d$, the finite-volume Gibbs distributions with $+$ or $-$ boundary conditions converge to infinite-volume Gibbs states:*

$$\langle\cdot\rangle^+_{\beta,h} = \lim_{n\to\infty}\langle\cdot\rangle^+_{\Lambda_n;\beta,h}\,, \qquad \langle\cdot\rangle^-_{\beta,h} = \lim_{n\to\infty}\langle\cdot\rangle^-_{\Lambda_n;\beta,h}\,. \tag{3.16}$$

The states $\langle\cdot\rangle^+_{\beta,h}$, $\langle\cdot\rangle^-_{\beta,h}$ do not depend on the sequence $(\Lambda_n)_{n\geq 1}$ and are both translation invariant.

The proof will be given later (on page 107), after introducing some important tools.

Remark 3.18. The previous theorem does not claim that $\langle\cdot\rangle^+_{\beta,h}$ and $\langle\cdot\rangle^-_{\beta,h}$ are distinct Gibbs states. Determining the set of values of the parameters β and h for which this is the case will be one of our main tasks in the remainder of this chapter. ◇

More generally, one can prove, albeit in a nonconstructive way, that any sequence of finite-volume Gibbs distributions admits converging subsequences.

Exercise 3.8. Let $(\eta_n)_{n\geq 1}$ be a sequence of boundary conditions and $\Lambda_n \uparrow \mathbb{Z}^d$. Prove that there exists an increasing sequence $(n_k)_{k\geq 1}$ of integers and a Gibbs state $\langle\cdot\rangle$ such that

$$\langle\cdot\rangle = \lim_{k\to\infty} \langle\cdot\rangle^{\eta_{n_k}}_{\Lambda_{n_k};\beta,h}$$

is well defined.

Another explicit example using the free boundary condition will be considered in Exercise 3.16.

3.5 Two Families of Local Functions

The construction of Gibbs states requires proving the existence of the limit

$$\lim_{n\to\infty} \langle f\rangle^{\eta_n}_{\Lambda_n;\beta,h}$$

for each local function f. Ideally, one would like to test convergence only on a restricted family of functions. The following lemma provides two particularly convenient such families, which will be especially well suited for the use of the correlation inequalities introduced in the next section. Define, for all $A \Subset \mathbb{Z}^d$,

$$\sigma_A \stackrel{\text{def}}{=} \prod_{j\in A} \sigma_j\,, \qquad n_A \stackrel{\text{def}}{=} \prod_{j\in A} n_j\,,$$

where $n_j \stackrel{\text{def}}{=} \frac{1}{2}(1+\sigma_j)$ is the **occupation variable** at j.

Lemma 3.19. *Let f be local. There exist real coefficients $(\hat{f}_A)_{A\subset\text{supp}(f)}$ and $(\tilde{f}_A)_{A\subset\text{supp}(f)}$ such that both of the following representations hold:*

$$f = \sum_{A\subset\text{supp}(f)} \hat{f}_A\sigma_A\,, \qquad f = \sum_{A\subset\text{supp}(f)} \tilde{f}_A n_A\,.$$

Proof The following orthogonality relation will be proved below: for all $B \Subset \mathbb{Z}^d$ and all configurations $\omega, \tilde{\omega}$,

$$2^{-|B|}\sum_{A\subset B} \sigma_A(\tilde{\omega})\sigma_A(\omega) = \mathbf{1}_{\{\omega_i=\tilde{\omega}_i,\,\forall i\in B\}}\,. \tag{3.17}$$

Applying (3.17) with $B = \text{supp}(f)$,

$$\begin{aligned}
f(\omega) &= \sum_{\omega'\in\Omega_{\text{supp}(f)}} f(\omega')\,\mathbf{1}_{\{\omega_i=\omega'_i\ \forall i\in\text{supp}(f)\}}\\
&= \sum_{\omega'\in\Omega_{\text{supp}(f)}} f(\omega')\,2^{-|\text{supp}(f)|} \sum_{A\subset\text{supp}(f)} \sigma_A(\omega)\sigma_A(\omega')\\
&= \sum_{A\subset\text{supp}(f)} \Big\{2^{-|\text{supp}(f)|} \sum_{\omega'\in\Omega_{\text{supp}(f)}} f(\omega')\sigma_A(\omega')\Big\}\sigma_A(\omega)\,.
\end{aligned}$$

This shows that the first identity holds with

$$\hat{f}_A = 2^{-|\mathrm{supp}(f)|} \sum_{\omega' \in \Omega_{\mathrm{supp}(f)}} f(\omega')\sigma_A(\omega').$$

Since $\sigma_A = \prod_{i\in A}(2n_i - 1)$, the second identity follows from the first one.

We now prove (3.17). Let us first assume that $\omega_i = \widetilde{\omega}_i$, for all $i \in B$. In that case, $\sigma_A(\widetilde{\omega})\sigma_A(\omega) = \prod_{i\in A} \widetilde{\omega}_i\omega_i = 1$, since $\widetilde{\omega}_i\omega_i = \omega_i^2 = 1$ for all $i \in A \subset B$. This implies (3.17). Assume then that there exists $i \in B$ such that $\omega_i \neq \widetilde{\omega}_i$ (and thus $\omega_i\widetilde{\omega}_i = -1$). Then,

$$\begin{aligned}\sum_{A\subset B} \sigma_A(\widetilde{\omega})\sigma_A(\omega) &= \sum_{A\subset B\setminus\{i\}} \left(\sigma_A(\widetilde{\omega})\sigma_A(\omega) + \sigma_{A\cup\{i\}}(\widetilde{\omega})\sigma_{A\cup\{i\}}(\omega)\right)\\ &= \sum_{A\subset B\setminus\{i\}} \left(\sigma_A(\widetilde{\omega})\sigma_A(\omega) + \omega_i\widetilde{\omega}_i\sigma_A(\widetilde{\omega})\sigma_A(\omega)\right)\\ &= \sum_{A\subset B\setminus\{i\}} \sigma_A(\widetilde{\omega})\sigma_A(\omega)\,(1 + \omega_i\widetilde{\omega}_i) = 0\,. \end{aligned}$$

□

Thanks to the above lemma and to linearity, checking convergence of $(\langle f\rangle^{\eta_n}_{\Lambda_n;\beta,h})_{n\geq 1}$ for all local functions can now be reduced to showing convergence of $(\langle\sigma_A\rangle^{\eta_n}_{\Lambda_n;\beta,h})_{n\geq 1}$ or $(\langle n_A\rangle^{\eta_n}_{\Lambda_n;\beta,h})_{n\geq 1}$ for all finite $A \Subset \mathbb{Z}^d$. This task will be greatly simplified once we have described some of the so-called *correlation inequalities* that hold for the Ising model.

3.6 Correlation Inequalities

Correlation inequalities are one of the major tools in the mathematical analysis of the Ising model. We will use them to construct $\langle\cdot\rangle^+_{\beta,h}$ and $\langle\cdot\rangle^-_{\beta,h}$, and to study many other properties.

The Ising model enjoys many such inequalities, but we will restrict our attention to the two most prominent ones: the GKS and FKG inequalities. Since the proofs are not particularly enlightening, they are postponed to the end of the chapter, in Section 3.8.

3.6.1 GKS Inequalities

As a motivation, consider the Ising model in a volume Λ, with $+$ boundary condition. First, the ferromagnetic nature of the model makes it likely that the $+$ boundary condition will favor a nonnegative magnetization inside the box, at least when $h \geq 0$. Therefore, if i is any point of Λ, it seems reasonable to expect that $h \geq 0$ implies

$$\langle\sigma_i\rangle^+_{\Lambda;\beta,h} \geq 0\,. \tag{3.18}$$

Similarly, knowing that the spin at some vertex j takes the value $+1$ should not decrease the probability of observing a $+$ spin at another given vertex i, that is, one would expect that

$$\mu^+_{\Lambda;\beta,h}(\sigma_i = 1 \mid \sigma_j = 1) \geq \mu^+_{\Lambda;\beta,h}(\sigma_i = 1),$$

which can equivalently be written

$$\mu^+_{\Lambda;\beta,h}(\sigma_i = 1, \sigma_j = 1) \geq \mu^+_{\Lambda;\beta,h}(\sigma_i = 1)\, \mu^+_{\Lambda;\beta,h}(\sigma_j = 1).$$

Since $\mathbf{1}_{\{\sigma_i=1\}} = \frac{1}{2}(\sigma_i + 1)$, this can also be expressed as

$$\langle \sigma_i \sigma_j \rangle^+_{\Lambda;\beta,h} \geq \langle \sigma_i \rangle^+_{\Lambda;\beta,h} \langle \sigma_j \rangle^+_{\Lambda;\beta,h}. \tag{3.19}$$

This is equivalent to asking whether σ_i and σ_j are **positively correlated** under $\mu^+_{\Lambda;\beta,h}$.

Inequalities (3.18) and (3.19) are actually true, and will be particular instances of the *GKS inequalities* (named after Griffiths, Kelly and Sherman) which hold in a more general setting.

Namely, let $\mathbf{J} = (J_{ij})$ be a collection of *nonnegative* real numbers J_{ij} indexed by pairs $\{i,j\} \in \mathscr{E}^{\mathrm{b}}_\Lambda$. Let also $\mathbf{h} = (h_i)$ be a collection of real numbers indexed by vertices of Λ. We write $\mathbf{h} \geq 0$ as a shortcut for $h_i \geq 0$ for all $i \in \Lambda$. We then write, for $\omega \in \Omega^\eta_\Lambda$,

$$\mathscr{H}_{\Lambda;\mathbf{J},\mathbf{h}}(\omega) \stackrel{\mathrm{def}}{=} -\sum_{\{i,j\}\in\mathscr{E}^{\mathrm{b}}_\Lambda} J_{ij}\sigma_i(\omega)\sigma_j(\omega) - \sum_{i\in\Lambda} h_i\sigma_i(\omega). \tag{3.20}$$

We denote the corresponding finite-volume Gibbs distribution by $\mu^\eta_{\Lambda;\mathbf{J},\mathbf{h}}$. Of course, we recover $\mathscr{H}_{\Lambda;\beta,h}$ and $\mu^\eta_{\Lambda;\beta,h}$ by setting $J_{ij} = \beta$ for all $\{i,j\} \in \mathscr{E}^{\mathrm{b}}_\Lambda$ and $h_i = h$ for all $i \in \Lambda$.

The GKS inequalities are mostly restricted to $+$, free and periodic boundary conditions and to nonnegative magnetic fields. They deal with expectations and covariances of random variables of the type σ_A, which is precisely what is needed for the study of the thermodynamic limit.

Theorem 3.20 (GKS inequalities). *Let* $\mathbf{J}, \mathbf{h}$ *be as above and* $\Lambda \Subset \mathbb{Z}^d$. *Assume that* $\mathbf{h} \geq 0$. *Then, for all* $A, B \subset \Lambda$,

$$\langle \sigma_A \rangle^+_{\Lambda;\mathbf{J},\mathbf{h}} \geq 0, \tag{3.21}$$

$$\langle \sigma_A \sigma_B \rangle^+_{\Lambda;\mathbf{J},\mathbf{h}} \geq \langle \sigma_A \rangle^+_{\Lambda;\mathbf{J},\mathbf{h}} \langle \sigma_B \rangle^+_{\Lambda;\mathbf{J},\mathbf{h}}. \tag{3.22}$$

These inequalities remain valid for $\langle \cdot \rangle^\varnothing_{\Lambda;\mathbf{J},\mathbf{h}}$ *and* $\langle \cdot \rangle^{\mathrm{per}}_{\Lambda;\mathbf{J},\mathbf{h}}$.

Exercise 3.9. Let $A \subset \Lambda \Subset \mathbb{Z}^d$. Under the assumptions of Theorem 3.20, prove that $\langle \sigma_A \rangle^+_{\Lambda;\mathbf{J},\mathbf{h}}$ is increasing in both $\mathbf{J}$ and $\mathbf{h}$.

3.6.2 FKG Inequality

The FKG inequality (named after Fortuin, Kasteleyn and Ginibre) states that increasing events are positively correlated.

The total order on the set $\{-1, 1\}$ induces a **partial order** on $\Omega : \omega \leq \omega'$ if and only if $\omega_i \leq \omega'_i$ for all $i \in \mathbb{Z}^d$. An event $E \subset \Omega$ is **increasing** if $\omega \in E$ and $\omega \leq \omega'$ implies $\omega' \in E$. If E and F are both increasing events depending on the spins inside Λ, then again, due to the ferromagnetic nature of the model, one can expect that the occurrence of an increasing event enhances the probability of another increasing event. That is, assuming that F has positive probability:

$$\mu^+_{\Lambda;\beta,h}(E \mid F) \geq \mu^+_{\Lambda;\beta,h}(E).$$

Multiplying by the probability of F, this inequality can be written as:

$$\mu^+_{\Lambda;\beta,h}(E \cap F) \geq \mu^+_{\Lambda;\beta,h}(E)\mu^+_{\Lambda;\beta,h}(F). \tag{3.23}$$

The precise result will be stated and proved in a more general setting, involving the expectation of nondecreasing local functions, of which $\mathbf{1}_E$ and $\mathbf{1}_F$ are particular instances.

A function $f : \Omega \to \mathbb{R}$ is **nondecreasing** if and only if $f(\omega) \leq f(\omega')$ for all $\omega \leq \omega'$.

Exercise 3.10. Prove that the following functions are nondecreasing: σ_i, n_i, n_A, $\sum_{i \in A} n_i - n_A$, for any $i \in \mathbb{Z}^d$, $A \Subset \mathbb{Z}^d$.

A particularly useful feature of the FKG inequality is its applicability for all possible boundary conditions and arbitrary (that is, not necessarily nonnegative) values of the magnetic field. It is also valid in the general setting presented in the last section, in which β and h are replaced by $\mathbf{J}$ and $\mathbf{h}$:

Theorem 3.21 (FKG inequality). *Let $\mathbf{J} = (J_{ij})_{i,j \in \mathbb{Z}^d}$ be a collection of nonnegative real numbers and let $\mathbf{h} = (h_i)_{i \in \mathbb{Z}^d}$ be a collection of arbitrary real numbers. Let $\Lambda \Subset \mathbb{Z}^d$ and # be some arbitrary boundary condition. Then, for any pair of nondecreasing functions f and g,*

$$\langle fg \rangle^{\#}_{\Lambda;\mathbf{J},\mathbf{h}} \geq \langle f \rangle^{\#}_{\Lambda;\mathbf{J},\mathbf{h}} \langle g \rangle^{\#}_{\Lambda;\mathbf{J},\mathbf{h}}. \tag{3.24}$$

Inequality (3.23) follows by taking $J_{ij} = \beta$ and $h_i = h$, and $f = \mathbf{1}_E$, $g = \mathbf{1}_F$. Note also that $\langle fg \rangle^{\eta}_{\Lambda;\mathbf{J},\mathbf{h}} \leq \langle f \rangle^{\eta}_{\Lambda;\mathbf{J},\mathbf{h}} \langle g \rangle^{\eta}_{\Lambda;\mathbf{J},\mathbf{h}}$ whenever f is nondecreasing and g is nonincreasing (simply apply (3.24) to f and $-g$).

Actually, (3.24) can be seen as a natural extension of the following elementary result: if f and g are two nondecreasing functions from $\mathbb{R}$ to $\mathbb{R}$ and μ is a probability measure on $\mathbb{R}$, then

$$\langle fg \rangle_\mu \geq \langle f \rangle_\mu \langle g \rangle_\mu .$$

Namely, it suffices to write

$$\langle fg\rangle_\mu - \langle f\rangle_\mu \langle g\rangle_\mu = \tfrac{1}{2}\int (f(x)-f(y))(g(x)-g(y))\mu(\mathrm{d}x)\mu(\mathrm{d}y)\,,$$

and to observe that $f(x)-f(y)$ *and* $g(x)-g(y)$ *have the same sign, since* f *and* g *are both nondecreasing.* ◇

3.6.3 Consequences

Many useful properties of finite-volume Gibbs distributions can be derived from the correlation inequalities of the previous section. The first is exactly the ingredient that will be needed for the study of the thermodynamic limit:

Lemma 3.22. *Let* f *be a nondecreasing function and* $\Lambda_1 \subset \Lambda_2 \Subset \mathbb{Z}^d$*. Then, for any* $\beta \geq 0$ *and* $h \in \mathbb{R}$*,*

$$\langle f\rangle^+_{\Lambda_1;\beta,h} \geq \langle f\rangle^+_{\Lambda_2;\beta,h}\,. \tag{3.25}$$

The same statement holds for the $-$ *boundary condition and a nonincreasing function* f*.*

Before turning to the proof, we need a spatial Markov property satisfied by $\mu^\eta_{\Lambda;\beta,h}$.

Exercise 3.11. Prove that, for all $\Delta \subset \Lambda \Subset \mathbb{Z}^d$ and all configurations $\eta \in \Omega$ and $\omega' \in \Omega^\eta_\Lambda$,

$$\mu^\eta_{\Lambda;\beta,h}\big(\cdot \mid \sigma_i = \omega'_i,\ \forall i \in \Lambda \setminus \Delta\big) = \mu^{\omega'}_{\Delta;\beta,h}(\cdot)\,. \tag{3.26}$$

The probability in the right-hand side of (3.26) *really only depends on* ω'_i *for* $i \in \partial^{\mathrm{ex}}\Delta$*, where* $\partial^{\mathrm{ex}}\Delta$ *is the* ***exterior boundary of*** $\boldsymbol{\Delta}$*, defined by*

$$\partial^{\mathrm{ex}}\Delta \overset{\mathrm{def}}{=} \{i \notin \Delta : \exists j \in \Delta, j \sim i\}\,.$$

This implies that

$$\mu^\eta_{\Lambda;\beta,h}\big(A \mid \sigma_i = \omega'_i,\ \forall i \in \Lambda \setminus \Delta\big) = \mu^\eta_{\Lambda;\beta,h}\big(A \mid \sigma_i = \omega'_i,\ \forall i \in \partial^{\mathrm{ex}}\Delta\big)\,,$$

for all events A *depending only on the spins located inside* Δ*. In this sense,* (3.26) *is indeed a spatial Markov property.* ◇

Proof of Lemma 3.22. It follows from (3.26) that

$$\langle f\rangle^+_{\Lambda_1;\beta,h} = \big\langle f \mid \sigma_i = 1,\ \forall i \in \Lambda_2 \setminus \Lambda_1\big\rangle^+_{\Lambda_2;\beta,h}\,.$$

The indicator $\mathbf{1}_{\{\sigma_i = 1,\ \forall i \in \Lambda_2\setminus\Lambda_1\}}$ being a nondecreasing function, the FKG inequality implies that

$$\langle f \rangle^+_{\Lambda_1;\beta,h} = \frac{\langle f \mathbf{1}_{\{\sigma_i=1,\,\forall i\in\Lambda_2\setminus\Lambda_1\}}\rangle^+_{\Lambda_2;\beta,h}}{\langle \mathbf{1}_{\{\sigma_i=1,\,\forall i\in\Lambda_2\setminus\Lambda_1\}}\rangle^+_{\Lambda_2;\beta,h}}$$
$$\geq \frac{\langle f\rangle^+_{\Lambda_2;\beta,h}\langle \mathbf{1}_{\{\sigma_i=1,\,\forall i\in\Lambda_2\setminus\Lambda_1\}}\rangle^+_{\Lambda_2;\beta,h}}{\langle \mathbf{1}_{\{\sigma_i=1,\,\forall i\in\Lambda_2\setminus\Lambda_1\}}\rangle^+_{\Lambda_2;\beta,h}} = \langle f \rangle^+_{\Lambda_2;\beta,h}. \qquad \square$$

Actually, some form of monotonicity with respect to the volume can also be established for the Gibbs distributions with free boundary condition:

Exercise 3.12. Using the GKS inequalities, prove that, for all $\beta, h \geq 0$,

$$\langle \sigma_A \rangle^+_{\Lambda_1;\beta,h} \geq \langle \sigma_A \rangle^+_{\Lambda_2;\beta,h}, \qquad \langle \sigma_A \rangle^\varnothing_{\Lambda_1;\beta,h} \leq \langle \sigma_A \rangle^\varnothing_{\Lambda_2;\beta,h},$$

for all $A \subset \Lambda_1 \subset \Lambda_2 \Subset \mathbb{Z}^d$.

The next lemma shows that the Gibbs distributions with $+$ and $-$ boundary conditions play an extremal role, in the sense that they maximally favor $+$, respectively $-$, spins.

Lemma 3.23. *Let f be an arbitrary nondecreasing function. Then, for any $\beta \geq 0$ and $h \in \mathbb{R}$,*

$$\langle f \rangle^-_{\Lambda;\beta,h} \leq \langle f \rangle^\eta_{\Lambda;\beta,h} \leq \langle f \rangle^+_{\Lambda;\beta,h},$$

for any boundary condition $\eta \in \Omega$ and any $\Lambda \Subset \mathbb{Z}^d$. Similarly, if f is a local function with supp$(f) \subset \Lambda$, *resp.* supp$(f) \subset V_N$, *then*

$$\langle f \rangle^-_{\Lambda;\beta,h} \leq \langle f \rangle^\varnothing_{\Lambda;\beta,h} \leq \langle f \rangle^+_{\Lambda;\beta,h},$$
$$\langle f \rangle^-_{V_{N-1};\beta,h} \leq \langle f \rangle^{\mathrm{per}}_{V_N;\beta,h} \leq \langle f \rangle^+_{V_{N-1};\beta,h}.$$

Proof Let $I(\omega) = \exp\{\beta \sum_{i\in\Lambda, j\notin\Lambda, i\sim j} \omega_i(1-\eta_j)\}$. First, observe that

$$\sum_{\omega\in\Omega^+_\Lambda} e^{-\mathscr{H}_{\Lambda;\beta,h}(\omega)} = \sum_{\omega\in\Omega^\eta_\Lambda} e^{-\mathscr{H}_{\Lambda;\beta,h}(\omega)} I(\omega),$$

and, for any nondecreasing f,

$$\sum_{\omega\in\Omega^+_\Lambda} e^{-\mathscr{H}_{\Lambda;\beta,h}(\omega)} f(\omega) \geq \sum_{\omega\in\Omega^\eta_\Lambda} e^{-\mathscr{H}_{\Lambda;\beta,h}(\omega)} I(\omega) f(\omega).$$

(The inequality is a consequence of our not assuming that supp$(f) \subset \Lambda$.) This implies that

$$\langle f \rangle^+_{\Lambda;\beta,h} = \frac{\sum_{\omega\in\Omega^+_\Lambda} e^{-\mathscr{H}_\Lambda(\omega)} f(\omega)}{\sum_{\omega\in\Omega^+_\Lambda} e^{-\mathscr{H}_\Lambda(\omega)}} \geq \frac{\sum_{\omega\in\Omega^\eta_\Lambda} e^{-\mathscr{H}_\Lambda(\omega)} I(\omega) f(\omega)}{\sum_{\omega\in\Omega^\eta_\Lambda} e^{-\mathscr{H}_\Lambda(\omega)} I(\omega)} = \frac{\langle If \rangle^\eta_{\Lambda;\beta,h}}{\langle I \rangle^\eta_{\Lambda;\beta,h}} \geq \langle f \rangle^\eta_{\Lambda;\beta,h},$$

where we applied the FKG inequality in the last step, making use of the fact that the function I is nondecreasing.

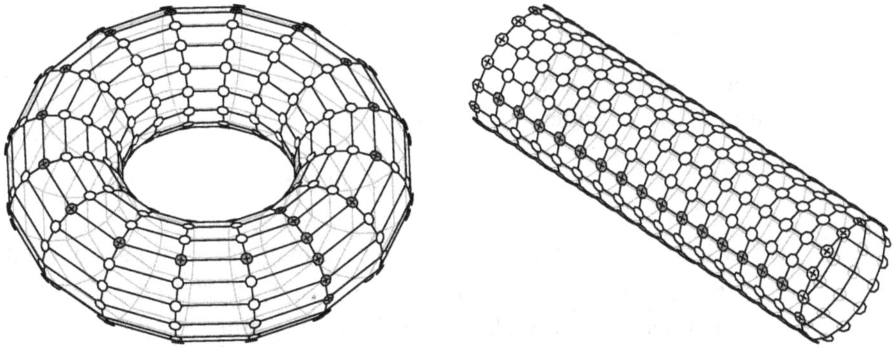

Figure 3.7. Left: the two-dimensional torus $\mathbb{T}_{16}$ with all spins along Σ_{16} forced to take the value $+1$. Right: opening the torus along the first "circle" of $+1$ yields an equivalent Ising model on a cylinder with $+$ boundary condition and all spins forced to take the value $+1$ along a line. Further opening the cylinder along the line of frozen $+$ spins yields an equivalent Ising model in the square $\{1, \dots, 15\}^2$ with $+$ boundary condition.

The proof for the free boundary condition is identical, using the nondecreasing function $I(\omega) = \exp\{\beta \sum_{\substack{i \in \Lambda, j \notin \Lambda \\ i \sim j}} \omega_i\}$.

Let us finally consider the Gibbs distribution with periodic boundary condition. In that case, we can argue as in the proof of Lemma 3.22, since, for any $\omega \in \Omega^+_{V_{N-1}}$ (considering $V_N = \{0, \dots, N\}^d$ as a subset of $\mathbb{Z}^d$),

$$\mu^{\mathrm{per}}_{V_N;\beta,h}\big(\omega|_{V_N} \bigm| \sigma_i = 1 \; \forall i \in \Sigma_N\big) = \mu^+_{V_{N-1};\beta,h}(\omega),$$

where $\Sigma_N \stackrel{\text{def}}{=} \big\{i = (i_1, \dots, i_d) \in V_N : \exists 1 \le k \le d \text{ such that } i_k = 0\big\}$ (see Figure 3.7) and the **restriction** of a configuration $\omega \in \Omega$ to a subset $S \subset \mathbb{Z}^d$ is defined by

$$\omega|_S \stackrel{\text{def}}{=} (\omega_i)_{i \in S}\,. \qquad \square$$

Exercise 3.13. Let $\eta, \omega \in \Omega$ be such that $\eta \le \omega$. Let f be a nondecreasing function. Show that, for any $\beta \ge 0$ and $h \in \mathbb{R}$,

$$\langle f \rangle^{\eta}_{\Lambda;\beta,h} \le \langle f \rangle^{\omega}_{\Lambda;\beta,h}\,,$$

for any $\Lambda \Subset \mathbb{Z}^d$. *Hint:* Adapt the argument in the proof of Lemma 3.23.

We can now prove the existence and translation invariance of $\langle \cdot \rangle^+_{\beta,h}$ and $\langle \cdot \rangle^-_{\beta,h}$.

Proof of Theorem 3.17. We consider the $+$ boundary condition. Let f be a local function. By Lemma 3.19 and linearity,

$$\langle f \rangle^+_{\Lambda_n;\beta,h} = \sum_{A \subset \mathrm{supp}(f)} \tilde{f}_A \langle n_A \rangle^+_{\Lambda_n;\beta,h}\,.$$

Since the functions n_A are nondecreasing, (3.25) implies that, for all A,

$$\langle n_A \rangle^+_{\Lambda_n;\beta,h} \geq \langle n_A \rangle^+_{\Lambda_{n+1};\beta,h}, \qquad \forall n \geq 1.$$

Being nonnegative, $\langle n_A \rangle^+_{\Lambda_n;\beta,h}$ thus converges as $n \to \infty$. It follows that $\langle f \rangle^+_{\Lambda_n;\beta,h}$ also has a limit, which we denote by

$$\langle f \rangle^+_{\beta,h} \stackrel{\text{def}}{=} \lim_{n\to\infty} \langle f \rangle^+_{\Lambda_n;\beta,h}.$$

Since it is obviously linear, positive and normalized, $\langle \cdot \rangle^+_{\beta,h}$ is a Gibbs state. We check now that it does not depend on the sequence $\Lambda_n \uparrow \mathbb{Z}^d$. Let $\Lambda^1_n \uparrow \mathbb{Z}^d$ and $\Lambda^2_n \uparrow \mathbb{Z}^d$ be two such sequences, and let us denote by $\langle \cdot \rangle^{+,1}_{\beta,h}$ and $\langle \cdot \rangle^{+,2}_{\beta,h}$ the corresponding limits. Since $\Lambda^1_n \uparrow \mathbb{Z}^d$ and $\Lambda^2_n \uparrow \mathbb{Z}^d$, we can always find a sequence $(\Delta_n)_{n\geq 1}$ such that, for all $k \geq 1$,

$$\Delta_{2k-1} \subset \{\Lambda^1_n : n \geq 1\}, \quad \Delta_{2k} \subset \{\Lambda^2_n : n \geq 1\}, \quad \Delta_k \subsetneq \Delta_{k+1}.$$

Of course, $\Delta_n \uparrow \mathbb{Z}^d$. Our previous considerations thus imply that $\lim_{k\to\infty} \langle f \rangle^+_{\Delta_k;\beta,h}$ exists, for every local function f. Moreover, since $(\langle f \rangle^+_{\Delta_{2k-1};\beta,h})_{k\geq 1}$ is a subsequence of $(\langle f \rangle^+_{\Lambda^1_n;\beta,h})_{n\geq 1}$ and $(\langle f \rangle^+_{\Delta_{2k};\beta,h})_{k\geq 1}$ is a subsequence of $(\langle f \rangle^+_{\Lambda^2_n;\beta,h})_{n\geq 1}$, we conclude that

$$\lim_{n\to\infty} \langle f \rangle^+_{\Lambda^1_n;\beta,h} = \lim_{k\to\infty} \langle f \rangle^+_{\Delta_k;\beta,h} = \lim_{n\to\infty} \langle f \rangle^+_{\Lambda^2_n;\beta,h},$$

for all local functions f. This shows that the state $\langle \cdot \rangle^+_{\beta,h}$ does not depend on the choice of the sequence $(\Lambda_n)_{n\geq 1}$.

We still have to prove translation invariance. Again, let f be a local function. For all $j \in \mathbb{Z}^d$, $f \circ \theta_j$ is also a local function and $\theta_{-j}\Lambda_n \uparrow \mathbb{Z}^d$ ($\theta_i \Lambda \stackrel{\text{def}}{=} \Lambda + i$). We thus have

$$\langle f \rangle^+_{\Lambda_n;\beta,h} \to \langle f \rangle^+_{\beta,h} \qquad \text{and} \qquad \langle f \circ \theta_j \rangle^+_{\theta_{-j}\Lambda_n;\beta,h} \to \langle f \circ \theta_j \rangle^+_{\beta,h}.$$

The conclusion follows, since $\langle f \circ \theta_j \rangle^+_{\theta_{-j}\Lambda_n;\beta,h} = \langle f \rangle^+_{\Lambda_n;\beta,h}$ (see Figure 3.8). □

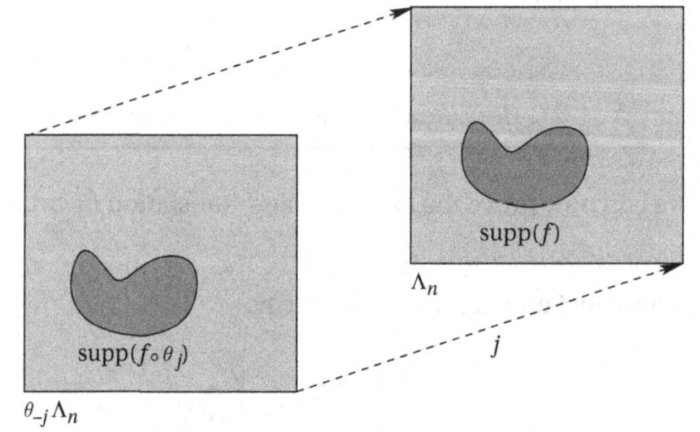

Figure 3.8. Proof of invariance under translation.

Exercise 3.14. Prove that $\langle\cdot\rangle^+_{\beta,h}$ and $\langle\cdot\rangle^-_{\beta,h}$ are also invariant under lattice rotations and reflections of $\mathbb{Z}^d$.

Exercise 3.15. Let $\beta \geq 0$ and $h \in \mathbb{R}$. Show that $\langle\cdot\rangle^+_{\beta,h}$ has **short-range correlations**, in the sense that, for all local functions f and g,

$$\lim_{\|i\|_1\to\infty} \langle f \cdot (g \circ \theta_i)\rangle^+_{\beta,h} = \langle f\rangle^+_{\beta,h}\langle g\rangle^+_{\beta,h}.$$

Hint: Use the FKG inequality to prove first the result with $f = n_A$ and $g = n_B$ for arbitrary $A, B \Subset \mathbb{Z}^d$.

With similar arguments, one can also construct Gibbs states using the free boundary condition:

Exercise 3.16. Prove that, for all $\beta \geq 0$, $h \in \mathbb{R}$ and any sequence $\Lambda_n \uparrow \mathbb{Z}^d$, the sequence $(\langle\cdot\rangle^{\varnothing}_{\Lambda_n;\beta,h})_{n\geq 1}$ converges to a Gibbs state $\langle\cdot\rangle^{\varnothing}_{\beta,h}$, independent of the sequence $(\Lambda_n)_{n\geq 1}$ chosen. Show that $\langle\cdot\rangle^{\varnothing}_{\beta,h}$ is translation invariant.

3.7 Phase Diagram

Now that we have seen that infinite-volume Gibbs states for a pair (β, h) can be constructed rigorously in various ways (for example, using $+$ or $-$ boundary conditions), the next problem is to determine whether these are the same Gibbs states, or whether there exist some values of the temperature and magnetic field for which the influence of the boundary condition survives in the thermodynamic limit, leading to multiple Gibbs states.

The answer to this question will be given in the next sections: it will depend on the dimension d and on the values of β and h. Contrary to what often happens in mathematics, the lack of uniqueness is not a defect of this approach, but is actually one of its main features: lack of uniqueness means that providing a complete microscopic description of the system (that is, the set of configurations and the Hamiltonian) as well as fixing all the relevant thermodynamic parameters (β and h) is not sufficient to completely determine the macroscopic behavior of the system.

Definition 3.24. If at least two distinct Gibbs states can be constructed for a pair (β, h), we say that there is a **first-order phase transition at (β, h).**

Later in this chapter (see Theorem 3.34), we will relate this *probabilistic* definition of a first-order phase transition to the *analytic* one associated with the pressure (Definition 3.8).

We can now turn to the main result of this chapter, which establishes the *phase diagram* of the Ising model, that is, the determination for each pair (β, h) of whether there is a unique or multiple Gibbs states. We gather the corresponding claims in the form of a theorem, the proof of which will be given in the remainder of the chapter (see Figure 3.9).

Theorem 3.25. *1. In any $d \geq 1$, when $h \neq 0$, there is a unique Gibbs state for all values of $\beta \in \mathbb{R}_{\geq 0}$.*
2. In $d = 1$, there is a unique Gibbs state at each $(\beta, h) \in \mathbb{R}_{\geq 0} \times \mathbb{R}$.
3. When $h = 0$ and $d \geq 2$, there exists $\beta_c = \beta_c(d) \in (0, \infty)$ such that:

- *when $\beta < \beta_c$, the Gibbs state at $(\beta, 0)$ is unique,*
- *when $\beta > \beta_c$, a first-order phase transition occurs at $(\beta, 0)$:*

$$\langle\cdot\rangle^+_{\beta,0} \neq \langle\cdot\rangle^-_{\beta,0}.$$

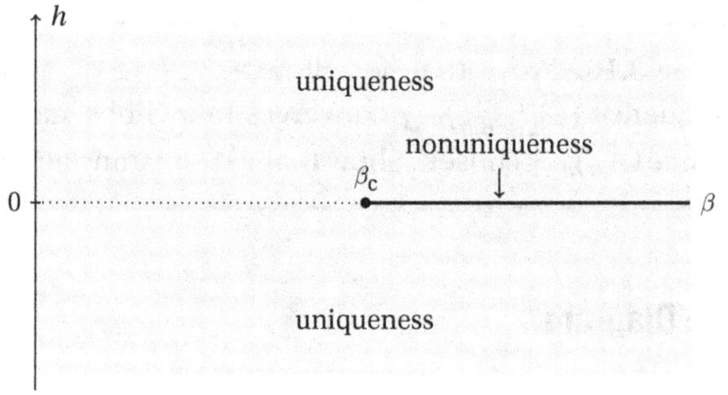

Figure 3.9. The phase diagram of the Ising model in $d \geq 2$. The line $\{(\beta, 0) : \beta > \beta_c\}$ is called the **coexistence line**. This diagram should be compared with the simulations of Figure 1.9.

The proof of Theorem 3.25 is quite long and is spread over several sections. The first item will be proved in Section 3.7.4. The second item was already proved in Section 3.3 (once the results there are combined with Theorem 3.34) and will be given an alternative proof in Section 3.7.3. The proof of the third item has two parts: the proof that $\beta_c < \infty$ is done in Section 3.7.2, while the proof that $\beta_c > 0$ is done in Section 3.7.3.

Remark 3.26. It can be proved that uniqueness holds also at $(\beta_c, 0)$, when $d \geq 2$, but the argument is beyond the scope of this book.[2] The phase transition occurring as β crosses β_c (at $h = 0$) is thus continuous. ◇

Remark 3.27. Although the above theorem claims the existence of at least two distinct Gibbs states when $d \geq 2$, $h = 0$ and $\beta > \beta_c$, it does not describe the *structure* of the set of Gibbs states associated with those values of (β, h). This is a much more difficult problem, to which we will return in Section 3.10.8. ◇

3.7.1 Two Criteria for (Non)Uniqueness

In this subsection, we establish a link between uniqueness of the Gibbs state, the average magnetization density and differentiability of the pressure. We use these quantities to formulate several equivalent characterizations of uniqueness of the Gibbs state, which play a crucial role in our determination of the phase diagram. Moreover, the second criterion provides the rigorous link between the analytic and probabilistic definitions of first-order phase transition introduced earlier.

A First Characterization of Uniqueness

The major role played by the states $\langle\cdot\rangle^+_{\beta,h}$ and $\langle\cdot\rangle^-_{\beta,h}$ is made clear by the following result.

Theorem 3.28. *Let $(\beta, h) \in \mathbb{R}_{\geq 0} \times \mathbb{R}$. The following statements are equivalent:*

1. *There is a unique Gibbs state at (β, h).*
2. $\langle\cdot\rangle^+_{\beta,h} = \langle\cdot\rangle^-_{\beta,h}$.
3. $\langle\sigma_0\rangle^+_{\beta,h} = \langle\sigma_0\rangle^-_{\beta,h}$.

Proof The implications 1⇒2⇒3 are trivial. Let us prove that 3⇒2. Take $\Lambda_n \uparrow \mathbb{Z}^d$ and $A \Subset \mathbb{Z}^d$. Since $\sum_{i\in A} n_i - n_A$ is nondecreasing (Exercise 3.10), Lemma 3.23 implies that, for all k,

$$\Big\langle \sum_{i\in A} n_i - n_A \Big\rangle^-_{\Lambda_k;\beta,h} \leq \Big\langle \sum_{i\in A} n_i - n_A \Big\rangle^+_{\Lambda_k;\beta,h}.$$

Using linearity, letting $k \to \infty$ and rearranging, we get

$$\sum_{i\in A} \big(\langle n_i\rangle^+_{\beta,h} - \langle n_i\rangle^-_{\beta,h}\big) \geq \langle n_A\rangle^+_{\beta,h} - \langle n_A\rangle^-_{\beta,h}.$$

If 3 holds, the left-hand side vanishes, since translation invariance then implies that

$$\langle n_i\rangle^+_{\beta,h} - \langle n_i\rangle^-_{\beta,h} = \langle n_0\rangle^+_{\beta,h} - \langle n_0\rangle^-_{\beta,h} = \tfrac{1}{2}\big(\langle\sigma_0\rangle^+_{\beta,h} - \langle\sigma_0\rangle^-_{\beta,h}\big) = 0.$$

But $\langle n_A\rangle^+_{\beta,h} \geq \langle n_A\rangle^-_{\beta,h}$ (again by Lemma 3.23), and so $\langle n_A\rangle^+_{\beta,h} = \langle n_A\rangle^-_{\beta,h}$. Together with Lemma 3.19, this implies that $\langle f\rangle^+_{\beta,h} = \langle f\rangle^-_{\beta,h}$ for every local function f. Therefore, 2 holds.

It only remains to prove that 2⇒1. Lemma 3.23 implies that any Gibbs state at (β, h), say $\langle\cdot\rangle_{\beta,h}$, is such that $\langle n_A\rangle^-_{\beta,h} \leq \langle n_A\rangle_{\beta,h} \leq \langle n_A\rangle^+_{\beta,h}$. If 2 holds, this implies $\langle n_A\rangle^-_{\beta,h} = \langle n_A\rangle_{\beta,h} = \langle n_A\rangle^+_{\beta,h}$. By Lemma 3.19, this extends to all local functions and, therefore, $\langle\cdot\rangle^-_{\beta,h} = \langle\cdot\rangle_{\beta,h} = \langle\cdot\rangle^+_{\beta,h}$. We conclude that 1 holds. □

Some Properties of the Magnetization Density

Remember that the average magnetization density in $\Lambda \Subset \mathbb{Z}^d$ with an arbitrary boundary condition # was defined by $m^\#_\Lambda(\beta, h) \stackrel{\text{def}}{=} \langle m_\Lambda\rangle^\#_{\Lambda;\beta,h}$. The uniqueness criterion developed in Theorem 3.28 is expressed in terms of the averages $\langle\sigma_0\rangle^+_{\beta,h}$ and $\langle\sigma_0\rangle^-_{\beta,h}$. It is natural to wonder whether these quantities are related to $m^+_\Lambda(\beta, h)$ and $m^-_\Lambda(\beta, h)$. The following result shows that they in fact coincide in the thermodynamic limit.

Proposition 3.29. *For any sequence $\Lambda \Uparrow \mathbb{Z}^d$, the limits*

$$m^+(\beta,h) \stackrel{\text{def}}{=} \lim_{\Lambda \Uparrow \mathbb{Z}^d} m^+_\Lambda(\beta,h), \quad m^-(\beta,h) \stackrel{\text{def}}{=} \lim_{\Lambda \Uparrow \mathbb{Z}^d} m^-_\Lambda(\beta,h)$$

exist and

$$m^+(\beta,h) = \langle \sigma_0 \rangle^+_{\beta,h}, \quad m^-(\beta,h) = \langle \sigma_0 \rangle^-_{\beta,h}.$$

Moreover, $h \mapsto m^+(\beta,h)$ is right-continuous, while $h \mapsto m^-(\beta,h)$ is left-continuous.

Remark 3.30. By Corollary 3.7, $m^+(\beta,h)$ and $m(\beta,h)$ are equal when $h \notin \mathfrak{B}_\beta$. Therefore, considering a sequence $h \downarrow 0$ in $\mathfrak{B}^c_\beta$,

$$m^*(\beta) = \lim_{h\downarrow 0} m(\beta,h) = \lim_{h\downarrow 0} m^+(\beta,h) = m^+(\beta,0) = \langle \sigma_0 \rangle^+_{\beta,0}.$$

Note also that, by Exercise 3.15,

$$\lim_{\|i\|_1\to\infty} \langle \sigma_0\sigma_i \rangle^+_{\beta,0} = \left(\langle \sigma_0 \rangle^+_{\beta,0}\right)^2 = m^*(\beta)^2, \qquad \forall \beta \geq 0.$$

This observation provides a convenient approach for its explicit computation in $d = 2$, which avoids having to work with a nonzero magnetic field. ◇

Proof Let $\Lambda_n \Uparrow \mathbb{Z}^d$. By the translation invariance of $\langle \cdot \rangle^+_{\beta,h}$ and by the monotonicity property (3.25),

$$\langle \sigma_0 \rangle^+_{\beta,h} = \langle m_{\Lambda_n} \rangle^+_{\beta,h} \leq \langle m_{\Lambda_n} \rangle^+_{\Lambda_n;\beta,h}.$$

This gives $\langle \sigma_0 \rangle^+_{\beta,h} \leq \liminf_n \langle m_{\Lambda_n} \rangle^+_{\Lambda_n;\beta,h}$. For the other bound, fix $k \geq 1$ and let $i \in \Lambda_n$. On the one hand, if $i + \mathrm{B}(k) \subset \Lambda_n$, (3.25) again gives

$$\langle \sigma_i \rangle^+_{\Lambda_n;\beta,h} \leq \langle \sigma_i \rangle^+_{i+\mathrm{B}(k);\beta,h} = \langle \sigma_0 \rangle^+_{\mathrm{B}(k);\beta,h}.$$

On the other hand, if $i + \mathrm{B}(k) \not\subset \Lambda_n$, then the box $i + \mathrm{B}(k)$ intersects $\partial^{\text{in}}\Lambda_n$. As a consequence,

$$\begin{aligned}\langle m_{\Lambda_n} \rangle^+_{\Lambda_n;\beta,h} &= \frac{1}{|\Lambda_n|} \sum_{\substack{i\in\Lambda_n:\\ i+\mathrm{B}(k)\subset\Lambda_n}} \langle \sigma_i \rangle^+_{\Lambda_n;\beta,h} + \frac{1}{|\Lambda_n|} \sum_{\substack{i\in\Lambda_n:\\ i+\mathrm{B}(k)\not\subset\Lambda_n}} \langle \sigma_i \rangle^+_{\Lambda_n;\beta,h} \\ &\leq \langle \sigma_0 \rangle^+_{\mathrm{B}(k);\beta,h} + 2\frac{|\mathrm{B}(k)||\partial^{\text{in}}\Lambda_n|}{|\Lambda_n|},\end{aligned}$$

since $|\langle \sigma_0 \rangle^+_{\Lambda_n;\beta,h}| \leq 1$. (Note that $\langle \sigma_0 \rangle^+_{\mathrm{B}(k);\beta,h}$ can be negative; this is the reason for the factor 2 in the last term). This implies that, for all $k \geq 1$, $\limsup_n \langle m_{\Lambda_n} \rangle^+_{\Lambda_n;\beta,h} \leq \langle \sigma_0 \rangle^+_{\mathrm{B}(k);\beta,h}$. Since $\lim_{k\to\infty} \langle \sigma_0 \rangle^+_{\mathrm{B}(k);\beta,h} = \langle \sigma_0 \rangle^+_{\beta,h}$, the desired result follows. The one-sided continuity of $m^+(\beta,h)$ and $m^-(\beta,h)$ will follow from Lemma 3.31 below. □

Lemma 3.31. *1. For all $\beta \geq 0$, $h \mapsto \langle\sigma_0\rangle^+_{\beta,h}$ is nondecreasing and right-continuous and $h \mapsto \langle\sigma_0\rangle^-_{\beta,h}$ is nondecreasing and left-continuous.*
2. For all $h \geq 0$, $\beta \mapsto \langle\sigma_0\rangle^+_{\beta,h}$ is nondecreasing and, for all $h \leq 0$, $\beta \mapsto \langle\sigma_0\rangle^-_{\beta,h}$ is nonincreasing.

Proof of Lemma 3.31. We prove the properties for $\langle\sigma_0\rangle^+_{\beta,h}$ (symmetry then allows us to deduce the corresponding properties for $\langle\sigma_0\rangle^-_{\beta,h}$).

1. Let $\Lambda \Subset \mathbb{Z}^d$. It follows from the FKG inequality that

$$\frac{\partial}{\partial h}\langle\sigma_0\rangle^+_{\Lambda;\beta,h} = \sum_{i\in\Lambda}\big(\langle\sigma_0\sigma_i\rangle^+_{\Lambda;\beta,h} - \langle\sigma_0\rangle^+_{\Lambda;\beta,h}\langle\sigma_i\rangle^+_{\Lambda;\beta,h}\big) \geq 0 .$$

So, at fixed Λ, $h \mapsto \langle\sigma_0\rangle^+_{\Lambda;\beta,h}$ is nondecreasing. This monotonicity clearly persists in the thermodynamic limit. Let then $(h_m)_{m\geq 1}$ be a sequence of real numbers such that $h_m \downarrow h$ and $(\Lambda_n)_{n\geq 1}$ be a sequence such that $\Lambda_n \uparrow \mathbb{Z}^d$. Lemma 3.22 implies that the double sequence $(\langle\sigma_0\rangle^+_{\Lambda_n;\beta,h_m})_{m,n\geq 1}$ is nonincreasing and bounded. Consequently, it follows from Lemma B.4 that

$$\begin{aligned}\lim_{m\to\infty}\langle\sigma_0\rangle^+_{\beta,h_m} &= \lim_{m\to\infty}\lim_{n\to\infty}\langle\sigma_0\rangle^+_{\Lambda_n;\beta,h_m}\\ &= \lim_{n\to\infty}\lim_{m\to\infty}\langle\sigma_0\rangle^+_{\Lambda_n;\beta,h_m} = \lim_{n\to\infty}\langle\sigma_0\rangle^+_{\Lambda_n;\beta,h} = \langle\sigma_0\rangle^+_{\beta,h} .\end{aligned}$$

The third identity relies on the fact that the finite-volume expectation $\langle\sigma_0\rangle^+_{\Lambda_n;\beta,h}$ is continuous in h.

2. Proceeding as before and using (3.22) with $A = \{0\}$ and $B = \{i,j\}$,

$$\frac{\partial}{\partial\beta}\langle\sigma_0\rangle^+_{\Lambda;\beta,h} = \sum_{\{i,j\}\in\mathscr{E}^{\mathrm{b}}_\Lambda}\big(\langle\sigma_0\sigma_i\sigma_j\rangle^+_{\Lambda;\beta,h} - \langle\sigma_0\rangle^+_{\Lambda;\beta,h}\langle\sigma_i\sigma_j\rangle^+_{\Lambda;\beta,h}\big) \geq 0 .$$

This monotonicity also clearly persists in the thermodynamic limit. □

Exercise 3.17. Let $A \Subset \mathbb{Z}^d$ and $h \geq 0$. Show that $\beta \mapsto \langle\sigma_A\rangle^\varnothing_{\beta,h}$ is left-continuous and $\beta \mapsto \langle\sigma_A\rangle^+_{\beta,h}$ is right-continuous.

Defining the Critical Inverse Temperature

Since $\langle\sigma_0\rangle^-_{\beta,0} = -\langle\sigma_0\rangle^+_{\beta,0}$ by symmetry, Theorem 3.28 and Remark 3.30 imply that, when $h = 0$, uniqueness is equivalent to $m^*(\beta) = 0$. Since Lemma 3.31 implies that $m^*(\beta) = \langle\sigma_0\rangle^+_{\beta,0}$ is monotone in β, we are led naturally to the following definition.

Definition 3.32. The **critical inverse temperature** is

$$\beta_c(d) \stackrel{\text{def}}{=} \inf\{\beta \geq 0 : m^*(\beta) > 0\} = \sup\{\beta \geq 0 : m^*(\beta) = 0\} . \qquad (3.27)$$

That is, $\beta_c(d)$ is the unique value of β such that $m^*(\beta) = 0$ if $\beta < \beta_c$, and $m^*(\beta) > 0$ if $\beta > \beta_c$. Of course, one still has to determine whether $\beta_c(d)$ is nontrivial, that is, whether $0 < \beta_c(d) < \infty$.

Remark 3.33. By translation invariance, $\langle\sigma_i\rangle^+_{\beta,0} = \langle\sigma_0\rangle^+_{\beta,0} = m^*(\beta)$ for all $i \in \mathbb{Z}^d$. This implies, using the FKG inequality, that

$$\langle\sigma_0\sigma_i\rangle^+_{\beta,0} \geq \langle\sigma_0\rangle^+_{\beta,0}\langle\sigma_i\rangle^+_{\beta,0} = m^*(\beta)^2 .$$

In particular,

$$\inf_{i\in\mathbb{Z}^d}\langle\sigma_0\sigma_i\rangle^+_{\beta,0} > 0, \qquad \forall\beta > \beta_c . \tag{3.28}$$

Such a behavior is referred to as **long-range order**. The presence of long-range order does not, however, imply that the random variables σ_i display strong correlations at large distances. Indeed, as follows from Exercise 3.15 (see also the more general statement in point 4 of Theorem 6.58), for any β,

$$\lim_{\|i\|_1\to\infty} \langle\sigma_0\sigma_i\rangle^+_{\beta,0} - \langle\sigma_0\rangle^+_{\beta,0}\langle\sigma_i\rangle^+_{\beta,0} = 0 ,$$

so that σ_0 and σ_i are always asymptotically (as $\|i\|_1 \to \infty$) uncorrelated.[3] ◇

A Second Characterization of Uniqueness

The following theorem providese the promised link between the two notions of first-order phase transition introduced in Definitions 3.8 and 3.24: nonuniqueness occurs at (β, h) if and only if the pressure fails to be differentiable in h at this point. The theorem also provides the extension of the relation (3.8) to values of h at which the pressure is not differentiable. In that case, we can rely on the convexity of the pressure, which we proved in Theorem 3.6, to conclude that its right- and left-derivatives with respect to h are always well defined.

Theorem 3.34. *The following identities hold for all values of $\beta \geq 0$ and $h \in \mathbb{R}$:*

$$\frac{\partial\psi}{\partial h^+}(\beta, h) = m^+(\beta, h), \qquad \frac{\partial\psi}{\partial h^-}(\beta, h) = m^-(\beta, h) .$$

In particular, $h \mapsto \psi(\beta, h)$ is differentiable at h if and only if there is a unique Gibbs state at (β, h).

Remark 3.35. Theorem 3.34 shows that the pressure is differentiable with respect to h precisely for those values of β and h at which there is a unique infinite-volume Gibbs state. We will see later (Exercise 6.33) that uniqueness of the infinite-volume Gibbs state also implies differentiability with respect to β. (Actually, although we will not prove it, the pressure of the Ising model on $\mathbb{Z}^d$ is *always* differentiable with respect to β.) ◇

Proof Remember that the set $\mathfrak{B}_\beta$ of points of nondifferentiability of the pressure is at most countable. Therefore, for each $h \in \mathbb{R}$, it is possible to find a sequence $h_k \downarrow h$ such that $h_k \notin \mathfrak{B}_\beta$ for all $k \geq 1$. It then follows from (3.9) that

$$\frac{\partial\psi}{\partial h^+}(\beta,h)=\lim_{h_k\downarrow h} m(\beta,h_k)=\lim_{h_k\downarrow h} m^+(\beta,h_k)=m^+(\beta,h),$$

since $m^+(\beta,h')=m(\beta,h')$ for all $h'\notin\mathfrak{B}_\beta$ (Corollary 3.7) and $m^+(\beta,h)$ is a right-continuous function of h (Proposition 3.29). Now, by symmetry,

$$\frac{\partial\psi}{\partial h^-}(\beta,h)=-\frac{\partial\psi}{\partial h^+}(\beta,-h)=-m^+(\beta,-h)=m^-(\beta,h).$$

As a consequence, we conclude that

$$\frac{\partial\psi}{\partial h}(\beta,h)\text{ exists}\quad\Leftrightarrow\quad m^+(\beta,h)=m^-(\beta,h).$$

The conclusion follows since, by Proposition 3.29 and Theorem 3.28,

$$m^+(\beta,h)=m^-(\beta,h)\quad\Leftrightarrow\quad\langle\sigma_0\rangle^+_{\beta,h}=\langle\sigma_0\rangle^-_{\beta,h}\quad\Leftrightarrow\quad\text{uniqueness at }(\beta,h).\ \square$$

In the following two sections, we prove item 3 of Theorem 3.25, which establishes, at $h=0$, distinct low- and high-temperature behaviors.

3.7.2 Spontaneous Symmetry Breaking at Low Temperatures

In this subsection, we prove that $\beta_c(d)<\infty$, for all $d\geq 2$. In order to do so, it is sufficient to show that, *uniformly in the size of* Λ,

$$\mu^+_{\Lambda;\beta,0}(\sigma_0=-1)\leq\delta(\beta),\tag{3.29}$$

where $\delta(\beta)\downarrow 0$ when $\beta\to\infty$. Indeed, this has the consequence that

$$\begin{aligned}\langle\sigma_0\rangle^+_{\Lambda;\beta,0}&=\mu^+_{\Lambda;\beta,0}(\sigma_0=+1)-\mu^+_{\Lambda;\beta,0}(\sigma_0=-1)\\&=1-2\mu^+_{\Lambda;\beta,0}(\sigma_0=-1)\\&\geq 1-2\delta(\beta).\end{aligned}$$

Therefore, if one fixes β large enough, so that $1-2\delta(\beta)>0$, and then takes the thermodynamic limit $\Lambda\uparrow\mathbb{Z}^d$, one deduces that

$$m^*(\beta)=\langle\sigma_0\rangle^+_{\beta,0}>0.\tag{3.30}$$

Using the characterization (3.27), this shows that $\beta_c<\infty$: a first-order phase transition indeed occurs at low temperatures.

The proof of (3.29) uses a key idea originally due to Peierls, today known as *Peierls' argument* and considered a cornerstone in the understanding of phase transitions. It consists in making the following idea rigorous.

When β is large, neighboring spins with different values make a high contribution to the total energy and are thus strongly suppressed. Therefore the contours, *which are the lines that separate regions of $+$ and $-$ spins, should be rare, and a typical configuration under $\mu^+_{\Lambda;\beta,0}$ should have the structure of a "sea" of* $+1$ *spins with small "islands" of $-$ spins (see Figure 3.10).* ◇

In other words, when β is large, typical configurations under $\mu^{+}_{\Lambda;\beta,0}$ are small *perturbations* of the ground state η^{+}, and these perturbations are realized by the contours of the configurations.

We will implement this strategy for the two-dimensional model and will see later how it can be extended to higher dimensions.

Low-Temperature Representation

Consider the two-dimensional Ising model in $\Lambda \Subset \mathbb{Z}^2$, with zero magnetic field and $+$ boundary condition. We fix some configuration $\omega \in \Omega^{+}_{\Lambda}$ and give a geometrical description of ω whose purpose is to account for the above-mentioned fact that a low temperature favors the alignment of nearest-neighbor spins. The starting point is thus to express the Hamiltonian in a way that emphasizes the role played by pairs of opposite spins:

$$\mathscr{H}_{\Lambda;\beta,0} = -\beta \sum_{\{i,j\}\in\mathscr{E}^{\mathrm{b}}_{\Lambda}} \sigma_i\sigma_j = -\beta|\mathscr{E}^{\mathrm{b}}_{\Lambda}| + \sum_{\{i,j\}\in\mathscr{E}^{\mathrm{b}}_{\Lambda}} \beta(1-\sigma_i\sigma_j).$$

The dependence on ω is only in the sum

$$\sum_{\{i,j\}\in\mathscr{E}^{\mathrm{b}}_{\Lambda}} \beta(1-\sigma_i\sigma_j) = \sum_{\substack{\{i,j\}\in\mathscr{E}^{\mathrm{b}}_{\Lambda}:\\ \sigma_i\neq\sigma_j}} 2\beta = 2\beta \cdot \#\big\{\{i,j\} \in \mathscr{E}^{\mathrm{b}}_{\Lambda} : \sigma_i \neq \sigma_j\big\}.$$

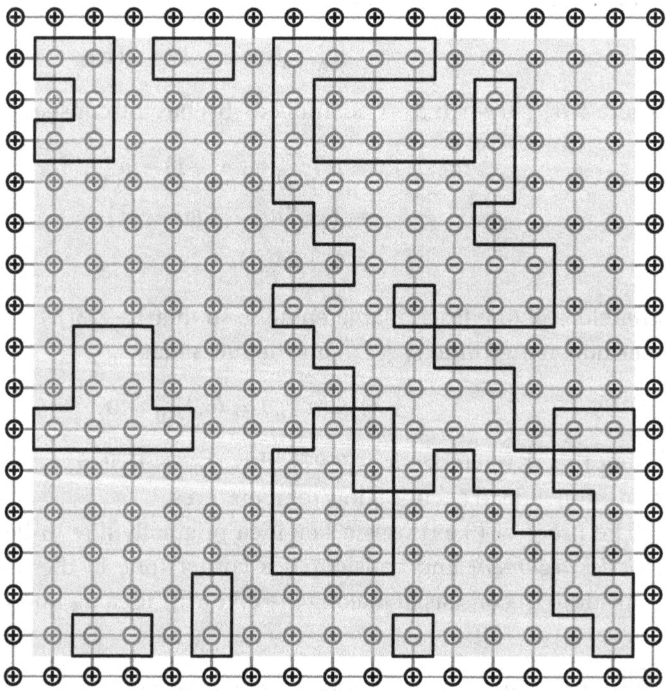

Figure 3.10. A configuration of the two-dimensional Ising model in a finite box Λ with $+$ boundary condition. At low temperature, the lines separating regions of $+$ and $-$ spins are expected to be short and sparse, leading to a positive magnetization in Λ (and thus the validity of (3.29)).

Let us associate with each vertex $i \in \mathbb{Z}^2$ the closed unit square centered at i:

$$\mathscr{S}_i \overset{\text{def}}{=} i + [-\tfrac{1}{2}, \tfrac{1}{2}]^2 . \tag{3.31}$$

The boundary (in the sense of the standard topology on $\mathbb{R}^2$) of $\mathscr{S}_i$, denoted by $\partial\mathscr{S}_i$, can be considered as being made of four edges connecting nearest neighbors of the *dual lattice*

$$\mathbb{Z}^2_* = \mathbb{Z}^2 + (\tfrac{1}{2}, \tfrac{1}{2}) \overset{\text{def}}{=} \{(i_1 + \tfrac{1}{2}, i_2 + \tfrac{1}{2}) : (i_1, i_2) \in \mathbb{Z}^2\} .$$

Notice that a given edge e of the original lattice intersects exactly one edge $e_\perp$ of the dual lattice. If we associate with a configuration $\omega \in \Omega^+_\Lambda$ the random set

$$\mathscr{M}(\omega) \overset{\text{def}}{=} \bigcup_{i \in \Lambda : \sigma_i(\omega) = -1} \mathscr{S}_i ,$$

then again $\partial\mathscr{M}(\omega)$ is made of edges of the dual lattice. Moreover, *each edge $e_\perp = \{i,j\}_\perp \subset \partial\mathscr{M}(\omega)$ separates two opposite spins: $\sigma_i(\omega) \neq \sigma_j(\omega)$.* One can therefore write

$$\mathscr{H}_{\Lambda;\beta,0}(\omega) = -\beta|\mathscr{E}^{\text{b}}_\Lambda| + 2\beta|\partial\mathscr{M}(\omega)| .$$

(Here, $|\partial\mathscr{M}(\omega)|$ denotes the number of edges contained in $\partial\mathscr{M}(\omega)$ or, equivalently, the total length of $\partial\mathscr{M}(\omega)$.) A configuration ω with its associated set $\partial\mathscr{M}(\omega)$ is represented in Figure 3.10.

We will now decompose $\partial\mathscr{M}(\omega)$ into disjoint components. For that, it is convenient to fix a deformation rule to decide how these components are defined. To this end, we first remark that each dual vertex of $\mathbb{Z}^2_*$ is adjacent to either 0, 2 or 4 edges of $\partial\mathscr{M}(\omega)$.[2] When this number is 4, we deform $\partial\mathscr{M}(\omega)$ using the following rule:

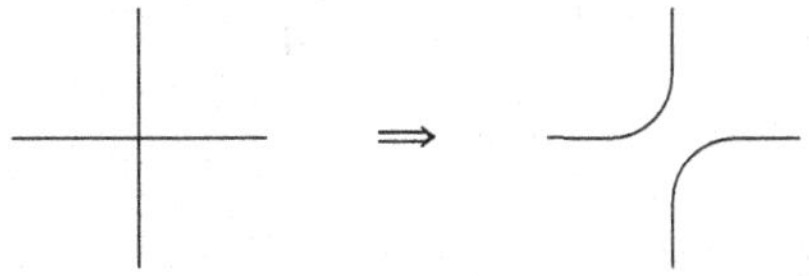

Figure 3.11. The deformation rule.

An application of this rule at all points at which the incidence number is 4 yields a decomposition of $\partial\mathscr{M}(\omega)$ into a set of disjoint closed simple paths on the dual lattice, as in Figure 3.12. In terms of dual edges,

$$\partial\mathscr{M}(\omega) = \gamma_1 \cup \cdots \cup \gamma_n .$$

Each path γ_i is called a **contour of ω**. Let $\Gamma(\omega) \overset{\text{def}}{=} \{\gamma_1, \dots, \gamma_n\}$ and define the **length** $|\gamma|$ of a contour $\gamma \in \Gamma(\omega)$ as the number of edges of the dual lattice that it contains. For example, in Figure 3.12, $|\gamma_5| = 14$.

[2] One way to show that is to consider a dual vertex $x \in \mathbb{Z}^2_*$ together with the four surrounding points of $\mathbb{Z}^2$, which we denote (in clockwise order) by i, j, k, l. Since $(\omega_i\omega_j)(\omega_j\omega_k)(\omega_k\omega_l)(\omega_l\omega_i) = \omega_i^2\omega_j^2\omega_k^2\omega_l^2 = 1$, the number of products equal to -1 in the leftmost expression is even. But such a product is equal to -1 precisely when the edge of a contour separates the corresponding spins.

Using the above notations, the energy of a configuration $\omega \in \Omega_\Lambda^+$ can be very simply expressed in terms of its contours:

$$\mathscr{H}_{\Lambda;\beta,0}(\omega) = -\beta|\mathscr{E}_\Lambda^{\mathrm{b}}| + 2\beta \sum_{\gamma\in\Gamma(\omega)} |\gamma| .$$

Consequently, the partition function in Λ with $+$ boundary condition can be written

$$\mathbf{Z}_{\Lambda;\beta,0}^{+} = e^{\beta|\mathscr{E}_\Lambda^{\mathrm{b}}|} \sum_{\omega\in\Omega_\Lambda^+} \prod_{\gamma\in\Gamma(\omega)} e^{-2\beta|\gamma|} . \tag{3.32}$$

Finally, the probability of $\omega \in \Omega_\Lambda^+$ can be expressed in terms of contours as

$$\mu_{\Lambda;\beta,0}^{+}(\omega) = \frac{e^{-\mathscr{H}_{\Lambda;\beta,0}(\omega)}}{\mathbf{Z}_{\Lambda;\beta,0}^{+}} = \frac{\prod_{\gamma\in\Gamma(\omega)} e^{-2\beta|\gamma|}}{\sum_{\omega}\prod_{\gamma\in\Gamma(\omega)} e^{-2\beta|\gamma|}} . \tag{3.33}$$

Remark 3.36. The above probability being a *ratio*, the terms $e^{\beta|\mathscr{E}_\Lambda^{\mathrm{b}}|}$ have canceled out. Therefore, having defined the Hamiltonian without the constant term $\beta|\mathscr{E}_\Lambda^{\mathrm{b}}|$ would have led to the same Gibbs distribution: the energy of a system can always be shifted by a constant without affecting the distribution. ◇

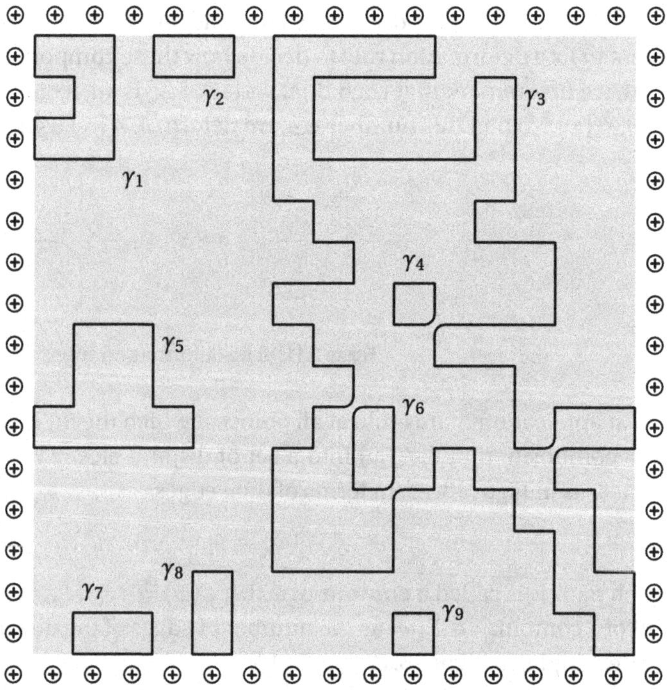

Figure 3.12. The contours (paths on the dual lattice) associated with the configuration of Figure 3.10. Together with the value of the spins on the boundary (+1 in the present case), the original configuration ω can be reconstructed in a unique manner.

Peierls' Argument

We consider the box $\mathrm{B}(n) = \{-n, \dots, n\}^2$. To study $\mu^+_{\mathrm{B}(n);\beta,0}(\sigma_0 = -1)$, we first observe that *any configuration* $\omega \in \Omega^+_{\mathrm{B}(n)}$ *such that* $\sigma_0(\omega) = -1$ *must possess at least one (actually, an odd number of) contours surrounding the origin.*

To make this statement precise, notice that each contour $\gamma \in \Gamma(\omega)$ is a bounded simple closed curve in $\mathbb{R}^2$ and therefore splits the plane into two regions, exactly one of which is bounded, called the **interior** of γ and denoted $\mathrm{Int}(\gamma)$. We can thus write

$$\mu^+_{\mathrm{B}(n);\beta,0}(\sigma_0 = -1) \leq \mu^+_{\mathrm{B}(n);\beta,0}\big(\exists \gamma_* \in \Gamma\colon \mathrm{Int}(\gamma_*) \ni 0\big) \leq \sum_{\gamma_*\colon \mathrm{Int}(\gamma_*)\ni 0} \mu^+_{\mathrm{B}(n);\beta,0}(\Gamma \ni \gamma_*).$$

Lemma 3.37. *For all* $\beta > 0$ *and any contour* γ_**,*

$$\mu^+_{\mathrm{B}(n);\beta,0}(\Gamma \ni \gamma_*) \leq e^{-2\beta|\gamma_*|}. \tag{3.34}$$

The bound (3.34) shows that the probability that a given contour appears in a configuration becomes small when β is large or when the contour is long. Later, we will refer to such a fact by saying that the ground state η^+ is *stable.*

Proof of Lemma 3.37. Using (3.33),

$$\begin{aligned}\mu^+_{\mathrm{B}(n);\beta,0}(\Gamma \ni \gamma_*) &= \sum_{\omega\colon \Gamma(\omega)\ni\gamma_*} \mu^+_{\mathrm{B}(n);\beta,0}(\omega) \\ &= e^{-2\beta|\gamma_*|}\frac{\sum_{\omega\colon \Gamma(\omega)\ni\gamma_*}\prod_{\gamma\in\Gamma(\omega)\setminus\{\gamma_*\}} e^{-2\beta|\gamma|}}{\sum_{\omega}\prod_{\gamma\in\Gamma(\omega)} e^{-2\beta|\gamma|}}.\end{aligned} \tag{3.35}$$

We will show that the ratio in (3.35) is bounded above by 1, by proving that the sum in the numerator is the same as the one in the denominator, but with an additional constraint. With each configuration ω with $\Gamma(\omega) \ni \gamma_*$ appearing in the sum of the numerator, we associate the configuration $\mathscr{E}_{\gamma_*}(\omega)$ obtained from ω by "removing γ_*". This can be done by simply flipping all spins in the interior of γ_*:

$$(\mathscr{E}_{\gamma_*}(\omega))_i \stackrel{\text{def}}{=} \begin{cases} -\omega_i & \text{if } i \in \mathrm{Int}(\gamma_*), \\ \omega_i & \text{otherwise.}\end{cases} \tag{3.36}$$

It is important to realize that $\mathscr{E}_{\gamma_*}(\omega)$ is the configuration whose set of contours is exactly $\Gamma(\omega) \setminus \{\gamma_*\}$. For instance, even if $\mathrm{Int}(\gamma_*)$ contains other contours (as γ_3 in Figure 3.12, which contains γ_4 and γ_6 in its interior), these continue to exist after flipping the spins. Let $\mathfrak{C}(\gamma_*)$ then be the set of configurations that can be obtained by removing γ_* from a configuration containing γ_*. We have

$$\sum_{\omega\colon \Gamma(\omega)\ni\gamma_*}\ \prod_{\gamma\in\Gamma(\omega)\setminus\{\gamma_*\}} e^{-2\beta|\gamma|} = \sum_{\omega'\in\mathfrak{C}(\gamma_*)}\ \prod_{\gamma'\in\Gamma(\omega')} e^{-2\beta|\gamma'|}.$$

But since the sum over $\omega' \in \mathfrak{C}(\gamma_*)$ is less than the sum over all $\omega' \in \Omega^+_{\mathrm{B}(n)}$, this shows that the ratio in (3.35) is indeed bounded above by 1. □

Each of the sums in the ratio in (3.35) *is typically exponentially large or small in* $|\mathrm{B}(n)|$. *We have proved that the ratio is nevertheless bounded above by* 1 *by flipping the spins of the configuration inside the contour* γ_*, *an operation that relied crucially on the symmetry of the model under a global spin flip.* ◇

Using (3.34), we bound the sum over all contours that surround the origin, by grouping them according to their lengths. Since the smallest contour surrounding the origin is made of 4 dual edges,

$$\mu^+_{\mathrm{B}(n);\beta,0}(\sigma_0 = -1) \leq \sum_{\gamma_*:\ \mathrm{Int}(\gamma_*)\ni 0} e^{-2\beta|\gamma_*|} \tag{3.37}$$

$$= \sum_{k\geq 4} \sum_{\substack{\gamma_*:\ \mathrm{Int}(\gamma_*)\ni 0 \\ |\gamma_*|=k}} e^{-2\beta|\gamma_*|}$$

$$= \sum_{k\geq 4} e^{-2\beta k}\, \#\left\{\gamma_* : \mathrm{Int}(\gamma_*) \ni 0,\ |\gamma_*| = k\right\}. \tag{3.38}$$

A contour of length k surrounding the origin necessarily contains a vertex of the set $\left\{(u - \frac{1}{2}, \frac{1}{2}) : u = 1, \ldots, [k/2]\right\}$. But the total number of contours of length k starting from a given vertex is at most $4 \cdot 3^{k-1}$. Indeed, there are 4 available directions for the first segment, then at most 3 for each of the remaining $k-1$ segments (since the contour does not use the same edge twice). Therefore,

$$\#\left\{\gamma_* : \mathrm{Int}(\gamma_*) \ni 0,\ |\gamma_*| = k\right\} \leq \tfrac{k}{2} \cdot 4 \cdot 3^{k-1}. \tag{3.39}$$

Gathering these estimates,

$$\mu^+_{\mathrm{B}(n);\beta,0}(\sigma_0 = -1) \leq \tfrac{2}{3} \sum_{k\geq 4} k 3^k e^{-2\beta k} \stackrel{\mathrm{def}}{=} \delta(\beta). \tag{3.40}$$

If β is large enough (so that $3e^{-2\beta} < 1$), then the series in (3.40) converges. Moreover, $\delta(\beta) \downarrow 0$ as $\beta \to \infty$. This proves (3.29), which concludes the proof that $\beta_c(2) < \infty$.

Before turning to the case $d \geq 3$, let us see what additional information about the low-temperature behavior of the two-dimensional Ising model can be extracted using the approach discussed above. The next exercise shows that, at sufficiently low temperatures, typical configurations in $\mathrm{B}(n)$, for the model with $+$ boundary condition, consist of a "sea" of $+$ spins with small islands of $-$ spins (the latter possibly containing "lakes" of $+$ spins, etc.). Namely, the largest contour in $\mathrm{B}(n)$ has a length of order $\log n$.

Exercise 3.18. Consider the two-dimensional Ising model.

1. Show that there exists $\beta_0 < \infty$ such that the following holds for all $\beta > \beta_0$. For any $c > 0$, there exists $K_0(c) < \infty$ such that, for all $K > K_0(c)$ and all n,

$$\mu^+_{\mathrm{B}(n);\beta,0}\left(\exists \gamma \in \Gamma \text{ with } |\gamma| \geq K \log n\right) \leq n^{-c}.$$

2. Show that, for all $\beta \geq 0$ and all $c > 0$, there exists $K_1(\beta, c) > 0$ and $n_0(c) < \infty$ such that, for all $K < K_1(\beta, c)$ and all $n \geq n_0(c)$,

$$\mu^+_{\mathrm{B}(n);\beta,0}(\exists \gamma \in \Gamma \text{ with } |\gamma| \geq K \log n) \geq 1 - e^{-n^{2-c}}.$$

Introducing a positive magnetic field h should only make the appearance of contours less likely, so it is natural to expect that the claims of the previous exercise still hold in that case.

Exercise 3.19. Extend the claims of Exercise 3.18 to the case $h > 0$. *Hint:* For the first claim, observe that the existence of a long contour implies the existence of a long path of $-$ spins, which is a decreasing event; then use the FKG inequality.

Extension to Larger Dimensions. It remains to show that a phase transition also occurs in the Ising model in dimensions $d \geq 3$. Adapting Peierls' argument to higher dimensions is possible, but the counting in (3.39) becomes a little trickier.

Exercise 3.20. Show that Peierls' estimate can be extended to $\mathbb{Z}^d$, $d \geq 3$. The combinatorial estimate on the sum of contours can be done using Lemma 3.38 below.

Nevertheless, we will analyze the model in $d \geq 3$ by following an alternative approach: using the natural embedding of $\mathbb{Z}^d$ into $\mathbb{Z}^{d+1}$ and the GKS inequalities, we will prove that $\beta_c(d)$ is nonincreasing in d.

The idea is elementary: one can build the Ising model on $\mathbb{Z}^{d+1}$ by considering a stack of Ising models on $\mathbb{Z}^d$ and adding interactions between neighboring spins living in successive layers. Then, the GKS inequalities tell us that adding these interactions does not decrease the magnetization and, thus, does not increase the inverse critical temperature. ◇

To simplify notation, we treat explicitly only the case $d = 3$; the extension to higher dimensions is straightforward. We will see $\mathbb{Z}^2$ as embedded in $\mathbb{Z}^3$. Therefore, we will temporarily use the following notations:

$$\mathrm{B}^3(n) \overset{\text{def}}{=} \{-n, \dots, n\}^3, \qquad \mathrm{B}^2(n) \overset{\text{def}}{=} \{-n, \dots, n\}^2.$$

We claim that

$$\langle \sigma_0 \rangle^+_{\mathrm{B}^3(n);\beta,0} \geq \langle \sigma_0 \rangle^+_{\mathrm{B}^2(n);\beta,0}.$$

Namely, consider the set of edges $\{i, j\}$ connecting two nearest-neighbor vertices $i = (i_1, i_2, i_3)$ and $j = (j_1, j_2, j_3)$ such that $i_3 = 0$ and $|j_3| = 1$. The two spins living at the endpoints of such an edge contribute to the total energy by an amount

$-\beta\sigma_i\sigma_j \equiv -J_{ij}\sigma_i\sigma_j$ (remember the Hamiltonian written as in (3.20)). Thanks to the GKS inequalities,

$$\frac{\partial}{\partial J_{ij}} \langle \sigma_0 \rangle^+_{B^3(n);\beta,0} = \langle \sigma_0\sigma_i\sigma_j \rangle^+_{B^3(n);\beta,0} - \langle \sigma_0 \rangle^+_{B^3(n);\beta,0} \langle \sigma_i\sigma_j \rangle^+_{B^3(n);\beta,0} \geq 0 .$$

We can therefore consider those edges, one after the other, and for each of them gradually decrease the interaction from its initial value $J_{ij} = \beta$ down to $J_{ij} = 0$. Denoting by $\mu^{+,0}_{B^3(n);\beta,0}$ the Gibbs distribution obtained after all those coupling constants J_{ij} have been brought down to zero, we obtain

$$\langle \sigma_0 \rangle^+_{B^3(n);\beta,0} \geq \langle \sigma_0 \rangle^{+,0}_{B^3(n);\beta,0} .$$

Observe that the spins contained in the layer $\{j_3 = 0\}$ interact now as if they were in a two-dimensional system, and so $\langle \sigma_0 \rangle^{+,0}_{B^3(n);\beta,0} = \langle \sigma_0 \rangle^+_{B^2(n);\beta,0}$. We therefore get

$$\lim_{n\to\infty} \langle \sigma_0 \rangle^+_{B^3(n);\beta,0} \geq \lim_{n\to\infty} \langle \sigma_0 \rangle^+_{B^2(n);\beta,0} .$$

Combined with (3.27), this inequality shows, in particular, that $\beta_c(3) \leq \beta_c(2)$. The existence of a first-order phase transition at low temperatures for the Ising model on $\mathbb{Z}^3$ thus follows from the already proven fact that $\beta_c(2) < \infty$.

Improved Bound. It is known that the inverse critical temperature of the two-dimensional Ising model equals

$$\beta_c(2) = \tfrac{1}{2}\text{arcsinh}(1) \cong 0.441 .$$

Obviously, not much care was taken, in our application of Peierls' argument, to optimize the resulting upper bound on $\beta_c(2)$. The following exercise shows how a slightly more careful application of the same ideas can lead to a rather decent upper bound (with a relative error of order 10%, compared to the exact value).

Exercise 3.21. 1. Check, using (3.40), that $\beta_c(2) < 0.88$.

2. The aim of this exercise is to improve this estimate to $\beta_c(2) < 0.493$. This will be done by showing that

$$\beta_c(2) \leq \tfrac{1}{2} \log \mu ,$$

where μ is the connectivity constant of $\mathbb{Z}^2$, defined by

$$\mu \stackrel{\text{def}}{=} \lim_{n\to\infty} \frac{1}{n} \log C_n ,$$

where C_n is the number of nearest-neighbor paths of length n starting at the origin and visiting each of its vertices at most once. It is known that $2.62 < \mu < 2.68$. *Hint:* Proceed similarly as in (3.37) and show that the ratio

$$\frac{\mu^+_{B(n);\beta,0}(\sigma_i = -1\ \forall i \in B(R))}{\mu^-_{B(n);\beta,0}(\sigma_i = -1\ \forall i \in B(R))} < 1 ,$$

uniformly in n, provided that $\beta > \frac{1}{2} \log \mu$ and that R is large enough.

3.7.3 Uniqueness at High Temperature

There exist several distinct methods to prove that there is a unique Gibbs state when the spins are weakly dependent, that is, at high temperatures. Two general approaches will be presented in Section 6.5. Here, we rely on a graphical representation, which is well adapted to a description of high-temperature correlations.

High-Temperature Representation. This representation relies on the following elementary identity. Since $\sigma_i\sigma_j$ takes only the two values ± 1,

$$e^{\beta\sigma_i\sigma_j} = \cosh(\beta) + \sigma_i\sigma_j \sinh(\beta) = \cosh(\beta)\left(1 + \tanh(\beta)\sigma_i\sigma_j\right). \tag{3.41}$$

Identity (3.41) can be used to rewrite the Boltzmann weight. For all $\Lambda \Subset \mathbb{Z}^d$ and $\omega \in \Omega_\Lambda^+$,

$$e^{-\mathscr{H}_{\Lambda;\beta,0}(\omega)} = \prod_{\{i,j\}\in\mathscr{E}_\Lambda^{\mathrm{b}}} e^{\beta\sigma_i(\omega)\sigma_j(\omega)} = \cosh(\beta)^{|\mathscr{E}_\Lambda^{\mathrm{b}}|} \prod_{\{i,j\}\in\mathscr{E}_\Lambda^{\mathrm{b}}} \left(1 + \tanh(\beta)\omega_i\omega_j\right), \tag{3.42}$$

where $\mathscr{E}_\Lambda^{\mathrm{b}}$ was defined in (3.2). We now expand the product over the edges.

Exercise 3.22. Show that, for any nonempty finite set $\mathscr{E}$,

$$\prod_{e\in\mathscr{E}}(1 + f(e)) = \sum_{E\subset\mathscr{E}} \prod_{e\in E} f(e). \tag{3.43}$$

Using (3.43) in (3.42) and changing the order of summation, we get

$$\mathbf{Z}_{\Lambda;\beta,0}^+ = \cosh(\beta)^{|\mathscr{E}_\Lambda^{\mathrm{b}}|} \sum_{E\subset\mathscr{E}_\Lambda^{\mathrm{b}}} \tanh(\beta)^{|E|} \underbrace{\sum_{\omega\in\Omega_\Lambda^+} \prod_{\{i,j\}\in E} \omega_i\omega_j}_{=\prod_{i\in\Lambda}\omega_i^{I(i,E)}},$$

where $I(i,E)$ is the incidence number: $I(i,E) \stackrel{\text{def}}{=} \#\{j \in \mathbb{Z}^d : \{i,j\} \in E\}$. Now, the summation over $\omega \in \Omega_\Lambda^+$ can be made separately for each vertex $i \in \Lambda$:

$$\sum_{\omega_i=\pm 1} \omega_i^{I(i,E)} = \begin{cases} 2 & \text{if } I(i,E) \text{ is even,} \\ 0 & \text{otherwise.} \end{cases} \tag{3.44}$$

We conclude that

$$\mathbf{Z}_{\Lambda;\beta,0}^+ = 2^{|\Lambda|} \cosh(\beta)^{|\mathscr{E}_\Lambda^{\mathrm{b}}|} \sum_{E\in\mathfrak{E}_\Lambda^{+;\mathrm{even}}} \tanh(\beta)^{|E|}, \tag{3.45}$$

where

$$\mathfrak{E}_\Lambda^{+;\mathrm{even}} \stackrel{\text{def}}{=} \left\{E \subset \mathscr{E}_\Lambda^{\mathrm{b}} : I(i,E) \text{ is even for all } i \in \Lambda\right\}.$$

When convenient, we will identify such sets of edges with the graph they induce.[3]

[3] The graph induced by a set E of edges is the graph having E as edges and the endpoints of the edges in E as vertices.

The expression (3.45) is called the *high-temperature representation* of the partition function. Proceeding in the same manner, we see that $\langle\sigma_0\rangle^+_{\Lambda;\beta,0}$ can be written

$$\langle\sigma_0\rangle^+_{\Lambda;\beta,0} = \big(\mathbf{Z}^+_{\Lambda;\beta,0}\big)^{-1} 2^{|\Lambda|}\cosh(\beta)^{|\mathscr{E}^{\mathrm{b}}_\Lambda|} \sum_{E\in\mathfrak{E}^{+;0}_\Lambda} \tanh(\beta)^{|E|}$$
$$= \frac{\sum_{E\in\mathfrak{E}^{+;0}_\Lambda}\tanh(\beta)^{|E|}}{\sum_{E\in\mathfrak{E}^{+;\mathrm{even}}_\Lambda}\tanh(\beta)^{|E|}}, \tag{3.46}$$

where

$$\mathfrak{E}^{+;0}_\Lambda \overset{\text{def}}{=} \big\{E\subset\mathscr{E}^{\mathrm{b}}_\Lambda : I(i,E) \text{ is even for all } i\in\Lambda\setminus\{0\}, \text{ but } I(0,E) \text{ is odd}\big\}.$$

Given $E\subset\mathscr{E}^{\mathrm{b}}_\Lambda$, we denote by $\Delta(E)$ the set of all edges of $\mathscr{E}^{\mathrm{b}}_\Lambda$ sharing no endpoint with an edge of E. Any collection of edges $E\in\mathfrak{E}^{+;0}_\Lambda$ can then be decomposed as $E = E_0\cup E'$, with $E_0\neq\varnothing$ the connected component of E containing 0, and $E'\in\mathfrak{E}^{+;\mathrm{even}}_\Lambda$ satisfying $E'\subset\Delta(E_0)$. Therefore,

$$\langle\sigma_0\rangle^+_{\Lambda;\beta,0} = \sum_{\substack{E_0\in\mathfrak{E}^{+;0}_\Lambda\\ \text{connected}, E_0\ni 0}} \tanh(\beta)^{|E_0|}\, \frac{\sum_{E'\in\mathfrak{E}^{+;\mathrm{even}}_\Lambda : E'\subset\Delta(E_0)}\tanh(\beta)^{|E'|}}{\sum_{E\in\mathfrak{E}^{+;\mathrm{even}}_\Lambda}\tanh(\beta)^{|E|}}. \tag{3.47}$$

Proof that $\beta_c(d) > 0$, for all d. Bounding the ratio in (3.47) by 1,

$$\langle\sigma_0\rangle^+_{\mathrm{B}(n);\beta,0} \le \sum_{\substack{E_0\in\mathfrak{E}^{+;0}_{\mathrm{B}(n)}\\ \text{connected}, E_0\ni 0}} \tanh(\beta)^{|E_0|}. \tag{3.48}$$

The sum can be bounded using the following lemma.

Lemma 3.38. *Let G be a connected graph with N edges. Starting from an arbitrary vertex of G, there exists a path in G crossing each edge of G exactly twice.*

Proof The proof proceeds by induction on N, observing that an arbitrary connected graph can always be built one edge at a time in such a way that all intermediate graphs are also connected. When $N = 1$, the result is trivial. Suppose that the result holds when $N = k$ and let $\pi = (\pi(1),\dots,\pi(2k))$ be one of the corresponding paths. We add a new edge to the graph, keeping it connected; this implies that at least one endpoint v of this edge belongs to the original graph. The desired path is obtained by following π until the first visit at v, then crossing the new edge once in each direction, and finally following the path π to its end. □

Using this lemma, we see that the number of graphs E_0 with ℓ edges contributing to (3.48) is bounded above by the number of paths of length 2ℓ starting from 0. The latter is certainly smaller than $(2d)^{2\ell}$ since each new edge can be taken in at most $2d$ different directions. On the other hand, E_0 necessarily connects 0 to $\mathrm{B}(n)^{\mathrm{c}}$: Indeed, $\sum_{i\in\mathbb{Z}^d} I(i,E_0) = 2|E_0|$ is even; since $I(0,E_0)$ is odd, there must be at least one vertex $i \neq 0$ with $I(i, E_0)$ odd; however, such a vertex cannot belong to $\mathrm{B}(n)$, since $I(i, E_0)$ is even for all $i \in \mathrm{B}(n) \setminus \{0\}$. We conclude that $|E_0| \geq n$, which yields, since $\tanh(\beta) \leq \beta$,

$$\langle \sigma_0 \rangle^+_{\mathrm{B}(n);\beta,0} \leq \sum_{\ell \geq n} (4d^2\beta)^\ell \leq e^{-cn}, \tag{3.49}$$

with $c = c(\beta, d) > 0$, for all $\beta < 1/(4d^2)$. In particular, $\langle \sigma_0 \rangle^+_{\beta,0} = 0$ for all $\beta < 1/(4d^2)$, which implies that $\beta_c(d) > 0$, that is, uniqueness at high temperatures, by Theorem 3.28 and the characterization (3.27).

Proof that $\beta_c(1) = +\infty$. Consider the Ising model in a one-dimensional box $\mathrm{B}(n)$ with $+$ boundary condition:

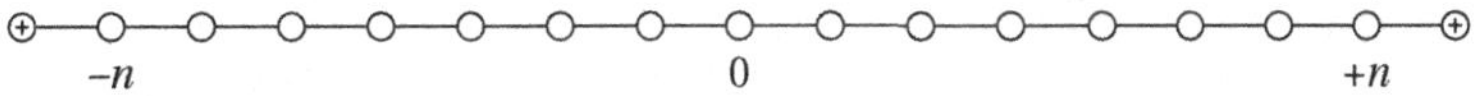

Due to the structure of $\mathbb{Z}$, there are only a few subgraphs of $E \subset \mathscr{E}^{\mathrm{b}}_{\mathrm{B}(n)}$ appearing in the ratio (3.46) and they are particularly simple. We first consider the denominator. Since the subgraphs appearing in the sum must be such that the incidence number of each $i \in \mathrm{B}(n)$ is either 0 or 2, $\mathfrak{C}^{+;\mathrm{even}}_{\mathrm{B}(n)}$ can contain only two graphs: the graph whose set of edges is $E = \varnothing$, as in the previous figure, and the one for which $E = \mathscr{E}^{\mathrm{b}}_{\mathrm{B}(n)}$:

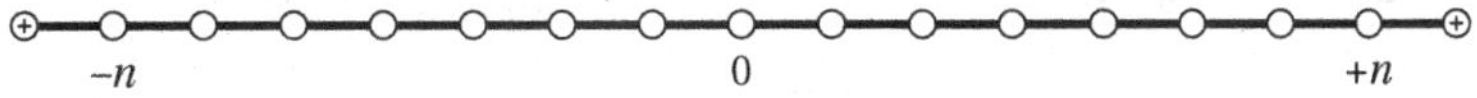

On the other hand, $\mathfrak{C}^{+;0}_{\mathrm{B}(n)}$ also reduces to two graphs, one composed of all edges with two nonnegative endpoints, and one composed of all edges with two nonpositive endpoints:

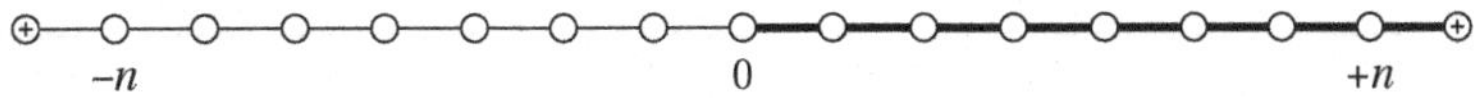

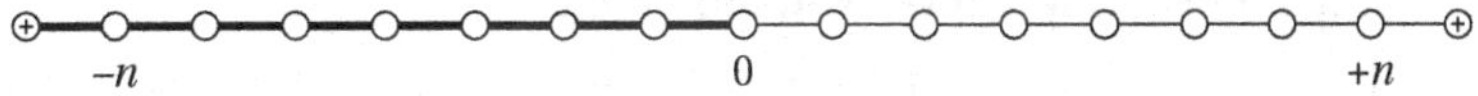

Consequently, (3.46) becomes

$$\langle \sigma_0 \rangle^+_{\mathrm{B}(n);\beta,0} = \frac{2\tanh(\beta)^{n+1}}{1 + \tanh(\beta)^{2(n+1)}},$$

which indeed tends to 0 when $n \to \infty$, for all $\beta < \infty$.

Exercise 3.23. Derive representations similar to (3.47) for $\langle \sigma_i \sigma_j \rangle^+_{\Lambda;\beta,0}$, $\mathbf{Z}^{\varnothing}_{\Lambda;\beta,0}$ and $\langle \sigma_i \sigma_j \rangle^{\varnothing}_{\Lambda;\beta,0}$.

The next exercise shows that the 2-point function decays exponentially when β is small enough.

Exercise 3.24. Using Exercise 3.23, prove that, for all β sufficiently small, there exists $c = c(\beta) > 0$ such that $\langle \sigma_i \sigma_j \rangle_{\beta,0} \leq e^{-c\|j-i\|_1}$, for all $i, j \in \mathbb{Z}^d$, where $\langle \cdot \rangle_{\beta,0}$ is the unique Gibbs state.

Note that, as shown in the next exercise, the decay of the 2-point function can never be faster than exponential (when $\beta \neq 0$):

Exercise 3.25. Using the GKS inequalities, prove that, in any dimension $d \geq 1$ and at any $\beta \geq 0$,

$$\langle \sigma_i \sigma_j \rangle^{+}_{\beta,0} \geq \langle \sigma_i \sigma_j \rangle^{\varnothing}_{\beta,0} \geq \langle \sigma_0 \sigma_{\|j-i\|_1} \rangle^{d=1}_{\Lambda_{i,j};\beta,0},$$

where the expectation in the right-hand side is with respect to the Gibbs distribution with free boundary condition in the box $\Lambda_{ij} = \{0, \dots, \|j-i\|_1\} \subset \mathbb{Z}$. Using Exercise 3.23, show that the 2-point function in the right-hand side is equal to $(\tanh \beta)^{\|j-i\|_1}$.

Remark 3.39. It is actually possible to prove that the exponential decay of $\langle \sigma_i \sigma_j \rangle_{\beta,0}$ and the exponential relaxation of $\langle \sigma_0 \rangle^{+}_{\mathrm{B}(n);\beta,0}$ toward $\langle \sigma_0 \rangle^{+}_{\beta,0}$ hold true for all $\beta < \beta_c(d)$.[3] ◇

Exercise 3.26. Use the high-temperature representation as an alternative way of computing the pressure of the one-dimensional Ising model with $h = 0$. Compare the expressions for $\psi^{+}_{\mathrm{B}(n)}$, $\psi^{\varnothing}_{\mathrm{B}(n)}$ and $\psi^{\mathrm{per}}_{\mathrm{B}(n)}$.

3.7.4 Uniqueness in Nonzero Magnetic Field

We are now left with the proof of item 1 of Theorem 3.25, which states that, when $h \neq 0$, the Gibbs state associated with (β, h) is always unique, regardless of the value of β. The proof will take us on a detour, using results from complex analysis, and will allow us to establish a very strong property of the pressure of the Ising model.

We will study the existence and properties of the pressure when h takes values in the complex domains

$$H^{+} \stackrel{\text{def}}{=} \{z \in \mathbb{C} \colon \mathfrak{Re}\, z > 0\},$$

$$H^{-} \stackrel{\text{def}}{=} \{z \in \mathbb{C} \colon \mathfrak{Re}\, z < 0\}.$$

Since the inverse temperature $\beta > 0$ will play no particular role in this section, we will omit it from the notations at some places. For example, we will write $\psi(h)$ rather than $\psi(\beta, h)$.

Theorem 3.40. *Let $\beta > 0$. As a function of the magnetic field h, the pressure of the Ising model in the thermodynamic limit, $\psi = \psi(h)$, can be extended from $\{h \in \mathbb{R} : h > 0\}$ (resp. $\{h \in \mathbb{R} : h < 0\}$) to an analytic function on the whole domain H^+ (resp. H^-). On H^+ and H^-, ψ can be computed using the thermodynamic limit with free boundary condition.*

This result of course implies that the complex derivative of ψ with respect to h exists on H^+ and H^-. Therefore, the real partial derivative $\frac{\partial \psi}{\partial h}$ exists at each real $h \neq 0$. By Theorem 3.34, this implies uniqueness of the Gibbs state for all $h \neq 0$, thus completing the proof of Theorem 3.25.

Remark 3.41. The GHS inequality, which is not discussed in this book, allows an alternative proof of the differentiability of the pressure when $h \neq 0$, avoiding complex analysis. Namely, the GHS inequality can be used to show that the magnetization $h \mapsto \langle \sigma_0 \rangle^+_{\beta,h}$ is concave, and hence continuous, on $\mathbb{R}_{\geq 0}$. This implies that its antiderivative (which is equal to ψ up to a constant) exists and is differentiable on $(0, +\infty)$. Of course, combined with Theorem 3.34, Theorem 3.40 implies the much stronger statement that $h \mapsto \langle \sigma_0 \rangle^+_{\beta,h}$ is real-analytic on $\{h < 0\}$ and $\{h > 0\}$. ◇

We have seen that, for real parameters, the thermodynamic limit of the pressure can be computed using an arbitrary boundary condition. When the magnetic field is complex, the boundary condition becomes a nuisance. It turns out that the free boundary condition is particularly convenient. We will therefore work in finite volumes $\Lambda \Subset \mathbb{Z}^d$ and study

$$\psi^{\varnothing}_{\Lambda}(h) = \frac{1}{|\Lambda|} \log \mathbf{Z}^{\varnothing}_{\Lambda;\beta,h} .$$

The existence and analyticity properties of the pressure are established by taking the thermodynamic limit $\Lambda \Uparrow \mathbb{Z}^d$ for this choice of boundary condition. The analytic function obtained is then the analytic continuation of the pressure to complex values of the field.[4]

On the one hand, when the magnetic field is real, $\mathbf{Z}^{\varnothing}_{\Lambda;\beta,h}$, being a finite linear combination of powers of $e^{\pm h}$, is real-analytic in h. Moreover, since

$$\mathbf{Z}^{\varnothing}_{\Lambda;\beta,h} > 0 \qquad \text{for all } h \in \mathbb{R}, \tag{3.50}$$

[4] We remind the reader of the following fact: if two functions analytic on a domain D coincide on a set $A \subset D$ which has an accumulation point in D, then these two functions are equal on D. Therefore, if it were possible to obtain another pressure $\widetilde{\psi}$ using a different boundary condition, analytic on H^+ and H^-, then, since this pressure coincides with the one obtained with free boundary conditions on the real axis, it must coincide with it on H^+ and H^-.

the pressure $\psi^{\varnothing}_{\Lambda}(\cdot)$ is also real-analytic in h. It is not true, however, that this real analyticity always holds in the thermodynamic limit $\Lambda \Uparrow \mathbb{Z}^d$. Indeed, we have seen (using Peierls' argument) that, at low temperature, the pressure is not even differentiable at $h = 0$.

On the other hand, since the Boltzmann weights are complex numbers when $h \in \mathbb{C}$, the partition function $\mathbf{Z}^{\varnothing}_{\Lambda;\beta,h}$ can very well vanish, leading to a problem even for the definition of the finite-volume pressure.

Fortunately, the celebrated *Lee-Yang Circle Theorem*, Theorem 3.43, will show that the partition function satisfies a remarkable property, analogous to (3.50), in suitable domains of the complex plane. This will allow us to control the analyticity of the pressure in the thermodynamic limit, as explained in the following result.

Theorem 3.42 (Lee-Yang). *Let $\beta \geq 0$. Let $D \subset \mathbb{C}$ be open, simply connected and such that $D \cap \mathbb{R}$ is an interval of $\mathbb{R}$. Assume that, for every finite volume $\Lambda \Subset \mathbb{Z}^d$,*

$$\mathbf{Z}^{\varnothing}_{\Lambda;\beta,h} \neq 0 \quad \forall h \in D. \tag{3.51}$$

Then, the pressure $h \mapsto \psi(h)$ admits an analytic continuation to D.

We know from the analysis done in the previous sections that the pressure is not differentiable at $h = 0$ when $\beta > \beta_c(d)$. When this happens, the previous theorem implies that there must exist a sequence $(h_k) \in \mathbb{C}$, tending to 0 and a sequence $\Lambda_k \Uparrow \mathbb{Z}^d$ such that $\mathbf{Z}^{\varnothing}_{\Lambda_k;\beta,h_k} = 0$ for all k. Therefore, even though the partition functions never vanish as long as h is real, complex zeros approach the point $h = 0$ in the thermodynamic limit. In this sense, although values of the magnetic field with a nonzero imaginary part may be experimentally meaningless,[4] the way the partition function behaves for such complex values of the magnetic field turns out to have profound physical consequences.

Proof of Theorem 3.42. (The precise statements of the few classical results of complex analysis needed in the proof below can be found in Appendix B.3.)

Let $\Lambda_n \Uparrow \mathbb{Z}^d$. Using (3.51), Theorem B.23 guarantees that one can find a function $h \mapsto \log \mathbf{Z}^{\varnothing}_{\Lambda_n;\beta,h}$ analytic on D and coinciding with the quantity studied in the rest of this chapter when $h \in D \cap \mathbb{R}$ (see Remark B.24 for the existence of a branch of the logarithm with this property). One can then define

$$g_n(h) \stackrel{\text{def}}{=} \exp\bigl(|\Lambda_n|^{-1} \log \mathbf{Z}^{\varnothing}_{\Lambda_n;\beta,h}\bigr),$$

which is also analytic on D. Now, when $h \in D \cap \mathbb{R}$, $g_n(h)$ coincides with $e^{\psi^{\varnothing}_{\Lambda_n}(h)}$, and Theorem 3.6 thus guarantees that, for such values of h, $g_n(h) \to g(h) \stackrel{\text{def}}{=} e^{\psi(h)}$ as $n \to \infty$, where ψ is the pressure of the Ising model in infinite volume.

The next observation is that the sequence (g_n) is locally uniformly bounded on D, since

$$
\begin{aligned}
|\mathbf{Z}^{\varnothing}_{\Lambda_n;\beta,h}| &\leq \sum_{\omega\in\Omega_{\Lambda_n}} \big|\exp\big(-\mathscr{H}^{\varnothing}_{\Lambda_n;\beta,h}(\omega)\big)\big| \\
&= \sum_{\omega\in\Omega_{\Lambda_n}} \exp\big(-\mathscr{H}^{\varnothing}_{\Lambda_n;\beta,\mathfrak{Re}\,h}(\omega)\big) \leq \exp\big((2d\beta + |\mathfrak{Re}\,h| + \log 2)|\Lambda_n|\big),
\end{aligned}
$$

and thus $|g_n(h)| = \exp(|\Lambda_n|^{-1}\log|\mathbf{Z}^{\varnothing}_{\Lambda_n;\beta,h}|) \leq \exp(2d\beta + |\mathfrak{Re}\,h| + \log 2)$ for all $h \in D$.

We are now in a position to apply Vitali's Convergence Theorem (Theorem B.25) in order to conclude that $(g_n)_{n\geq 1}$ converges locally uniformly, on D, to an analytic function g.

Moreover, since $g_n(h) \neq 0$ for all $h \in D$ and all $n \geq 1$, Hurwitz Theorem (Theorem B.26) implies that g has no zeros on D. Indeed, the other possibility (that is, $g \equiv 0$ on D) is incompatible with the fact that $g = e^{\psi} > 0$ on $D \cap \mathbb{R}$.

Since g does not vanish on D, it follows from Theorem B.23 that the latter admits an analytic logarithm in D. However, choosing again the branch that is real on $D \cap \mathbb{R}$, the function $\log g$ coincides with the pressure of the Ising model on the real axis, which proves the theorem. □

To prove Theorem 3.40 using Theorem 3.42, we still have to show

Theorem 3.43 (Lee–Yang Circle Theorem). *Condition* (3.51) *is satisfied when* $D = H^+$ *and when* $D = H^-$.

The proof given below will involve working with the variable

$$
z \stackrel{\text{def}}{=} e^{-2h}
$$

rather than h. But $h \in H^+$ if and only if $z \in \mathbb{U}$, where $\mathbb{U}$ is the open unit disk

$$
\mathbb{U} \stackrel{\text{def}}{=} \{z \in \mathbb{C} : |z| < 1\}.
$$

Therefore, Theorem 3.43 implies that all zeros of $\mathbf{Z}^{\varnothing}_{\Lambda_n;\beta,h}$ (seen as a function of z) lie on the unit circle. This explains the origin of the name given to the above result.

Proof When $\beta = 0$, the claim is trivial. We therefore assume from now on that $\beta > 0$. It will be convenient to consider the model as defined on a subgraph of $\mathbb{Z}^d$ with no isolated vertices, that is, to consider the model on a graph (V, E), where E is a finite set of edges between nearest neighbors of $\mathbb{Z}^d$ and where V is the set of all endpoints of edges in E. It will be assumed that the interactions among the spins on V appearing in the Hamiltonian are only between spins at vertices connected by an edge of E.

As we already said, the partition function with the free boundary condition in V is a finite linear combination of powers of $e^{\pm h}$. We now express it as a polynomial in the variable $z = e^{-2h}$. Namely,

$$\mathbf{Z}^{\varnothing}_{V;\beta,h} = \sum_{\omega\in\Omega_V} \prod_{\{i,j\}\in E} e^{\beta\sigma_i(\omega)\sigma_j(\omega)} \prod_{i\in V} e^{h\sigma_i(\omega)}$$
$$= e^{\beta|E|+h|V|} \sum_{\omega\in\Omega_V} \prod_{\{i,j\}\in E} e^{\beta(\sigma_i(\omega)\sigma_j(\omega)-1)} \prod_{i\in V} e^{h(\sigma_i(\omega)-1)} .$$

A configuration $\omega \in \Omega_V$ can always be identified with the set $X = X(\omega) \subset V$ defined by $X(\omega) \stackrel{\text{def}}{=} \{i \in V : \sigma_i(\omega) = -1\}$. We can therefore write

$$\sum_{\omega\in\Omega_V} \prod_{\{i,j\}\in E} e^{\beta(\sigma_i(\omega)\sigma_j(\omega)-1)} \prod_{i\in V} e^{h(\sigma_i(\omega)-1)} = \sum_{X\subset V} a_E(X) z^{|X|} \stackrel{\text{def}}{=} \mathscr{P}_E(z) ,$$

where $a_E(\varnothing) = a_E(V) \stackrel{\text{def}}{=} 1$ and, in all other cases,

$$a_E(X) \stackrel{\text{def}}{=} \prod_{\substack{\{i,j\}\in E\\ i\in X, j\in V\setminus X}} e^{-2\beta} .$$

Observe that these coefficients satisfy $0 \leq a_E(X) \leq 1$. Observing that $\mathbf{Z}^{\varnothing}_{V;\beta,h} = e^{\beta|E|+h|V|}\, \mathscr{P}_E(z)$, in order to show that $\mathbf{Z}^{\varnothing}_{V;\beta,h} \neq 0$ for all $h \in H^+$, it suffices to prove that $\mathscr{P}_E(z)$ does not vanish on $\mathbb{U}$.

The next step is to turn the one-variable but high-degree polynomial $\mathscr{P}_E$ into a many-variables but degree-one (in each variable) polynomial: let $\mathbf{z}_V = (z_i)_{i\in V} \in \mathbb{C}^V$ and consider the polynomial

$$\hat{\mathscr{P}}_E(\mathbf{z}_V) \stackrel{\text{def}}{=} \sum_{X\subset V} a_E(X) \prod_{i\in X} z_i .$$

Of course, the original polynomial $\mathscr{P}_E(z)$ is recovered by taking $z_i = z$ for all $i \in V$. We will show that

$$|z_i| < 1, \forall i \in V \implies \hat{\mathscr{P}}_E(\mathbf{z}_V) \neq 0 . \tag{3.52}$$

The proof proceeds by induction on the cardinality of E. We first check that (3.52) holds when E consists of a single edge $\{i,j\}$. In that case, since $a_E(\{i\}) = a_E(\{j\}) = e^{-2\beta}$,

$$\hat{\mathscr{P}}_E(\mathbf{z}_{\{i,j\}}) = z_i z_j + e^{-2\beta}(z_i + z_j) + 1 .$$

Therefore, $\hat{\mathscr{P}}_E(\mathbf{z}_{\{i,j\}}) = 0$ if and only if

$$z_i = -\frac{e^{-2\beta} z_j + 1}{z_j + e^{-2\beta}} .$$

Using the fact that $0 \leq e^{-2\beta} < 1$, it is easy to check (see Exercise 3.27 below) that the Möbius transformation $z \mapsto -(e^{-2\beta}z + 1)/(z + e^{-2\beta})$ interchanges the interior and the exterior of $\mathbb{U}$. This implies that if $|z_j| < 1$, then $|z_i| > 1$, so that $\hat{\mathscr{P}}_E(z_i, z_j)$ never vanishes when both $|z_i|, |z_j| < 1$.

Let us now assume that (3.52) holds for (V, E) and let $b = \{i,j\}$ be an edge of $\mathbb{Z}^d$ not contained in E. We want to show that (3.52) still holds for the graph $(V \cup \{i,j\}, E \cup \{b\})$.

There are three cases to consider, depending on whether $V \cap \{i,j\}$ is empty, contains one vertex, or contains two vertices.

Case 1: $V \cap \{i,j\} = \varnothing$. In this case, the sum over $X \subset V \cup \{i,j\}$ can be split into two independent sums, over $X_1 \subset V$ and $X_2 \subset \{i,j\}$, giving

$$\hat{\mathscr{P}}_{E\cup\{b\}}(\mathbf{z}_{V\cup\{i,j\}}) = \hat{\mathscr{P}}_E(\mathbf{z}_V)\hat{\mathscr{P}}_{\{b\}}(\mathbf{z}_{\{i,j\}}). \tag{3.53}$$

Since neither of the polynomials on the right-hand side vanishes (by the induction hypothesis) when $|z_k| < 1$ for all $k \in V \cup \{i,j\}$, the same must be true of the polynomial on the left-hand side.

Case 2: $V \cap \{i,j\} = \{i\}$. The main idea here is to add the new edge b in two steps. First, we add to E a "virtual" edge $b' = \{i',j\}$, where i' is a virtual vertex not present in V, and then identify i' with i, by a procedure called Asano contraction:

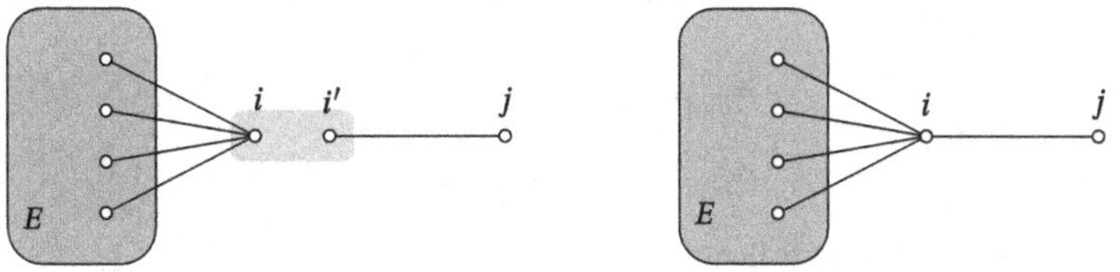

On the one hand, since $V \cap \{i',j\} = \varnothing$, we are back to Case 1: the polynomial $\hat{\mathscr{P}}_{E\cup\{b'\}}(\mathbf{z}_{V\cup\{i',j\}})$ can be factorized as in (3.53) and, by the induction hypothesis, we conclude that it cannot vanish when all its variables have modulus smaller than 1.

On the other hand, the sum over $X \subset V \cup \{i',j\}$ in $\hat{\mathscr{P}}_{E\cup\{b'\}}(\mathbf{z}_{V\cup\{i',j\}})$ can be split depending on $X \cap \{i',i\}$ being $\{i,i'\}$, $\{i'\}$, $\{i\}$ or $\varnothing$, giving

$$\hat{\mathscr{P}}_{E\cup\{b'\}}(\mathbf{z}_{V\cup\{i',j\}}) = \hat{\mathscr{P}}^{-,-}z_i z_{i'} + \hat{\mathscr{P}}^{+,-}z_{i'} + \hat{\mathscr{P}}^{-,+}z_i + \hat{\mathscr{P}}^{+,+},$$

where $\hat{\mathscr{P}}^{+,+}$, $\hat{\mathscr{P}}^{+,-}$, $\hat{\mathscr{P}}^{-,+}$ and $\hat{\mathscr{P}}^{-,-}$ are polynomials in the remaining variables: z_j and z_k, $k \in V \setminus \{i\}$.

The **Asano contraction of** $\hat{\mathscr{P}}_{E\cup\{b'\}}(\mathbf{z}_{V\cup\{i',j\}})$ is defined as the polynomial

$$\hat{\mathscr{P}}^{-,-}z_i + \hat{\mathscr{P}}^{+,+}.$$

It turns out that this polynomial coincides with $\hat{\mathscr{P}}_{E\cup\{b\}}(\mathbf{z}_{V\cup\{j\}})$.

Lemma 3.44. $\hat{\mathscr{P}}_{E\cup\{b\}}(\mathbf{z}_{V\cup\{j\}}) = \hat{\mathscr{P}}^{-,-}z_i + \hat{\mathscr{P}}^{+,+}$.

Proof Let $\tilde{V} \stackrel{\text{def}}{=} (V \setminus \{i\}) \cup \{j\}$. For $\sigma_1, \sigma_2 \in \{-,+\}$, the polynomials $\hat{\mathscr{P}}^{\sigma_1,\sigma_2}$ are explicitly given by

$$\hat{\mathscr{P}}^{\sigma_1,\sigma_2} = \sum_{X\subset\tilde{V}} a^{\sigma_1,\sigma_2}_{E\cup\{b'\}}(X)\prod_{k\in X} z_k,$$

with

$$a^{-,-}_{E\cup\{b'\}}(X) \stackrel{\text{def}}{=} \big(\mathbf{1}_{\{X\ni j\}} + \mathbf{1}_{\{X\not\ni j\}}e^{-2\beta}\big)a_E(X\cup\{i\}),$$
$$a^{+,-}_{E\cup\{b'\}}(X) \stackrel{\text{def}}{=} \big(\mathbf{1}_{\{X\ni j\}} + \mathbf{1}_{\{X\not\ni j\}}e^{-2\beta}\big)a_E(X),$$
$$a^{-,+}_{E\cup\{b'\}}(X) \stackrel{\text{def}}{=} \big(\mathbf{1}_{\{X\not\ni j\}} + \mathbf{1}_{\{X\ni j\}}e^{-2\beta}\big)a_E(X\cup\{i\}),$$
$$a^{+,+}_{E\cup\{b'\}}(X) \stackrel{\text{def}}{=} \big(\mathbf{1}_{\{X\not\ni j\}} + \mathbf{1}_{\{X\ni j\}}e^{-2\beta}\big)a_E(X).$$

Doing a similar decomposition for the polynomial $\hat{\mathscr{P}}_{E\cup\{b\}}(\mathbf{z}_{V\cup\{j\}})$, we get

$$\hat{\mathscr{P}}_{E\cup\{b\}}(\mathbf{z}_{V\cup\{j\}}) = \hat{\mathscr{P}}^{-}z_i + \hat{\mathscr{P}}^{+},$$

where, for $\sigma \in \{-,+\}$, we have introduced

$$\hat{\mathscr{P}}^{\sigma} \stackrel{\text{def}}{=} \sum_{X\subset\tilde{V}} a^{\sigma}_{E\cup\{b\}}(X)\prod_{k\in X} z_k,$$

with

$$a^{-}_{E\cup\{b\}}(X) \stackrel{\text{def}}{=} \big(\mathbf{1}_{\{X\ni j\}} + \mathbf{1}_{\{X\not\ni j\}}e^{-2\beta}\big)a_E(X\cup\{i\}),$$
$$a^{+}_{E\cup\{b\}}(X) \stackrel{\text{def}}{=} \big(\mathbf{1}_{\{X\not\ni j\}} + \mathbf{1}_{\{X\ni j\}}e^{-2\beta}\big)a_E(X).$$

The conclusion follows. □

Since we have seen that the polynomial $\hat{\mathscr{P}}_{E\cup\{b'\}}(\mathbf{z}_{V\cup\{i',j\}})$ does not vanish when all its variables have modulus smaller than 1, it suffices to show that its Asano contraction also cannot vanish when all its variables have modulus smaller than 1.

Let us fix the variables z_k, $k \in V\setminus\{i\}$, and z_j so that they all belong to $\mathbb{U}$. By Case 1, we know that, in this situation, $\hat{\mathscr{P}}_{E\cup\{b'\}}(\mathbf{z}_{E\cup\{i',j\}})$ cannot vanish when z_i and $z_{i'}$ also both belong to $\mathbb{U}$. By taking $z_i = z_{i'} = z$, we conclude that

$$z \mapsto \hat{\mathscr{P}}^{-,-}z^2 + (\hat{\mathscr{P}}^{-,+} + \hat{\mathscr{P}}^{+,-})z + \hat{\mathscr{P}}^{+,+}$$

cannot have zeros of modulus smaller than 1. In particular, the product of its two roots has modulus 1 or larger. But the latter implies that $|\hat{\mathscr{P}}^{+,+}| \geq |\hat{\mathscr{P}}^{-,-}|$ and, thus, $z \mapsto \hat{\mathscr{P}}^{-,-}z + \hat{\mathscr{P}}^{+,+}$ cannot vanish when $|z| < 1$.

Case 3: $V \cap \{i,j\} = \{i,j\}$**.** This case is treated in a very similar way, so we only sketch the argument and leave the details as an exercise to the reader.

Adding a virtual edge $b'' = \{i',j'\}$ yields a polynomial $\hat{\mathscr{P}}_{E\cup\{b''\}}(\mathbf{z}_{V\cup\{i',j'\}})$ satisfying (3.52) by Case 1. We then proceed as above and apply two consecutive Asano contractions: the first to identify the variables $z_{j'}$ and z_j, the second to identify the variables $z_{i'}$ and z_i. □

Remark 3.45. The reader might have noticed that the proof given above does not depend on the structure of the graph inherited from the Hamiltonian of the model. Moreover, the fact that the interaction is the same between each pair of nearest-neighbor spins was not used: the coupling constant β used for all edges could be

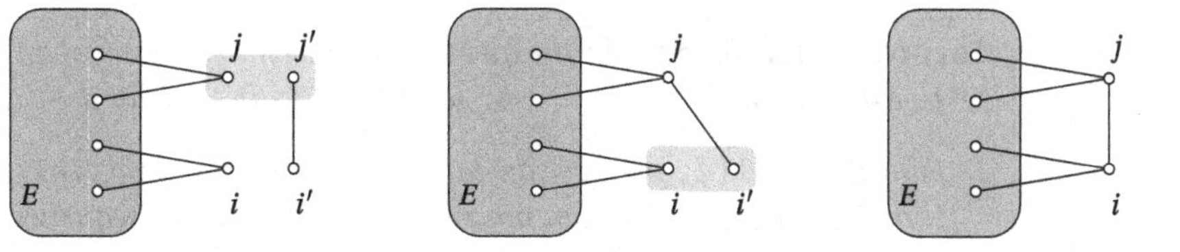

Figure 3.13. A picture of Case 3: We first add a virtual edge $\{i',j'\}$ to E, then identify first j and j', and then i and i'.

replaced by couplings J_{ij} varying from edge to edge. Therefore, the Circle Theorem and its consequence, Theorem 3.42, can be adapted to obtain analyticity of the pressure in more general settings. ◇

Exercise 3.27. Let $\varphi(z) \stackrel{\text{def}}{=} \frac{\alpha z+1}{\alpha+z}$, where $0 \le \alpha < 1$. Show that $\partial\mathbb{U}$ is invariant under φ, and that φ maps the interior of $\mathbb{U}$ onto its exterior and vice versa.

Exercise 3.28. Using the explicit formula (3.10) for the pressure of the one-dimensional Ising model, determine the location of its singularities as a function of the (complex) magnetic field h. What happens as β tends to infinity?

The next exercise provides an alternative approach to the analyticity of the pressure in a smaller open part of the complex plane, still containing $\mathbb{R} \setminus \{0\}$.

Exercise 3.29. Assume that $\mathfrak{Re}\, h > 0$. Observe that, by considering two independent copies of the system with magnetic field h and $\bar{h}$, one can write

$$|\mathbf{Z}^{\varnothing}_{\Lambda;\beta,h}|^2 = \sum_{\omega,\omega'} \exp\Big\{\beta \sum_{\{i,j\}\in\mathscr{E}_\Lambda} (\omega_i\omega_j + \omega'_i\omega'_j) + \sum_{i\in\Lambda}(h\omega_i + \bar{h}\omega'_i)\Big\}.$$

We define $\theta_i \in \{0, \pi/2, \pi, 3\pi/2\}$, $i \in \Lambda$, by $\cos\theta_i \stackrel{\text{def}}{=} \frac{1}{2}(\omega_i+\omega'_i)$ and $\sin\theta_i \stackrel{\text{def}}{=} \frac{1}{2}(\omega_i-\omega'_i)$. Show that, after changing to these variables and expanding the exponential, one obtains

$$|\mathbf{Z}^{\varnothing}_{\Lambda;\beta,h}|^2 = \sum_{(\theta_i)_{i\in\Lambda}} \sum_{\substack{\mathbf{m}=(m_i)_{i\in\Lambda}\\ m_i\in\{0,1,2,3\}}} \widehat{\alpha}_{\mathbf{m}} e^{\mathrm{i}\sum_{i\in\Lambda} m_i\theta_i} = 4^{|\Lambda|}\,\widehat{\alpha}_{(0,\dots,0)},$$

with coefficients $\widehat{\alpha}_{\mathbf{m}}$ nonnegative and nondecreasing in $\mathfrak{Re}\, h + \mathfrak{Im}\, h$ and in $\mathfrak{Re}\, h - \mathfrak{Im}\, h$. Conclude that $|\mathbf{Z}^{\varnothing}_{\Lambda;\beta,h}| > 0$ when $\mathfrak{Re}\, h > |\,\mathfrak{Im}\, h|$.

3.7.5 Summary of What Has Been Proved

In this brief subsection, we summarize the main results that have been derived. First, we emphasize that the main features of the discussion in Section 1.4.3 have been fully recovered (compare, in particular, with Figure 1.11):

Theorem 3.46. *Let $\beta_c(d)$ be the inverse critical temperature of the Ising model on $\mathbb{Z}^d$ (we have seen that $\beta_c(1) = +\infty$, while $0 < \beta_c(d) < \infty$ for $d \geq 2$).*

1. *For all $\beta < \beta_c(d)$, the average magnetization density $m(\beta, h)$ is well defined (and independent of the boundary condition and of the sequence of boxes used in its definition) for all $h \in \mathbb{R}$. It is an odd, nondecreasing, continuous function of h; in particular, $m(\beta, 0) = 0$.*
2. *For all $\beta > \beta_c(d)$, the average magnetization density $m(\beta, h)$ is well defined (and independent of the boundary condition and of the sequence of boxes used in its definition) for all $h \in \mathbb{R} \setminus \{0\}$. It is an odd, nondecreasing function of h, which is continuous everywhere except at $h = 0$, where*

$$\lim_{h \downarrow 0} m(\beta, h) = m^+(\beta, h) > 0, \quad \lim_{h \uparrow 0} m(\beta, h) = m^-(\beta, h) < 0.$$

In particular, the spontaneous magnetization satisfies

$$m^*(\beta) = 0 \text{ when } \beta < \beta_c(d), \qquad m^*(\beta) > 0 \text{ when } \beta > \beta_c(d).$$

Remark 3.47. As has already been mentioned, it is known that $m^*(\beta_c) = 0$. By Exercise 3.17, this implies that the function $\beta \mapsto m^*(\beta)$ is continuous at β_c. ◇

Remark 3.48. It follows from the above that, when $h = 0$, the spontaneous magnetization $m^*(\beta)$ allows one to distinguish the ordered regime (in which $m^*(\beta) > 0$) from the disordered regime (in which $m^*(\beta) = 0$). A function with this property is said to be an **order parameter**. ◇

Proof of Theorem 3.46. On the one hand, we know from Theorem 3.43 that, for all $\beta \geq 0$, the pressure $\psi(\beta, h)$ is differentiable with respect to h at all $h \neq 0$. On the other hand, point 3 of Theorem 3.25 and Theorem 3.34 imply that the function $h \mapsto \psi(\beta, h)$ is differentiable at $h = 0$ when $\beta < \beta_c(d)$, but is not differentiable at $h = 0$ when $\beta > \beta_c(d)$. This implies that $\mathfrak{B}_\beta = \varnothing$ when $h \neq 0$ or $\beta < \beta_c(d)$, and that $\mathfrak{B}_\beta = \{0\}$ when $h = 0$ and $\beta > \beta_c(d)$.

By Corollary 3.7, the above implies that $m(\beta, h)$ is well defined and independent of the boundary condition whenever $h \neq 0$ or $\beta < \beta_c(d)$. This shows, in particular, that $m(\beta, h) = m^+(\beta, h)$ for all $h > 0$.

The claim that $m(\beta, h)$ is an odd, nondecreasing function of h that is continuous for all $h \notin \mathfrak{B}_\beta$ follows from symmetry and Corollary 3.7. □

We have also seen that the Gibbs states provide a satisfactory description of the model in the thermodynamic limit. These states give a first glimpse of the way models in infinite volume will be described later in the book. The states $\langle \cdot \rangle^+_{\beta,h}$ and $\langle \cdot \rangle^-_{\beta,h}$, constructed with $+$ and $-$ boundary conditions respectively, were instrumental in characterizing the uniqueness regime. Much more will be said on these states, in particular in Chapter 6.

3.8 Proof of the Correlation Inequalities

3.8.1 Proof of the GKS Inequalities

Although the GKS inequalities (3.21) and (3.22) are already more than we need to study the (nearest-neighbor) Ising model, we will prove them in an even more general setting.

Let $\Lambda \Subset \mathbb{Z}^d$ and let $\mathbf{K} = (K_C)_{C\subset\Lambda}$ be a family of real numbers, called **coupling constants**. Consider the following probability distribution on Ω_Λ:

$$\nu_{\Lambda;\mathbf{K}}(\omega) \stackrel{\text{def}}{=} \frac{1}{\mathbf{Z}_{\Lambda;\mathbf{K}}} \exp\Big\{\sum_{C\subset\Lambda} K_C\omega_C\Big\},$$

where $\omega_C \stackrel{\text{def}}{=} \prod_{i\in C}\omega_i$ and $\mathbf{Z}_{\Lambda;\mathbf{K}}$ is the associated partition function. The Gibbs distributions $\mu^+_{\Lambda;\mathbf{J},\mathbf{h}}$, $\mu^\varnothing_{\Lambda;\mathbf{J},\mathbf{h}}$ and $\mu^{\text{per}}_{\Lambda;\mathbf{J},\mathbf{h}}$ can all be written in this form, with $K_C \geq 0\, \forall C \subset \Lambda$, if $\mathbf{h} \geq 0$. For example, $\mu^+_{\Lambda;\beta,h} = \nu_{\Lambda;\mathbf{K}}$ once

$$K_C = \begin{cases} h + \beta\#\{j \notin \Lambda : j \sim i\} & \text{if } C = \{i\} \subset \Lambda, \\ \beta & \text{if } C = \{i,j\} \subset \Lambda,\, i \sim j, \\ 0 & \text{otherwise.} \end{cases}$$

Exercise 3.30. Check that $\mu^\varnothing_{\Lambda;\beta,h}$ and $\mu^{\text{per}}_{\Lambda;\beta,h}$ can also be written in this form for a suitable choice of the coefficients $\mathbf{K}$, and that these coefficients can all be taken nonnegative if $\mathbf{h} \geq 0$.

We can now state the following generalization of Theorem 3.20.

Theorem 3.49. *Let* $\mathbf{K} = (K_C)_{C\subset\Lambda}$ *be such that* $K_C \geq 0$ *for all* $C \subset \Lambda$. *Then, for any* $A, B \subset \Lambda$,

$$\langle\sigma_A\rangle_{\Lambda;\mathbf{K}} \geq 0, \tag{3.54}$$

$$\langle\sigma_A\sigma_B\rangle_{\Lambda;\mathbf{K}} \geq \langle\sigma_A\rangle_{\Lambda;\mathbf{K}}\langle\sigma_B\rangle_{\Lambda;\mathbf{K}}. \tag{3.55}$$

Proof Clearly, $\mathbf{Z}_{\Lambda;\mathbf{K}} > 0$. We can thus focus on the numerators. Expanding the exponentials as Taylor series as $e^{K_C\omega_C} = \sum_{n_C\geq 0} \frac{1}{n_C!} K_C^{n_C}\omega_C^{n_C}$, we can write

$$\begin{aligned} \mathbf{Z}_{\Lambda;\mathbf{K}}\langle\sigma_A\rangle_{\Lambda;\mathbf{K}} &= \sum_\omega \omega_A \prod_{C\subset\Lambda} e^{K_C\omega_C} \\ &= \sum_{\substack{(n_C)_{C\subset\Lambda}\\ n_C\geq 0}} \prod_{C\subset\Lambda} \frac{K_C^{n_C}}{n_C!} \sum_\omega \omega_A \prod_{C\subset\Lambda} \omega_C^{n_C}. \end{aligned} \tag{3.56}$$

We rewrite $\omega_A \prod_{C\subset\Lambda} \omega_C^{n_C} = \prod_{i\in\Lambda} \omega_i^{m_i}$, where $m_i = \mathbf{1}_{\{i\in A\}} + \sum_{C\subset\Lambda, C\ni i} n_C$. Upon summation, since

$$\sum_{\omega_i=\pm 1} \omega_i^{m_i} = \begin{cases} 2 & \text{if } m_i \text{ is even,} \\ 0 & \text{if } m_i \text{ is odd,} \end{cases}$$

it follows that

$$\sum_{\omega} \prod_{i\in\Lambda} \omega_i^{m_i} = \prod_{i\in\Lambda} \sum_{\omega_i=\pm 1} \omega_i^{m_i} \geq 0.$$

This establishes (3.54). To prove (3.55), we duplicate the system. That is, we consider the product probability distribution $\nu_{\Lambda;\mathbf{K}} \otimes \nu_{\Lambda;\mathbf{K}}$ on $\Omega_\Lambda \times \Omega_\Lambda$ defined by

$$\nu_{\Lambda;\mathbf{K}} \otimes \nu_{\Lambda;\mathbf{K}}(\omega, \omega') \stackrel{\text{def}}{=} \nu_{\Lambda;\mathbf{K}}(\omega)\nu_{\Lambda;\mathbf{K}}(\omega').$$

If we define $\sigma_i(\omega, \omega') \stackrel{\text{def}}{=} \omega_i$ and $\sigma_i'(\omega, \omega') \stackrel{\text{def}}{=} \omega_i'$, then

$$\langle \sigma_A \sigma_B \rangle_{\Lambda;\mathbf{K}} - \langle \sigma_A \rangle_{\Lambda;\mathbf{K}} \langle \sigma_B \rangle_{\Lambda;\mathbf{K}} = \langle \sigma_A(\sigma_B - \sigma_B') \rangle_{\nu_{\Lambda;\mathbf{K}} \otimes \nu_{\Lambda;\mathbf{K}}}.$$

The problem is thus reduced to proving the nonnegativity of

$$(\mathbf{Z}_{\Lambda;\mathbf{K}})^2 \langle \sigma_A(\sigma_B - \sigma_B') \rangle_{\nu_{\Lambda;\mathbf{K}} \otimes \nu_{\Lambda;\mathbf{K}}} = \sum_{\omega,\omega'} \omega_A(\omega_B - \omega_B') \prod_{C\subset\Lambda} e^{K_C(\omega_C + \omega_C')}.$$

Introducing the variables $\omega_i'' \stackrel{\text{def}}{=} \omega_i \omega_i' = \omega_i'/\omega_i$,

$$\begin{aligned} \sum_{\omega,\omega'} \omega_A(\omega_B - \omega_B') \prod_{C\subset\Lambda} e^{K_C(\omega_C + \omega_C')} &= \sum_{\omega,\omega''} \omega_A \omega_B (1 - \omega_B'') \prod_{C\subset\Lambda} e^{K_C(1+\omega_C'')\omega_C} \\ &= \sum_{\omega''} (1 - \omega_B'') \sum_{\omega} \omega_A \omega_B \prod_{C\subset\Lambda} e^{K_C(1+\omega_C'')\omega_C}. \end{aligned}$$

Since $1 - \omega_B'' \geq 0$, (3.55) follows by treating this last sum over ω (for each fixed ω'') as the one in (3.56), working with coupling constants $K_C(1 + \omega_C'') \geq 0$. □

Exercise 3.31. Let $\mathbf{K} = (K_C)_{C\subset\Lambda}$ and $\mathbf{K}' = (K_C')_{C\subset\Lambda}$ be such that $K_C \geq |K_C'|$ (in particular, $K_C \geq 0$), for all $C \subset \Lambda$. Show that, for any $A, B \subset \Lambda$,

$$\langle \sigma_A \rangle_{\Lambda;\mathbf{K}} \geq \langle \sigma_A \rangle_{\Lambda;\mathbf{K}'}.$$

Hint: Apply a variant of the argument used to prove (3.55).

3.8.2 Proof of the FKG Inequality

We provide here a very general and short proof of the FKG inequality. The interested reader can find an alternative proof in Section 3.10.3, based on Markov chain techniques, which one might find more intuitive.

Our aim is to show that, for a finite volume $\Lambda \Subset \mathbb{Z}^d$ and two nondecreasing functions $f, g\colon \Omega \to \mathbb{R}$,

$$\langle fg\rangle^{\eta}_{\Lambda;\mathbf{J},\mathbf{h}} \geq \langle f\rangle^{\eta}_{\Lambda;\mathbf{J},\mathbf{h}} \langle g\rangle^{\eta}_{\Lambda;\mathbf{J},\mathbf{h}} . \tag{3.57}$$

Again, we will prove a result that is more general than required. Remember that the order we use on Ω_Λ is the following: $\omega \leq \omega'$ if and only if $\omega_i \leq \omega'_i$ for all $i \in \Lambda$. We also define, for $\omega = (\omega_i)_{i\in\Lambda}$ and $\omega' = (\omega'_i)_{i\in\Lambda}$,

$$\omega \wedge \omega' \stackrel{\text{def}}{=} (\omega_i \wedge \omega'_i)_{i\in\Lambda} ,$$

$$\omega \vee \omega' \stackrel{\text{def}}{=} (\omega_i \vee \omega'_i)_{i\in\Lambda} .$$

As explained below, (3.57) is a consequence of the following general result.

Theorem 3.50. *Let $\mu = \bigotimes_{i\in\Lambda} \mu_i$ be a product measure on Ω_Λ. Let $f_1, \dots, f_4\colon \Omega_\Lambda \to \mathbb{R}$ be nonnegative functions on Ω_Λ such that*

$$f_1(\omega) f_2(\omega') \leq f_3(\omega \wedge \omega') f_4(\omega \vee \omega'), \qquad \forall \omega, \omega' \in \Omega_\Lambda. \tag{3.58}$$

Then

$$\langle f_1\rangle_\mu \langle f_2\rangle_\mu \leq \langle f_3\rangle_\mu \langle f_4\rangle_\mu . \tag{3.59}$$

Before turning to the proof of this result, let us explain why it implies (3.57). With no loss of generality, we can assume that f and g depend only on the values of the configuration inside Λ and that both are nonnegative.[5] For $i \in \Lambda$, $s \in \{\pm 1\}$, let

$$\mu_i(s) \stackrel{\text{def}}{=} e^{hs + s\sum_{j\notin\Lambda, j\sim i} J_{ij}\eta_j} .$$

We have

$$\langle f\rangle^{\eta}_{\Lambda;\mathbf{J},\mathbf{h}} = \sum_{\omega\in\Omega_\Lambda} f(\omega) p(\omega) \mu(\omega) = \langle fp\rangle_\mu ,$$

where

$$p(\omega) \stackrel{\text{def}}{=} \frac{\exp\{\sum_{\{i,j\}\in\mathscr{E}_\Lambda} J_{ij}\omega_i\omega_j\}}{\mathbf{Z}^{\eta}_{\Lambda;\mathbf{J},\mathbf{h}}} .$$

Let $f_1 = pf$, $f_2 = pg$, $f_3 = p$, $f_4 = pfg$. If (3.58) holds for this choice, then (3.59) holds, and so (3.57) is proved. To check (3.58), we must verify that

$$p(\omega)p(\omega') \leq p(\omega \vee \omega')p(\omega \wedge \omega') .$$

But this is true since

$$\omega_i\omega_j + \omega'_i\omega'_j \leq (\omega_i \vee \omega'_i)(\omega_j \vee \omega'_j) + (\omega_i \wedge \omega'_i)(\omega_j \wedge \omega'_j) .$$

[5] Indeed, if these hypotheses are not verified, we can redefine $f(\omega)$, for $\omega \in \Omega_\Lambda$, by $f(\omega\eta|_{\Lambda^c}) - \min_{\omega'} f(\omega'\eta|_{\Lambda^c})$, where $\omega\eta|_{\Lambda^c}$ is the configuration that coincides with ω on Λ and with η on Λ^c. The same can be done with g. Note that this does not affect the covariance of f and g.

Indeed, the inequality is obvious if both terms in the right-hand side are equal to 1. Let us therefore assume that at least one of them is equal to -1. This cannot happen if both $\omega_i \neq \omega_i'$ and $\omega_j \neq \omega_j'$. Without loss of generality, we can thus suppose that $\omega_i = \omega_i'$. In that case, the right-hand side equals

$$\omega_i\{(\omega_j \vee \omega_j') + (\omega_j \wedge \omega_j')\} = \omega_i(\omega_j + \omega_j') = \omega_i\omega_j + \omega_i'\omega_j'.$$

Remark 3.51. As the reader can easily check, the proof below does not rely on the fact that the spins take their values in $\{\pm 1\}$; it actually holds for arbitrary real-valued spins. ◇

Proof of Theorem 3.50. For some fixed $i \in \Lambda$, any configuration $\omega \in \Omega_\Lambda$ can be identified with the pair $(\tilde{\omega}, \omega_i)$, where $\tilde{\omega} \in \Omega_{\Lambda\setminus\{i\}}$. We will show that

$$f_1(\omega)f_2(\omega') \leq f_3(\omega \wedge \omega')f_4(\omega \vee \omega') \tag{3.60}$$

implies

$$\tilde{f}_1(\tilde{\omega})\tilde{f}_2(\tilde{\omega}') \leq \tilde{f}_3(\tilde{\omega} \wedge \tilde{\omega}')\tilde{f}_4(\tilde{\omega} \vee \tilde{\omega}'), \tag{3.61}$$

where (for $k = 1,2,3,4$) $\tilde{f}_k(\tilde{\omega}) \stackrel{\text{def}}{=} \langle f_k(\tilde{\omega}, \cdot)\rangle_{\mu_i} = \sum_{v=\pm 1} f_k(\tilde{\omega}, v)\mu_i(v)$. Using this observation $|\Lambda|$ times yields the desired result.

The left-hand side of (3.61) can be written

$$\begin{aligned}\langle f_1(\tilde{\omega}, u)f_2(\tilde{\omega}', v)\rangle_{\mu_i\otimes\mu_i} &= \langle \mathbf{1}_{\{u=v\}} f_1(\tilde{\omega}, u)f_2(\tilde{\omega}', v)\rangle_{\mu_i\otimes\mu_i} \\ &\quad + \langle \mathbf{1}_{\{u<v\}}(f_1(\tilde{\omega}, u)f_2(\tilde{\omega}', v) + f_1(\tilde{\omega}, v)f_2(\tilde{\omega}', u))\rangle_{\mu_i\otimes\mu_i}.\end{aligned}$$

Similarly, the right-hand side of (3.61) can be written

$$\begin{aligned}\langle f_3(\tilde{\omega} \wedge \tilde{\omega}', u)f_4(\tilde{\omega} \vee \tilde{\omega}', v)\rangle_{\mu_i\otimes\mu_i} &= \langle \mathbf{1}_{\{u=v\}} f_3(\tilde{\omega} \wedge \tilde{\omega}', u)f_4(\tilde{\omega} \vee \tilde{\omega}', v)\rangle_{\mu_i\otimes\mu_i} \\ + \langle \mathbf{1}_{\{u<v\}}(f_3(\tilde{\omega} \wedge \tilde{\omega}', u)&f_4(\tilde{\omega} \vee \tilde{\omega}', v) + f_3(\tilde{\omega} \wedge \tilde{\omega}', v)f_4(\tilde{\omega} \vee \tilde{\omega}', u))\rangle_{\mu_i\otimes\mu_i}.\end{aligned}$$

We thus obtain

$$\begin{aligned}&\tilde{f}_3(\tilde{\omega} \wedge \tilde{\omega}')\tilde{f}_4(\tilde{\omega} \vee \tilde{\omega}') - \tilde{f}_1(\tilde{\omega})\tilde{f}_2(\tilde{\omega}') \\ &\quad = \langle \mathbf{1}_{\{u=v\}}(f_3(\tilde{\omega} \wedge \tilde{\omega}', u)f_4(\tilde{\omega} \vee \tilde{\omega}', v) - f_1(\tilde{\omega}, u)f_2(\tilde{\omega}', v))\rangle_{\mu_n\otimes\mu_n} \\ &\qquad + \langle \mathbf{1}_{\{u<v\}}(C + D - A - B)\rangle_{\mu_n\otimes\mu_n},\end{aligned} \tag{3.62}$$

where we have introduced $A \stackrel{\text{def}}{=} f_1(\tilde{\omega}, u)f_2(\tilde{\omega}', v)$, $B \stackrel{\text{def}}{=} f_1(\tilde{\omega}, v)f_2(\tilde{\omega}', u)$, $C \stackrel{\text{def}}{=} f_3(\tilde{\omega} \wedge \tilde{\omega}', u)f_4(\tilde{\omega} \vee \tilde{\omega}', v)$ and $D \stackrel{\text{def}}{=} f_3(\tilde{\omega} \wedge \tilde{\omega}', v)f_4(\tilde{\omega} \vee \tilde{\omega}', u)$.

The first term in the right-hand side of (3.62) is nonnegative thanks to inequality (3.60). The desired claim (3.61) will thus follow if we can show that $A + B \leq C + D$.

Observe first that (3.60) implies that $A \leq C$, $B \leq C$ and

$$\begin{aligned}AB &= f_1(\tilde{\omega}, u)f_2(\tilde{\omega}', u)f_1(\tilde{\omega}, v)f_2(\tilde{\omega}', v) \\ &\leq f_3(\tilde{\omega} \wedge \tilde{\omega}', u)f_4(\tilde{\omega} \vee \tilde{\omega}', u)f_3(\tilde{\omega} \wedge \tilde{\omega}', v)f_4(\tilde{\omega} \vee \tilde{\omega}', v) = CD.\end{aligned}$$

On the one hand, if $C = 0$, then $A = B = 0$ and the inequality $A + B \leq C + D$ is obvious. On the other hand, when $C \neq 0$, the inequality follows from

$$(C + D - A - B)/C \geq 1 + AB/C^2 - (A + B)/C = (1 - A/C)(1 - B/C) \geq 0. \quad \square$$

3.9 Bibliographical References

The Ising model is probably the most studied model in statistical physics and, as such, is discussed in countless books and review articles. An old, but very good, general discussion in the spirit of what is done here is [146]. We list some references for the material presented in the chapter.

Pressure. The notion of convergence in the sense of van Hove (formulated in a slightly different, but equivalent way) was first introduced in [345].

In the context of lattice spin systems, the existence and the basic properties of the thermodynamic limit for the pressure were first established by Griffiths [145] and Gallavotti and Miracle-Solé [128]. The proofs given in this chapter (Theorem 3.6 and Exercise 3.3) can be extended to cover a very wide class of models, possibly with interactions of infinite range. See the books by Ruelle [289] and Simon [308] for additional results and information.

The computation of the pressure of the one-dimensional (nearest-neighbor) Ising model (Theorem 3.9) was the main result of Ising's PhD thesis and was published in [175]. It relied on some simple combinatorics in order to compute the generating function $\sum_N \mathbf{Z}_{V_N;\beta,h} s^N$, from which Ising then extracted the value of the partition function. The transfer matrix computation seems to be due to Kramers and Wannier [200].

The first computation of the pressure of the two-dimensional Ising model without magnetic field, whose result is stated at the end of Section 3.3, was achieved in a groundbreaking work by Onsager [259]. Extensions of the computations to nonzero magnetic fields in two dimensions, or to higher dimensions, have not been found despite much effort.

Gibbs States. The notion of Gibbs state as used in this chapter (rather than the more general version discussed in Chapter 6) was commonly used in the 1960s and 1970s; see, for example, the early review by Gallavotti [127].

Correlation Inequalities and Applications. The first version of the GKS inequalities was obtained by Griffiths [142]; in the form stated in Theorem 3.49 they are due to Kelly and Sherman [188]. These inequalities admit important generalizations to more general single-spin spaces; see, for example, [139, 310]. The proof of the GKS inequalities given in Section 3.8 is due to Ginibre [138].

The FKG inequality was first established by Fortuin, Kasteleyn and Ginibre [110]. The proof given in Section 3.8.2 is due to Ahlswede and Daykin [2]; our presentation is inspired by [10]. The alternative proof presented in Section 3.10.3 was found by Holley [163]; see also [132, 225].

The applications of the correlation inequalities given in Section 3.6 are part of the folklore and are spread out over many papers. A good early reference is [146]. Exercise 3.15 is adapted from [229].

The uniqueness criteria given in Theorems 3.28 and 3.34 are due to Lebowitz and Martin-Löf [219]. The other claims concerning the magnetization density are again part of the folklore.

Peierls' Argument. The geometric proof described in Section 3.7.2 is due to Peierls [266]; see also [80, 144]. This argument has become central in the rigorous analysis of first-order phase transitions and is at the basis of the Pirogov–Sinai theory, a far-reaching generalization which is the main topic of Chapter 7.

The approach described in Exercise 3.21 is inspired by [198]. The bounds $2.625622 < \mu < 2.679193$ on the connectivity constant have been taken from [182] and [277] respectively. Numerically, the best estimate at the time of writing seems to be $\mu \cong 2.63815853032790(3)$ [180].

High-Temperature Representation. The high-temperature representation, which is described in Section 3.7.3, was introduced by van der Waerden in [340].

The proof of uniqueness based on the high-temperature expansion is again part of the folklore. There are many alternative ways of establishing uniqueness at high enough temperature, among which are Dobrushin's Uniqueness Theorem (discussed in Section 6.5.2), the cluster expansion (discussed in Section 6.5.4) and disagreement percolation (see, for example, [132]). These can be used to extract additional information, such as analyticity of the pressure, exponential decay of correlations, exponential convergence of the finite-volume expectations of local functions, etc.; see [86] for a discussion of the remarkable additional properties that hold at sufficiently high temperatures.

Uniqueness in Nonzero Magnetic Field. Theorems 3.40, 3.42 and 3.43 are due to Lee and Yang and appeared first in [220, 353]. The Asano contraction method used in the proof of the last theorem was introduced by Asano in [13]; see also [290]. For a rather extensive bibliography on this topic and various extensions, see [33].

Although we do not discuss this in the text, it is possible to derive various properties of interest from the Lee–Yang Theorem, such as exponential decay of truncated correlation functions (for example, $\langle \sigma_0 \sigma_i \rangle_{\beta,h} - \langle \sigma_0 \rangle_{\beta,h} \langle \sigma_i \rangle_{\beta,h}$) at all β when $h \neq 0$ [95], as well as analyticity in h of correlation functions [216]. See also [120, 121].

Another route to the proof of uniqueness at nonzero magnetic field is through the GHS inequality. The latter was first proved by Griffiths, Hurst and Sherman in [143]. It states that the Ising model with magnetic field $\mathbf{h} = (h_i)_{i\in\Lambda}$ satisfies

$$\frac{\partial^2}{\partial h_i \partial h_j} \langle \sigma_k \rangle^{\varnothing}_{\Lambda;\beta,\mathbf{h}} \leq 0 \,,$$

for all $\Lambda \Subset \mathbb{Z}^d$ and $i, j, k \in \Lambda$, provided that $h_\ell \geq 0$ for all $\ell \in \Lambda$. Taking $h_i = h$ for all i, it implies in particular that the magnetization density $m(\beta, h)$ is concave (in particular, continuous) as a function of $h \geq 0$.

The alternative argument given in Exercise 3.29 is adapted from a more general approach by Dunlop [96].

3.10 Complements and Further Reading

3.10.1 Kramers–Wannier Duality

In this section we present an argument, proposed by Kramers and Wannier [200], which suggests that the critical inverse temperature of the Ising model on $\mathbb{Z}^2$ is equal to

$$\beta_c(2) = \tfrac{1}{2}\log(1+\sqrt{2}). \tag{3.63}$$

The starting point is the representation of the partition function with + boundary condition in terms of contours in (3.32):

$$\mathbf{Z}^+_{\mathrm{B}(n);\beta,0} = e^{\beta|\mathscr{E}^{\mathrm{b}}_{\mathrm{B}(n)}|} \sum_{\omega\in\Omega^+_{\mathrm{B}(n)}} \prod_{\gamma\in\Gamma(\omega)} e^{-2\beta|\gamma|}. \tag{3.64}$$

Let $\mathrm{B}(n)^* = \{-n-\frac{1}{2}, -n+\frac{1}{2}, \dots, n-\frac{1}{2}, n+\frac{1}{2}\}^2 \subset \mathbb{Z}^2_*$ be the box dual to $\mathrm{B}(n)$. From Exercise 3.23, we have the high-temperature representation

$$\mathbf{Z}^{\varnothing}_{\mathrm{B}(n)^*;\beta^*,0} = 2^{|\mathrm{B}(n)^*|}\cosh(\beta^*)^{|\mathscr{E}_{\mathrm{B}(n)^*}|} \sum_{E\in\mathfrak{C}^{\mathrm{even}}_{\mathrm{B}(n)^*}} \tanh(\beta^*)^{|E|}. \tag{3.65}$$

We now identify each set $E \in \mathfrak{C}^{\mathrm{even}}_{\mathrm{B}(n)^*}$ with the edges of the contours of a unique configuration $\omega \in \Omega^+_{\mathrm{B}(n)}$:

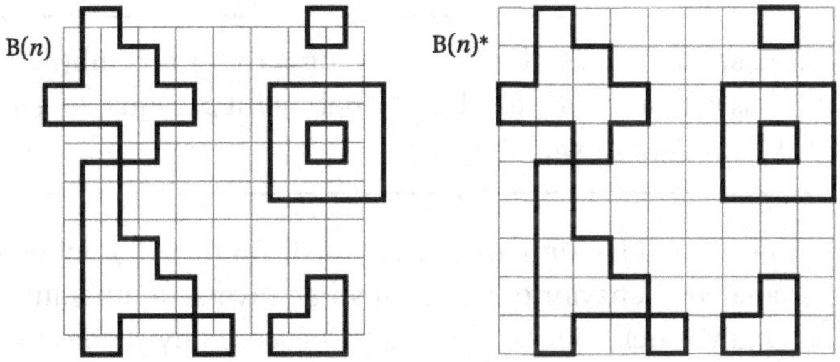

Lemma 3.52. *Let $E \in \mathscr{E}_{\mathrm{B}(n)^*}$. Then $E \in \mathfrak{C}^{\mathrm{even}}_{\mathrm{B}(n)^*}$ if and only if E coincides with the edges of the contours of a configuration $\omega \in \Omega^+_{\mathrm{B}(n)}$.*

Proof If $E \in \mathfrak{C}^{\mathrm{even}}_{\mathrm{B}(n)^*}$, then applying the rounding operation of Figure 3.11 yields a set of disjoint closed loops which are the contours of the configuration $\omega \in \Omega^+_{\mathrm{B}(n)}$ defined by

$$\omega_i \stackrel{\text{def}}{=} (-1)^{\#\{\text{loops surrounding } i\}}, \quad i \in \mathrm{B}(n).$$

Conversely, we have already seen in footnote 2, page 117, that the set of edges of the contours of a configuration $\omega \in \Omega^{+}_{\mathrm{B}(n)}$ belongs to $\mathfrak{E}^{\text{even}}_{\mathrm{B}(n)^*}$. □

It follows from the previous lemma that

$$\sum_{E \in \mathfrak{E}^{\text{even}}_{\mathrm{B}(n)^*}} \tanh(\beta^*)^{|E|} = \sum_{\omega \in \Omega^{+}_{\mathrm{B}(n)}} \prod_{\gamma \in \Gamma(\omega)} \tanh(\beta^*)^{|\gamma|}.$$

Therefore, if β^* satisfies

$$\tanh(\beta^*) = e^{-2\beta}, \tag{3.66}$$

we obtain the identity

$$2^{-|\mathrm{B}(n)^*|} \cosh(\beta^*)^{-|\mathscr{E}_{\mathrm{B}(n)^*}|} \mathbf{Z}^{\varnothing}_{\mathrm{B}(n)^*;\beta^*,0} = e^{-\beta|\mathscr{E}^{\mathrm{b}}_{\mathrm{B}(n)}|} \mathbf{Z}^{+}_{\mathrm{B}(n);\beta,0}. \tag{3.67}$$

When $n \to \infty$,

$$\frac{|\mathrm{B}(n)^*|}{|\mathrm{B}(n)|} \to 1, \quad \frac{|\mathscr{E}_{\mathrm{B}(n)^*}|}{|\mathrm{B}(n)|} \to 2, \quad \frac{|\mathscr{E}^{\mathrm{b}}_{\mathrm{B}(n)}|}{|\mathrm{B}(n)|} \to 2.$$

We thus obtain, by Theorem 3.6,

$$\psi(\beta, 0) = \psi(\beta^*, 0) - \log \sinh(2\beta^*). \tag{3.68}$$

The meaning of (3.68) is that the pressure is essentially invariant under the transformation

$$\beta \mapsto \beta^* = \operatorname{arctanh}(e^{-2\beta}), \tag{3.69}$$

which interchanges the low and high temperatures, as can be verified in the following exercise.

Exercise 3.32. Show that the mapping $\phi : x \mapsto \operatorname{arctanh}(e^{-2x})$ is an involution ($\phi \circ \phi = \mathrm{id}$) with a unique fixed (self-dual) point β_{sd} equal to $\frac{1}{2}\log(1+\sqrt{2})$. Moreover, $\phi([0, \beta_{\mathrm{sd}})) = (\beta_{\mathrm{sd}}, \infty]$.

Since ϕ and log sinh are both analytic on $(0, \infty)$, it follows from (3.68) that any nonanalytic behavior of $\psi(\cdot, 0)$ at some inverse temperature β must also imply a nonanalytic behavior at $\beta^* = \phi(\beta)$. Consequently, if one assumes that the pressure $\psi(\cdot, 0)$

1. is nonanalytic at β_c,
2. is analytic everywhere else,

then β_c must coincide with β_{sd}. This leads to the conjecture (3.63).

That the inverse critical temperature of the Ising model on $\mathbb{Z}^2$ actually coincides with the self-dual point of this transformation follows from the exact expression for the pressure derived by Onsager. There exists in fact a variety of ways to prove that this is the correct value for β_c in the two-dimensional Ising model, relying on the self-duality of the model, but avoiding exact computations; see, for example, [350]. Extension to other planar graphs is possible; see [70] and references therein.

The duality relation (3.67) and various generalizations have found numerous other uses in the rigorous analysis of the two-dimensional Ising model. The book by Gruber, Hintermann and Merlini [154] discusses duality in considerably more detail and in a more general framework.

3.10.2 Mean-Field Bounds

Let $\psi_\beta^{\mathrm{CW}}(h)$, $m_\beta^{\mathrm{CW}}(h)$ and $\beta_c^{\mathrm{CW}} \stackrel{\mathrm{def}}{=} (2d)^{-1}$ be the pressure, magnetization and critical inverse temperature of the Curie–Weiss model associated with the d-dimensional Ising model (remember the dependence on d in the Hamiltonian (2.2)). The following theorem, due to Thompson [330, 332], shows that these quantities provide rigorous bounds on the corresponding quantities for the Ising model on $\mathbb{Z}^d$. References to additional results pertaining to the relations between a model on $\mathbb{Z}^d$ and its mean-field approximation can be found in Section 2.5.4.

Theorem 3.53. *The following holds for the Ising model on $\mathbb{Z}^d$, $d \geq 1$:*

1. $\psi(\beta, h) \geq \psi_\beta^{\mathrm{CW}}(h)$, *for all* $\beta \geq 0$ *and all* $h \in \mathbb{R}$;
2. $\langle \sigma_0 \rangle_{\beta,h}^{+} \leq m_\beta^{\mathrm{CW}}(h)$, *for all* $\beta \geq 0$ *and all* $h \geq 0$;
3. $\beta_c(d) \geq \beta_c^{\mathrm{CW}}$, *for all* $d \geq 1$.

Proof *1.* Since the pressures are even functions of h, we can assume that $h \geq 0$. We start by decomposing the Hamiltonian with periodic boundary condition:

$$\mathscr{H}_{V_n;\beta,h}^{\mathrm{per}} \stackrel{\mathrm{def}}{=} -\beta \sum_{\{i,j\} \in \mathscr{E}_{V_n}^{\mathrm{per}}} \sigma_i \sigma_j - h \sum_{i \in V_n} \sigma_i = \mathscr{H}_{V_n;\beta,h}^{\mathrm{per},0} + \mathscr{H}_{V_n;\beta,h}^{\mathrm{per},1},$$

where

$$\mathscr{H}_{V_n;\beta,h}^{\mathrm{per},0} \stackrel{\mathrm{def}}{=} d\beta |V_n| m^2 - (h + 2d\beta m) \sum_{i \in V_n} \sigma_i,$$

$$\mathscr{H}_{V_n;\beta,h}^{\mathrm{per},1} \stackrel{\mathrm{def}}{=} -\beta \sum_{\{i,j\} \in \mathscr{E}_{V_n}^{\mathrm{per}}} (\sigma_i - m)(\sigma_j - m),$$

where $m \in \mathbb{R}$ will be chosen later. We can then rewrite the corresponding partition function as

$$\begin{aligned}
\mathbf{Z}_{V_n;\beta,h}^{\mathrm{per}} &\stackrel{\mathrm{def}}{=} \sum_{\omega \in \Omega_{V_n}} \exp\bigl(-\mathscr{H}_{V_n;\beta,h}^{\mathrm{per}}(\omega)\bigr) \\
&= \sum_{\omega \in \Omega_{V_n}} \exp\bigl(-\mathscr{H}_{V_n;\beta,h}^{\mathrm{per},1}(\omega)\bigr) \exp\bigl(-\mathscr{H}_{V_n;\beta,h}^{\mathrm{per},0}(\omega)\bigr) \\
&= \mathbf{Z}_{V_n;\beta,h}^{\mathrm{per},0} \bigl\langle \exp\bigl(-\mathscr{H}_{V_n;\beta,h}^{\mathrm{per},1}\bigr)\bigr\rangle_{V_n;\beta,h}^{\mathrm{per},0},
\end{aligned}$$

where we have introduced the Gibbs distribution

$$\mu^{\mathrm{per},0}_{V_n;\beta,h}(\omega) \stackrel{\text{def}}{=} \frac{\exp\bigl(-\mathscr{H}^{\mathrm{per},0}_{V_n;\beta,h}(\omega)\bigr)}{\mathbf{Z}^{\mathrm{per},0}_{V_n;\beta,h}}, \qquad \text{with} \quad \mathbf{Z}^{\mathrm{per},0}_{V_n;\beta,h} \stackrel{\text{def}}{=} \sum_{\omega\in\Omega_{V_n}} \exp\bigl(-\mathscr{H}^{\mathrm{per},0}_{V_n;\beta,h}(\omega)\bigr).$$

By Jensen's inequality,

$$\mathbf{Z}^{\mathrm{per}}_{V_n;\beta,h} \geq \mathbf{Z}^{\mathrm{per},0}_{V_n;\beta,h} \exp\bigl(-\langle \mathscr{H}^{\mathrm{per},1}_{V_n;\beta,h}\rangle^{\mathrm{per},0}_{V_n;\beta,h}\bigr).$$

Observe that

$$\begin{aligned}\langle \mathscr{H}^{\mathrm{per},1}_{V_n;\beta,h}\rangle^{\mathrm{per},0}_{V_n;\beta,h} &= -\beta \sum_{\{i,j\}\in\mathscr{E}^{\mathrm{per}}_{V_n}} \bigl(\langle\sigma_i\rangle^{\mathrm{per},0}_{V_n;\beta,h} - m\bigr)\bigl(\langle\sigma_j\rangle^{\mathrm{per},0}_{V_n;\beta,h} - m\bigr) \\ &= -\beta d|V_n|\bigl(m - \langle\sigma_0\rangle^{\mathrm{per},0}_{V_n;\beta,h}\bigr)^2.\end{aligned}$$

Since

$$\langle\sigma_0\rangle^{\mathrm{per},0}_{V_n;\beta,h} = \tanh(2d\beta m + h),$$

choosing m to be the largest solution to

$$m = \tanh(2d\beta m + h)$$

we get $\langle \mathscr{H}^{\mathrm{per},1}_{V_n;\beta,h}\rangle^{\mathrm{per},0}_{V_n;\beta,h} = 0$ and, therefore,

$$\mathbf{Z}^{\mathrm{per}}_{V_n;\beta,h} \geq \mathbf{Z}^{\mathrm{per},0}_{V_n;\beta,h} = e^{-d\beta m^2|V_n|} 2^{|V_n|} \cosh(2d\beta m + h)^{|V_n|}.$$

The conclusion follows (just compare with the expression in Exercise 2.4).

2. Let $\Lambda = \mathrm{B}(n)$, with $n \geq 1$, and let $i \sim 0$ be any nearest neighbor of the origin. Let $\langle\cdot\rangle^{+,1}_{\Lambda;\beta,h}$ denote the expectation with respect to the Gibbs distribution in Λ with no interaction between the two vertices 0 and i. Then, using (3.41),

$$\begin{aligned}\langle\sigma_0\rangle^{+}_{\Lambda;\beta,h} &= \frac{\sum_{\omega\in\Omega^+_\Lambda} \omega_0 \exp\bigl\{\beta\sum_{\{j,k\}\in\mathscr{E}^{\mathrm{b}}_\Lambda\setminus\{0,i\}} \omega_j\omega_k\bigr\}\bigl(1+\omega_0\omega_i\tanh\beta\bigr)}{\sum_{\omega\in\Omega^+_\Lambda} \exp\bigl\{\beta\sum_{\{j,k\}\in\mathscr{E}^{\mathrm{b}}_\Lambda\setminus\{0,i\}} \omega_j\omega_k\bigr\}\bigl(1+\omega_0\omega_i\tanh\beta\bigr)} \\ &= \frac{\langle\sigma_0\rangle^{+,1}_{\Lambda;\beta,h} + \langle\sigma_i\rangle^{+,1}_{\Lambda;\beta,h}\tanh\beta}{1+\langle\sigma_0\sigma_i\rangle^{+,1}_{\Lambda;\beta,h}\tanh\beta} \\ &\leq \frac{\langle\sigma_0\rangle^{+,1}_{\Lambda;\beta,h} + \langle\sigma_i\rangle^{+,1}_{\Lambda;\beta,h}\tanh\beta}{1+\langle\sigma_0\rangle^{+,1}_{\Lambda;\beta,h}\langle\sigma_i\rangle^{+,1}_{\Lambda;\beta,h}\tanh\beta},\end{aligned} \tag{3.70}$$

where we used the GKS inequality. Now, observe that, for any $x \geq 0$, $a \in [0,1]$ and $b \in [-1,1]$,

$$\frac{b + a\tanh(x)}{1 + ba\tanh(x)} \leq \frac{b+\tanh(ax)}{1+b\tanh(ax)}. \tag{3.71}$$

Indeed, $y \mapsto (b+y)/(1+by)$ is increasing in $y \geq 0$, and $\tanh(ax) \geq a\tanh(x)$ (by concavity). Applying (3.71) to (3.70), we get

$$\langle\sigma_0\rangle^{+}_{\Lambda;\beta,h} \leq \frac{\langle\sigma_0\rangle^{+,1}_{\Lambda;\beta,h} + \tanh\bigl(\beta\langle\sigma_i\rangle^{+,1}_{\Lambda;\beta,h}\bigr)}{1+\langle\sigma_0\rangle^{+,1}_{\Lambda;\beta,h}\tanh\bigl(\beta\langle\sigma_i\rangle^{+,1}_{\Lambda;\beta,h}\bigr)}.$$

But, since $(\tanh(x)+\tanh(y)) \,/\, (1+\tanh(x)\tanh(y)) = \tanh(x+y)$, this gives

$$\langle\sigma_0\rangle^+_{\Lambda;\beta,h} \le \tanh\bigl\{\operatorname{arctanh}\bigl(\langle\sigma_0\rangle^{+,1}_{\Lambda;\beta,h}\bigr) + \beta\langle\sigma_i\rangle^{+,1}_{\Lambda;\beta,h}\bigr\},$$

which can be rewritten as

$$\operatorname{arctanh}\bigl(\langle\sigma_0\rangle^+_{\Lambda;\beta,h}\bigr) \le \operatorname{arctanh}\bigl(\langle\sigma_0\rangle^{+,1}_{\Lambda;\beta,h}\bigr) + \beta\langle\sigma_i\rangle^{+,1}_{\Lambda;\beta,h}.$$

Finally, by GKS inequalities, $\langle\sigma_i\rangle^{+,1}_{\Lambda;\beta,h} \le \langle\sigma_i\rangle^+_{\Lambda;\beta,h}$, so that

$$\operatorname{arctanh}\bigl(\langle\sigma_0\rangle^+_{\Lambda;\beta,h}\bigr) \le \operatorname{arctanh}\bigl(\langle\sigma_0\rangle^{+,1}_{\Lambda;\beta,h}\bigr) + \beta\langle\sigma_i\rangle^{+}_{\Lambda;\beta,h}. \tag{3.72}$$

Clearly, one can iterate (3.72), removing all edges between 0 and its nearest neighbors, one at a time. This yields

$$\operatorname{arctanh}\bigl(\langle\sigma_0\rangle^+_{\Lambda;\beta,h}\bigr) \le \operatorname{arctanh}\bigl(\langle\sigma_0\rangle^{\varnothing}_{\{0\};\beta,h}\bigr) + \beta\sum_{i\sim 0}\langle\sigma_i\rangle^{+}_{\Lambda;\beta,h}.$$

Of course, $\langle\sigma_0\rangle^{\varnothing}_{\{0\};\beta,h} = \tanh(h)$. Therefore,

$$\operatorname{arctanh}\bigl(\langle\sigma_0\rangle^+_{\Lambda;\beta,h}\bigr) \le h + \beta\sum_{i\sim 0}\langle\sigma_i\rangle^{+}_{\Lambda;\beta,h},$$

that is,

$$\langle\sigma_0\rangle^+_{\Lambda;\beta,h} \le \tanh\Bigl(h + \beta\sum_{i\sim 0}\langle\sigma_i\rangle^{+}_{\Lambda;\beta,h}\Bigr).$$

We can now let $\Lambda \uparrow \mathbb{Z}^d$ and use the fact that $\langle\sigma_i\rangle^+_{\beta,h} = \langle\sigma_0\rangle^+_{\beta,h}$ for all i to obtain the desired bound:

$$\langle\sigma_0\rangle^+_{\beta,h} \le \tanh\bigl(h + 2d\beta\langle\sigma_0\rangle^+_{\beta,h}\bigr).$$

From this we conclude that $\langle\sigma_0\rangle^+_{\beta,h} \le m^{\mathrm{CW}}_\beta(h)$.

3. When $\beta < \beta_c^{\mathrm{CW}}$, the previous item implies that $\langle\sigma_0\rangle^+_{\beta,0} \le m^{\mathrm{CW}}_\beta(0) = 0$. This implies $\beta < \beta_c(d)$, which proves the claim. □

3.10.3 An Alternative Proof of the FKG Inequality

Here, we provide an alternative proof of the FKG inequality. Although possibly less general and somewhat longer than the one provided in Section 3.8.2, we believe that it has the undeniable advantage of being more enlightening. It relies on some basic knowledge of discrete-time finite-state Markov chains, as given, for example, in the book [156].

The Gibbs Sampler. Let $\Lambda \Subset \mathbb{Z}^d$ and let μ be some probability distribution on $\Omega_\Lambda = \{-1,1\}^\Lambda$ satisfying $\mu(\omega) > 0$ for all $\omega \in \Omega_\Lambda$.

We construct a discrete-time Markov chain $(X_n)_{n\ge 0}$ on Ω_Λ as follows: given that $X_n = \omega \in \Omega_\Lambda$, the value of X_{n+1}, say ω', is sampled using the following algorithm:

1. Sample a number u according to the uniform distribution on $[0,1]$ (independently of all other sources of randomness).

2. Sample a vertex $i \in \Lambda$ with uniform distribution (independently of all other sources of randomness).
3. Set $\omega'_j = \omega_j$ for all $j \neq i$.
4. Set

$$\omega'_i = \begin{cases} +1 & \text{if } u \leq \mu\big(\sigma_i = 1 \;\big|\; \sigma_j = \omega_j \;\forall j \neq i\big), \\ -1 & \text{otherwise.} \end{cases}$$

In other words, there are no transitions between two configurations differing at more than one vertex; moreover, given two configurations $\omega, \omega' \in \Omega_\Lambda$ differing at a single vertex $i \in \Lambda$, the transition probability from ω to ω' is given by

$$p(\omega \to \omega') = \frac{1}{|\Lambda|}\,\mu\big(\sigma_i = \omega'_i \;\big|\; \sigma_j = \omega_j \;\forall j \neq i\big) = \frac{1}{|\Lambda|}\,\frac{\mu(\omega')}{\mu(\omega) + \mu(\omega')}\,.$$

Observe that the Markov chain $(X_n)_{n\geq 0}$ is irreducible (since one can move between two arbitrary configurations by changing one spin at a time, each such transition occurring with positive probability) and aperiodic (since $p(\omega \to \omega) > 0$). Therefore the distribution of X_n converges almost surely towards the unique stationary distribution. We claim that the latter is given by μ. Indeed, $(X_n)_{n\geq 0}$ is reversible with respect to μ: if $\omega, \omega' \in \Omega_\Lambda$ are two configurations differing only at one vertex, then

$$\mu(\omega)p(\omega \to \omega') = \frac{1}{|\Lambda|}\,\frac{\mu(\omega)\mu(\omega')}{\mu(\omega) + \mu(\omega')} = \mu(\omega')p(\omega' \to \omega)\,.$$

Monotone Coupling. Let us now consider two probability distributions μ and $\tilde{\mu}$ on Ω_Λ. As above, we assume that $\mu(\omega) > 0$ and $\tilde{\mu}(\omega) > 0$. Moreover, we assume that

$$\mu\big(\sigma_i = 1 \;\big|\; \sigma_j = \omega_j \;\forall j \neq i\big) \leq \tilde{\mu}\big(\sigma_i = 1 \;\big|\; \sigma_j = \tilde{\omega}_j \;\forall j \neq i\big)\,, \tag{3.73}$$

for all $\omega, \tilde{\omega} \in \Omega_\Lambda$ such that $\tilde{\omega} \geq \omega$.

Let us denote by $(X_n)_{n\geq 0}$ and $(\tilde{X}_n)_{n\geq 0}$ the Markov chains on Ω_Λ associated with μ and $\tilde{\mu}$, as described above. We are going to define the **monotone coupling** of these two Markov chains. The coupling is defined by the previous construction, but using, at each step of the process, the *same* $u \in [0, 1]$ and $i \in \Lambda$ for both chains. The important observation is that

$$\tilde{X}_n \geq X_n \implies \tilde{X}_{n+1} \geq X_{n+1}\,.$$

Indeed, let us denote by i the vertex that has been selected at this step. In order to violate the inequality $\tilde{X}_{n+1} \geq X_{n+1}$, it is necessary that $\sigma_i(X_{n+1}) = 1$ and $\sigma_i(\tilde{X}_{n+1}) = -1$. But this is impossible, since, for the former to be true, one needs to have $u \leq \mu(\sigma_i = 1 \,|\, \sigma_j = \sigma_j(X_n) \;\forall j \neq i)$, which, by (3.73), would imply that $u \leq \tilde{\mu}(\sigma_i = 1 \,|\, \sigma_j = \sigma_j(\tilde{X}_n) \;\forall j \neq i)$ and, thus, $\sigma_i(\tilde{X}_{n+1}) = 1$.

Stochastic Domination. Let μ and $\tilde{\mu}$ be as above. It is now very easy to prove that, for every nondecreasing function f,

$$\langle f \rangle_{\tilde{\mu}} \geq \langle f \rangle_\mu\,. \tag{3.74}$$

In that case, we say that $\tilde{\mu}$ **stochastically dominates** μ.

Let us consider the two monotonically coupled Markov chains, as described above, with initial values $X_0 = \eta^- \equiv -1$ and $\tilde{X}_0 = \eta^+ \equiv 1$. We denote by $\mathbb{P}$ the distribution of the coupled Markov chains. Now, since these chains converge, respectively, to μ and $\tilde{\mu}$, we can write

$$\langle f\rangle_{\tilde{\mu}} - \langle f\rangle_{\mu} = \lim_{n\to\infty} \sum_{\eta,\tilde{\eta}\in\Omega_\Lambda} \{f(\tilde{\eta}) - f(\eta)\}\, \mathbb{P}\big(X_n = \eta, \tilde{X}_n = \tilde{\eta}\big).$$

Moreover, by monotonicity of the coupling,

$$\mathbb{P}\big(\tilde{X}_n \geq X_n, \text{ for all } n \geq 0\big) = 1.$$

We can thus restrict the summation to pairs $\tilde{\eta} \geq \eta$:

$$\langle f\rangle_{\tilde{\mu}} - \langle f\rangle_{\mu} = \lim_{n\to\infty} \sum_{\substack{\eta,\tilde{\eta}\in\Omega_\Lambda \\ \tilde{\eta}\geq\eta}} \{f(\tilde{\eta}) - f(\eta)\}\, \mathbb{P}\big(X_n = \eta, \tilde{X}_n = \tilde{\eta}\big).$$

(3.74) follows since $\tilde{\eta} \geq \eta$ implies that $f(\tilde{\eta}) - f(\eta) \geq 0$.

Proof of the FKG Inequality. We can now prove the FKG inequality for the Ising model on $\mathbb{Z}^d$. Let $\Lambda \Subset \mathbb{Z}^d$, $\eta \in \Omega$, $\beta \geq 0$ and $h \in \mathbb{R}$. We want to prove that

$$\langle fg\rangle^{\eta}_{\Lambda;\beta,h} \geq \langle f\rangle^{\eta}_{\Lambda;\beta,h} \langle g\rangle^{\eta}_{\Lambda;\beta,h}, \tag{3.75}$$

for all nondecreasing functions f and g. Note that we can, and will, assume that $g(\tau) > 0$ for all $\tau \in \Omega^{\eta}_{\Lambda}$, since adding a constant to g does not affect (3.75). We can thus consider the following two probability distributions on Ω_Λ:

$$\mu(\omega) \stackrel{\text{def}}{=} \mu^{\eta}_{\Lambda;\beta,h}(\omega\eta), \qquad \tilde{\mu}(\omega) \stackrel{\text{def}}{=} \frac{g(\omega\eta)}{\langle g\rangle^{\eta}_{\Lambda;\beta,h}} \mu^{\eta}_{\Lambda;\beta,h}(\omega\eta),$$

where, given $\omega \in \Omega_\Lambda$, $\omega\eta$ denotes the configuration coinciding with ω in Λ and with η outside Λ. Clearly $\mu(\omega) > 0$ and $\tilde{\mu}(\omega) > 0$ for all $\omega \in \Omega_\Lambda$; (3.75) can then be rewritten as

$$\langle f\rangle_{\tilde{\mu}} \geq \langle f\rangle_{\mu}.$$

Since this is exactly (3.74), it is sufficient to prove that (3.73) holds for these two distributions.

Observe first that, since g is nondecreasing,

$$\begin{aligned}
\tilde{\mu}\big(\sigma_i = 1 \mid \sigma_j = \tilde{\omega}_j(n)\ \forall j \neq i\big) &= \frac{\mu((+1)\tilde{\omega})g((+1)\tilde{\omega})}{\mu((+1)\tilde{\omega})g((+1)\tilde{\omega}) + \mu((-1)\tilde{\omega})g((-1)\tilde{\omega})} \\
&= \left\{1 + \frac{\mu((-1)\tilde{\omega})\, g((-1)\tilde{\omega})}{\mu((+1)\tilde{\omega})\, g((+1)\tilde{\omega})}\right\}^{-1} \\
&\geq \left\{1 + \frac{\mu((-1)\tilde{\omega})}{\mu((+1)\tilde{\omega})}\right\}^{-1},
\end{aligned}$$

where $(+1)\tilde{\omega}$, resp. $(-1)\tilde{\omega}$, is the configuration given by $\tilde{\omega}$ at vertices different from i and by $+1$, resp. -1, at i.

Now,

$$\frac{\mu((-1)\tilde{\omega})}{\mu((+1)\tilde{\omega})} = \frac{\mu^{\eta}_{\Lambda;\beta,h}((-1)\tilde{\omega}\eta)}{\mu^{\eta}_{\Lambda;\beta,h}((+1)\tilde{\omega}\eta)} = \exp\Bigl(-2\beta\sum_{i\sim j}(\tilde{\omega}\eta)_j - 2h\Bigr)$$

is a nonincreasing function of $\tilde{\omega}$. It follows that, for any $\omega \in \Omega_\Lambda$ such that $\tilde{\omega} \geq \omega$,

$$\begin{aligned}\tilde{\mu}(\sigma_i = 1 \mid \sigma_j = \tilde{\omega}_j(n)\ \forall j \neq i) &\geq \Bigl\{1 + \frac{\mu((-1)\omega)}{\mu((+1)\omega)}\Bigr\}^{-1}\\ &= \mu\bigl(\sigma_i = 1 \bigm| \sigma_j = \omega_j(n)\ \forall j \neq i\bigr),\end{aligned}$$

and (3.73), and thus (3.75), follows.

3.10.4 Transfer Matrix and Markov Chains

In Section 3.3, we described how the pressure of the one-dimensional Ising model could be determined using the transfer matrix. Readers familiar with Markov chains might have noted certain obvious similarities. In this complement, we explain how these tools can be related and what additional information can be extracted.

Let A be the transfer matrix of the one-dimensional Ising model, defined in (3.11). For simplicity, let us denote by $\mathbf{Z}_n^{s,s'} \equiv \mathbf{Z}^{\eta^{s,s'}}_{\Lambda_n;\beta,h}$, $s, s' \in \{\pm 1\}$, the partition function of the model on $\Lambda_n = \{1, \dots, n\}$, with boundary condition $\eta^{s,s'}$ given by $\eta_i^{s,s'} = s$ if $i \leq 0$ and $\eta_i^{s,s'} = s'$ if $i > 0$.

Proceeding as in Section 3.3, the transfer matrix can be related to the partition function $\mathbf{Z}_n^{s,s'}$ in the following way: for all $n \geq 1$,

$$\mathbf{Z}_n^{s,s'} = \bigl(A^{n+1}\bigr)_{s,s'} .$$

Let $\lambda > 0$ be the largest of the two eigenvalues of A. We denote by φ, respectively φ^*, the right-eigenvector, respectively left-eigenvector, associated with λ: $A\varphi = \lambda\varphi$, $\varphi^* A = \lambda\varphi^*$. We assume that these eigenvectors satisfy the following normalization assumption: $\varphi \cdot \varphi^* = 1$. All these quantities can be computed explicitly, but we will not need the resulting expressions here. Notice however that, either by an explicit computation or by the Perron–Frobenius Theorem [45, Theorem 1.1], all components of φ and φ^* are positive.

We now define a new matrix $\Pi = (\pi_{s,s'})_{s,s'=\pm 1}$ by

$$\pi_{s,s'} \stackrel{\text{def}}{=} \frac{\varphi_{s'}}{\lambda \varphi_s} A_{s,s'} .$$

Π is the transition matrix of an irreducible, aperiodic Markov chain. Indeed, for $s \in \{\pm 1\}$,

$$\sum_{s' \in \{\pm 1\}} \pi_{s,s'} = \frac{1}{\lambda\varphi_s} \sum_{s' \in \{\pm 1\}} A_{s,s'}\varphi_{s'} = \frac{1}{\lambda\varphi_s}\bigl(A\varphi\bigr)_s = 1 .$$

Irreducibility and aperiodicity follow from the positivity of $\pi_{s,s'}$ for all $s, s' \in \{\pm 1\}$.

Being irreducible, Π possesses a unique stationary distribution ν, given by

$$\nu(\{s\}) = \varphi_s \varphi_s^* , \qquad s \in \{\pm 1\} .$$

Indeed, $\nu(\{1\}) + \nu(\{-1\}) = 1$, by our normalization assumption, and

$$(\nu\Pi)(\{s'\}) = \sum_{s\in\{\pm1\}} \nu(\{s\})\,\pi_{s,s'} = \frac{1}{\lambda}\varphi_{s'} \sum_{s\in\{\pm1\}} \varphi_s^* A_{s,s'} = \varphi_{s'}\varphi_{s'}^* = \nu(\{s'\}),$$

since $\varphi^* A = \lambda\varphi^*$.

The probability distribution ν on $\{\pm1\}$ provides the distribution of σ_0 under the infinite-volume Gibbs state. Indeed, denoting by $\mu^{s,s'}_{\mathrm{B}(n);\beta,h}$ the Gibbs distribution on $\mathrm{B}(n) = \{-n,\dots,n\}$ with boundary condition $\eta^{s,s'}$, the probability that $\sigma_0 = s_0$ is given by

$$\mu^{s,s'}_{\mathrm{B}(n);\beta,h}(\sigma_0 = s_0) = \frac{\mathbf{Z}_n^{s,s_0}\mathbf{Z}_n^{s_0,s'}}{\mathbf{Z}^{s,s'}_{2n+1}} = \frac{\left(A^{n+1}\right)_{s,s_0}\left(A^{n+1}\right)_{s_0,s'}}{\left(A^{2n+2}\right)_{s,s'}} .$$

Now, as can be checked, for any $s, s' \in \{\pm1\}$,

$$\left(A^n\right)_{s,s'} = \lambda^n \frac{\varphi_s}{\varphi_{s'}} \left(\Pi^n\right)_{s,s'} ,$$

which gives, after substitution in the above expression,

$$\mu^{s,s'}_{\mathrm{B}(n);\beta,h}(\sigma_0 = s_0) = \frac{\left(\Pi^n\right)_{s,s_0}\left(\Pi^n\right)_{s_0,s'}}{\left(\Pi^{2n+2}\right)_{s,s'}} .$$

Since the Markov chain is irreducible and aperiodic, $\lim_{n\to\infty}(\Pi^n)_{s,s'} = \nu(\{s'\})$ for all $s, s' \in \{\pm1\}$. We conclude that

$$\lim_{n\to\infty} \mu^{s,s'}_{\mathrm{B}(n);\beta,h}(\sigma_0 = s_0) = \frac{\nu(\{s_0\})\nu(\{s'\})}{\nu(\{s'\})} = \nu(\{s_0\}) .$$

One can check similarly that the joint distribution of any finite collection $(\sigma_i)_{a\le i\le b}$ of spins is given by

$$\lim_{n\to\infty} \mu^{s,s'}_{\mathrm{B}(n);\beta,h}\big(\sigma_k = s_k\,, \forall a \le k \le b\big) = \nu(\{s_a\}) \prod_{k=a}^{b-1} \pi_{s_k,s_{k+1}} .$$

3.10.5 Ising Antiferromagnet

The **Ising antiferromagnet** is a model whose neighboring spins tend to point in *opposite* directions, this effect becoming stronger at lower temperatures. It therefore does not exhibit spontaneous magnetization.

We only consider the antiferromagnet in the absence of a magnetic field. This model can be thought of as an Ising model with negative coupling constants:

$$\mathscr{H}^{\mathrm{anti}}_{\Lambda;\beta}(\omega) \overset{\mathrm{def}}{=} \beta \sum_{\{i,j\}\in\mathscr{E}^{\mathrm{b}}_\Lambda} \sigma_i(\omega)\sigma_j(\omega) . \tag{3.76}$$

Let a vertex $i = (i_1, \dots, i_d) \in \mathbb{Z}^d$ be called **even** (resp. **odd**) if $i_1 + \cdots + i_d$ is even (resp. odd). Consider the transformation $\tau_{\text{even}} \colon \Omega \to \Omega$ defined by

$$(\tau_{\text{even}}\omega)_i \stackrel{\text{def}}{=} \begin{cases} +\omega_i & \text{if } i \text{ is even,} \\ -\omega_i & \text{otherwise.} \end{cases}$$

One can then define $\tau_{\text{odd}} \colon \Omega \to \Omega$ by

$$(\tau_{\text{odd}}\omega)_i \stackrel{\text{def}}{=} -(\tau_{\text{even}}\omega)_i\,, \qquad i \in \mathbb{Z}^d\,.$$

Not surprisingly, the main features of this model can be derived from the results obtained for the Ising model:

Exercise 3.33. Observing that

$$\mathscr{H}^{\text{anti}}_{\Lambda;\beta}(\omega) = \mathscr{H}_{\Lambda;\beta}(\tau_{\text{even}}\omega)\,,$$

use the results obtained in this chapter to show that, when $\beta > \beta_c(d)$, two distinct Gibbs states can be constructed, $\langle\cdot\rangle^{\text{even}}_\beta$ and $\langle\cdot\rangle^{\text{odd}}_\beta$. Describe the typical configurations under these two states.

Let us just emphasize that the trick used in the previous exercise to reduce the analysis to the ferromagnetic case relies in an essential way on the fact that the lattice $\mathbb{Z}^d$ is **bipartite**; that is, one can color each of its vertices in either black or white in such a way that no neighboring vertices have the same color. On a nonbipartite lattice, or in the presence of a magnetic field, the behavior of the antiferromagnet is much more complicated; some aspects will be discussed in Exercises 7.5 and 7.7.

3.10.6 Random-Cluster and Random-Current Representations

In this chapter, we chose an approach to the Ising model that we deemed best suited to the generalization to other models done in the rest of the book. In particular, we barely touched on the topics of *geometrical representations*: we only introduced the low- and high-temperature representations in Sections 3.7.2 and 3.7.3, in the course of our analysis of the phase diagram. In this section, we briefly introduce two other graphical representations that have played and continue to play a central role in the mathematical analysis of the Ising model, the *random-cluster* and *random-current representations.*

Good references to the random-cluster representation can be found in the review paper [132] by Georgii, Häggström and Maes, and the books by Grimmett [150] and Werner [350]. The lecture notes [91] by Duminil-Copin provide a good introduction to several graphical representations, including random-cluster and random-current. In addition to these, graphical representations of correlation functions in terms of interacting random paths (an example being the high-temperature representation of Section 3.7.3) are also very important tools; a thorough discussion can be found in the book [102] by Fernández, Fröhlich and Sokal.

The Random-Cluster Representation. This representation was introduced by Fortuin and Kasteleyn [109]. Besides playing an instrumental role in many mathematical investigations of the Ising model, it also provides a deep link with other classical models, in particular the q-state Potts model and the Bernoulli bond percolation process. Moreover, this representation is the basis of numerical algorithms, first introduced by Swendsen and Wang [323], which are very efficient at sampling from such Gibbs distributions.

The starting point is similar to what was done to derive the high-temperature representation of the model: we expand in a suitable way the Boltzmann weight. Here, we write

$$e^{\beta\sigma_i\sigma_j} = e^{-\beta} + (e^{\beta} - e^{-\beta})\mathbf{1}_{\{\sigma_i=\sigma_j\}} = e^{\beta}\big((1-p_\beta) + p_\beta \mathbf{1}_{\{\sigma_i=\sigma_j\}}\big),$$

where we have introduced $p_\beta \stackrel{\text{def}}{=} 1 - e^{-2\beta} \in [0,1]$.

Let $\Lambda \Subset \mathbb{Z}^d$. Using the above notations, we obtain, after expanding the product (remember Exercise 3.22),

$$\prod_{\{i,j\}\in\mathscr{E}^{\mathrm{b}}_\Lambda} e^{\beta\sigma_i\sigma_j} = e^{\beta|\mathscr{E}^{\mathrm{b}}_\Lambda|} \sum_{E\subset\mathscr{E}^{\mathrm{b}}_\Lambda} p_\beta^{|E|}(1-p_\beta)^{|\mathscr{E}^{\mathrm{b}}_\Lambda\setminus E|} \prod_{\{i,j\}\in E} \mathbf{1}_{\{\sigma_i=\sigma_j\}}.$$

The partition function $\mathbf{Z}^{+}_{\Lambda;\beta,0}$ can thus be expressed as

$$\begin{aligned}\mathbf{Z}^{+}_{\Lambda;\beta,0} &= e^{\beta|\mathscr{E}^{\mathrm{b}}_\Lambda|} \sum_{E\subset\mathscr{E}^{\mathrm{b}}_\Lambda} p_\beta^{|E|}(1-p_\beta)^{|\mathscr{E}^{\mathrm{b}}_\Lambda\setminus E|} \sum_{\omega\in\Omega^{+}_\Lambda} \prod_{\{i,j\}\in E} \mathbf{1}_{\{\sigma_i(\omega)=\sigma_j(\omega)\}} \\ &= e^{\beta|\mathscr{E}^{\mathrm{b}}_\Lambda|} \sum_{E\subset\mathscr{E}^{\mathrm{b}}_\Lambda} p_\beta^{|E|}(1-p_\beta)^{|\mathscr{E}^{\mathrm{b}}_\Lambda\setminus E|}\, 2^{N^{\mathrm{w}}_\Lambda(E)},\end{aligned}$$

where $N^{\mathrm{w}}_\Lambda(E)$ denotes the number of connected components (usually called **clusters** in this context) of the graph $(\mathbb{Z}^d, E\cup\mathscr{E}_{\mathbb{Z}^d\setminus\Lambda})$ (in other words, the graph obtained by considering all vertices of $\mathbb{Z}^d$ and all edges of $\mathbb{Z}^d$ that either belong to E or do not intersect the box Λ). Indeed, in the sum over $\omega\in\Omega^{+}_\Lambda$, the only configurations contributing are those in which all spins belonging to the same cluster agree.

The **FK-percolation process** in Λ with **wired boundary condition** is the probability distribution on the set $\mathscr{P}(\mathscr{E}^{\mathrm{b}}_\Lambda)$ of all subsets of $\mathscr{E}^{\mathrm{b}}_\Lambda$ assigning to a subset of edges $E\subset\mathscr{E}^{\mathrm{b}}_\Lambda$ the probability

$$\nu^{\mathrm{FK,w}}_{\Lambda;p_\beta,2}(E) \stackrel{\text{def}}{=} \frac{p_\beta^{|E|}(1-p_\beta)^{|\mathscr{E}^{\mathrm{b}}_\Lambda\setminus E|}\, 2^{N^{\mathrm{w}}_\Lambda(E)}}{\sum_{E'\subset\mathscr{E}^{\mathrm{b}}_\Lambda} p_\beta^{|E'|}(1-p_\beta)^{|\mathscr{E}^{\mathrm{b}}_\Lambda\setminus E'|}\, 2^{N^{\mathrm{w}}_\Lambda(E')}}.$$

Remark 3.54. Observe that, by replacing the factor 2 in the above expression by 1, the distribution $\nu^{\mathrm{FK,w}}_{\Lambda;p_\beta,2}$ reduces to the Bernoulli bond percolation process on $\mathscr{E}^{\mathrm{b}}_\Lambda$, in which each edge of $\mathscr{E}_{\mathbb{Z}^d}$ belongs to E with probability p_β, independently from the other edges. Similarly, the random-cluster representation of the q-state Potts model is obtained by replacing the factor 2 by q. In this sense, the FK-percolation process provides a one-parameter family of models interpolating between Bernoulli percolation, Ising and Potts models. ◇

For $A, B \subset \mathbb{Z}^d$, let us write $\{A \leftrightarrow B\}$ for the event that there exists a cluster intersecting both A and B.

Exercise 3.34. Proceeding as above, check the following identities: for any $i, j \in \Lambda \Subset \mathbb{Z}^d$,

$$\langle \sigma_i \rangle^+_{\Lambda;\beta,0} = \nu^{\mathrm{FK,w}}_{\Lambda;p_\beta,2}(i \leftrightarrow \partial^{\mathrm{ex}}\Lambda), \qquad \langle \sigma_i \sigma_j \rangle^+_{\Lambda;\beta,0} = \nu^{\mathrm{FK,w}}_{\Lambda;p_\beta,2}(i \leftrightarrow j).$$

One feature that makes the random-cluster representation particularly useful, as it makes it possible to successfully import many ideas and techniques developed for Bernoulli bond percolation, is the availability of an FKG inequality. Let $\Lambda \Subset \mathbb{Z}^d$ and consider the partial order on $\mathscr{P}(\mathscr{E}^{\mathrm{b}}_\Lambda)$ given by $E \leq E'$ if and only if $E \subset E'$.

Exercise 3.35. Show, using Theorem 3.50, that

$$\nu^{\mathrm{FK,w}}_{\Lambda;p_\beta,2}(\mathscr{A} \cap \mathscr{B}) \geq \nu^{\mathrm{FK,w}}_{\Lambda;p_\beta,2}(\mathscr{A})\, \nu^{\mathrm{FK,w}}_{\Lambda;p_\beta,2}(\mathscr{B}),$$

for all pairs $\mathscr{A}, \mathscr{B}$ of nondecreasing events on $\mathscr{P}(\mathscr{E}^{\mathrm{b}}_\Lambda)$.

As an immediate application, one can prove the existence of the thermodynamic limit.

Exercise 3.36. Show that, for every local increasing event $\mathscr{A}$,

$$\lim_{\Lambda \uparrow \mathbb{Z}^d} \nu^{\mathrm{FK,w}}_{\Lambda;p_\beta,2}(\mathscr{A})$$

exists. *Hint:* Proceed as in the proof of Theorem 3.17.

As already mentioned in Remark 3.15 and as will be explained in more detail in Chapter 6, it follows from the previous exercise and the Riesz–Markov–Kakutani Representation Theorem that one can define a probability measure $\nu^{\mathrm{FK,w}}_{p_\beta,2}$ on $\mathscr{E}$ such that

$$\nu^{\mathrm{FK,w}}_{p_\beta,2}(\mathscr{A}) = \lim_{\Lambda \uparrow \mathbb{Z}^d} \nu^{\mathrm{FK,w}}_{\Lambda;p_\beta,2}(\mathscr{A}),$$

for all local events $\mathscr{A}$. A simple but remarkable observation is that the statements of Exercise 3.34 still hold under $\nu^{\mathrm{FK,w}}_{p_\beta,2}$. In particular,

$$\langle \sigma_0 \rangle^+_{\beta,0} = \nu^{\mathrm{FK,w}}_{p_\beta,2}(0 \leftrightarrow \infty), \tag{3.77}$$

where $\{0 \leftrightarrow \infty\} \stackrel{\mathrm{def}}{=} \bigcap_n \{0 \leftrightarrow \partial^{\mathrm{ex}}\mathrm{B}(n)\}$ corresponds to the event that there exists an infinite path of disjoint open edges starting from 0 (or, equivalently, that the cluster containing 0 has infinite cardinality). Since Theorem 3.28 shows that the existence of a first-order phase transition at inverse temperature β (and magnetic field $h = 0$) is equivalent to $\langle\sigma_0\rangle^+_{\beta,0} > 0$, the above relation implies that the latter is also equivalent to **percolation** in the associated FK-percolation process. This observation provides new insights into the phase transition we have studied in this chapter and provides the basis for a geometrical analysis of the Ising model using methods inherited from percolation theory.

Exercise 3.37. Prove the identity (3.77).

The Random-Current Representation. Also of great importance in the mathematical analysis of the Ising model, with many fundamental applications, this representation had already been introduced in [143], but its true power was realized by Aizenman [4].

Once again, the strategy is to expand the Boltzmann weight in a suitable way, then expand the product over pairs of neighbors, and finally sum explicitly over the spins. For the first step, we simply expand the exponential as a Taylor series:

$$e^{\beta\sigma_i\sigma_j} = \sum_{n=0}^{\infty} \frac{\beta^n}{n!}(\sigma_i\sigma_j)^n .$$

We then get, writing $\mathbf{n} = (n_e)_{e\in\mathscr{E}^{\mathrm{b}}_\Lambda}$ for a collection of nonnegative integers,

$$\prod_{\{i,j\}\in\mathscr{E}^{\mathrm{b}}_\Lambda} e^{\beta\sigma_i\sigma_j} = \sum_{\mathbf{n}} \Big\{\prod_{e\in\mathscr{E}^{\mathrm{b}}_\Lambda} \frac{\beta^{n_e}}{n_e!}\Big\} \prod_{\{i,j\}\in\mathscr{E}^{\mathrm{b}}_\Lambda} (\sigma_i\sigma_j)^{n_{\{i,j\}}} .$$

The partition function $\mathbf{Z}^+_{\Lambda;\beta,0}$ can thus be expressed as

$$\begin{aligned}\mathbf{Z}^+_{\Lambda;\beta,0} &= \sum_{\mathbf{n}} \Big\{\prod_{e\in\mathscr{E}^{\mathrm{b}}_\Lambda} \frac{\beta^{n_e}}{n_e!}\Big\} \sum_{\omega\in\Omega^+_\Lambda} \prod_{\{i,j\}\in\mathscr{E}^{\mathrm{b}}_\Lambda} (\sigma_i(\omega)\sigma_j(\omega))^{n_{\{i,j\}}} \\ &= \sum_{\mathbf{n}} \Big\{\prod_{e\in\mathscr{E}^{\mathrm{b}}_\Lambda} \frac{\beta^{n_e}}{n_e!}\Big\} \prod_{i\in\Lambda} \sum_{\omega_i=\pm1} \omega_i^{\hat{I}(i,\mathbf{n})} ,\end{aligned}$$

where $\hat{I}(i,\mathbf{n}) \stackrel{\mathrm{def}}{=} \sum_{j:\, j\sim i} n_{\{i,j\}}$. Since

$$\sum_{\omega_i=\pm1} \omega_i^m = \begin{cases} 2 & \text{if } m \text{ is even,} \\ 0 & \text{if } m \text{ is odd,} \end{cases}$$

we conclude that

$$\mathbf{Z}^+_{\Lambda;\beta,0} = 2^{|\Lambda|} \sum_{\mathbf{n}:\, \partial_\Lambda\mathbf{n}=\varnothing} \prod_{e\in\mathscr{E}^{\mathrm{b}}_\Lambda} \frac{\beta^{n_e}}{n_e!} = 2^{|\Lambda|} e^{\beta|\mathscr{E}^{\mathrm{b}}_\Lambda|}\, \mathbb{P}^+_{\Lambda;\beta}(\partial_\Lambda\mathbf{n} = \varnothing),$$

where $\partial_\Lambda\mathbf{n} \stackrel{\mathrm{def}}{=} \big\{i \in \Lambda : \hat{I}(i,\mathbf{n}) \text{ is odd}\big\}$ and, under the probability distribution $\mathbb{P}^+_{\Lambda;\beta}$, $\mathbf{n} = (n_e)_{e\in\mathscr{E}^{\mathrm{b}}_\Lambda}$ is a collection of independent random variables, each one distributed according to the Poisson distribution of parameter β. We will call $\mathbf{n}$ a **current configuration** in Λ.

In the same way, one easily derives similar representations for arbitrary correlation functions.

Exercise 3.38. Derive the following identity: for all $A \subset \Lambda \Subset \mathbb{Z}^d$,

$$\langle \sigma_A \rangle^+_{\Lambda;\beta,0} = \frac{\mathbb{P}^+_{\Lambda;\beta}(\partial_\Lambda \mathbf{n} = A)}{\mathbb{P}^+_{\Lambda;\beta}(\partial_\Lambda \mathbf{n} = \varnothing)}.$$

The power of the random-current representation, however, lies in the fact that it also allows a probabilistic interpretation of *truncated correlations* in terms of various geometric events. The crucial result is the following lemma, which deals with a distribution on *pairs* of current configurations $\mathbb{P}^{+(2)}_{\Lambda;\beta}(\mathbf{n}^1, \mathbf{n}^2) \stackrel{\text{def}}{=} \mathbb{P}^+_{\Lambda;\beta}(\mathbf{n}^1)\mathbb{P}^+_{\Lambda;\beta}(\mathbf{n}^2)$. Let us denote by $i \stackrel{\mathbf{n}}{\longleftrightarrow} \partial^{\text{ex}}\Lambda$ the event that there is a path connecting i to $\partial^{\text{ex}}\Lambda$ along which $\mathbf{n}$ takes only positive values.

Lemma 3.55 (Switching Lemma). *Let $\Lambda \Subset \mathbb{Z}^d$, $A \subset \Lambda$, $i \in \Lambda$ and $\mathscr{I}$ a set of current configurations in Λ. Then,*

$$\begin{aligned}
&\mathbb{P}^{+(2)}_{\Lambda;\beta}(\partial_\Lambda \mathbf{n}^1 = A, \partial_\Lambda \mathbf{n}^2 = \{i\}, \mathbf{n}^1 + \mathbf{n}^2 \in \mathscr{I}) \\
&\quad = \mathbb{P}^{+(2)}_{\Lambda;\beta}(\partial_\Lambda \mathbf{n}^1 = A \,\triangle\, \{i\}, \partial_\Lambda \mathbf{n}^2 = \varnothing, \mathbf{n}^1 + \mathbf{n}^2 \in \mathscr{I}, i \stackrel{\mathbf{n}^1+\mathbf{n}^2}{\longleftrightarrow} \partial^{\text{ex}}\Lambda).
\end{aligned} \tag{3.78}$$

Proof We use the following notations:

$$\mathsf{w}(\mathbf{n}) \stackrel{\text{def}}{=} \prod_{e \in \mathscr{E}_\Lambda} \frac{\beta^{n_e}}{n_e!}$$

and, for two current configurations satisfying $\mathbf{n} \le \mathbf{m}$ (that is, $n_e \le m_e$, $\forall e \in \mathscr{E}^{\text{b}}_\Lambda$),

$$\binom{\mathbf{m}}{\mathbf{n}} \stackrel{\text{def}}{=} \prod_{e \in \mathscr{E}^{\text{b}}_\Lambda} \binom{m_e}{n_e}.$$

We are going to change variables from the pair $(\mathbf{n}^1, \mathbf{n}^2)$ to the pair $(\mathbf{m}, \mathbf{n})$, where $\mathbf{m} = \mathbf{n}^1 + \mathbf{n}^2$ and $\mathbf{n} = \mathbf{n}^2$. Since $\partial_\Lambda(\mathbf{n}^1 + \mathbf{n}^2) = \partial_\Lambda \mathbf{n}^1 \,\triangle\, \partial_\Lambda \mathbf{n}^2$, $\mathbf{n} \le \mathbf{m}$ and

$$\mathsf{w}(\mathbf{n}^1)\mathsf{w}(\mathbf{n}^2) = \binom{\mathbf{n}^1 + \mathbf{n}^2}{\mathbf{n}^2} \mathsf{w}(\mathbf{n}^1 + \mathbf{n}^2) = \binom{\mathbf{m}}{\mathbf{n}} \mathsf{w}(\mathbf{m}),$$

we can write

$$\sum_{\substack{\partial_\Lambda \mathbf{n}^1 = A \\ \partial_\Lambda \mathbf{n}^2 = \{i\} \\ \mathbf{n}^1 + \mathbf{n}^2 \in \mathscr{I}}} \mathsf{w}(\mathbf{n}^1)\mathsf{w}(\mathbf{n}^2) = \sum_{\substack{\partial_\Lambda \mathbf{m} = A \triangle \{i\} \\ \mathbf{m} \in \mathscr{I}}} \mathsf{w}(\mathbf{m}) \sum_{\substack{\mathbf{n} \le \mathbf{m} \\ \partial_\Lambda \mathbf{n} = \{i\}}} \binom{\mathbf{m}}{\mathbf{n}}. \tag{3.79}$$

The first observation is that $i \overset{\mathbf{m}}{\not\longleftrightarrow} \partial^{\mathrm{ex}}\Lambda \implies i \overset{\mathbf{n}}{\not\longleftrightarrow} \partial^{\mathrm{ex}}\Lambda$, since $\mathbf{n} \leq \mathbf{m}$. Consequently,

$$\sum_{\substack{\mathbf{n}\leq\mathbf{m}\\ \partial_\Lambda\mathbf{n}=\{i\}}} \binom{\mathbf{m}}{\mathbf{n}} = 0, \quad \text{when } i \overset{\mathbf{m}}{\not\longleftrightarrow} \partial^{\mathrm{ex}}\Lambda, \tag{3.80}$$

since $i \overset{\mathbf{n}}{\longleftrightarrow} \partial^{\mathrm{ex}}\Lambda$ whenever $\partial_\Lambda\mathbf{n} = \{i\}$. Let us therefore assume that $i \overset{\mathbf{m}}{\longleftrightarrow} \partial^{\mathrm{ex}}\Lambda$, which allows us to use the following lemma, which will be proven below.

Lemma 3.56. *Let* $\mathbf{m}$ *be a current configuration in* $\Lambda \Subset \mathbb{Z}^d$ *and* $C, D \subset \Lambda$. *If there exists a current configuration* $\mathbf{k}$ *such that* $\mathbf{k} \leq \mathbf{m}$ *and* $\partial_\Lambda\mathbf{k} = C$, *then*

$$\sum_{\substack{\mathbf{n}\leq\mathbf{m}\\ \partial_\Lambda\mathbf{n}=D}} \binom{\mathbf{m}}{\mathbf{n}} = \sum_{\substack{\mathbf{n}\leq\mathbf{m}\\ \partial_\Lambda\mathbf{n}=C\triangle D}} \binom{\mathbf{m}}{\mathbf{n}}. \tag{3.81}$$

An application of this lemma with $C = D = \{i\}$ yields

$$\sum_{\substack{\mathbf{n}\leq\mathbf{m}\\ \partial_\Lambda\mathbf{n}=\{i\}}} \binom{\mathbf{m}}{\mathbf{n}} = \sum_{\substack{\mathbf{n}\leq\mathbf{m}\\ \partial_\Lambda\mathbf{n}=\varnothing}} \binom{\mathbf{m}}{\mathbf{n}}, \quad \text{when } i \overset{\mathbf{m}}{\longleftrightarrow} \partial^{\mathrm{ex}}\Lambda. \tag{3.82}$$

Using (3.80) and (3.82) in (3.79), and returning to the variables $\mathbf{n}^1 = \mathbf{m} - \mathbf{n}$ and $\mathbf{n}^2 = \mathbf{n}$, we get

$$\begin{aligned}\sum_{\substack{\partial_\Lambda\mathbf{n}^1=A\\ \partial_\Lambda\mathbf{n}^2=\{i,j\}\\ \mathbf{n}^1+\mathbf{n}^2\in\mathscr{I}}} \mathsf{w}(\mathbf{n}^1)\mathsf{w}(\mathbf{n}^2) &= \sum_{\substack{\partial_\Lambda\mathbf{m}=A\triangle\{i,j\}\\ \mathbf{m}\in\mathscr{I}\\ i\overset{\mathbf{m}}{\longleftrightarrow}j}} \mathsf{w}(\mathbf{m}) \sum_{\substack{\mathbf{n}\leq\mathbf{m}\\ \partial_\Lambda\mathbf{n}=\varnothing}} \binom{\mathbf{m}}{\mathbf{n}}\\ &= \sum_{\substack{\partial_\Lambda\mathbf{n}^1=A\triangle\{i,j\}\\ \partial_\Lambda\mathbf{n}^2=\varnothing\\ \mathbf{n}^1+\mathbf{n}^2\in\mathscr{I}}} \mathsf{w}(\mathbf{n}^1)\mathsf{w}(\mathbf{n}^2)\,\mathbf{1}_{\{i\overset{\mathbf{n}^1+\mathbf{n}^2}{\longleftrightarrow}\partial^{\mathrm{ex}}\Lambda\}},\end{aligned}$$

and the proof is complete. □

Proof of Lemma 3.56. Let us associate the configuration $\mathbf{m}$ with the graph $G_\mathbf{m}$ with vertices $\Lambda \cup \partial^{\mathrm{ex}}\Lambda$ and with m_e edges between the endpoints of each edge $e \in \mathscr{E}^{\mathrm{b}}_\Lambda$. By assumption, $G_\mathbf{m}$ possesses a subgraph $G_\mathbf{k}$ with $\partial_\Lambda G_\mathbf{k} = C$, where $\partial_\Lambda G_\mathbf{k}$ is the set of vertices of Λ belonging to an odd number of edges.

The left-hand side of (3.81) is equal to the number of subgraphs G of $G_\mathbf{m}$ satisfying $\partial_\Lambda G = D$, while the right-hand side counts the number of subgraphs G of $G_\mathbf{m}$ satisfying $\partial_\Lambda G = C \triangle D$. But the application $G \mapsto G \triangle G_\mathbf{k}$ defines a bijection between these two families of graphs, since $\partial_\Lambda(G \triangle G_\mathbf{k}) = \partial_\Lambda G \triangle \partial_\Lambda G_\mathbf{k}$ and $(G \triangle G_\mathbf{k}) \triangle G_\mathbf{k} = G$. □

As one simple application of the Switching Lemma, let us derive a probabilistic representation for the truncated 2-point function.

Lemma 3.57. *For all distinct $i, j \in \Lambda \Subset \mathbb{Z}^d$,*

$$\langle \sigma_i ; \sigma_j \rangle^+_{\Lambda;\beta,0} = \frac{\mathbb{P}^{+(2)}_{\Lambda;\beta}(\partial_\Lambda \mathbf{n}^1 = \{i,j\}, \partial_\Lambda \mathbf{n}^2 = \varnothing, i \overset{\mathbf{n}^1+\mathbf{n}^2}{\not\longleftrightarrow} \partial^{\mathrm{ex}}\Lambda)}{\mathbb{P}^{+(2)}_{\Lambda;\beta}(\partial_\Lambda \mathbf{n}^1 = \varnothing, \partial_\Lambda \mathbf{n}^2 = \varnothing)}. \tag{3.83}$$

Proof Using the representation of Exercise 3.38,

$$\begin{aligned}\langle \sigma_i ; \sigma_j \rangle^+_{\Lambda;\beta,0} &= \frac{\mathbb{P}_{\Lambda;\beta}(\partial_\Lambda \mathbf{n} = \{i,j\})}{\mathbb{P}_{\Lambda;\beta}(\partial_\Lambda \mathbf{n} = \varnothing)} - \frac{\mathbb{P}_{\Lambda;\beta}(\partial_\Lambda \mathbf{n} = \{i\})\,\mathbb{P}_{\Lambda;\beta}(\partial_\Lambda \mathbf{n} = \{j\})}{\mathbb{P}_{\Lambda;\beta}(\partial_\Lambda \mathbf{n} = \varnothing)\,\mathbb{P}_{\Lambda;\beta}(\partial_\Lambda \mathbf{n} = \varnothing)} \\ &= \frac{\mathbb{P}^{+(2)}_{\Lambda;\beta}(\partial_\Lambda \mathbf{n}^1 = \{i,j\}, \partial_\Lambda \mathbf{n}^2 = \varnothing) - \mathbb{P}^{+(2)}_{\Lambda;\beta}(\partial_\Lambda \mathbf{n}^1 = \{i\}, \partial_\Lambda \mathbf{n}^2 = \{j\})}{\mathbb{P}^{+(2)}_{\Lambda;\beta}(\partial_\Lambda \mathbf{n}^1 = \varnothing, \partial_\Lambda \mathbf{n}^2 = \varnothing)}.\end{aligned}$$

Since the Switching Lemma implies that

$$\mathbb{P}^{+(2)}_{\Lambda;\beta}(\partial_\Lambda \mathbf{n}^1 = \{i\}, \partial_\Lambda \mathbf{n}^2 = \{j\}) = \mathbb{P}^{+(2)}_{\Lambda;\beta}(\partial_\Lambda \mathbf{n}^1 = \{i,j\}, \partial_\Lambda \mathbf{n}^2 = \varnothing, i \overset{\mathbf{n}^1+\mathbf{n}^2}{\longleftrightarrow} \partial^{\mathrm{ex}}\Lambda),$$

we can cancel terms in the numerator and the conclusion follows. □

Observe that (3.83) implies that $\langle \sigma_i ; \sigma_j \rangle^+_{\Lambda;\beta,0} \geq 0$, which is a particular instance of the GKS (or FKG) inequalities. However, having such a probabilistic representation also opens up the possibility of proving nontrivial lower and upper bounds.

Among the numerous fundamental applications of the random-current representation, let us mention the proof that $m^*(\beta_c(d)) = 0$ in all dimensions $d \geq 2$ [3, 7, 8], the proof that, for all $\beta < \beta_c(d)$ and all $d \geq 1$, there exists $c = c(\beta, d) > 0$ such that $\langle \sigma_0 \sigma_i \rangle^+_{\beta,0} \leq e^{-c\|i\|_2}$ [5], the fact that $\langle \sigma_0 \sigma_i \rangle_{\beta_c(d),0} \simeq c_d \|i\|_2^{2-d}$ in all large enough dimensions [292] and the determination of the sign of all Ursell functions in [306]. Additional information can be found in the references given above.

3.10.7 Non-Translation-Invariant Gibbs States and Interfaces

In this subsection, we briefly discuss the existence or absence of non-translation-invariant Gibbs states describing coexistence of phases. The first proof of the existence of non-translation-invariant Gibbs states in the Ising model on $\mathbb{Z}^d$, $d \geq 3$, at sufficiently low temperatures, is due to Dobrushin [81]; the much simpler argument we provide below is due to van Beijeren [338].

We require the parameters of the model to be such that the system is in the non-uniqueness regime. So, for the rest of the section, we always assume that $d \geq 2$, $h = 0$ and $\beta > \beta_c(d)$.

A natural way to try to induce spatial coexistence of the $+$ and $-$ phases in a system is to use nonhomogeneous boundary conditions. Let us therefore consider the **Dobrushin boundary condition** η^{Dob}, defined by (see Figure 3.14)

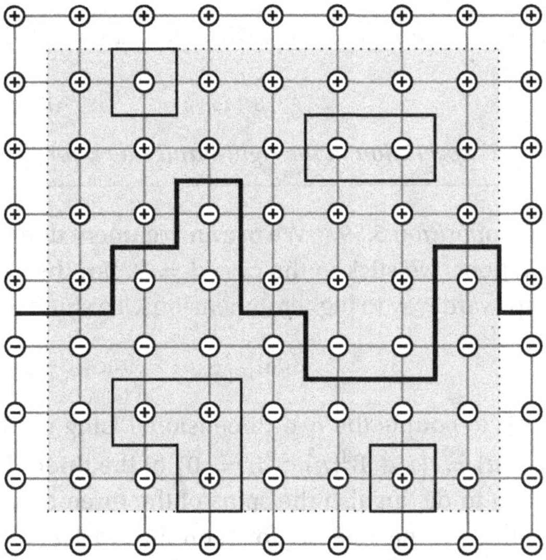

Figure 3.14. In $d = 2$, with the Dobrushin boundary condition, a configuration always has a unique open contour (called interface in the text, the thickest line on the figure) connecting the two vertical sides of the box.

$$\eta_i^{\mathrm{Dob}} \stackrel{\mathrm{def}}{=} \begin{cases} +1 & \text{if } i = (i_1, \dots, i_d) \text{ with } i_d \geq 0, \\ -1 & \text{otherwise.} \end{cases}$$

Let us then define the sequence of boxes to be used for the rest of the section, more suited to the use of the Dobrushin boundary condition:

$$\Lambda^d(n) \stackrel{\mathrm{def}}{=} \{i \in \mathbb{Z}^d : -n \leq i_j \leq n \text{ if } 1 \leq j < d,\ -n \leq i_d \leq n-1\}.$$

If $i = (i_1, i_2, \dots, i_{d-1}, i_d) \in \mathbb{Z}^d$, we denote by $\bar{i} = (i_1, i_2, \dots, i_{d-1}, -1 - i_d) \in \mathbb{Z}^d$ its reflection through the plane $\{x \in \mathbb{R}^d : x_d = -\frac{1}{2}\}$.

The nonhomogeneity of the Dobrushin boundary condition can be shown to have a significant effect in higher dimensions:

Theorem 3.58. *Assume $d \geq 3$. Then, for all $\beta > \beta_c(d-1)$, there exists a sequence of integers $n_k \uparrow \infty$ along which*

$$\langle \cdot \rangle_{\beta,0}^{\mathrm{Dob}} \stackrel{\mathrm{def}}{=} \lim_{k \to \infty} \langle \cdot \rangle_{\Lambda^d(n_k);\beta,0}^{\mathrm{Dob}}$$

is a well-defined Gibbs state that satisfies

$$\langle \sigma_0 \rangle_{\beta,0}^{\mathrm{Dob}} > 0 > \langle \sigma_{\bar{0}} \rangle_{\beta,0}^{\mathrm{Dob}}.$$

In particular, $\langle \cdot \rangle_{\beta,0}^{\mathrm{Dob}}$ is not invariant under vertical translations.

The states constructed in the previous theorem are usually called **Dobrushin states**. The proof of this result relies on the following key inequality:

Proposition 3.59. *Let $d \geq 2$. Then, for all $i \in \mathrm{B}^d(n)$ such that $i_d = 0$,*

$$\langle \sigma_i \rangle^{\mathrm{Dob}}_{\mathrm{B}^d(n);\beta,0} \geq \langle \sigma_i \rangle^{+}_{\mathrm{B}^{d-1}(n);\beta,0}, \tag{3.84}$$

where the expectation in the right-hand side is for the Ising model in $\mathbb{Z}^{d-1}$.

Proof of Proposition 3.59. We use an argument due to van Beijeren [338]. To simplify notations, we stick to the case $d = 3$, but the argument can be adapted in a straightforward way to higher dimensions. To show that

$$\langle \sigma_0 \rangle^{\mathrm{Dob}}_{\mathrm{B}^3(n);\beta,0} \geq \langle \sigma_0 \rangle^{+}_{\mathrm{B}^2(n);\beta,0}, \tag{3.85}$$

the idea is to couple the two-dimensional Ising model in the box $\mathrm{B}^2(n)$ with the layer $\mathrm{B}^{3,0}(n) \stackrel{\text{def}}{=} \{i \in \mathrm{B}^3(n) : i_3 = 0\}$ of the three-dimensional model. It will be convenient to distinguish the spins of the three-dimensional model and those of the two-dimensional one. We thus continue to denote by σ_i the former, but we write τ_i for the latter. We then introduce new random variables. For all $i \in \mathrm{B}^{3,+}(n) \stackrel{\text{def}}{=} \{i \in \mathrm{B}^3(n) : i_3 > 0\}$, we set

$$s_i \stackrel{\text{def}}{=} \tfrac{1}{2}(\sigma_i + \sigma_{\underline{i}}), \qquad t_i \stackrel{\text{def}}{=} \tfrac{1}{2}(\sigma_i - \sigma_{\underline{i}}),$$

where , for $i = (i_1, i_2, i_3)$, we have set $\underline{i} \stackrel{\text{def}}{=} (i_1, i_2, -i_3)$. Moreover, for all $i \in \mathrm{B}^{3,0}(n)$, we set

$$s_i \stackrel{\text{def}}{=} \tfrac{1}{2}(\sigma_i + \tau_i), \qquad t_i \stackrel{\text{def}}{=} \tfrac{1}{2}(\sigma_i - \tau_i).$$

These random variables are $\{-1, 0, 1\}$-valued and satisfy the constraint

$$s_i = 0 \Leftrightarrow t_i \neq 0, \qquad \forall i \in \mathrm{B}^{3,+}(n) \cup \mathrm{B}^{3,0}(n). \tag{3.86}$$

Observe now that (3.85) is equivalent to

$$\langle t_0 \rangle \geq 0, \tag{3.87}$$

where the expectation is with respect to $\mu^{\mathrm{Dob}}_{\mathrm{B}^3(n);\beta,0} \otimes \mu^{+}_{\mathrm{B}^2(n);\beta,0}$. The conclusion thus follows from Exercise 3.39 below. □

Equation (3.87) is actually a particular instance of a set of GKS-type inequalities, originally studied by Percus.

Exercise 3.39. Prove (3.87). *Hint:* Expand the numerator of $\langle t_0 \rangle$ according to the realization of $A = \{i \in \mathrm{B}^{3,+}(n) \cup \mathrm{B}^{3,0}(n) : s_i = 0\}$. Observe that, once A is fixed, there remains exactly one nontrivial $\{-1, 1\}$-valued variable at each vertex. Verify that you can then apply the usual GKS inequalities to show that each term of the sum is nonnegative (you will have to check that the resulting Hamiltonian has the proper form).

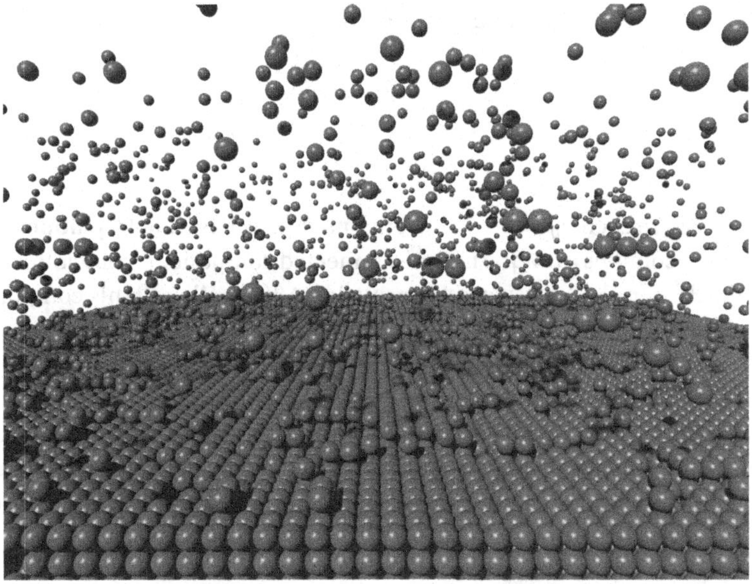

Figure 3.15. A typical configuration of the low-temperature three-dimensional Ising model with Dobrushin boundary condition. For convenience, in this picture, $-$ spins are represented by balls and $+$ spins by empty space. The interface is a perfect plane with only local defects.

Proof of Theorem 3.58. The construction of $\langle\cdot\rangle^{\mathrm{Dob}}_{\beta,0}$ along some subsequence $\Lambda^d(n_k)$ can be done as in Exercise 3.8. Observe that, by symmetry,

$$\langle\sigma_0\rangle^{\mathrm{Dob}}_{\Lambda^d(n_k);\beta,0} = -\langle\sigma_{\bar{0}}\rangle^{\mathrm{Dob}}_{\Lambda^d(n_k);\beta,0}, \tag{3.88}$$

which gives, after $k \to \infty$,

$$\langle\sigma_0\rangle^{\mathrm{Dob}}_{\beta,0} = -\langle\sigma_{\bar{0}}\rangle^{\mathrm{Dob}}_{\beta,0}. \tag{3.89}$$

Observe that, by the FKG inequality, applying a magnetic field $h \uparrow \infty$ on the spins living in $\mathrm{B}^d(n_k) \setminus \Lambda^d(n_k)$ yields

$$\langle\sigma_0\rangle^{\mathrm{Dob}}_{\Lambda^d(n_k);\beta,0} \geq \langle\sigma_0\rangle^{\mathrm{Dob}}_{\mathrm{B}^d(n_k);\beta,0}.$$

Using (3.84), we deduce that

$$\langle\sigma_0\rangle^{\mathrm{Dob}}_{\Lambda^d(n_k);\beta,0} \geq \langle\sigma_0\rangle^{+}_{\mathrm{B}^{d-1}(n_k);\beta,0}.$$

The limit $k \to \infty$ of the right-hand side converges to the spontaneous magnetization of the $(d-1)$-dimensional Ising model, which is positive when $\beta > \beta_c(d-1)$. The claim thus follows from (3.89). □

The Interface. Whether non-translation-invariant infinite-volume Gibbs states exist in $d \geq 3$ is in fact closely related to the behavior of the *macroscopic interface* induced by the Dobrushin boundary condition.

Let $\omega \in \Omega^{\text{Dob}}_{\Lambda^d(n)}$ and consider the set

$$\mathscr{B}(\omega) \stackrel{\text{def}}{=} \bigcup_{\substack{\{i,j\}\in\mathscr{E}_{\mathbb{Z}^d}\\ \omega_i\neq\omega_j}} \pi_{ij}\,,$$

where each $\pi_{ij} \stackrel{\text{def}}{=} \mathscr{S}_i \cap \mathscr{S}_j$ (remember (3.31)) is called a **plaquette**. By construction, $\mathscr{B}$ contains a unique infinite connected component (coinciding with the plane $\{x \in \mathbb{R}^d : x_d = -\frac{1}{2}\}$ everywhere outside $\Lambda^d(n)$). We call this component the **interface** and denote it by $\Gamma = \Gamma(\omega)$.

It turns out that, in $d \geq 3$, Γ is *rigid* at low temperature: in typical configurations, Γ coincides with $\{x_d = -\frac{1}{2}\}$ apart from local defects; see Figure 3.15. This can be quantified very precisely using cluster expansion techniques, as was done in Dobrushin's original work [81]. The much simpler description given below provides substantially less information, but still allows one to prove localization of Γ in a weaker sense.

Theorem 3.60. *Assume that $d \geq 3$. There exists $c'(\beta) > 0$ satisfying $\lim_{\beta\to\infty} c'(\beta) = 0$ such that, uniformly in n and in $i \in \{j \in \Lambda^d(n) : j_d = 0\}$,*

$$\mu^{\text{Dob}}_{\Lambda^d(n);\beta,0}(\Gamma \supset \pi_{i\bar{i}}) \geq 1 - c'(\beta)\,.$$

Proof of Theorem 3.60. We first decompose

$$\langle\sigma_i\sigma_{\bar{i}}\rangle^{\text{Dob}}_{\Lambda^d(n);\beta,0} = \langle\sigma_i\sigma_{\bar{i}}\mathbf{1}_{\{\Gamma\supset\pi_{i\bar{i}}\}}\rangle^{\text{Dob}}_{\Lambda^d(n);\beta,0} + \langle\sigma_i\sigma_{\bar{i}}\mathbf{1}_{\{\Gamma\not\supset\pi_{i\bar{i}}\}}\rangle^{\text{Dob}}_{\Lambda^d(n);\beta,0}\,. \tag{3.90}$$

On the one hand, $\sigma_i\sigma_{\bar{i}} = -1$ whenever $\Gamma \supset \pi_{i\bar{i}}$. On the other hand, when $\Gamma \not\supset \pi_{i\bar{i}}$, i and $\bar{i}$ belong to the same (random) component of $\Lambda^d(n) \setminus \Gamma$, with constant (either $+$ or $-$) boundary condition. More precisely, with a fixed interface Γ we associate a partition of $\Lambda^d(n)$ into connected regions $D_1, \dots, D_k$. The Dobrushin boundary condition, together with Γ, induces a well-defined constant boundary condition $\#_i$ on each region D_i, either $+$ or $-$. $\Gamma \not\supset \pi_{i\bar{i}}$ means that the edge $\{i, \bar{i}\}$ is contained inside one of these components, say D_*. We can therefore write

$$\langle\sigma_i\sigma_{\bar{i}}\mathbf{1}_{\{\Gamma\not\supset\pi_{i\bar{i}}\}}\rangle^{\text{Dob}}_{\Lambda^d(n);\beta,0} = \sum_{\Gamma\not\supset\pi_{i\bar{i}}} \langle\sigma_i\sigma_{\bar{i}}\rangle^{\#_*}_{D_*;\beta,0}\, \mu^{\text{Dob}}_{\Lambda^d(n);\beta,0}(\Gamma(\omega) = \Gamma)\,. \tag{3.91}$$

Assume that $\#_* = +$. Then the GKS inequalities (see Exercise 3.12) imply that

$$\langle\sigma_i\sigma_{\bar{i}}\rangle^{+}_{D_*;\beta,0} \geq \langle\sigma_i\sigma_{\bar{i}}\rangle^{+}_{\Lambda^d(n);\beta,0} \geq \langle\sigma_i\sigma_{\bar{i}}\rangle^{+}_{\beta,0}\,.$$

When $\#_* = -$, the same holds since, by symmetry, $\langle\sigma_i\sigma_{\bar{i}}\rangle^{+}_{D_*;\beta,0} = \langle\sigma_i\sigma_{\bar{i}}\rangle^{-}_{D_*;\beta,0}$ and $\langle\sigma_i\sigma_{\bar{i}}\rangle^{+}_{\beta,0} = \langle\sigma_i\sigma_{\bar{i}}\rangle^{-}_{\beta,0}$. Thus,

$$\langle\sigma_i\sigma_{\bar{i}}\mathbf{1}_{\{\Gamma\not\supset\pi_{i\bar{i}}\}}\rangle^{\text{Dob}}_{\Lambda^d(n);\beta,0} \geq \langle\sigma_i\sigma_{\bar{i}}\rangle^{+}_{\beta,0}\, \mu^{\text{Dob}}_{\Lambda^d(n);\beta,0}(\Gamma \not\supset \pi_{i\bar{i}})\,.$$

Collecting the above and rearranging the terms, we get

$$\mu^{\text{Dob}}_{\Lambda^d(n);\beta,0}(\Gamma \supset \pi_{i\bar{i}}) \geq 1 - \frac{1 + \langle \sigma_i \sigma_{\bar{i}} \rangle^{\text{Dob}}_{\Lambda^d(n);\beta,0}}{1 + \langle \sigma_i \sigma_{\bar{i}} \rangle^{+}_{\beta,0}} .$$

Let us consider the numerator in the right-hand side. Using Jensen's inequality, we can write

$$\langle \sigma_i \sigma_{\bar{i}} \rangle^{\text{Dob}}_{\Lambda^d(n);\beta,0} = 1 - \tfrac{1}{2}\langle (\sigma_i - \sigma_{\bar{i}})^2 \rangle^{\text{Dob}}_{\Lambda^d(n);\beta,0} \leq 1 - \tfrac{1}{2}\big(\langle \sigma_i - \sigma_{\bar{i}} \rangle^{\text{Dob}}_{\Lambda^d(n);\beta,0}\big)^2 .$$

But $\langle \sigma_{\bar{i}} \rangle^{\text{Dob}}_{\Lambda^d(n);\beta,0} = -\langle \sigma_i \rangle^{\text{Dob}}_{\Lambda^d(n);\beta,0}$ and so, by Proposition 3.59,

$$\langle \sigma_i - \sigma_{\bar{i}} \rangle^{\text{Dob}}_{\Lambda^d(n);\beta,0} = 2\langle \sigma_i \rangle^{\text{Dob}}_{\Lambda^d(n);\beta,0} \geq 2\langle \sigma_0 \rangle^{+}_{\beta,0;d-1} ,$$

from which we conclude that

$$\mu^{\text{Dob}}_{\Lambda^d(n);\beta,0}(\Gamma \supset \pi_{i\bar{i}}) \geq 1 - 2\frac{1 - \big(\langle \sigma_0 \rangle^{+}_{\beta,0;d-1}\big)^2}{1 + \langle \sigma_i \sigma_{\bar{i}} \rangle^{+}_{\beta,0}} \geq 1 - 2\frac{1 - \big(\langle \sigma_0 \rangle^{+}_{\beta,0;d-1}\big)^2}{1 + \big(\langle \sigma_0 \rangle^{+}_{\beta,0}\big)^2} .$$

This lower bound is uniform in n and i and converges to 1 as $\beta \to \infty$. □

Of course, Theorem 3.59 only shows the existence of non-translation-invariant Gibbs states when $\beta > \beta_c(d-1)$, and one might wonder what happens for values of β in the remaining interval $[\beta_c(d-1), \beta_c(d))$. It turns out that this problem is still open. The conjectured behavior, however, is as follows:[5]

- When $d = 3$, there should exist a value $\beta_R \in [\beta_c(2), \beta_c(3))$ such that the existence of Gibbs states which are not translation invariant holds for all $\beta > \beta_R$, but not for $\beta < \beta_R$. At β_R, the system is said to undergo a **roughening transition**. At this transition the interface is supposed to lose its rigidity and to start having unbounded fluctuations.[6]
- When $d \geq 4$, Dobrushin's non-translation-invariant Gibbs states are believed to exist (with a rigid interface) for all $\beta > \beta_c(d)$.

Two-Dimensional Model. The behavior of the interface in two dimensions is very different and, from a mathematical point of view, a rather detailed and complete picture is available.

Consider again a configuration $\omega \in \Omega^{\text{Dob}}_{\Lambda_n}$ and, in particular, the associated interface Γ. Let us denote by ω_Γ the configuration in $\Omega^{\text{Dob}}_{\Lambda_n}$ for which $\mathscr{B}(\omega_\Gamma) = \{\Gamma\}$. We can then define the upper and lower "envelopes" $\Gamma^{\pm} : \mathbb{Z} \to \mathbb{Z}$ of Γ by

$$\Gamma^+(i) \stackrel{\text{def}}{=} \max\{j \in \mathbb{Z} : \sigma_{(i,j)}(\omega_\Gamma) = -1\} + 1 ,$$

$$\Gamma^-(i) \stackrel{\text{def}}{=} \min\{j \in \mathbb{Z} : \sigma_{(i,j)}(\omega_\Gamma) = +1\} - 1 .$$

Note that $\Gamma^+(i) > \Gamma^-(i)$ for all $i \in \mathbb{Z}$. One can show [60] that, with probability close to 1, Γ^- and Γ^+ remain very close to each other: there exists $K = K(\beta) < \infty$ such that, with probability tending to 1 as $n \to \infty$,

$$\max_{i\in\mathbb{Z}} |\Gamma^+(i) - \Gamma^-(i)| \leq K \log n . \tag{3.92}$$

Let us now introduce the diffusively rescaled profiles $\hat{\Gamma}^{\pm}\colon [-1,1] \to \mathbb{R}$. Given $y = (y_1,\dots,y_d) \in \mathbb{R}^d$, let us write $\lfloor y \rfloor \stackrel{\text{def}}{=} (\lfloor y_1 \rfloor,\dots,\lfloor y_d \rfloor)$. We then set, for any $x \in [-1,1]$,

$$\hat{\Gamma}^{+}(x) = \frac{1}{\sqrt{n}}\Gamma^{+}(\lfloor nx \rfloor),$$

and similarly for Γ^{-}. Observe that, thanks to (3.92), we know that

$$\lim_{n\to\infty} \mu^{\text{Dob}}_{\Lambda_n}\big(\sup_{x\in[-1,1]} |\hat{\Gamma}^{+}(x) - \hat{\Gamma}^{-}(x)| \le \epsilon \big) = 1, \text{ for all } \epsilon > 0.$$

Since the interface Γ is squeezed between Γ^{+} and Γ^{-}, studying the limiting behavior of $\hat{\Gamma}^{+}$ suffices to understand the asymptotic behavior of the interface under diffusive scaling. This is the content of the next theorem, first proved by Higuchi [161] for large enough values of β and then extended to all $\beta > \beta_c(2)$ by Greenberg and Ioffe [141].

Figure 3.16. A typical configuration of the low-temperature two-dimensional Ising model with Dobrushin boundary condition. Once properly rescaled, the interface between the two phases converges weakly to a Brownian bridge process.

Theorem 3.61. *For all $\beta > \beta_c(2)$, there exists $\kappa_\beta \in (0,\infty)$ such that $\hat{\Gamma}^{+}$ converges weakly to a Brownian bridge on $[-1,1]$ with diffusivity constant κ_β.*

(The Brownian bridge is a Brownian motion $(B_t)_{t\in[-1,1]}$ starting at 0 at $t = -1$ and conditioned to be at 0 at $t = +1$; see [251].) It is also possible [141] to express the diffusivity constant κ_β in terms of the physically relevant quantity, the surface tension, but this is beyond the scope of this book.

Theorem 3.61 shows that, in contrast to what happens in higher dimensions, the interface of the two-dimensional Ising model is never rigid (except in the trivial case $\beta = +\infty$); see Figure 3.16. Moreover, in a finite box Λ_n, Γ undergoes vertical fluctuations of order $\sqrt{n}$. A consequence of this delocalization of the interface is the following: when n becomes very large, the behavior of the system near the center of the box Λ_n will be typical of either the $+$ phase (if Γ has wandered far away below the origin) or the $-$ phase (if Γ has wandered far away above the origin), and the probability of each of these two alternatives converges to $\frac{1}{2}$ as $n \to \infty$. In particular, in this case, the infinite-volume Gibbs state resulting from the Dobrushin boundary condition is translation invariant and given by $\frac{1}{2}\mu^+_{\beta,0} + \frac{1}{2}\mu^-_{\beta,0}$. More details and far-reaching generalizations are discussed in Section 3.10.8.

3.10.8 Gibbs States and Local Behavior in Large Finite Systems

When introducing the notion of Gibbs states in Section 3.4, we motivated the definition by saying that the latter should lead to an interpretation of Gibbs states as providing approximate descriptions of all possible local behaviors in large finite systems, the quality of this approximation improving with the distance to the system's boundary. It turns out that, in the two-dimensional Ising model, this heuristic discussion can be made precise and rigorous.

Let us consider an arbitrary finite subset $\Lambda \Subset \mathbb{Z}^d$ and an arbitrary boundary condition $\eta \in \Omega$. We are interested in describing the local behavior of the Gibbs distribution $\mu^\eta_{\Lambda;\beta,0}$ in the vicinity of a point $i \in \Lambda$. Since Λ and η are arbitrary, there is no loss of generality in assuming that $i = 0$.

The Case of Pure Boundary Conditions. Let us first consider the simpler case of constant boundary conditions, which we assume to be $+$ for the sake of concreteness. We know from the definition of $\langle\cdot\rangle^+_{\beta,0}$ that, for any local function f,

$$\langle f\rangle^+_{\Lambda;\beta,0} \to \langle f\rangle^+_{\beta,0} \quad \text{as } \Lambda \uparrow \mathbb{Z}^d .$$

We now state a result, first proved by Bricmont, Lebowitz and Pfister [49], that says that $\langle\cdot\rangle^+_{\beta,0}$ actually provides an approximation for the finite-volume expectation $\langle\cdot\rangle^+_{\Lambda;\beta,0}$ with an error exponentially small in the distance from the support of f to the boundary of the box. Let

$$R \stackrel{\text{def}}{=} \max\{n : \mathrm{B}(n) \subset \Lambda\}$$

denote the distance from the origin to the boundary of Λ and let $r \stackrel{\text{def}}{=} \lfloor R/2 \rfloor$.

Theorem 3.62 (Exponential relaxation). *Assume that $\beta > \beta_c(2)$. There exists $c_1 = c_1(\beta) > 0$ such that the following holds. Let $\Lambda \Subset \mathbb{Z}^2$. Then, uniformly in all functions f with* $\mathrm{supp}(f) \subset \mathrm{B}(r)$,

$$|\langle f\rangle^+_{\Lambda;\beta,0} - \langle f\rangle^+_{\beta,0}| \le \tfrac{1}{c_1}\|f\|_\infty e^{-c_1 n}.$$

The same holds for the − boundary condition.

This fully vindicates the statement that the Gibbs state $\langle\cdot\rangle^+_{\beta,0}$ (resp. $\langle\cdot\rangle^-_{\beta,0}$) provides an accurate description of the local behavior of any finite-volume Gibbs distribution with + (resp. −) boundary condition, in regions of size proportional to the distance to the boundary of the system.

The Case of General Boundary Conditions. Let us now turn to the case of a Gibbs distribution with an arbitrary boundary condition η, which is much more delicate. For $\Lambda \Subset \mathbb{Z}^d$, take R as before, but this time define r as follows: fix some small $\epsilon \in (0, 1/2)$ and set

$$r \stackrel{\text{def}}{=} \lfloor R^{1/2-\epsilon} \rfloor . \tag{3.93}$$

A **circuit** is a set of distinct vertices $(t_0, t_1, \ldots, t_k)$ of $\mathbb{Z}^2$ with the property that $\|t_m - t_{m-1}\|_\infty = 1$, for all $1 \le m \le k$, and $\|t_k - t_0\|_\infty = 1$. Let $\mathscr{C}_\epsilon$ be the event

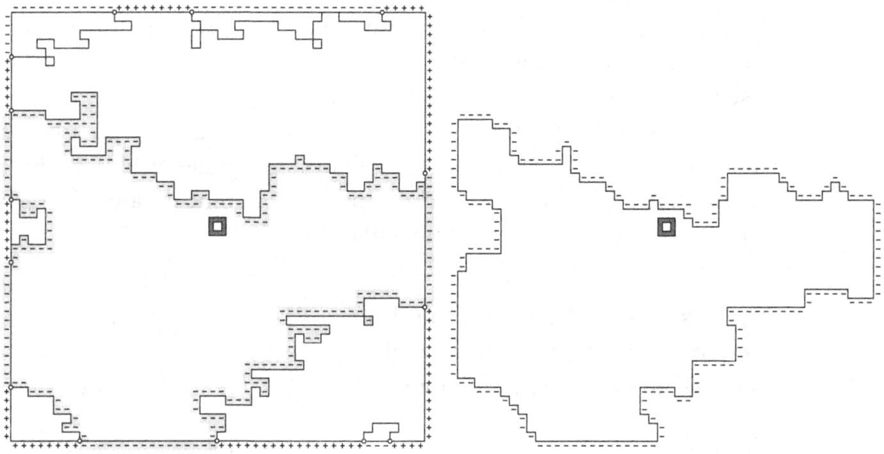

Figure 3.17. A (square) box Λ with a nonconstant boundary condition. The boundary condition induces open Peierls contours inside the system. With probability close to 1, none of them intersect the box $\mathrm{B}(2r)$ located in the middle (represented by the dark square). Left: a realization of the open Peierls contours. The event $\mathscr{C}_\epsilon^-$ occurs. The relevant − spins, the value of which is forced by the realization of the open contours, are indicated (and shaded). Right: the induced random box with − boundary condition. The box $\mathrm{B}(r)$, represented by the small white square in the middle, is located at a distance at least r from the boundary of this box.

that there is a circuit surrounding $\mathrm{B}(2r)$ in $\Lambda \cup \partial^{\mathrm{ex}}\Lambda$ and along which the spins take a constant value. We decompose $\mathscr{C}_\epsilon = \mathscr{C}_\epsilon^+ \cup \mathscr{C}_\epsilon^-$ according to the sign of the spins along the outermost such circuit. The main observation is that the event $\mathscr{C}_\epsilon$ is typical when $\beta > \beta_c(2)$, a fact first proved by Coquille and Velenik [73].

Theorem 3.63. *Assume that* $\beta > \beta_c(2)$. *For all* $\epsilon > 0$, *there exists* $c_2 = c_2(\beta, \epsilon) > 0$ *such that*

$$\mu^\eta_{\Lambda;\beta,0}(\mathscr{C}_\epsilon) \geq 1 - c_2 R^{-\epsilon},$$

uniformly in $\Lambda \Subset \mathbb{Z}^2$ *and* $\eta \in \Omega$.

Therefore, neglecting an event of probability at most $c_2 R^{-\epsilon}$, we can assume that one of the events $\mathscr{C}_\epsilon^+$ or $\mathscr{C}_\epsilon^-$ occurs. For definiteness, let us consider the latter case and let us denote by π the corresponding outermost circuit. Observe now that, conditionally on $\mathscr{C}_\epsilon^-$ and π, any function f with $\mathrm{supp}(f) \subset \mathrm{B}(r)$ finds itself in a box (delimited by π) with $-$ boundary condition (see Figure 3.17). Moreover, its support is at a distance at least r from the boundary of this box. It thus follows from Theorem 3.62 that its (conditional) expectation is closely approximated by $\langle f\rangle^-_{\beta,0}$. This leads [73] to the following generalization of Theorem 3.62.

Theorem 3.64. *Assume that* $\beta > \beta_c(2)$. *There exist constants* $\alpha = \alpha(\Lambda, \eta, \beta)$ *and* $c_3 = c_3(\beta)$ *such that, uniformly in functions* f *with* $\mathrm{supp}(f) \subset \mathrm{B}(r)$, *one has*

$$\left|\langle f\rangle^\eta_{\Lambda;\beta,0} - \left(\alpha \langle f\rangle^+_{\beta,0} + (1-\alpha)\langle f\rangle^-_{\beta,0}\right)\right| \leq c_3 \|f\|_\infty R^{-\epsilon}. \tag{3.94}$$

The coefficients α and $1-\alpha$ in (3.94) are given by

$$\alpha = \mu^\eta_{\Lambda;\beta,0}(\mathscr{C}_\epsilon^+ \,|\, \mathscr{C}_\epsilon), \qquad 1-\alpha = \mu^\eta_{\Lambda;\beta,0}(\mathscr{C}_\epsilon^- \,|\, \mathscr{C}_\epsilon),$$

that is, by the probabilities that the box $\mathrm{B}(R)$ (in which one measures f) finds itself deep inside a $+$, resp. $-$, region (conditionally on the typical event $\mathscr{C}_\epsilon$).

Again, the statement (3.94) fully vindicates the interpretation of Gibbs states as describing all possible local behaviors of any finite-volume system, with an accuracy improving with the distance to the system's boundary. This requires however, in general, that the size of the observation window be chosen small compared to the square root of the distance to the boundary. We will explain the reason for this restriction at the end of the section.

Note that (3.94) also shows that, in the two-dimensional Ising model, the only possible local behaviors are those corresponding to the $+$ and $-$ phases, since the approximation is stated in terms of the two Gibbs states $\langle\cdot\rangle^+_{\beta,0}$ and $\langle\cdot\rangle^-_{\beta,0}$. In other words, looking at local properties of the system, one will see behavior typical of the $+$ phase with probability close to α, and of the $-$ phase with probability close to $1-\alpha$. Actually this can be made a little more precise, as we explain now.

What Are the Possible Gibbs States? Let us consider a sequence of boundary conditions $(\eta_n)_{n\geq 1}$ and a sequence of boxes $\Lambda_n \uparrow \mathbb{Z}^2$. We assume that the corresponding sequence of Gibbs distributions $(\mu^{\eta_n}_{\Lambda_n;\beta,0})_{n\geq 1}$ converges to some Gibbs state $\langle\cdot\rangle$. Then, applying (3.94) with $f = \sigma_0$, we conclude that

$$\lim_{n\to\infty}\Big|\langle\sigma_0\rangle^{\eta_n}_{\Lambda_n;\beta,0} - \big(\alpha_n \langle\sigma_0\rangle^{+}_{\beta,0} + (1-\alpha_n)\langle\sigma_0\rangle^{-}_{\beta,0}\big)\Big| = 0 .$$

Since, by assumption, $\lim_{n\to\infty}\langle\sigma_0\rangle^{\eta_n}_{\Lambda_n;\beta,0} = \langle\sigma_0\rangle$, this implies the existence of

$$\alpha \stackrel{\text{def}}{=} \lim_{n\to\infty}\alpha_n = \frac{\langle\sigma_0\rangle - \langle\sigma_0\rangle^{-}_{\beta,0}}{\langle\sigma_0\rangle^{+}_{\beta,0} - \langle\sigma_0\rangle^{-}_{\beta,0}} .$$

Applying again (3.94) to arbitrary local functions, we conclude that

$$\langle\cdot\rangle = \lim_{n\to\infty}\langle\cdot\rangle^{\eta_n}_{\Lambda_n;\beta,0} = \alpha\,\langle\cdot\rangle^{+}_{\beta,0} + (1-\alpha)\,\langle\cdot\rangle^{-}_{\beta,0} ,$$

and thus all possible Gibbs states are convex combinations of the Gibbs states $\langle\cdot\rangle^{+}_{\beta,0}$ and $\langle\cdot\rangle^{-}_{\beta,0}$. This is the **Aizenman–Higuchi Theorem**, originally derived by Aizenman and Higuchi [160] directly for infinite-volume states; see also [135] for a self-contained, somewhat simpler and more general argument.

A more general formulation of the previous derivation will be presented in Chapter 6, once we have introduced the notion of infinite-volume Gibbs measures.

As we have seen in Section 3.10.7, when $d \geq 3$ and β is large enough, there exist Gibbs states that are not translation invariant. In particular, this implies that the Aizenman–Higuchi Theorem does not extend to this setting. Nevertheless, it can be proved that all *translation-invariant* Gibbs states of the Ising model on $\mathbb{Z}^d$, $d \geq 3$ are convex combinations of $\langle\cdot\rangle^{+}_{\beta,h}$ and $\langle\cdot\rangle^{-}_{\beta,h}$. This result is due to Bodineau [27], who completed earlier analyses started by Gallavotti and Miracle-Solé [129] and by Lebowitz [218].

Why This Constraint on the Size of the Observation Window? In the case of pure boundary conditions, it was possible to take an observation window with a radius proportional to the distance to the boundary. We now explain why one cannot, in general, improve Theorem 3.64 to larger windows. Let us thus consider an observation window $\mathrm{B}(r)$, with now an arbitrary radius r.

The reason is to be found in the probability of observing some "pathological" behavior in our finite system. Namely, we have seen above that, typically, the event $\mathscr{C}_\epsilon$ is realized. It turns out that, for some choices of the boundary condition η, the probability of not observing $\mathscr{C}_\epsilon$ is really of order $r/\sqrt{R}$ and thus small only when $r \ll \sqrt{R}$.

As a simple example, consider the box $\Lambda = \mathrm{B}(n)$ with the Dobrushin boundary condition η^{Dob}, as introduced in Section 3.10.7. As explained there, in that case the open Peierls contour has fluctuations of order $\sqrt{n}$ and its scaling limit is a Brownian bridge. This implies that the probability that this contour intersects $\mathrm{B}(2r)$ is indeed of order $r/\sqrt{n}$; note that, when this occurs, the event $\mathscr{C}_\epsilon$ becomes impossible.

Remark 3.65. In the uniqueness regime, quantitative estimates are easier to obtain. Consider an Ising model either at $\beta < \beta_c(2)$ and $h = 0$, or at $h \neq 0$ and arbitrary β, and let $\langle\cdot\rangle_{\beta,h}$ denote the associated (unique) Gibbs state. Then it can be shown [49, 95] that there is again exponential relaxation: there exists a constant $c_4 = c_4(\beta, h)$ such that

$$\left|\langle f\rangle^{\eta}_{\Lambda;\beta,h} - \langle f\rangle_{\beta,h}\right| \leq c\,\|f\|_\infty\, e^{-R/c_4}\,,$$

uniformly in functions f satisfying $\mathrm{supp}(f) \subset \mathrm{B}(r)$, $r = \lfloor R/2 \rfloor$. ◇

3.10.9 Absence of Analytic Continuation of the Pressure

From the point of view of complex analysis, the properties of the pressure of the Ising model that we have obtained raise natural questions, which will turn out to have physical relevance, as explained in Chapter 4, in particular in the discussion of Section 4.12.3. Since we are interested in fixing the temperature and studying the analyticity properties with respect to the magnetic field, in this section, we will denote the pressure by

$$h \mapsto \psi_\beta(h)\,.$$

For the sake of concreteness, let us consider only positive fields (by the identity $\psi_\beta(-h) = \psi_\beta(h)$, everything we say here admits an equivalent for negative fields). Although the pressure was first shown to exist on the real axis, we have seen in Theorem 3.42 that it can actually be extended to the whole half-plane $H^+ = \{\mathfrak{Re}\, h > 0\}$ as an analytic function $\psi_\beta \colon H^+ \to \mathbb{C}$. We will also see in Section 5.7.1 how to obtain the coefficients of the expansion of $\psi_\beta(h)$ in the variable e^{-2h}, with the latter being convergent for all $h \in H^+$. Unfortunately, these results do not provide any information on the behavior of the pressure on the boundary of H^+, $\partial H^+ \stackrel{\text{def}}{=} \{\mathfrak{Re}\, h = 0\}$. In function-theoretic terms, the most natural question is whether ψ_β can be *analytically continued* outside H^+. We will thus distinguish two scenarios.

Scenario 1: Analytic Continuation Is Possible. Analytic continuation means that there exist a strictly larger domain $H' \supset H^+$ and an analytic map $\widetilde{\psi}_\beta \colon H' \to \mathbb{C}$, which coincides with ψ_β on H^+, as depicted in Figure 3.18. This scenario is seen, for example, in the one-dimensional Ising model: the exact solution (3.10) guarantees that ψ_β can be continued analytically through $h = 0$, at all temperatures. Of course, since it can be defined as an analytic function on the whole real line, the analytic continuation $\widetilde{\psi}_\beta$ obtained when crossing $h = 0$ is nothing but the usual pressure: for $h < 0$, $\widetilde{\psi}_\beta(h) = \psi_\beta(h)$ (see Figure 3.4).

Another example where analytic continuation is possible is provided by the Curie–Weiss model. Indeed, we have already seen in Exercise 2.4 that, starting from

$$\psi^{\mathrm{CW}}_\beta(h) = \max_m\{hm - f^{\mathrm{CW}}_\beta(m)\}\,,$$

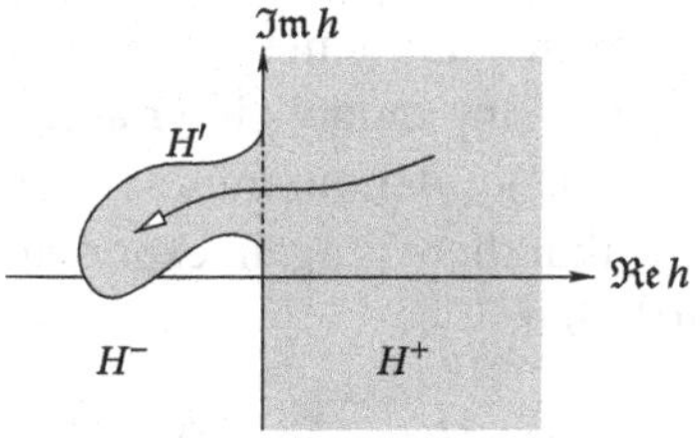

Figure 3.18. In Scenario 1, there exists an interval of the imaginary axis through which the pressure can be continued analytically.

a more explicit expression can be obtained for the pressure:

$$\psi_\beta^{\mathrm{CW}}(h) = -\frac{\beta m_\beta^{\mathrm{CW}}(h)^2}{2} + \log\cosh\bigl(\beta m_\beta^{\mathrm{CW}}(h) + h\bigr) + \log 2\,.$$

Although this function is not differentiable at $h = 0$ when β is large, it possesses an analytic continuation across $h = 0$. Namely, remember that $m_\beta^{\mathrm{CW}}(h)$ is the largest solution (in m) of the mean-field equation

$$\tanh(\beta m + h) = m\,. \tag{3.95}$$

A look at Figure 2.4 shows that the map $h \mapsto m_\beta^{\mathrm{CW}}(h)$, well defined for $h > 0$, can obviously be continued analytically through $h = 0$, to small negative values of h, see Figure 3.19. The continuation $\widetilde{m}_\beta^{\mathrm{CW}}(\cdot)$, for small $h < 0$, is still a solution of (3.95), but corresponds only to a *local* maximum of $m \mapsto hm - f_\beta^{\mathrm{CW}}(m)$, and thus does not represent the equilibrium value of the magnetization.

As a consequence, the pressure $\psi_\beta(h)$ can also be continued analytically through $h = 0$, $h \mapsto \widetilde{\psi}_\beta^{\mathrm{CW}}(h)$, as depicted in Figure 3.19.

Remark 3.66. If the analytic continuation can be made to reach the negative real axis $\{h \in \mathbb{R} : h < 0\}$, as in Figure 3.19, then the analytically continued pressure at such (physically relevant) values of $h < 0$ can acquire an imaginary part, and some (nonrigorous) theories predict that this imaginary component should be related to the lifetime of the corresponding metastable state. See [206]. $\diamond$

Scenario 2: Analytic Continuation Is Blocked by the Presence of Singularities. In the second scenario, there exists no analytic continuation across the imaginary axis. This happens when the singularities form a dense subset of the imaginary axis, see Figure 3.20. In such a case, $\{\mathfrak{Re}\,h = 0\}$ is called a **natural boundary** for ψ_β.

Which Scenario Occurs in the Ising Model on $\mathbb{Z}^d$, $d \geq 2$? With the exception of the (trivial) one-dimensional case, the results concerning the possibility of analytically continuing the pressure of the Ising model across ∂H^+ are largely incomplete.

In the supercritical regime $\beta < \beta_c(d)$, the pressure is differentiable at $h = 0$ and analytic continuation is expected to be possible, through any point of the imaginary

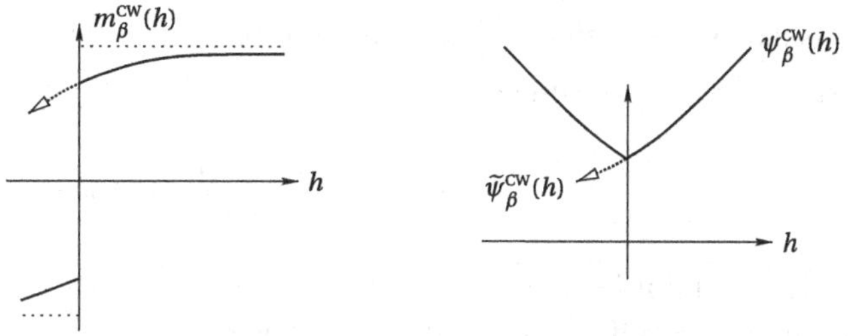

Figure 3.19. Left: the analytic continuation (dotted) of the magnetization of the Curie–Weiss model, across $h = 0$, along $h \to 0^+$. Right: the corresponding analytic continuation of the pressure. When $\beta > \beta_c$, the analytic continuation *differs* from the values of the true pressure for small $h < 0$: $\widetilde{\psi}_\beta^{\mathrm{CW}}(h) < \psi_\beta^{\mathrm{CW}}(h)$.

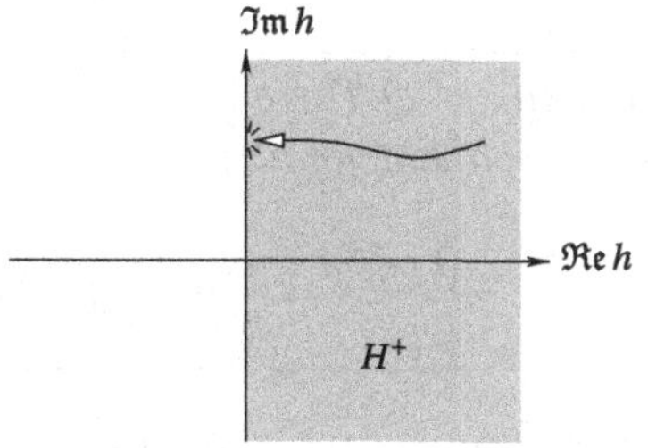

Figure 3.20. In Scenario 2, $\{\mathfrak{Re}\, h = 0\}$ is a natural boundary of the pressure: any path crossing the imaginary axis "hits" a singularity, which prevents analytic continuation.

axis. (Analyticity at $h = 0$ for the two-dimensional Ising model at any $\beta < \beta_c$ is established in [231].) For sufficiently large temperatures, a proof will be provided in Chapter 5 using the cluster expansion technique (see Exercise 5.8).

In the subcritical regime $\beta > \beta_c(d)$, the only rigorous contribution remains the study of Isakov [174], who considered the d-dimensional Ising model ($d \geq 2$) at low temperature and studied the high-order derivatives of the pressure at $h = 0$. Before stating his result, note that Theorem 3.42 allows one to use Cauchy's formula to obtain that, for all $h_0 \in H^+$,

$$\frac{\mathrm{d}^k \psi_\beta}{\mathrm{d}h^k}(h_0) = \frac{k!}{2\pi \mathrm{i}} \oint_\gamma \frac{\psi_\beta(z)}{(z - h_0)^{k+1}} \,\mathrm{d}z\,, \tag{3.96}$$

where γ is a smooth simple closed curve contained in H^+, surrounding h_0, oriented counterclockwise. Choosing γ as the circle of radius $|\,\mathfrak{Re}\, h_0|/2$ centered at h_0, we get the upper bound

$$\left| \frac{\mathrm{d}^k \psi_\beta}{\mathrm{d}h^k}(h_0) \right| \leq C^k k!\,. \tag{3.97}$$

The constant C being proportional to $1/|\,\mathfrak{Re}\, h_0|$, this upper bound provides no information on the behavior near the imaginary axis.

Isakov showed that, for all k, the kth one-sided derivative at $h = 0$,[6] $\frac{\mathrm{d}^k\psi_\beta}{\mathrm{d}h_+^k}(0)$, exists, is finite and equals

$$\frac{\mathrm{d}^k\psi_\beta}{\mathrm{d}h_+^k}(0) = \lim_{h_0\downarrow 0} \frac{\mathrm{d}^k\psi_\beta}{\mathrm{d}h^k}(h_0),$$

where the limit $h_0 \downarrow 0$ is taken along the real axis. This implies that the pressure, although not differentiable at $h = 0$, has right-derivatives of all orders at $h = 0$. Therefore, the Taylor series for the pressure at $h = 0$ exists:

$$a_0 + a_1 h + a_2 h^2 + a_3 h^3 + \cdots, \qquad \text{where } a_k = \frac{1}{k!}\frac{\mathrm{d}^k\psi_\beta}{\mathrm{d}h_+^k}(0). \tag{3.98}$$

But Isakov also obtained the following remarkable result:

Theorem 3.67. *($d \geq 2$) There exist $\beta_0 < \infty$ and $0 < A < B < \infty$, both depending on β, such that, for all $\beta \geq \beta_0$, as $k \to \infty$,*

$$A^k k!^{\frac{d}{d-1}} \leq \lim_{h_0\downarrow 0}\left|\frac{\mathrm{d}^k\psi_\beta}{\mathrm{d}h^k}(h_0)\right| \leq B^k k!^{\frac{d}{d-1}}. \tag{3.99}$$

Since $\frac{d}{d-1} > 1$, (3.99) shows that the high-order derivatives at $0 \in \partial H^+$ diverge much faster than inside H^+, as seen in (3.97). This implies in particular that the series (3.98) *diverges* for all $h \neq 0$, and therefore does not represent the function in the neighborhood of 0.[7] In other words, the pressure has a singularity at $h = 0$ and there exists no analytic continuation of ψ_β through the transition point. We will study this phenomenon in a simple toy model in Exercise 4.16.

Although this result has only been established at very low temperature, it is expected to hold for all $\beta > \beta_c$. Observe that, since $e^{(h+2\pi k\mathrm{i})\sigma_j} = e^{h\sigma_j}$, the pressure is periodic in the imaginary direction, with period 2π. The singularity at $h = 0$ therefore implies the presence of singularities at each of the points $2\pi k\mathrm{i} \in \partial H^+$.

Isakov's result was later extended to other models (see the references at the end of Section 4.12.3). But the problem of determining whether there exists some analytic continuation *around* the singularity at $h = 0$, across some interval on the imaginary axis as on Figure 3.18, is still open.

3.10.10 Metastable Behavior in Finite Systems

As explained in Section 3.10.9, the spontaneous magnetization of the Ising model at low temperatures cannot be analytically continued from negative values of h to positive values of h. Of course, this only applies in the thermodynamic limit, since the magnetization is an analytic function in a finite system. It is thus of interest

6 For $k = 1$, the one-sided derivative is the same as encountered earlier in the chapter: $\frac{\mathrm{d}\psi_\beta}{\mathrm{d}h^+}(0)$. For $k \geq 2$, the kth one-sided derivative is defined by induction.

to understand what happens, in finite systems, to the $-$ phase when h becomes positive.

To discuss this issue, let us consider the low-temperature d-dimensional Ising model in the box $\mathrm{B}(n)$ with a magnetic field h and $-$ boundary condition. When $h \leq 0$ and β is large enough, typical configurations are given by small perturbations of the ground state η^-: they consist of a large sea of $-$ spins with small islands of $+$ spins (see Exercises 3.18 and 3.19). The $-$ boundary condition is said to be **stable** in $\mathrm{B}(n)$. The situation is more interesting when $h > 0$. To get some insight, let us consider the two configurations $\omega^-, \omega^+ \in \Omega^-_{\mathrm{B}(n)}$, in which all the spins in $\mathrm{B}(n)$ take the value -1, resp. $+1$. Then, $\mathscr{H}_{\mathrm{B}(n);\beta,h}(\omega^-) - \mathscr{H}_{\mathrm{B}(n);\beta,h}(\omega^+) = -2\beta|\partial^{\mathrm{ex}}\mathrm{B}(n)| + 2h|\mathrm{B}(n)|$. We thus see that ω^- and ω^+ have the same energy if and only if

$$h = \beta\frac{|\partial^{\mathrm{ex}}\mathrm{B}(n)|}{|\mathrm{B}(n)|} = \frac{2d\beta}{|\mathrm{B}(n)|^{1/d}}\,.$$

We would thus expect that, provided that $h > 0$ satisfies

$$h|\mathrm{B}(n)|^{1/d} < 2d\beta\,, \tag{3.100}$$

the $-$ boundary condition should remain stable in $\mathrm{B}(n)$ even though there is a positive magnetic field, in the sense that typical low-temperature configurations should be small perturbations of ω^-, as on the left of Figure 3.21. In contrast, when h satisfies

$$h|\mathrm{B}(n)|^{1/d} > 2d\beta\,, \tag{3.101}$$

one would expect the $+$ phase to invade the box, with only a narrow layer of $-$ phase along the boundary of $\mathrm{B}(n)$, as on the right of Figure 3.21. In this case, the $-$ boundary condition is **unstable**.

Of course, the previous argument is very rough, taking into account only constant configurations inside $\mathrm{B}(n)$, and one should expect the above claims to be valid only for extremely low temperatures. Nevertheless, in a more careful analysis [296], Schonmann and Shlosman showed that the above remains qualitatively true for the two-dimensional Ising model at any $\beta > \beta_c(2)$: there exists $c = c(\beta) \in (0, \infty)$ such that the $-$ boundary condition is stable as long as $h < c|\mathrm{B}(n)|^{-1/2}$, while it becomes unstable when $h > c|\mathrm{B}(n)|^{-1/2}$. In the latter case, the macroscopic shape of the region occupied by the $+$ phase can be characterized precisely (showing, in particular, that macroscopic regions remain occupied by the $-$ phase near the four corners of $\mathrm{B}(n)$, as long as h is not too large). In particular, these results show that the magnetization at the center of the box satisfies, for large n and small $|h|$,

$$\langle\sigma_0\rangle^-_{\mathrm{B}(n);\beta,h} \cong \begin{cases} -m^* & \text{if } h < c|\mathrm{B}(n)|^{-1/2}\,, \\ +m^* & \text{if } h > c|\mathrm{B}(n)|^{-1/2}\,. \end{cases}$$

In this sense, the negative-h magnetization can be "continued" into the positive-h region, but only as long as $h < c|\mathrm{B}(n)|^{-1/2}$. The fact that the size of the latter interval vanishes as $n \to \infty$ explains why the above discussion does not contradict the absence of analytic continuation in the thermodynamic limit.

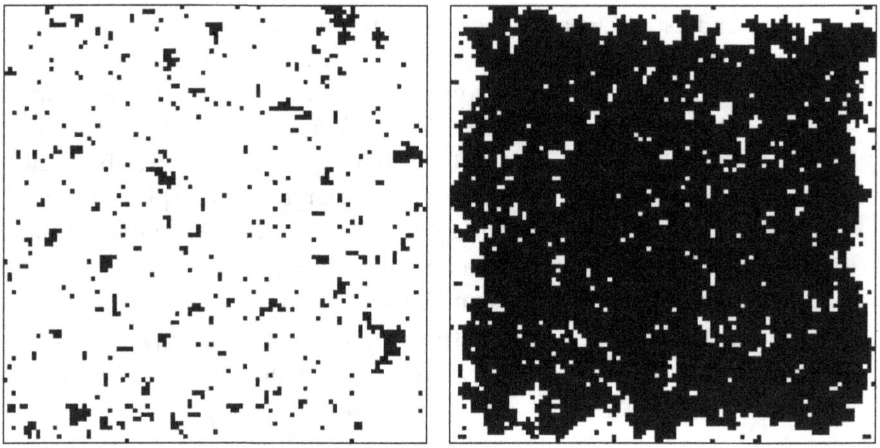

Figure 3.21. Typical low-temperature configurations of the two-dimensional Ising model in a box of sidelength 100 with $-$ boundary condition and magnetic field $h > 0$. Left: for small positive values of h (depending on β and the size of the box), typical configurations are small perturbations of the ground state η^-, even though the $-$ phase is thermodynamically unstable, the cost of creating a droplet of $+$ phase being too large. Right: for larger values of h, the $+$ phase invades the box, while the unstable $-$ phase is restricted to a layer along the boundary, where it is stabilized by the boundary condition. Partial information on the size of this layer (in a slightly different geometrical setting) can be found in [346].

3.10.11 Critical Phenomena

As explained in this chapter, a first-order phase transition occurs at each point of the line $\{(\beta, h) \in \mathbb{R}_{\geq 0} \times \mathbb{R} : \beta > \beta_c(d),\, h = 0\}$, where $\beta_c(d) \in (0, \infty)$ for all $d \geq 2$. One of the manifestations of these first-order phase transitions is the discontinuity of the magnetization density at $h = 0$:

$$\lim_{h \downarrow 0} \{m(\beta, h) - m(\beta, -h)\} = 2m^*(\beta) > 0$$

for all $\beta > \beta_c(d)$. It can be shown [7, 8, 352] (for $d = 2$, remember (1.51) and Figure 1.10) that $\beta \mapsto m^*(\beta)$ is decreasing and vanishes continuously as $\beta \downarrow \beta_c$. Therefore, since $m^*(\beta) = 0$ for all $\beta \leq \beta_c(d)$, the magnetization density $m(\beta, h)$ (and thus the pressure) cannot be analytic at the point $(\beta_c(d), 0)$. The corresponding phase transition, however, is not of first order anymore: it is said to be **continuous** and the point $(\beta_c(d), 0)$ is said to be a **critical point**.

As we have already mentioned in Section 2.5.3, the behavior of a system at a critical point displays remarkable features. In particular, many quantities of interest have singular behavior, with qualitative features that depend only on rough properties of the model, such as its spatial dimensionality, its symmetries and the short- or long-range nature of its interactions. Models can then be distributed into large families with the same critical behavior, known as **universality classes**.

Among the characteristic features that are used to determine the universality class to which a model belongs, an important role is played by the critical exponents. The definitions of several of these have been given for the Curie–Weiss model in Section 2.5.3 and can be used also for the Ising model (using the corresponding quantities). For the Ising model, we have gathered these exponents in Table 3.1:

Table 3.1. Some critical exponents of the Ising model.

	$d=2$	$d=3$	$d \geq 4$
α	0	0.110(1)	0
b	1/8	0.3265(3)	1/2
γ	7/4	1.2372(5)	1
δ	15	4.789(2)	3

The exponents given are rigorously only known to hold when $d = 2$ [59, 259, 352] and when $d \geq 4$ [4, 7, 9, 318]. The values given for $d = 3$ are taken from the review [267]; much more precise estimates are now available [194].

Observe that the exponents become independent of the dimension as soon as $d \geq 4$. The dimension $d_u \stackrel{\text{def}}{=} 4$ is known as the **upper critical dimension**. Above d_u, the exponents take the same values as in the Curie–Weiss model (see Section 2.5.3), in line with the interpretation of the mean-field approximation as the limit of the model as $d \to \infty$ (see Section 2.5.4). Such a behavior is expected to be general, but with a value of d_u depending on the universality class.

At a heuristic level, the core reason for this universality can be traced back to the divergence of the **correlation length** at the critical point. This measures the range over which spins are strongly correlated. In the Ising model, the correlation length ξ is such that

$$\langle \sigma_0 ; \sigma_i \rangle^+_{\beta,h} \stackrel{\text{def}}{=} \langle \sigma_0 \sigma_i \rangle^+_{\beta,h} - \langle \sigma_0 \rangle^+_{\beta,h} \langle \sigma_i \rangle^+_{\beta,h} \sim e^{-\|i\|_2/\xi},$$

for all i for which $\|i\|_2$ is large enough. More precisely,

$$\xi(\beta, h)(\mathbf{n}) \stackrel{\text{def}}{=} \lim_{k\to\infty} \frac{-k}{\log \langle \sigma_0 ; \sigma_{[k\mathbf{n}]} \rangle^+_{\beta,h}},$$

where $\mathbf{n}$ is a unit vector in $\mathbb{R}^d$ and we have written $[x] \stackrel{\text{def}}{=} (\lfloor x(1) \rfloor, \dots, \lfloor x(d) \rfloor)$ for any $x = (x(1), \dots, x(d)) \in \mathbb{R}^d$.

In the Ising model, it is expected that the correlation length is finite (in all directions) for all $(\beta, h) \neq (\beta_c(d), 0)$. This has been proved when $d = 2$ [239]; in higher dimensions, this is only known when either $\beta < \beta_c(d)$ [5] or when β is large enough (we will prove it in Theorem 5.16), while it is known to diverge as $\beta \uparrow \beta_c(d)$ [238].

Under the assumption that there is only one relevant length scale close to the critical point, the divergence of the correlation length implies the absence of any characteristic length scale at the critical point: at this point, the system is expected to be invariant under a change of scale. Based on such ideas, physicists have developed a nonrigorous, but powerful, framework in which this picture can be substantiated and which allows the approximate determination of the critical behavior: the **renormalization group**.

Let us briefly describe the idea in a simple case. We define a mapping $T\colon \Omega \to \Omega$ as follows: given $\omega \in \Omega$, $\omega' \stackrel{\text{def}}{=} T(\omega)$ is defined by

$$\omega'_i \stackrel{\text{def}}{=} \begin{cases} +1 & \text{if } \sum_{j\in 3i+\mathrm{B}(1)} \omega_j > 0\,, \\ -1 & \text{if } \sum_{j\in 3i+\mathrm{B}(1)} \omega_j < 0\,. \end{cases}$$

In other words, we partition $\mathbb{Z}^d$ into cubic blocks of sidelength 3, and replace the 3^d spins in each of these blocks by a single spin, equal to $+1$ if the magnetization in the block is positive, and to -1 otherwise. This transformation is called a **majority transformation**.

One can then iterate this transformation. Figure 3.22 shows the first two iterations starting from three different initial configurations, corresponding to the two-dimensional Ising model at $h = 0$ and at three values of β: slightly subcritical ($\beta < \beta_c(2)$), critical ($\beta = \beta_c(2)$) and slightly supercritical ($\beta > \beta_c(2)$). At first sight, it looks as though the transformation corresponds to decreasing β in the first case, keeping it critical in the second and increasing it in the third. Of course, the situation cannot be that simple: the probability distribution describing the transformed configuration clearly does *not* correspond to an Ising model anymore. Nevertheless, it might correspond to a model with additional interactions. One could then consider the action of this transformation in the space of all Hamiltonians. The idea is then the following: this transformation has two stable fixed points corresponding to infinite and zero temperatures, which attracts all initial states with $\beta < \beta_c(d)$, respectively $\beta > \beta_c(d)$. In addition it has an unstable fixed point corresponding to the critical point. This can, heuristically, be understood in terms of the correlation length: since each application of the transformation corresponds roughly to a zoom by a factor 3, the correlation length is divided by 3 at each step. As the number of iterations grows, the correlation length converges to 0, which corresponds to $\beta = 0$ or $\beta = \infty$, *except* when it was initially equal to infinity, in which case it remains infinite; this case corresponds to the critical point. An analysis of the behavior of the transformation close to the unstable fixed point then provides information on the critical behavior of the original system.

These ideas are compelling but, at least in this naive form, the above procedure is known to be problematic from a mathematical point of view; see [343] for a detailed discussion or the comments in Section 6.14.2. Nevertheless, more sophisticated versions do allow physicists to obtain remarkably accurate estimates of critical exponents. Moreover, the *philosophy* of the renormalization group has played a key role in several rigorous investigations (even outside the realm of critical phenomena).

From a rigorous point of view, the analysis of critical systems is usually done using alternative approaches, limited to rather specific classes of models and mostly in two situations: systems above their upper critical dimensions and two-dimensional systems. Since research in these fields is still very actively developing, we will not discuss them any further. Instead, we list several good sources where these topics are discussed at length; these should be quite accessible if the reader is familiar with the content of the present book.

A first approach to critical phenomena in lattice spin systems and (Euclidean) quantum field theory, based on random walk (or random surfaces) representations, is given in considerable detail in the monograph [102] by Fernández, Fröhlich and Sokal; it provides a thorough discussion of scaling limits, inequalities for critical exponents, the validity of mean-field exponents above the upper critical dimension, etc.

A second approach is described in the books by Brydges [57] and Mastropietro [234]. It is based on a rigorous implementation of a version of the renormalization

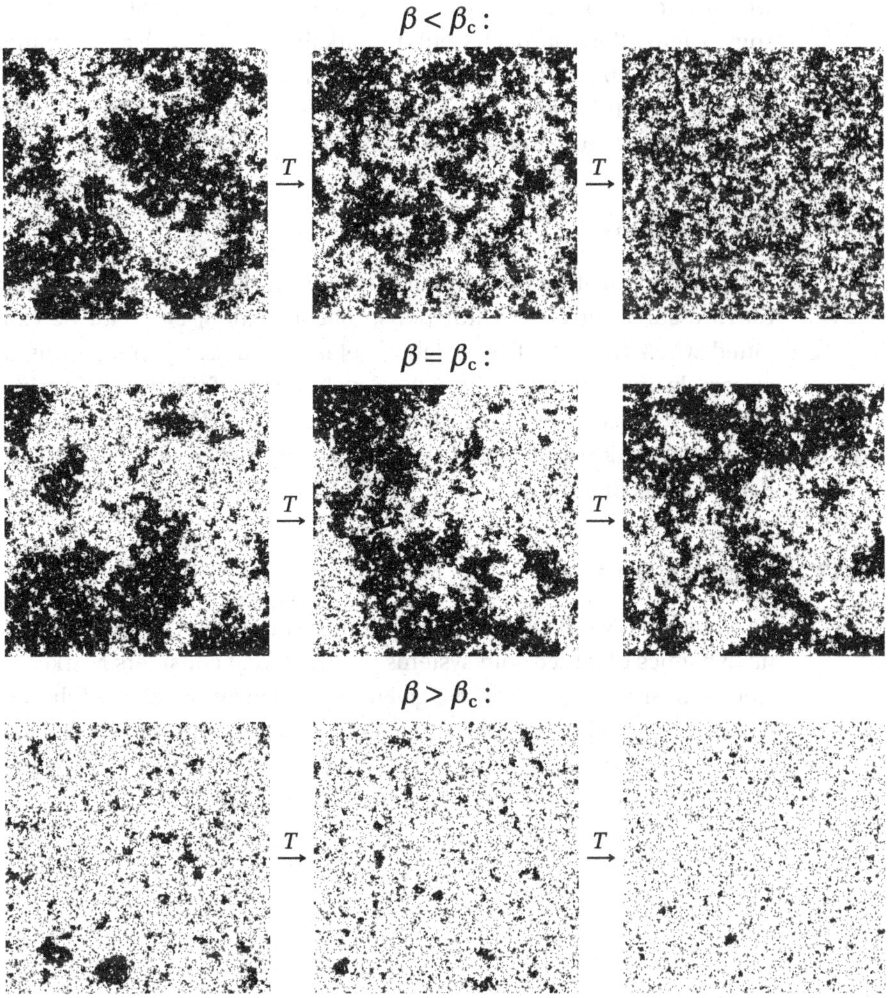

Figure 3.22. Two iterations of the majority transformation at different temperatures.

group. These books cover both the perturbative and nonperturbative renormalization group approaches from the functional-integral point of view and cover a broad spectrum of applications.

A third approach is presented in the book [315] by Slade. This provides an introduction to the lace expansion, a powerful tool allowing one to obtain precise information on the critical behavior of systems above their upper critical dimension, at least for quantities admitting representation in terms of self-interacting random paths.

A fourth approach, at the base of many of the recent developments of this field, is based on the Schramm–Löwner evolution (SLE). This approach to critical phenomena is restricted to two-dimensional systems, but yields extremely detailed and complete information when it is applicable. An introduction to SLE can be found in the book [210] and in lecture notes by Werner [349] and Lawler [208]. Combined with discrete complex analytic methods and specific graphical representations of spin systems, this approach yields remarkable results, such as the conformal invariance of the scaling limit, explicit expressions for the critical exponents, etc. Good references on this topic are the books by Werner [350] and Duminil-Copin [91, 92], as well as the lecture notes by Duminil-Copin and Smirnov [94].

3.10.12 Exact Solution

A remarkable feature of the planar Ising model is that many quantities of interest (pressure, correlation functions, magnetization, etc.) can be explicitly computed when $h = 0$. The insights yielded by these computations have had an extremely important impact on the development of the theory of critical phenomena. There exist today many different approaches. The interested reader can find more information on this topic in the books by McCoy and Wu [239], Baxter [17] or Palmer [261], for example.

3.10.13 Stochastic Dynamics

Another topic we have only barely touched upon is the analysis of the stochastic dynamics of lattice spin systems. In these, one considers Markov chains on Ω, whose invariant measures are given by the corresponding Gibbs measures. We made use of such a dynamics in Section 3.10.3 in the simplest case of the finite-volume Ising model. The book [225] by Liggett and the lecture notes [232] by Martinelli provide good introductions to this topic.

4 Liquid–Vapor Equilibrium

In this chapter, we develop a rigorous theory of the liquid–vapor equilibrium. In particular, we will provide a version of the van der Waals–Maxwell theory of condensation, which was briefly presented in Section 1.1.6. For that, we will study the lattice gas and construct the two main thermodynamic quantities associated with it, namely the *free energy* and the *pressure*, under general two-body interactions. These will be studied for different types of microscopic interactions:

1. When particles do not interact with each other except through exclusion, the thermodynamic quantities can be computed explicitly (as in the ideal gas). This will be done in Section 4.7, where the **hard-core gas** will be studied in detail.
2. In the **nearest-neighbor** gas (Section 4.8), only neighboring particles attract each other, and a direct link with the Ising model, as introduced in Chapter 3, can be made. A satisfactory qualitative thermodynamic description of the condensation phenomenon will then be obtained by importing results from Chapter 3.
3. The **van der Waals gas** (Section 4.9) is the mean-field version of the lattice gas, and is a reformulation in this language of the Curie–Weiss model of Chapter 2. As we will see, this model displays a number of unphysical properties. Nevertheless, it turns out that Maxwell's construction appears naturally as a consequence of the Legendre transform.
4. Finally, we consider **Kać interactions** (Section 4.10), in which a small parameter $\gamma > 0$ is used to tune the range of the interaction. By sending the range of the interaction to infinity, in the so-called **van der Waals limit**, we will make a bridge between the two previous models, restoring the correct behavior of the thermodynamic potentials, and put Maxwell's construction on rigorous grounds.

In contrast to most other chapters, this one focuses more on the study of the *thermodynamic potentials*, free energy and pressure, rather than on the Gibbs distribution and its sensitivity to boundary conditions. Typical configurations under the relevant Gibbs distributions will nevertheless be briefly discussed in Sections 4.6 and 4.12.1.

Remark 4.1. In order to ease the physical interpretation of the results obtained, we will adopt, in this chapter, the convention used in physics: namely, we will keep

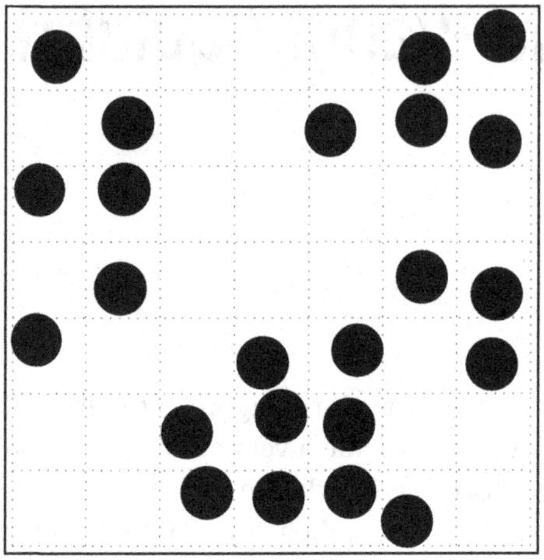

Figure 4.1. In the lattice gas approximation, the vessel is divided into imaginary cells, such that each cell can contain at most one particle.

the inverse temperature β outside the Hamiltonian, and will add a multiplicative constant $\frac{1}{\beta}$ in front of the free energy and pressure. $\diamond$

We will rely on some results on real convex functions; these are collected in Appendix B.2.

4.1 Lattice Gas Approximation

The lattice gas was introduced informally in Chapter 1, Section 1.2.4. Consider a gas contained in a vessel. We will build a model based on van der Waals' two main assumptions concerning the interactions between the particles that compose the gas: [1]

- *repulsion*: at short distances, particles interact in a repulsive way (as small, impenetrable spheres);
- *attraction*: attractive forces act at larger distances.

In order to avoid the many technicalities inherent to the continuum (and because this book is about lattice models), we will introduce a natural discretization. Although it might appear as a significant departure from reality, we will see that it leads to satisfactory results and allows a good qualitative understanding of the corresponding phenomena.

In the **lattice gas approximation**, the vessel is partitioned into imaginary microscopic cubic cells of sidelength 1 (in some suitable units), and it is assumed that *each cell can be either empty or occupied by exactly one particle*; see Figure 4.1. Since it prevents particles from overlapping, this assumption embodies the short-range repulsive part of the interaction. Each cell is identified with a vertex $i \in \Lambda$, where

Λ is some finite subset of $\mathbb{Z}^d$. As an additional simplification, we only keep track of the cells that are occupied, and not of the exact position of each particle inside its cell.

Turning to the attractive part of the interaction, we assume that a pair of particles in cells i and j contributes an amount $-K(i,j) \le 0$ to the total energy, where $K(i,j)$ decreases to zero when $\|j-i\|_2 \to \infty$. We also assume that $K(\cdot,\cdot)$ is **translation invariant**: $K(i,j) = K(0,j-i)$, and **symmetric**: $K(0,-j) = K(0,j)$. The total interaction energy of the system is thus

$$-\sum_{\substack{\{i,j\}\subset\Lambda\\ i \text{ and } j \text{ occupied}}} K(i,j)\,.$$

Later, some specific choices for $K(i,j)$ will be considered.

It is natural to assign to every cell $i \in \Lambda$ its **occupation number** (notice that in Chapter 3 the corresponding random variables were denoted n_i):

$$\eta_i \stackrel{\text{def}}{=} \begin{cases} 1 & \text{if } i \text{ contains a particle,}\\ 0 & \text{otherwise.}\end{cases}$$

A **configuration** of the lattice gas in the vessel is therefore given by the set of occupation numbers, $\eta = (\eta_i)_{i\in\Lambda}$, and is thus an element of $\{0,1\}^\Lambda$. Using occupation numbers, one can consider the interaction between pairs of cells i and j, $-K(i,j)\eta_i\eta_j$, which can be nonzero only if i and j both contain a particle.

Definition 4.2. Let $\Lambda \Subset \mathbb{Z}^d$, $\eta \in \{0,1\}^\Lambda$. The **Hamiltonian of the lattice gas in Λ** is

$$\mathscr{H}_{\Lambda;K}(\eta) \stackrel{\text{def}}{=} -\sum_{\{i,j\}\subset\Lambda} K(i,j)\eta_i\eta_j. \tag{4.1}$$

We will actually be mostly interested in systems with **finite-range** interactions, that is, those for which

$$r \stackrel{\text{def}}{=} \inf\{R \ge 0 : K(i,j) = 0 \text{ if } \|j-i\|_2 > R\} < \infty\,.$$

More generally, in order to have a well-defined thermodynamic limit, we will need to assume that the maximal interaction between a particle and the rest of the system is bounded. In our case, this condition can be written

$$\kappa \stackrel{\text{def}}{=} \sup_{i\in\mathbb{Z}^d}\sum_{j\ne i} K(i,j) = \sum_{j\ne 0} K(0,j) < \infty\,,$$

since $-\kappa$ represents the interaction of a particle with the rest of an infinite system in which each other cell contains a particle.

The **number of particles** in a configuration η can be expressed as

$$N_\Lambda(\eta) \stackrel{\text{def}}{=} \sum_{i\in\Lambda}\eta_i\,,$$

and the **empirical density** is defined by

$$\rho_\Lambda \stackrel{\text{def}}{=} \frac{N_\Lambda}{|\Lambda|}.$$

When studying the lattice gas in a large vessel, we will assume either that the number of particles is fixed (describing a fluid confined to some hermetically sealed container), or that this number can fluctuate (the system can exchange particles with an external reservoir). As explained in Chapter 1, these two descriptions of the gas are called, respectively, **canonical** and **grand canonical**. They will both be associated with a thermodynamic potential (respectively, the free energy and the pressure), which will contain the relevant information about the thermodynamic behavior of the system.

Remark 4.3. Except in our discussion in Section 4.12.1, we will only consider the free boundary condition in this chapter. The reason for this is that we are mostly interested in thermodynamic potentials, which turn out to be insensitive to the chosen boundary condition, for precisely the same reason as in Chapter 3. ◇

4.2 Canonical Ensemble and Free Energy

In the canonical ensemble (Section 1.2.2), the number of particles is fixed.

Definition 4.4. Let $\Lambda \Subset \mathbb{Z}^d$, $N \in \{0, 1, 2, \dots, |\Lambda|\}$. The **canonical Gibbs distribution at inverse temperature β** is the probability distribution on $\{0,1\}^\Lambda$ defined by

$$\nu_{\Lambda;\beta,N}(\eta) \stackrel{\text{def}}{=} \frac{\exp(-\beta\mathscr{H}_{\Lambda;K}(\eta))}{\mathbf{Q}_{\Lambda;\beta,N}} \mathbf{1}_{\{N_\Lambda(\eta)=N\}}, \tag{4.2}$$

where the **canonical partition function** is defined by

$$\mathbf{Q}_{\Lambda;\beta,N} \stackrel{\text{def}}{=} \sum_{\substack{\eta\in\{0,1\}^\Lambda: \\ N_\Lambda(\eta)=N}} \exp(-\beta\mathscr{H}_{\Lambda;K}(\eta)). \tag{4.3}$$

The thermodynamic potential describing an infinite system of fixed density at equilibrium is the *free energy*. It is convenient to first define the **free energy in a finite region** Λ as a function of a continuous parameter $\rho \in [0,1]$. For that, assume first that ρ is such that $\rho|\Lambda| \in \{0, 1, \dots, |\Lambda|\}$ and let

$$f_{\Lambda;\beta}(\rho) \stackrel{\text{def}}{=} \frac{-1}{\beta|\Lambda|} \log \mathbf{Q}_{\Lambda;\beta,\rho|\Lambda|}. \tag{4.4}$$

This defines a function on $\{0, \frac{1}{|\Lambda|}, \frac{2}{|\Lambda|}, \dots, \frac{|\Lambda|-1}{|\Lambda|}, 1\}$, which can be extended to a continuous function on $[0,1]$ by interpolating linearly on each interval $[\frac{k}{|\Lambda|}, \frac{k+1}{|\Lambda|}]$.

When taking the thermodynamic limit $\Lambda \Uparrow \mathbb{Z}^d$ (for a definition, see page 86) in the canonical ensemble, the number of particles will increase with the size of the system, $N \to \infty$, but the density of particles will remain constant. To simplify, we will not consider the thermodynamic limit along general sequences Λ_n that converge in the sense of van Hove, but rather use everywhere sequences of **parallelepipeds**, that is, sets of the form $\big([a_1, b_1]\times[a_2, b_2]\times\cdots\times[a_d, b_d]\big)\cap\mathbb{Z}^d$. Arguments similar to those used in Section 3.2.2 can be used to remove this restriction. We denote by $\mathscr{R}$ the collection of all parallelepipeds.

Theorem 4.5. *Let $\mathscr{R} \ni \Lambda_n \Uparrow \mathbb{Z}^d$. Let $\rho \in [0,1]$ and $N_n \in \mathbb{N}$ be such that $\frac{N_n}{|\Lambda_n|} \to \rho$. The limit*

$$f_\beta(\rho) \stackrel{\text{def}}{=} \lim_{n\to\infty} f_{\Lambda_n;\beta}(\rho) \tag{4.5}$$

*exists, does not depend on the choice of the sequences $(\Lambda_n)_{n\geq1}$ and $(N_n)_{n\geq1}$, and is called the **free energy**. Moreover, the convergence is uniform on compact subsets of $(0,1)$ and $\rho \mapsto f_\beta(\rho)$ is convex and continuous.*

To prove Theorem 4.5, the first ingredient is the following basic property of the canonical partition function:

Lemma 4.6. *Consider two disjoint regions $\Lambda, \Lambda' \Subset \mathbb{Z}^d$. If $N \leq |\Lambda|$ and $N' \leq |\Lambda'|$, then*

$$\mathbf{Q}_{\Lambda\cup\Lambda';\beta,N+N'} \geq \mathbf{Q}_{\Lambda;\beta,N}\mathbf{Q}_{\Lambda';\beta,N'} \,. \tag{4.6}$$

Proof In $\mathbf{Q}_{\Lambda\cup\Lambda';\beta,N+N'}$, we obtain a lower bound by keeping only the configurations in which Λ contains N particles and Λ' contains N' particles. Moreover, since $K(i,j) \geq 0$, we can ignore the interactions between pairs of particles at vertices $i \in \Lambda, j \in \Lambda'$. After summing separately over the configurations in Λ and Λ', we get (4.6). □

The second ingredient is the following "continuity" property of the partition function with respect to the number of particles in the vessel:

Lemma 4.7. *Let $\Lambda \Subset \mathbb{Z}^d$ and $N \in \{0,1,\dots,|\Lambda|-1\}$. Then,*

$$\tfrac{|\Lambda|-N}{N+1}\mathbf{Q}_{\Lambda;\beta,N} \leq \mathbf{Q}_{\Lambda;\beta,N+1} \leq e^{\beta\kappa}\tfrac{|\Lambda|-N}{N+1}\mathbf{Q}_{\Lambda;\beta,N} \,. \tag{4.7}$$

Proof Observe that

$$\begin{aligned}\mathbf{Q}_{\Lambda;\beta,N+1} &= \sum_{\substack{\eta\in\{0,1\}^\Lambda:\\ N_\Lambda(\eta)=N+1}} \exp\big(-\beta\mathscr{H}_{\Lambda;K}(\eta)\big)\\ &= \frac{1}{N+1}\sum_{\substack{\eta\in\{0,1\}^\Lambda:\\ N_\Lambda(\eta)=N}}\ \sum_{\substack{\eta'\in\{0,1\}^\Lambda:\\ \eta'_i\geq\eta_i,\forall i\\ N_\Lambda(\eta')=N+1}} \exp\big(-\beta\mathscr{H}_{\Lambda;K}(\eta')\big)\,.\end{aligned}$$

Since $\mathscr{H}_{\Lambda;K}(\eta) - \kappa \le \mathscr{H}_{\Lambda;K}(\eta') \le \mathscr{H}_{\Lambda;K}(\eta)$ and since there are exactly $|\Lambda| - N$ terms in the sum over η', this proves (4.7). □

Exercise 4.1. Using Lemma 4.7, show that, for all $\epsilon > 0$, when Λ is large, $\mathbf{Q}_{\Lambda;\beta,N} \le \epsilon \mathbf{Q}_{\Lambda;\beta,N+1}$ if $\frac{N}{|\Lambda|}$ is sufficiently small, and $\mathbf{Q}_{\Lambda;\beta,N+1} \le \epsilon \mathbf{Q}_{\Lambda;\beta,N}$ if $\frac{N}{|\Lambda|}$ is sufficiently close to 1.

Proof of Theorem 4.5. For simplicity, we do not include β in the notation of the partition functions. Let $\rho \in [0,1]$. We will first take for N_n the particular sequence $N_n = \lceil \rho |\Lambda_n| \rceil$, and show the existence of the limit

$$f_\beta(\rho) = \lim_{n\to\infty} \frac{-1}{\beta |\Lambda_n|} \log \mathbf{Q}_{\Lambda_n;\lceil \rho|\Lambda_n|\rceil} . \tag{4.8}$$

The boundary cases $\rho = 0$, $\rho = 1$ can be computed explicitly:

$$f_\beta(0) = 0 , \qquad f_\beta(1) = -\tfrac{\kappa}{2} . \tag{4.9}$$

For intermediate densities, we use a subadditivity argument. For convenience, we write (4.7) as follows:

$$c^{-1}\mathbf{Q}_{\Lambda;N} \le \mathbf{Q}_{\Lambda;N+1} \le c\,\mathbf{Q}_{\Lambda;N} , \tag{4.10}$$

for some $c > 1$ that can be chosen uniformly if $\frac{N}{|\Lambda|}$ belongs to some closed interval $[a,b] \subset (0,1)$.

Let $\rho \in (0,1)$. For all disjoint $\Lambda', \Lambda'' \in \mathscr{R}$, with $\Lambda = \Lambda' \cup \Lambda'' \in \mathscr{R}$, we have $\lceil \rho|\Lambda|\rceil \ge \lceil \rho|\Lambda'|\rceil + \lceil \rho|\Lambda''|\rceil - 2$. Therefore, applying (4.10) twice, followed by (4.6),

$$\mathbf{Q}_{\Lambda;\lceil\rho|\Lambda|\rceil} \ge c^{-2}\mathbf{Q}_{\Lambda';\lceil\rho|\Lambda'|\rceil}\mathbf{Q}_{\Lambda'';\lceil\rho|\Lambda''|\rceil} .$$

It follows that the numbers $a(\Lambda) \stackrel{\text{def}}{=} -\log(c^{-2}\mathbf{Q}_{\Lambda;\lceil\rho|\Lambda|\rceil})$ enjoy the following subadditivity property:

$$a(\Lambda' \cup \Lambda'') \le a(\Lambda') + a(\Lambda'') .$$

Moreover, these numbers are translation invariant: $a(\Lambda + i) = \Lambda$. This implies (see Lemma B.6) that

$$\lim_{n\to\infty} \frac{a(\Lambda_n)}{|\Lambda_n|} \quad \text{exists and equals} \quad \inf_{\Lambda\in\mathscr{R}} \frac{a(\Lambda)}{|\Lambda|} .$$

This shows the existence of the limit in (4.8) and also provides the following useful upper bound, valid for all $\Lambda \in \mathscr{R}$:

$$\mathbf{Q}_{\Lambda;\lceil\rho|\Lambda|\rceil} \le c^2 e^{-\beta f_\beta(\rho)|\Lambda|} . \tag{4.11}$$

Remark 4.8. Before going further, let us derive some simple bounds on $f_\beta(\rho)$, which we will need later. First, we can bound the energy of each configuration η appearing in $\mathbf{Q}_{\Lambda;\lceil\rho|\Lambda|\rceil}$ (everywhere below, $\Lambda \in \mathscr{R}$):

$$-\beta\mathscr{H}_{\Lambda;K}(\eta) = \tfrac{1}{2}\beta \sum_{i\in\Lambda} \eta_i \sum_{\substack{j\in\Lambda\\ j\ne i}} K(i,j)\eta_j \le \tfrac{1}{2}\beta\kappa\lceil\rho|\Lambda|\rceil ,$$

which gives

$$\mathbf{Q}_{\Lambda;\lceil\rho|\Lambda|\rceil} \le e^{\frac{1}{2}\beta\kappa\lceil\rho|\Lambda|\rceil}\binom{|\Lambda|}{\lceil\rho|\Lambda|\rceil}. \tag{4.12}$$

Approximating the combinatorial factor using Stirling's formula as in (B.2), we can write, when $|\Lambda|$, N and $|\Lambda|-N$ are large,

$$\binom{|\Lambda|}{N} = \frac{1+o(1)}{\sqrt{2\pi\frac{N}{|\Lambda|}(1-\frac{N}{|\Lambda|})}}\Big\{\big(\tfrac{N}{|\Lambda|}\big)^{\frac{N}{|\Lambda|}}\big(1-\tfrac{N}{|\Lambda|}\big)^{1-\frac{N}{|\Lambda|}}\Big\}^{-|\Lambda|}. \tag{4.13}$$

Therefore, letting

$$s^{\text{l.g.}}(\rho) \stackrel{\text{def}}{=} -\rho\log\rho - (1-\rho)\log(1-\rho), \tag{4.14}$$

using (4.12) and taking the thermodynamic limit, we obtain

$$-\tfrac{1}{2}\kappa\rho - \tfrac{1}{\beta}s^{\text{l.g.}}(\rho) \le f_\beta(\rho) \le 0. \tag{4.15}$$

This bound implies in particular that $f_\beta(\rho)$ is finite, since $\kappa<\infty$. One can also use $-\beta\mathscr{H}_{\Lambda;K}(\eta)\ge 0$, which gives

$$f_\beta(\rho) \le -\tfrac{1}{\beta}s^{\text{l.g.}}(\rho). \tag{4.16}$$

Alternatively, one can bound the partition function from below by keeping a single configuration,

$$\mathbf{Q}_{\Lambda;\lceil\rho|\Lambda|\rceil} \ge e^{-\beta\mathscr{H}_{\Lambda;K}(\eta_*)}. \tag{4.17}$$

Exercise 4.2. Show that there is a configuration η_*, contributing to $\mathbf{Q}_{\Lambda;\lceil\rho|\Lambda|\rceil}$, such that

$$\mathscr{H}_{\Lambda;K}(\eta_*) = -\tfrac{1}{2}\kappa\rho|\Lambda| + o(|\Lambda|). \tag{4.18}$$

Using (4.18) in (4.17) gives

$$f_\beta(\rho) \le -\tfrac{1}{2}\kappa\rho. \tag{4.19}$$

◇

Let us now assume that $(N_n)_{n\ge1}$ is an arbitrary sequence satisfying $\frac{N_n}{|\Lambda_n|}\to\rho$. We can again use (4.10) repeatedly and get, for large n,

$$c^{-|N_n-\lceil\rho|\Lambda_n|\rceil|}\mathbf{Q}_{\Lambda_n;\lceil\rho|\Lambda_n|\rceil} \le \mathbf{Q}_{\Lambda_n;N_n} \le c^{|N_n-\lceil\rho|\Lambda_n|\rceil|}\mathbf{Q}_{\Lambda_n;\lceil\rho|\Lambda_n|\rceil}.$$

Since $\frac{N_n-\lceil\rho|\Lambda_n|\rceil}{|\Lambda_n|}\to 0$, this shows that the limit in (4.5) exists and coincides with the one in (4.8).

Let $I=[a,b]\subset(0,1)$. Using (4.7) for $\frac{N}{|\Lambda|},\frac{N+1}{|\Lambda|}\in I$,

$$\big|f_{\Lambda;\beta}\big(\tfrac{N+1}{|\Lambda|}\big) - f_{\Lambda;\beta}\big(\tfrac{N}{|\Lambda|}\big)\big| \le \tfrac{1}{|\Lambda|}\big\{\kappa + \tfrac{1}{\beta}\sup_{\rho\in I}\log(\tfrac{1-\rho}{\rho})\big\}.$$

From this, one easily deduces the existence of $C=C(\beta,I)>0$ such that, for all $\Lambda\Subset\mathbb{Z}^d$,

$$|f_{\Lambda;\beta}(\rho) - f_{\Lambda;\beta}(\rho')| \leq C|\rho - \rho'|, \qquad \forall \rho, \rho' \in I. \tag{4.20}$$

Combined with the already established pointwise convergence, (4.20) implies uniform convergence on I. Moreover, the limiting function $f_\beta(\rho)$ is continuous (actually, C-Lipschitz) on I. Using (4.15)-(4.19) yields $\lim_{\rho\downarrow 0} f_\beta(\rho) = 0$ and $\lim_{\rho\uparrow 1} f_\beta(\rho) = -\frac{\kappa}{2}$, which by (4.9) guarantees continuity at 0 and 1.

To show that f_β is convex, we fix $\rho_1, \rho_2 \in (0,1)$ and consider the sequence of cubes $D_k = \{1,2,3,\dots,2^k\}^d$. For each k, D_{k+1} is the union of 2^d translates of D_k, denoted $D_k^{(1)},\dots,D_k^{(2^d)}$. We split these boxes into two groups, each subgroup containing $2^d/2$ boxes. Putting $\lceil \rho_1 |D_k| \rceil$ particles in each box of the first group and $\lceil \rho_2 |D_k| \rceil$ particles in each box of the second group, and using translation invariance,

$$\mathbf{Q}_{D_{k+1};\lceil \frac{\rho_1+\rho_2}{2}|D_{k+1}|\rceil} \geq c^{-2^d} \{\mathbf{Q}_{D_k;\lceil \rho_1 |D_k|\rceil}\}^{2^d/2} \{\mathbf{Q}_{D_k;\lceil \rho_2 |D_k|\rceil}\}^{2^d/2}.$$

This implies, after letting $k \to \infty$,

$$f_\beta\left(\tfrac{\rho_1+\rho_2}{2}\right) \leq \tfrac{1}{2}\{f_\beta(\rho_1) + f_\beta(\rho_2)\}. \tag{4.21}$$

Convexity of $f_\beta(\rho)$ thus follows from its continuity (see Proposition B.9). □

4.3 Grand Canonical Ensemble and Pressure

In the grand canonical ensemble (Section 1.2.3), the system can exchange particles with an external reservoir of fixed chemical potential μ (and inverse temperature β).

Remark 4.9. In this chapter, the letter μ always denotes the chemical potential, *not* a probability measure. ◇

Definition 4.10. Let $\mu \in \mathbb{R}$. The **grand canonical Gibbs distribution** at inverse temperature β is the probability distribution on $\{0,1\}^\Lambda$ defined by

$$\nu_{\Lambda;\beta,\mu}(\eta) \stackrel{\text{def}}{=} \frac{\exp(-\beta\{\mathscr{H}_{\Lambda;K}(\eta) - \mu N_\Lambda(\eta)\})}{\Theta_{\Lambda;\beta,\mu}}, \tag{4.22}$$

where the **grand canonical partition function** is defined by

$$\Theta_{\Lambda;\beta,\mu} \stackrel{\text{def}}{=} \sum_{\eta\in\{0,1\}^\Lambda} \exp(-\beta\{\mathscr{H}_{\Lambda;K}(\eta) - \beta N_\Lambda(\eta)\}). \tag{4.23}$$

By summing over the possible number of particles, one gets the following simple relation between the canonical and grand canonical partition functions:

$$\Theta_{\Lambda;\beta,\mu} = \sum_{N=0}^{|\Lambda|} e^{\beta\mu N} \mathbf{Q}_{\Lambda;\beta,N}. \tag{4.24}$$

Exercise 4.3. Let $\Lambda \Subset \mathbb{Z}^d$, and let $f, g\colon \{0,1\}^\Lambda \to \mathbb{R}$ be two nondecreasing functions. Using Theorem 3.50, prove that f and g are positively correlated: For all $\beta, \mu \in \mathbb{R}$,

$$\mathrm{Cov}_{\Lambda;\beta,\mu}(f,g) \geq 0\,,$$

where $\mathrm{Cov}_{\Lambda;\beta,\mu}$ denotes the covariance under $\nu_{\Lambda;\beta,\mu}$.

The thermodynamic potential describing an infinite system at equilibrium with a reservoir of particles at fixed chemical potential is the *pressure*. The **pressure in a finite volume** $\Lambda \Subset \mathbb{Z}^d$ is defined as

$$p_{\Lambda;\beta}(\mu) \stackrel{\text{def}}{=} \frac{1}{\beta|\Lambda|} \log \Theta_{\Lambda;\beta,\mu}\,, \qquad \mu \in \mathbb{R}\,.$$

Observe that the derivative of this quantity yields the average density of particles under $\nu_{\Lambda;\beta,\mu}$:

$$\frac{\partial p_{\Lambda;\beta}}{\partial \mu} = \left\langle \frac{N_\Lambda}{|\Lambda|} \right\rangle_{\Lambda;\beta,\mu}. \tag{4.25}$$

We thus see that tuning the chemical potential allows one to control the average number of particles in the system. In particular, as discussed in Exercise 4.6 below, large negative values of μ result in a dilute (gas) phase, while large positive values of μ yield a dense (liquid) phase.

Theorem 4.11. *Let $\mathscr{R} \ni \Lambda_n \Uparrow \mathbb{Z}^d$. For all $\mu \in \mathbb{R}$, the limit*

$$p_\beta(\mu) \stackrel{\text{def}}{=} \lim_{n\to\infty} p_{\Lambda_n;\beta}(\mu) \tag{4.26}$$

*exists and does not depend on the choice of the sequence $(\Lambda_n)_{n\geq 1}$; it is called the **pressure**. Moreover, $\mu \mapsto p_\beta(\mu)$ is convex and continuous.*

Since p_β is convex, its derivative

$$\rho_\beta(\mu) \stackrel{\text{def}}{=} \frac{\partial p_\beta}{\partial \mu} \tag{4.27}$$

exists everywhere except possibly on a countable set of points (Theorem B.12) and will be called the **average (grand canonical) density**. The one-sided derivatives $\frac{\partial p_\beta}{\partial \mu^+}$ and $\frac{\partial p_\beta}{\partial \mu^-}$, are well defined at each μ. Theorem B.12, together with (4.25), also guarantees that, when $\rho_\beta(\mu)$ exists,

$$\rho_\beta(\mu) = \lim_{n\to\infty} \left\langle \frac{N_{\Lambda_n}}{|\Lambda_n|} \right\rangle_{\Lambda_n;\beta,\mu}.$$

The existence of the limit in (4.26) will be seen to be a consequence of the existence of the free energy (see below), but it can also be proved directly:

Exercise 4.4. Prove the existence of the pressure in Theorem 4.11 using the method suggested in Exercise 3.3.

Theorem 4.11 leaves open the possibility that the pressure has affine pieces, along which an increase of the chemical potential μ would not result in an increase of the average density $\rho_\beta(\mu)$. This turns out to be impossible:

Theorem 4.12. *$\mu \mapsto p_\beta(\mu)$ is strictly convex and increasing.*

Proof Differentiating (4.25) once again,

$$\frac{\partial^2 p_{\Lambda;\beta}}{\partial \mu^2} = \frac{\beta}{|\Lambda|} \mathrm{Var}_{\Lambda;\beta,\mu}(N_\Lambda),$$

where $\mathrm{Var}_{\Lambda;\beta,\mu}$ denotes the variance under $\nu_{\Lambda;\beta,\mu}$. Let us first observe that there exists $c > 0$, depending on β, κ and μ, such that

$$c < \nu_{\Lambda;\beta,\mu}\big(\eta_i = 1 \mid \eta_j = m_j, \forall j \in \Lambda \setminus \{i\}\big) < 1 - c, \qquad \forall i \in \Lambda, \tag{4.28}$$

for all choices of $m_j \in \{0, 1\}, j \in \Lambda \setminus \{i\}$. In particular, $\mathrm{Var}_{\Lambda;\beta,\mu}(\eta_i) \geq c(1 - c)$ for all $i \in \Lambda$. Moreover, Exercise 4.3 guarantees that $\mathrm{Cov}_{\Lambda;\beta,\mu}(\eta_i, \eta_j) \geq 0$, so

$$\mathrm{Var}_{\Lambda;\beta,\mu}(N_\Lambda) = \sum_{i\in\Lambda} \mathrm{Var}_{\Lambda;\beta,\mu}(\eta_i) + \sum_{\substack{i,j\in\Lambda \\ i\neq j}} \mathrm{Cov}_{\Lambda;\beta,\mu}(\eta_i, \eta_j) \geq c(1 - c)\,|\Lambda|\,.$$

That p_β is increasing and strictly convex follows from the fact that $\frac{\partial^2 p_{\Lambda;\beta}}{\partial \mu^2} \geq \beta c(1 - c) > 0$, uniformly in $\Lambda \Subset \mathbb{Z}^d$ (see Exercise B.5). □

Exercise 4.5. Find the constant c in (4.28).

Exercise 4.6. Assuming that $p_\beta(\mu)$ exists, show that

$$\lim_{\mu\to-\infty} p_\beta(\mu) = 0\,, \qquad \lim_{\mu\to+\infty} \frac{p_\beta(\mu)}{\mu} = 1\,. \tag{4.29}$$

Conclude that

$$\lim_{\mu\to-\infty} \rho_\beta(\mu) = 0\,, \qquad \lim_{\mu\to+\infty} \rho_\beta(\mu) = 1\,. \tag{4.30}$$

4.4 Equivalence of Ensembles

During our brief discussion of thermodynamics in Chapter 1, we saw that the entropy was related to the other thermodynamic potentials through Legendre transforms. We now check that similar relations between the free energy and pressure hold for the lattice gas.

Theorem 4.13. ***Equivalence of ensembles at the level of potentials*** *holds for the general lattice gas. That is, the free energy and pressure are each other's Legendre transform:*

$$f_\beta(\rho) = \sup_{\mu\in\mathbb{R}}\{\mu\rho - p_\beta(\mu)\} \qquad \forall \rho \in [0,1], \tag{4.31}$$

$$p_\beta(\mu) = \sup_{\rho\in[0,1]}\{\rho\mu - f_\beta(\rho)\} \qquad \forall \mu \in \mathbb{R}. \tag{4.32}$$

Since each can be obtained from the other by a Legendre transform, f_β and p_β contain the same information about the system, and either of them can be used to study the thermodynamic behavior of the lattice gas.

Proof We use (4.24). Since the $|\Lambda_n| + 1$ terms of that sum are all nonnegative,

$$\max_N\{e^{\beta\mu N}\mathbf{Q}_{\Lambda_n;\beta,N}\} \le \Theta_{\Lambda_n;\beta,\mu} \le (|\Lambda_n| + 1)\max_N\{e^{\beta\mu N}\mathbf{Q}_{\Lambda_n;\beta,N}\}.$$

By Exercise 4.1, we see that the maximum over N is attained for values of $\frac{N}{|\Lambda|}$ bounded away from 0 or 1. For those N, one can use (4.11):

$$\begin{aligned} e^{\beta\mu N}\mathbf{Q}_{\Lambda_n;\beta,N} &\le c^2\exp\bigl(\beta\{\mu\tfrac{N}{|\Lambda_n|} - f_\beta(\tfrac{N}{|\Lambda_n|})\}|\Lambda_n|\bigr)\\ &\le c^2\exp\bigl(\beta\sup_\rho\{\mu\rho - f_\beta(\rho)\}|\Lambda_n|\bigr). \end{aligned}$$

This gives

$$\limsup_{n\to\infty}\frac{1}{\beta|\Lambda_n|}\log\Theta_{\Lambda_n;\beta,\mu} \le \sup_\rho\{\rho\mu - f_\beta(\rho)\}.$$

For the lower bound, we first use the continuity of $\rho \mapsto \rho\mu - f_\beta(\rho)$, and consider some $\rho_* \in [0,1]$ for which $\sup_\rho\{\rho\mu - f_\beta(\rho)\} = \rho_*\mu - f_\beta(\rho_*)$. Let $\epsilon > 0$, and n be large enough to ensure that $\mathbf{Q}_{\Lambda_n;\beta,\lceil\rho_*|\Lambda_n|\rceil} \ge e^{-\beta f_\beta(\rho_*)|\Lambda_n| - \beta\epsilon|\Lambda_n|}$. Then, taking $N = \lceil\rho_*|\Lambda_n|\rceil$,

$$\begin{aligned} \max_N\{e^{\beta\mu N}\mathbf{Q}_{\Lambda_n;\beta,N}\} &\ge e^{\beta\mu\lceil\rho_*|\Lambda_n|\rceil}\mathbf{Q}_{\Lambda_n;\beta,\lceil\rho_*|\Lambda_n|\rceil}\\ &\ge \exp\Bigl(\beta\bigl\{\tfrac{\lceil\rho_*|\Lambda_n|\rceil}{|\Lambda_n|}\mu - f_\beta(\rho_*)\bigr\}|\Lambda_n| - \beta\epsilon|\Lambda_n|\Bigr), \end{aligned}$$

which gives

$$\liminf_{n\to\infty}\frac{1}{\beta|\Lambda_n|}\log\Theta_{\Lambda_n;\beta,\mu} \ge \rho_*\mu - f_\beta(\rho_*) - \epsilon.$$

Since this holds for all $\epsilon > 0$, (4.32) (and thereby the existence of the pressure) is proved. Then, $f_\beta(\rho)$ being convex and continuous, it coincides with the Legendre transform of its Legendre transform (Theorem B.19[1]). This proves (4.31). □

[1] To apply that theorem, one needs to define $f_\beta(\rho) \stackrel{\text{def}}{=} +\infty$ for all $\rho \notin [0,1]$, so that $f_\beta : \mathbb{R} \to \mathbb{R} \cup \{\infty\}$ is convex and lower semi-continuous.

The equivalence of ensembles allows us to derive a further smoothness property for the free energy.

Corollary 4.14. *The free energy f_β is differentiable everywhere on* $(0, 1)$.

Proof If there existed a point ρ_* at which f_β were not differentiable, Theorem B.20 would imply that p_β is affine on some interval; this would contradict the claim of Theorem 4.12. □

4.5 An Overview of the Rest of the Chapter

The existence of the free energy and pressure and the equivalence of ensembles, proved in the previous sections, hold under quite general assumptions (in our case: $\kappa < \infty$). In Section 4.6, we will see how these can be used to derive general properties of the canonical and grand canonical Gibbs distributions (similarly to what was done in earlier chapters for the Curie–Weiss and Ising models). Namely, the first concerns the typical density of particles, $\frac{N_\Lambda}{|\Lambda|}$, under the grand canonical distribution $\nu_{\Lambda;\beta,\mu}$, and the second concerns the geometrical properties of configurations under the canonical distribution $\nu_{\Lambda;\beta,N}$.

The remainder of this chapter is devoted to the study of particular cases. Our main concern will be to determine under which conditions *phase transitions* can occur at low temperature. For each of the models considered, we will study the qualitative properties of the free energy $f_\beta(\rho)$ and of the pressure $p_\beta(\mu)$. We will also express the pressure as a function of the density $\rho \in (0, 1)$ and of the volume per particle $v \stackrel{\text{def}}{=} \rho^{-1}$, yielding two functions $\rho \mapsto \widetilde{p}_\beta(\rho)$, $v \mapsto \widehat{p}_\beta(v)$. Since the latter are considered at a fixed value of β, they are **isotherms of the pressure**.

A salient feature of the occurrence of phase transitions, in the canonical lattice gas, is the *condensation phenomenon*, that is, the coexistence of macroscopic regions with different densities, gas and liquid. Although a complete description of this phenomenon is outside the scope of this book, some aspects of the problem will be described in the complements at the end of the chapter (Section 4.12.1, see Figure 4.23).

4.6 Concentration and Typical Configurations

In this section, still under general assumptions (we only assume that $\kappa < \infty$), we use the existence of the free energy and pressure to derive properties of the Gibbs distributions.

4.6.1 Typical Densities

In the grand canonical ensemble, the number of particles in Λ, N_Λ, can fluctuate, and we expect $\frac{N_\Lambda}{|\Lambda|}$ to concentrate around its average value, given by (4.25):

$$\left\langle \frac{N_\Lambda}{|\Lambda|} \right\rangle_{\Lambda;\beta,\mu} = \frac{\partial p_{\Lambda;\beta}}{\partial \mu}.$$

The next result characterizes the typical values of the density under $\nu_{\Lambda_n;\beta,\mu}$ as minimizers of a suitable function (compare with the similar results obtained in Section 2.2 in the context of mean-field models).

Theorem 4.15. *Let $\mathscr{R} \ni \Lambda_n \Uparrow \mathbb{Z}^d$ and let $J \subset [0,1]$ be an interval. Then,*

$$\lim_{n\to\infty} \frac{1}{|\Lambda_n|} \log \nu_{\Lambda_n;\beta,\mu}\Big(\frac{N_{\Lambda_n}}{|\Lambda_n|} \in J\Big) = -\min_{\rho\in J} I_{\beta,\mu}(\rho),$$

where

$$I_{\beta,\mu}(\rho) \stackrel{\text{def}}{=} \beta\big\{(f_\beta(\rho) - \mu\rho) - \min_{\rho'\in[0,1]} (f_\beta(\rho') - \mu\rho')\big\} \tag{4.33}$$

is called the ***rate function.***

Proof The proof follows the same steps as the proof of the equivalence of ensembles. Using the same decomposition as in (4.24),

$$\nu_{\Lambda;\beta,\mu}\big(\tfrac{N_\Lambda}{|\Lambda|} \in J\big) = \frac{1}{\Theta_{\Lambda;\beta,\mu}} \sum_{\substack{0\le N\le|\Lambda|:\\ N/|\Lambda|\in J}} e^{\beta\mu N} \mathbf{Q}_{\Lambda;\beta,N}.$$

The denominator $\Theta_{\Lambda;\beta,\mu}$ is treated using Theorem 4.5. For the numerator,

$$\max_{N:\, N/|\Lambda|\in J} \{e^{\beta\mu N}\mathbf{Q}_{\Lambda;\beta,N}\} \le \sum_{\substack{0\le N\le|\Lambda|:\\ N/|\Lambda|\in J}} e^{\beta\mu N}\mathbf{Q}_{\Lambda;\beta,N} \le (|\Lambda|+1) \max_{N:\, N/|\Lambda|\in J} \{e^{\beta\mu N}\mathbf{Q}_{\Lambda;\beta,N}\},$$

and we can proceed as in the proof of Theorem 4.13. □

Consider the set of minimizers of $I_{\beta,\mu} \ge 0$:

$$\mathscr{M}_{\beta,\mu} \stackrel{\text{def}}{=} \big\{\rho \in [0,1] \,:\, I_{\beta,\mu}(\rho) = 0\big\}.$$

By continuity of $I_{\beta,\mu}$, $\mathscr{M}_{\beta,\mu}$ is closed. Since f_β is convex, so is $I_{\beta,\mu}$. Therefore, $\mathscr{M}_{\beta,\mu}$ is either a singleton, or a closed interval:

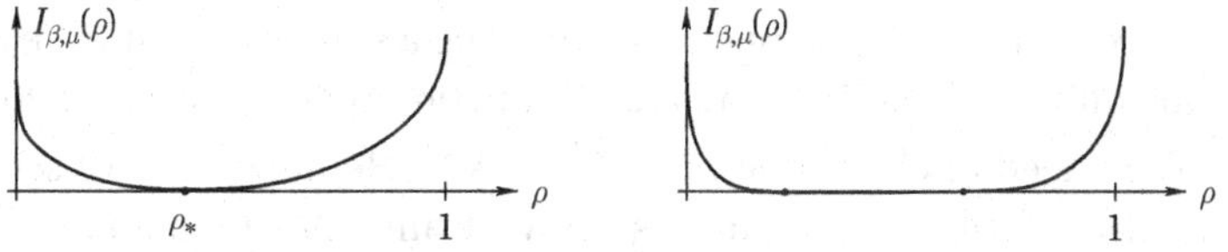

Figure 4.2. Depending on (β,μ), the minimizers of the rate function form either a singleton, $\mathscr{M}_{\beta,\mu} = \{\rho_*\}$ (left), or a closed interval (right).

Remark 4.16. Most of the plots given in this chapter were made to illustrate important features of the functions under consideration; in order to better emphasize the functions, we have often decided to accentuate them. Nevertheless, the qualitative properties have been preserved. Only those for the hard-core gas, and some of those for the van der Waals model, are drawn from an expression computed rigorously. ◇

Theorem 4.15 thus says that, in a grand canonical system with chemical potential μ, the particle density $\frac{N_\Lambda}{|\Lambda|}$ concentrates on $\mathcal{M}_{\beta,\mu}$ in the following sense: for any open set $G \subset [0,1]$, with $G \supset \mathcal{M}_{\beta,\mu}$, we have that

$$\text{as } \Lambda \Uparrow \mathbb{Z}^d, \quad \nu_{\Lambda;\beta,\mu}\Big(\frac{N_\Lambda}{|\Lambda|} \in G\Big) \to 1 \quad \text{exponentially fast in } |\Lambda|.$$

Indeed, for any closed interval $J \subset [0,1] \setminus \mathcal{M}_{\beta,\mu}$, we have $\lambda \stackrel{\text{def}}{=} \min_{\rho\in J} I_{\beta,\mu}(\rho) > 0$ and, for any large enough box Λ,

$$\nu_{\Lambda;\beta,\mu}\Big(\frac{N_\Lambda}{|\Lambda|} \in J\Big) \leq e^{-\frac{\lambda}{2}|\Lambda|}.$$

Densities outside $\mathcal{M}_{\beta,\mu}$ are therefore very atypical in large systems. Of course, it does not follow from the above theorem that all values in the set $\mathcal{M}_{\beta,\mu}$ are equally likely. Investigating this question requires a much more delicate analysis, taking into account *surface effects*; see the complements to this chapter (Section 4.12.1) for a discussion.

Remark 4.17. Using the equivalence of ensembles, one can also express $\mathcal{M}_{\beta,\mu}$ as

$$\mathcal{M}_{\beta,\mu} = \big\{\rho \in [0,1] \colon p_\beta(\mu) = \mu\rho - f_\beta(\rho)\big\}.$$ ◇

When the pressure is differentiable (that is, in the absence of a first-order phase transition), one knows exactly at which value the density concentrates:

Proposition 4.18. *Assume that p_β is differentiable at μ. Then, under $\nu_{\Lambda;\beta,\mu}$, the density $\frac{N_\Lambda}{|\Lambda|}$ concentrates on $\rho_\beta(\mu) \stackrel{\text{def}}{=} \frac{\partial p_\beta}{\partial \mu}$: for all $\epsilon > 0$, as $\Lambda \Uparrow \mathbb{Z}^d$,*

$$\nu_{\Lambda;\beta,\mu}\Big(\Big|\frac{N_\Lambda}{|\Lambda|} - \rho_\beta(\mu)\Big| \geq \epsilon\Big) \to 0, \qquad \textit{exponentially fast in } |\Lambda|.$$

Proof If $\frac{\partial p_\beta}{\partial \mu}$ exists, then $\mathcal{M}_{\beta,\mu}$ must be a singleton (if it were an interval, f_β would be affine on that interval, a contradiction with Theorem B.20): $\mathcal{M}_{\beta,\mu} = \{\rho_*\}$. We only need to check that $\rho_* = \frac{\partial p_\beta}{\partial \mu}$. Using Remark 4.17, we see that ρ_* must satisfy $p_\beta(\mu) = \mu\rho_* - f_\beta(\rho_*)$. It thus follows from (4.32) that, for all $\epsilon > 0$,

$$p_\beta(\mu+\epsilon) - p_\beta(\mu) \geq \big\{(\mu+\epsilon)\rho_* - f_\beta(\rho_*)\big\} - p_\beta(\mu) = \epsilon\rho_*$$

and

$$p_\beta(\mu) - p_\beta(\mu-\epsilon) \leq p_\beta(\mu) - \big\{(\mu-\epsilon)\rho_* - f_\beta(\rho_*)\big\} = \epsilon\rho_*,$$

which, dividing by ϵ and letting $\epsilon \downarrow 0$, gives $\frac{\partial p_\beta}{\partial \mu} = \rho_*$, proving the claim. □

4.6.2 Strict Convexity and Spatial Homogeneity

In this section we describe typical configurations of particles in the canonical ensemble, by looking at how the density can vary from one point to another. More precisely, we consider the canonical Gibbs distribution $\nu_{\Lambda;\beta,N}$ in a large box Λ, and assume that the density $\frac{N}{|\Lambda|} \simeq \rho$ belongs to some interval I on which the free energy is *strictly convex*:

$$f_\beta(\lambda\rho_1 + (1-\lambda)\rho_2) < \lambda f_\beta(\rho_1) + (1-\lambda)f_\beta(\rho_2),$$

for all $0 < \lambda < 1$ and all $\rho_1 < \rho_2$ in I. We will show that, under such conditions, the system is *homogeneous*: with high probability under $\nu_{\Lambda;\beta,N}$, all macroscopic sub-boxes of Λ have the same density ρ. (We have proved a similar claim in the microcanonical ensemble, in Section 1.3.1.)

We consider the thermodynamic limit along a sequence $\mathscr{R} \ni \Lambda \Uparrow \mathbb{Z}^d$ and, for each $0 < \alpha < 1$ and each Λ, consider a collection $\mathscr{D}_\alpha(\Lambda) \subset \mathscr{R}$ of subsets $\Lambda' \subset \Lambda$ with the property that $\frac{|\Lambda'|}{|\Lambda|} \to \alpha$ when $\Lambda \Uparrow \mathbb{Z}^d$.

Theorem 4.19. *Let $\mathscr{R} \ni \Lambda \Uparrow \mathbb{Z}^d$. Assume that $\frac{N}{|\Lambda|} \to \rho \in (0,1)$ and that f_β is strictly convex in a neighborhood of ρ. Fix $0 < \alpha < 1$. Then, for all small $\epsilon > 0$, as $\Lambda \Uparrow \mathbb{Z}^d$,*

$$\nu_{\Lambda;\beta,N}\Big(\exists \Lambda' \in \mathscr{D}_\alpha(\Lambda) \text{ such that} \Big|\frac{N_{\Lambda'}}{|\Lambda'|} - \rho\Big| \geq \epsilon\Big) \to 0, \tag{4.34}$$

exponentially fast in $|\Lambda|$.

Proof Together with $\mathscr{R} \ni \Lambda \Uparrow \mathbb{Z}^d$, we consider $N \to \infty$ such that $\frac{N}{|\Lambda|} \to \rho \in (0,1)$. Fix $\epsilon > 0$ and some $\delta > 0$ (which will be fixed later), and cover the set $[0,1] \setminus (\rho - \epsilon, \rho + \epsilon)$ with closed intervals J_k, $k = 1, \dots, m$, of sizes $\leq \delta$, all at distance at least ϵ from ρ. We can assume that $m \leq 2/\delta$. We will first show that there exists $b_0 > 0$ such that, when Λ is large enough,

$$\nu_{\Lambda;\beta,N}\Big(\frac{N_{\Lambda'}}{|\Lambda'|} \in J_k\Big) \leq |\Lambda| e^{-b_0|\Lambda|}, \tag{4.35}$$

for all $\Lambda' \in \mathscr{D}_\alpha(\Lambda)$ and all k. Since there are at most $|\Lambda|$ sub-boxes $\Lambda' \in \mathscr{D}_\alpha(\Lambda)$ and since the number of intervals J_k is bounded, the main claim will then follow.

Consider some J_k. For definiteness, we assume that $\min J_k > \rho$ (the other case is treated similarly). We can of course assume that ρ, α and J_k are such that $\{N_{\Lambda'}/|\Lambda'| \in J_k\} \neq \varnothing$. First, decompose

$$\nu_{\Lambda;\beta,N}\Big(\frac{N_{\Lambda'}}{|\Lambda'|} \in J_k\Big) = \sum_{N' :\, \frac{N'}{|\Lambda'|} \in J_k} \nu_{\Lambda;\beta,N}(N_{\Lambda'} = N'). \tag{4.36}$$

Let $\Lambda'' \stackrel{\text{def}}{=} \Lambda \setminus \Lambda'$. If the interaction has finite range (see Exercise 4.7 below for the general case), then, for all configurations $\eta \in \{0,1\}^\Lambda$,

$$\mathscr{H}_{\Lambda;K}(\eta) = \mathscr{H}_{\Lambda';K}(\eta|_{\Lambda'}) + \mathscr{H}_{\Lambda'';K}(\eta|_{\Lambda''}) + O(|\partial\Lambda|), \tag{4.37}$$

where, as usual, we denote by $\eta|_\Delta$ the restriction of $\eta \in \{0,1\}^\Lambda$ to $\Delta \subset \Lambda$. Therefore, letting $N'' \stackrel{\text{def}}{=} N - N'$,

$$\nu_{\Lambda;\beta,N}(N_{\Lambda'} = N') \le e^{O(|\partial\Lambda|)} \frac{\mathbf{Q}_{\Lambda';\beta,N'}\mathbf{Q}_{\Lambda'';\beta,N''}}{\mathbf{Q}_{\Lambda;\beta,N}}.$$

For the denominator, we use

$$\lim_{\Lambda\Uparrow\mathbb{Z}^d} \frac{1}{\beta|\Lambda|} \log \mathbf{Q}_{\Lambda;\beta,N} = -f_\beta(\rho).$$

Let $N'_{\text{min}} = \min\{N' : N'/|\Lambda'| \in J_k\}$, $N''_{\text{max}} = N - N'_{\text{min}}$. Using Lemma 4.7 repeatedly,

$$\mathbf{Q}_{\Lambda';\beta,N'} \le \Big[e^{\beta\kappa}\Big(\frac{1-N'_{\text{min}}/|\Lambda'|}{N'_{\text{min}}/|\Lambda'|} \vee 1\Big)\Big]^{N'-N'_{\text{min}}} \mathbf{Q}_{\Lambda';\beta,N'_{\text{min}}},$$

$$\mathbf{Q}_{\Lambda'';\beta,N''} \le \Big(\frac{(N''_{\text{max}}+1)/|\Lambda''|}{1-N''_{\text{max}}/|\Lambda''|} \vee 1\Big)^{N''_{\text{max}}-N''} \mathbf{Q}_{\Lambda'';\beta,N'_{\text{max}}}.$$

As $\Lambda \Uparrow \mathbb{Z}^d$, we have

$$\frac{N'_{\text{min}}}{|\Lambda'|} \to \rho^k_{\text{min}} \stackrel{\text{def}}{=} \min J_k, \qquad \frac{N''_{\text{max}}}{|\Lambda''|} \to \rho^k_{\text{max}},$$

where ρ^k_{max} satisfies

$$\alpha\rho^k_{\text{min}} + (1-\alpha)\rho^k_{\text{max}} = \rho.$$

Therefore, since $\Lambda' \in \mathscr{R}$ and $|\Lambda'|/|\Lambda| \to \alpha$,

$$\lim_{\Lambda\Uparrow\mathbb{Z}^d} \frac{1}{\beta|\Lambda|} \log \mathbf{Q}_{\Lambda';\beta,N'_{\text{min}}} = -\alpha f_\beta(\rho^k_{\text{min}}).$$

Observe that Λ'' is *not* a parallelepiped, but we can use Lemma 4.6 as follows:

$$\mathbf{Q}_{\Lambda'';\beta,N''_{\text{max}}} \le \frac{\mathbf{Q}_{\Lambda;\beta,\tilde{N}}}{\mathbf{Q}_{\Lambda';\beta,\tilde{N}'}}, \tag{4.38}$$

where $\tilde{N}' \stackrel{\text{def}}{=} \lfloor \frac{N''_{\text{max}}}{|\Lambda''|}|\Lambda'| \rfloor$, $\tilde{N} \stackrel{\text{def}}{=} N''_{\text{max}} + \tilde{N}'$. Now, $\frac{\tilde{N}}{|\Lambda|} \to \rho^k_{\text{max}}$, and $\frac{\tilde{N}'}{|\Lambda'|} \to \rho^k_{\text{max}}$. Therefore,

$$\lim_{\Lambda\Uparrow\mathbb{Z}^d} \frac{1}{\beta|\Lambda|} \log \frac{\mathbf{Q}_{\Lambda;\beta,\tilde{N}}}{\mathbf{Q}_{\Lambda';\beta,\tilde{N}'}} = -(1-\alpha) f_\beta(\rho^k_{\text{max}}).$$

We have thus proved that

$$\limsup_{\Lambda\Uparrow\mathbb{Z}^d} \frac{1}{\beta|\Lambda|} \log \max_{N' : \frac{N'}{|\Lambda'|} \in J_k} \nu_{\Lambda;\beta,N}(N_{\Lambda'} = N')$$
$$\le M_k\delta - \big\{\alpha f_\beta(\rho^k_{\text{min}}) + (1-\alpha) f_\beta(\rho^k_{\text{max}}) - f_\beta(\rho)\big\},$$

where

$$M_k \stackrel{\text{def}}{=} \kappa + \log\Big(\frac{1-\rho^k_{\text{min}}}{\rho^k_{\text{min}}} \vee 1\Big) + \log\Big(\frac{\rho^k_{\text{max}}}{1-\rho^k_{\text{max}}} \vee 1\Big).$$

Observe that M_k is bounded uniformly in k. Namely, there exists $0 < \epsilon' < \epsilon$ (depending on ρ and α) such that $\rho^k_{\text{max}} < \rho - \epsilon' < \rho + \epsilon < \rho^k_{\text{min}}$ for all k. Moreover, by the strict convexity of f_β in the neighborhood of ρ, there exists some $b_0 > 0$ such that

$$\min_{1\leq k\leq m} \big\{\alpha f_\beta(\rho^k_{\text{min}}) + (1-\alpha) f_\beta(\rho^k_{\text{max}}) - f_\beta(\rho)\big\} \geq 2b_0 > 0,$$

uniformly in m. One can thus take δ small enough that $M_k\delta \leq b_0$. The sum in (4.36) contains at most $|\Lambda|$ terms, which proves (4.35) for large enough Λ. □

Exercise 4.7. Show that, when the interaction is not of finite range (but assuming $\kappa < \infty$), (4.37) becomes

$$\mathscr{H}_{\Lambda;K}(\eta) = \mathscr{H}_{\Lambda';K}(\eta|_{\Lambda'}) + \mathscr{H}_{\Lambda'';K}(\eta|_{\Lambda''}) + o(|\Lambda|),$$

so that the rest of the proof remains unchanged.

4.7 Hard-Core Lattice Gas

Let us see what happens when

$$K(i,j) = 0$$

for all pairs i, j. This model, already considered in Chapter 1, is called the **hard-core lattice gas**, since the only interaction between the particles is the constraint of having at most one of them at each vertex. Due to the lack of an attractive part in its Hamiltonian, this model will not present a particularly interesting behavior, but it remains a good starting point, since its thermodynamic potentials can be computed explicitly.

When $K \equiv 0$, the canonical partition function becomes a purely combinatorial quantity, counting the configurations $\eta \in \{0,1\}^\Lambda$ with $N_\Lambda(\eta) = N$:

$$\mathbf{Q}^{\text{hard}}_{\Lambda;N} = \binom{|\Lambda|}{N}.$$

Since $\binom{|\Lambda|}{0} = \binom{|\Lambda|}{|\Lambda|} = 1$, we get $f^{\text{hard}}_\beta(0) = f^{\text{hard}}_\beta(1) = 0$. For intermediate densities, $0 < \rho < 1$, we use again (4.13)-(4.14), and obtain (see Figure 4.3)

$$f^{\text{hard}}_\beta(\rho) = -\tfrac{1}{\beta} s^{\text{l.g.}}(\rho). \tag{4.39}$$

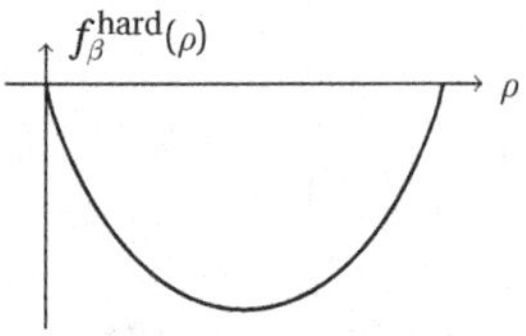

Figure 4.3. The free energy of the hard-core lattice gas: strictly convex at all temperatures.

The pressure can also be computed explicitly (see Figure 4.4), using either the equivalence of ensembles, or simply (4.24):

$$\Theta_{\Lambda;\mu}^{\mathrm{hard}} = \sum_{N=0}^{|\Lambda|} \binom{|\Lambda|}{N} e^{\beta\mu N} = (1+e^{\beta\mu})^{|\Lambda|},$$

which yields

$$p_\beta^{\mathrm{hard}}(\mu) = \frac{1}{\beta}\log(1+e^{\beta\mu}). \tag{4.40}$$

The expressions obtained for $f_\beta^{\mathrm{hard}}(\rho)$ and $p_\beta^{\mathrm{hard}}(\mu)$ imply that these functions are analytic. We will now see how to express the pressure as a function of ρ rather than μ. To this end, one must answer the following question: can one realize a chosen average density ρ by suitably tuning μ?

Observe that the average density of particles,

$$\rho_\beta^{\mathrm{hard}}(\mu) = \frac{\partial p_\beta^{\mathrm{hard}}}{\partial\mu} = \frac{e^{\beta\mu}}{1+e^{\beta\mu}}, \tag{4.41}$$

is smooth for all values of μ: when μ increases from $-\infty$ to $+\infty$, the density of the hard-core gas increases from 0 to 1 without discontinuities and exhibits no phase transition (see Figure 4.4). This absence of condensation is of course due to the lack of attraction between the particles.

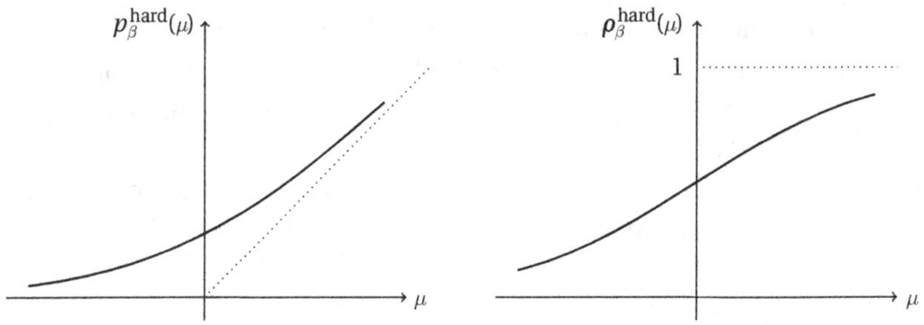

Figure 4.4. The pressure (left) and average density (right) of the hard-core lattice gas.

By Proposition 4.18, we also know that the density of particles in a large grand canonical system, $\frac{N_\Lambda}{|\Lambda|}$, concentrates on $\rho_\beta^{\mathrm{hard}}(\mu)$. Since $\rho_\beta^{\mathrm{hard}}(\mu)$ is increasing in μ, the equation

$$\rho_\beta^{\text{hard}}(\mu) = \rho \tag{4.42}$$

has a unique solution in μ, for each fixed $\rho \in (0,1)$. This solution can of course be given explicitly:

$$\mu_\beta^{\text{hard}}(\rho) = \tfrac{1}{\beta} \log \tfrac{\rho}{1-\rho} .$$

Therefore, densities $\rho \in (0,1)$ are in one-to-one correspondence with chemical potentials $\mu \in \mathbb{R}$. This bijection allows the pressure to be expressed as a function of the density:

$$\widetilde{p}_\beta^{\text{hard}}(\rho) \stackrel{\text{def}}{=} p_\beta^{\text{hard}}(\mu_\beta^{\text{hard}}(\rho)) = -\tfrac{1}{\beta}\log(1-\rho) .$$

At low densities, $\log(1-\rho) \simeq -\rho$, which allows recovery of the qualitative behavior provided by the equation of state of the ideal gas:

$$\beta \widetilde{p}_\beta^{\text{hard}}(\rho) = \rho + O(\rho^2) \qquad (\rho \text{ small}) .$$

In terms of the **volume per particle**, $v = \rho^{-1}$,

$$\widehat{p}_\beta^{\text{hard}}(v) \stackrel{\text{def}}{=} \widetilde{p}_\beta^{\text{hard}}(v^{-1}) = -\tfrac{1}{\beta}\log\left(1 - \tfrac{1}{v}\right) .$$

Remark 4.20. When v is large, $-\log(1 - \frac{1}{v}) \simeq \frac{1}{v}$, and the above provides an approximation to the *ideal gas law* (1.14), with $R = 1$,

$$pv = RT . \qquad \diamond$$

4.7.1 Parenthesis: Equivalence of Ensembles at the Level of Measures

Consider the canonical hard-core lattice gas along a sequence $\Lambda \Uparrow \mathbb{Z}^d$, $N \to \infty$, with $\frac{N}{|\Lambda|} \to \rho$. What can be said about the distribution of particles in a smaller subsystem $\Delta \subset \Lambda$, whose size remains fixed as $\Lambda \Uparrow \mathbb{Z}^d$?

Although the density of particles in Λ is fixed, close to ρ, the number of particles in Δ can fluctuate. We therefore expect to obtain, when $\Lambda \Uparrow \mathbb{Z}^d$, some distribution of the grand canonical type inside Δ, with a chemical potential μ to be determined; not surprisingly, it will be exactly the one obtained earlier through the relation (4.42).

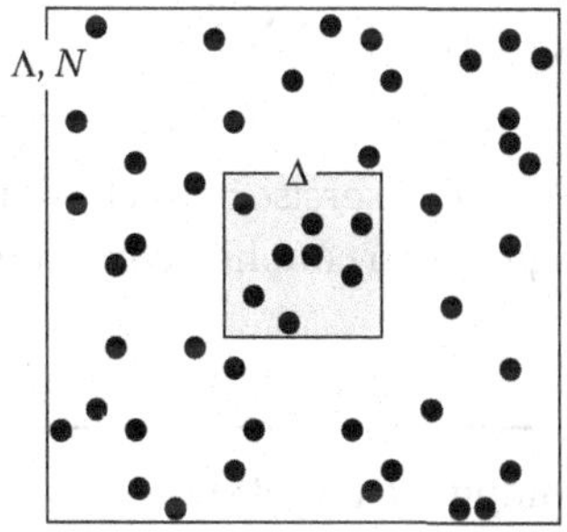

Figure 4.5. In a large system Λ with a fixed number of particles N, what can be said about the probability distribution describing a smaller subsystem $\Delta \subset \Lambda$?

Proposition 4.21. *Let $\Delta \Subset \mathbb{Z}^d$, $\Lambda \Uparrow \mathbb{Z}^d$, and assume that $\frac{N}{|\Lambda|} \to \rho \in (0,1)$. Let $\eta_\Delta \in \{0,1\}^\Delta$ be any configuration of particles in Δ. Then, as $\Lambda \Uparrow \mathbb{Z}^d$,*

$$\nu_{\Lambda;\beta,N}\big(\{\eta : \eta|_\Delta = \eta_\Delta\}\big) \longrightarrow \nu_{\Delta;\beta,\mu}(\eta_\Delta), \tag{4.43}$$

where μ is the unique solution to (4.42).

Relation (4.43) is the simplest instance of **equivalence of ensembles at the level of measures**. A similar statement holds much more generally, at least away from phase transitions, but is substantially harder to establish.[2]

Proof The proof is a direct application of Stirling's formula: if $M = N_\Delta(\eta_\Delta)$,

$$\nu_{\Lambda;\beta,N}\big(\{\eta : \eta|_\Delta = \eta_\Delta\}\big) = \frac{\binom{|\Lambda|-|\Delta|}{N-M}}{\binom{|\Lambda|}{N}} = (1+o(1))\Big(1-\frac{N}{|\Lambda|}\Big)^{|\Delta|}\Big(\frac{N}{|\Lambda|-N}\Big)^{M}.$$

Since $\frac{N}{|\Lambda|} \to \rho$ and since (4.42) can be written as $\frac{\rho}{1-\rho} = e^{\beta\mu_\beta^{\text{hard}}(\rho)}$,

$$\Big(\frac{N}{|\Lambda|-N}\Big)^{M} \longrightarrow \Big(\frac{\rho}{1-\rho}\Big)^{M} = \exp\{\beta\mu_\beta^{\text{hard}}(\rho)M\},$$

and

$$\Big(1-\frac{N}{|\Lambda|}\Big)^{|\Delta|} \longrightarrow (1-\rho)^{|\Delta|} = \frac{1}{(1+e^{\beta\mu_\beta^{\text{hard}}(\rho)})^{|\Delta|}} = \frac{1}{\Theta_{\Delta;\beta,\mu_\beta^{\text{hard}}(\rho)}}. \qquad \square$$

4.8 Nearest-Neighbor Lattice Gas

In this section, we take further advantage of the binary nature of the lattice gas to link it precisely to the Ising ferromagnet. The occupation numbers $\eta_i \in \{0,1\}$ of the lattice gas can be mapped to Ising spins $\omega_i \in \{-1,+1\}$ by

$$\eta_i \mapsto \omega_i \overset{\text{def}}{=} 2\eta_i - 1. \tag{4.44}$$

Expressed in terms of the Ising spins, (4.1) becomes

$$\beta\{\mathscr{H}_{\Lambda;K}(\eta) - \mu N_\Lambda(\eta)\} = -\tfrac{\beta}{4}\sum_{\{i,j\}\subset\Lambda} K(i,j)\omega_i\omega_j - \tfrac{\beta}{4}(\kappa+2\mu)\sum_{i\in\Lambda}\omega_i - \beta(\tfrac{\mu}{2}+\tfrac{\kappa}{8})|\Lambda| + b_\Lambda,$$

where $b_\Lambda = o(|\Lambda|)$ (see exercise below). We thus see that the lattice gas is linked to an Ising ferromagnet with coupling constants $J_{ij} = \frac{\beta}{4}K(i,j) \geq 0$ and magnetic field $h' = \frac{\beta}{4}(\kappa + 2\mu)$.

Exercise 4.8. Compute b_Λ, and show that

$$\lim_{\Lambda \Uparrow \mathbb{Z}^d} \frac{|b_\Lambda|}{|\Lambda|} = 0.$$

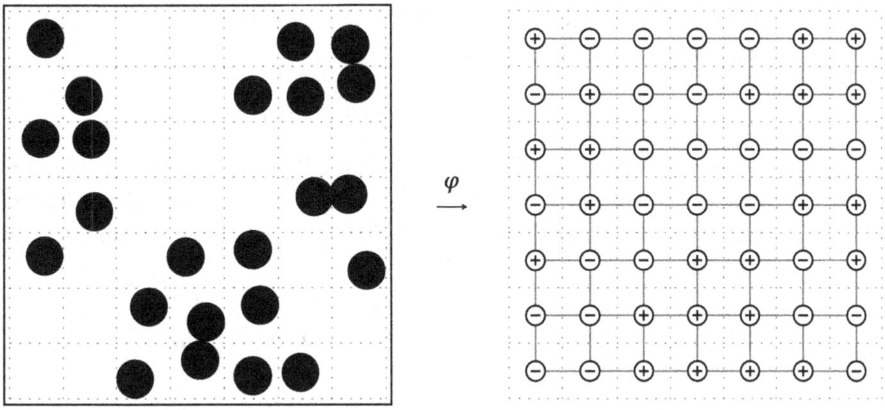

Figure 4.6. Left: a configuration of the lattice gas, in which each cell i is either occupied by a particle, or empty. Right: the occupation variables $\eta_i \in \{0,1\}$ are mapped to spin variables $\omega_i \in \{\pm 1\}$ using the mapping (4.44).

In order to take advantage of the results obtained in Chapter 3, in this subsection we restrict our attention to the **nearest-neighbor lattice gas**, for which

$$K(i,j) \stackrel{\text{def}}{=} \mathbf{1}_{\{i\sim j\}} . \tag{4.45}$$

In this case, $\kappa = 2d$.

4.8.1 Pressure

The parameters (β, μ) of the grand canonical lattice gas are related to those of the nearest-neighbor Ising model, (β', h'), by the relations

$$\beta' = \tfrac{1}{4}\beta , \qquad h' = \tfrac{\beta}{4}(\kappa + 2\mu) .$$

By Exercise 4.8, for all $\epsilon > 0$, one can take n sufficiently large that

$$e^{-\epsilon|\mathrm{B}(n)|} e^{\beta(\frac{\mu}{2}+\frac{\kappa}{8})|\mathrm{B}(n)|} \mathbf{Z}^{\varnothing}_{\mathrm{B}(n);\beta',h'} \le \Theta_{\mathrm{B}(n);\beta,\mu} \le e^{\epsilon|\mathrm{B}(n)|} e^{\beta(\frac{\mu}{2}+\frac{\kappa}{8})|\mathrm{B}(n)|} \mathbf{Z}^{\varnothing}_{\mathrm{B}(n);\beta',h'} .$$

We thus get, after taking the limits $n \to \infty$ and $\epsilon \downarrow 0$,

$$\beta p_\beta(\mu) = \psi_{\beta'}(h') + \tfrac{\beta\mu}{2} + \tfrac{\beta\kappa}{8} . \tag{4.46}$$

We can now extract qualitative information from the Ising model and translate it into the lattice gas language.

For instance, we know from Theorem 3.9 that, in $d = 1$, $h' \mapsto \psi_{\beta'}(h')$ is everywhere analytic in h' (at all temperatures). This implies that the corresponding lattice gas has no phase transition, $\mu \mapsto p_\beta(\mu)$ being analytic everywhere. In fact, the exact solution of Theorem 3.9, together with (4.46), yields an explicit expression for $p_\beta(\mu)$.

When $d \ge 2$, it follows from the Lee-Yang Circle Theorem and Theorem 3.40 that $\psi_{\beta'}$ is analytic at least outside $h' = 0$. Therefore, to $h' = 0$ corresponds the unique value of the chemical potential at which the lattice gas can exhibit a first-order phase transition, namely:

$$\mu_* \stackrel{\text{def}}{=} -\tfrac{1}{2}\kappa\,.$$

We also know, from Theorem 3.34 and Peierls' argument, that there exists an inverse critical temperature $\beta_c(d) \in (0,\infty)$ such that a first-order phase transition does occur whenever $\beta' > \beta_c(d)$; see (3.27). We gather these results in the following:

Theorem 4.22. *Let $\mu \mapsto p_\beta(\mu)$ denote the pressure of the nearest-neighbor lattice gas.*

1. *When $d = 1$, p_β is analytic everywhere.*
2. *When $d \geq 2$, p_β is analytic everywhere on $\{\mu : \mu \neq \mu_*\}$. Moreover, letting*
$$\beta_c^{\text{l.g.}} = \beta_c^{\text{l.g.}}(d) \stackrel{\text{def}}{=} \tfrac{1}{4}\beta_c(d)\,,$$
p_β is differentiable at μ_* if $\beta < \beta_c^{\text{l.g.}}$, but nondifferentiable at μ_* if $\beta > \beta_c^{\text{l.g.}}$.

In particular, at all temperatures, the density of particles $\mu \mapsto \rho_\beta(\mu) = \frac{\partial p_\beta}{\partial \mu}$ exists (and is analytic) everywhere outside μ_*. Using (4.46), this can be related directly to the infinite-volume magnetization $m_{\beta'}$ of the Ising model:

$$\rho_\beta(\mu) = \frac{\partial p_\beta}{\partial \mu} = \frac{1}{\beta}\frac{\partial \psi_{\beta'}}{\partial h'}\frac{\partial h'}{\partial \mu} + \frac{1}{2} = \frac{1 + m_{\beta'}(h')}{2}\,.$$

(We write $m_{\beta'}(h')$ rather than $m(\beta', h')$, since we are mainly interested in the dependence on h'.) We call $(-\infty, \mu_*)$ the **gas branch** of the pressure, and $(\mu_*, +\infty)$ the **liquid branch**. Although the pressure is not differentiable at μ_* when $\beta > \beta_c^{\text{l.g.}}$, convexity guarantees that its one-sided derivatives are well defined and given by

$$\rho_l \stackrel{\text{def}}{=} \frac{\partial p_\beta}{\partial \mu^+}\Big|_{\mu_*} = \frac{1 + m_{\beta'}^*}{2}\,, \qquad \rho_g \stackrel{\text{def}}{=} \frac{\partial p_\beta}{\partial \mu^-}\Big|_{\mu_*} = \frac{1 - m_{\beta'}^*}{2}\,.$$

At μ_*, the grand canonical system becomes sensitive to the boundary condition, and the density is only guaranteed to satisfy $\rho_g \leq \rho \leq \rho_l$. The reader can actually take a look back at the pictures of Figure 1.9 for the typical configurations of the lattice gas at low and high temperature. Observe that the densities ρ_g, ρ_l always satisfy

$$\rho_g + \rho_l = 1\,. \tag{4.47}$$

(Let us mention that this property is really a consequence of the hidden spin-flip symmetry of the underlying Ising model, and does not hold in general lattice gases.) The pressure and the density therefore have the qualitative behavior displayed in Figure 4.7. (We remind the reader that the graphs shown in this section are only qualitative; their purpose is to emphasize the main features observed in the nearest-neighbor lattice gas.)

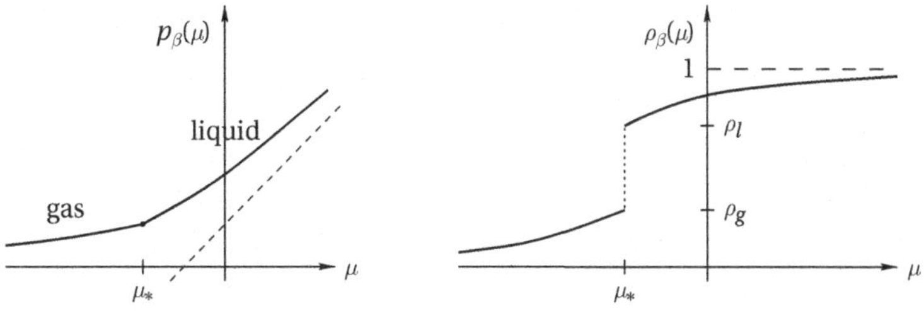

Figure 4.7. The pressure (left) and density (right) of the nearest-neighbor lattice gas, exhibiting a first-order phase transition when $\beta > \beta_c^{\text{l.g.}}$. The function $\mu \mapsto p_\beta(\mu)$ is analytic everywhere, except at μ_*, at which the one-sided derivatives differ, implying a jump in density, corresponding to the change from a gas of density ρ_g to a liquid of density ρ_l.

4.8.2 Free Energy

In the canonical ensemble, we can obtain the main qualitative properties of the free energy using the fact that it is the Legendre transform of the pressure and applying Theorem B.20.

At high temperature, $\beta < \beta_c^{\text{l.g.}}$, the pressure is differentiable everywhere and the free energy is therefore strictly convex; see Figure 4.8. By Theorem 4.19, this implies that the typical configurations under the canonical Gibbs distribution are always spatially homogeneous, at all densities.

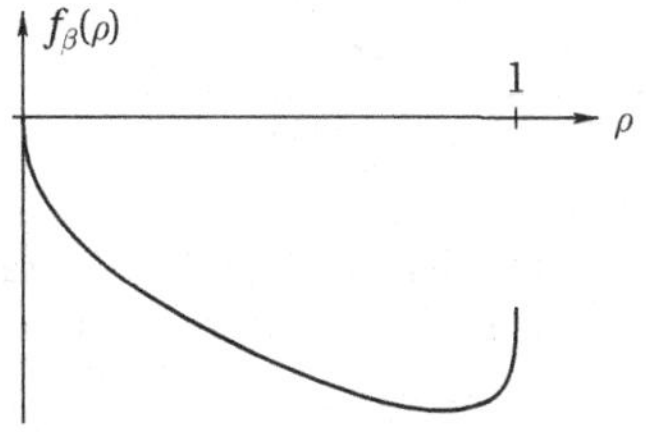

Figure 4.8. The free energy of the nearest-neighbor lattice gas is strictly convex when $\beta < \beta_c^{\text{l.g.}}$.

At low temperature, when $\beta > \beta_c^{\text{l.g.}}$, p_β is not differentiable at μ_* and, again by Theorem B.20, f_β is affine on the interval $[\rho_g, \rho_l]$, called the **coexistence plateau**. As for the pressure, we refer to $(0, \rho_g)$ as the **gas branch**, and to $(\rho_l, 1)$ as the **liquid branch**; see Figure 4.9.

Exercise 4.9. Show that when $\beta > \beta_c^{\text{l.g.}}$, f_β is analytic on the gas and liquid branches. *Hint:* Use Theorem 4.22, the strict convexity of the pressure, and the Implicit Function Theorem.

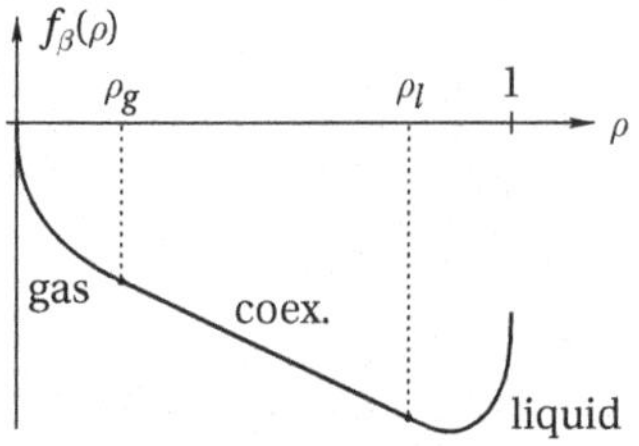

Figure 4.9. The free energy of the nearest-neighbor lattice gas, when $\beta > \beta_c^{\text{l.g.}}$, is analytic everywhere (see Exercise 4.9) except at ρ_g and ρ_l. On the coexistence plateau $[\rho_g, \rho_l]$, both gas and liquid are present in the system, in various proportions: there is *coexistence* and *phase separation.*

4.8.3 Typical Densities

The typical density $\frac{N_\Lambda}{|\Lambda|}$ under the grand canonical Gibbs distribution $\nu_{\Lambda;\beta,\mu}$ can be characterized using the analysis following Theorem 4.15; see Figure 4.10.

When $\beta \leq \beta_c^{\text{l.g.}}$ and for all $\mu \in \mathbb{R}$, the rate function $I_{\beta,\mu}$ is strictly convex and has a unique minimizer, $\mathscr{M}_{\beta,\mu} = \{\partial p_\beta / \partial \mu\}$, at which the density concentrates. The scenario is similar if $\beta > \beta_c^{\text{l.g.}}$ and $\mu \neq \mu_*$.

When $\beta > \beta_c^{\text{l.g.}}$ and $\mu = \mu_*$, the rate function attains its minima on the coexistence plateau: $\mathscr{M}_{\beta,\mu} = [\rho_g, \rho_l]$.

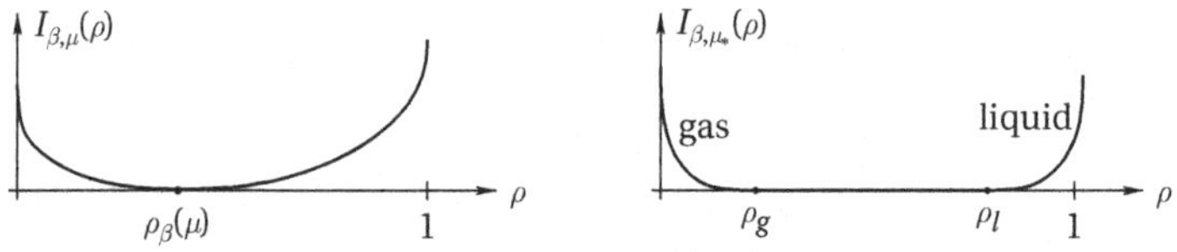

Figure 4.10. Values of the density at which the rate function does not attain its minimum are very unlikely to be observed in a large system distributed according to $\nu_{\Lambda;\beta,\mu}$. When $\beta < \beta_c^{\text{l.g.}}$, or when $\beta > \beta_c^{\text{l.g.}}$ and $\mu \neq \mu_*$ (left), $I_{\beta,\mu}(\rho)$ has a unique minimum at $\rho_\beta(\mu)$. When $\beta > \beta_c^{\text{l.g.}}$ and $\mu = \mu_*$ (right), $I_{\beta,\mu_*}(\rho)$ is minimal on the whole coexistence plateau.

Theorem 4.15 does not provide any information on the typical densities when $\mu = \mu_*$, beyond concentration on the coexistence plateau, and a more detailed analysis is necessary; this will be discussed in Section 4.12.1.

4.8.4 Pressure as a Function of ρ and v

Let us now express the pressure in terms of either the density ρ or the volume per particle $v = \rho^{-1}$.

When $\beta \leq \beta_c^{\text{l.g.}}$, p_β is differentiable for all values of μ and $\rho_\beta(\mu) = \frac{\partial p_\beta}{\partial \mu}$ is continuous (Theorem B.12) and increasing. Remember from Exercise 4.6 that $\rho_\beta(\mu) \to 0$

as $\mu \to -\infty$, and $\rho_\beta(\mu) \to 1$ as $\mu \to +\infty$. Therefore, one can proceed as for the hard-core gas: for any $\rho \in (0,1)$, the equation

$$\rho_\beta(\mu) = \rho \tag{4.48}$$

has a unique solution, which we denote $\mu_\beta^{\text{l.g.}}(\rho)$. We can thus define

$$\widetilde{p}_\beta(\rho) \stackrel{\text{def}}{=} p_\beta(\mu_\beta^{\text{l.g.}}(\rho)), \quad \rho \in (0,1). \tag{4.49}$$

When $\beta > \beta_c^{\text{l.g.}}$, existence of a solution to (4.48) is not guaranteed for all $\rho \in (0,1)$, because of the jump of $\rho_\beta(\mu)$ at μ_* (see Figure 4.7). In fact, inversion is only possible when $\rho < \rho_g$ or $\rho > \rho_l$; in this case, we also denote the inverse by $\mu_\beta^{\text{l.g.}}(\rho)$. To extend this function to a well-defined $\mu_\beta^{\text{l.g.}} : (0,1) \to \mathbb{R}$, we set

$$\mu_\beta^{\text{l.g.}}(\rho) \stackrel{\text{def}}{=} \mu_* \qquad \forall \rho \in [\rho_g, \rho_l]. \tag{4.50}$$

The reason for defining the inverse that way on the coexistence plateau is that, for finite systems, the average particle density in Λ, which is equal to $\frac{\partial p_{\Lambda;\beta}(\mu)}{\partial \mu}$, is increasing and differentiable (in fact, analytic) as a function of μ. In particular, with any density $\rho \in (0,1)$ is associated a unique value $\mu_{\Lambda;\beta}(\rho)$ of the chemical potential. The latter function being increasing, it is clear that it converges for all $\rho \in (0,1)$, as $\Lambda \Uparrow \mathbb{Z}^d$, to the function $\mu_\beta^{\text{l.g.}}(\rho)$ defined above. ◇

We can then define $\widetilde{p}_\beta(\rho)$ as in (4.49). Its qualitative behavior is sketched in Figure 4.11.

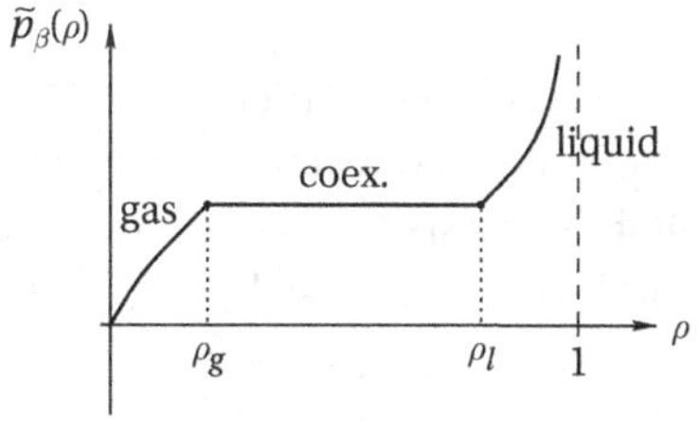

Figure 4.11. The pressure of the nearest-neighbor lattice gas at low temperature, as a function of the density $\rho \in (0,1)$.

Remark 4.23. In Section 5.7.2, we will see that the nearest-neighbor lattice gas also presents the ideal gas behavior at small densities:

$$\beta\widetilde{p}_\beta(\rho) = \rho + O(\rho^2), \qquad (\rho \text{ small}).$$

Using the *cluster expansion technique*, we will see in Theorem 5.12 that the coefficients of the **virial expansion** can actually be computed, yielding the exact higher-order corrections to the pressure at low density:

$$\beta\widetilde{p}_\beta(\rho) = \rho + b_2\rho^2 + b_3\rho^3 + \cdots \qquad (\rho \text{ small}).$$ ◇

Exercise 4.10. Show that, when $\beta > \beta_c^{\text{l.g.}}$, $\rho \mapsto \widetilde{p}_\beta(\rho)$ is analytic on the gas and liquid branches.

Finally, we can also express the pressure as a function of the volume per particle, $v = \rho^{-1}$,

$$\widehat{p}_\beta(v) \stackrel{\text{def}}{=} \widetilde{p}_\beta(v^{-1}), \qquad v \in (1, \infty),$$

to obtain the qualitative behavior of the isotherms in a form directly comparable to the van der Waals–Maxwell theory. The sketch of a typical low-temperature isotherm is given in Figure 4.12, where we have set $v_l \stackrel{\text{def}}{=} \rho_l^{-1}$ and $v_g \stackrel{\text{def}}{=} \rho_g^{-1}$.

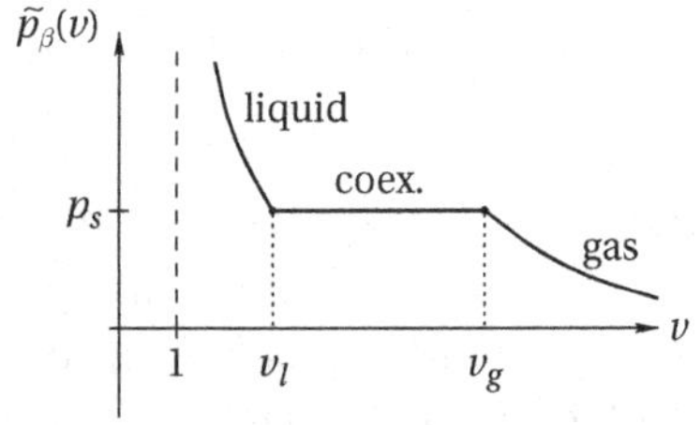

Figure 4.12. The pressure of the nearest-neighbor lattice gas at low temperature, as a function of the volume per particle $v > 1$. The value $v = 1$ plays the same role as $v = b$ in van der Waals' isotherms (Figure 1.4). The **saturation pressure** is given by $p_s = p_\beta(\mu_*)$.

4.9 Van der Waals Lattice Gas

In this section, we consider a lattice gas that does not fit in the general framework described earlier, but which will be important from the point of view of the van der Waals–Maxwell theory, especially in the next section.

Consider a lattice gas in a vessel $\Lambda \Subset \mathbb{Z}^d$, in which the interaction between the particles at vertices $i, j \in \Lambda$ is given by

$$K(i,j) \stackrel{\text{def}}{=} \frac{1}{|\Lambda|}. \tag{4.51}$$

This type of interaction is not physical, since the contribution to the total energy from a pair of particles depends on the size of the region Λ in which they live: it becomes of infinite range and tends to zero when $|\Lambda| \to \infty$. Nevertheless, the sum over the pairs of particles can be expressed as

$$\begin{aligned}\sum_{\{i,j\}\subset\Lambda} K(i,j)\eta_i\eta_j &= \frac{1}{2|\Lambda|}\sum_{i\in\Lambda}\sum_{\substack{j\in\Lambda\\ j\neq i}} \eta_i\eta_j \\ &= \frac{1}{2|\Lambda|}\sum_{i\in\Lambda}\eta_i\Big(\sum_{j\in\Lambda}\eta_j - \eta_i\Big) = \tfrac{1}{2}\rho_\Lambda^2|\Lambda| - \tfrac{1}{2}\rho_\Lambda\,,\end{aligned}$$

where $\rho_\Lambda \stackrel{\text{def}}{=} \frac{N_\Lambda}{|\Lambda|}$ is the empirical density. Since it is bounded, the second term $-\frac{1}{2}\rho_\Lambda$ does not contribute on the macroscopic scale and will be neglected.

Therefore, although not physically realistic, interactions of the form (4.51) lead to a model in which the square of the density appears explicitly in the Hamiltonian. In this sense, it can be considered as a microscopic toy model that embodies the main assumption made by van der Waals and discussed in Chapter 1. This model, with Hamiltonian

$$\mathscr{H}^{\text{vW}}_{\Lambda;\mu} \stackrel{\text{def}}{=} -\tfrac{1}{2}\rho_\Lambda^2|\Lambda| - \mu\rho_\Lambda|\Lambda|,$$

will be called the **van der Waals model.**

We encountered the same interaction (formulated in the spin language) when considering the Curie–Weiss model in Section 2.1 (remember in particular the Curie–Weiss Hamiltonian (2.4)). Therefore, this model could also be called the **mean-field** or **Curie–Weiss** lattice gas. A large part of the rest of this section will be the translation of the discussion of Section 2.1 into the lattice gas language. We nevertheless discuss a new important feature: its link with Maxwell's construction.

Let us denote the canonical and grand canonical partition functions of the van der Waals model by $\mathbf{Q}^{\text{vW}}_{\Lambda;\beta,N}$, respectively $\Theta^{\text{vW}}_{\Lambda;\beta,\mu}$, and consider the associated free energy and pressure:

$$f^{\text{vW}}_\beta(\rho) \stackrel{\text{def}}{=} \lim_{n\to\infty} \frac{-1}{\beta|\Lambda_n|}\log \mathbf{Q}^{\text{vW}}_{\Lambda_n;\beta,\lceil\rho|\Lambda_n|\rceil}, \qquad \rho\in[0,1],$$

$$p^{\text{vW}}_\beta(\mu) \stackrel{\text{def}}{=} \lim_{n\to\infty} \frac{1}{\beta|\Lambda_n|}\log \Theta^{\text{vW}}_{\Lambda_n;\beta,\mu}, \qquad \mu\in\mathbb{R}.$$

The dependence of $K(i,j)$ on Λ prevents us from using Theorems 4.5 and 4.11 to show the existence of these limits. However, it is not difficult to compute the latter explicitly. Remember the definition of $s^{\text{l.g.}}(\rho)$ in (4.14).

Theorem 4.24. *The above limits exist, and are given by*

$$f^{\text{vW}}_\beta(\rho) = -\tfrac{1}{2}\rho^2 - \tfrac{1}{\beta}s^{\text{l.g.}}(\rho), \tag{4.52}$$

$$p^{\text{vW}}_\beta(\mu) = \sup_{\rho\in[0,1]}\{\mu\rho - f^{\text{vW}}_\beta(\rho)\}. \tag{4.53}$$

Exactly as we already saw in (2.5), the free energy splits into an **energy** term $-\frac{1}{2}\rho^2$ and an **entropy** term $-\frac{1}{\beta}s^{\text{l.g.}}(\rho)$.

Proof The simple structure of the Hamiltonian yields

$$\mathbf{Q}^{\text{vW}}_{\Lambda;\beta,N} = \binom{|\Lambda|}{N} e^{\frac{1}{2}\beta\rho_\Lambda^2|\Lambda|}.$$

We then use (4.13) and get (4.52). For the pressure, we use a decomposition like the one in (4.24) and proceed as in the proof of (4.32). □

Of course, the properties of f_β^{vW} and p_β^{vW} can also be derived directly from those of the Curie–Weiss model. Therefore, parts of the material presented below have already been presented, in a different form, in Chapter 2.

By (4.53), p_β^{vW} is the Legendre transform of f_β^{vW}, but the converse is only true when f_β^{vW} is convex.

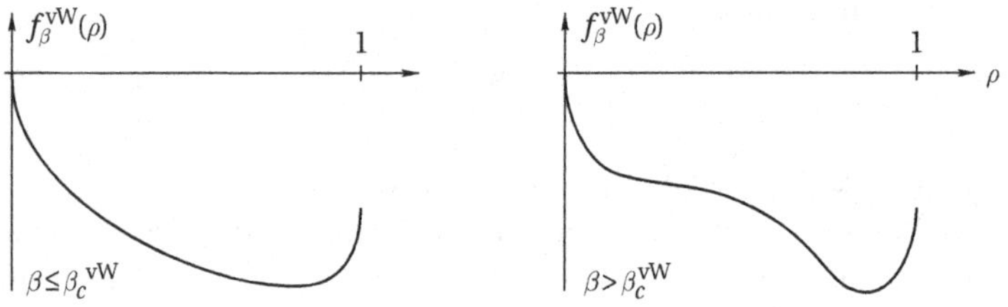

Figure 4.13. The free energy (4.52) of the van der Waals model at high (left) and low (right) temperatures.

4.9.1 (Non)Convexity of the Free Energy

Since $f_\beta^{\mathrm{vW}}(\rho)$ is the sum of a concave energy term and a convex entropy term, its convexity is not clear a priori. But an elementary computation shows that

$$\frac{\partial^2 f_\beta^{\mathrm{vW}}}{\partial \rho^2} \geq 0 \quad \forall \rho \in (0,1) \qquad \text{if and only if} \qquad \beta \leq \beta_c^{\mathrm{vW}}, \tag{4.54}$$

where the **critical inverse temperature** is

$$\beta_c^{\mathrm{vW}} \stackrel{\text{def}}{=} 4.$$

We can thus determine exactly when f_β^{vW} is the Legendre transform of p_β^{vW}:

1. When $\beta \leq \beta_c^{\mathrm{vW}}$, f_β^{vW} is convex and, since the Legendre transform is an involution on convex lower semi-continuous functions (Theorem B.19), this means that

$$f_\beta^{\mathrm{vW}}(\rho) = \sup_{\mu \in \mathbb{R}} \{\rho\mu - p_\beta^{\mathrm{vW}}(\mu)\}.$$

   Therefore, *equivalence of ensembles holds at high temperature.*
2. When $\beta > \beta_c^{\mathrm{vW}}$, f_β^{vW} is nonconvex and therefore cannot be the Legendre transform of p_β^{vW} (see Exercise B.6): there exist values of ρ for which

$$f_\beta^{\mathrm{vW}}(\rho) \neq \sup_{\mu \in \mathbb{R}} \{\rho\mu - p_\beta^{\mathrm{vW}}(\mu)\}. \tag{4.55}$$

   Therefore, *equivalence of ensembles does not hold at low temperature.*

The reader might wonder whether physical significance can be attached to the Legendre transform of the pressure, namely the right-hand side of (4.55). In fact, since the pressure is the Legendre transform of the free energy (by (4.53)), its Legendre transform is given by (see Theorem B.17)

$$\sup_{\mu\in\mathbb{R}}\{\rho\mu - p_\beta^{\mathrm{vW}}(\mu)\} = \mathrm{CE}f_\beta^{\mathrm{vW}}(\rho), \tag{4.56}$$

where $\mathrm{CE}f_\beta^{\mathrm{vW}}(\rho)$ is the **convex envelope**[2] **of** f_β^{vW}, defined by

$$\mathrm{CE}f_\beta^{\mathrm{vW}} \stackrel{\mathrm{def}}{=} \text{largest convex function } g \text{ such that } g \le f_\beta^{\mathrm{vW}} . \tag{4.57}$$

The regime $\beta > \beta_c^{\mathrm{vW}}$ thus corresponds to $\mathrm{CE}f_\beta^{\mathrm{vW}} \neq f_\beta^{\mathrm{vW}}$ (see Figure 4.14). In the next section, we will relate this to Maxwell's construction.

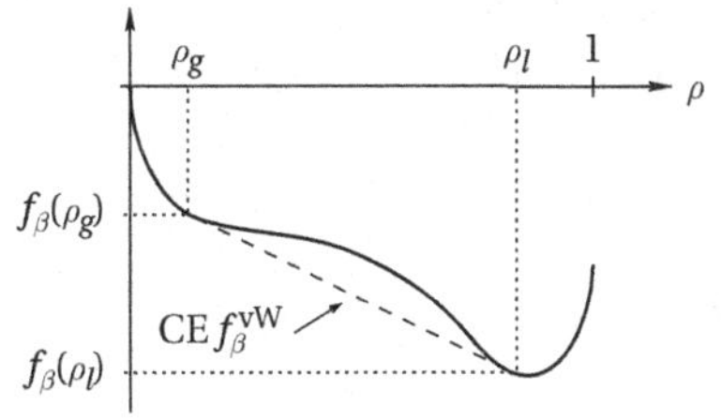

Figure 4.14. At low temperature, the free energy of the van der Waals model differs from its convex envelope. The points ρ_g and ρ_l will be identified below.

The nonconvexity observed at low temperature in the van der Waals model is due to the fact that the geometry of the system plays no role: any pair of particles interacts in the same way, no matter how distant. Therefore, not surprisingly, we end up with the same conclusions as in van der Waals' theory when making the homogeneity assumption.

Convexity is known, since Chapter 1, to be a consequence of the variational principles satisfied by the fundamental functions of thermostatics. For systems with finite-range interactions, it appeared in the proof of Theorem 4.5. In the present setting, the argument can be formulated as follows. A system with density $\rho \in (\rho_g, \rho_l)$ living in Λ can always be split into two subsystems: a first one, with volume $|\Lambda_1| = \alpha|\Lambda|$ and density ρ_g and a second one with volume $|\Lambda_2| = (1-\alpha)|\Lambda|$ and density ρ_l, where α is chosen such that the overall density is unchanged: $\alpha\rho_g + (1-\alpha)\rho_l = \rho$. Indeed, the free energy density associated with these two systems is $\alpha f(\rho_g) + (1-\alpha)f(\rho_l)$, which is smaller than the free energy density of the original system: $\alpha f(\rho_g) + (1-\alpha)f(\rho_l) \le f(\alpha\rho_g + (1-\alpha)\rho_l) = f(\rho)$.

The reason this does not occur in the van der Waals model is that it is impossible to split the system into two pieces in such a way that the energy of interaction between the two subsystems is negligible (that is, is $o(|\Lambda|)$). This peculiarity, ultimately due to the long-range nature of the interactions, explains the unphysical features of these

[2] A more precise definition can be found in Appendix B.2.3.

systems, such as the nonconvexity of the free energy. In models with short-range interactions, such a splitting is indeed possible, and the spatial coexistence of gas and liquid phases occurs at the phase transition. ◇

4.9.2 An Expression for the Pressure: Maxwell's Construction

We have already seen in (4.56) that the Legendre transform of the pressure is given by $\mathrm{CE} f_\beta^{\mathrm{vW}}$. At low temperature, $\mathrm{CE} f_\beta^{\mathrm{vW}}$ is affine on a segment (Figure 4.14), and by Theorem B.20 this implies that p_β^{vW} has a point of nondifferentiability. We make this analysis more explicit below.

Using (4.53), the analysis of the pressure $p_\beta^{\mathrm{vW}}(\mu)$ for a fixed μ can be done through the study of the maxima of the function $\rho \mapsto \mu\rho - f_\beta^{\mathrm{vW}}(\rho)$. Since f_β^{vW} is differentiable, these can be found by solving

$$\frac{\partial f_\beta^{\mathrm{vW}}}{\partial \rho} = \mu\,, \tag{4.58}$$

which can be written as

$$\theta(\rho) = \beta(\rho + \mu)\,, \tag{4.59}$$

where $\theta(\rho) \stackrel{\mathrm{def}}{=} \log \frac{\rho}{1-\rho}$. We fix μ and make a qualitative analysis of the solutions (in ρ) to (4.59). Observe that $\theta(\rho) \to -\infty$ when $\rho \downarrow 0$, and $\theta(\rho) \to +\infty$ when $\rho \uparrow 1$. The graph of $\theta(\rho)$ therefore intersects the straight line $\rho \mapsto \beta(\rho + \mu)$ at least once, for all $\mu \in \mathbb{R}$ and $\beta > 0$; see Figure 4.15. However, the graph of $\theta(\rho)$ may intersect that line more than once. Actually, since $\theta'(\rho) \geq \theta'(\frac{1}{2}) = 4 = \beta_c^{\mathrm{vW}}$, we see that this intersection is unique when $\beta \leq \beta_c^{\mathrm{vW}}$, but not necessarily so when $\beta > \beta_c^{\mathrm{vW}}$.

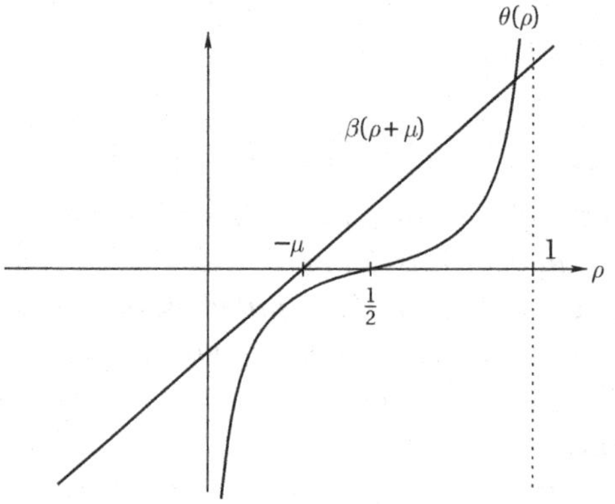

Figure 4.15. Solving (4.59).

Van der Waals Pressure When $\beta \leq \beta_c^{\mathrm{vW}}$

When $\beta \leq \beta_c^{\mathrm{vW}}$, the unique solution to (4.59), denoted $\mu \mapsto \rho_\beta^{\mathrm{vW}}(\mu)$, is differentiable (analytic in fact) with respect to μ, and the pressure is given by

$$p_\beta^{\mathrm{vW}}(\mu) = \mu\rho_\beta^{\mathrm{vW}}(\mu) - f_\beta^{\mathrm{vW}}(\rho_\beta^{\mathrm{vW}}(\mu))\,. \tag{4.60}$$

Since $\rho_\beta^{\text{vW}}(\mu) = \rho$ can be inverted to obtain $\mu_\beta^{\text{vW}}(\rho)$, we can express the pressure as a function of the density; from (4.58),

$$\widetilde{p}_\beta^{\text{vW}}(\rho) \stackrel{\text{def}}{=} p_\beta^{\text{vW}}(\mu_\beta(\rho)) = \frac{\partial f_\beta^{\text{vW}}}{\partial \rho}\rho - f_\beta^{\text{vW}}(\rho) = -\tfrac{1}{2}\rho^2 - \tfrac{1}{\beta}\log(1-\rho)\,. \tag{4.61}$$

Exercise 4.11. Check (4.61).

Again, at low densities, (4.61) reduces to the equation of state for the ideal gas (1.14):

$$\beta\widetilde{p}_\beta^{\text{vW}}(\rho) = \rho + O(\rho^2)\,.$$

As a function of $v = \rho^{-1}$ (see Figure 4.19),

$$\widehat{p}_\beta^{\text{vW}}(v) = -\frac{1}{2v^2} - \frac{1}{\beta}\log\Bigl(1 - \frac{1}{v}\Bigr)\,. \tag{4.62}$$

Once more, when v is large, $\widehat{p}_\beta^{\text{vW}}(v)$ is well approximated by the solution to

$$\Bigl(p + \frac{1}{2v^2}\Bigr)v = \beta^{-1}\,,$$

which is essentially van der Waals' expression (1.23), with $a = \frac{1}{2}$. (For large v, one could also replace v by $v-1$.)

Van der Waals Pressure When $\beta > \beta_c^{\text{vW}}$

When $\beta > \beta_c^{\text{vW}}$, the pressure is also of the form (4.60), but the solution to (4.59) may not be unique. In that case, one must select those that correspond to a maxima of $\rho \mapsto \mu\rho - f_\beta(\rho)$. This can be made visually transparent by defining a new variable:

$$x \stackrel{\text{def}}{=} \rho - \tfrac{1}{2}\,.$$

(This change of variable symmetrizes the problem using a variable better suited than ρ. The analysis then reduces to the one done for the Curie–Weiss model, in Section 2.3.) After rearranging the terms, we are thus looking for the points $x \in [-\frac{1}{2}, \frac{1}{2}]$ that maximize the function

$$x \mapsto \varphi_\mu(x) \stackrel{\text{def}}{=} (\mu + \tfrac{1}{2})x - g_\beta(x)\,,$$

where $g_\beta(x) \stackrel{\text{def}}{=} -\frac{1}{2}x^2 - \frac{1}{\beta}s_\beta(\frac{1}{2} + x)$, and where we have ignored a term that depends on μ but not on x. The advantage of working with the variable x is that $g_\beta(-x) = g_\beta(x)$. This shows that

$$\mu_*^{\text{vW}} \stackrel{\text{def}}{=} -\tfrac{1}{2}$$

is the only value of μ for which φ_μ is symmetric and has two distinct maximizers. For all other values of μ this maximizer is unique; see Figure 4.16.

When μ increases from $\mu < \mu_*^{\text{vW}}$ to $\mu > \mu_*^{\text{vW}}$, the unique maximizer of $x \mapsto \varphi_\mu(x)$ jumps discontinuously from a value $x_g \stackrel{\text{def}}{=} \rho_g - \frac{1}{2} < 0$ to a value $x_l \stackrel{\text{def}}{=} \rho_l - \frac{1}{2} > 0$.

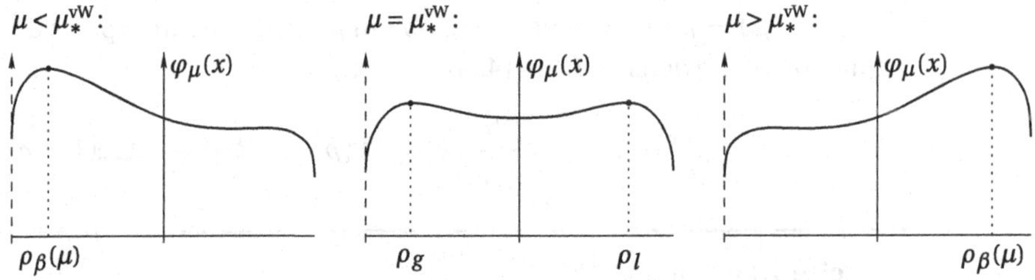

Figure 4.16. The maximizers of φ_μ.

We conclude that $\rho \mapsto \mu\rho - f_\beta^{\rm vW}(\rho)$ has two distinct maximizers when $\mu = \mu_*^{\rm vW}$: ρ_g and $\rho_l = 1 - \rho_g$. Moreover,

$$p_\beta^{\rm vW}(\mu_*^{\rm vW}) = \mu_*^{\rm vW}\rho_g - f_\beta(\rho_g) = \mu_*^{\rm vW}\rho_l - f_\beta(\rho_l). \tag{4.63}$$

When $\mu \neq \mu_*^{\rm vW}$, the maximizer is unique; we continue denoting it by $\rho_\beta^{\rm vW}(\mu)$. Of course, $\rho_\beta^{\rm vW}(\mu) < \rho_g$ when $\mu < \mu_*^{\rm vW}$, $\rho_\beta^{\rm vW}(\mu) > \rho_l$ when $\mu > \mu_*^{\rm vW}$, and

$$\rho_g = \lim_{\mu \uparrow \mu_*^{\rm vW}} \rho_\beta^{\rm vW}(\mu), \quad \rho_l = \lim_{\mu \downarrow \mu_*^{\rm vW}} \rho_\beta^{\rm vW}(\mu).$$

Since we have, for $\mu \neq \mu_*^{\rm vW}$,

$$\frac{\partial p_\beta^{\rm vW}}{\partial \mu} = \frac{\partial}{\partial \mu}\{\mu\rho_\beta^{\rm vW}(\mu) - f_\beta^{\rm vW}(\rho_\beta^{\rm vW}(\mu))\} = \rho_\mu^{\rm vW}(\mu)$$

and since $p_\beta^{\rm vW}$ is convex (being a Legendre transform), Theorem B.12 then gives

$$\frac{\partial p_\beta^{\rm vW}}{\partial \mu^-}\bigg|_{\mu_*^{\rm vW}} = \rho_g < \rho_l = \frac{\partial p_\beta^{\rm vW}}{\partial \mu^+}\bigg|_{\mu_*^{\rm vW}}.$$

Remark 4.25. By Theorem B.20, the lack of differentiability of $p_\beta^{\rm vW}$ at $\mu_*^{\rm vW}$ implies that its Legendre transform is affine on $[\rho_g, \rho_l]$ and $\mu_*^{\rm vW}$ gives the slope of $\mathrm{CE} f_\beta^{\rm vW}$ on $[\rho_g, \rho_l]$. Using the second and third terms in (4.63), we get

$$f_\beta^{\rm vW}(\rho_l) - f_\beta^{\rm vW}(\rho_g) = \mu_*^{\rm vW}(\rho_l - \rho_g) = \frac{\partial f_\beta^{\rm vW}}{\partial \rho}\bigg|_{\rho_g}(\rho_l - \rho_g).$$

This is clearly seen on the graph of Figure 4.14. ◇

Let us now complete the description of the pressure in terms of the variables ρ and v. Since $\rho_\beta^{\rm vW}(\mu)$ behaves discontinuously at $\mu_*^{\rm vW}$, we define its inverse as we did earlier (see page 201). For $\rho < \rho_g$ or $\rho > \rho_l$, $\rho_\beta(\mu) = \rho$ has an inverse which we again denote $\mu_\beta^{\rm vW}(\rho)$ and, for those densities, the pressure is obtained as in (4.61). Thus,

$$\widetilde{\mu}_\beta(\rho) \stackrel{\rm def}{=} \begin{cases} \mu_\beta^{\rm vW}(\rho) & \text{if } \rho \in (0, \rho_g), \\ \mu_*^{\rm vW} & \text{if } \rho \in [\rho_g, \rho_l], \\ \mu_\beta^{\rm vW}(\rho) & \text{if } \rho \in (\rho_l, 1). \end{cases}$$

As a function of the density, $\widetilde{p}_\beta^{\rm vW}(\rho) \stackrel{\rm def}{=} p_\beta(\widetilde{\mu}_\beta^{\rm vW}(\rho))$ then takes the following form (see Figure 4.17).

$$\widetilde{p}_\beta^{\rm vW}(\rho) = \begin{cases} -\frac{1}{2}\rho^2 - \frac{1}{\beta}\log(1-\rho) & \text{if } \rho \in (0, \rho_g), \\ p_\beta^{\rm vW}(\mu_*^{\rm vW}) & \text{if } \rho \in [\rho_g, \rho_l], \\ -\frac{1}{2}\rho^2 - \frac{1}{\beta}\log(1-\rho) & \text{if } \rho \in (\rho_l, 1). \end{cases} \tag{4.64}$$

Remark 4.26. On the gas branch, at any temperature, we can use the Taylor expansion for $\log(1-\rho)$, and get

$$\beta\widetilde{p}_\beta^{\rm vW}(\rho) = \rho + \tfrac{1}{2}(1-\beta)\rho^2 + \tfrac{1}{3}\rho^3 + \tfrac{1}{4}\rho^4 + \cdots.$$

The above series, which in fact converges for all complex ρ inside the unit disk, is called the **virial expansion** for the pressure. It provides high-order corrections to the equation of the ideal gas. ◇

Finally, we can express the pressure as a function of the volume per particle, $v = \rho^{-1}$. At low temperature $\widehat{p}_\beta^{\rm vW}(v)$ presents a striking difference with its high-temperature counterpart, computed earlier. Namely, at low temperature, the expression (4.62) has to be replaced by a *constant* on the coexistence plateau $[v_l, v_g]$, where $v_l = \rho_l^{-1}$, $v_g = \rho_g^{-1}$. Quite remarkably, this constant is the same as the one provided by Maxwell's construction; see Figure 4.18.

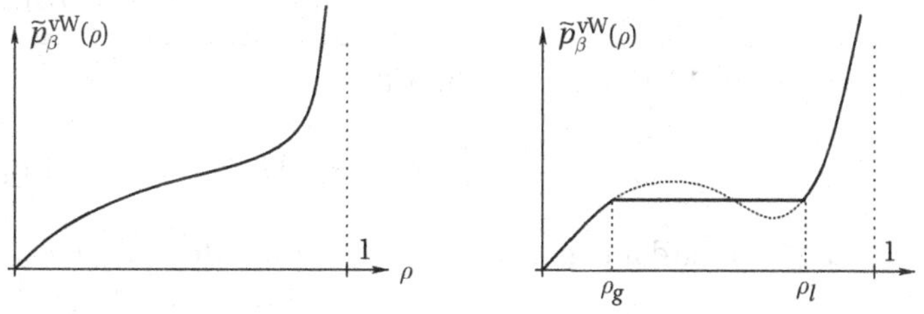

Figure 4.17. The pressure of the van der Waals model as a function of the density. Left: the regime $\beta \le \beta_c^{\rm vW}$; Right: $\beta > \beta_c^{\rm vW}$.

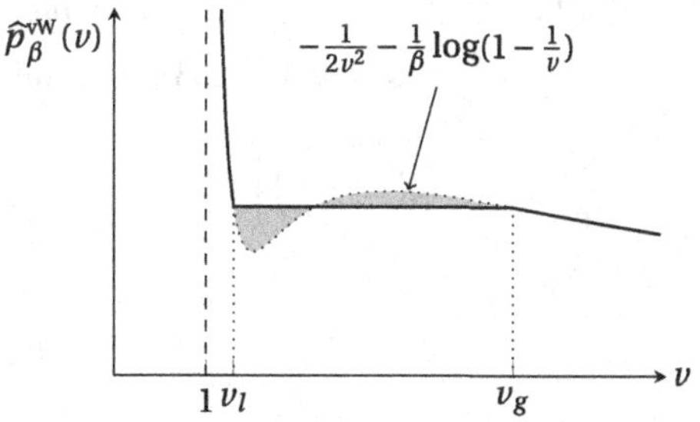

Figure 4.18. The pressure of the van der Waals model at low temperature as a function of the volume per particle $v > 1$, obtained by applying the *equal area rule* (Maxwell construction) to an everywhere smooth function: the two shaded areas are equal. The coexistence plateau is at a height given by the saturation pressure $p_\beta^{\rm vW}(\mu_*^{\rm vW})$.

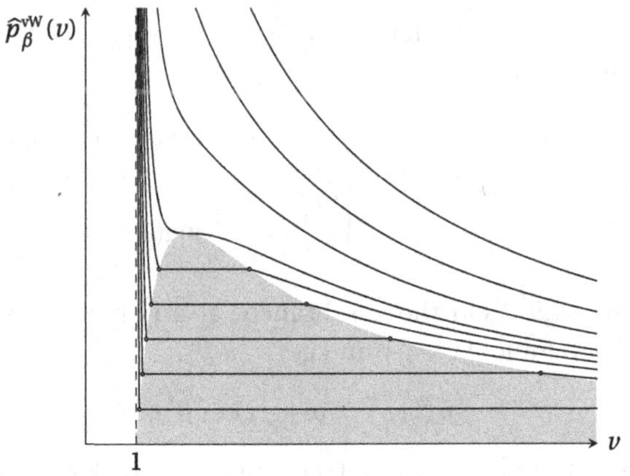

Figure 4.19. The pressure of the van der Waals-Maxwell model as a function of $v > 1$. The top four curves represent isotherms for values of $\beta \le \beta_c^{\mathrm{vW}}$ (see (4.62)) and are smooth everywhere; the fourth one is the critical isotherm. The remaining curves correspond to values $\beta > \beta_c^{\mathrm{vW}}$ and include a coexistence plateau due to Maxwell's construction. The shaded region represents the values of the parameters (v, p) located on a coexistence plateau. These plots were obtained from (4.65).

Theorem 4.27 (Maxwell's Construction). *When* $\beta > \beta_c^{\mathrm{vW}}$,

$$\widehat{p}^{\,\mathrm{vW}}_{\beta}(v) = \mathrm{MC}\Big\{-\frac{1}{2v^2} - \frac{1}{\beta}\log\Big(1 - \frac{1}{v}\Big)\Big\}. \tag{4.65}$$

Proof As in (1.24), we must show that

$$\int_{v_l}^{v_g} \Big\{-\frac{1}{2v^2} - \frac{1}{\beta}\log\Big(1 - \frac{1}{v}\Big)\Big\}\,\mathrm{d}v = p_\beta(\mu_*^{\mathrm{vW}})(v_g - v_l). \tag{4.66}$$

A straightforward integration shows that this integral equals, after rearrangement,

$$\begin{aligned}\tfrac{1}{2}\Big[\tfrac{1}{v}\Big]_{v_l}^{v_g} - \tfrac{1}{\beta}\Big[(v-1)\log(v-1) - v\log v\Big]_{v_l}^{v_g} &= \tfrac{1}{2}\Big\{\tfrac{1}{v_g} - \tfrac{1}{v_l}\Big\} - \tfrac{1}{\beta}\Big[-vs(\tfrac{1}{v})\Big]_{v_l}^{v_g} \\ &= -\tfrac{1}{2}\tfrac{v_g - v_l}{v_l v_g} - \tfrac{1}{\beta}\Big\{v_l s(\tfrac{1}{v_l}) - v_g s(\tfrac{1}{v_g})\Big\}.\end{aligned}$$

We then use the fact that $-\frac{1}{2} = \mu_*^{\mathrm{vW}}$, $\frac{1}{v_l v_g} = \rho_l\rho_g = \rho_g(1-\rho_g)$, $s(\frac{1}{v_l}) = s(\rho_l) = s(1-\rho_l) = s(\rho_g) = s(\frac{1}{v_g})$, as well as (4.63), to obtain

$$\begin{aligned}\int_{v_l}^{v_g} \Big\{-\frac{1}{2v^2} - \frac{1}{\beta}\log\Big(1 - \frac{1}{v}\Big)\Big\}\,\mathrm{d}v &= (v_g - v_l)\big\{\mu_*^{\mathrm{vW}}\rho_g + \tfrac{1}{2}\rho_g^2 + \tfrac{1}{\beta}s(\rho_g)\big\} \\ &= (v_g - v_l)\big\{\mu_*^{\mathrm{vW}}\rho_g - f_\beta(\rho_g)\big\} \\ &= (v_g - v_l)p_\beta^{\mathrm{vW}}(\mu_*^{\mathrm{vW}}).\end{aligned}$$

□

4.10 Kać Interactions and the Van der Waals Limit

Although Maxwell's construction was obtained rigorously in the previous section, it was only proved to occur in a model with nonphysical interactions. It therefore remains to be understood whether the van der Waals–Maxwell theory can be given a precise meaning in the framework of equilibrium statistical mechanics, but starting from *finite-range* interactions. This will be done in this section.

Consider a lattice gas in a large vessel, with a fixed density. We have seen how van der Waals' main simplifying hypothesis could be realized in a model (the van der Waals model), in which a quadratic term ρ_Λ^2 appears in the Hamiltonian, as a consequence of the nonlocal structure of the interaction.

As explained earlier, the *local* density of a real gas at fixed overall density ρ can undergo large fluctuations, in particular in the coexistence regime. This makes it possible to observe the true physical phenomenon of interest: condensation.

Instead of making homogeneity assumptions on the density, we introduce a class of interactions that allows one to compute the energy in large but finite regions, whatever the density of particles is (in that region). To this end, we will consider some initial pair interaction, given by some function φ, and then *scale it* in an appropriate way:

Definition 4.28. Let $\varphi\colon \mathbb{R}^d \to \mathbb{R}_{\geq 0}$ be a Riemann-integrable function with compact support, satisfying $\varphi(-x) = \varphi(x)$ and

$$\int \varphi(x)\,\mathrm{d}x = 1\,. \tag{4.67}$$

The **Kać interaction associated with φ with scaling parameter** $\gamma > 0$ is

$$K_\gamma(i,j) \stackrel{\text{def}}{=} \gamma^d \varphi(\gamma(j-i))\,.$$

We will mostly be interested in small values of γ, that is, when the interaction of a particle at $i \in \mathbb{Z}^d$ is essentially the same with all the other particles located in a neighborhood of i of diameter γ^{-1}. The smaller γ, the more particles interact, but the less a pair $\{i,j\}$ contributes to the total energy. In this respect, the interaction K_γ is similar, at a microscopic scale, to the van der Waals model. Nevertheless, because φ is compactly supported, K_γ *always has a finite range of order* γ^{-1}. Let us denote the maximal interaction of a particle with the rest of the system by $\kappa_\gamma = \sum_{j \neq 0} K_\gamma(0,j)$.

Exercise 4.12. Show, using (4.67), that

$$\lim_{\gamma \downarrow 0} \kappa_\gamma = 1\,. \tag{4.68}$$

A possible choice for φ is

$$\varphi(x) \stackrel{\text{def}}{=} \begin{cases} 2^{-d} & \text{if } \|x\|_\infty \leq 1\,, \\ 0 & \text{if } \|x\|_\infty > 1\,. \end{cases} \tag{4.69}$$

The scaling of this function, for some $0 < \gamma < 1$, is depicted in Figure 4.20.

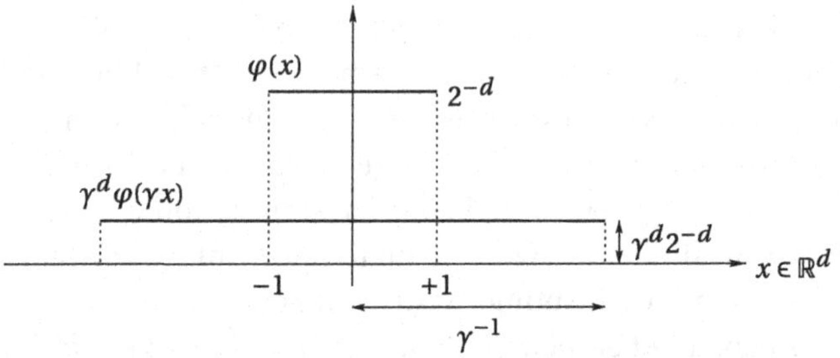

Figure 4.20. The function φ in (4.69)

The canonical and grand canonical partition functions associated with $\mathscr{H}_{\Lambda;K_\gamma}$ will be denoted $\mathbf{Q}_{\Lambda;\gamma,\beta,N}$, respectively $\Theta_{\Lambda;\gamma,\beta,\mu}$. The free energy and pressure will be denoted $f_{\Lambda;\gamma,\beta}(\rho)$, respectively $p_{\Lambda;\gamma,\beta}(\mu)$.

We consider a two-step limiting procedure: first, we take the thermodynamic limit at a fixed positive value of γ (below, $\Lambda \Uparrow \mathbb{Z}^d$ actually means using $\mathscr{R} \ni \Lambda_n \Uparrow \mathbb{Z}^d$):

$$f_{\gamma,\beta}(\rho) = \lim_{\Lambda \Uparrow \mathbb{Z}^d} f_{\Lambda;\gamma,\beta}(\rho), \qquad p_{\gamma,\beta}(\mu) = \lim_{\Lambda \Uparrow \mathbb{Z}^d} p_{\Lambda;\gamma,\beta}(\mu).$$

The existence of these limits is guaranteed by Theorems 4.5 and 4.11. In the second step, we let the parameter γ tend to 0.

Definition 4.29. The limit $\gamma \downarrow 0$ is called the **van der Waals limit.**[3] When they exist, we denote the limits by

$$f_{0^+,\beta}(\rho) \stackrel{\text{def}}{=} \lim_{\gamma \downarrow 0} f_{\gamma,\beta}(\rho), \qquad p_{0^+,\beta}(\mu) \stackrel{\text{def}}{=} \lim_{\gamma \downarrow 0} p_{\gamma,\beta}(\mu).$$

Remark 4.30. To summarize, the relevant limiting procedure, in the study of Kać interactions, is $\lim_{\gamma \downarrow 0} \lim_{\Lambda \Uparrow \mathbb{Z}^d}\{\dots\}$. Observe that if the limits are taken in the other order, $\lim_{\Lambda \Uparrow \mathbb{Z}^d} \lim_{\gamma \downarrow 0}\{\dots\}$, this yields the hard-core model of Section 4.7. Indeed, for all fixed $i, j \in \mathbb{Z}^d$,

$$\lim_{\gamma \downarrow 0} K_\gamma(i,j) = 0. \tag{4.70}$$

Therefore,

$$\lim_{\Lambda \Uparrow \mathbb{Z}^d} \lim_{\gamma \downarrow 0} f_{\Lambda;\gamma,\beta}(\rho) = \lim_{\Lambda \Uparrow \mathbb{Z}^d} f^{\text{hard}}_{\Lambda;\beta}(\rho) = f^{\text{hard}}_{\beta}(\rho) = -\tfrac{1}{\beta} s^{\text{l.g.}}(\rho),$$
$$\lim_{\Lambda \Uparrow \mathbb{Z}^d} \lim_{\gamma \downarrow 0} p_{\Lambda;\gamma,\beta}(\mu) = \lim_{\Lambda \Uparrow \mathbb{Z}^d} p^{\text{hard}}_{\Lambda;\beta}(\mu) = p^{\text{hard}}_{\beta}(\mu) = \tfrac{1}{\beta} \log(1 + e^{\beta\mu}),$$

[3] This limit is also called the **mean-field, Kać** or **Lebowitz–Penrose limit.**

as seen in (4.39) and (4.40). Therefore, taking the limits in that order does not lead to interesting phenomena. ◇

4.10.1 Van der Waals Limit of the Thermodynamic Potentials

When $\gamma \downarrow 0$, Kać interactions become, loosely speaking, infinitely weak and of infinite range. We therefore expect $p_{0^+,\beta}$ and $f_{0^+,\beta}$ to be related to the thermodynamic potentials of the van der Waals model, in some sense.

Notice also that, since $p_{\gamma,\beta}$ and $f_{\gamma,\beta}$ are convex, their limits as $\gamma \downarrow 0$ must also be convex. Since we know that nonconvexity does occur in the van der Waals model at low temperature, some new feature is to be expected.

Theorem 4.31 (Van der Waals limit of Kać interactions). *For all* $\beta > 0$,

$$f_{0^+,\beta}(\rho) = \mathrm{CE} f_\beta^{\mathrm{vW}}(\rho), \qquad \forall \rho \in [0,1], \tag{4.71}$$

$$p_{0^+,\beta}(\mu) = p_\beta^{\mathrm{vW}}(\mu), \qquad \forall \mu \in \mathbb{R}. \tag{4.72}$$

The most remarkable feature of this result is that the limiting behavior of Kać interactions is described by the van der Waals model, *but with a free energy having the correct convexity property.* (We remind the reader that the convex envelope did not appear naturally in the van der Waals model, but only when considering the Legendre transform of the pressure in (4.56).) Remember that

$$f_\beta^{\mathrm{vW}}(\rho) = -\tfrac{1}{2}\rho^2 - \tfrac{1}{\beta} s^{\mathrm{l.g.}}(\rho).$$

The double limiting procedure $\lim_{\gamma \downarrow 0} \lim_{\Lambda \Uparrow \mathbb{Z}^d}\{\cdot\}$ thus leads to two main features. The first is the appearance of a quadratic term in the free energy, $-\frac{1}{2}\rho^2$, without it having been introduced artificially in the Hamiltonian. Here, as will be seen, it stems from the interaction of a particle with the rest of the system, which provides a nonvanishing contribution even in the limit $\gamma \downarrow 0$. The second new feature is of course the geometric modification of f_β^{vW} by the convex envelope. This modification is nontrivial at low temperature, since it leads to the appearance of an *affine portion* on the graph of the free energy, as seen earlier; see Figure 4.21.

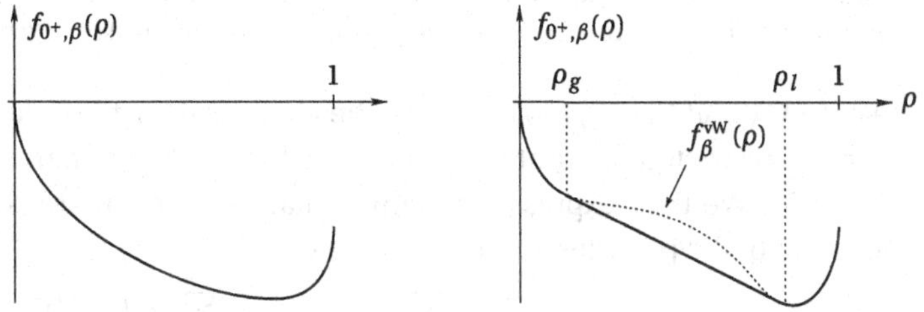

Figure 4.21. The van der Waals limit of the free energy. Left: $\beta \le \beta_c^{\mathrm{vW}}$, right: $\beta > \beta_c^{\mathrm{vW}}$

Proof of (4.71)
Since the temperature plays no role in the proof below, we will usually omit β from the notations.

As the mechanism of the proof will show, the appearance of the convex envelope in (4.71) is precisely due to the fact that, for finite-range interactions, the system is free to let the density of particles vary from place to place.

We thus use an intermediate scale, $\ell \in \mathbb{N}$, which we assume to be large, of the form $\ell = 2^p$, but smaller than the scale of the interaction, $\ell \ll \gamma^{-1}$. In the end, we will consecutively take the limits $\Lambda \Uparrow \mathbb{Z}^d$, then $\gamma \downarrow 0$ and finally $\ell \uparrow \infty$.

By Theorem 4.5, the free energy can be computed using a sequence of cubic boxes $\Lambda \Uparrow \mathbb{Z}^d$ whose sidelength is always a multiple of ℓ. We therefore consider a partition of $\mathbb{Z}^d$ into cubes $\Lambda^{(\alpha)}$, $\alpha = 1, 2, \dots$, of sidelength ℓ (see Figure 4.22). For simplicity, we can assume that $\Lambda^{(1)}$ always contains the origin, that Λ is a cube of sidelength $\ell 2^n$, given by the union of M cubes of the partition and centered on the cube $\Lambda^{(1)}$, and denote these cubes by $\Lambda^{(1)}, \dots, \Lambda^{(M)}$.

We fix ρ and take $N = \lceil \rho |\Lambda| \rceil$. The starting point is to consider all possible arrangements of the N particles in the boxes $\Lambda^{(\alpha)}$, by writing

$$\mathbf{Q}_{\Lambda;\gamma,N} = \sum_{\substack{N_1,\dots,N_M:\\ N_1+\dots+N_M=N}} \mathbf{Q}_{\Lambda;\gamma,N}(N_1, \dots, N_M), \tag{4.73}$$

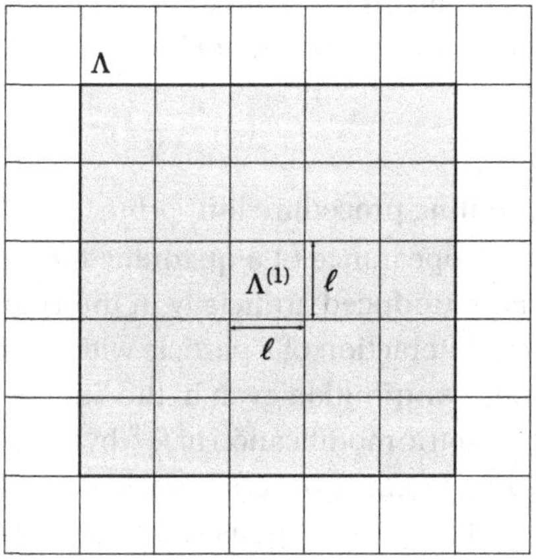

Figure 4.22. $\mathbb{Z}^d$ is partitioned into cubes $\Lambda^{(\alpha)}$, $\alpha = 1, 2, \dots$, of sidelength ℓ. The box Λ is assumed to be built from boxes of the partition, and centered on the box $\Lambda^{(1)}$.

where $\mathbf{Q}_{\Lambda;\gamma,N}(N_1, \dots, N_M)$ denotes the canonical partition function in which the sum is restricted to configurations in which $\Lambda^{(\alpha)}$ contains N_α particles, $\alpha = 1, \dots, M$. We then express the Hamiltonian in a way that takes into account the location of the particles among the boxes $\Lambda^{(\alpha)}$:

$$\mathscr{H}_{\Lambda;K_\gamma} = \sum_{\alpha} \mathscr{H}_{\Lambda^{(\alpha)};K_\gamma} + \sum_{\substack{\{\alpha,\alpha'\}\\ \alpha\neq\alpha'}} I_\gamma(\alpha, \alpha'),$$

where $I_\gamma(\alpha,\alpha')$ represents the interactions between the particles in $\Lambda^{(\alpha)}$ and those in $\Lambda^{(\alpha')}$:

$$I_\gamma(\alpha,\alpha') = -\sum_{i\in\Lambda^{(\alpha)}}\sum_{j\in\Lambda^{(\alpha')}} K_\gamma(i,j)\eta_i\eta_j .$$

By defining

$$\overline{K}_\gamma(\alpha,\alpha') \stackrel{\text{def}}{=} \max_{i\in\Lambda^{(\alpha)},j\in\Lambda^{(\alpha')}} K_\gamma(i,j),$$

$$\underline{K}_\gamma(\alpha,\alpha') \stackrel{\text{def}}{=} \min_{i\in\Lambda^{(\alpha)},j\in\Lambda^{(\alpha')}} K_\gamma(i,j),$$

we have

$$-\overline{K}_\gamma(\alpha,\alpha')N_{\Lambda^{(\alpha)}}N_{\Lambda^{(\alpha')}} \le I_\gamma(\alpha,\alpha') \le -\underline{K}_\gamma(\alpha,\alpha')N_{\Lambda^{(\alpha)}}N_{\Lambda^{(\alpha')}} . \tag{4.74}$$

Upper Bound. Since $f_\gamma(\rho)$ is convex, its limit as $\gamma \downarrow 0$ is also convex, and therefore continuous (Exercise B.3 and Proposition B.9). (Continuity at 0 and 1 follows from (4.9), (4.15), (4.16) and (4.19).) It is thus sufficient to prove an upper bound for (4.71) for densities ρ belonging to a dense subset of $(0,1)$. So, let us fix a dyadic density, of the form $\rho = \frac{k}{2^m}$, $0 < k < 2^m$. By construction, when ℓ is large enough, $N_* \stackrel{\text{def}}{=} \lceil \rho|\Lambda^{(\alpha)}|\rceil = \rho|\Lambda^{(\alpha)}|$ for each α. We thus get a lower bound on the partition function by keeping only the configurations in which each box $\Lambda^{(\alpha)}$ contains exactly N_* particles:

$$\mathbf{Q}_{\Lambda;\gamma,N} \ge \mathbf{Q}_{\Lambda;\gamma,N}(N_*,\ldots,N_*) . \tag{4.75}$$

Using (4.74) and performing separately the sums over the configurations of N_* particles in each box $\Lambda^{(\alpha)}$,

$$\begin{aligned}\mathbf{Q}_{\Lambda;\gamma,N}(N_*,\ldots,N_*) &\ge \{\mathbf{Q}_{\Lambda^{(1)};\gamma,N_*}\}^M \prod_{\substack{\{\alpha,\alpha'\}\\ \alpha\neq\alpha'}} e^{\beta \underline{K}_\gamma(\alpha,\alpha')N_*^2}\\ &= \{\mathbf{Q}_{\Lambda^{(1)};\gamma,N_*}\}^M \exp\Bigl(\tfrac12\beta N_*^2 \sum_{\alpha=1}^{M}\sum_{\substack{\alpha'=1\\(\alpha'\neq\alpha)}}^{M} \underline{K}_\gamma(\alpha,\alpha')\Bigr) .\end{aligned}$$

Now,

$$\sum_{\substack{\alpha'=1\\(\alpha'\neq\alpha)}}^{M} \underline{K}_\gamma(\alpha,\alpha') = \sum_{\substack{\alpha'\ge1\\(\alpha'\neq\alpha)}} \underline{K}_\gamma(\alpha,\alpha') - \sum_{\substack{\alpha'\ge1\\(\Lambda^{(\alpha')}\not\subset\Lambda)}} \underline{K}_\gamma(\alpha,\alpha') . \tag{4.76}$$

By translation invariance, the first sum on the right-hand side does not depend on α, which can thus be assumed to be equal to 1. Then, since K_γ has finite range, say R_γ, the second sum is (finite and) nonzero only if $\Lambda^{(\alpha)}$ is at distance at most R_γ from Λ^{c}. It thus represents a boundary term, of order $(R_\gamma/\ell)^d|\partial^{\mathrm{ex}}\Lambda|$. Since we take the thermodynamic limit *before* the limit $\gamma \downarrow 0$, we get, after letting $\Lambda \Uparrow \mathbb{Z}^d$ (along that specific sequence of cubes),

$$f_\gamma(\rho) \le \frac{-1}{\beta|\Lambda^{(1)}|} \log \mathbf{Q}_{\Lambda^{(1)};\gamma,N_*} - \frac{\rho^2}{2}|\Lambda^{(1)}| \sum_{\alpha'>1} \underline{K}_\gamma(1,\alpha') . \tag{4.77}$$

Exercise 4.13. For all fixed $\ell \in \mathbb{N}$,

$$\lim_{\gamma\downarrow 0} |\Lambda^{(1)}| \sum_{\alpha'>1} \underline{K}_\gamma(1,\alpha') = \int \varphi(x)\,dx = \lim_{\gamma\downarrow 0} |\Lambda^{(1)}| \sum_{\alpha'>1} \overline{K}_\gamma(1,\alpha') . \tag{4.78}$$

We can now compute the van der Waals limit. By (4.70), we know that $\lim_{\gamma\downarrow 0} \mathbf{Q}_{\Lambda;\gamma,N_*} = \mathbf{Q}^{\text{hard}}_{\Lambda;N_*}$ and, since we assumed that $\int \varphi(x)\,dx = 1$, (4.77) and (4.78) yield

$$\limsup_{\gamma\downarrow 0} f_\gamma(\rho) \le f^{\text{hard}}_{\Lambda^{(1)}}(\rho) - \tfrac{1}{2}\rho^2 .$$

Taking $\ell \to \infty$ gives

$$\limsup_{\gamma\downarrow 0} f_\gamma(\rho) \le -\tfrac{1}{2}\rho^2 - \tfrac{1}{\beta} s^{\text{l.g.}}(\rho) = f^{\text{vW}}_\beta(\rho) .$$

This bound holds for all dyadic $\rho \in (0,1)$. Since $f_\gamma(\rho)$ is convex, $\limsup_{\gamma\downarrow 0} f_\gamma(\rho)$ also is; in particular, it is continuous. This implies that this last upper bound holds for all $\rho \in (0,1)$, and, using again the convexity of $\limsup_{\gamma\downarrow 0} f_\gamma(\rho)$,

$$\limsup_{\gamma\downarrow 0} f_\gamma(\rho) \le \text{CE} f^{\text{vW}}_\beta(\rho) .$$

Lower Bound. We start by bounding (4.73) as follows:

$$\mathbf{Q}_{\Lambda;\gamma,N} \le \mathscr{N}(N;M) \max_{\substack{N_1,\dots,N_M:\\ N_1+\dots+N_M=N}} \mathbf{Q}_{\Lambda;\gamma,N}(N_1,\dots,N_M) ,$$

where $\mathscr{N}(N;M)$ is the number of M-tuples $(N_1,\dots,N_M)$ with $N_1+\dots+N_M = N$.

Exercise 4.14. Show that, for all $\rho \in (0,1)$,

$$\lim_{\ell\to\infty} \lim_{\Lambda\Uparrow\mathbb{Z}^d} \frac{1}{|\Lambda|} \log \mathscr{N}(\lceil \rho|\Lambda| \rceil; M) = 0 . \tag{4.79}$$

Then, using again (4.74),

$$\mathbf{Q}_{\Lambda;\gamma,N}(N_1,\dots,N_M) \le \Big\{ \prod_\alpha \mathbf{Q}_{\Lambda^{(\alpha)};\gamma,N_\alpha} \Big\} \prod_{\substack{\{\alpha,\alpha'\}\\ \alpha\neq\alpha'}} e^{\beta \overline{K}_\gamma(\alpha,\alpha') N_\alpha N_{\alpha'}} .$$

For the first product, we write

$$\mathbf{Q}_{\Lambda^{(\alpha)};\gamma,N_\alpha} = \exp\big(-\beta f_{\Lambda^{(1)};\gamma}\big(\tfrac{N_\alpha}{|\Lambda^{(\alpha)}|}\big)|\Lambda^{(1)}|\big) .$$

For the second, we use $N_\alpha N_{\alpha'} \le \frac{1}{2}(N_\alpha^2 + N_{\alpha'}^2)$ and the same argument given after (4.76) to obtain

$$\prod_{\substack{\{\alpha,\alpha'\}\\ \alpha\neq\alpha'}} e^{\beta \overline{K}_\gamma(\alpha,\alpha')N_\alpha N_{\alpha'}} \le e^{c|\partial^{\rm ex}\Lambda|}\exp\Bigl(\tfrac{1}{2}\beta\overline{\kappa}_\gamma|\Lambda^{(1)}|\sum_{\alpha=1}^{M}\bigl(\tfrac{N_\alpha}{|\Lambda^{(\alpha)}|}\bigr)^2\Bigr),$$

where c depends on γ and ℓ, and

$$\overline{\kappa}_\gamma \stackrel{\text{def}}{=} |\Lambda^{(1)}|\sum_{\alpha'>1}\overline{K}_\gamma(1,\alpha').$$

By Lemma 4.13, $\overline{\kappa}_\gamma \to 1$ as $\gamma \downarrow 0$. If we define $g_{\Lambda^{(1)},\gamma}(\rho) \stackrel{\text{def}}{=} -\frac{1}{2}\overline{\kappa}_\gamma\rho^2 + f_{\Lambda^{(1)};\gamma}(\rho)$, then

$$\begin{aligned}\frac{-1}{\beta|\Lambda|}\log \mathbf{Q}_{\Lambda;\gamma,N} \ge{}& \frac{-1}{\beta|\Lambda|}\log \mathscr{N}(N;M)\\ &+ \min_{\substack{N_1,\dots,N_M:\\ N_1+\cdots+N_M=N}} \frac{1}{M}\sum_{\alpha=1}^{M} g_{\Lambda^{(1)};\gamma}\bigl(\tfrac{N_\alpha}{|\Lambda^{(\alpha)}|}\bigr) - \frac{c|\partial^{\rm ex}\Lambda|}{\beta|\Lambda|}.\end{aligned}$$

Now, for each M-tuple $(N_1,\dots,N_M)$ above,

$$\begin{aligned}\frac{1}{M}\sum_{\alpha=1}^{M} g_{\Lambda^{(1)};\gamma}\bigl(\tfrac{N_\alpha}{|\Lambda^{(\alpha)}|}\bigr) &\ge \frac{1}{M}\sum_{\alpha=1}^{M} \mathrm{CE}\, g_{\Lambda^{(1)};\gamma}\bigl(\tfrac{N_\alpha}{|\Lambda^{(\alpha)}|}\bigr)\\ &\ge \mathrm{CE}\, g_{\Lambda^{(1)};\gamma}\Bigl(\tfrac{1}{M}\sum_{\alpha=1}^{M}\tfrac{N_\alpha}{|\Lambda^{(\alpha)}|}\Bigr) = \mathrm{CE}\, g_{\Lambda^{(1)};\gamma}\bigl(\tfrac{N}{|\Lambda|}\bigr).\end{aligned}$$

For the first inequality, we used that $g \ge \mathrm{CE}\, g$, and then that $\mathrm{CE}\, g$ is convex (see Exercise B.2). Since $\frac{N}{|\Lambda|} \to \rho$,

$$f_\gamma(\rho) \ge \liminf_{\Lambda\Uparrow\mathbb{Z}^d} \frac{-1}{\beta|\Lambda|}\log \mathscr{N}(N;M) + \mathrm{CE}\, g_{\Lambda^{(1)};\gamma}(\rho).$$

By (4.79), the first term on the right-hand side will vanish once we take the limit $\ell \to \infty$. To take the van der Waals limit in the second term, we first observe that

$$\lim_{\gamma\downarrow 0} g_{\Lambda^{(1)};\gamma}(\rho) = -\tfrac{1}{2}\rho^2 + f^{\rm hard}_{\Lambda^{(1)}}(\rho), \tag{4.80}$$

uniformly in $\rho \in (0,1)$, and rely on the following:

Exercise 4.15. Let $f_n\colon [a,b] \to \mathbb{R}$ be a sequence of functions converging uniformly to $f\colon [a,b] \to \mathbb{R}$. Then $\mathrm{CE} f_n$ converges uniformly to $\mathrm{CE} f$.

We therefore get, for all $\rho \in [a,b] \subset [0,1]$,

$$\liminf_{\gamma\downarrow 0} f_\gamma(\rho) \ge \liminf_{\Lambda\Uparrow\mathbb{Z}^d} \frac{-1}{\beta|\Lambda|}\log \mathscr{N}(N;M) + \mathrm{CE}\bigl\{-\tfrac{\rho^2}{2} + f^{\rm hard}_{\Lambda^{(1)}}(\rho)\bigr\}.$$

We finally take the limit $\ell \to \infty$, which yields

$$\liminf_{\gamma \downarrow 0} f_\gamma(\rho) \geq \mathrm{CE} f_\beta^{\mathrm{vW}}(\rho).$$

This concludes the proof of (4.71). □

Proof of (4.72)
For a fixed $\mu \in \mathbb{R}$, we take $\gamma \downarrow 0$ on both sides of

$$p_{\gamma,\beta}(\mu) = \sup_{\rho \in [0,1]} \{\rho\mu - f_{\gamma,\beta}(\rho)\}.$$

We first show that the right-hand side converges to $\sup_{\rho\in[0,1]}\{\rho\mu - f_{0^+,\beta}(\rho)\}$.

Since they are fixed, let us omit β and μ from the notations. Let $F_\gamma(\rho) \stackrel{\text{def}}{=} \rho\mu - f_\gamma(\rho)$ and $F_{0^+}(\rho) \stackrel{\text{def}}{=} \lim_{\gamma\downarrow 0} F_\gamma(\rho) = \rho\mu - f_{0^+}(\rho)$. As we already know, F_{0^+} attains its maximum away from 0 and 1. Therefore we can take $\delta > 0$ small enough so that $\sup_\rho F_{0^+}(\rho) = \sup_{\rho\in K} F_{0^+}(\rho)$, where $K = [\delta, 1-\delta]$. Observe that, because $\rho \mapsto F_{0^+}(\rho)$ is concave and by our choice of δ, $\partial^+ F_{0^+}(\delta) > 0$ and $\partial^- F_{0^+}(1-\delta) < 0$; by Theorem B.12, this implies that $\partial^+ F_\gamma(\delta) > 0$, $\partial^- F_\gamma(1-\delta) < 0$, for all small enough $\gamma > 0$. Since the family $(f_\gamma)_{\gamma>0}$ is bounded by (4.15)-(4.19), the convergence $f_\gamma \to f_{0^+}$ is uniform on K (Lemma B.10). This implies

$$\lim_{\gamma\downarrow 0} \sup_{\rho\in K} F_\gamma(\rho) = \sup_{\rho\in K} F_{0^+}(\rho),$$

which implies what we wanted. Therefore, in terms of the Legendre transform $(\cdot)^*$, we have obtained $p_{0^+} = (f_{0^+})^* = (\mathrm{CE} f^{\mathrm{vW}})^*$. But, by Corollary B.18, $(\mathrm{CE} f^{\mathrm{vW}})^* = (f^{\mathrm{vW}})^* = p^{\mathrm{vW}}$. □

4.11 Bibliographical References

Lattice models of gases have been studied since the early stages of statistical mechanics; Boltzmann, in particular, considered similar approximations as a computational device. See the book by Gallavotti [130] and references therein.

Thermodynamic Potentials and Equivalence of Ensembles. The construction of the thermodynamic potentials and the derivation of their general convexity properties are classical and can be found in several sources. A very general approach can be found in the important work of Lanford [205]. A classical reference for the existence of the pressure is the book by Ruelle [289]. See also the more recent book by Presutti [279], in which the equivalence of ensembles is proved at the level of thermodynamic potentials. More on the equivalence of ensembles at the level of measures can be found in Section 6.14.1.

Van der Waals Lattice Gas. The van der Waals model studied in Section 4.9 incorporates the main assumptions made by van der Waals [339] about the interactions of a gas of particles, which we had already discussed in Chapter 1. As already mentioned in Chapter 2, the Curie–Weiss version of the van der Waals lattice gas model was introduced independently by many people, including Temperley [328], Husimi [167] and Kać [183]. Our treatment of its pressure, as a function of the variables ρ and v, was taken from Dorlas' book [90].

Ising Lattice Gas. The mapping between the Ising model and the lattice gas first appeared explicitly in [220].

Kać Limit. The first justification of Maxwell's construction based on the van der Waals limit was given by Kać, Uhlenbeck and Hemmer [184] for a one-dimensional gas of hard rods. The result was then substantially generalized by Lebowitz and Penrose [215]; our proof of Theorem 4.31 follows essentially theirs. A general reference covering much more material on systems with Kać interactions is Presutti's book [279]. More bibliographical references on Kać interactions will be given in the complements.

4.12 Complements and Further Reading

4.12.1 The Phase Separation Phenomenon

In the current chapter, we have provided a satisfactory description of the condensation phenomenon in terms of the thermodynamic potentials, but we have not discussed what really happens during condensation, as observed in typical configurations. In this section, we provide a brief description of what can be said about this problem from a mathematical point of view. To keep the discussion as simple as possible, only the nearest-neighbor lattice gas will be considered, although much of what follows can be extended to general finite-range ferromagnetic interactions. Detailed information and more references can be found in the review [28].

Consider the nearest-neighbor lattice gas in $\mathrm{B}(n) \subset \mathbb{Z}^d$, $d \geq 2$. We have seen in Theorem 4.15 that, in the grand canonical ensemble with $\mu = \mu_*$, the typical values of the density lie in the interval $[\rho_g, \rho_l]$, on which the rate function $I_{\beta,\mu}$ vanishes. In particular, this result does not allow us to discriminate between the various possible values of the density in this interval. There is a reason for that: in this regime, the average density in the box is in fact very sensitive to the boundary condition and thus cannot be derived using only the thermodynamic limit of the pressure and of the free energy.

Similarly, in the canonical ensemble with $\rho \in [\rho_g, \rho_l]$, the lack of strict convexity of the free energy prevents us from using Theorem 4.19 to determine the typical values of the local density in various subsets of the box $\Lambda = \mathrm{B}(n)$ for such values of ρ. As we will see, in this regime, typical values of the *local* density are still given by ρ_g and ρ_l. However, on a macroscopic scale, typical configurations are not homogeneous anymore, but exhibit *phase coexistence*. Namely, in order to satisfy the constraint that the overall density in Λ is ρ, the system reacts by *spatially segregating* the gas and liquid phases: for example, if the boundary condition favors the gas phase, there is spontaneous creation of a droplet of liquid phase surrounded by the gas phase, as depicted in Figure 4.23.

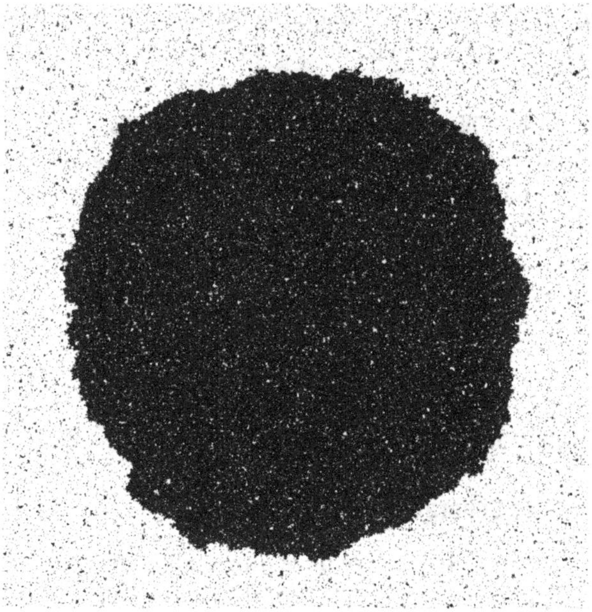

Figure 4.23. A typical configuration of the nearest-neighbor lattice gas in a box of size 500×500 in the canonical ensemble at inverse temperature $\beta = 2$ (with a boundary condition favoring the gas phase). The simulation was made by fixing the density to the value $\rho = \frac{1}{2} \in (\rho_g, \rho_l)$. Clearly, spatial homogeneity in the sense of (4.34) is no longer true; phase separation occurs. In a typical configuration, a macroscopic liquid droplet of density ρ_l appears, immersed in a gas phase of density ρ_g. This droplet's shape is described, asymptotically, by a Wulff crystal (see below).

In the thermodynamic limit, the droplet's macroscopic shape becomes deterministic, with microscopic fluctuations. Namely, let us denote the droplet by $V \subset \Lambda$. The following occurs with a probability tending to 1 as $\Lambda \uparrow \mathbb{Z}^d$:

1. Up to microscopic corrections, its *volume* is given by $|V| = \frac{\rho-\rho_g}{\rho_l-\rho_g}|\Lambda|$. In this way, the average density in Λ is indeed ρ, since

$$\rho_l|V| + \rho_g(|\Lambda| - |V|) \simeq \rho|\Lambda|.$$

2. The *shape* of the droplet converges to a deterministic shape characterized as a minimizer of a surface functional involving the *surface free energy*, to be defined below.

These two statements will be given a more precise form in Theorem 4.33 below. We will see that the macroscopic geometry of the regions occupied by the gas and liquid phases depends strongly on the boundary condition and thus, again, cannot be deduced using only the thermodynamic limit of the pressure and the free energy. In order to go further, we need to go beyond such bulk quantities, and consider corrections coming from surface effects.

Surface Corrections to the Pressure

In order to discuss corrections to the pressure, we need to consider nontrivial boundary conditions. Similarly to what we did in Chapter 3, we consider the Hamiltonian

$$\mathscr{H}_{\Lambda;\mu_*}(\eta) \stackrel{\text{def}}{=} -\sum_{\{i,j\}\in\mathscr{E}^{\text{b}}_{\Lambda}} \eta_i\eta_j - \mu_*\sum_{i\in\Lambda}\eta_i\,,$$

defined on infinite configurations $\eta \in \Omega \stackrel{\text{def}}{=} \{0,1\}^{\mathbb{Z}^d}$. Let

$$\Omega^1_\Lambda \stackrel{\text{def}}{=} \{\eta\in\Omega : \eta_i = 1 \text{ for all } i\notin\Lambda\}\,, \qquad \Omega^0_\Lambda \stackrel{\text{def}}{=} \{\eta\in\Omega : \eta_i = 0 \text{ for all } i\notin\Lambda\}\,.$$

We denote by $\nu^1_{\Lambda;\beta,\mu^*}$ and $\nu^0_{\Lambda;\beta,\mu^*}$ the corresponding Gibbs distributions in Λ:

$$\nu^1_{\Lambda;\beta,\mu^*}(\eta) \stackrel{\text{def}}{=} \frac{e^{-\mathscr{H}_{\Lambda;\mu_*}(\eta)}}{\Theta^1_{\Lambda;\beta,\mu_*}}\mathbf{1}_{\{\eta\in\Omega^1_\Lambda\}}\,, \qquad \nu^0_{\Lambda;\beta,\mu^*}(\eta) \stackrel{\text{def}}{=} \frac{e^{-\mathscr{H}_{\Lambda;\mu_*}(\eta)}}{\Theta^0_{\Lambda;\beta,\mu_*}}\mathbf{1}_{\{\eta\in\Omega^0_\Lambda\}}\,,$$

where $\Theta^1_{\Lambda;\beta,\mu_*}$ and $\Theta^0_{\Lambda;\beta,\mu_*}$ are the associated partition functions.

Using the mapping between the lattice gas and the Ising model, it is easy to check that these two probability measures correspond exactly to the Gibbs distribution of the Ising model, at inverse temperature $\beta/4$ and with magnetic field $h = 0$, in Λ with $+$, respectively $-$, boundary condition. It thus follows from the analysis in Chapter 3 that, when $\beta > \beta_c^{\text{l.g.}}$, typical configurations under $\nu^1_{\Lambda;\beta,\mu^*}$ have a (homogeneous) density larger than $1/2$, while typical configurations under $\nu^0_{\Lambda;\beta,\mu^*}$ have a (homogeneous) density smaller than $1/2$. They thus describe, respectively, the liquid and gas phases.

Let us now turn to the corresponding finite-volume pressures:

$$p^1_{\Lambda;\beta} \stackrel{\text{def}}{=} \frac{1}{\beta|\Lambda|}\log\Theta^1_{\Lambda;\beta,\mu_*}\,, \qquad p^0_{\Lambda;\beta} \stackrel{\text{def}}{=} \frac{1}{\beta|\Lambda|}\log\Theta^0_{\Lambda;\beta,\mu_*}\,.$$

(For simplicity, we do not indicate μ_* in the notation for the pressures.) As usual, since the boundary condition plays no role in the definition of the thermodynamic pressure, $p^1_{\Lambda;\beta}$ and $p^0_{\Lambda;\beta}$ both converge to p_β in the thermodynamic limit, which implies in particular that

$$p^1_{\Lambda;\beta}|\Lambda| = p_\beta|\Lambda| + o(|\Lambda|)\,,$$

and similarly for $p^0_{\Lambda;\beta}$. In fact, it follows from the proof of Theorem 3.6 that the error $o(|\Lambda|)$ is in fact a *boundary term*, that is, it is $O(|\partial^{\text{in}}\Lambda|)$. It should thus not come as a surprise that this error term depends in general on the choice of the boundary condition. We therefore expect a more accurate description of the following type:

$$\begin{aligned} p^1_{\Lambda;\beta}|\Lambda| &= p_\beta|\Lambda| - \tau^1_\beta|\partial^{\rm in}\Lambda| + o(|\partial^{\rm in}\Lambda|), \\ p^0_{\Lambda;\beta}|\Lambda| &= p_\beta|\Lambda| - \tau^0_\beta|\partial^{\rm in}\Lambda| + o(|\partial^{\rm in}\Lambda|). \end{aligned} \tag{4.81}$$

Here, $-\tau^1_\beta$ (resp. $-\tau^0_\beta$) should be interpreted as the contribution *per unit of area* to $p^1_{\Lambda;\beta}|\Lambda|$ (resp. $p^0_{\Lambda;\beta}|\Lambda|$), resulting from the interaction between the phase contained inside Λ and the boundary of the box. (The negative signs are introduced to respect the conventions, making τ^1_β and τ^0_β nonnegative.)

Surface Tension. Up to now, we have only considered the correction to the pressure in cases in which the boundary condition typically induces homogeneous configurations inside the box. We now consider what happens when the boundary condition induces the presence of a macroscopic interface. The surface tension measures the free energy (per unit of area) associated with an interface. It is given by a function $\tau_\beta(\cdot)$ defined on $\{\mathbf{n} \in \mathbb{R}^d : \|\mathbf{n}\|_2 = 1\}$, where $\mathbf{n}$ represents the direction perpendicular to the interface.

In order to induce the presence of a macroscopic interface, we proceed as in Section 3.10.7. Let us fix a direction $\mathbf{n} \in \mathbb{R}^d$ and define, for each $i \in \mathbb{Z}^d$,

$$\eta^{\mathbf{n}}_i \stackrel{\text{def}}{=} \begin{cases} 0 & \text{if } i \cdot \mathbf{n} \geq 0, \\ 1 & \text{otherwise,} \end{cases}$$

where $i \cdot \mathbf{n}$ denotes the scalar product on $\mathbb{R}^d$. This boundary condition is illustrated in Figure 4.24; it is a natural generalization of Dobrushin's boundary condition,

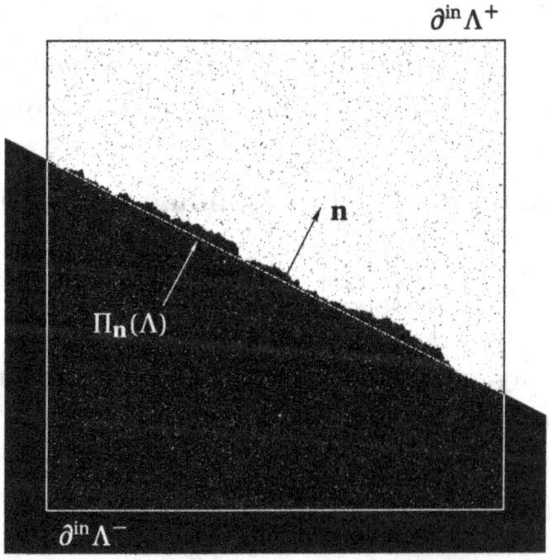

Figure 4.24. A picture representing the construction of the surface tension $\tau_\beta(\mathbf{n})$, by fixing a boundary condition in which all cells below (resp. above) the plane $\{x \cdot \mathbf{n} = 0\}$ are occupied by particles (resp. vacant).

which was introduced in Section 3.10.7. As explained there, the boundary condition $\eta^{\mathbf{n}}$ leads to the presence of an interface, separating the lower part of the box (filled with liquid) from its upper half (filled with gas).

Let us now extract the contribution to the pressure $p^{\mathbf{n}}_{\Lambda;\beta}$ due to this interface.

Let $\Pi_{\mathbf{n}}(\Lambda) \stackrel{\text{def}}{=} \{x \in [-n,n]^d : x \cdot \mathbf{n} = 0\}$ be the intersection of the hyperplane orthogonal to $\mathbf{n}$ with the box (seen as a subset of $\mathbb{R}^d$). In the discussion below, we assume that $\mathbf{n}$ has its last coordinate positive. This will allow us to refer to the parts of Λ located *above* and *below* $\Pi_{\mathbf{n}}(\Lambda)$.

With the boundary condition $\eta^{\mathbf{n}}$, two contributions to the pressure should come from $\partial^{\text{in}}\Lambda$: one coming from the contact of the liquid with the lower part of the boundary of the box (below $\Pi_{\mathbf{n}}(\Lambda)$), denoted $-\tau^1_\beta|\partial^{\text{in}}\Lambda^-|$, and the other coming from the contact of the gas with the upper part of the boundary of the box (above $\Pi_{\mathbf{n}}(\Lambda)$), denoted $-\tau^0_\beta|\partial^{\text{in}}\Lambda^+|$.

Finally, the third contribution to the pressure should come from the *interface* that crosses the box, whose existence is forced by the choice of the boundary condition; it should depend on $\mathbf{n}$ and be proportional to $|\Pi_{\mathbf{n}}(\Lambda)|$ (that is, the area of $\Pi_{\mathbf{n}}(\Lambda)$). The decomposition of the pressure into its volume and surface contributions should therefore be

$$p^{\mathbf{n}}_{\Lambda;\beta}|\Lambda| = p_\beta|\Lambda| - \tau^0_\beta|\partial^{\text{in}}\Lambda^+| - \tau^1_\beta|\partial^{\text{in}}\Lambda^-| - \tau_\beta(\mathbf{n})|\Pi_{\mathbf{n}}(\Lambda)| + o(|\partial^{\text{in}}\Lambda|)\,. \tag{4.82}$$

Using (4.82), (4.81) and the fact that $|\partial^{\text{in}}\Lambda| = 2|\partial^{\text{in}}\Lambda^{\pm}|$, we get

$$\begin{aligned}\tau_\beta(\mathbf{n})|\Pi_{\mathbf{n}}(\Lambda)| &= -p^{\mathbf{n}}_{\Lambda;\beta}|\Lambda| + \tfrac{1}{2}(p^0_{\Lambda;\beta} + p^1_{\Lambda;\beta})|\Lambda| + o(|\partial^{\text{in}}\Lambda|)\\ &= -\frac{1}{\beta}\log\frac{\Theta^{\mathbf{n}}_{\Lambda;\beta,\mu_*}}{\left(\Theta^0_{\Lambda;\beta,\mu_*}\Theta^1_{\Lambda;\beta,\mu_*}\right)^{1/2}} + o(|\partial^{\text{in}}\Lambda|)\,.\end{aligned}$$

This then leads to the following natural definition.

Definition 4.32. Let $\mathbf{n}$ be a unit vector in $\mathbb{R}^d$. The **surface tension** per unit area, orthogonally to the direction $\mathbf{n}$, is defined by

$$\tau_\beta(\mathbf{n}) \stackrel{\text{def}}{=} -\lim_{k\to\infty}\frac{1}{\beta|\Pi_{\mathbf{n}}(\mathrm{B}(k))|}\log\frac{\Theta^{\mathbf{n}}_{\mathrm{B}(k);\beta,\mu_*}}{\left(\Theta^0_{\mathrm{B}(k);\beta,\mu_*}\Theta^1_{\mathrm{B}(k);\beta,\mu_*}\right)^{1/2}}\,.$$

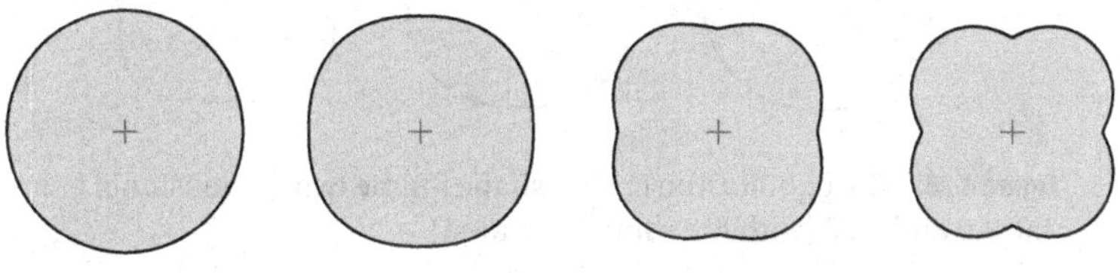

Figure 4.25. The surface tension of the two-dimensional nearest-neighbor lattice gas (as a function of the direction) for $\beta = 2.0, 4.0, 8.0$ and 16.0 (the scale differs for each value of β).

The existence of the above limit can be proved using a subadditivity argument. We refer to [244] for a proof in a more general setup.

The surface tension has a number of important properties, the main one, for our purposes, being the following: for all $\mathbf{n}$,

$$\tau_\beta(\mathbf{n}) > 0 \text{ if and only if } \beta > \beta_c^{\text{l.g.}} \,. \tag{4.83}$$

The proof can be found in [53] and [217]. More information on the surface tension can be found in the review [271].

Equilibrium Crystal Shapes

We can then define a functional on subsets $\mathsf{V} \subset \mathbb{R}^d$ with a smooth boundary (sufficiently smooth, say, to have a well-defined exterior unit normal $\mathbf{n}_x$ at almost all $x \in \partial\mathsf{V}$):

$$\mathscr{W}(\mathsf{V}) \stackrel{\text{def}}{=} \int_{\partial\mathsf{V}} \tau_\beta(\mathbf{n}_x)\mathrm{d}S_x \,,$$

where $\mathrm{d}S_x$ represents the infinitesimal surface element of $\partial\mathsf{V}$ at x.

We can now get back to the problem of describing the droplet mentioned at the beginning of the section. Let $\Lambda = \mathrm{B}(n)$ with n large. Below, we also identify Λ with the subset of $\mathbb{R}^d$ given as the union of all closed unit cubes centered at the vertices of Λ.

Consider now the following variational problem:[4]

Minimize $\mathscr{W}(\mathsf{V})$ *among all* $\mathsf{V} \subset \Lambda$ *whose volume equals* $|\mathsf{V}| = \frac{\rho-\rho_g}{\rho_l-\rho_g}|\Lambda|$.

It can be shown that, up to translations, the solution to this problem is unique; we denote it by $\mathsf{V}_* = \mathsf{V}_*(\beta,\rho)$. As a matter of fact, it can be given explicitly. Consider the **Wulff shape** or **equilibrium crystal shape** associated with $\tau_\beta(\cdot)$, defined by (see Figure 4.26)

$$\mathsf{V}_* = \mathsf{V}_*(\beta) \stackrel{\text{def}}{=} \left\{x \in \mathbb{R}^d \,:\, x \cdot \mathbf{n} \le \tau_\beta(\mathbf{n}) \text{ for every unit vector } \mathbf{n} \in \mathbb{R}^d\right\}.$$

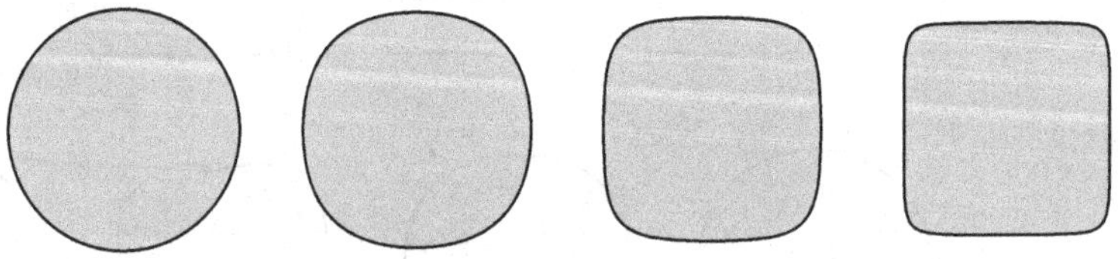

Figure 4.26. The equilibrium crystal shape for the two-dimensional Ising lattice gas at $\beta = 2.0$, 4.0, 8.0 and 16.0 (with the same fixed area).

[4] For this to make full sense, we should impose some regularity on the class of sets V involved, but we will abstain from discussing these issues here and refer to [28] for more information.

A solution to the variational problem is then given by an appropriate dilation of V_* (together with translations), provided that it is not too large to fit inside Λ:

$$\mathsf{V}_* = \frac{\rho - \rho_g}{\rho_l - \rho_g} |\Lambda| \frac{\mathsf{V}_*}{|\mathsf{V}_*|}.$$

In the general case, the solution is also obtained starting from V_*, but with some modifications; see [28].

Let us now state the result that characterizes the separation of phases. To keep things simple, and because the most precise results have been obtained in this context, we only discuss the two-dimensional case.

It will be convenient to describe the configurations using the Peierls contours introduced in Section 3.7.2; in the lattice gas language, the latter separate empty vertices from those containing a particle.

Let us state a precise result, choosing a specific boundary condition:

Theorem 4.33. *Consider the two-dimensional nearest-neighbor lattice gas in the box $\Lambda = \mathrm{B}(n)$ with 0 boundary condition. Assume that $\beta > \beta_c^{\mathrm{l.g.}}$, fix some $\rho \in (\rho_g, \rho_l)$ and set $N = \lfloor \rho |\Lambda| \rfloor$. Then,*

$$\lim_{n \to \infty} \nu^0_{\Lambda;\beta,N}(\mathscr{D}) = 1,$$

where $\mathscr{D}$ is the event defined as follows. There exist constants c_1, c_2, depending only on β, such that

- *all contours, except one which we denote by γ_0, have diameter at most $c_1 \log n$;*
- *the contour γ_0 has macroscopic size; it is closely approximated by a translate of V_*:*

$$\min_{x \in [-1,1]^2} n^{-1} d_{\mathbb{H}}(\gamma_0, x + \partial \mathsf{V}_*) \le c_2 n^{-1/4} (\log n)^{1/2},$$

where $d_{\mathbb{H}}(A; B)$ denotes the Hausdorff distance between A and B.

It can also be shown that the local density is given by ρ_g outside γ_0 and by ρ_l inside.

Together with the analysis of the thermodynamic potentials developed earlier in this chapter, this theory provides a satisfactory description of the condensation phenomenon at equilibrium.

The first result on phase separation is due to Minlos and Sinai [248], who showed that, at very low temperature, a unique large contour appears in Λ, whose shape is close to a square. (Their result actually holds in all $d \ge 2$.) For the two-dimensional model, the proper understanding of the role played by the surface tension, and the description of the scaling limit of this contour as the Wulff shape, was first achieved by Dobrushin, Kotecký and Shlosman [79]. A different proof was then obtained by Pfister [270]. Building on the latter work, Ioffe managed to extend the proof to all $\beta > \beta_c(2)$ [170, 171]; the version stated above is due to Ioffe and Schonmann [173].

Extensions to the Ising model in higher dimensions have been obtained by Bodineau [26] (at sufficiently low temperatures) and Cerf and Pisztora [64] (for all $\beta > \beta_c(d)$). A detailed analysis of the effect of boundary conditions on the equilibrium crystal shape is given in [29, 272]. Further relevant references and historical notes can be found in the review paper [28]. Similar results have been obtained for models with Kać interactions; see, for example, [18].

4.12.2 Kać Interactions When γ Is Small but Fixed

We saw in Theorem 4.31 that, at low temperature, the limiting free energy and pressure obtained via the van der Waals limit exhibit the characteristic features of a first-order phase transition. Let us make a few comments concerning what happens when studying these thermodynamic potentials when $\gamma > 0$ is small but fixed, not necessarily tending to 0.

First of all, observe that, *regardless of the dimension of the system*, the functions obtained in the van der Waals limit are all given by some transformation of the *same* function $f_\beta^{\mathrm{vW}}(\rho)$, that is, in the van der Waals limit $\gamma \downarrow 0$, the system loses its dependence on the dimension. This leads us to an important remark.

Namely, when $d=1$, since $\gamma > 0$ corresponds to a potential with finite range interactions, the associated pressure $\mu \mapsto p_{\gamma,\beta}(\mu)$ is differentiable at all temperatures, as will be seen in Exercise 6.34 (in fact, it is even analytic [289, Theorem 5.6.2]). Therefore, by Theorem B.20, $\rho \mapsto f_{\gamma,\beta}(\rho)$ is always strictly convex when $\beta > \beta_c^{\mathrm{vW}}$, for all $\gamma > 0$, and only becomes affine on the coexistence plateau *in the limit* $\gamma \downarrow 0$. This is of course not in contradiction with Theorem 4.31, since a convex but non-strictly convex function can be uniformly approximated arbitrarily well by a strictly convex function. Therefore when $\beta > \beta_c^{\mathrm{vW}}$, when $d = 1$, the functions obtained in the van der Waals limit, $f_{0^+,\beta}(\rho)$ and $p_{0^+,\beta}(\mu)$, present nonanalytic behaviors that are not representative of what happens when $\gamma > 0$, however small γ might be.

When $d \geq 2$, the situation is different, since we know that the Ising model exhibits a phase transition at sufficiently low temperature. One is thus naturally led to ask about the behavior of $f_{\gamma,\beta}$ and $p_{\gamma,\beta}$ for *small but fixed* values of $\gamma > 0$. This problem is not trivial since the interaction between two particles at fixed vertices (for example nearest neighbors) becomes small when γ is small.

The next theorem answers this question; it follows from the original works of Cassandro and Presutti [62] and Bovier and Zahradník [39], who introduced a convenient notion of contours for the Ising ferromagnet with Kać interactions.

Theorem 4.34. *($d \geq 2$) Let $\beta > \beta_c^{\mathrm{vW}}$. There exists $\gamma_0 = \gamma_0(\beta) > 0$ such that, for all $0 < \gamma < \gamma_0$, $p_{\gamma,\beta}(\mu)$ is nondifferentiable at $\mu_{*,\gamma} \overset{\mathrm{def}}{=} -\kappa_\gamma/2$ and, as a consequence, $f_{\gamma,\beta}(\rho)$ is affine on $[\rho_{g,\gamma}, \rho_{l,\gamma}]$, where*

$$\rho_{g,\gamma} = \frac{\partial p_{\gamma,\beta}}{\partial \mu^-}\Big|_{\mu_{*,\gamma}} < \frac{\partial p_{\gamma,\beta}}{\partial \mu^+}\Big|_{\mu_{*,\gamma}} = \rho_{l,\gamma}\,.$$

Moreover, as $\gamma \downarrow 0$, $\rho_{g,\gamma} \to \rho_g$ and $\rho_{l,\gamma} \to \rho_l$, where ρ_g and ρ_l are the endpoints of the coexistence plateau of the van der Waals model.

Models with Kać interactions at small values of γ can be considered as *perturbations* of the mean-field behavior observed in the limit $\gamma \downarrow 0$. This allows, for instance, comparison of the expectation of local observables with their mean-field counterparts, and extraction of useful information to study the model. This method was used by Lebowitz, Mazel and Presutti in [214] to provide one of the very few rigorous proofs of occurrence of a phase transition in the continuum.

For a much more detailed description of systems with Kać interactions, we refer the reader to the book by Presutti [279]. More comments on the case $\gamma > 0$ are made in the next section.

4.12.3 Condensation, Metastability and the Analytic Structure of the Isotherms

As we already said, from its very beginning, one of the central issues of statistical mechanics was to provide an explanation for the phase transitions observed in gases and liquids. In particular, the condensation phenomenon was used as a test to decide whether the theory of Boltzmann and Gibbs provided a sufficient structure on which phase transitions could be firmly understood. It was not even clear, at that time, whether a detailed study of the partition function could lead to a single function describing two distinct states, gas and liquid, or whether some additional hypotheses had to be made in order to allow for their coexistence.

The results obtained in this chapter, in particular those concerning the condensation phenomenon at low temperature, provide a satisfactory answer. Since condensation has been, historically, one of the cornerstones in the development of statistical mechanics, we will end this chapter with comments regarding some of the first attempts made at describing condensation rigorously.

Mayer's Conjecture. The first notable attempt at obtaining a theoretical explanation of the condensation phenomenon, starting only from the partition function, was initiated by Mayer in the 1930s. In a series of papers with several coauthors, Mayer developed a theory to study the pressure of a model of particles in the continuum with pairwise interactions. Although not completely rigorous, his theory also firmly established the basis of the method invented by Ursell [337], known nowadays as the *cluster expansion* (see Chapter 5). We will not enter into too much detail, but rather sketch the argument he proposed for the mathematical description of condensation.

In [236], Mayer provided an expression for the coefficients a_n of the expansion of the pressure as a function of the **fugacity** $z \stackrel{\text{def}}{=} e^{\beta\mu}$:

$$\beta p_\beta(\mu) = a_1 z + a_2 z^2 + a_3 z^3 + \cdots . \tag{4.84}$$

Mayer's argument then proceeded as follows. The series should converge at least for small values of z, which corresponds to large negative values of μ, that is, to a dilute phase. But if the series converges when $0 \leq z < r_0$, and z is allowed to take complex values, then it defines a function, analytic in the disk $\{z \in \mathbb{C} : |z| < r_0\}$.

With a function describing the gas phase for small values of z at hand, Mayer associated the condensation phenomenon with *the first singularity encountered when continuing the pressure analytically along the real axis, from small to large values of z*. Let us assume that one such singularity is indeed encountered and denote it by z_s; see Figure 4.27. This way of defining the condensation point (the same characterization can be used when using other variables, such as ρ or v) would later be referred to as **Mayer's conjecture**.

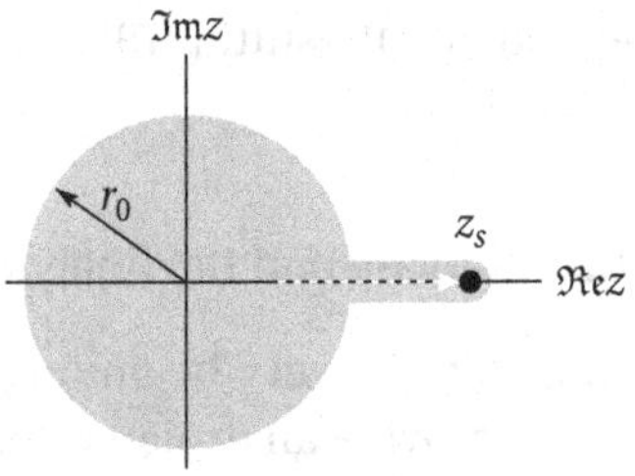

Figure 4.27. The determination of the condensation point according to Mayer: find the first singularity of the function defined by the series $a_1 z + a_2 z^2 + a_3 z^3 + \cdots$, encountered along the positive real axis.

At that time, the question of whether Mayer's method could really describe the condensation phenomenon was debated (see [36]). One reason for that was that Mayer obtained (4.84) under several radical assumptions, one of them being that *the particles of the system are sufficiently far apart*, which is equivalent to assuming that the system is in a dilute (gas) phase. Therefore, the equation of state given by the series had no reason a priori to be able to describe at the same time the dense (liquid) phase. This indicates that, in order for the full equation of state to be given, some other argument should be used to yield a second function describing the dense phase (large values of z). The two functions should then be combined, possibly using a thermodynamic argument similar to the Maxwell construction, in order for equilibrium to be described.

But another question was raised. If a singularity z_s is indeed found on the positive real axis, closest to the origin, does it necessarily describe the true condensation point? Here, van der Waals and Maxwell's theory provides the immediate counterexample showing that condensation is not necessarily related to a singularity. Namely, consider the pressure of the van der Waals model at low temperature, computed in Section 4.9.2. On the gas branch, $\rho \in (0, \rho_g)$, we obtained

$$\widetilde{p}_\beta^{\mathrm{vW}}(\rho) = -\tfrac{1}{2}\rho^2 - \tfrac{1}{\beta}\log(1-\rho),$$

and Taylor expanded $\log(1-\rho)$ (Remark 4.26) to obtain the virial expansion

$$\beta \widetilde{p}_\beta^{\mathrm{vW}}(\rho) = \rho + \tfrac{1}{2}(1-\beta)\rho^2 + \tfrac{1}{3}\rho^3 + \tfrac{1}{4}\rho^4 + \cdots.$$

The radius of convergence of the latter series is equal to 1: it defines an analytic function on the unit disk $\{\rho \in \mathbb{C} : |\rho| < 1\}$ and, as ρ increases from $\rho = 0$, the first singularity along the positive real axis is encountered at $\rho = 1$, *not* at $\rho = \rho_g < 1$. This shows that, for this model, Mayer's way of determining the condensation point fails: the absence of a complex singularity at ρ_g makes it impossible to determine the position of the condensation point only from the knowledge of the values of the pressure on the gas branch.

Since the van der Waals–Maxwell theory remained of central importance at the time, theoretical physicists had good reasons to believe that its way of describing isotherms, by patching different branches together, was generic and should be a consequence of the first principles of statistical mechanics. It was therefore taken for granted at that time in the physics community that analytic continuations such as the one observed in van der Waals' theory were always possible. They were actually even given some importance, due to their relation to another important physical phenomenon.

Metastable States. Consider the isotherm of Figure 1.5. What is the significance of the part of the van der Waals isotherm $p(v)$ that is left out *after* Maxwell's construction, that is, for $v \in [v_l, v_g]$?

From a mathematical point of view, $p(v)$ provides of course the unique *analytic continuation* from one of the branches of MC $p(v)$ to the other, along the paths $v \uparrow v_l$ and $v \downarrow v_g$. From an experimental point of view, an interesting observation can be made, which we have not described yet. Namely, if the experiment is done with care, it is actually possible to drive a real gas slowly along the path $v \downarrow v_g$, across v_g, without it starting to condense. The state obtained has a pressure larger than the saturation pressure, and is called a **supersaturated vapor**. It is not an equilibrium state, but what is called a **metastable gas state**. Such a state can have a very long lifetime, but a sufficiently strong external perturbation abruptly drives the system away from it, resulting in a mixture of equilibrium gas and liquid phases at the saturation pressure. Similarly, it is possible to observe a metastable liquid phase, the so-called **superheated liquid**, by slowly increasing v beyond v_l starting from the liquid phase.

Since metastable states are observed in the laboratory but are not equilibrium states in the sense of statistical mechanics, theoretical physicists considered analytic continuation as a way of at least defining their pressure [206].

Moving back to the case considered earlier, the analytic continuation of $\tilde{p}_\beta(\rho)$ through ρ_g coming from the gas branch, as represented by the dotted line below, would therefore provide the pressure of a supersaturated vapor:

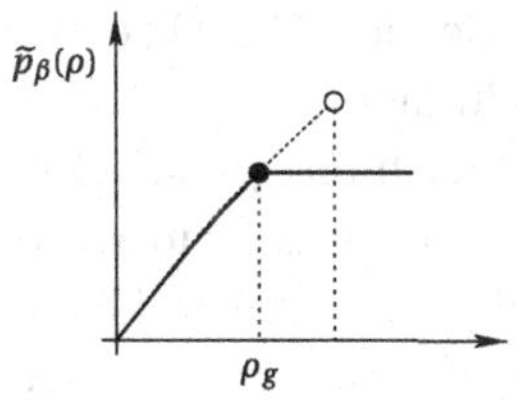

Singularity and the Droplet Mechanism. However, the true analytic structure of isotherms would later prove to be very different. In the 1960s–1970s, an argument of a completely different nature suggested that the branches of an isotherm were separated by *singularities* preventing an analytic continuation. The argument was based on the use of the *droplet model*, whose pressure mimics the pressure of a single droplet of fluid immersed in a gas. This model was introduced for the first time by Andreev [11] and studied more systematically by Fisher [106] (see Exercise 4.16 below).

The striking feature suggested by this toy model was that the actual condensation phenomenon, namely the appearance of large stable droplets of liquid, was responsible for the presence of a singularity of the pressure at the condensation point.

These predictions were confirmed rigorously in the celebrated work [174] of Isakov, which we already mentioned in Section 3.10.9, Theorem 3.67. Isakov implemented rigorously the mechanism suggested by Andreev and Fisher, by giving a detailed study of the large contours (representing droplets) in the low-temperature Ising model ($d \geq 2$), as a function of the magnetic field. In essence, he showed that the coexistence of both phases, at $h = 0$, was responsible for the peculiar behavior of the derivatives of the pressure seen in (3.99).

When translated into the nearest-neighbor lattice gas language, Isakov's analysis implies that, when β is sufficiently large, all the thermodynamic potentials considered in Section 4.8 have singularities blocking analytic continuation at their transition points. For instance p_β, which is analytic on the branches $(-\infty, \mu_*)$ and (μ_*, ∞) by Theorem 4.22, has a singularity at μ_* that forbids analytic continuations along the paths $\mu \uparrow \mu_*$ and $\mu \downarrow \mu_*$. This can also be shown to prevent the existence analytic continuations of f_β and $\widetilde{p}_\beta$ along $\rho \uparrow \rho_g$ or $\rho \downarrow \rho_l$, or of $\widehat{p}_\beta$ along $v \uparrow v_l$ or $v \downarrow v_g$, (see [112]).

These results strongly support Mayer's conjecture, at least for discrete spin systems with finite-range interactions: the condensation point can in principle be detected by studying a single branch up to its first singularity. They also definitely rule out the possibility of studying metastability by means of analytic continuation (see the bibliographical references given below).

Moreover, Isakov's result indicates that the global structure of the isotherms in "real" systems is more complex, since the branches of the isotherms are represented by functions that cannot be united into one single analytic function. In particular, the pressure of a model with short-range interactions *is not obtained by applying some Maxwell-type construction to a smooth function.* This sharp contrast with the van der Waals model comes from the fact that separation of phases (as briefly described in Section 4.12.1) occurs in systems with finite-range interactions, but not in mean-field models.

We mention further bibliographical references related to the topics discussed above. Interesting papers related to Mayer's conjecture include the papers of Kahn and Uhlenbeck [185], and Born and Fuchs [36]. In [204], Lanford and Ruelle ruled out the possibility of analytic continuation of the pressure, using an argument involving the variational principle for Gibbs states. This variational principle will be described later in Chapter 6.

Prior to the work of Isakov, various attempts had been made at describing metastability via analytic continuation for several toy models. These include papers of Schulman and coauthors [253, 280, 281].

In [114], the analysis of Isakov was generalized to the class of two-phase models with finite-range interactions considered in Pirogov–Sinai theory (Chapter 7). The link between the absence of analytic continuation for finite-range models and the mean-field behavior of the van der Waals model was clarified in [113], where a Kać potential with a magnetic field was considered, and the disappearance of its singularity in the van der Waals limit was analyzed in detail.

Additional information on the nonanalytic aspects of thermodynamic potentials at first-order phase transitions can be found in the review of Pfister [275] or in [112].

It is widely accepted, nowadays, that metastability is a dynamical phenomenon that does not enter the framework of equilibrium statistical mechanics. An important contribution to the understanding of metastability from such a point of view can be found in [297]. A modern presentation of metastability, from the point of view of stochastic dynamics, can be found in the books by Olivieri and Vares [258] and Bovier and den Hollander [38].

Exercise 4.16. Consider, for $d \geq 2$,

$$\psi_\beta(h) \stackrel{\text{def}}{=} \sum_{n \geq 1} e^{-\beta 2d n^{(d-1)/d}} e^{-hn} \,,$$

which is a version of the droplet toy model considered by Fisher [106], formulated in the spin language. (The sum is to be interpreted as the pressure of a cubic droplet of $-$ spins immersed in a sea of $+$ spins, centered at the origin; if the droplet has volume n, $-\beta 2dn^{(d-1)/d}$ represents its surface energy, and $-hn$ the energy due to the effect of the magnetic field on its volume.) Verify that ψ_β is analytic in $H^+ \stackrel{\text{def}}{=} \{\mathfrak{Re}\, h > 0\}$. Then, show that ψ_β has no analytic continuation across $h = 0$ (along $h \downarrow 0$), by computing the limits

$$\lim_{h \downarrow 0} \frac{\mathrm{d}^k \psi_\beta}{\mathrm{d}h^k}\bigg|_h$$

and showing that they have the same behavior as that described in (3.99).

---

5 Cluster Expansion

5.1 Introduction

The cluster expansion is a powerful tool in the rigorous study of statistical mechanics. It was introduced by Mayer during the early stages of the study of the phenomenon of condensation and remains widely used nowadays. In particular, it remains at the core of the implementation of many renormalization arguments in mathematical physics, yielding rigorous results that no other methods have yet been able to provide.

Simply stated, the cluster expansion provides a method for studying the logarithm of a partition function. We will use it in various situations, for instance to obtain new analyticity results for the Ising model in the thermodynamic limit.

In a first application, we will obtain new results on the pressure $h \mapsto \psi_\beta(h)$, completing those of Chapter 3. There, we saw that the pressure is analytic in the half-space $\{\mathfrak{Re}\, h > 0\}$, but the techniques we used did not provide further quantitative information. Here we will use the cluster expansion to compute the coefficients of the expansion of $\psi_\beta(h) - h$, in terms of the variable $z = e^{-2h}$:

$$\psi_\beta(h) - h = a_0 + a_1 z + a_2 z^2 + a_3 z^3 + \cdots, \qquad (\mathfrak{Re}\, h \text{ large}).$$

In our second application, we will fix $h = 0$, and study the analyticity of $\beta \mapsto \psi_\beta(0)$. We will obtain, when β is sufficiently *small*, an expansion in terms of the variable $z = \tanh(\beta)$,

$$\psi_\beta(0) - d \log(\cosh \beta) = b_0 + b_1 z + b_2 z^2 + b_3 z^3 + \cdots, \qquad (\beta \text{ small}).$$

One might hope that this series converges for all $\beta < \beta_c(d)$. Unfortunately, the method developed in this chapter will guarantee analyticity only when $\beta < \beta_0$, where $\beta_0 = \beta_0(d)$ is some number, strictly smaller than the critical value $\beta_c(d)$. We will call a regime such as $\beta < \beta_0$ a regime of *very high temperature* to distinguish it from the *high-temperature* regime $\beta < \beta_c$ used in earlier chapters.

Similarly, we will also obtain, when β is sufficiently *large*, an expansion in terms of the variable $z = e^{-2\beta}$,

$$\psi_\beta(0) - \beta d = c_0 + c_1 z + c_2 z^2 + c_3 z^3 + \cdots, \qquad (\beta \text{ large}).$$

Once again, this series will be guaranteed to converge at very low temperature, that is, for all $\beta > \beta_0'$, where $\beta_0' > \beta_c(d)$.

Of course, the cluster expansion is not limited to the study of the pressure in these different regimes and we will show how it can be used to extract additional information on other quantities of interest. In particular, at very low temperatures, we will derive a series expansion for the spontaneous magnetization and prove exponential decay of the truncated 2-point function.

We hope that this sample of applications will convince the reader that the cluster expansion is a versatile tool that, even though applicable only in restricted regions of the space of parameters, provides precious information often unavailable when using other techniques.

Remark 5.1. The cluster expansion will also be used in other parts of this book. We will use it in Chapter 6 to derive uniqueness of the infinite-volume Gibbs state at sufficiently high temperatures for a rather large class of models, and it will play a central role in the Pirogov–Sinai theory described in Chapter 7. ◇

5.2 Polymer Models

The cluster expansion applies when the model under consideration has a partition function that can be written in a particular form, already encountered earlier in the book. For instance, remember from Section 3.7.2 that the configurations of the Ising model at low temperature were conveniently described using extended geometric objects, the *contours*, rather than the individual spins; namely (see (3.32)):

1. each configuration was set in one-to-one correspondence with a family of *pairwise disjoint* contours;
2. once expressed in terms of contours, the Boltzmann weight split into a *product of weights* associated with the contours.

Relying on this geometric representation, Peierls' argument allowed us to prove positivity of the spontaneous magnetization at sufficiently low temperature.

Later, when studying the Ising model at high temperature, a different representation of the partition function was used. Although the objects involved were of a different nature (especially in higher dimensions, see (3.45)), they also satisfied some geometric compatibility condition, namely that of being pairwise disjoint. Moreover, the Boltzmann weight again factorized as a product of the weights associated with these objects.

The description of a system in terms of geometric objects (rather than the original microscopic components, such as Ising spins) turns out to be common in equilibrium statistical mechanics; the resulting class of models, usually called *polymer models*, is precisely the one for which the cluster expansion will be developed. The corresponding partition functions often have a common structure that can be exploited to provide, under suitable hypotheses, detailed information on their logarithm.

Consider a finite set Γ, the elements of which are called **polymers** and usually denoted by $\gamma \in \Gamma$. In specific situations, polymers can be complicated objects, but in this abstract setting we need only two main ingredients:

1. With each polymer $\gamma \in \Gamma$ is associated a **weight** (or **activity**) $\mathrm{w}(\gamma)$, which can be a real or complex number.
2. The interaction between polymers is **pairwise** and is encoded in a function $\delta : \Gamma \times \Gamma \to \mathbb{R}$, which is assumed to be symmetric (that is, $\delta(\gamma, \gamma') = \delta(\gamma', \gamma)$) and to satisfy the following two conditions:

$$\delta(\gamma, \gamma) = 0, \qquad \forall \gamma \in \Gamma, \tag{5.1}$$

$$|\delta(\gamma, \gamma')| \leq 1, \qquad \forall \gamma, \gamma' \in \Gamma. \tag{5.2}$$

Definition 5.2. The **(polymer) partition function** is defined by

$$\Xi \overset{\text{def}}{=} \sum_{\Gamma' \subset \Gamma} \Big\{ \prod_{\gamma \in \Gamma'} \mathrm{w}(\gamma) \Big\} \Big\{ \prod_{\{\gamma, \gamma'\} \subset \Gamma'} \delta(\gamma, \gamma') \Big\}, \tag{5.3}$$

where the sum is over all finite subsets of Γ.

Of course, each pair $\{\gamma, \gamma'\}$ appears only once in the product. We allow $\Gamma' = \varnothing$, in which case the products are, as usual, defined to be 1.

The polymers will always be geometric objects of finite size living on $\mathbb{Z}^d$ (or, possibly, on the dual lattice) and their interaction will be related to pairwise geometric compatibility conditions between the polymers; these conditions will usually be *local*, that is, the compatibility of two polymers can be checked by inspecting their "neighborhood" on $\mathbb{Z}^d$.

5.3 Formal Expansion

The cluster expansion provides an explicit expansion for $\log \Xi$, in the form of a series. To obtain the coefficients of this expansion, we will perform a sequence of operations on Ξ, leading to an expression of the form

$$\Xi = \exp(\cdots).$$

As a first step, the sum over $\Gamma' \subset \Gamma$ can be decomposed according to the number $|\Gamma'|$ of polymers contained in Γ':

$$\sum_{\Gamma' \subset \Gamma} (\cdots) = 1 + \sum_{n \geq 1} \sum_{\substack{\Gamma' \subset \Gamma: \\ |\Gamma'| = n}} (\cdots).$$

For convenience, we now transform the second sum over $\Gamma' \subset \Gamma$ into a sum over *ordered* n-tuples. So let $G_n = (V_n, E_n)$ be the complete graph on $V_n = \{1, 2, \ldots, n\}$. That is, G_n is the simple undirected graph in which there is precisely one edge

$\{i,j\} \in E_n$ for each pair of distinct vertices $i,j \in V_n$ (see Figure 2.1). We can then write

$$\Xi = 1 + \sum_{n\geq 1} \frac{1}{n!} \sum_{\gamma_1} \cdots \sum_{\gamma_n} \Big\{ \prod_{i\in V_n} \mathrm{w}(\gamma_i) \Big\} \Big\{ \prod_{\{i,j\}\in E_n} \delta(\gamma_i,\gamma_j) \Big\}. \tag{5.4}$$

Notice that, the sum being now over all ordered n-tuples $(\gamma_1,\ldots,\gamma_n) \in \Gamma^{V_n}$, we had to introduce a factor $\frac{1}{n!}$ to avoid overcounting. Observe that only collections in which all polymers $\gamma_1,\ldots,\gamma_n$ are distinct contribute to the sum, since $\delta(\gamma_i,\gamma_j) = 0$ whenever $\gamma_i = \gamma_j$. This means that only a finite number of terms, in this sum over $n \geq 1$, are nonzero.

The next step is the following: rather than working with a sum over the n-tuples of Γ^{V_n}, we will work with a sum over suitable subgraphs of G_n. We write $G \subset G_n$ to indicate that G is a subgraph of G_n with the same set of vertices V_n and with a set of edges that is a subset of E_n. Given a graph $G = (V, E)$, we will often write $i \in G$, respectively $e \in G$, instead of $i \in V$, respectively $e \in E$.

Subgraphs of G_n can be introduced if one uses the "$+1 - 1$" trick to expand the product containing the interactions between the polymers (see Exercise 3.22). Letting

$$\zeta(\gamma,\gamma') \stackrel{\text{def}}{=} \delta(\gamma,\gamma') - 1,$$

we get

$$\prod_{\{i,j\}\in E_n} \delta(\gamma_i,\gamma_j) = \prod_{\{i,j\}\in E_n} (1+\zeta(\gamma_i,\gamma_j)) = \sum_{E\subset E_n} \prod_{\{i,j\}\in E} \zeta(\gamma_i,\gamma_j).$$

Since a set $E \subset E_n$ can be put in one-to-one correspondence with the subgraph $G \subset G_n$ defined by $G \stackrel{\text{def}}{=} (V_n, E)$, we can interpret the sum over $E \subset E_n$ as a sum over $G \subset G_n$. We thus obtain

$$\begin{aligned}\Xi &= 1 + \sum_{n\geq 1} \frac{1}{n!} \sum_{G\subset G_n} \sum_{\gamma_1} \cdots \sum_{\gamma_n} \Big\{ \prod_{i\in V_n} \mathrm{w}(\gamma_i) \Big\} \Big\{ \prod_{\{i,j\}\in E} \zeta(\gamma_i,\gamma_j) \Big\} \\ &= 1 + \sum_{n\geq 1} \frac{1}{n!} \sum_{G\subset G_n} Q[G], \end{aligned} \tag{5.5}$$

where we have introduced, for a graph $G = (V, E)$,

$$Q[G] \stackrel{\text{def}}{=} \sum_{\gamma_1} \cdots \sum_{\gamma_{|V|}} \Big\{ \prod_{i\in V} \mathrm{w}(\gamma_i) \Big\} \Big\{ \prod_{\{i,j\}\in E} \zeta(\gamma_i,\gamma_j) \Big\}.$$

Let us now define the **Ursell functions** φ on ordered families $(\gamma_1,\ldots,\gamma_m)$ by $\varphi(\gamma_1) \stackrel{\text{def}}{=} 1$, when $m = 1$, and

$$\varphi(\gamma_1,\ldots,\gamma_m) \stackrel{\text{def}}{=} \frac{1}{m!} \sum_{\substack{G\subset G_m \\ \text{connected}}} \prod_{\{i,j\}\in G} \zeta(\gamma_i,\gamma_j),$$

when $m \geq 2$.

Proposition 5.3.

$$\Xi = \exp\Big(\sum_{m\geq 1}\sum_{\gamma_1}\cdots\sum_{\gamma_m}\varphi(\gamma_1,\dots,\gamma_m)\prod_{i\in V_m}\mathrm{w}(\gamma_i)\Big). \tag{5.6}$$

Observe that, even if Ξ is a finite sum, the resulting series in (5.6) is infinite, since a given polymer can appear several times in the same collection, without the Ursell function necessarily vanishing. In the next section, we will state conditions that ensure that the series is actually absolutely convergent, which will justify the rearrangements done in the proof below. For the time being, however, we are only interested in the structure of its coefficients, so the series in (5.6) should (temporarily) only be considered as formal.

Proof of Proposition 5.3. Notice that, if $G'_1,\dots,G'_k$ represent the (maximal) connected components of G, then

$$Q[G] = \prod_{r=1}^{k} Q[G'_r].$$

Now, observe that $Q[G] = Q[G']$ whenever G and G' are isomorphic.[1] One can thus replace the vertex set V'_i of G'_i by $\{1,\dots,m_i\}$, where $m_i = |V'_i|$. Therefore,

$$\begin{aligned}\sum_{G\subset G_n} Q[G] &= \sum_{k=1}^{n}\sum_{\substack{G\subset G_n\\ G=(G'_1,\dots,G'_k)}}\prod_{r=1}^{k} Q[G'_r]\\ &= \sum_{k=1}^{n}\frac{1}{k!}\sum_{\substack{m_1,\dots,m_k\\ m_1+\cdots+m_k=n}}\frac{n!}{m_1!\cdots m_k!}\sum_{\substack{G'_1\subset G_{m_1}\\ \text{connected}}}\cdots\sum_{\substack{G'_k\subset G_{m_k}\\ \text{connected}}}\prod_{r=1}^{k} Q[G'_r], \end{aligned}\tag{5.7}$$

where, in the second identity, the coefficient $n!/(m_1!\cdots m_k!)$ takes into account the number of ways of partitioning V_n into k disjoint subsets of respective cardinalities $m_1,\dots,m_k\geq 1$. Observe that, at least formally,

$$\sum_{n\geq 1}\sum_{k=1}^{n}\sum_{\substack{m_1,\dots,m_k\\ m_1+\cdots+m_k=n}}(\cdots) = \sum_{k\geq 1}\sum_{n\geq k}\sum_{\substack{m_1,\dots,m_k\\ m_1+\cdots+m_k=n}}(\cdots) = \sum_{k\geq 1}\sum_{m_1,\dots,m_k}(\cdots), \tag{5.8}$$

which leads to

$$\begin{aligned}\Xi &= 1+\sum_{k\geq 1}\frac{1}{k!}\sum_{m_1,\dots,m_k}\prod_{r=1}^{k}\Big\{\frac{1}{m_r!}\sum_{\substack{G'_r\subset G_{m_r}\\ \text{connected}}} Q[G'_r]\Big\}\\ &= 1+\sum_{k\geq 1}\frac{1}{k!}\Big(\sum_{m\geq 1}\sum_{\gamma_1}\cdots\sum_{\gamma_m}\varphi(\gamma_1,\dots,\gamma_m)\prod_{j=1}^{m}\mathrm{w}(\gamma_j)\Big)^k,\end{aligned}$$

which is (5.6). □

[1] Two graphs $G=(V,E)$ and $G=(V',E')$ are **isomorphic** if there exists a bijection $f\colon V\to V'$ such that an edge $e=\{x,y\}$ belongs to E if and only if $e'=\{f(x),f(y)\}$ belongs to E'.

Let us emphasize a delicate point ignored in the above computation. In a first step, in (5.5), Ξ *was written with the help of a sum* $\sum_{n\geq 1} a_n$, *where* n *indexes the size of the complete graph* G_n, *and where the* a_n *are all equal to zero when* n *is sufficiently large. In a second step (see* (5.7)*), each* a_n *was decomposed as* $a_n = \sum_{k=1}^{n} b_{k,n}$, *where the index* k *denotes the number of connected components of the subgraph* $G \subset G_n$. *Since* a_n *vanishes for large* n, *this means that important cancellations occur among the* $b_{k,n}$ *(when summed over* k*). The main formal computation that requires justification was done in* (5.8), *when we interchanged the summations over* n *and* k:

$$\sum_{n\geq 1} a_n = \sum_{n\geq 1}\sum_{k=1}^{n} b_{k,n} = \sum_{k\geq 1}\sum_{n\geq k} b_{k,n}\,.$$

Namely, the interchange is allowed only if each of the series $\sum_{n\geq k} b_{k,n}$ *is known to converge, and this is not guaranteed in general.* ◇

Proving that Ξ has a well-defined logarithm implies in particular that $\Xi \neq 0$. As we saw when studying uniqueness in the Ising model, the absence of zeros of the partition function on a complex domain (for each Λ along a sequence $\Lambda \Uparrow \mathbb{Z}^d$) in fact entails uniqueness of the infinite-volume Gibbs measure of this model. This indicates that guaranteeing the absolute convergence of the series for $\log \Xi$ is non-trivial in general, and the latter will usually hold only for some restricted range of values of the parameters of the underlying model.

5.4 A Condition Ensuring Convergence

We now impose conditions on the weights which ensure that the series in (5.6) converges absolutely:

$$\sum_{k\geq 1}\sum_{\gamma_1}\cdots\sum_{\gamma_k} |\varphi(\gamma_1,\ldots,\gamma_k)| \prod_{i=1}^{k} |\mathrm{w}(\gamma_i)| < \infty\,. \tag{5.9}$$

The main ingredient is the following:

Theorem 5.4. *Assume that* (5.2) *holds and that there exists* $a\colon \Gamma \to \mathbb{R}_{>0}$ *such that, for each* $\gamma_* \in \Gamma$,

$$\sum_{\gamma} |\mathrm{w}(\gamma)| e^{a(\gamma)} |\zeta(\gamma,\gamma_*)| \leq a(\gamma_*)\,. \tag{5.10}$$

Then, for all $\gamma_1 \in \Gamma$,

$$1 + \sum_{k\geq 2} k \sum_{\gamma_2}\cdots\sum_{\gamma_k} |\varphi(\gamma_1,\gamma_2,\ldots,\gamma_k)| \prod_{j=2}^{k} |\mathrm{w}(\gamma_j)| \leq e^{a(\gamma_1)}\,. \tag{5.11}$$

In particular, (5.9) *holds.*

Remark 5.5. In this chapter, we always assume that $|\Gamma| < \infty$. Nevertheless, this restriction is not necessary. When it is not imposed, in addition to (5.10), one has to require that

$$\sum_{\gamma} |\mathrm{w}(\gamma)| e^{a(\gamma)} < \infty . \qquad \diamond$$

The series in (5.10) *should remind the reader of those considered when implementing Peierls' argument, such as* (3.37). *Actually, verifying that these conditions hold in a specific situation usually amounts to a similar energy–entropy argument.* $\diamond$

Exercise 5.1. Verify that (5.11) implies (5.9).

Proof of Theorem 5.4. We fix $\gamma_1 \in \Gamma$ and show that, for all $N \geq 2$,

$$1 + \sum_{k=2}^{N} k \sum_{\gamma_2} \cdots \sum_{\gamma_k} |\varphi(\gamma_1, \gamma_2, \dots, \gamma_k)| \prod_{j=2}^{k} |\mathrm{w}(\gamma_j)| \leq e^{a(\gamma_1)} . \tag{5.12}$$

Clearly, letting $N \to \infty$ in (5.12) yields (5.11). The proof of (5.12) is done by induction over N.

For $N = 2$, the only connected graph $G \subset G_2$ is the one with one edge connecting 1 and 2, and so $\varphi(\gamma_1, \gamma_2) = \frac{1}{2!}\zeta(\gamma_1, \gamma_2)$. Therefore, the left-hand side of (5.12) is

$$1 + 2\sum_{\gamma_2} |\varphi(\gamma_1, \gamma_2)||\mathrm{w}(\gamma_2)| = 1 + \sum_{\gamma_2} |\zeta(\gamma_1, \gamma_2)||\mathrm{w}(\gamma_2)| \leq e^{a(\gamma_1)} ,$$

where we used $1 \leq e^{a(\gamma_2)}$, (5.10) and $1 + x \leq e^x$. This proves (5.12) for $N = 2$. We now show that if (5.12) holds for N, then it also holds for $N + 1$.

To do that, consider the left-hand side of (5.12) with $N + 1$ in place of N, take some $k \leq N + 1$, and consider any connected graph $G \subset G_k$ appearing in the sum defining $\varphi(\gamma_1, \gamma_2, \dots, \gamma_k)$. Let E' denote the nonempty set of edges of G with an endpoint at 1. The graph G', obtained from G by removing 1 together with each edge of E', splits into a set of connected components $G'_1, \dots, G'_l$.

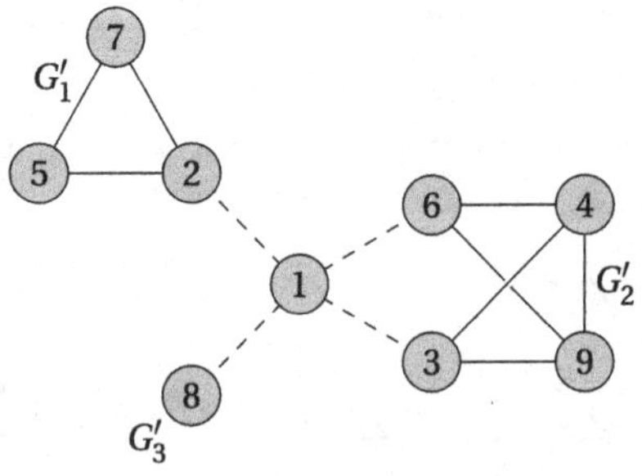

We can thus see G as obtained by (i) partitioning the set $\{2,3,\dots,k\}$ into subsets $V_1',\dots,V_l'$, $l \le k-1$, (ii) associating with each V_i' a connected graph G_i', and (iii) connecting 1 in all possible ways to at least one point in each connected component V_i'. Accordingly,

$$\varphi(\gamma_1,\gamma_2,\dots,\gamma_k) = \tag{5.13}$$
$$\frac{1}{k!}\sum_{l=1}^{k-1}\frac{1}{l!}\sum_{V_1',\dots,V_l'}\prod_{i=1}^{l}\Big\{\sum_{\substack{G_i':\ V(G_i')=V_i'\\ \text{connected}}}\ \prod_{\{i',j'\}\in G_i'}\zeta(\gamma_{i'},\gamma_{j'})\Big\}\Big\{\sum_{\substack{K_i\subset V_i'\\ K_i\neq\varnothing}}\prod_{j'\in K_i}\zeta(\gamma_1,\gamma_{j'})\Big\}.$$

The next step is to specify the number of points in each V_i'. If $|V_i'| = m_i$,

$$\sum_{\substack{G_i':\ V(G_i')=V_i'\\ \text{connected}}}\ \prod_{\{i',j'\}\in G_i'}\zeta(\gamma_{i'},\gamma_{j'}) = m_i!\,\varphi\big((\gamma_{j'})_{j'\in V_i'}\big).$$

Moreover,

$$\sum_{\substack{K_i\subset V_i'\\ K_i\neq\varnothing}}\prod_{j'\in K_i}\zeta(\gamma_1,\gamma_{j'}) = \Big\{\prod_{j'\in V_i'}\big(1+\zeta(\gamma_1,\gamma_{j'})\big)\Big\} - 1. \tag{5.14}$$

Exercise 5.2. Assuming $|1+\alpha_k| \le 1$ for all $k \ge 1$, show that

$$\Big|\prod_{k=1}^{n}(1+\alpha_k) - 1\Big| \le \sum_{k=1}^{n}|\alpha_k|. \tag{5.15}$$

Since (5.2) guarantees that $|1+\zeta| \le 1$, (5.14) and (5.15) yield

$$\Big|\sum_{\substack{K_i\subset V_i'\\ K_i\neq\varnothing}}\prod_{j'\in K_i}\zeta(\gamma_1,\gamma_{j'})\Big| \le \sum_{j'\in V_i'}|\zeta(\gamma_1,\gamma_{j'})|.$$

We now use (5.13) to bound the sum on the left-hand side of (5.12) (with $N+1$ in place of N). The sum over the sets V_i' will be made as in the proof of Proposition 5.3: the number of partitions of $\{2,3,\dots,k\}$ into $(V_1',\dots,V_l')$, with $|V_i'| = m_i$, $m_1+\dots+m_l = k-1$, is equal to $\frac{(k-1)!}{m_1!\cdots m_l!}$. But, since the summands are nonnegative, we can bound

$$\sum_{k=2}^{N+1}\sum_{l=1}^{k-1}\sum_{\substack{m_1,\dots,m_l:\\ m_1+\dots+m_l=k-1}}(\cdots) = \sum_{l=1}^{N}\sum_{k=l+1}^{N+1}\sum_{\substack{m_1,\dots,m_l:\\ m_1+\dots+m_l=k-1}}(\cdots)$$
$$\le \sum_{l=1}^{N}\sum_{m_1=1}^{N}\cdots\sum_{m_l=1}^{N}(\cdots),$$

which leaves us with

$$\sum_{k=2}^{N+1} k \sum_{\gamma_2} \cdots \sum_{\gamma_k} |\varphi(\gamma_1, \gamma_2, \ldots, \gamma_k)| \prod_{j=2}^{k} |\mathrm{w}(\gamma_j)|$$

$$\leq \sum_{l \geq 1} \frac{1}{l!} \prod_{i=1}^{l} \Big\{ \sum_{m_i=1}^{N} \sum_{\gamma'_1} \cdots \sum_{\gamma'_{m_i}} |\varphi(\gamma'_1, \ldots, \gamma'_{m_i})| \prod_{j'=1}^{m_i} |\mathrm{w}(\gamma'_{j'})| \sum_{j'=1}^{m_i} |\zeta(\gamma_1, \gamma'_{j'})| \Big\}. \tag{5.16}$$

Lemma 5.6. *If* (5.12) *holds, then, for all* $\gamma_* \in \Gamma$,

$$\sum_{k=1}^{N} \sum_{\gamma_1} \cdots \sum_{\gamma_k} \Big\{ \sum_{i=1}^{k} |\zeta(\gamma_*, \gamma_i)| \Big\} |\varphi(\gamma_1, \ldots, \gamma_k)| \prod_{j=1}^{k} |\mathrm{w}(\gamma_j)| \leq a(\gamma_*). \tag{5.17}$$

Proof We fix $\gamma_* \in \Gamma$, and multiply both sides of (5.12) by $|\zeta(\gamma_*, \gamma_1)| \cdot |\mathrm{w}(\gamma_1)|$, and sum over γ_1. Using (5.10), the right-hand side of the expression obtained can be bounded by $a(\gamma_*)$, whereas the left-hand side becomes

$$\sum_{k=1}^{N} k \sum_{\gamma_1} \cdots \sum_{\gamma_k} |\zeta(\gamma_*, \gamma_1)| |\varphi(\gamma_1, \ldots, \gamma_k)| \prod_{j=1}^{k} |\mathrm{w}(\gamma_j)| .$$

But clearly, for all $i \in \{2, \ldots, k\}$,

$$\sum_{\gamma_1} \cdots \sum_{\gamma_k} |\zeta(\gamma_*, \gamma_1)| |\varphi(\gamma_1, \ldots, \gamma_k)| \prod_{j=1}^{k} |\mathrm{w}(\gamma_j)|$$

$$= \sum_{\gamma_1} \cdots \sum_{\gamma_k} |\zeta(\gamma_*, \gamma_i)| |\varphi(\gamma_1, \ldots, \gamma_k)| \prod_{j=1}^{k} |\mathrm{w}(\gamma_j)| ,$$

which proves the claim. □

Using (5.17), we can bound (5.16) by $\sum_{l \geq 1} \frac{1}{l!} a(\gamma_1)^l = e^{a(\gamma_1)} - 1$. This concludes the proof of Theorem 5.4. □

The determination of a suitable function $a(\gamma)$ for (5.10) will depend on the problem considered. As we will see in the applications below, $a(\gamma)$ will usually be naturally related to some measure of the size of γ.

Example 5.7. As the most elementary application of the previous lemma, let us consider the expansion of $\log(1 + z)$ for small $|z|$.

The function $1 + z$ can be seen as a particularly simple example of polymer partition function: one with a single polymer, $\Gamma = \{\gamma\}$, and a weight function $\mathrm{w}(\gamma) = z$. Indeed, in this case, there are only two terms in the right-hand side of (5.3) ($\Gamma' = \varnothing$ and $\Gamma' = \{\gamma\}$) and the partition function reduces to

$$\Xi = 1 + z .$$

Theorem 5.4 guarantees that the cluster expansion for Ξ converges whenever (5.10) is satisfied:

$$|z|e^a \leq a,$$

where $a > 0$ is a constant we can choose. Since the function $a \mapsto ae^{-a}$ is maximal when $a = 1$, the best possible choice for a is $a = 1$ and the condition for convergence becomes

$$|z| \leq e^{-1}.$$

Theorem 5.4 then guarantees convergence of the cluster expansion for $\log \Xi$ for all such values of z: by (5.6),

$$\log(1+z) = \log \Xi = \sum_{m \geq 1} \varphi_m z^m,$$

where we have introduced

$$\varphi_m \stackrel{\text{def}}{=} \varphi \underbrace{(\gamma, \ldots, \gamma)}_{m \text{ copies}} = \frac{1}{m!} \sum_{k=0}^{\binom{m}{2}} (-1)^k \, |\mathscr{G}_{m,k}|,$$

and $\mathscr{G}_{m,k}$ is the set of all connected subgraphs of the complete graph G_m with m vertices and k edges.

It is instructive to compare the above result with the classical Taylor expansion:

$$\log(1+z) = \sum_{m \geq 1} \frac{(-1)^{m-1}}{m} z^m.$$

First, we see that the condition in Theorem 5.4 is not optimal, since the latter series actually converges whenever $|z| < 1$. Moreover, identifying the coefficients of z^n in both expansions, we obtain the following nontrivial combinatorial identity:

$$\sum_{k=0}^{\binom{m}{2}} (-1)^k |\mathscr{G}_{m,k}| = (-1)^{m-1}(m-1)!. \qquad \diamond$$

5.5 When the Weights Depend on a Parameter

The convergence of the cluster expansion is very often used to prove analyticity of the pressure in the thermodynamic limit. So let us assume that the weights of the polymers depend on some complex parameter:

$$z \mapsto \mathrm{w}_z(\gamma), \qquad z \in D,$$

where D is a domain of $\mathbb{C}$. When each weight depends smoothly (for example, analytically) on z, it can be useful to determine whether this smoothness extends to $\log \Xi$.

Theorem 5.8. *Assume that* $z \mapsto \mathrm{w}_z(\gamma)$ *is analytic on D, for each* $\gamma \in \Gamma$, *and that there exists a real weight* $\overline{\mathrm{w}}(\gamma) \geq 0$ *such that*

$$\sup_{z \in D} |\mathrm{w}_z(\gamma)| \leq \overline{\mathrm{w}}(\gamma), \quad \forall \gamma \in \Gamma, \tag{5.18}$$

and such that (5.10) *holds with* $\overline{\mathrm{w}}(\gamma)$ *in place of* $\mathrm{w}(\gamma)$. *Then,* (5.6) *and* (5.9) *hold with* $\mathrm{w}_z(\gamma)$ *in place of* $\mathrm{w}(\gamma)$, *and* $z \mapsto \log \Xi$ *is analytic on D.*

Proof Let us write the expansion as $\log \Xi = \sum_{n \geq 1} f_n(z)$, where

$$f_n(z) \stackrel{\text{def}}{=} \sum_{\gamma_1} \cdots \sum_{\gamma_n} \varphi(\gamma_1, \dots, \gamma_n) \prod_{i=1}^{n} \mathrm{w}_z(\gamma_i).$$

Since $|\Gamma| < \infty$, f_n is a sum containing only a finite number of terms; it is therefore analytic in D. If we can verify that the series $\sum_n f_n$ is uniformly convergent on compact sets $K \subset D$, Theorem B.27 will imply that it represents an analytic function on D. We therefore compute

$$\begin{aligned}
\sup_{z \in K} \Big| \sum_{n \geq 1} f_n(z) - \sum_{n=1}^{N} f_n(z) \Big| &\leq \sup_{z \in K} \sum_{n > N} |f_n(z)| \\
&\leq \sum_{n > N} \sup_{z \in K} |f_n(z)| \\
&\leq \sum_{n > N} \sum_{\gamma_1} \cdots \sum_{\gamma_n} |\varphi(\gamma_1, \dots, \gamma_n)| \prod_{i=1}^{n} \overline{\mathrm{w}}(\gamma_i).
\end{aligned} \tag{5.19}$$

By our assumptions, Theorem 5.4 implies that (5.9) holds, with $\overline{\mathrm{w}}(\cdot)$ in place of $|\mathrm{w}(\cdot)|$. This implies that (5.19) goes to zero when $N \to \infty$. The fact that (5.11) holds is immediate. □

5.6 The Case of Hard-Core Interactions

Up to now, we have considered fairly general interactions. But often in practice, and in all cases treated in this book, δ takes the particularly simple form of a **hard-core** interaction, that is,

$$\delta(\gamma, \gamma') \in \{0, 1\} \qquad \text{for all } \gamma, \gamma' \in \Gamma.$$

In such a case, two polymers γ and γ' will be said to be **compatible** if $\delta(\gamma, \gamma') = 1$ and **incompatible** if $\delta(\gamma, \gamma') = 0$. Obviously, only collections of pairwise compatible polymers yield a nonzero contribution to the partition function Ξ in (5.4).

Let us now turn to the series (5.6) for $\log \Xi$. We say that a collection $\{\gamma_1, \dots, \gamma_n\}$ is **decomposable** if it is possible to express it as a disjoint union of two nonempty sets, in such a way that each γ_i in the first set is compatible with each γ_j in the second. It follows immediately from the definition of the Ursell functions that

$$\varphi(\gamma_1, \dots, \gamma_n) = 0 \quad \text{if } \{\gamma_1, \dots, \gamma_n\} \text{ is decomposable.}$$

In particular, the nonzero contributions to $\log \Xi$ in (5.6) therefore come from the *nondecomposable* collections. An unordered, nondecomposable collection $X = \{\gamma_1, \dots, \gamma_n\}$ is called a **cluster**. Note that X is actually a *multiset*, that is, the same polymer can appear multiple times. We denote by $n_X(\gamma)$ the number of times the polymer $\gamma \in \Gamma$ appears in X. We can write

$$\log \Xi = \sum_{n\geq 1} \sum_{\gamma_1} \cdots \sum_{\gamma_n} \varphi(\gamma_1, \dots, \gamma_n) \prod_{i=1}^{n} \mathrm{w}(\gamma_i) = \sum_{X} \Psi(X),$$

where the sum is over all clusters of polymers in Γ and, for a cluster $X = \{\tilde{\gamma}_1, \dots, \tilde{\gamma}_n\}$,

$$\Psi(X) \stackrel{\text{def}}{=} \Big\{\prod_{\gamma\in\Gamma} \frac{1}{n_X(\gamma)!}\Big\} \Big\{ \sum_{\substack{G\subset G_n \\ \text{connected}}} \prod_{\{i,j\}\in G} \zeta(\tilde{\gamma}_i, \tilde{\gamma}_j)\Big\} \prod_{i=1}^{n} \mathrm{w}(\tilde{\gamma}_i). \tag{5.20}$$

Indeed, given a cluster $X = \{\tilde{\gamma}_1, \dots, \tilde{\gamma}_n\}$, there are $\frac{n!}{\prod_{\gamma\in\Gamma} n_X(\gamma)!}$ distinct ways of assigning the polymers $\tilde{\gamma}_1, \dots, \tilde{\gamma}_n$ to the summation variables $\gamma_1, \dots, \gamma_n$ above.

5.7 Applications

The cluster expansion can be applied in many situations. Our main systematic use of it will be in Chapter 7, when developing the Pirogov–Sinai theory. We will also use it to obtain a uniqueness criterion for infinite-volume Gibbs measures, in Section 6.5.4.

Before that, we apply it in various ways to the Ising model (and to the corresponding nearest-neighbor lattice gas). We will see that to different regions of the phase diagram correspond different well-suited polymer models. The cluster expansion can then be used to extract useful information on the model for parameters in these regions.

When checking Condition (5.10), we will see that the regions in which the cluster expansion converges for those polymer models are all far from the point $(\beta, h) = (\beta_c, 0)$:

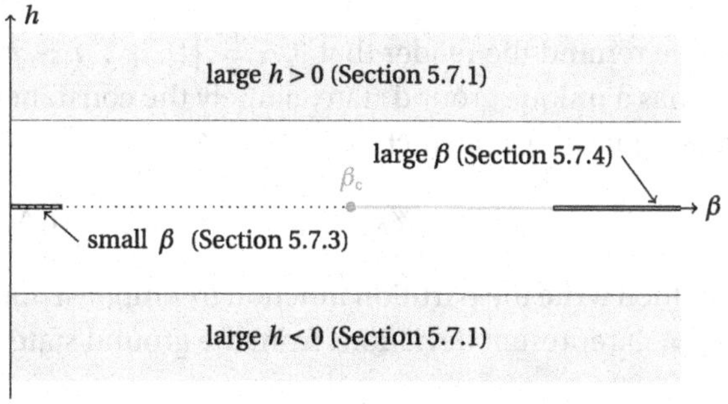

5.7.1 Ising Model in a Strong Magnetic Field

Consider the Ising model with a complex magnetic field $h \in \mathbb{C}$, at an arbitrary inverse temperature $\beta \geq 0$. The Lee–Yang Circle Theorem proved in Chapter 3 yields existence and analyticity of the pressure in the half-planes $H^+ \stackrel{\text{def}}{=} \{h \in \mathbb{C} : \Re e\, h > 0\}$ and $H^- \stackrel{\text{def}}{=} \{h \in \mathbb{C} : \Re e\, h < 0\}$. Here, we will use the cluster expansion to obtain a weaker result, namely that analyticity holds in the regions $\{h \in \mathbb{C} : \Re e\, h > x_0 > 0\}$ and $\{h \in \mathbb{C} : \Re e\, h < -x_0 < 0\}$ (see below for the value of x_0). Although these regions are proper subsets of the half-planes H^+ and H^-, the convergent expansion provides a wealth of additional information on the pressure in these regions, not provided by the Lee–Yang approach.

We will consider the case $\Re e\, h > 0$. As seen in Chapter 3, when $h \in \mathbb{R}$, the pressure does not depend on the boundary condition used in the thermodynamic limit, and we can thus choose the most convenient one. In this section, this turns out to be the $+$ boundary condition. The first step is to define a polymer model that is well suited for the analysis of the Ising model with a large magnetic field.

When the magnetic field $h > 0$ is large, there is a very strong incentive for spins to take the value $+1$. It is therefore natural to describe configurations by only keeping track of the negative spins. ◇

We emphasize the role of the negative spins by writing the Hamiltonian as follows:

$$\begin{aligned}\mathscr{H}_{\Lambda;\beta,h} &= -\beta \sum_{\{i,j\}\in\mathscr{E}_\Lambda^{\mathrm{b}}} \sigma_i\sigma_j - h\sum_{i\in\Lambda}\sigma_i \\ &= -\beta|\mathscr{E}_\Lambda^{\mathrm{b}}| - h|\Lambda| - \beta \sum_{\{i,j\}\in\mathscr{E}_\Lambda^{\mathrm{b}}} (\sigma_i\sigma_j - 1) - h\sum_{i\in\Lambda}(\sigma_i - 1).\end{aligned} \tag{5.21}$$

Let $\omega \in \Omega_\Lambda^+$. Introducing the set

$$\Lambda^-(\omega) \stackrel{\text{def}}{=} \{i \in \Lambda : \omega_i = -1\}, \tag{5.22}$$

we can write

$$\mathscr{H}_{\Lambda;\beta,h}(\omega) = -\beta|\mathscr{E}_\Lambda^{\mathrm{b}}| - h|\Lambda| + 2\beta|\partial_e\Lambda^-(\omega)| + 2h|\Lambda^-(\omega)|,$$

where we remind the reader that $\partial_e A \stackrel{\text{def}}{=} \big\{\{i,j\} : i \sim j,\, i \in A,\, j \notin A\big\}$. Notice that $\mathscr{H}_{\Lambda;\beta,h}$ has a unique ground state, namely the constant configuration η^+ (in which all spins equal $+1$), for which $\Lambda^-(\eta^+) = \varnothing$ and

$$\mathscr{H}_{\Lambda;\beta,h}(\eta^+) = -\beta|\mathscr{E}_\Lambda^{\mathrm{b}}| - h|\Lambda|.$$

We can then write the partition function by emphasizing that configurations ω with $\Lambda^-(\omega) \neq \varnothing$ represent deviations from the ground state:

$$\mathbf{Z}^{+}_{\Lambda;\beta,h} = e^{\beta|\mathscr{E}^{\mathrm{b}}_{\Lambda}|+h|\Lambda|} \sum_{\Lambda^- \subset \Lambda} e^{-2\beta|\partial_e \Lambda^-|-2h|\Lambda^-|}$$
$$= e^{\beta|\mathscr{E}^{\mathrm{b}}_{\Lambda}|+h|\Lambda|} \Big\{ 1 + \sum_{\substack{\Lambda^- \subset \Lambda: \\ \Lambda^- \neq \varnothing}} e^{-2\beta|\partial_e \Lambda^-|-2h|\Lambda^-|} \Big\}.$$

Let us declare two vertices $i, j \in \Lambda^-$ to be connected if $d_1(i,j) \stackrel{\text{def}}{=} \|j-i\|_1 = 1$. We can then decompose Λ^- into maximal connected components (see Figure 5.1):

$$\Lambda^- = S_1 \cup \cdots \cup S_n .$$

By definition, $d_1(S_i, S_j) \stackrel{\text{def}}{=} \inf\{d_1(k,l) : k \in S_i, l \in S_j\} > 1$ if $i \neq j$. The components S_i play the role of the polymers in the present application. Since $|\partial_e \Lambda^-| = \sum_{i=1}^n |\partial_e S_i|$ and $|\Lambda^-| = \sum_{i=1}^n |S_i|$, we can write

$$\mathbf{Z}^{+}_{\Lambda;\beta,h} = e^{\beta|\mathscr{E}^{\mathrm{b}}_{\Lambda}|+h|\Lambda|}\, \Xi^{\mathrm{LF}}_{\Lambda;\beta,h}, \tag{5.23}$$

where the **large-field polymer partition function** is

$$\Xi^{\mathrm{LF}}_{\Lambda;\beta,h} \stackrel{\text{def}}{=} 1 + \sum_{n\geq 1} \frac{1}{n!} \sum_{S_1 \subset \Lambda} \cdots \sum_{S_n \subset \Lambda} \Big\{ \prod_{i=1}^n \mathrm{w}_h(S_i) \Big\} \Big\{ \prod_{1\leq i<j\leq n} \delta(S_i, S_j) \Big\}. \tag{5.24}$$

Each sum $\sum_{S_i \subset \Lambda}$ is over nonempty connected subsets of Λ (from now on, all sets denoted by the letter S, with or without a subscript, will be considered as nonempty and connected), the weights are

$$\mathrm{w}_h(S_i) \stackrel{\text{def}}{=} e^{-2\beta|\partial_e S_i|-2h|S_i|},$$

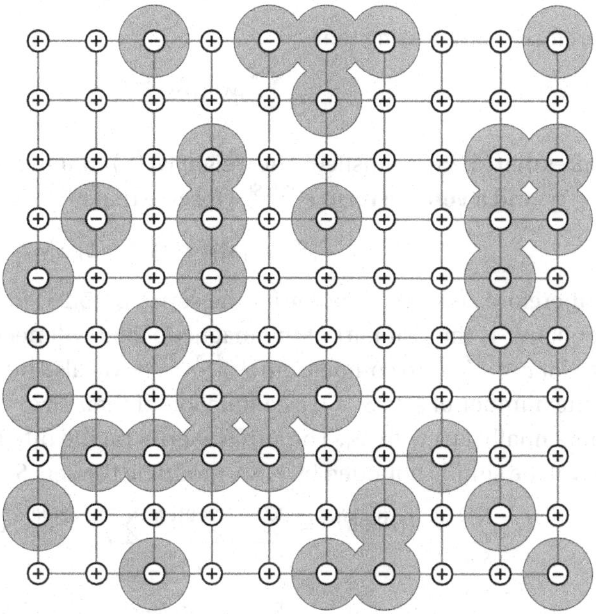

Figure 5.1. A configuration of the Ising model. Each connected component of the shaded area delimits one of the polymers $S_1, \ldots, S_{17}$.

and the interactions are of hard-core type:

$$\delta(S_i, S_j) \stackrel{\text{def}}{=} \begin{cases} 1 & \text{if } d_1(S_i, S_j) > 1, \\ 0 & \text{otherwise.} \end{cases}$$

We will now show that there exists a function $a(S) \geq 0$ such that (5.10) holds when $\mathfrak{Re}\, h$ is taken sufficiently large. In the present context, this condition becomes

$$\forall S_* \subset \Lambda, \quad \sum_{S \subset \Lambda} |w_h(S)| e^{a(S)} |\zeta(S, S_*)| \leq a(S_*), \tag{5.25}$$

where we remind the reader that $\zeta(S, S_*) \stackrel{\text{def}}{=} \delta(S, S_*) - 1$. Observe that $\zeta(S, S_*) \neq 0$ if and only if $S \cap [S_*]_1 \neq \varnothing$, where

$$[S_*]_1 \stackrel{\text{def}}{=} \left\{ j \in \mathbb{Z}^d : d_1(j, S_*) \leq 1 \right\}.$$

Therefore, the sum in (5.25) can be bounded by

$$\sum_{S \subset \Lambda} |w_h(S)| e^{a(S)} |\zeta(S, S_*)| \leq |[S_*]_1| \max_{j \in [S_*]_1} \sum_{S \ni j} |w_h(S)| e^{a(S)},$$

where now the sum over $S \ni j$ is an infinite sum over all finite connected subsets of $\mathbb{Z}^d$ that contain the point j. Let us define, for all S,

$$a(S) \stackrel{\text{def}}{=} |[S]_1|.$$

Since both the weights and $a(\cdot)$ are invariant under translations,

$$\max_{j \in [S_*]_1} \sum_{S \ni j} |w_h(S)| e^{|[S]_1|} = \sum_{S \ni 0} |w_h(S)| e^{|[S]_1|}.$$

Therefore, guaranteeing that

$$\sum_{S \ni 0} |w_h(S)| e^{|[S]_1|} \leq 1 \tag{5.26}$$

ensures that (5.25) is satisfied. The weight $w_h(S)$ contains two terms: a **surface term** $e^{-2\beta|\partial_e S|}$ and a **volume term** $e^{-2h|S|}$. Observe that $e^{|[S]_1|}$ is also a volume term, since

$$|S| \leq |[S]_1| \leq (2d+1)|S|.$$

We therefore see that, in order for the series in (5.26) to converge and be smaller or equal to 1, the real part of the magnetic field will need to be taken sufficiently large for $e^{-2\,\mathfrak{Re}\, h|S|}$ to compensate $e^{(2d+1)|S|}$. It will also be necessary to compensate for the number of sets $S \ni 0$ as a function of their size, since the latter also grows exponentially fast with $|S|$. The surface term, on the other hand, will be of no help and will be simply bounded by 1. So, grouping the sets $S \ni 0$ by size,

$$\begin{aligned} \sum_{S \ni 0} |w_h(S)| e^{|[S]_1|} &= \sum_{k \geq 1} e^{-2k\,\mathfrak{Re}\, h} \sum_{\substack{S \ni 0 \\ |S| = k}} e^{-2\beta|\partial_e S|} e^{|[S]_1|} \\ &\leq \sum_{k \geq 1} e^{-(2\,\mathfrak{Re}\, h - 2d - 1)k} \# \{S \ni 0 : |S| = k\}. \end{aligned}$$

Exercise 5.3. Using Lemma 3.38, show that

$$\#\{S \ni 0 : |S| = k\} \le (2d)^{2k}. \tag{5.27}$$

Using (5.27), we get

$$\sum_{S \ni 0} |\mathrm{w}_h(S)| e^{|[S]_1|} \le \eta(\Re\mathrm{e}\, h, d), \tag{5.28}$$

where $\eta(x, d) \stackrel{\text{def}}{=} \sum_{k\ge 1} e^{-(2x-2d-1-2\log(2d))k}$. If we define

$$x_0 = x_0(d) \stackrel{\text{def}}{=} \inf\{x > 0 : \eta(x, d) \le 1\}$$

and let

$$H^+_{x_0} \stackrel{\text{def}}{=} \{h \in \mathbb{C} : \Re\mathrm{e}\, h > x_0\},$$

then, for all $h \in H^+_{x_0}$, the cluster expansion

$$\log \Xi^{\mathrm{LF}}_{\Lambda;\beta,h} = \sum_{n\ge 1} \sum_{S_1 \subset \Lambda} \cdots \sum_{S_n \subset \Lambda} \varphi(S_1, \dots, S_n) \prod_{i=1}^{n} \mathrm{w}_h(S_i)$$

converges absolutely. As seen in Section 5.6, the contributions to the expansion come from the *clusters* $X = \{S_1, \dots, S_n\}$. We define the **support** of $X = \{S_1, \dots, S_n\}$ by $\overline{X} \stackrel{\text{def}}{=} S_1 \cup \cdots \cup S_n$. With these notations,

$$\sum_{n\ge 1} \sum_{S_1 \subset \Lambda} \cdots \sum_{S_n \subset \Lambda} \varphi(S_1, \dots, S_n) \prod_{i=1}^{n} \mathrm{w}_h(S_i) = \sum_{X:\, \overline{X} \subset \Lambda} \Psi(X),$$

where $\Psi(\cdot)$ was defined in (5.20).

We will now see how to use this to extract the volume and surface contributions to the pressure in Λ. First, notice that, when defining x_0 above, we have actually guaranteed that the sum in (5.28) converges even if the sum is over *all* connected subsets $S \ni 0$ (not only over those contained in Λ). This allows us to bound series containing clusters of all sizes whose support includes a given vertex: using Theorem 5.4 for the terms $n \ge 2$,

$$\begin{aligned}\sum_{X:\, \overline{X} \ni i} |\Psi(X)| &\le \sum_{n \ge 1} n \sum_{S_1 \ni i} \sum_{S_2} \cdots \sum_{S_n} |\varphi(S_1, \dots, S_n)| \prod_{k=1}^{n} |\mathrm{w}_h(S_k)| \\ &\le \sum_{S_1 \ni i} |\mathrm{w}_h(S_1)| e^{|[S_1]_1|} \le \eta(\Re\mathrm{e}\, h, d) \le 1. \end{aligned} \tag{5.29}$$

We can then rearrange the terms of the cluster expansion in Λ as follows. Since $\frac{1}{|\overline{X}|} \sum_{i \in \Lambda} \mathbf{1}_{\{\overline{X} \ni i\}} = 1$ for any $\overline{X} \subset \Lambda$,

$$\sum_{X:\overline{X}\subset\Lambda}\Psi(X)=\sum_{i\in\Lambda}\sum_{\substack{X:\\ i\in\overline{X}\subset\Lambda}}\frac{1}{|\overline{X}|}\Psi(X)$$
$$=\sum_{i\in\Lambda}\Big\{\sum_{\substack{X:\\ i\in\overline{X}}}\frac{1}{|\overline{X}|}\Psi(X)-\sum_{\substack{X:\\ i\in\overline{X}\not\subset\Lambda}}\frac{1}{|\overline{X}|}\Psi(X)\Big\}. \tag{5.30}$$

The difference between the two series is well defined, since both are absolutely convergent. Notice that both of them contain clusters of unbounded sizes. By translation invariance, the first sum over X in the right-hand side of (5.30) does not depend on i, and thus yields a constant contribution. The second sum is a boundary term. Indeed, whenever $i\in\overline{X}\not\subset\Lambda$, there must exist at least one component $S_k\in X$ that intersects the boundary of Λ: $\overline{X}\cap\partial^{\mathrm{ex}}\Lambda\neq\varnothing$. Therefore, using (5.29) for the second inequality,

$$\Big|\sum_{i\in\Lambda}\sum_{\substack{X:\\ i\in\overline{X}\not\subset\Lambda}}\frac{1}{|\overline{X}|}\Psi(X)\Big|\leq|\partial^{\mathrm{ex}}\Lambda|\max_{j\in\partial^{\mathrm{ex}}\Lambda}\sum_{X:\overline{X}\ni j}|\Psi(X)|\leq|\partial^{\mathrm{ex}}\Lambda|.$$

We thus obtain

$$\frac{1}{|\Lambda|}\log\mathbf{Z}^{+}_{\Lambda;\beta,h}=\beta\frac{|\mathscr{E}^{\mathrm{b}}_{\Lambda}|}{|\Lambda|}+h+\sum_{X:\overline{X}\ni0}\frac{1}{|\overline{X}|}\Psi(X)+\frac{O(|\partial^{\mathrm{ex}}\Lambda|)}{|\Lambda|}. \tag{5.31}$$

We now fix $h\in H^{+}_{x_0}$ and take the thermodynamic limit in (5.31) along the sequence of boxes $\mathrm{B}(n)$. In this limit, the boundary term vanishes and $|\mathscr{E}^{\mathrm{b}}_{\mathrm{B}(n)}|/|\mathrm{B}(n)|\to d$, yielding

$$\psi_\beta(h)=\beta d+h+\sum_{X:\overline{X}\ni0}\frac{1}{|\overline{X}|}\Psi(X),\qquad \mathfrak{Re}\,h>x_0. \tag{5.32}$$

Remark 5.9. Inserting (5.32) into (5.31), we can write, for any fixed region Λ,

$$\mathbf{Z}^{+}_{\Lambda;\beta,h}=e^{\psi_\beta(h)|\Lambda|+O(|\partial^{\mathrm{ex}}\Lambda|)}, \tag{5.33}$$

which provides a direct access to the finite-volume corrections to the pressure (the boundary term can of course be written down explicitly, as was done above). Thus, the cluster expansion provides a tool to study systematically *finite-size effects*, at least in perturbative regimes. This plays a particularly important role when extracting information about thermodynamic behavior from (finite-volume) numerical simulations. Such a decomposition will also be used repeatedly in Chapter 7. ◇

The cluster expansion of the pressure, in (5.32), describes the contributions to the pressure when $\mathfrak{Re}\,h$ is large. Namely, the term $\beta d+h$ corresponds to the **energy density** of the ground state η^{+}:

$$\lim_{n\to\infty}\frac{-\mathscr{H}_{\mathrm{B}(n);\beta,h}(\eta^{+})}{|\mathrm{B}(n)|}=\beta d+h.$$

The contributions due to the excitations away from η^{+} are added successively by considering terms of the series associated with larger and larger clusters. The contribution of a cluster $X=\{S_1,\dots,S_n\}$ is of order $e^{-2h(|S_1|+\cdots+|S_n|)}$. Thanks to the absolute summability of the series (5.32), we can regroup all terms coming from clusters contributing to the same order e^{-2nh}, $n \geq 1$. In this way, we obtain an absolutely convergent series for $\psi_\beta(h) - \beta d - h$ in the variable e^{-2h}.

Lemma 5.10. *When $h \in H^+_{x_0}$, the pressure of the Ising model on $\mathbb{Z}^d$ satisfies, with $z = e^{-2h}$,*

$$\psi_\beta(h) - \beta d - h = a_1 z + a_2 z^2 + a_3 z^3 + \cdots, \tag{5.34}$$

where

$$a_1 = e^{-4d\beta},$$
$$a_2 = de^{-(8d-4)\beta} - (\tfrac{1}{2} + d)e^{-8d\beta}.$$

Proof As pointed out above, the contribution of a cluster $X = \{S_1, \dots, S_n\}$ is of order $e^{-2h(|S_1|+\cdots+|S_n|)}$. The associated combinatorial factor can be read from (5.20) and (5.32), namely

$$\underbrace{\frac{1}{|\overline{X}|}}_{A} \underbrace{\prod_{S\in\Gamma} \frac{1}{n_X(S)!}}_{B} \underbrace{\Big\{ \sum_{\substack{G\subset G_n\\ \text{connected}}} \prod_{\{i,j\}\in G} \zeta(S_i, S_j)\Big\}}_{C} \prod_{i=1}^{n} \mathrm{w}_h(S_i),$$

where Γ is here the set of all connected components in $\mathbb{Z}^d$. The only cluster contributing to a_1 is the cluster composed of the single polymer $\{0\}$. In this case, $A = 1$, $B = 1$ and $C = 1$; this yields the first coefficient since $\partial_e\{0\} = 2d$.

There are two types of clusters contributing to a_2: the clusters composed of a single polymer of size 2 containing 0, and the clusters made of two polymers of size 1, at least one of which is $\{0\}$.

Let us first consider the former: there are exactly $2d$ polymers of size 2 containing the origin and, for each such polymer S, $\partial_e S = 2(2d-1)$, $A = \frac{1}{2}$, $B = 1$ and $C = 1$; this yields a contribution

$$2d \cdot \frac{1}{2} \cdot e^{-2(2d-1)2\beta} = de^{-(8d-4)\beta}.$$

Let us now turn to the clusters made up of two polymers of size 1, at least one of which is $\{0\}$. The first possibility is that both polymers are $\{0\}$, and therefore $A = 1$, $B = \frac{1}{2}$ and $C = -1$; this yields the contribution

$$-\tfrac{1}{2}\, e^{-2\cdot 4d\beta} = -\tfrac{1}{2}\, e^{-8d\beta}.$$

The second possibility is that $X = \{S_1, S_2\} = \{\{0\}, \{i\}\}$, with $i \sim 0$. There are $2d$ ways of choosing i, and for each of those, $A = \frac{1}{2}$, $B = 1$ and $C = -1$; we thus obtain a contribution of

$$2d \cdot (-\tfrac{1}{2}) \cdot e^{-2\cdot 4d\beta} = -de^{-8d\beta}. \qquad \square$$

It is of course possible to compute the coefficients to arbitrary order, but the computations become tricky when the order gets large.

Exercise 5.4. Show that

$$a_3 = ((2d+1)d + \tfrac{1}{3})e^{-12d\beta} - 4d^2 e^{-(12d-4)\beta} + d(2d-1)e^{-(12d-8)\beta}\,.$$

Remark 5.11. The expansion obtained in (5.34) converges when $|z| < e^{-2x_0}$. Remember that the Lee–Yang Theorem (Theorem 3.43) implies analyticity of the pressure (as a function of z) in the whole open unit disk $\mathbb{U} = \{z \in \mathbb{C} : |z| < 1\}$. By the uniqueness of the Taylor coefficients, this means that the series in (5.34) converges not only for $\mathfrak{Re}\, h > x_0$, but for all $\mathfrak{Re}\, h > 0$. ◇

Using the $+$ boundary condition was quite convenient, but the same analysis could have been done with any other boundary condition, with slight changes, and would have led to the same expansion (5.32), only the boundary term in (5.30) being affected by the choice of boundary condition.

Exercise 5.5. Prove that last statement. What changes must be made if one uses nonconstant boundary conditions? Conclude that, when $|\mathfrak{Re}\, h|$ is large, the thermodynamic limit for the pressure exists for arbitrary boundary conditions.

To summarize, we have seen that considering the Ising model with $\mathfrak{Re}\, h > 0$ large allows one to see the regions of $-$ spins as perturbations of the ground state η^+. These perturbations are under control whenever the cluster expansion converges. This led us to a series expansion for the pressure of the model in the variable e^{-2h}.

5.7.2 Virial Expansion for the Lattice Gas

(In order to get motivation and notation for the material in this section, we strongly recommend the reader to have a look at Chapter 4.)

We have seen, in Section 4.8, that the pressure $p_\beta(\mu)$ of the nearest-neighbor lattice gas is analytic everywhere except at μ_*, where it has a discontinuous derivative if the temperature is sufficiently low. To express the pressure as a function of the particle density $\rho \in [0,1]$, we inverted the relation $\rho = \frac{\partial p_\beta}{\partial \mu}$, to obtain $\mu_\beta = \mu_\beta(\rho)$, and defined $\widetilde{p}_\beta(\rho) \stackrel{\text{def}}{=} p_\beta(\mu_\beta(\rho))$. The latter function was shown to be analytic on the gas branch $(0, \rho_g)$, constant on the coexistence plateau $[\rho_g, \rho_l]$, and again analytic on the liquid branch $(\rho_l, 1)$ (see Figure 4.11 and Exercise 4.10).

In this section, we go one step further. We consider the behavior of the model on the gas branch, for small values of the density, and obtain a representation of $\widetilde{p}_\beta$ as a convergent series, called the **virial expansion**:

$$\beta \widetilde{p}_\beta(\rho) = b_1\rho + b_2\rho^2 + b_3\rho^3 + \cdots \qquad (\rho \text{ small}),$$

with (in principle) explicit expressions for the **virial coefficients** b_k, $k \geq 1$.

The canonical lattice gas at low density corresponds, in the grand canonical ensemble, to large negative values of the chemical potential μ (remember Exercise 4.6). We have also seen in Section 4.8 that the nearest-neighbor lattice gas can be mapped, via $\eta_i \mapsto 2\eta_i - 1$, to the Ising model with an inverse temperature $\beta' = \frac{1}{4}\beta$ and magnetic field $h' = \frac{\beta}{2}(2d + \mu)$; in particular, their pressures are related by

$$\beta p_\beta(\mu) = \psi_{\beta'}(h') + \tfrac{\beta\mu}{2} + \tfrac{\beta\kappa}{8} . \tag{5.35}$$

Since a large negative chemical potential corresponds to a large negative magnetic field, we can derive the virial expansion from the results for the Ising model at large values of $\mathfrak{Re}\, h$ which were obtained in the previous section. Namely, using the symmetry $\psi_\beta(-h) = \psi_\beta(h)$ and using the expansion (5.34), in terms of the variable $z' = e^{2h'}$, with $\mathfrak{Re}\, h' < -x_0$:

$$\psi_{\beta'}(h') = \beta' d - h' + a_1 z' + a_2 z'^2 + a_3 z'^3 + \cdots . \tag{5.36}$$

(Remember that each a_n should be used with β' instead of β.) This gives

$$\beta p_\beta(\mu) = \sum_{n\geq 1} a_n z'^n , \tag{5.37}$$

which is called the **Mayer expansion.** The Mayer series is absolutely convergent and can therefore be differentiated term by term with respect to μ. Since $\frac{\partial z'^n}{\partial \mu} = n\beta z'^n$, this yields

$$\rho = \frac{\partial p_\beta}{\partial \mu} = \sum_{n\geq 1} n a_n z'^n \stackrel{\text{def}}{=} \sum_{n\geq 1} \tilde{a}_n z'^n \stackrel{\text{def}}{=} \phi(z') .$$

We will obtain the virial expansion by inverting this last expression, obtaining $z' = \phi^{-1}(\rho)$, and injecting the result into (5.37). Since $\frac{\mathrm{d}\phi}{\mathrm{d}z}(0) = a_1 = e^{-4d\beta} > 0$, the analytic Implicit Function Theorem (Theorem B.28) implies that ϕ can indeed be inverted on a small disk $D \subset \mathbb{C}$ centered at the origin, and the inverse is analytic on that disk. We write the Taylor expansion of the inverse as $\phi^{-1}(\rho) = \sum_k c_k \rho^k$. Assuming that the coefficients c_k are known (they will be computed below), we can write down the virial expansion. Namely,

$$\begin{aligned}
\beta \widetilde{p}_\beta(\rho) &= \sum_{n\geq 1} a_n \{\phi^{-1}(\rho)\}^n \\
&= \sum_{n\geq 1} a_n \sum_{k_1\geq 1} \cdots \sum_{k_n\geq 1} \prod_{i=1}^{n} c_{k_i} \rho^{k_i} \\
&= \sum_{n\geq 1} a_n \sum_{m\geq n} \sum_{\substack{k_1,\dots,k_n\geq 1\\ k_1+\cdots+k_n=m}} \prod_{i=1}^{n} c_{k_i} \rho^{k_i} = \sum_{m\geq 1} \tilde{b}_m \rho^m ,
\end{aligned}$$

where

$$\tilde{b}_m \stackrel{\text{def}}{=} \sum_{n=1}^{m} a_n \sum_{\substack{k_1,\dots,k_n\geq 1\\ k_1+\cdots+k_n=m}} \prod_{i=1}^{n} c_{k_i} .$$

A similar computation can be used in the following exercise.

Exercise 5.6 (Computing the Taylor coefficients of an inverse function). Let $\phi(z) = \sum_{k\geq 1} \tilde{a}_k z^k$ be convergent in a neighborhood of $z = 0$, with $\tilde{a}_1 \neq 0$ (in particular, ϕ is invertible in a neighborhood of $z = 0$). Write its compositional inverse ϕ^{-1} as $\phi^{-1}(z) = \sum_{k\geq 1} c_k z^k$. Show that $c_1 = \tilde{a}^{-1}$ and that, for all $m \geq 2$,

$$\sum_{n=1}^{m} \tilde{a}_n \sum_{\substack{k_1,\dots,k_n\geq 1\\ k_1+\dots+k_n=m}} \prod_{i=1}^{n} c_{k_i} = 0\,.$$

Using this, compute the first few coefficients of ϕ^{-1}:

$$c_2 = -\frac{\tilde{a}_2}{\tilde{a}_1^3}\,, \quad c_3 = 2\frac{\tilde{a}_2^2}{\tilde{a}_1^5} - \frac{\tilde{a}_3}{\tilde{a}_1^4}\,, \qquad \text{etc.}$$

As can be verified, using the coefficients a_k computed in Lemma 5.10,

$$\tilde{b}_1 = 1\,, \quad \tilde{b}_2 = -\frac{a_2}{a_1^2} = \tfrac{1}{2} + d - de^{\beta}\,, \qquad \text{etc.}$$

We have thus shown

Theorem 5.12. *At low densities, the pressure of the nearest-neighbor lattice gas satisfies*

$$\beta\tilde{p}_\beta(\rho) = \rho + (\tfrac{1}{2} + d - de^{\beta})\rho^2 + O(\rho^3)\,.$$

5.7.3 Ising Model at High Temperature ($h = 0$)

In this section, we consider again the pressure of the Ising model but in another regime: $h = 0$ and $\beta \ll 1$. Here, thermal fluctuations are so strong that the spins behave nearly independently from each other.

We choose the free boundary condition, as it is the most convenient one in the high-temperature regime. Proceeding as in Section 3.7.3, we express the partition function as in Exercise 3.23:

$$\mathbf{Z}^{\varnothing}_{\Lambda;\beta,0} = 2^{|\Lambda|}(\cosh\beta)^{|\mathscr{E}_\Lambda|} \sum_{E\in\mathfrak{E}^{\mathrm{even}}_\Lambda} (\tanh\beta)^{|E|}\,, \tag{5.38}$$

where the sum is over all subsets of edges $E \subset \mathscr{E}_\Lambda$ such that the number of edges of E incident to each vertex $i \in \Lambda$ is even.

Each set $E \in \mathfrak{E}^{\mathrm{even}}_\Lambda$ can be identified with a graph, by simply considering it together with the endpoints of each of its edges. This graph can be decomposed into (maximal) connected components, which play the role of polymers. In terms of edges, this decomposition can be written $E = E_1 \cup \dots \cup E_n$, so that we obtain

$$\mathbf{Z}^{\varnothing}_{\Lambda;\beta,0} = 2^{|\Lambda|}(\cosh\beta)^{|\mathscr{E}_\Lambda|}\, \Xi^{\mathrm{HT}}_{\Lambda;\beta,0}\,,$$

with

$$\Xi^{\mathrm{HT}}_{\Lambda;\beta,0} \stackrel{\mathrm{def}}{=} 1 + \sum_{n\geq 1} \frac{1}{n!} \sum_{E_1 \subset \mathscr{E}_\Lambda} \cdots \sum_{E_n \subset \mathscr{E}_\Lambda} \Big\{ \prod_{i=1}^{n} (\tanh \beta)^{|E_i|} \Big\} \prod_{1\leq i<j\leq n} \delta(E_i, E_j),$$

where $E_i \in \mathfrak{E}^{\mathrm{even}}_\Lambda$, for all i, and

$$\delta(E_i, E_j) \stackrel{\mathrm{def}}{=} \begin{cases} 1 & \text{if } E_i \text{ and } E_j \text{ have no vertex in common,} \\ 0 & \text{otherwise,} \end{cases}$$

is again of hard-core type.

The above representation is well suited to the high-temperature regime, since the weight $(\tanh\beta)^{|E_i|}$*, associated with a polymer* E_i*, decays fast when* β *is small.* ◇

Proceeding as in Section 5.7.1, we can show that the conditions for the convergence of the cluster expansion are satisfied when β is sufficiently small, thus proving that the pressure behaves analytically at high temperature:

Theorem 5.13. *There exists* $r_0 > 0$ *such that* $\beta \mapsto \psi_\beta(0)$ *is analytic in the disk* $\{\beta \in \mathbb{C} : |\beta| < r_0\}$.

Exercise 5.7. Prove Theorem 5.13, and compute the first few terms of the expansion of $\psi_\beta(0) - d\log(\cosh\beta) - \log 2$ as a power series in the variable $z = \tanh\beta$.

Remark 5.14. Even though we have only considered analyticity of the pressure as a function of β here, it is possible to extract a lot of additional information on the model in this regime. In Section 6.5.4, we will use a variant of the above approach to prove uniqueness of the infinite-volume Gibbs measure at all sufficiently high temperatures, for a large class of models. ◇

It follows from the results of Chapter 3 that $h \mapsto \psi_\beta(h)$ is continuously differentiable at $h = 0$ when $\beta < \beta_c(d)$. In the next exercise, the reader is asked to adapt the high-temperature representation to show that it is in fact analytic in a neighborhood of $h = 0$, at least when β is sufficiently small.

Exercise 5.8. Show that there exists $\beta_0 = \beta_0(d) > 0$ such that, for all $0 \leq \beta \leq \beta_0$, $h \mapsto \psi_\beta(h)$ is analytic at $h = 0$.

5.7.4 Ising Model at Low Temperature ($h = 0$)

We now consider the Ising model on $\mathbb{Z}^d$, $d \geq 2$, at very low temperature and in the absence of a magnetic field. Our goals, in this regime, are (i) to establish analyticity of the pressure $\beta \mapsto \psi_\beta(0)$, (ii) to derive an explicit series expansion for the magnetization and (iii) to prove exponential decay of the truncated 2-point correlation function $\langle \sigma_i; \sigma_j \rangle^+_{\beta,0}$ as $\|i - j\|_2 \to \infty$.

We know from Section 3.7.2 that, when $h = 0$ and β is large, the relevant objects for the description of configurations are the *contours* separating the regions of $+$ and $-$ spins. In dimension 2, we used the deformation rule of Figure 3.11. Since that deformation was specific to $d = 2$, we will here define contours in a slightly different manner. The description will be used in any dimension $d \geq 2$, in a large box Λ, with either $+$ or $-$ boundary condition.

We again write the Hamiltonian in a way that emphasizes the role played by pairs of neighboring spins with opposite signs:

$$\mathscr{H}_{\Lambda;\beta,0} = -\beta|\mathscr{E}^{\mathrm{b}}_\Lambda| - \beta \sum_{\{i,j\}\in\mathscr{E}^{\mathrm{b}}_\Lambda} (\sigma_i\sigma_j - 1). \tag{5.39}$$

We consider the $+$ boundary condition in a region $\Lambda \Subset \mathbb{Z}^d$. Given $\omega \in \Omega^+_\Lambda$, we use again $\Lambda^-(\omega)$ to denote the set of vertices i at which $\omega_i = -1$. Rather than $\Lambda^-(\omega)$ itself (which was relevant when considering a large magnetic field), we will be interested only in its boundary.

We associate with each $i \in \mathbb{Z}^d$ the closed unit cube of $\mathbb{R}^d$ centered at i:

$$\mathscr{S}_i \stackrel{\text{def}}{=} i + [-\tfrac{1}{2}, \tfrac{1}{2}]^d,$$

and let

$$\mathscr{M}(\omega) \stackrel{\text{def}}{=} \bigcup_{i\in\Lambda^-(\omega)} \mathscr{S}_i. \tag{5.40}$$

We can then consider the (maximal) connected components of $\partial\mathscr{M}(\omega)$ (here, the boundary ∂ is in the sense of the Euclidean topology of $\mathbb{R}^d$):

$$\Gamma'(\omega) \stackrel{\text{def}}{=} \{\gamma_1, \dots, \gamma_n\}.$$

Each γ_i is called a **contour** of ω. (Note that when $d = 2$, this notion slightly differs from the one used in Chapter 3.) In $d = 2$ (see Figure 3.10), contours can be identified with connected sets of dual edges. In higher dimensions, contours are connected sets of *plaquettes*, which are the $(d-1)$-dimensional faces of the d-dimensional hypercubes $\mathscr{S}_i$, $i \in \mathbb{Z}^d$. The number of plaquettes contained in γ_i will be denoted $|\gamma_i|$. Observe that there is a one-to-one mapping between the plaquettes of $\partial\mathscr{M}(\omega)$ and the edges of $\partial_e\Lambda^-(\omega)$ (associating with a plaquette the unique edge crossing it), and so $|\partial_e\Lambda^-(\omega)| = \sum_{i=1}^n |\gamma_i|$. We can thus write

$$\mathbf{Z}^+_{\Lambda;\beta,0} = e^{\beta|\mathscr{E}^{\mathrm{b}}_\Lambda|} \sum_{\omega\in\Omega^+_\Lambda} \prod_{\gamma\in\Gamma'(\omega)} \mathrm{w}_\beta(\gamma),$$

where

$$\mathrm{w}_\beta(\gamma) \stackrel{\text{def}}{=} e^{-2\beta|\gamma|}. \tag{5.41}$$

The final step is to transform the summation over ω into a summation over families of contours. To this end, we introduce a few notions. Let $\Gamma_\Lambda \stackrel{\text{def}}{=} \{\gamma \in \Gamma'(\omega) : \omega \in \Omega_\Lambda^+\}$ denote the set of all possible contours in Λ.

A collection of contours $\Gamma' \subset \Gamma_\Lambda$ is **admissible** if there exists a configuration $\omega \in \Omega_\Lambda^+$ such that $\Gamma'(\omega) = \Gamma'$. We say that $\Lambda \subset \mathbb{Z}^d$ is **c-connected** if $\mathbb{R}^d \setminus \bigcup_{i\in\Lambda} \mathscr{S}_i$ is a connected subset of $\mathbb{R}^d$.

Exercise 5.9. Assuming that Λ is c-connected, show that a collection $\Gamma' = \{\gamma_1, \dots, \gamma_n\} \subset \Gamma_\Lambda$ is admissible if and only if its contours are pairwise disjoint: $\gamma_i \cap \gamma_j = \varnothing$ for all $i \neq j$. Why is this not necessarily true when Λ is not c-connected?

Therefore, provided that Λ is c-connected,

$$\mathbf{Z}^+_{\Lambda;\beta,0} = e^{\beta|\mathscr{E}^{\mathrm{b}}_\Lambda|}\, \Xi^{\mathrm{LT}}_{\Lambda;\beta,0}\,, \tag{5.42}$$

where

$$\begin{aligned}\Xi^{\mathrm{LT}}_{\Lambda;\beta,0} &\stackrel{\text{def}}{=} \sum_{\substack{\Gamma'\subset\Gamma_\Lambda\\ \text{admiss.}}} \prod_{\gamma\in\Gamma'} \mathrm{w}_\beta(\gamma)\\ &= 1 + \sum_{n\geq 1} \frac{1}{n!} \sum_{\gamma_1\in\Gamma_\Lambda} \cdots \sum_{\gamma_n\in\Gamma_\Lambda} \Bigl\{\prod_{i=1}^n \mathrm{w}_\beta(\gamma_i)\Bigr\} \prod_{1\leq i<j\leq n} \delta(\gamma_i, \gamma_j)\,,\end{aligned}$$

with interactions that are once again of hard-core type:

$$\delta(\gamma_i, \gamma_j) \stackrel{\text{def}}{=} \begin{cases} 1 & \text{if } \gamma_i \cap \gamma_j = \varnothing\,, \\ 0 & \text{otherwise.}\end{cases} \tag{5.43}$$

The above representation of the Ising model is adapted to the low-temperature regime, since the weight $\mathrm{w}_\beta(\gamma_i) = e^{-2\beta|\gamma_i|}$ associated with γ_i decays fast when β is large. This observation was, of course, at the core of Peierls' argument. ◇

For the rest of the section, we will allow β to take complex values. We first verify that (5.10) holds with $a(\gamma) \stackrel{\text{def}}{=} |\gamma|$.

Exercise 5.10. Prove that there exists $x_0 = x_0(d) > 0$ such that, for all β satisfying $\mathfrak{Re}\,\beta > x_0$,

$$\sum_\gamma |\mathrm{w}_\beta(\gamma)| e^{|\gamma|} < \infty\,, \tag{5.44}$$

and, for each $\gamma_* \in \Gamma_\Lambda$,

$$\sum_{\gamma} |\mathrm{w}_\beta(\gamma)|\, e^{|\gamma|}\, |\zeta(\gamma,\gamma_*)| \leq |\gamma_*|\,. \tag{5.45}$$

Hint: Use Lemma 3.38 to count the number of contours γ whose support contains a fixed point.

Pressure

Observe that, when $d=2$, the analyticity of $\beta \mapsto \psi_\beta(0)$ for large β can be deduced directly from the analyticity at small values of β (Theorem 5.13 above), using the duality transformation described in Section 3.10.1. However, there is no analogous transformation in $d \geq 3$.

We leave it as an exercise to provide the details of the proof of the following result:

Theorem 5.15. *($d \geq 2$). There exists $x_0 = x_0(d) > 0$ such that $\beta \mapsto \psi_\beta(0)$ is analytic on $\{\beta \in \mathbb{C} : \Re\mathfrak{e}\,\beta > x_0\}$. Moreover,*

$$\psi_\beta(0) = \beta d + e^{-4d\beta} + d e^{-4(2d-1)\beta} + O(e^{-8d\beta})\,.$$

Magnetization and Decay of the Truncated 2-Point Function

We now move on to the study of correlation functions at low temperature.

Let $\Lambda \Subset \mathbb{Z}^d$ be c-connected, and $A \subset \Lambda$. Remembering that $\sigma_A \stackrel{\text{def}}{=} \prod_{i\in A}\sigma_i$, we will express the correlation function

$$\langle \sigma_A \rangle^+_{\Lambda;\beta,0} = \sum_{\omega\in\Omega^+_\Lambda} \sigma_A(\omega) \frac{e^{-\mathscr{H}_{\Lambda;\beta,0}(\omega)}}{\mathbf{Z}^+_{\Lambda;\beta,0}}$$

in a form suitable for an analysis based on the cluster expansion. The denominator, $\mathbf{Z}^+_{\Lambda;\beta,0}$, can be expressed using (5.42). To do the same for the numerator, we start with

$$\sum_{\omega\in\Omega^+_\Lambda} \sigma_A(\omega) e^{-\mathscr{H}_{\Lambda;\beta,0}(\omega)} = e^{\beta|\mathscr{E}^{\mathrm{b}}_\Lambda|} \sum_{\omega\in\Omega^+_\Lambda} \sigma_A(\omega) \prod_{\gamma\in\Gamma'(\omega)} \mathrm{w}_\beta(\gamma)\,.$$

Let $\omega \in \Omega^+_\Lambda$ and let $\gamma \in \Gamma'(\omega)$ be one of its contours. Consider the configuration $\omega^\gamma \in \Omega^+_\Lambda$ which has γ as its unique contour: $\Gamma'(\omega^\gamma) = \{\gamma\}$. The **interior** of γ is defined by (see Figure 5.2)

$$\operatorname{Int}\gamma \stackrel{\text{def}}{=} \{i \in \Lambda : \omega^\gamma_i = -1\} = \Lambda^-(\omega^\gamma)\,.$$

The important observation is that, for any $\omega \in \Omega^+_\Lambda$,

$$\omega_i = (-1)^{\#\{\gamma\in\Gamma'(\omega):\ i\in\operatorname{Int}\gamma\}}\,,$$

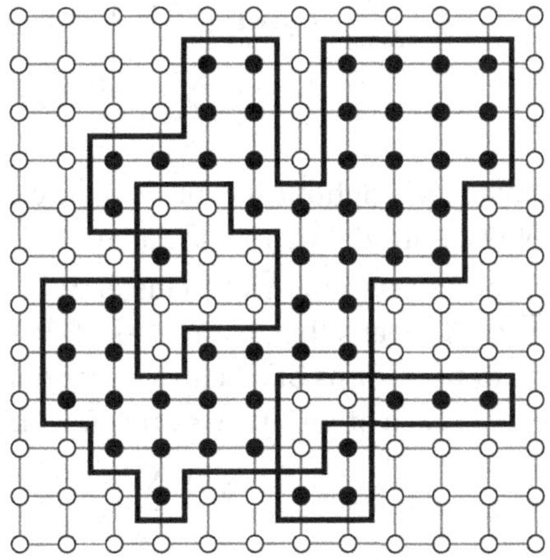

Figure 5.2. The interior of a (here two-dimensional) contour: the interior is the set of all black vertices.

that is, the sign of the spin at the vertex i is equal to $+1$ if and only if there is an even number of contours **surrounding** i (in the sense that i belongs to their interior). It follows from this observation that

$$\sigma_A(\omega) = (-1)^{\sum_{i\in A} \#\{\gamma\in\Gamma'(\omega)\,:\, i\in\operatorname{Int}\gamma\}} = \prod_{\gamma\in\Gamma'(\omega)} (-1)^{\#\{i\in A\,:\, i\in\operatorname{Int}\gamma\}}.$$

We therefore get

$$\sigma_A(\omega) \prod_{\gamma\in\Gamma'(\omega)} \mathrm{w}_\beta(\gamma) = \prod_{\gamma\in\Gamma'(\omega)} \mathrm{w}_\beta^A(\gamma),$$

where

$$\mathrm{w}_\beta^A(\gamma) \stackrel{\text{def}}{=} (-1)^{\#\{i\in A\,:\, i\in\operatorname{Int}\gamma\}}\mathrm{w}_\beta(\gamma).$$

We conclude that

$$\langle\sigma_A\rangle^+_{\Lambda;\beta,0} = \frac{\sum_{\Gamma'\subset\Gamma_\Lambda,\text{admiss.}} \prod_{\gamma\in\Gamma'} \mathrm{w}_\beta^A(\gamma)}{\sum_{\Gamma'\subset\Gamma_\Lambda,\text{admiss.}} \prod_{\gamma\in\Gamma'} \mathrm{w}_\beta(\gamma)} \equiv \frac{\Xi^{\mathrm{LT},A}_{\Lambda;\beta,0}}{\Xi^{\mathrm{LT}}_{\Lambda;\beta,0}}. \tag{5.46}$$

Now, the polymer partition functions in the numerator and denominator in (5.46) differ only in the weight associated with contours that surround vertices in A. When $\mathfrak{Re}\,\beta > x_0$ (see Exercise 5.10), the cluster expansion for $\log \Xi^{\mathrm{LT}}_{\Lambda;\beta,0}$ converges and, since $|\mathrm{w}_\beta^A(\gamma)| = |\mathrm{w}_\beta(\gamma)|$ for all γ, the same holds for $\log \Xi^{\mathrm{LT},A}_{\Lambda;\beta,0}$. We thus obtain

$$\begin{aligned}\langle\sigma_A\rangle^+_{\Lambda;\beta,0} &= \exp\{\log \Xi^{\mathrm{LT},A}_{\Lambda;\beta,0} - \log \Xi^{\mathrm{LT}}_{\Lambda;\beta,0}\}\\ &= \exp\Big\{\sum_{X:\,\overline{X}\subset\Lambda} \Psi^A_\beta(X) - \sum_{X:\,\overline{X}\subset\Lambda} \Psi_\beta(X)\Big\},\end{aligned}$$

where the sums in the rightmost expression are over clusters of contours in Λ and $\Psi_\beta(X)$ and $\Psi^A_\beta(X)$ are defined as in (5.20) with weights w given by w_β and w^A_β respectively, and the support $\overline{X}$ of a cluster $X = \{\gamma_1, \dots, \gamma_n\}$ is defined as $\bigcup_{k=1}^n \gamma_k$ (of course, $\overline{X} \subset \Lambda$ means that, as subsets of $\mathbb{R}^d$, $\overline{X} \subset \bigcup_{i\in\Lambda} \mathscr{S}_i$). In particular, the contributions to both sums of all clusters containing no contour γ surrounding a vertex of A cancel each other, and we are left with

$$\langle\sigma_A\rangle^+_{\Lambda;\beta,0} = \exp\Big\{\sum_{\substack{X\sim A:\\ \overline{X}\subset\Lambda}} (\Psi^A_\beta(X) - \Psi_\beta(X))\Big\},$$

where $X \sim A$ means that X contains at least one contour γ such that $A \cap \mathrm{Int}\,\gamma \neq \varnothing$. We leave it as an exercise to show that one can let $\Lambda \uparrow \mathbb{Z}^d$ in the above expression:

Exercise 5.11. $(d \geq 2)$ Prove that

$$\langle\sigma_A\rangle^+_{\beta,0} = \exp\Big\{\sum_{X\sim A} (\Psi^A_\beta(X) - \Psi_\beta(X))\Big\}, \tag{5.47}$$

provided that $\mathfrak{Re}\,\beta$ is sufficiently large.

We now turn to two applications of this formula.

Magnetization at Very Low Temperatures. In Section 3.7.2, we used Peierls' argument to obtain a lower bound on $\langle\sigma_0\rangle^+_{\beta,0}$ that tends to 1 as $\beta \to \infty$. We can use the cluster expansion to obtain an explicit expansion in $e^{-2\beta}$ for $\langle\sigma_0\rangle^+_{\beta,0}$, valid for large enough values of β. Namely, an application of (5.47) with $A = \{0\}$ yields

$$\langle\sigma_0\rangle^+_{\beta,0} = \exp\Big\{\sum_{X\sim\{0\}} (\Psi^{\{0\}}_\beta(X) - \Psi_\beta(X))\Big\}, \tag{5.48}$$

where the condition $X \sim \{0\}$ now reduces to the requirement that at least one of the contours γ in X surrounds 0. It is then a simple exercise, proceeding as in the previous sections, to obtain the desired expansion.

Exercise 5.12. $(d \geq 2)$ Prove that, for all sufficiently large values of β,

$$\langle\sigma_0\rangle^+_{\beta,0} = 1 - 2e^{-4d\beta} - 4de^{-(8d-4)\beta} + O(e^{-8d\beta}).$$

Decay of the Truncated 2-Point Function. As we saw in Exercises 3.23 and 3.24, the correlations of the Ising model decay exponentially fast at sufficiently high temperature (small β)

$$\langle\sigma_i\sigma_j\rangle_{\beta,0} \leq e^{-c_{HT}(\beta)\|j-i\|_1}, \quad \forall i,j \in \mathbb{Z}^d.$$

In contrast, we know that at low temperature, $\beta > \beta_c$, the correlations do not decay anymore since, by the GKS inequalities, uniformly in i and j,

$$\langle \sigma_i \sigma_j \rangle^+_{\beta,0} \geq \langle \sigma_i \rangle^+_{\beta,0} \langle \sigma_j \rangle^+_{\beta,0} = (\langle \sigma_0 \rangle^+_{\beta,0})^2 > 0 \,. \tag{5.49}$$

Here, we will study the **truncated 2-point function**, which is the name usually given in physics to the covariance between the random variables σ_i and σ_j, in the Gibbs state $\langle \cdot \rangle^+_{\beta,0}$:

$$\langle \sigma_i ; \sigma_j \rangle^+_{\beta,0} \stackrel{\text{def}}{=} \langle \sigma_i \sigma_j \rangle^+_{\beta,0} - \langle \sigma_i \rangle^+_{\beta,0} \langle \sigma_j \rangle^+_{\beta,0} \,.$$

Theorem 5.16. *($d \geq 2$) There exist $0 < \beta_0 < \infty$, $c > 0$ and $C < \infty$ such that, for all $\beta \geq \beta_0$,*

$$0 \leq \langle \sigma_i ; \sigma_j \rangle^+_{\beta,0} \leq C e^{-c\beta \|j-i\|_1} \,, \quad \forall i,j \in \mathbb{Z}^d \,. \tag{5.50}$$

This result shows that, at least at low enough temperatures, the correlation length of the Ising model on $\mathbb{Z}^d$, $d \geq 2$, is finite (and actually tends to 0 as $\beta \uparrow \infty$). In particular, the spins are only weakly correlated, even though there is long-range order.

Proof The first inequality is just (5.49), so we only prove the second one. Let us write $\widetilde{\Psi}^A_\beta(X) \stackrel{\text{def}}{=} \Psi^A_\beta(X) - \Psi_\beta(X)$. On the one hand, by (5.48),

$$\langle \sigma_i \rangle^+_{\beta,0} \langle \sigma_j \rangle^+_{\beta,0} = \exp\Big\{ \sum_{X \sim \{i\}} \widetilde{\Psi}^{\{i\}}_\beta(X) + \sum_{X \sim \{j\}} \widetilde{\Psi}^{\{j\}}_\beta(X) \Big\} .$$

On the other hand, by the general formula (5.47),

$$\langle \sigma_i \sigma_j \rangle^+_{\beta,0} = \exp\Big\{ \sum_{X \sim \{i,j\}} \widetilde{\Psi}^{\{i,j\}}_\beta(X) \Big\} .$$

Clusters $X \sim \{i,j\}$ can be split into three disjoint classes:

$$\mathscr{C}_i \stackrel{\text{def}}{=} \big\{ X : X \sim \{i\} \text{ but } X \not\sim \{j\} \big\}, \qquad \mathscr{C}_j \stackrel{\text{def}}{=} \big\{ X : X \sim \{j\} \text{ but } X \not\sim \{i\} \big\},$$
$$\mathscr{C}_{i,j} \stackrel{\text{def}}{=} \big\{ X : X \sim \{i\} \text{ and } X \sim \{j\} \big\} .$$

Observe now that $\Psi^{\{i,j\}}_\beta(X) = \Psi^{\{i\}}_\beta(X)$ for all $X \in \mathscr{C}_i$, and $\Psi^{\{i,j\}}_\beta(X) = \Psi^{\{j\}}_\beta(X)$ for all $X \in \mathscr{C}_j$. This implies that

$$\begin{aligned}
\langle \sigma_i \sigma_j \rangle^+_{\beta,0} &= \exp\Big\{ \sum_{X \sim \{i\}} \widetilde{\Psi}^{\{i\}}_\beta(X) + \sum_{X \sim \{j\}} \widetilde{\Psi}^{\{j\}}_\beta(X) \\
&\qquad + \sum_{X \in \mathscr{C}_{i,j}} \big(\widetilde{\Psi}^{\{i,j\}}_\beta(X) - \widetilde{\Psi}^{\{i\}}_\beta(X) - \widetilde{\Psi}^{\{j\}}_\beta(X) \big) \Big\} \\
&= \langle \sigma_i \rangle^+_{\beta,0} \langle \sigma_j \rangle^+_{\beta,0} \exp\Big\{ \sum_{X \in \mathscr{C}_{i,j}} \big(\widetilde{\Psi}^{\{i,j\}}_\beta(X) - \widetilde{\Psi}^{\{i\}}_\beta(X) - \widetilde{\Psi}^{\{j\}}_\beta(X) \big) \Big\} .
\end{aligned}$$

Now, for all A, $|\widetilde{\Psi}^A_\beta(X)| \leq 2|\Psi_\beta(X)|$, and therefore

$$\langle\sigma_i\sigma_j\rangle^+_{\beta,0} \leq \langle\sigma_i\rangle^+_{\beta,0}\langle\sigma_j\rangle^+_{\beta,0} \exp\Bigl\{6 \sum_{X\in\mathscr{C}_{i,j}} |\Psi_\beta(X)|\Bigr\}.$$

The conclusion will thus follow once we prove that

$$\sum_{X\in\mathscr{C}_{i,j}} |\Psi_\beta(X)| \leq C' e^{-c\beta\|j-i\|_1}, \qquad \forall i,j \in \mathbb{Z}^d,$$

for some constants $c > 0$ and $C' < \infty$. To prove this claim, assume $\beta \geq 2x_0$. By (5.29), for any vertex $v \in \mathbb{R}^d$, since $|\overline{X}| \leq \sum_{\gamma\in X} |\gamma|$,

$$\sum_{X:\,\overline{X}\ni v} |\Psi_\beta(X)|e^{\beta|\overline{X}|} \leq \sum_{X:\,\overline{X}\ni v} |\Psi_{\beta/2}(X)| \leq 1.$$

This implies that, for any $R > 0$,

$$\sum_{\substack{X:\\ \overline{X}\ni v,\,|\overline{X}|\geq R}} |\Psi_\beta(X)| \leq e^{-\beta R} \sum_{X:\,\overline{X}\ni v} |\Psi_\beta(X)|e^{\beta|\overline{X}|} \leq e^{-\beta R}.$$

Therefore, since each $X \in \mathscr{C}_{i,j}$ satisfies $|\overline{X}| \geq \|j - i\|_1$,

$$\begin{aligned}\sum_{X\in\mathscr{C}_{i,j}} |\Psi_\beta(X)| &\leq \sum_{R\geq\|j-i\|_1} R^d \sum_{\substack{X:\\ \overline{X}\ni v,\,|\overline{X}|=R}} |\Psi_\beta(X)| \\ &\leq \sum_{R\geq\|j-i\|_1} R^d e^{-\beta R} \leq C' e^{-c\beta\|j-i\|_1},\end{aligned}$$

uniformly in $i,j \in \mathbb{Z}^d$, for some $c = c(d) > 0$ and all β large enough. □

5.8 Bibliographical References

The cluster expansion is one of the oldest tools of statistical mechanics. As already mentioned in Section 4.12.3, Mayer [236] started using it systematically in his analysis leading to the coefficients of the virial expansion of the pressure of a real gas. Groeneveld [153] was one of the first to provide a rigorous proof of its convergence.

Nowadays, there exist various approaches to the problem of convergence of the expansion, all leading more or less to the same conclusions. Adopting one is essentially a matter of personal taste. The proof of convergence we gave in Section 5.4 was taken from Ueltschi [335], since it is pretty straightforward and keeps the combinatorics elementary.

Some standard references on the subject include the following papers. Polymer models were introduced for the first time by Gruber and Kunz [155]. Kotecký and Preiss [196] gave the first inductive proof of the convergence of the cluster expansion, similar to the one used in Theorem 5.4. An interesting alternative approach, where the expansion is obtained as the result of a multi-variable Taylor expansion, was proposed by Dobrushin in [82]. A pedagogical description of the tree-graph approach that originated with the work of Penrose [269] can be found in Pfister [270]; see also the paper of Fernández and Procacci [103], where several of these methods are compared.

6 Infinite-Volume Gibbs Measures

In this chapter, we give an introduction to the theory of Gibbs measures, which describes the properties of infinite systems at equilibrium. We will not cover all the aspects of the theory, but instead present the most important ideas and results in the simplest possible setting; the Ising model being a guiding example throughout the chapter.

Remark 6.1. Due to the rather abstract nature of this theory, it will be necessary to resort to some notions from measure theory that were not necessary in the previous chapters. From the probabilistic point of view, we will use extensively the fundamental notion of *conditional expectation,* central in the description of Gibbs measures. The reader familiar with these subjects (some parts of which are briefly presented in Appendix B, Sections B.5 and B.8) will certainly feel more comfortable. Certain topological notions will also be used, but will be presented from scratch in the chapter. Nevertheless, we emphasize that, although of great importance in the understanding of the mathematical framework of statistical mechanics, *a detailed understanding of this chapter is not required for the rest of the book.* ◇

Some Models to Which the Theory Applies. The theory of Gibbs measures presented in this chapter is general and applies to a wide range of models. Although the description of the equilibrium properties of these models will always follow the standard prescription of equilibrium statistical mechanics, what distinguishes the models is their *microscopic* specificities. That is, in our context: (i) the possible values of a spin at a given vertex of $\mathbb{Z}^d$, and (ii) the interactions between spins contained in a finite region $\Lambda \Subset \mathbb{Z}^d$.

A model is thus defined by first considering the set Ω_0, called the **single-spin space**, which describes all the possible states of one spin. The **spin configurations** on a (possible infinite) subset $S \subset \mathbb{Z}^d$ are defined as in Chapter 3:

$$\Omega_S \stackrel{\text{def}}{=} \Omega_0^S = \left\{(\omega_i)_{i\in S} \,:\, \omega_i \in \Omega_0 \,\forall i \in S\right\} .$$

When $S = \mathbb{Z}^d$, we simply write $\Omega \equiv \Omega_{\mathbb{Z}^d}$. Then, for each finite subset $\Lambda \Subset \mathbb{Z}^d$, the energy of a configuration in Λ is determined by a **Hamiltonian**

$$\mathscr{H}_\Lambda : \Omega \to \mathbb{R} .$$

We list some of the examples that will be used as illustrations throughout the chapter.

- For the **Ising model,**

$$\Omega_0 = \{+1, -1\}.$$

The nearest-neighbor version studied in Chapter 3 corresponds to

$$\mathscr{H}_\Lambda(\omega) = -\beta \sum_{\{i,j\}\in\mathscr{E}^{\mathrm{b}}_\Lambda} \omega_i\omega_j - h\sum_{i\in\Lambda}\omega_i,$$

where we remind the reader that $\mathscr{E}^{\mathrm{b}}_\Lambda$ is the set of nearest-neighbor edges of $\mathbb{Z}^d$ with at least one endpoint in Λ, see (3.2). We will also consider a long-range version of this model:

$$\mathscr{H}_\Lambda(\omega) = -\sum_{\{i,j\}\cap\Lambda\neq\varnothing} J_{ij}\omega_i\omega_j - h\sum_{i\in\Lambda}\omega_i,$$

where $J_{ij} \to 0$ (sufficiently fast) when $\|j-i\|_1 \to \infty$.

- For the q**-state Potts model,** where $q \geq 2$ is an integer, we set

$$\Omega_0 = \{0, 1, 2, \dots, q-1\},$$
$$\mathscr{H}_\Lambda(\omega) = -\beta \sum_{\{i,j\}\in\mathscr{E}^{\mathrm{b}}_\Lambda} \delta_{\omega_i,\omega_j}.$$

- For the **Blume–Capel model,**

$$\Omega_0 = \{+1, 0, -1\},$$
$$\mathscr{H}_\Lambda(\omega) = -\beta \sum_{\{i,j\}\in\mathscr{E}^{\mathrm{b}}_\Lambda} (\omega_i-\omega_j)^2 - h\sum_{i\in\Lambda}\omega_i - \lambda\sum_{i\in\Lambda}\omega_i^2.$$

- The **XY model** is an example with an uncountable single-spin space,

$$\Omega_0 = \left\{x \in \mathbb{R}^2 : \|x\|_2 = 1\right\},$$

and Hamiltonian

$$\mathscr{H}_\Lambda(\omega) = -\beta \sum_{\{i,j\}\in\mathscr{E}^{\mathrm{b}}_\Lambda} \omega_i\cdot\omega_j,$$

where $\omega_i \cdot \omega_j$ denotes the scalar product.

All the models above have a common property: their single-spin space is *compact* (see below). Models with noncompact single-spin spaces present additional interesting difficulties which will not be discussed in this chapter. One important case, the **Gaussian Free Field** for which $\Omega_0 = \mathbb{R}$, will be studied separately in Chapter 8.

About the Point of View Adopted in this Chapter. Describing the above models in infinite volume will require a fair amount of mathematical tools. For simplicity, *we will only provide the details of the theory for models whose spins take their values in* $\{\pm1\}$; the set of configurations is thus the same as in Chapter 3: $\Omega = \{\pm1\}^{\mathbb{Z}^d}$.

Even with this simplification, we will face most of the mathematical difficulties that are unavoidable when attempting to describe infinite systems at equilibrium. It will however allow us to provide elementary proofs, in several cases, and to somewhat reduce the overall amount of abstraction (and notation) required.

Let us stress that the set $\{\pm 1\}$ has been chosen for convenience, but that it could be replaced by any finite set; our discussion (including the proofs) applies essentially verbatim also in that setting. In fact, all the results presented here remain valid, modulo some minor changes, for any model whose spins take their values in a compact set. At the end of the chapter, in Section 6.10, we will mention the few differences that appear in this more general situation.

So, from now on, and until the end of the chapter, unless explicitly stipulated otherwise, Ω_0 *will be* $\{\pm 1\}$, *and*

$$\Omega_\Lambda = \{\pm 1\}^\Lambda, \quad \Omega = \{\pm 1\}^{\mathbb{Z}^d}.$$

Outline of the Chapter

The probabilistic framework used to decribe infinite systems on the lattice will be presented in Section 6.2, together with a motivation for the notion of *specification*, central to the definition of infinite-volume Gibbs measures. After introducing the necessary topological notions, the existence of Gibbs measures will be proved in Section 6.4. Several uniqueness criteria, among which is Dobrushin's condition of weak dependence, will be described in Section 6.5. Gibbs measures enjoying symmetries will be described rapidly in Section 6.6; translation invariance, which plays a special role, will be described in Section 6.7. In Section 6.8, the convex structure of the set of Gibbs measures will be described, as well as the decomposition of any Gibbs measure into a convex combination of extremal elements and the latter's remarkable properties. In Section 6.9, we will present the variational principle, which provides an alternative description of translation-invariant Gibbs measures, in more thermodynamical terms. In Section 6.10, we will sketch the changes necessary in order to describe infinite systems whose spins take infinitely many values; these being considered at several places in the rest of the book. In Section 6.11, we will give a criterion for nonuniqueness involving the nondifferentiability of the pressure, which will be used later in the book. The remaining sections are complements to the chapter.

6.1 The Problem with Infinite Systems

Let us recall the approach used in Chapter 3. By considering for example the + boundary condition, we started in a finite volume $\Lambda \Subset \mathbb{Z}^d$, and defined the Gibbs distribution unambiguously by

$$\mu^+_{\Lambda;\beta,h}(\omega) = \frac{e^{-\mathscr{H}_{\Lambda;\beta,h}(\omega)}}{\mathbf{Z}^+_{\Lambda;\beta,h}}, \quad \omega \in \Omega^+_\Lambda.$$

Then, to describe the Ising model on the infinite lattice, we introduced the thermodynamic limit. We considered a sequence of subsets $\Lambda_n \uparrow \mathbb{Z}^d$ and showed, for each local function f, existence of the limit

$$\langle f \rangle^+_{\beta,h} = \lim_{n\to\infty} \langle f \rangle^+_{\Lambda_n;\beta,h}.$$

This defined a linear functional $\langle \cdot \rangle^+_{\beta,h}$ on local functions, which was called an *infinite-volume Gibbs state.*

This procedure was sufficient for us to determine the phase diagram of the Ising model (Section 3.7), but leaves several natural questions open. For instance, we know that

$$\lim_{n\to\infty} \mu^+_{\Lambda_n;\beta,h}(\sigma_0 = -1) = \lim_{n\to\infty} \tfrac{1}{2}\big(1 - \langle \sigma_0 \rangle^+_{\Lambda_n;\beta,h}\big) = \tfrac{1}{2}\big(1 - \langle \sigma_0 \rangle^+_{\beta,h}\big)$$

exists. This raises the question whether this limit represents the probability that $\sigma_0 = -1$ under some *infinite-volume* probability measure $\mu^+_{\beta,h}$:

$$\mu^+_{\beta,h}(\sigma_0 = -1) = \tfrac{1}{2}\big(1 - \langle \sigma_0 \rangle^+_{\beta,h}\big). \tag{6.1}$$

In infinite volume, neither the Hamiltonian nor the partition function is well-defined. Moreover, it is easy to check that each individual configuration would have to have probability zero. Therefore, extending the definition of a Gibbs distribution to the uncountable set of configurations Ω requires a different approach, involving the methods of measure theory.

6.2 Events and Probability Measures on Ω

As we said above, it is easy to construct a probability distribution on a finite set such as Ω_Λ, since this can be done by specifying the probability of each configuration. Another convenient consequence of the finiteness of Ω_Λ is that the set of events associated with Ω_Λ is naturally identified with the collection $\mathscr{P}(\Omega_\Lambda)$ of all subsets with Ω_Λ. The set of probability distributions on the finite measurable space $(\Omega_\Lambda, \mathscr{P}(\Omega_\Lambda))$ is denoted simply $\mathscr{M}_1(\Omega_\Lambda)$.

Notation 6.2. *In this chapter, it will often be convenient to add a subscript to configurations to specify explicitly the domain in which they are defined. For example, elements of Ω_Λ will usually be denoted $\omega_\Lambda, \eta_\Lambda$, etc.*

Given $S \subset \mathbb{Z}^d$ and a configuration ω defined on a set larger than S, we will also write ω_S to denote the restriction of ω to S, $(\omega_i)_{i\in S}$. We will also often decompose a configuration $\omega_S \in \Omega_S$ as a concatenation: $\omega_S = \omega_\Lambda \omega_{S\setminus\Lambda}$.

These notations should not be confused with the notation in Chapter 3, where σ_Λ was used to denote the product of all spins in Λ, while the restriction of ω to Λ was written $\omega|_\Lambda$.

We first define the natural collection of *events* on Ω, based on the notion of *cylinder*. The restriction of $\omega \in \Omega$ to $S \subset \mathbb{Z}^d$, ω_S, can be expressed using the **projection map** $\Pi_S \colon \Omega \to \Omega_S$:

$$\Pi_S(\omega) \stackrel{\text{def}}{=} \omega_S .$$

In particular, with this notation, given $A \in \mathscr{P}(\Omega_\Lambda)$, the event that "$A$ occurs in Λ" can be written $\Pi_\Lambda^{-1}(A) = \{\omega \in \Omega : \omega_\Lambda \in A\}$.

For each $\Lambda \Subset \mathbb{Z}^d$, consider the set

$$\mathscr{C}(\Lambda) \stackrel{\text{def}}{=} \left\{\Pi_\Lambda^{-1}(A) : A \in \mathscr{P}(\Omega_\Lambda)\right\}$$

of all events on Ω that depend only on the spins located inside Λ. Each event $C \in \mathscr{C}(\Lambda)$ is called a **cylinder (with base Λ)**. For example, $\{\omega_0 = -1\}$, the event containing all configurations ω for which $\omega_0 = -1$, is a cylinder with base $\Lambda = \{0\}$.

Exercise 6.1. Show that $\mathscr{C}(\Lambda)$ has the structure of an **algebra**: (i) $\varnothing \in \mathscr{C}(\Lambda)$, (ii) $A \in \mathscr{C}(\Lambda)$ implies $A^c \in \mathscr{C}(\Lambda)$, and (iii) $A, B \in \mathscr{C}(\Lambda)$ implies $A \cup B \in \mathscr{C}(\Lambda)$.

For any $S \subset \mathbb{Z}^d$ (possibly infinite), consider the collection

$$\mathscr{C}_S \stackrel{\text{def}}{=} \bigcup_{\Lambda \Subset S} \mathscr{C}(\Lambda)$$

of all **local events** in S, that is, all events that depend on finitely many spins, all located in S.

Exercise 6.2. Check that, for all $S \subset \mathbb{Z}^d$, $\mathscr{C}_S$ contains countably many events and that it has the structure of an algebra. *Hint:* First, show that $\mathscr{C}(\Lambda) \subset \mathscr{C}(\Lambda')$ whenever $\Lambda \subset \Lambda'$.

The **σ-algebra generated by cylinders with base contained in S** is denoted by

$$\mathscr{F}_S \stackrel{\text{def}}{=} \sigma(\mathscr{C}_S)$$

and consists of all the events that depend only on the spins inside S. When $S = \mathbb{Z}^d$, we simply write

$$\mathscr{C} \equiv \mathscr{C}_{\mathbb{Z}^d}, \quad \mathscr{F} \equiv \sigma(\mathscr{C}).$$

The cylinders $\mathscr{C}$ should be considered as the algebra of *local* events. Although generated from these local events, the σ-algebra $\mathscr{F}$ automatically contains *macroscopic* events, that is, events that depend on the system as a whole (a precise definition of macroscopic events will be given in Section 6.8.1). For example, the event

$$\left\{\omega \in \Omega : \limsup_{n\to\infty} \frac{1}{|\mathrm{B}(n)|} \sum_{i\in\mathrm{B}(n)} \omega_i > 0\right\} = \bigcup_{k\geq 1} \bigcap_{n\geq 1} \bigcup_{m\geq n} \left\{\frac{1}{|\mathrm{B}(m)|} \sum_{i\in\mathrm{B}(m)} \omega_i \geq \tfrac{1}{k}\right\}$$

belongs to $\mathscr{F}$ (and is obviously not local). The importance of macroscopic events will be emphasized in Section 6.8.

The reader might wonder whether there are interesting events that do not belong to $\mathscr{F}$. As a matter of fact, all events we will need can be described explicitly *in terms of the individual spins in S, using (possibly infinite) unions and intersections. Those are all in $\mathscr{F}$.* ◇

The set of probability measures on $(\Omega, \mathscr{F})$ will be denoted $\mathscr{M}_1(\Omega, \mathscr{F})$, or simply $\mathscr{M}_1(\Omega)$ when no ambiguity is possible. The elements of $\mathscr{M}_1(\Omega)$ will usually be denoted μ or ν.

A function $g \colon \Omega \to \mathbb{R}$ is **measurable with respect to** $\mathscr{F}_S$ (or simply $\mathscr{F}_S$**-measurable**) if $g^{-1}(I) \in \mathscr{F}_S$ for all Borel sets $I \subset \mathbb{R}$. Intuitively, such a function should be a *function* of the spins living in S:

Lemma 6.3. *A function $g \colon \Omega \to \mathbb{R}$ is $\mathscr{F}_S$-measurable if and only if there exists $\varphi \colon \Omega_S \to \mathbb{R}$ such that*

$$g(\omega) = \varphi(\omega_S).$$

Proof Let g be $\mathscr{F}_S$-measurable. On Ω_S, consider the set of cylinder events $\mathscr{C}'_S$, and $\mathscr{F}'_S = \sigma(\mathscr{C}'_S)$. If $\Pi_S \colon \Omega \to \Omega_S$ denotes the projection map, we have $\Pi_S^{-1}(C') \in \mathscr{F}_S$ for all $C' \in \mathscr{C}'_S$. This implies that $\mathscr{F}_S$ is generated by Π_S: $\mathscr{F}_S = \sigma(\Pi_S)$ (see Section B.5.2). Therefore, by Lemma B.38, there exists $\varphi \colon \Omega \to \Omega_S$ such that $g = \varphi \circ \Pi_S$. Conversely, if g is of this form, then clearly $g^{-1}(I) \in \mathscr{F}_S$ for each Borel set $I \subset \mathbb{R}$ so that g is $\mathscr{F}_S$-measurable. □

Remember that $f \colon \Omega \to \mathbb{R}$ is **local** if it only depends on a finite number of spins: there exists $\Lambda \Subset \mathbb{Z}^d$ such that $f(\omega) = f(\omega')$ as soon as $\omega_\Lambda = \omega'_\Lambda$. By Lemma 6.3, this is equivalent to saying that f is $\mathscr{F}_\Lambda$-measurable. In fact, since the spins take finitely many values, a local function can only take finitely many values and can therefore be expressed as a finite linear combination of indicators of cylinders. Since, for each of the latter, $f^{-1}(I) \in \mathscr{C} \subset \mathscr{F}$, local functions are always measurable. In the sequel, all the functions $f \colon \Omega \to \mathbb{R}$ that we will consider will be assumed to be measurable.

Notation 6.4. *In Chapter 3, we denoted the expectation of a function f under a probability measure μ by $\langle f \rangle_\mu$. For the rest of this chapter, it will be convenient to also use the following equivalent notations: $\int f \, d\mu$, or $\mu(f)$.*

States vs. Probability Measures

Remember from Section 3.4 that a *state* is a normalized positive linear functional $f \mapsto \langle f \rangle$ acting on local functions. Observe that a state can be associated with each probability measure $\mu \in \mathscr{M}_1(\Omega)$ by setting, for all local functions f,

$$\langle f \rangle \stackrel{\text{def}}{=} \mu(f).$$

It turns out that all states are of this form:

Theorem 6.5. *For every state $\langle \cdot \rangle$, there exists a unique probability measure $\mu \in \mathcal{M}_1(\Omega)$ such that $\langle f \rangle = \mu(f)$ for every local function $f: \Omega \to \mathbb{R}$.*

This result is a particular case of the Riesz–Markov–Kakutani Representation Theorem. Its proof requires a few tools that will be presented later, and can be found in Section 6.12.

Two Infinite-Volume Measures for the Ising Model

Using Theorem 6.5, we can associate a probability measure to each Gibbs state of the Ising model. In particular, let us denote by $\mu^+_{\beta,h}$ (resp. $\mu^-_{\beta,h}$) the measure associated with $\langle \cdot \rangle^+_{\beta,h}$ (resp. $\langle \cdot \rangle^-_{\beta,h}$). For these measures, relations such as (6.1) hold. A lot will be learned about these measures throughout the chapter.

For the time being, one should remember that the construction of $\mu^+_{\beta,h}$ and $\mu^-_{\beta,h}$ was based on the thermodynamic limit, which was used to define the states $\langle \cdot \rangle^+_{\beta,h}$ and $\langle \cdot \rangle^-_{\beta,h}$. Our aim, in the following sections, is to present a way of defining measures *directly* on the infinite lattice, without involving any limiting procedure. As we will see, this alternative approach presents a number of substantial advantages.

Why Not Simply Use Kolmogorov's Extension Theorem?

In probability theory, the standard approach to construct infinite collections of dependent random variables relies on Kolmogorov's Extension Theorem, in which the strategy is to define a measure by requiring it to satisfy a set of local conditions. In our case, these conditions should depend on the microscopic description of the system under consideration, which is encoded in its Hamiltonian. We briefly outline this approach and explain why it does not solve the problem we are interested in.

Given $\mu \in \mathcal{M}_1(\Omega)$ and $\Lambda \Subset \mathbb{Z}^d$, the **marginal distribution of μ on Λ** is the probability distribution $\mu|_\Lambda \in \mathcal{M}_1(\Omega_\Lambda)$ defined by

$$\mu|_\Lambda \stackrel{\text{def}}{=} \mu \circ \Pi_\Lambda^{-1}. \tag{6.2}$$

In other words, $\mu|_\Lambda$ is the only distribution in $\mathcal{M}_1(\Omega_\Lambda)$ such that, for all $A \in \mathcal{P}(\Omega_\Lambda)$, $\mu|_\Lambda(A) = \mu(\{\omega \in \Omega : \omega_\Lambda \in A\})$. By construction, the marginals satisfy

$$\mu|_\Delta = \mu|_\Lambda \circ (\Pi^\Lambda_\Delta)^{-1}, \quad \forall \Delta \subset \Lambda \Subset \mathbb{Z}^d, \tag{6.3}$$

where $\Pi^\Lambda_\Delta: \Omega_\Lambda \to \Omega_\Delta$ is the canonical projection defined by $\Pi^\Lambda_\Delta \stackrel{\text{def}}{=} \Pi_\Delta \circ \Pi_\Lambda^{-1}$.

It turns out that a measure $\mu \in \mathcal{M}_1(\Omega)$ is entirely characterized by its marginals $\mu|_\Lambda$, $\Lambda \Subset \mathbb{Z}^d$, but more is true: given any collection of probability distributions $\{\mu_\Lambda\}_{\Lambda \Subset \mathbb{Z}^d}$, with $\mu_\Lambda \in \mathcal{M}_1(\Omega_\Lambda)$ for all Λ, which satisfies a compatibility condition of the type (6.3), there exists a unique probability measure $\mu \in \mathcal{M}_1(\Omega)$ admitting them as marginals. This is the content of the following famous theorem:

Theorem 6.6 (Kolmogorov's Extension Theorem). *Let* $\{\mu_\Lambda\}_{\Lambda \Subset \mathbb{Z}^d}$, $\mu_\Lambda \in \mathcal{M}_1(\Omega_\Lambda)$, *be* ***consistent*** *in the sense that*

$$\text{for all } \Lambda \Subset \mathbb{Z}^d : \quad \mu_\Delta = \mu_\Lambda \circ (\Pi^\Lambda_\Delta)^{-1}, \quad \forall \Delta \subset \Lambda . \tag{6.4}$$

Then there exists a unique $\mu \in \mathcal{M}_1(\Omega)$ *such that* $\mu|_\Lambda = \mu_\Lambda$ *for all* $\Lambda \Subset \mathbb{Z}^d$.

Proof See Section 6.12. □

Theorem 6.6 yields an efficient way of constructing a measure in $\mathcal{M}_1(\Omega)$, provided that one can define the desired collection $\{\mu_\Lambda\}_{\Lambda \Subset \mathbb{Z}^d}$ of candidates for its marginals. An important such application is the construction of the product measure, that is, of an **independent field**; in our setting, this covers for example the case of the Ising model at infinite temperature, $\beta = 0$.

Exercise 6.3. (Construction of a product measure on $(\Omega, \mathcal{F})$) For each $i \in \mathbb{Z}^d$, let ρ_i be a probability distribution on $\{\pm 1\}$ and let, for all $\Lambda \Subset \mathbb{Z}^d$,

$$\mu_\Lambda(\omega_\Lambda) \stackrel{\text{def}}{=} \prod_{j \in \Lambda} \rho_j(\omega_j), \qquad \omega_\Lambda \in \Omega_\Lambda .$$

Check that $\{\mu_\Lambda\}_{\Lambda \Subset \mathbb{Z}^d}$ is consistent. The resulting measure on $(\Omega, \mathcal{F})$, whose existence is guaranteed by Theorem 6.6, is denoted $\rho^{\mathbb{Z}^d}$.

If one tries to use Theorem 6.6 to construct infinite-volume measures for the Ising model on $\mathbb{Z}^d$, we face a difficulty. Namely, the Boltzmann weight allows one to define finite-volume Gibbs distributions in terms of the underlying Hamiltonian. However, as we will explain now, *in general, there is no way to express the marginals associated with an infinite-volume Gibbs measure without making explicit reference to the latter.*

Indeed, let us consider the simplest case of the marginal distribution of the spin at the origin, σ_0, and let us assume that $d \geq 2$ and $h = 0$. Of course, σ_0 follows a Bernoulli distribution (with values in $\{\pm 1\}$) for some parameter $p \in [0, 1]$. The only thing that needs to be determined is the value of p. However, we already know from the results in Chapter 3 that, for all large enough values of β, the average value of σ_0, and thus the relevant value of p, depends on the chosen Gibbs state. However, *all these states correspond to the same Hamiltonian and the same values of the parameters β and h.* This means that it is impossible to determine p from a knowledge of the Hamiltonian and the parameters β and h: one needs to know the macroscopic state the system is in, which is precisely what we are trying to construct. This shows that Kolmogorov's Extension Theorem is doomed to fail for the construction of the Ising model in infinite volume. [1]

Exercise 6.4. Consider $\{\mu^{\varnothing}_{\Lambda}\}_{\Lambda \Subset \mathbb{Z}^d}$, where $\mu^{\varnothing}_{\Lambda}$ is the Gibbs distribution associated with the two-dimensional Ising model in Λ, with free boundary condition, at parameters $\beta > 0$ and $h = 0$. Show that the family obtained is not consistent.

6.2.1 The DLR Approach

A key observation, made by Dobrushin, Lanford and Ruelle, is that if one considers *conditional probabilities* rather than marginals, then one is led to a different consistency condition, much better suited to our needs. Before stating this condition precisely (see Lemma 6.7 below), we explain it at an elementary level, using the Ising model and the notations of Chapter 3.

Consider $\Delta \subset \Lambda \Subset \mathbb{Z}^d$ and a boundary condition $\eta \in \Omega$:

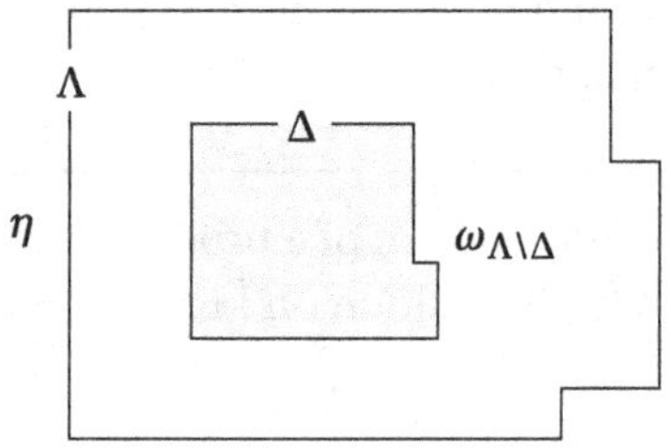

The Ising model in Λ with boundary condition η is described by $\mu^{\eta}_{\Lambda;\beta,h}$. Let f be a local function depending only on the variables ω_j, $j \in \Delta$, and consider the expectation of f under $\mu^{\eta}_{\Lambda;\beta,h}$. Since f only depends on the spins located inside Δ, this expectation can be computed by first fixing the values of the spins in $\Lambda \setminus \Delta$. As we already saw in Exercise 3.11, $\mu^{\eta}_{\Lambda;\beta,h}$, conditioned on $\omega_{\Lambda\setminus\Delta}$, is equivalent to the Gibbs distribution on Δ with boundary condition $\omega_{\Lambda\setminus\Delta}\eta_{\Lambda^c}$ outside Δ. Therefore,

$$\begin{aligned}\langle f\rangle^{\eta}_{\Lambda;\beta,h} &= \sum_{\omega_{\Lambda\setminus\Delta}} \langle f\,\mathbf{1}_{\{\omega_{\Lambda\setminus\Delta}\text{ outside }\Delta\}}\rangle^{\eta}_{\Lambda;\beta,h} \\ &= \sum_{\omega_{\Lambda\setminus\Delta}} \langle f\rangle^{\omega_{\Lambda\setminus\Delta}\eta_{\Lambda^c}}_{\Delta;\beta,h}\, \mu^{\eta}_{\Lambda;\beta,h}(\omega_{\Lambda\setminus\Delta}\text{ outside }\Delta)\,. \end{aligned} \tag{6.5}$$

(Notice a slight abuse of notation in the last line.) A particular instance of (6.5) is when f is the indicator of some event A occurring in Δ, in which case

$$\mu^{\eta}_{\Lambda;\beta,h}(A) = \sum_{\omega_{\Lambda\setminus\Delta}} \mu^{\omega_{\Lambda\setminus\Delta}\eta_{\Lambda^c}}_{\Delta;\beta,h}(A)\, \mu^{\eta}_{\Lambda;\beta,h}(\omega_{\Lambda\setminus\Delta}\text{ outside }\Delta)\,. \tag{6.6}$$

The above discussion expresses the idea of Dobrushin, Lanford and Ruelle: the relation (6.5), or its second equivalent version (6.6), can be interpreted as a consistency relation between the Gibbs distributions in Λ and Δ. We can formulate (6.5) in a more precise way:

Lemma 6.7. *For all $\Delta \subset \Lambda \Subset \mathbb{Z}^d$ and all bounded measurable $f : \Omega \to \mathbb{R}$,*

$$\langle f \rangle^{\eta}_{\Lambda;\beta,h} = \big\langle \langle f \rangle^{\cdot}_{\Delta;\beta,h} \big\rangle^{\eta}_{\Lambda;\beta,h}, \qquad \forall \eta \in \Omega. \tag{6.7}$$

Proof of Lemma 6.7. To lighten the notations, we omit any mention of the dependence on β and h. Each $\omega \in \Omega^{\eta}_{\Lambda}$ is of the form $\omega = \omega_{\Lambda}\eta_{\Lambda^c}$, with $\omega_{\Lambda} \in \Omega_{\Lambda}$. Therefore,

$$\big\langle \langle f \rangle^{\cdot}_{\Delta} \big\rangle^{\eta}_{\Lambda} = \sum_{\omega_{\Lambda}} \langle f \rangle^{\omega_{\Lambda}\eta_{\Lambda^c}}_{\Delta} \frac{e^{-\mathscr{H}_{\Lambda}(\omega_{\Lambda}\eta_{\Lambda^c})}}{\mathbf{Z}^{\eta}_{\Lambda}}. \tag{6.8}$$

In the same way,

$$\langle f \rangle^{\omega_{\Lambda}\eta_{\Lambda^c}}_{\Delta} = \sum_{\omega'_{\Delta}} f(\omega'_{\Delta}\omega_{\Lambda\setminus\Delta}\eta_{\Lambda^c}) \frac{e^{-\mathscr{H}_{\Delta}(\omega'_{\Delta}\omega_{\Lambda\setminus\Delta}\eta_{\Lambda^c})}}{\mathbf{Z}^{\omega_{\Lambda}\eta_{\Lambda^c}}_{\Delta}}. \tag{6.9}$$

In (6.8), we decompose $\omega_{\Lambda} = \omega_{\Delta}\omega_{\Lambda\setminus\Delta}$, and sum separately over $\omega_{\Lambda\setminus\Delta}$ and ω_{Δ}. Observe that

$$\begin{aligned}\mathscr{H}_{\Lambda}(\omega_{\Delta}\omega_{\Lambda\setminus\Delta}\eta_{\Lambda^c}) - \mathscr{H}_{\Delta}(\omega_{\Delta}\omega_{\Lambda\setminus\Delta}\eta_{\Lambda^c}) = \\ \mathscr{H}_{\Lambda}(\omega'_{\Delta}\omega_{\Lambda\setminus\Delta}\eta_{\Lambda^c}) - \mathscr{H}_{\Delta}(\omega'_{\Delta}\omega_{\Lambda\setminus\Delta}\eta_{\Lambda^c}).\end{aligned} \tag{6.10}$$

Indeed, the difference on each side represents the interactions among the spins inside $\Lambda \setminus \Delta$, and between these spins and those outside Λ, and so does not depend on ω_{Δ} or ω'_{Δ}. Therefore, plugging (6.9) into (6.8), using (6.10), rearranging and calling $\omega'_{\Delta}\omega_{\Lambda\setminus\Delta} \equiv \omega'_{\Lambda}$, we get

$$\begin{aligned}\big\langle \langle f \rangle^{\cdot}_{\Delta} \big\rangle^{\eta}_{\Lambda} &= \sum_{\omega_{\Lambda\setminus\Delta}} \sum_{\omega'_{\Delta}} f(\omega'_{\Delta}\omega_{\Lambda\setminus\Delta}\eta_{\Lambda^c}) \frac{e^{-\mathscr{H}_{\Lambda}(\omega'_{\Delta}\omega_{\Lambda\setminus\Delta}\eta_{\Lambda^c})}}{\mathbf{Z}^{\eta}_{\Lambda}} \underbrace{\frac{\sum_{\omega_{\Delta}} e^{-\mathscr{H}_{\Delta}(\omega_{\Delta}\omega_{\Lambda\setminus\Delta}\eta_{\Lambda^c})}}{\mathbf{Z}^{\omega_{\Lambda}\eta_{\Lambda^c}}_{\Delta}}}_{=1} \\ &= \sum_{\omega'_{\Lambda}} f(\omega'_{\Lambda}\eta_{\Lambda^c}) \frac{e^{-\mathscr{H}_{\Lambda}(\omega'_{\Lambda}\eta_{\Lambda^c})}}{\mathbf{Z}^{\eta}_{\Lambda}} \\ &= \langle f \rangle^{\eta}_{\Lambda}.\end{aligned}$$

□

Remark 6.8. The proof given above does not depend on the details of the Ising Hamiltonian, but rather on the property (6.10), which will be used again later. ◇

We now explain why (6.7) leads to a natural characterization of *infinite-volume Gibbs states*, more general than the one introduced in Chapter 3.

First observe that, since we are considering the Ising model in which the interactions are only between nearest neighbors, the function $\omega \mapsto \langle f \rangle^{\omega}_{\Delta;\beta,h}$ is local (it depends only on those ω_i for which $i \in \partial^{\mathrm{ex}}\Delta$). So, *if* the distributions $\langle \cdot \rangle^{\eta}_{\Lambda;\beta,h}$ converge to a Gibbs state $\langle \cdot \rangle$ when $\Lambda \uparrow \mathbb{Z}^d$, in the sense of Definition 3.14, then we can take the thermodynamic limit on both sides of (6.7), obtaining

$$\langle f \rangle = \big\langle \langle f \rangle^{\cdot}_{\Delta;\beta,h} \big\rangle, \tag{6.11}$$

for all $\Delta \Subset \mathbb{Z}^d$ and all local functions f. We conclude that (6.11) must be satisfied by all states $\langle\cdot\rangle$ obtained as limits. But this can also be used to characterize states *without reference to limits.* Namely, we could extend the notion of infinite-volume Gibbs state by saying that *a state* $\langle\cdot\rangle$ *(not necessarily obtained as a limit) is an infinite-volume Gibbs state for the Ising model at* (β, h) *if* (6.11) *holds for every* $\Delta \Subset \mathbb{Z}^d$ *and all local functions* f. This new characterization has mathematical advantages that will become clear later.

If one identifies a Gibbs state $\langle\cdot\rangle$ with the corresponding measure μ given in Theorem 6.5, then μ should satisfy the infinite-volume version of (6.6): by taking $f = \mathbf{1}_A$, for some local event A, (6.11) becomes

$$\mu(A) = \int \mu^{\omega}_{\Delta;\beta,h}(A)\mu(\mathrm{d}\omega)\,. \tag{6.12}$$

Once again, we can use (6.12) as a set of conditions that *define* those measures that describe the Ising model in infinite volume. We say that $\mu \in \mathcal{M}_1(\Omega)$ *is a Gibbs measure for the parameters* (β, h) *if* (6.12) *holds for all* $\Delta \Subset \mathbb{Z}^d$ *and all local events* A. An important feature of this point of view is that it characterizes probability measures directly on the infinite lattice $\mathbb{Z}^d$, without assuming that they have been obtained from a limiting procedure.

6.3 Specifications and Measures

We will formulate the DLR approach introduced in the previous section in a more precise and more general way. The theory will apply to a large class of models, containing the Ising model as a particular case. It will also include models with a more complex structure, for example with long-range interactions or interactions between larger collections of spins.

We will proceed in two steps. First, we will generalize the consistency relation (6.7) by introducing the notion of *specification.*

In our discussion of the Ising model, the starting ingredient was the family of finite-volume Gibbs distributions $\{\mu^{\cdot}_{\Lambda;\beta,h}(\cdot)\}_{\Lambda\Subset\mathbb{Z}^d}$, whose main features we gather as follows:

1. For a fixed boundary condition ω, $\mu^{\omega}_{\Lambda;\beta,h}(\cdot)$ is a probability distribution on $(\Omega^{\omega}_{\Lambda}, \mathcal{P}(\Omega^{\omega}_{\Lambda}))$. It can however also be seen as a probability measure on $(\Omega, \mathcal{F})$ by letting, for all $A \in \mathcal{F}$,

$$\mu^{\omega}_{\Lambda;\beta,h}(A) \stackrel{\text{def}}{=} \sum_{\tau_\Lambda \in \Omega_\Lambda} \mu^{\omega}_{\Lambda;\beta,h}(\tau_\Lambda\omega_{\Lambda^c})\mathbf{1}_A(\tau_\Lambda\omega_{\Lambda^c})\,. \tag{6.13}$$

   In particular,

$$\forall B \in \mathcal{F}_{\Lambda^c}\,, \quad \mu^{\omega}_{\Lambda;\beta,h}(B) = \mathbf{1}_B(\omega)\,. \tag{6.14}$$

2. For a fixed $A \in \mathcal{F}$, $\mu^{\omega}_{\Lambda;\beta,h}(A)$ is entirely determined by ω_{Λ^c} (actually, even by $\omega_{\partial^{\text{ex}}\Lambda}$). In particular, $\omega \mapsto \mu^{\omega}_{\Lambda;\beta,h}(A)$ is $\mathcal{F}_{\Lambda^c}$-measurable.
3. When considering regions $\Delta \subset \Lambda \Subset \mathbb{Z}^d$, the consistency condition (6.7) is satisfied.

The maps $\mu^{\cdot}_{\Lambda;\beta,h}(\cdot)$ depend of course on the specific form of the Hamiltonian of the Ising model, but the three properties above can in fact be introduced without reference to any particular Hamiltonian. In a fixed volume, we start by incorporating the first two features in a general definition:

Definition 6.9. Let $\Lambda \Subset \mathbb{Z}^d$. A **probability kernel from $\mathscr{F}_{\Lambda^c}$ to $\mathscr{F}$** is a map $\pi_\Lambda : \mathscr{F} \times \Omega \to [0,1]$ with the following properties:

- For each $\omega \in \Omega$, $\pi_\Lambda(\cdot \mid \omega)$ is a probability measure on $(\Omega, \mathscr{F})$.
- For each $A \in \mathscr{F}$, $\pi_\Lambda(A \mid \cdot)$ is $\mathscr{F}_{\Lambda^c}$-measurable.

If, moreover,

$$\pi_\Lambda(B \mid \omega) = \mathbf{1}_B(\omega), \qquad \forall B \in \mathscr{F}_{\Lambda^c} \tag{6.15}$$

for all $\omega \in \Omega$, π_Λ is said to be **proper**.

Note that, if π_Λ is a proper probability kernel from $\mathscr{F}_{\Lambda^c}$ to $\mathscr{F}$, then the probability measure $\pi_\Lambda(\cdot \mid \omega)$ is concentrated on the set Ω^ω_Λ. Indeed, for any $\omega \in \Omega$,

$$\pi_\Lambda(\Omega^\omega_\Lambda \mid \omega) = \mathbf{1}_{\Omega^\omega_\Lambda}(\omega) = 1\,, \tag{6.16}$$

since $\Omega^\omega_\Lambda \in \mathscr{F}_{\Lambda^c}$. For this reason, we will call ω the **boundary condition** of $\pi_\Lambda(\cdot \mid \omega)$. Our first example of a proper probability kernel was thus $(A, \omega) \mapsto \mu^\omega_{\Lambda;\beta,h}(A)$, defined in (6.13).

For a fixed boundary condition ω, a bounded measurable function $f \colon \Omega \to \mathbb{R}$ can be integrated with respect to $\pi_\Lambda(\cdot \mid \omega)$. We denote by $\pi_\Lambda f$ the $\mathscr{F}_{\Lambda^c}$-measurable function defined by

$$\pi_\Lambda f(\omega) \stackrel{\text{def}}{=} \int f(\eta)\pi_\Lambda(\mathrm{d}\eta \mid \omega)\,.$$

Although this integral notation is convenient, our assumption on the finiteness of Ω_0 implies that most of the integrals that will appear in this chapter are actually finite sums. Indeed, we will always work with proper probability kernels, and the observation (6.16) implies that the measure $\pi_\Lambda(\cdot \mid \omega)$ is entirely characterized by the probability it associates with the configurations in the finite set Ω^ω_Λ. In particular, we can verify that π_Λ is proper if and only if it is of the form (6.13). Namely, using (6.16), one can compute the probability of any event $A \in \mathscr{F}$ by summing over the configurations in Ω^ω_Λ:

$$\pi_\Lambda(A \mid \omega) = \sum_{\eta \in \Omega^\omega_\Lambda} \pi_\Lambda(\{\eta\} \mid \omega)\mathbf{1}_A(\eta)\,.$$

Since each $\eta \in \Omega^\omega_\Lambda$ is of the form $\eta = \eta_\Lambda \omega_{\Lambda^c}$, this sum can equivalently be expressed as

$$\pi_\Lambda(A \mid \omega) = \sum_{\eta_\Lambda \in \Omega_\Lambda} \pi_\Lambda(\{\eta_\Lambda \omega_{\Lambda^c}\} \mid \omega)\mathbf{1}_A(\eta_\Lambda \omega_{\Lambda^c})\,.$$

In the sequel, *all kernels π_Λ to be considered will be proper*, which, by the above discussion, means that π_Λ is entirely defined by the numbers $\pi_\Lambda(\{\eta_\Lambda \omega_{\Lambda^c}\}) \mid \omega)$. To lighten the notations, we will abbreviate

$$\pi_\Lambda(\{\eta_\Lambda\omega_{\Lambda^c}\}\,|\,\omega) \equiv \pi_\Lambda(\eta_\Lambda\,|\,\omega)\,.$$

These sums will be used constantly throughout the chapter. We summarize this discussion in the following statement.

Lemma 6.10. *If π_Λ is proper, then, for all $\omega \in \Omega$,*

$$\pi_\Lambda(A\,|\,\omega) = \sum_{\eta_\Lambda\in\Omega_\Lambda} \pi_\Lambda(\eta_\Lambda\,|\,\omega)\mathbf{1}_A(\eta_\Lambda\omega_{\Lambda^c})\,, \qquad \forall A \in \mathscr{F}\,, \tag{6.17}$$

and, for any bounded measurable function $f\colon \Omega \to \mathbb{R}$,

$$\pi_\Lambda f(\omega) = \sum_{\eta_\Lambda\in\Omega_\Lambda} \pi_\Lambda(\eta_\Lambda\,|\,\omega)f(\eta_\Lambda\omega_{\Lambda^c})\,. \tag{6.18}$$

In order to describe an infinite system on $\mathbb{Z}^d$, we will actually need a *family* of proper probability kernels, $\{\pi_\Lambda\}_{\Lambda\Subset\mathbb{Z}^d}$, satisfying consistency relations of the type (6.6)–(6.7). These consistency relations are conveniently expressed in terms of the **composition of kernels**: given π_Λ and π_Δ, set

$$\pi_\Lambda\pi_\Delta(A\,|\,\eta) \stackrel{\text{def}}{=} \int \pi_\Delta(A\,|\,\omega)\pi_\Lambda(\mathrm{d}\omega\,|\,\eta)\,.$$

Exercise 6.5. Let $\Delta \subset \Lambda \Subset \mathbb{Z}^d$. Show that $\pi_\Lambda\pi_\Delta$ is a proper probability kernel from $\mathscr{F}_{\Lambda^c}$ to $\mathscr{F}$.

In these terms, the generalization of (6.6) can be stated as follows.

Definition 6.11. A **specification** is a family $\pi = \{\pi_\Lambda\}_{\Lambda\Subset\mathbb{Z}^d}$ of proper probability kernels that is **consistent**, in the sense that

$$\pi_\Lambda\pi_\Delta = \pi_\Lambda, \quad \forall \Delta \subset \Lambda \Subset \mathbb{Z}^d\,.$$

In order to formulate an analogue of (6.12) for probability kernels, it is natural to define, for every kernel π_Λ and every $\mu \in \mathscr{M}_1(\Omega)$, the probability measure $\mu\pi_\Lambda \in \mathscr{M}_1(\Omega)$ via

$$\mu\pi_\Lambda(A) \stackrel{\text{def}}{=} \int \pi_\Lambda(A\,|\,\omega)\,\mu(\mathrm{d}\omega), \quad A \in \mathscr{F}\,. \tag{6.19}$$

Exercise 6.6. Show that, for every bounded measurable function f, every measure $\mu \in \mathscr{M}_1(\Omega)$ and every kernel π_Λ, $\mu\pi_\Lambda(f) = \mu(\pi_\Lambda f)$. *Hint:* Start with $f = \mathbf{1}_A$.

With a specification at hand, we can now introduce the central definition of this chapter. Expression (6.20) below is the generalization of (6.12).

Definition 6.12. Let $\pi = \{\pi_\Lambda\}_{\Lambda \Subset \mathbb{Z}^d}$ be a specification. A measure $\mu \in \mathcal{M}_1(\Omega)$ is said to be **compatible with** (or **specified by**) π if

$$\mu = \mu\pi_\Lambda, \quad \forall \Lambda \Subset \mathbb{Z}^d. \tag{6.20}$$

The set of measures compatible with π (if any) is denoted by $\mathscr{G}(\pi)$.

The above characterization raises several questions, which we will investigate in quite some generality in the rest of this chapter.

- *Existence.* Is there always at least one measure μ satisfying (6.20)? This problem will be tackled in Section 6.4.
- *Uniqueness.* Can there be several such measures? The uniqueness problem will be considered in Section 6.5, where we will introduce a condition on a specification π which guarantees that $\mathscr{G}(\pi)$ contains exactly one probability measure: $|\mathscr{G}(\pi)| = 1$.
- *Comparison with the former approach.* We will also consider the important question of comparing the approach based on Definition 6.12 with the approach used in Chapter 3, in which infinite-volume states were obtained as the thermodynamic limits of finite-volume ones. We will see that Definition 6.12 yields, in general, a strictly larger set of measures than those produced by the approach via the thermodynamic limit (proof of Theorem 6.26 and Example 6.64). Nevertheless, all the relevant (in a sense to be discussed later) measures in $\mathscr{G}(\pi)$ can in fact be obtained using the latter approach (Section 6.8).

When the specification does not involve interactions between the spins, these questions can be answered easily:

Exercise 6.7. For each $i \in \mathbb{Z}^d$, let ρ_i be a probability distribution on $\{\pm 1\}$. For each $\Lambda \Subset \mathbb{Z}^d$, define the product distribution ρ^Λ on Ω_Λ by

$$\rho^\Lambda(\omega_\Lambda) \stackrel{\text{def}}{=} \prod_{i\in\Lambda} \rho_i(\omega_i).$$

For $\tau_\Lambda \in \Omega_\Lambda$ and $\eta \in \Omega$, let

$$\pi_\Lambda(\tau_\Lambda \mid \eta) \stackrel{\text{def}}{=} \rho^\Lambda(\tau_\Lambda). \tag{6.21}$$

1. Show that $\pi = \{\pi_\Lambda\}_{\Lambda \Subset \mathbb{Z}^d}$ is a specification.
2. Show that the product measure $\rho^{\mathbb{Z}^d}$ (remember Exercise 6.3) is the unique probability measure specified by π: $\mathscr{G}(\pi) = \{\rho^{\mathbb{Z}^d}\}$.

In the previous exercise, establishing existence and uniqueness for a probability measure compatible with the specification (6.21) is straightforward, thanks to the independence of the spins. In the next sections, we will introduce a general procedure for constructing specifications corresponding to systems of interacting spins and we will see that existence/uniqueness can be derived for abstract specifications under fairly general assumptions. (Establishing nonuniqueness, on the other hand, usually requires a case-by-case study.)

6.3.1 Kernels vs. Conditional Probabilities

Before continuing, we emphasize the important relation existing between a specification and the measures it specifies (if any). We first verify the following simple property:

Lemma 6.13. *Assume that π_Λ is proper. Then, for all $A \in \mathscr{F}$ and all $B \in \mathscr{F}_{\Lambda^c}$,*

$$\pi_\Lambda(A \cap B \,|\, \cdot) = \pi_\Lambda(A \,|\, \cdot)\mathbf{1}_B(\cdot)\,. \tag{6.22}$$

Proof Assume first that $\omega \in B$. Then, since the kernel is proper, B has probability 1 under $\pi_\Lambda(\cdot \,|\, \omega)$: $\pi_\Lambda(B \,|\, \omega) = \mathbf{1}_B(\omega) = 1$. Therefore

$$\pi_\Lambda(A \cap B \,|\, \omega) = \pi_\Lambda(A \,|\, \omega) - \pi_\Lambda(A \cap B^c \,|\, \omega) = \pi_\Lambda(A \,|\, \omega) = \pi_\Lambda(A \,|\, \omega)\mathbf{1}_B(\omega)\,.$$

Similarly, if $\omega \notin B$, $\pi_\Lambda(B \,|\, \omega) = 0$ and thus

$$\pi_\Lambda(A \cap B \,|\, \omega) = 0 = \pi_\Lambda(A \,|\, \omega)\mathbf{1}_B(\omega)\,. \qquad \square$$

Now, observe that if $\mu \in \mathscr{G}(\pi)$, then (6.22) implies that, for all $A \in \mathscr{F}_\Lambda$ and $B \in \mathscr{F}_{\Lambda^c}$,

$$\int_B \pi_\Lambda(A \,|\, \omega)\mu(d\omega) = \int \pi_\Lambda(A \cap B \,|\, \omega)\mu(d\omega) = \mu\pi_\Lambda(A \cap B) = \mu(A \cap B)\,.$$

But, by definition of the conditional probability,

$$\mu(A \cap B) = \int_B \mu(A \,|\, \mathscr{F}_{\Lambda^c})(\omega)\mu(d\omega)\,.$$

By the almost-sure uniqueness of the conditional expectation (Lemma B.50), we thus see that

$$\mu(A \,|\, \mathscr{F}_{\Lambda^c})(\cdot) = \pi_\Lambda(A \,|\, \cdot), \qquad \mu\text{-almost surely.} \tag{6.23}$$

Since $A \mapsto \pi_\Lambda(A \,|\, \omega)$ is a measure for *each* ω, we thus see that π_Λ *provides a regular conditional distribution for μ, when conditioned with respect to $\mathscr{F}_{\Lambda^c}$*. On the other hand, if (6.23) holds, then, for all $\Lambda \Subset \mathbb{Z}^d$ and all $A \in \mathscr{F}$,

$$\mu\pi_\Lambda(A) = \int \pi_\Lambda(A \,|\, \omega)\mu(d\omega) = \int \mu(A \,|\, \mathscr{F}_{\Lambda^c})\mu(d\omega) = \mu(A),$$

and so $\mu \in \mathscr{G}(\pi)$. We have thus shown that *a measure μ is compatible with a specification $\pi = \{\pi_\Lambda\}_{\Lambda \Subset \mathbb{Z}^d}$ if and only if each kernel π_Λ provides a regular version of $\mu(\cdot \,|\, \mathscr{F}_{\Lambda^c})$.*

6.3.2 Gibbsian Specifications

Before moving on to the existence problem, we introduce the class of specifications representative of the models studied in this book.

The Ising Hamiltonian $\mathscr{H}_{\Lambda;\beta,h}$ (see (3.1)) contains two sums: the first one is over pairs of nearest neighbors $\{i,j\} \in \mathscr{E}^{\mathrm{b}}_{\Lambda}$, the second one is over single vertices $i \in \Lambda$. It thus contains interactions among pairs, and singletons. This structure can be generalized, including interactions among spins on sets of larger (albeit finite) cardinality. ◇

We will define a Hamiltonian by defining the energy of a configuration on each subset $B \Subset \mathbb{Z}^d$, via the notion of *potential.*

Definition 6.14. If, for each finite $B \Subset \mathbb{Z}^d$, $\Phi_B \colon \Omega \to \mathbb{R}$ is $\mathscr{F}_B$-measurable, then the collection $\Phi = \{\Phi_B\}_{B \Subset \mathbb{Z}^d}$ is called a **potential**. The **Hamiltonian** in the box $\Lambda \Subset \mathbb{Z}^d$ associated with the potential Φ is defined by

$$\mathscr{H}_{\Lambda;\Phi}(\omega) \stackrel{\text{def}}{=} \sum_{\substack{B \Subset \mathbb{Z}^d: \\ B \cap \Lambda \neq \varnothing}} \Phi_B(\omega), \qquad \forall \omega \in \Omega. \tag{6.24}$$

Since the sum (6.24) can a priori contain infinitely many terms, we must guarantee that it converges. Let

$$r(\Phi) \stackrel{\text{def}}{=} \inf\bigl\{R > 0 \,:\, \Phi_B \equiv 0 \text{ for all } B \text{ with } \operatorname{diam}(B) > R\bigr\},$$

If $r(\Phi) < \infty$, Φ has **finite range** and $\mathscr{H}_{\Lambda;\Phi}$ is well defined. If $r(\Phi) = \infty$, Φ has **infinite range** and, for the Hamiltonian to be well defined, we will assume that Φ is **absolutely summable** in the sense that

$$\sum_{\substack{B \Subset \mathbb{Z}^d \\ B \ni i}} \|\Phi_B\|_\infty < \infty, \qquad \forall i \in \mathbb{Z}^d \tag{6.25}$$

(remember that $\|f\|_\infty \stackrel{\text{def}}{=} \sup_\omega |f(\omega)|$), which ensures that the interaction of a spin with the rest of the system is always bounded, and therefore that $\|\mathscr{H}_{\Lambda;\Phi}\|_\infty < \infty$.

We now present a few examples of models discussed in this book with the corresponding potentials.

- The (nearest-neighbor) **Ising model** on $\mathbb{Z}^d$ can be recovered from the potential

$$\Phi_B(\omega) = \begin{cases} -\beta\omega_i\omega_j & \text{if } B = \{i,j\}\,, i \sim j\,, \\ -h\omega_i & \text{if } B = \{i\}\,, \\ 0 & \text{otherwise.} \end{cases} \tag{6.26}$$

Observe that the corresponding specification describes a model at specific values of its parameters: in the present case, we get a different specification for each choice of the parameters β and h.

One can introduce an infinite-range version of the Ising model, by introducing a collection $\{J_{ij}\}_{i,j\in\mathbb{Z}^d}$ of real numbers and setting

$$\Phi_B(\omega) = \begin{cases} -J_{ij}\omega_i\omega_j & \text{if } B = \{i,j\}, \\ -h\omega_i & \text{if } B = \{i\}, \\ 0 & \text{otherwise.} \end{cases} \tag{6.27}$$

- The (nearest-neighbor) q-**state Potts model** corresponds to the potential

$$\Phi_B(\omega) = \begin{cases} -\beta\delta_{\omega_i,\omega_j} & \text{if } B = \{i,j\}, i \sim j, \\ 0 & \text{otherwise.} \end{cases} \tag{6.28}$$

- The (nearest-neighbor) **Blume–Capel model** is characterized by the potential

$$\Phi_B(\omega) = \begin{cases} \beta(\omega_i - \omega_j)^2 & \text{if } B = \{i,j\}, i \sim j, \\ -h\omega_i - \lambda\omega_i^2 & \text{if } B = \{i\}, \\ 0 & \text{otherwise.} \end{cases} \tag{6.29}$$

This model will be studied in Chapter 7.

Exercise 6.8. If $J_{ij} = \|j - i\|_\infty^{-\alpha}$, determine the values of $\alpha > 0$ (depending on the dimension) for which (6.27) is absolutely summable.

In the above examples, the parameters of each model have been introduced according to different sets B. Sometimes, one might want the inverse temperature to be introduced separately, so as to appear as a multiplicative constant in front of the Hamiltonian. This amounts to considering an absolutely summable potential $\Phi = \{\Phi_B\}_{B\Subset\mathbb{Z}^d}$, and then multiplying it by β: $\beta\Phi \equiv \{\beta\Phi_B\}_{B\Subset\mathbb{Z}^d}$.

We now proceed to define a specification $\pi^\Phi = \{\pi_\Lambda^\Phi\}_{\Lambda\Subset\mathbb{Z}^d}$ such that $\pi_\Lambda^\Phi(\cdot\,|\,\omega)$ gives to each configuration $\tau_\Lambda\omega_{\Lambda^c}$ a probability proportional to the Boltzmann weight prescribed by equilibrium statistical mechanics:

$$\pi_\Lambda^\Phi(\tau_\Lambda|\omega) \stackrel{\text{def}}{=} \frac{1}{\mathbf{Z}_{\Lambda;\Phi}^\omega} e^{-\mathscr{H}_{\Lambda;\Phi}(\tau_\Lambda\omega_{\Lambda^c})}, \tag{6.30}$$

where we have written explicitly the dependence on ω_{Λ^c}, and where the partition function $\mathbf{Z}_{\Lambda;\Phi}^\omega$ is given by

$$\mathbf{Z}_{\Lambda;\Phi}^\omega \stackrel{\text{def}}{=} \sum_{\tau_\Lambda\in\Omega_\Lambda} \exp(-\mathscr{H}_{\Lambda;\Phi}(\tau_\Lambda\omega_{\Lambda^c})). \tag{6.31}$$

Lemma 6.15. *$\pi^\Phi = \{\pi_\Lambda^\Phi\}_{\Lambda\Subset\mathbb{Z}^d}$ is a specification.*

Proof To lighten the notation, let us omit Φ everywhere from the notation. It will also help to change momentarily the way we denote partition functions, namely, in this proof, we will write

$$\mathbf{Z}_\Lambda(\omega_{\Lambda^c}) \equiv \mathbf{Z}^{\omega}_{\Lambda;\Phi}\,.$$

The fact that each π_Λ defines a proper kernel follows by what was said earlier, so it remains to verify consistency. We fix $\Delta \subset \Lambda \Subset \mathbb{Z}^d$, and show that $\pi_\Lambda \pi_\Delta = \pi_\Lambda$. The proof follows the same steps as the one of Lemma 6.7. Using Lemma 6.10,

$$\begin{aligned}\pi_\Lambda\pi_\Delta(A\,|\,\omega) &= \sum_{\tau_\Lambda} \pi_\Lambda(\tau_\Lambda\,|\,\omega)\pi_\Delta(A\,|\,\tau_\Lambda\,\omega_{\Lambda^c})\\ &= \sum_{\tau_\Lambda}\sum_{\eta_\Delta} \mathbf{1}_A(\eta_\Delta\,\tau_{\Lambda\setminus\Delta}\omega_{\Lambda^c})\pi_\Lambda(\tau_\Lambda\,|\,\omega)\pi_\Delta(\eta_\Delta\,|\,\tau_{\Lambda\setminus\Delta}\omega_{\Lambda^c})\,.\end{aligned}$$

We split the first sum in two, writing $\tau_\Lambda = \tau'_\Delta\tau''_{\Lambda\setminus\Delta}$. Using the definition of the kernels π_Λ and π_Δ, the above becomes

$$\sum_{\tau''_{\Lambda\setminus\Delta}}\sum_{\eta_\Delta} \mathbf{1}_A(\eta_\Delta\,\tau''_{\Lambda\setminus\Delta}\omega_{\Lambda^c})\frac{e^{-\mathscr{H}_\Delta(\eta_\Delta\tau''_{\Lambda\setminus\Delta}\omega_{\Lambda^c})}}{\mathbf{Z}_\Lambda(\omega_{\Lambda^c})\mathbf{Z}_\Delta(\tau''_{\Lambda\setminus\Delta}\omega_{\Lambda^c})}\sum_{\tau'_\Delta} e^{-\mathscr{H}_\Lambda(\tau'_\Delta\tau''_{\Lambda\setminus\Delta}\omega_{\Lambda^c})}\,.$$

But, exactly as in (6.10),

$$\mathscr{H}_\Lambda(\tau'_\Delta\tau''_{\Lambda\setminus\Delta}\omega_{\Lambda^c}) - \mathscr{H}_\Delta(\tau'_\Delta\tau''_{\Lambda\setminus\Delta}\omega_{\Lambda^c}) = \mathscr{H}_\Lambda(\eta_\Delta\tau''_{\Lambda\setminus\Delta}\omega_{\Lambda^c}) + \mathscr{H}_\Delta(\eta_\Delta\tau''_{\Lambda\setminus\Delta}\omega_{\Lambda^c})\,,$$

which gives

$$\sum_{\tau'_\Delta} e^{-\mathscr{H}_\Lambda(\tau'_\Delta\tau''_{\Lambda\setminus\Delta}\omega_{\Lambda^c})} = \mathbf{Z}_\Delta(\tau''_{\Lambda\setminus\Delta}\omega_{\Lambda^c})e^{-\mathscr{H}_\Lambda(\eta_\Delta\tau''_{\Lambda\setminus\Delta}\omega_{\Lambda^c})}e^{-\mathscr{H}_\Delta(\eta_\Delta\tau''_{\Lambda\setminus\Delta}\omega_{\Lambda^c})}\,.$$

Inserting this in the above expression, and renaming $\eta_\Delta\tau''_{\Lambda\setminus\Delta} \equiv \eta'_\Lambda$, we get

$$\pi_\Lambda\pi_\Delta(A\,|\,\omega) = \sum_{\eta'_\Lambda} \mathbf{1}_A(\eta'_\Lambda\omega_{\Lambda^c})\frac{e^{-\mathscr{H}_\Lambda(\eta'_\Lambda\omega_{\Lambda^c})}}{\mathbf{Z}_\Lambda(\omega_{\Lambda^c})} = \pi_\Lambda(A\,|\,\omega)\,. \qquad \square$$

We can now state the general definition of a Gibbs measure.

Definition 6.16. The specification π^Φ associated with a potential Φ is said to be **Gibbsian**. A probability measure μ compatible with the Gibbsian specification π^Φ is said to be an **infinite-volume Gibbs measure** (or simply a **Gibbs measure**) associated with the potential Φ.

It is customary to use the abbreviation $\mathscr{G}(\Phi) \equiv \mathscr{G}(\pi^\Phi)$. Actually, when the potential is parametrized by a few variables, we will write them rather than Φ. For example, in the case of the (nearest-neighbor) Ising model, whose specification depends on β and h, we will simply write $\mathscr{G}(\beta, h)$.

Remark 6.17. Notice that different potentials can lead to the same specification. For example, in the case of the Ising model, one could as well have considered the potential

$$\tilde{\Phi}_B(\omega) = \begin{cases} -\beta\omega_i\omega_j - \frac{h}{2d}(\omega_i + \omega_j) & \text{if } B = \{i,j\}\,, i \sim j\,, \\ 0 & \text{otherwise.} \end{cases}$$

Since they give rise to the same Hamiltonian, up to a term depending only on ω_{Λ^c}, these potentials also give rise to the same specification. They thus describe precisely the same physics. For this reason, they are said to be **physically equivalent**. ◇

When introducing a model, it is often quite convenient, instead of giving the corresponding potential $\{\Phi_B\}_{B \Subset \mathbb{Z}^d}$, to provide its **formal Hamiltonian**

$$\mathscr{H}(\omega) \stackrel{\text{def}}{=} \sum_{B \Subset \mathbb{Z}^d} \Phi_B(\omega)\,.$$

Of course, this notation is purely formal and does not specify a well-defined function on Ω. It is however possible to read from $\mathscr{H}$ the corresponding potential (up to physical equivalence).

As an example, the effective Hamiltonian of the Ising model on $\mathbb{Z}^d$ may be denoted by

$$-\beta \sum_{\{i,j\} \in \mathscr{E}_{\mathbb{Z}^d}} \sigma_i\sigma_j - h \sum_{i \in \mathbb{Z}^d} \sigma_i\,.$$

In view of what we saw in Chapter 3, the following is a natural definition of phase transition, in terms of *nonuniqueness* of the Gibbs measure:

Definition 6.18. If $\mathscr{G}(\Phi)$ contains at least two distinct Gibbs measures, $|\mathscr{G}(\Phi)| > 1$, we say that there is a **first-order phase transition** for the potential Φ.

6.4 Existence

Going back to the case of a general specification, we now turn to the problem of determining conditions that ensure the existence of at least one measure compatible with a given specification. As in many existence proofs in analysis and probability theory, this will be based on a *compactness* argument, and thus requires that we introduce a few topological notions. We will take advantage of the fact that the spins take values in a finite set to provide elementary proofs.

The approach is similar to the construction of Gibbs states in Chapter 3. We fix an arbitrary boundary condition $\omega \in \Omega$ and consider the sequence $(\mu_n)_{n\geq 1} \subset \mathscr{M}_1(\Omega)$ defined by

$$\mu_n(\cdot) \stackrel{\text{def}}{=} \pi_{B(n)}(\cdot \mid \omega), \tag{6.32}$$

where, as usual, $B(n) = \{-n, \ldots, n\}^d$. To study this sequence, we will first introduce a suitable notion of convergence for sequences of measures (Definition 6.23). This will make $\mathscr{M}_1(\Omega)$ *sequentially compact*; in particular, there always exists $\mu \in \mathscr{M}_1(\Omega)$ and a subsequence of $(\mu_n)_{n\geq 1}$, say $(\mu_{n_k})_{k\geq 1}$, such that $(\mu_{n_k})_{k\geq 1}$ converges to μ (Theorem 6.24). To guarantee that $\mu \in \mathscr{G}(\pi)$, we will impose a natural condition on π, called *quasilocality*.

6.4.1 Convergence on Ω

We first introduce a **topology** on Ω, that is, a notion of convergence for sequences of configurations.

Definition 6.19. A sequence $\omega^{(n)} \in \Omega$ **converges** to $\omega \in \Omega$ if

$$\lim_{n\to\infty} \omega_j^{(n)} = \omega_j, \qquad \forall j \in \mathbb{Z}^d.$$

We then write $\omega^{(n)} \to \omega$.

Since $\{\pm 1\}$ is a finite set, this convergence can be reformulated as follows: $\omega^{(n)} \to \omega$ if and only if, for all N, there exists n_0 such that

$$\omega^{(n)}_{B(N)} = \omega_{B(N)} \qquad \text{for all } n \geq n_0\,.$$

The notion of neighborhood in this topology should thus be understood as follows: two configurations are close to each other if they coincide on a large region containing the origin. The following exercise shows that this topology is *metrizable.*

Exercise 6.9. For $\omega, \eta \in \Omega$, let

$$d(\omega, \eta) \stackrel{\text{def}}{=} \sum_{i\in\mathbb{Z}^d} 2^{-\|i\|_\infty} \mathbf{1}_{\{\omega_i \neq \eta_i\}}\,. \tag{6.33}$$

Show that $d(\cdot,\cdot)$ is a distance on Ω, and that $\omega^{(n)} \to \omega^*$ if and only if $d(\omega^{(n)}, \omega^*) \to 0$.

Another consequence of the finiteness of the spin space is that Ω is compact in the topology just introduced:

Proposition 6.20 (Compactness of Ω). *With the above notion of convergence, Ω is* ***sequentially compact:*** *for every sequence $(\omega^{(n)})_{n\geq 1} \subset \Omega$, there exists $\omega^* \in \Omega$ and a subsequence $(\omega^{(n_k)})_{k\geq 1}$ such that $\omega^{(n_k)} \to \omega^*$ when $k \to \infty$.*

Proof We use a standard diagonalization argument. Consider $(\omega^{(n)})_{n\geq 1} \subset \Omega$ and let $i_1, i_2, \ldots$ be an arbitrary enumeration of $\mathbb{Z}^d$. Then $(\omega^{(n)}_{i_1})_{n\geq 1}$ is a sequence in

$\{\pm 1\}$, from which we can extract a subsequence $(\omega_{i_1}^{(n_{1,j})})_{j\geq 1}$ that converges (in fact, it can be taken constant). We then consider $(\omega_{i_2}^{(n_{1,j})})_{j\geq 1}$, from which we extract a converging subsequence $(\omega_{i_2}^{(n_{2,j})})_{j\geq 1}$, etc., until we have, for each k, a converging subsequence $(\omega_{i_k}^{(n_{k,j})})_{j\geq 1}$. Let $\omega^* \in \Omega$ be defined by

$$\omega_{i_k}^* \stackrel{\text{def}}{=} \lim_{j\to\infty} \omega_{i_k}^{(n_{k,j})}, \qquad \forall k \geq 1\,.$$

Now, the diagonal subsequence $(\omega^{(n_{j,j})})_{j\geq 1}$ is a subsequence of $(\omega^{(n)})_{n\geq 1}$ and satisfies $\omega^{(n_{j,j})} \to \omega^*$ as $j \to \infty$. □

We can now define a function $f\colon \Omega \to \mathbb{R}$ to be **continuous** if $\omega^{(n)} \to \omega$ implies $f(\omega^{(n)}) \to f(\omega)$. The set of continuous functions on Ω is denoted by $C(\Omega)$.

Exercise 6.10. Show that each $f \in C(\Omega)$ is measurable. *Hint:* First show that $\mathscr{C} \subset \{\text{open sets}\} \subset \mathscr{F}$, where the open sets are those associated with the topology defined above.

We say that f is **uniformly continuous** (see Appendix B.4) if

$$\forall \epsilon > 0, \text{ there exists } \delta > 0 \text{ such that } d(\omega, \eta) \leq \delta \text{ implies } |f(\omega) - f(\eta)| \leq \epsilon.$$

Exercise 6.11. Using Proposition 6.20, give a direct proof of the following facts: if f is continuous, it is also uniformly continuous, bounded, and it attains its supremum and its infimum.

Local functions are clearly continuous (since they do not depend on remote spins); they are in fact *dense* in $C(\Omega)$:[2]

Lemma 6.21. *$f \in C(\Omega)$ if and only if it is* **quasilocal**, *that is, if and only if there exists a sequence of local functions $(g_n)_{n\geq 1}$ such that $\|g_n - f\|_\infty \to 0$.*

Proof Let $f\colon \Omega \to \mathbb{R}$ be continuous. Fix some $\epsilon > 0$. Since f is also uniformly continuous, there exists some $\Lambda \Subset \mathbb{Z}^d$ such that $|f(\omega) - f(\eta)| \leq \epsilon$ for any pair η and ω coinciding on Λ. Therefore, if one chooses some arbitrary $\widetilde{\omega} \in \Omega$ and introduces the local function $g(\omega) \stackrel{\text{def}}{=} f(\omega_\Lambda \widetilde{\omega}_{\Lambda^c})$, we have that $|f(\omega) - g(\omega)| \leq \epsilon \ \forall \omega \in \Omega$. Conversely, let $(g_n)_{n\geq 1}$ be a sequence of local functions such that $\|g_n - f\|_\infty \to 0$. Fix $\epsilon > 0$ and let n be such that $\|g_n - f\|_\infty \leq \epsilon$. Since g_n is uniformly continuous, let $\delta > 0$ be such that $d(\omega, \eta) \leq \delta$ implies $|g_n(\omega) - g_n(\eta)| \leq \epsilon$. For each such pair ω, η we also have

$$|f(\omega) - f(\eta)| \leq |f(\omega) - g_n(\omega)| + |g_n(\omega) - g_n(\eta)| + |g_n(\eta) - f(\eta)| \leq 3\epsilon\,.$$

Since this can be done for all $\epsilon > 0$, we have shown that $f \in C(\Omega)$. □

We will often use the fact that probability measures on $(\Omega, \mathscr{F})$ are uniquely determined by their action on cylinders, or by the value they associate with the expectation of local or continuous functions.

Lemma 6.22. *If $\mu, \nu \in \mathscr{M}_1(\Omega)$, then the following are equivalent:*

1. $\mu = \nu$.
2. $\mu(C) = \nu(C)$ *for all cylinders* $C \in \mathscr{C}$.
3. $\mu(g) = \nu(g)$ *for all local functions* g.
4. $\mu(f) = \nu(f)$ *for all* $f \in C(\Omega)$.

Proof 1⇒2 is trivial, and 2⇒1 is a consequence of the Uniqueness Theorem for measures (Corollary B.37). 2⇔3 is immediate, since the indicator of a cylinder is a local function. 3⇒4: Let $f \in C(\Omega)$ and let $(g_n)_{n\geq 1}$ be a sequence of local functions such that $\|g_n - f\|_\infty \to 0$ (Lemma 6.21). This implies $|\mu(g_n) - \mu(f)| \leq \|g_n - f\|_\infty \to 0$. Similarly, $|\nu(g_n) - \nu(f)| \to 0$. Therefore,

$$\mu(f) = \lim_{n\to\infty} \mu(g_n) = \lim_{n\to\infty} \nu(g_n) = \nu(f).$$

Finally, 4⇒3 holds because local functions are continuous. □

6.4.2 Convergence on $\mathscr{M}_1(\Omega)$

The topology on $\mathscr{M}_1(\Omega)$ will be the following:

Definition 6.23. A sequence $(\mu_n)_{n\geq 1} \subset \mathscr{M}_1(\Omega)$ **converges to** $\mu \in \mathscr{M}_1(\Omega)$ if

$$\lim_{n\to\infty} \mu_n(C) = \mu(C), \quad \text{for all cylinders } C \in \mathscr{C}.$$

We then write $\mu_n \Rightarrow \mu$.

The fact that the convergence of a sequence of measures is tested on local events (the cylinders) should remind the reader of the convergence encountered in Chapter 3 (Definition 3.14), where a similar notion of convergence was introduced to define Gibbs states.

Before continuing, we let the reader check the following equivalent characterizations of convergence on $\mathscr{M}_1(\Omega)$.

Exercise 6.12. Show the equivalence between:

1. $\mu_n \Rightarrow \mu$.
2. $\mu_n(f) \to \mu(f)$ for all local functions f.
3. $\mu_n(f) \to \mu(f)$ for all $f \in C(\Omega)$.
4. $\rho(\mu_n, \mu) \to 0$, where we defined, for all $\mu, \nu \in \mathscr{M}_1(\Omega)$, the distance

$$\rho(\mu, \nu) \stackrel{\text{def}}{=} \sup_{k\geq 1} \frac{1}{k} \max_{C\in\mathscr{C}(\mathrm{B}(k))} |\mu(C) - \nu(C)|.$$

Theorem 6.24 (Compactness of $\mathscr{M}_1(\Omega)$). *With the above notion of convergence, $\mathscr{M}_1(\Omega)$ is sequentially compact: for every sequence $(\mu_n)_{n\geq 1} \subset \mathscr{M}_1(\Omega)$, there exist $\mu \in \mathscr{M}_1(\Omega)$ and a subsequence $(\mu_{n_k})_{k\geq 1}$ such that $\mu_{n_k} \Rightarrow \mu$ when $k \to \infty$.*

Since the proof of this result is similar, in spirit, to the one used in the proof of the compactness of Ω, we postpone it to Section 6.12.

6.4.3 Existence and Quasilocality

We will see below that the following condition on a specification π guarantees that $\mathscr{G}(\pi) \neq \varnothing$.

Definition 6.25. A specification $\pi = \{\pi_\Lambda\}_{\Lambda \Subset \mathbb{Z}^d}$ is **quasilocal** if each kernel π_Λ is continuous with respect to its boundary condition. That is, if for all $C \in \mathscr{C}$, $\omega \mapsto \pi_\Lambda(C \mid \omega)$ is continuous.

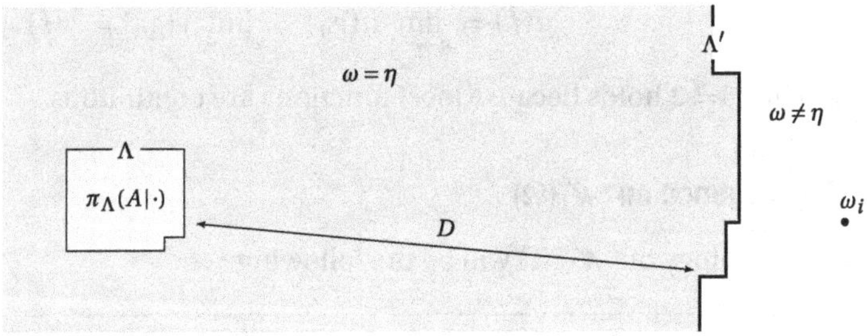

Figure 6.1. Understanding quasilocality: when D is large, $\pi_\Lambda(A \mid \omega)$ depends weakly on the values of ω_i for all i at distance larger than D from Λ (assuming all closer spins are fixed). In other words, for all $\epsilon > 0$, if ω and η coincide on a sufficiently large region $\Lambda' \supset \Lambda$, then $|\pi_\Lambda(A \mid \omega) - \pi_\Lambda(A \mid \eta)| \leq \epsilon$.

The next exercise shows that quasilocal specifications map continuous (and, in particular, local) functions to continuous functions.

Exercise 6.13. Let $\pi = \{\pi_\Lambda\}_{\Lambda \Subset \mathbb{Z}^d}$ be quasilocal and fix some $\Lambda \Subset \mathbb{Z}^d$. Show that $f \in C(\Omega)$ implies $\pi_\Lambda f \in C(\Omega)$. (This property is sometimes referred to as the **Feller property**.)

We can now state the main existence theorem.

Theorem 6.26. *If $\pi = \{\pi_\Lambda\}_{\Lambda \Subset \mathbb{Z}^d}$ is quasilocal, then $\mathscr{G}(\pi) \neq \varnothing$.*

Proof Fix an arbitrary $\omega \in \Omega$ and let $\mu_n(\cdot) \stackrel{\text{def}}{=} \pi_{\mathrm{B}(n)}(\cdot \mid \omega)$. (One could also choose a different ω for each n.) Observe that, by the consistency assumption of the kernels forming π, we have that, once n is so large that $\mathrm{B}(n) \supset \Lambda$,

$$\mu_n \pi_\Lambda = \pi_{B(n)} \pi_\Lambda(\cdot \,|\, \omega) = \pi_{B(n)}(\cdot \,|\, \omega) = \mu_n . \tag{6.34}$$

By Theorem 6.24, there exist $\mu \in \mathscr{M}_1(\Omega)$ and a subsequence $(\mu_{n_k})_{k\geq 1}$ such that $\mu_{n_k} \Rightarrow \mu$ as $k \to \infty$. We prove that $\mu \in \mathscr{G}(\pi)$. Fix $f \in C(\Omega)$, $\Lambda \Subset \mathbb{Z}^d$. Since π is quasilocal, Exercise 6.13 shows that $\pi_\Lambda f \in C(\Omega)$. Therefore,

$$\mu\pi_\Lambda(f) = \mu(\pi_\Lambda f) = \lim_{k\to\infty} \mu_{n_k}(\pi_\Lambda f) = \lim_{k\to\infty} \mu_{n_k}\pi_\Lambda(f) = \lim_{k\to\infty} \mu_{n_k}(f) = \mu(f) .$$

We used Exercise 6.6 for the first and third identities. The fourth identity follows from (6.34). By Lemma 6.22, we conclude that $\mu\pi_\Lambda = \mu$. Since this holds for all $\Lambda \Subset \mathbb{Z}^d$, this shows that $\mu \in \mathscr{G}(\pi)$. □

Since ω should be interpreted as a boundary condition, a Gibbs measure μ constructed as in the above proof,

$$\pi_{B(n_k)}(\cdot \,|\, \omega) \Rightarrow \mu ,$$

is said to be **prepared with the boundary condition ω**. A priori, μ can depend on the chosen boundary condition, and should therefore be denoted by μ^ω. A fundamental question, of course, is to determine whether uniqueness holds, or whether under certain conditions there exist distinct boundary conditions ω, ω' for which $\mu^\omega \neq \mu^{\omega'}$.

Adapting the proof of the above theorem allows us to obtain the following topological property of $\mathscr{G}(\pi)$ (completing the proof is left as an exercise):

Lemma 6.27. *Let π be a quasilocal specification. Then, $\mathscr{G}(\pi)$ is a closed subset of $\mathscr{M}_1(\Omega)$.*

A class of quasilocal specifications of central importance is provided by the Gibbsian specifications:

Lemma 6.28. *If Φ is absolutely summable, then π^Φ is quasilocal.*

Proof Fix $\Lambda \Subset \mathbb{Z}^d$. Let ω be fixed, and ω' be another configuration that coincides with ω on a region $\Delta \supset \Lambda$. Let $\tau_\Lambda \in \Omega_\Lambda$. We can write

$$\left|\pi^\Phi_\Lambda(\tau_\Lambda \,|\, \omega) - \pi^\Phi_\Lambda(\tau_\Lambda \,|\, \omega')\right| = \left|\int_0^1 \left\{\frac{\mathrm{d}}{\mathrm{d}t}\frac{e^{-h_t(\tau_\Lambda)}}{z_t}\right\}\mathrm{d}t\right| , \tag{6.35}$$

where we have set, for $0 \leq t \leq 1$, $h_t(\tau_\Lambda) \stackrel{\text{def}}{=} t\mathscr{H}_{\Lambda;\Phi}(\tau_\Lambda \omega_{\Lambda^c}) + (1-t)\mathscr{H}_{\Lambda;\Phi}(\tau_\Lambda \omega'_{\Lambda^c})$ and $z_t \stackrel{\text{def}}{=} \sum_{\tau_\Lambda} e^{-h_t(\tau_\Lambda)}$. As can be easily verified,

$$\begin{aligned}\left|\frac{\mathrm{d}}{\mathrm{d}t}\frac{e^{-h_t(\tau_\Lambda)}}{z_t}\right| &\leq 2 \max_{\eta_\Lambda \in \Omega_\Lambda} \left|\mathscr{H}_{\Lambda;\Phi}(\eta_\Lambda \omega_{\Lambda^c}) - \mathscr{H}_{\Lambda;\Phi}(\eta_\Lambda \omega'_{\Lambda^c})\right| \\ &\leq 4|\Lambda| \max_{i\in\Lambda} \sum_{\substack{B \Subset \mathbb{Z}^d, B \ni i \\ \mathrm{diam}(B) \geq D}} \|\Phi_B\|_\infty ,\end{aligned}$$

where D is the distance between Λ and Δ^{c}. Due to the absolute summability of Φ, this last series goes to 0 when $D \to \infty$. As a consequence, $\pi_\Lambda^\Phi(\tau_\Lambda \mid \cdot)$ is continuous at ω. This implies that $\pi_\Lambda^\Phi(C \mid \cdot)$ is continuous for all $C \in \mathscr{C}$. □

Lemma 6.28 provides an efficient solution to the problem of constructing quasilocal specifications. Coupled with Theorem 6.26, it provides a general approach to the construction of Gibbs measures.[3]

In Chapter 3, we also considered other types of boundary conditions, namely free and periodic. It is not difficult to show, arguing similarly as in the proof of Theorem 6.26, that these also lead to Gibbs measures:

Exercise 6.14. Use the finite-volume Gibbs distributions of the Ising model with free boundary condition, $\mu^{\varnothing}_{\Lambda;\beta,h}$, and the thermodynamic limit to construct a measure $\mu^{\varnothing}_{\beta,h}$. Show that $\mu^{\varnothing}_{\beta,h} \in \mathscr{G}(\beta,h)$.

The following exercise shows that existence is not guaranteed in the absence of quasilocality.[4]

Exercise 6.15. Let η^- denote the configuration in which all spins are -1, and let $\eta^{-,i}$ denote the configuration in which all spins are -1, except at the vertex i, at which it is $+1$. For $\Lambda \Subset \mathbb{Z}^d$, let

$$\pi_\Lambda(A \mid \omega) \overset{\text{def}}{=} \begin{cases} \frac{1}{|\Lambda|}\sum_{i\in\Lambda} \mathbf{1}_A(\eta^{-,i}) & \text{if } \omega_{\Lambda^{\mathrm{c}}} = \eta^-_{\Lambda^{\mathrm{c}}}, \\ \mathbf{1}_A(\eta^-_\Lambda \omega_{\Lambda^{\mathrm{c}}}) & \text{otherwise.} \end{cases}$$

Show that $\pi = \{\pi_\Lambda\}_{\Lambda \Subset \mathbb{Z}^d}$ is a specification and explain why it describes a system consisting of a single + spin, located anywhere on $\mathbb{Z}^d$, in a sea of − spins. Show that π is not quasilocal and that $\mathscr{G}(\pi) = \varnothing$. *Hint:* Let $N^+(\omega)$ denote the number of vertices $i \in \mathbb{Z}^d$ at which $\omega_i = +1$. Assume $\mu \in \mathscr{G}(\pi)$, and show that $\mu(\{N^+ = 0\} \cup \{N^+ = 1\} \cup \{N^+ \geq 2\}) = 0$, which gives $\mu(\Omega) = 0$.

6.5 Uniqueness

Now that we have a way of ensuring that $\mathscr{G}(\pi)$ contains at least one measure, we describe further conditions on π which ensure that this measure is actually unique. As will be seen later, the measure, when it is unique, inherits several useful properties.

Remark 6.29. We continue using Ising spins, but emphasize, however, that all statements and proofs in this section remain valid for any finite single-spin space. This matters, since, in contrast to most results in this chapter, some of the statements below are not of a qualitative nature, but involve quantitative criteria.

The point is, then, that these criteria still apply verbatim to this more general setting. ◇

6.5.1 Uniqueness vs. Sensitivity to Boundary Conditions

The following result shows that when (and only when) there is a unique Gibbs measure, the system enjoys a very strong form of lack of sensitivity to boundary condition: any sequence of finite-volume Gibbs distributions converges.

Lemma 6.30. *The following are equivalent.*

1. *Uniqueness holds:* $\mathscr{G}(\pi) = \{\mu\}$.
2. *For all* ω, *all* $\Lambda_n \uparrow \mathbb{Z}^d$ *and all local functions* f,

$$\pi_{\Lambda_n} f(\omega) \to \mu(f). \tag{6.36}$$

The convergence *for all* ω is essential here. We will see later, in Section 6.8.2, that convergence can also be guaranteed to occur in other important situations, but only for suitable sets of boundary conditions.

Proof Fix some boundary condition ω. Remember from the proof of Theorem 6.26 that, from any sequence $(\pi_{\Lambda_n}(\cdot\,|\,\omega))_{n\geq 1}$, one can extract a subsequence converging to some element of $\mathscr{G}(\pi)$. If $\mathscr{G}(\pi) = \{\mu\}$, all these subsequences must have the same limit μ. Therefore, the sequence itself converges to μ.

On the other hand, if $\mu, \nu \in \mathscr{G}(\pi)$, then one can write, for all local functions f,

$$\begin{aligned}|\mu(f) - \nu(f)| &= \big|\mu\pi_\Lambda(f) - \nu\pi_\Lambda(f)\big| \\ &= \Big|\int \{\pi_\Lambda f(\omega) - \pi_\Lambda f(\eta)\}\mu(d\omega)\nu(d\eta)\Big| \\ &\leq \int \big|\pi_\Lambda f(\omega) - \pi_\Lambda f(\eta)\big|\mu(d\omega)\nu(d\eta)\,.\end{aligned}$$

Since $|\pi_\Lambda f(\cdot)| \leq \|f\|_\infty$, we can use dominated convergence and (6.36) to conclude that $\mu(f) = \nu(f)$. Since this holds for all local functions, it follows that $\mu = \nu$. □

6.5.2 Dobrushin's Uniqueness Theorem

Our first uniqueness criterion will be formulated in terms of the one-vertex kernels $\pi_{\{i\}}(\cdot\,|\,\omega)$, which for simplicity will be denoted by π_i. Each $\pi_i(\cdot\,|\,\omega)$ should be considered as a distribution for the spin at vertex i, with boundary condition ω. We will measure the dependence of $\pi_i(\cdot\,|\,\omega)$ on the value of the boundary condition ω at other vertices. We will measure the proximity between two such distributions using the **total variation distance** (see Section B.10):

$$\|\pi_i(\cdot\,|\,\omega) - \pi_i(\cdot\,|\,\omega')\|_{TV} \stackrel{\text{def}}{=} \sum_{\omega_i=\pm 1} \big|\pi_i(\omega_i\,|\,\omega) - \pi_i(\omega_i\,|\,\omega')\big|\,.$$

We can then introduce

$$c_{ij}(\pi) \stackrel{\text{def}}{=} \sup_{\substack{\omega,\omega' \in \Omega: \\ \omega_k = \omega'_k \,\forall k \neq j}} \|\pi_i(\cdot \,|\, \omega) - \pi_i(\cdot \,|\, \omega')\|_{TV}$$

and

$$c(\pi) \stackrel{\text{def}}{=} \sup_{i \in \mathbb{Z}^d} \sum_{j \in \mathbb{Z}^d} c_{ij}(\pi) .$$

Theorem 6.31. *Let π be a quasilocal specification satisfying* ***Dobrushin's condition of weak dependence:***

$$c(\pi) < 1 . \tag{6.37}$$

Then the probability measure specified by π is unique: $|\mathscr{G}(\pi)| = 1$.

Before starting the proof, we need to introduce a few notions. Define the **oscillation** of $f: \Omega \to \mathbb{R}$ at $i \in \mathbb{Z}^d$ by

$$\delta_i(f) \stackrel{\text{def}}{=} \sup_{\substack{\omega,\eta \in \Omega \\ \omega_k = \eta_k \,\forall k \neq i}} |f(\omega) - f(\eta)| . \tag{6.38}$$

The oscillation enables us to quantify the variation of $f(\omega)$ when one changes ω into another configuration by successive spin flips. Namely, if $\omega_{\Lambda^c} = \eta_{\Lambda^c}$, then

$$|f(\omega) - f(\eta)| \leq \sum_{i \in \Lambda} \delta_i(f) . \tag{6.39}$$

It is thus natural to define the **total oscillation** of f by

$$\Delta(f) \stackrel{\text{def}}{=} \sum_{i \in \mathbb{Z}^d} \delta_i(f) . \tag{6.40}$$

We denote the space of functions with finite total oscillation by $\mathscr{O}(\Omega)$. All local functions have finite total oscillation; by Lemma 6.21, this implies that $\mathscr{O}(\Omega)$ is dense in $C(\Omega)$. Nevertheless,

Exercise 6.16. Show that $\mathscr{O}(\Omega) \not\subset C(\Omega)$ and $\mathscr{O}(\Omega) \not\supset C(\Omega)$.

Intuitively, $\Delta(f)$ measures how far f is from being a constant. This is made clear in the following lemma. Letting $C_{\mathscr{O}}(\Omega) \stackrel{\text{def}}{=} C(\Omega) \cap \mathscr{O}(\Omega)$, we have:

Lemma 6.32. *Let $f \in C_{\mathscr{O}}(\Omega)$. Then $\Delta(f) \geq \sup f - \inf f$.*

Proof Let $f \in C_{\mathscr{O}}(\Omega)$. By Exercise 6.11, f attains its supremum and its infimum. In particular, there exist, for all $\epsilon > 0$, two configurations ω^1, ω^2 such that $\omega^1_{\Lambda^c} = \omega^2_{\Lambda^c}$

for some sufficiently large box Λ, and such that $\sup f \leq f(\omega^1)+\epsilon$, $\inf f \geq f(\omega^2)-\epsilon$. Then, using (6.39),

$$\sup f - \inf f \leq f(\omega^1) - f(\omega^2) + 2\epsilon \leq \sum_{i\in\Lambda} \delta_i(f) + 2\epsilon \leq \Delta(f) + 2\epsilon\,. \qquad \square$$

Using Lemma 6.32, we can always write

$$|\mu(f) - \nu(f)| \leq \Delta(f)\,, \qquad \forall f \in C_{\mathscr{O}}(\Omega)\,. \tag{6.41}$$

Proposition 6.33. *Assume* (6.37). *Let* $\mu, \nu \in \mathscr{G}(\pi)$ *be such that*

$$|\mu(f) - \nu(f)| \leq \alpha\Delta(f)\,, \qquad \forall f \in C_{\mathscr{O}}(\Omega)\,, \tag{6.42}$$

for some constant $\alpha \leq 1$. *Then,*

$$|\mu(f) - \nu(f)| \leq c(\pi)\alpha\Delta(f)\,, \qquad \forall f \in C_{\mathscr{O}}(\Omega)\,. \tag{6.43}$$

Assuming, for the moment, the validity of this proposition, we can easily conclude the proof of Theorem 6.31.

Proof of Theorem 6.31. Let $\mu, \nu \in \mathscr{G}(\pi)$ and let f be a local function. (6.41) shows that (6.42) holds with $\alpha = 1$. Since $c(\pi) < 1$, we can apply Proposition 6.33 repeatedly:

$$\begin{aligned} |\mu(f) - \nu(f)| \leq \Delta(f) &\implies |\mu(f) - \nu(f)| \leq c(\pi)\Delta(f) \\ &\implies |\mu(f) - \nu(f)| \leq c(\pi)^2\Delta(f) \\ &\implies |\mu(f) - \nu(f)| \leq c(\pi)^n\Delta(f)\,, \quad \forall n \geq 0\,. \end{aligned}$$

Since $\Delta(f) < \infty$ and $c(\pi) < 1$, taking $n \to \infty$ leads to $\mu(f) = \nu(f)$. By Lemma 6.22, $\mu = \nu$. $\square$

The proof of Proposition 6.33 relies on a technical estimate:

Lemma 6.34. *Let* $f \in C_{\mathscr{O}}(\Omega)$. *Then,* $\delta_j(\pi_j f) = 0$ *for all* j, *and, for any* $i \neq j$,

$$\delta_i(\pi_j f) \leq \delta_i(f) + c_{ji}(\pi)\delta_j(f)\,. \tag{6.44}$$

The content of this lemma can be given an intuitive meaning, as follows. If f is constant, then $\delta_i(f) = 0$ for all $i \in \mathbb{Z}^d$. If f is nonconstant, each oscillation $\delta_i(f)$ can be seen as a quantity of dust present at i, measuring how far f is from being constant: the less dust, the closer f is to a constant function. With this interpretation, the map $f \mapsto \pi_j f$ can be interpreted as a dusting off at vertex j. Namely, before the dusting at j, the oscillation at any given point i is $\delta_i(f)$. After the dusting at j, Lemma 6.34 says that the amount of dust at j becomes zero ($\delta_j(\pi_j f) = 0$) and that the total amount of dust at every other point $i \neq j$ is incremented, at most, by a fraction $c_{ji}(\pi)$ of the dust present at j before the dusting. For this reason, Lemma 6.34 is often called the ***dusting lemma****.* ◇

Proof of Lemma 6.34. If $i = j$, then $\delta_j(\pi_j f) = 0$ (remember that the function $\pi_j f$ is $\mathscr{F}_{\{j\}^c}$-measurable). Let us thus assume that $i \neq j$. Let ω, ω' be two configurations that agree everywhere outside i. We write

$$\pi_j f(\omega) - \pi_j f(\omega') = \sum_{\eta_j=\pm 1} \{\pi_j(\eta_j \,|\, \omega) f(\eta_j \omega_{\{j\}^c}) - \pi_j(\eta_j \,|\, \omega') f(\eta_j \omega'_{\{j\}^c})\}$$
$$= \sum_{\eta_j=\pm 1} \{\pi_j(\eta_j \,|\, \omega) \tilde{f}(\eta_j \omega_{\{j\}^c}) - \pi_j(\eta_j \,|\, \omega') \tilde{f}(\eta \omega'_{\{j\}^c})\},$$

where $\tilde{f}(\cdot) \stackrel{\text{def}}{=} f(\cdot) - m$, for some constant m to be chosen later. We add and subtract $\pi_j(\eta_j \,|\, \omega) \tilde{f}(\eta_j \omega'_{\{j\}^c})$ from each term of the last sum and use

$$|\tilde{f}(\eta_j \omega_{\{j\}^c}) - \tilde{f}(\eta_j \omega'_{\{j\}^c})| = |f(\eta_j \omega_{\{j\}^c}) - f(\eta_j \omega'_{\{j\}^c})| \leq \delta_i(f),$$
$$\sum_{\eta_j=\pm 1} |\pi_j(\eta_j \,|\, \omega) - \pi_j(\eta_j \,|\, \omega')| = \|\pi_j(\cdot \,|\, \omega) - \pi_j(\cdot \,|\, \omega')\|_{TV} \leq c_{ji}(\pi).$$

Since $\sum_{\eta_j} \pi_j(\eta_j \,|\, \omega) = 1$,

$$\delta_i(\pi_j f) \leq \delta_i(f) + c_{ji}(\pi) \max_{\eta_j} |\tilde{f}(\eta_j \omega'_{\{j\}^c})|.$$

Choosing $m = f((+1)_j \omega'_{\{j\}^c})$, we have $\max_{\eta_j} |\tilde{f}(\eta_j \omega'_{\{j\}^c})| \leq \delta_j(f)$ and (6.44) follows. □

Proof of Proposition 6.33. Fix an arbitrary total order on $\mathbb{Z}^d$, denoted $\succ$, in which the smallest element is the origin. We first prove that, when (6.42) holds, one has, for all $i \in \mathbb{Z}^d$,

$$|\mu(f) - \nu(f)| \leq c(\pi)\alpha \sum_{k \prec i} \delta_k(f) + \alpha \sum_{k \succeq i} \delta_k(f), \qquad \forall f \in C_{\mathscr{O}}(\Omega). \tag{6.45}$$

When $i = 0$, the first sum is empty and the claim reduces to our assumption (6.42). Let us thus assume that (6.45) has been proved for i.

Observe that, for all k, $\pi_k f \in C_{\mathscr{O}}(\Omega)$. Indeed, on the one hand, $\pi_k f$ is continuous since π is quasilocal. On the other hand, by (6.44),

$$\Delta(\pi_k f) = \sum_j \delta_j(\pi_k f) \leq \sum_j \delta_j(f) + c(\pi)\delta_k(f) < \infty.$$

Using (6.45) with f replaced by $\pi_i f$, and since $\delta_i(\pi_i f) = 0$,

$$|\mu(f) - \nu(f)| = |\mu(\pi_i f) - \nu(\pi_i f)| \leq c(\pi)\alpha \sum_{k \prec i} \delta_k(\pi_i f) + \alpha \sum_{k \succ i} \delta_k(\pi_i f).$$

Using again (6.44),

$$|\mu(f) - \nu(f)| \leq c(\pi)\alpha \sum_{k \prec i} \delta_k(f) + \alpha \sum_{k \succeq i} \delta_k(f)$$
$$+ \alpha \delta_i(f) c(\pi) \sum_{k \prec i} c_{ik}(\pi) + \alpha \delta_i(f) \sum_{k \succ i} c_{ik}(\pi).$$

Now, observe that, since $c(\pi) < 1$,

$$c(\pi)\sum_{k\prec i} c_{ik}(\pi) + \sum_{k\succ i} c_{ik}(\pi) \le \sum_{k\in\mathbb{Z}^d} c_{ik}(\pi) = c(\pi),$$

which yields

$$\begin{aligned}|\mu(f) - \nu(f)| &\le c(\pi)\alpha \sum_{k\prec i} \delta_k(f) + \alpha \sum_{k\succ i} a_k\delta_k(f) + c(\pi)\alpha\delta_i(f)\\ &= c(\pi)\alpha \sum_{k\preceq i} \delta_k(f) + \alpha \sum_{k\succ i} \delta_k(f).\end{aligned}$$

This shows that (6.45) holds for all $i \in \mathbb{Z}^d$. Since $\sum_k \delta_k(f) = \Delta(f) < \infty$, (6.43) now follows by letting i increase to infinity (with respect to $\succ$) in (6.42). □

6.5.3 Application to Gibbsian Specifications

Theorem 6.31 is very general. We will now apply it to several Gibbsian specifications. We will start with regimes in which the Gibbs measure is unique despite possibly strong interactions between the spins.

Exercise 6.17. Consider the Ising model ($d \ge 1$) with a magnetic field $h > 0$ and arbitrary inverse temperature β. Use Theorem 6.35 to show that $|\mathscr{G}(\beta, h)| = 1$ for all large enough h. (Contrast this result with the corresponding one obtained in Theorem 3.25, where it was shown that uniqueness holds for all $h \ne 0$.)

Exercise 6.18. Consider the Blume–Capel model in $d \ge 1$ (see (6.29)).

1. Consider first $(\lambda, h) = (0, 0)$, and give a range of values of β for which uniqueness holds.
2. Then, fix $(\lambda, h) = t\mathbf{e}$, with $t > 0$. Show that if $\mathbf{e} \in \mathbb{S}^1$ points in any direction different from $(1, 0)$, $(-1, 1)$ or $(-1, -1)$, then, for all $\beta > 0$, the Gibbs measure is unique as soon as t is sufficiently large. (See Figure 6.2.)

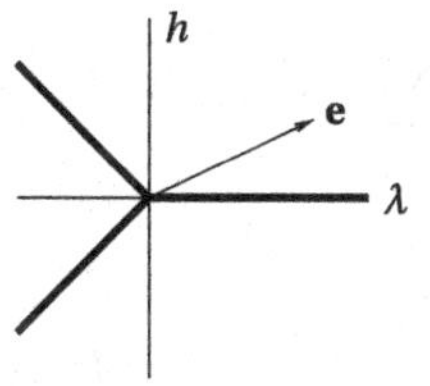

Figure 6.2. The Blume–Capel model with parameters $(\lambda, h) = t\mathbf{e}$ has a unique Gibbs measure when $t > 0$ is large enough, and when $\mathbf{e}$ points to any direction distinct from those indicated by the bold line.

Exercise 6.19. The (nearest-neighbor) **Potts antiferromagnet** on $\mathbb{Z}^d$ at inverse temperature $\beta \geq 0$ has single-spin space $\Omega_0 \stackrel{\text{def}}{=} \{0,\dots,q-1\}$ and is associated with the potential

$$\Phi_B(\omega) = \begin{cases} +\beta\delta_{\omega_i,\omega_j} & \text{if } B = \{i,j\}\,, i \sim j, \\ 0 & \text{otherwise.} \end{cases}$$

Show that this model has a unique Gibbs measure for all $\beta \in \mathbb{R}_{\geq 0}$, provided that $q > 6d$.

Let us now formulate the criterion of Theorem 6.31 in a form better suited to the treatment of weak interactions. Let $\delta(f) \stackrel{\text{def}}{=} \sup_{\eta',\eta''} |f(\eta') - f(\eta'')|$.

Theorem 6.35. *Assume that $\Phi = \{\Phi_B\}_{B \Subset \mathbb{Z}^d}$ is absolutely summable and satisfies*

$$\sup_{i\in\mathbb{Z}^d} \sum_{j\neq i} \sum_{B\supset\{i,j\}} \delta(\Phi_B) < 1\,. \tag{6.46}$$

Then Dobrushin's condition of weak dependence is satisfied, and therefore there is a unique Gibbs measure specified by π^Φ.

Proof Fix some $i \in \mathbb{Z}^d$, and let ω and ω' coincide everywhere except at some vertex $j \neq i$. Starting as in the proof of Lemma 6.28,

$$\|\pi_i^\Phi(\cdot\,|\,\omega) - \pi_i^\Phi(\cdot\,|\,\omega')\|_{TV} \leq \int_0^1 \Bigl\{\sum_{\eta_i}\Bigl|\frac{\mathrm{d}\nu_t(\eta_i)}{\mathrm{d}t}\Bigr|\Bigr\}\mathrm{d}t,$$

where, for $0 \leq t \leq 1$, $\nu_t(\eta_i) \stackrel{\text{def}}{=} \frac{e^{-h_t(\eta_i)}}{z_t}$, with

$$h_t(\eta_i) \stackrel{\text{def}}{=} t\mathscr{H}_{\{i\};\Phi}(\eta_i\omega_{\{i\}^c}) + (1-t)\mathscr{H}_{\{i\};\Phi}(\eta_i\omega'_{\{i\}^c}),$$

and $z_t \stackrel{\text{def}}{=} \sum_{\eta_i} e^{-h_t(\eta_i)}$. A straightforward computation shows that

$$\frac{\mathrm{d}\nu_t(\eta_i)}{\mathrm{d}t} = \bigl\{\Delta\mathscr{H}_i - E_{\nu_t}[\Delta\mathscr{H}_i]\bigr\}\nu_t(\eta_i),$$

where $\Delta\mathscr{H}_i(\eta_i) \stackrel{\text{def}}{=} \mathscr{H}_{\{i\};\Phi}(\eta_i\omega'_{\{i\}^c}) - \mathscr{H}_{\{i\};\Phi}(\eta_i\omega_{\{i\}^c})$. We therefore have

$$\begin{aligned}\sum_{\eta_i}\Bigl|\frac{\mathrm{d}\nu_t(\eta_i)}{\mathrm{d}t}\Bigr| &= E_{\nu_t}\Bigl[\bigl|\Delta\mathscr{H}_i - E_{\nu_t}[\Delta\mathscr{H}_i]\bigr|\Bigr] \\ &\leq E_{\nu_t}\Bigl[\bigl(\Delta\mathscr{H}_i - E_{\nu_t}[\Delta\mathscr{H}_i]\bigr)^2\Bigr]^{1/2} \\ &\leq E_{\nu_t}\bigl[(\Delta\mathscr{H}_i - m)^2\bigr]^{1/2},\end{aligned}$$

where the first inequality follows from the Cauchy-Schwartz inequality, and we introduced an arbitrary number $m \in \mathbb{R}$ (remember that $m \mapsto E[(X-m)^2]$ is minimal when $m = E[X]$). Choosing $m = (\max\Delta\mathscr{H}_i + \min\Delta\mathscr{H}_i)/2$, we have

$$|\Delta\mathscr{H}_i - m| \leq \tfrac{1}{2}\max_{\eta_i,\eta'_i}\bigl|\Delta\mathscr{H}_i(\eta_i) - \Delta\mathscr{H}_i(\eta'_i)\bigr|\,.$$

Each $\Delta\mathscr{H}_i(\cdot)$ contains a sum over sets $B \ni i$, and notice that those sets B which do not contain j do not contribute. We can thus restrict to those sets $B \supset \{i,j\}$ and get

$$\begin{aligned}
&|\Delta\mathscr{H}_i(\eta_i) - \Delta\mathscr{H}_i(\eta_i')| \\
&\quad \leq \sum_{B\supset\{i,j\}} \left\{ |\Phi_B(\eta_i\omega'_{\{i\}^c}) - \Phi_B(\eta_i'\omega'_{\{i\}^c})| + |\Phi_B(\eta_i\omega_{\{i\}^c}) - \Phi_B(\eta_i'\omega_{\{i\}^c})| \right\} \\
&\quad \leq 2\sum_{B\supset\{i,j\}} \delta(\Phi_B)\,.
\end{aligned}$$

This proves (6.46). □

Let us give a simple example of application of the above criterion.

Example 6.36. Consider first the nearest-neighbor Ising model with $h = 0$ on $\mathbb{Z}^d$, whose potential was given in (6.26). The only sets B that contribute to the sum (6.46) are the nearest neighbors $B = \{i,j\}$, $i \sim j$, for which $\delta(\Phi_B) = 2\beta$. Therefore (6.46) reads, since each i has $2d$ neighbors,

$$2\beta \cdot 2d < 1\,.$$

In $d = 1$, this means that uniqueness holds when $\beta < \frac{1}{4}$, although we know from the results of Chapter 3 that uniqueness holds at *all* temperatures. In $d = 2$, the above guarantees uniqueness when $\beta < \frac{1}{8} = 0.125$, which should be compared with the exact range, known to be $\beta \leq \beta_c(2) = 0.4406\ldots$ ◇

We will actually see in Corollary 6.41 that, for finite-range models, uniqueness holds at all temperatures when $d = 1$.

More generally, the above criterion allows one to prove uniqueness at sufficiently high temperature for a wide class of models. A slight rewriting of the condition makes the application more immediate. If one changes the order of summation in the double sum in (6.46), the latter becomes

$$\sum_{j\neq i}\sum_{B\supset\{i,j\}} \delta(\Phi_B) = \sum_{B\ni i}(|B|-1)\delta(\Phi_B)\,. \tag{6.47}$$

We can thus state a general, easily applicable high-temperature uniqueness result. Remember that the inverse temperature β can always be associated with a potential $\Phi \stackrel{\text{def}}{=} \{\Phi_B\}_{B\Subset\mathbb{Z}^d}$, by multiplication: $\beta\Phi \stackrel{\text{def}}{=} \{\beta\Phi_B\}_{B\Subset\mathbb{Z}^d}$.

Corollary 6.37. *Let $\Phi = \{\Phi_B\}_{B\Subset\mathbb{Z}^d}$ be an absolutely summable potential satisfying*

$$b \stackrel{\text{def}}{=} \sup_{i\in\mathbb{Z}^d}\sum_{B\ni i}(|B|-1)\|\Phi_B\|_\infty < \infty\,, \tag{6.48}$$

and let $\beta_0 \stackrel{\text{def}}{=} \frac{1}{2b}$. Then, for all $\beta < \beta_0$, there is a unique measure compatible with $\pi^{\beta\Phi}$.

Proof It suffices to use (6.47) in Theorem 6.35, with $\delta(\Phi_B) \leq 2\|\Phi_B\|_\infty$. □

Exercise 6.20. Consider the long-range Ising model introduced in (6.27), with $h = 0$ and $J_{ij} = \|j-i\|_\infty^{-\alpha}$. Find a range of values of $\alpha > 0$ (depending on the dimension) for which (6.48) holds, and deduce a range of values of $0 < \beta < \infty$ for which uniqueness holds.

In dimension 1, the previous exercise guarantees that uniqueness holds at sufficiently high temperature whenever $\alpha > 1$. We will prove later that, when $\alpha > 2$, uniqueness actually holds for all positive temperatures; see Example 6.42.

6.5.4 Uniqueness at High Temperature via Cluster Expansion

In this section, we consider an alternative approach, relying on the cluster expansion, to establish uniqueness of the Gibbs measure at sufficiently high temperature. We have seen in Lemma 6.30 that $\mathscr{G}(\beta\Phi) = \{\mu\}$ if and only if

$$\pi_{\Lambda_n}^{\beta\Phi} f(\omega) \to \mu(f), \quad \forall \omega \in \Omega, \tag{6.49}$$

for every local function f. Here we provide a direct way of proving such a convergence.

Theorem 6.38. *Assume that* $\Phi = \{\Phi_B\}_{B \Subset \mathbb{Z}^d}$ *satisfies*

$$\sup_{i\in\mathbb{Z}^d} \sum_{B\ni i} \|\Phi_B\|_\infty e^{4|B|} < \infty . \tag{6.50}$$

Then, there exists $0 < \beta_1 < \infty$ *such that, for all* $\beta \leq \beta_1$, (6.49) *holds. As a consequence:* $\mathscr{G}(\beta\Phi) = \{\mu\}$. *Moreover, when* Φ *has finite range, the convergence in* (6.49) *is exponential: for all sufficiently large* Λ,

$$\left|\pi_\Lambda^{\beta\Phi} f(\omega) - \mu(f)\right| \leq D\|f\|_\infty e^{-Cd(\mathrm{supp}(f),\Lambda^c)}, \qquad \forall \omega \in \Omega, \tag{6.51}$$

where $C > 0$ *and* D *depend on* Φ.

Since $\mu(f) = \int \pi_\Lambda^{\beta\Phi}(f|\omega)\mu(d\omega)$, for any $\mu \in \mathscr{G}(\beta\Phi)$, Theorem 6.38 is a consequence of the following proposition, whose proof relies on the cluster expansion and provides an explicit expression for $\mu(f)$. In order not to delve here into the technicalities of the cluster expansion, we postpone this proof to the end of Section 6.12.

Proposition 6.39. *If* (6.50) *holds, then there exists* $0 < \beta_1 < \infty$ *such that, for all* $\beta \leq \beta_1$, *the following holds. Fix some* ω. *For every local function* f, *there exists* $c(f)$ *(independent of* ω*) such that*

$$\lim_{\Lambda\uparrow\mathbb{Z}^d} \pi_\Lambda^{\beta\Phi} f(\omega) = c(f) . \tag{6.52}$$

Moreover, if Φ *has finite range,*

$$\left|\pi_\Lambda^{\beta\Phi} f(\omega) - c(f)\right| \leq D\|f\|_\infty e^{-Cd(\mathrm{supp}(f),\Lambda^c)}, \tag{6.53}$$

where $C > 0$ *and* D *depend on* Φ.

6.5.5 Uniqueness in One Dimension

In one dimension, the criterion (6.46) implies uniqueness for any model with absolutely convergent potential, but only at sufficiently high temperatures (small values of β).

We know that the nearest-neighbor Ising model on $\mathbb{Z}$ has a unique Gibbs measure at *all* temperatures, so the proof given above, relying on Dobrushin's condition of weak dependence, ignores some important features of one-dimensional systems. We now establish another criterion, of less general applicability, but providing considerably stronger results when $d = 1$.

Theorem 6.40. *Let Φ be an absolutely summable potential such that*

$$D \stackrel{\text{def}}{=} \sup_n \sup_{\substack{\omega_{B(n)} \\ \eta_{B(n)^c}, \eta'_{B(n)^c}}} \left| \mathscr{H}_{B(n);\Phi}(\omega_{B(n)}\eta_{B(n)^c}) - \mathscr{H}_{B(n);\Phi}(\omega_{B(n)}\eta'_{B(n)^c}) \right| < \infty. \tag{6.54}$$

Then there is a unique Gibbs measure compatible with π^{Φ}.

Since

$$\left| \mathscr{H}_{B(n);\Phi}(\omega_{B(n)}\eta_{B(n)^c}) - \mathscr{H}_{B(n);\Phi}(\omega_{B(n)}\eta'_{B(n)^c}) \right| \leq 2 \sum_{\substack{A \cap B(n) \neq \varnothing \\ A \cap B(n)^c \neq \varnothing}} \|\Phi_A\|_\infty, \tag{6.55}$$

condition (6.54) will be satisfied for one-dimensional systems in which the interaction between the inside and the outside of any interval $B(n) = \{-n, \dots, n\}$ is uniformly bounded in n, meaning that *boundary effects are negligible.*

The sum on the right-hand side of (6.55) is of course finite when Φ has finite range, which allows a general uniqueness result for one-dimensional systems to be stated:

Corollary 6.41. *($d = 1$) If Φ is any finite-range potential, then $|\mathscr{G}(\Phi)| = 1$.*

However, the sum on the right-hand side of (6.55) can contain infinitely many terms, as long as these decay sufficiently fast, as the next example shows.

Example 6.42. Consider the one-dimensional long-range Ising model (6.27), with

$$J_{ij} = |j - i|^{-(2+\epsilon)},$$

with $\epsilon > 0$. Using (6.55),

$$\begin{aligned}
\left| \mathscr{H}_{B(n);\beta\Phi}(\omega_{B(n)}\eta_{B(n)^c}) - \mathscr{H}_{B(n);\beta\Phi}(\omega_{B(n)}\eta'_{B(n)^c}) \right| &\leq 2 \sum_{i \in B(n)} \sum_{j \in B(n)^c} \frac{\beta}{|j - i|^{2+\epsilon}} \\
&\leq 2\beta \sum_{k \geq 1} \sum_{\substack{i \in B(n): \\ d(i, B(n)^c) = k}} \sum_{r \geq k} \frac{1}{r^{2+\epsilon}} \\
&\leq 2\beta c_\epsilon \sum_{k \geq 1} \frac{1}{k^{1+\epsilon}} < \infty.
\end{aligned}$$

Theorem 6.40 implies uniqueness for all finite values of $\beta \geq 0$ whenever $\epsilon > 0$. Remarkably, this is a sharp result, as it can be shown that uniqueness fails at large values of β whenever $\epsilon \leq 0$.[5] ◇

Since we do not yet have all the necessary tools, we postpone the proof of Theorem 6.40 to the end of Section 6.8.4 (p. 319). It will rely on the following ingredient:

Lemma 6.43. *Let D be defined as in* (6.54). *Then, for all* $\omega, \eta \in \Omega$ *and all cylinders* $C \in \mathscr{C}$, *for all large enough n,*

$$e^{-2D}\pi^{\Phi}_{\mathrm{B}(n)}(C\,|\,\eta) \leq \pi^{\Phi}_{\mathrm{B}(n)}(C\,|\,\omega) \leq e^{2D}\pi^{\Phi}_{\mathrm{B}(n)}(C\,|\,\eta)\,. \tag{6.56}$$

Proof Using (6.54) in the Boltzmann weight, we obtain

$$e^{-D}e^{-\mathscr{H}_{\mathrm{B}(n);\Phi}(\tau_{\mathrm{B}(n)}\eta_{\mathrm{B}(n)^c})} \leq e^{-\mathscr{H}_{\mathrm{B}(n);\Phi}(\tau_{\mathrm{B}(n)}\omega_{\mathrm{B}(n)^c})} \leq e^{D}e^{-\mathscr{H}_{\mathrm{B}(n);\Phi}(\tau_{\mathrm{B}(n)}\eta_{\mathrm{B}(n)^c})}\,.$$

This yields $\mathbf{Z}^{\omega}_{\mathrm{B}(n);\Phi} \leq \mathbf{Z}^{\eta}_{\mathrm{B}(n);\Phi}e^{D}$. Thus, $\pi^{\Phi}_{\mathrm{B}(n)}(\tau_{\mathrm{B}(n)}\,|\,\omega) \geq e^{-2D}\pi^{\Phi}_{\mathrm{B}(n)}(\tau_{\mathrm{B}(n)}\,|\,\eta)$. Let $C \in \mathscr{C}$. If n is large enough for $\mathrm{B}(n)$ to contain the base of C, then $\mathbf{1}_C(\tau_{\mathrm{B}(n)}\omega_{\mathrm{B}(n)^c}) = \mathbf{1}_C(\tau_{\mathrm{B}(n)}\eta_{\mathrm{B}(n)^c})$. Therefore,

$$\begin{aligned}\pi^{\Phi}_{\mathrm{B}(n)}(C\,|\,\omega) &= \sum_{\tau_{\mathrm{B}(n)}}\pi^{\Phi}_{\mathrm{B}(n)}(\tau_{\mathrm{B}(n)}\,|\,\omega)\mathbf{1}_C(\tau_{\mathrm{B}(n)}\omega_{\mathrm{B}(n)^c})\\ &\geq e^{-2D}\sum_{\tau_{\mathrm{B}(n)}}\pi^{\Phi}_{\mathrm{B}(n)}(\tau_{\mathrm{B}(n)}\,|\,\eta)\mathbf{1}_C(\tau_{\mathrm{B}(n)}\eta_{\mathrm{B}(n)^c}) = e^{-2D}\pi^{\Phi}_{\mathrm{B}(n)}(C\,|\,\eta)\,.\end{aligned}$$ □

6.6 Symmetries

In this section, we study how the presence of symmetries in a specification π can extend to the measures in $\mathscr{G}(\pi)$. We do not assume that the single-spin space Ω_0 is necessarily $\{\pm 1\}$.

We are interested in the action of a group $(\mathsf{G},\cdot)$ on the set of configurations Ω. That is, we consider a family $(\tau_{\mathsf{g}})_{\mathsf{g}\in\mathsf{G}}$ of maps $\tau_{\mathsf{g}}\colon \Omega \to \Omega$ such that

1. $(\tau_{\mathsf{g}_1} \circ \tau_{\mathsf{g}_2})\omega = \tau_{\mathsf{g}_1\cdot\mathsf{g}_2}\omega$ for all $\mathsf{g}_1, \mathsf{g}_2 \in \mathsf{G}$, and
2. $\tau_{\mathsf{e}}\omega = \omega$ for all $\omega \in \Omega$, where e is the neutral element of G.

Note that $\tau_{\mathsf{g}}^{-1} = \tau_{\mathsf{g}^{-1}}$ for all $\mathsf{g} \in \mathsf{G}$. The action of the group can be extended to functions and measures. For all $\mathsf{g} \in \mathsf{G}$, all functions $f\colon \Omega \to \mathbb{R}$ and all $\mu \in \mathscr{M}_1(\Omega)$, we define

$$\tau_{\mathsf{g}}f(\omega) \stackrel{\text{def}}{=} f(\tau_{\mathsf{g}}^{-1}\omega)\,, \qquad \tau_{\mathsf{g}}\mu(A) \stackrel{\text{def}}{=} \mu(\tau_{\mathsf{g}}^{-1}A)\,,$$

for all $\omega \in \Omega$ and all $A \in \mathscr{F}$. Of course, we then have $\tau_{\mathsf{g}}\mu(f) = \mu(\tau_{\mathsf{g}}^{-1}f)$, for all integrable functions f.

We will use τ_{g} to act on a specification $\pi = \{\pi_\Lambda\}_{\Lambda \Subset \mathbb{Z}^d}$ and turn it into a new specification $\tau_{\mathsf{g}}\pi = \{\tau_{\mathsf{g}}\pi_\Lambda\}_{\Lambda\Subset\mathbb{Z}^d}$. We will mainly consider two types of transformations, *internal* and *spatial*.

1. An **internal transformation** starts with a group G acting on the single-spin space Ω_0. The action of G is then extended to Ω by setting, for all $g \in G$ and all $\omega \in \Omega$,

$$(\tau_g\omega)_i \stackrel{\text{def}}{=} \tau_g\omega_i, \quad \forall i \in \mathbb{Z}^d .$$

(We use the same notation for the action on both Ω_0 and Ω as this will never lead to ambiguity.) The action of τ_g on a kernel π_Λ is defined by

$$(\tau_g\pi)_\Lambda(A \mid \omega) \stackrel{\text{def}}{=} \pi_\Lambda(\tau_g^{-1}A \mid \tau_g^{-1}\omega) . \tag{6.57}$$

2. In the case of a **spatial transformation**, we start with a group G acting on $\mathbb{Z}^d$ and we extend its action to Ω by setting, for all $g \in G$ and all $\omega \in \Omega$,

$$(\tau_g\omega)_i \stackrel{\text{def}}{=} \omega_{\tau_g^{-1}i}, \quad \forall i \in \mathbb{Z}^d .$$

Basic examples of spatial transformations are: translations, rotations and reflections. The action of τ_g on π_Λ is defined by

$$(\tau_g\pi)_\Lambda(A \mid \omega) \stackrel{\text{def}}{=} \pi_{\tau_g^{-1}\Lambda}(\tau_g^{-1}A \mid \tau_g^{-1}\omega) . \tag{6.58}$$

A **general transformation** is then a composition of these two types of transformations. To simplify the exposition, for the rest of this section, we focus on internal transformations. However, everything can be extended in a straightforward way to the other cases. Invariance under translations will play an important role in the rest of the book. For that reason, we will describe this type of spatial transformation in more detail in Section 6.7, together with translation-invariant specifications and Gibbs measures.

So, until the end of this section, we assume that the actions τ_g are associated with an internal transformation group.

Definition 6.44. π is **G-invariant** if $(\tau_g\pi)_\Lambda = \pi_\Lambda$ for all $\Lambda \Subset \mathbb{Z}^d$ and all $g \in G$.

The most important example is that of a Gibbsian specification associated with a potential that is invariant under the action of G. Namely, consider an absolutely summable potential $\Phi = \{\Phi_A\}_{A \Subset \mathbb{Z}^d}$, and let us assume that $\tau_g\Phi_A = \Phi_A$ for all $A \Subset \mathbb{Z}^d$ and all $g \in G$. It then follows that, for all $\Lambda \Subset \mathbb{Z}^d$ and all $g \in G$,

$$\mathscr{H}_\Lambda(\omega) = \mathscr{H}_\Lambda(\tau_g\omega), \quad \forall \omega \in \Omega .$$

As a consequence, the associated specification is G-invariant: $\tau_g\pi^\Phi = \pi^\Phi$ for all $g \in G$. Let us mention a few specific examples.

- **The Ising Model with $h = 0$.** In this case, the internal symmetry group is given by the cyclic group Z_2, that is, the group with two elements: the neutral element e and the **spin flip** f which acts on Ω_0 via $\tau_f\omega_0 = -\omega_0$. As already discussed in Chapter 3, the Hamiltonian is invariant under the global spin flip,

$$\mathscr{H}_{\Lambda;\beta,0}(\omega) = \mathscr{H}_{\Lambda;\beta,0}(\tau_f\omega) ,$$

and the specification of the Ising model with $h = 0$ is therefore invariant under the action of Z_2. When $h \neq 0$, this is of course no longer true.

- **The Potts Model.** In this case, the internal symmetry group is S_q, the group of all permutations on the set $\Omega_0 = \{0, \dots, q-1\}$. It is immediate to check that the potential defining the Potts model (see (6.28)) is S_q-invariant.
- **The Blume–Capel Model with $h = 0$.** As in the Ising model, the potential is invariant under the action of Z_2, the spin flip acting again on $\Omega_0 = \{-1, 0, 1\}$ via $\tau_{\mathsf{f}}\omega_0 = -\omega_0$. That is, the model is invariant under the interchange of $+$ and $-$ spins (leaving the 0 spins unchanged).

At the end of the chapter, we will also consider models in which the spin space is not a finite set.

6.6.1 Measures Compatible with a G-Invariant Specification

Theorem 6.45. *Let G be an internal transformation group and π be a G-invariant specification. Then, $\mathscr{G}(\pi)$ is preserved by G:*

$$\mu \in \mathscr{G}(\pi) \Rightarrow \tau_{\mathsf{g}}\mu \in \mathscr{G}(\pi), \quad \forall \mathsf{g} \in \mathsf{G}.$$

Proof Let $\mathsf{g} \in \mathsf{G}$, $\Lambda \Subset \mathbb{Z}^d$, $\omega \in \Omega$ and $A \in \mathscr{F}$. Since $(\tau_{\mathsf{g}}\mu)(f) = \mu(f \circ \tau_{\mathsf{g}})$,

$$\begin{aligned}(\tau_{\mathsf{g}}\mu)\pi_\Lambda(A) &= \int \pi_\Lambda(A \,|\, \tau_{\mathsf{g}}\omega)\,\mu(\mathrm{d}\omega) \\ &= \int \pi_\Lambda(\tau_{\mathsf{g}}^{-1}A \,|\, \omega)\,\mu(\mathrm{d}\omega) = \mu\pi_\Lambda(\tau_{\mathsf{g}}^{-1}A) = \mu(\tau_{\mathsf{g}}^{-1}A) = \tau_{\mathsf{g}}\mu(A).\end{aligned}$$

It follows that $\tau_{\mathsf{g}}\mu \in \mathscr{G}(\pi)$. □

The above result does not necessarily mean that $\tau_{\mathsf{g}}\mu = \mu$ for all $\mathsf{g} \in \mathsf{G}$, but this property is of course verified when uniqueness holds: in this case, the unique Gibbs measure inherits all the symmetries of the Hamiltonian.

Corollary 6.46. *Assume that $\mathscr{G}(\pi) = \{\mu\}$. If π is G-invariant, then μ is G-invariant: $\tau_{\mathsf{g}}\mu = \mu$ for all $\mathsf{g} \in \mathsf{G}$.*

However, when there are multiple measures compatible with a given specification, it can happen that some of these measures are not G-invariant.

Definition 6.47. Let π be G-invariant. If there exists $\mu \in \mathscr{G}(\pi)$ for which $\tau_{\mathsf{g}}\mu \neq \mu$, the associated symmetry is said to be **spontaneously broken** under μ.

"Spontaneous" is used here to distinguish this phenomenon from an ***explicit symmetry breaking****. The latter occurs, for example, when one introduces a nonzero*

magnetic field h in the Ising model, thereby deliberately destroying the symmetry present when $h = 0$. ◇

Example 6.48. We have seen that, when $h = 0$, the interactions of the Ising model treat $+$ and $-$ spins in a completely symmetric way: $\mathscr{H}_{\Lambda;\beta,0}(\tau_{\mathsf{f}}\omega) = \mathscr{H}_{\Lambda;\beta,0}(\omega)$, where f denotes the global spin flip. Nevertheless, when $d \geq 2$ and $\beta > \beta_c(d)$, we know that the associated Gibbs measures $\mu^+_{\beta,0} \neq \mu^-_{\beta,0}$ are not invariant under a global spin flip, since $\langle\sigma_0\rangle^+_{\beta,0} > 0 > \langle\sigma_0\rangle^-_{\beta,0}$: the symmetry is spontaneously broken. We nevertheless have that $\tau_{\mathsf{f}}\mu^+_{\beta;0} = \mu^-_{\beta;0}$, in complete accordance with the claim of Theorem 6.45. ◇

6.7 Translation-Invariant Gibbs Measures

The theory of Gibbs measures often becomes simpler once restricted to translation-invariant measures. We will see for instance in Section 6.9 that, in this framework, Gibbs measures can be characterized in an alternative way, allowing us to establish a close relation between the DLR formalism and thermostatics.

Translations on $\mathbb{Z}^d$ are a particular type of spatial transformation group, as described in the previous section. Remember from Chapter 3 (see (3.15)) that the **translation by** $j \in \mathbb{Z}^d$, denoted $\theta_j \colon \mathbb{Z}^d \to \mathbb{Z}^d$, is defined by

$$\theta_j i \stackrel{\text{def}}{=} i + j,$$

and can be seen as an action of $\mathbb{Z}^d$ on itself. Notice that $\theta_j^{-1} = \theta_{-j}$.

Definition 6.49. $\mu \in \mathscr{M}_1(\Omega)$ is **translation invariant** if $\theta_j\mu = \mu$ for all $j \in \mathbb{Z}^d$.

Example 6.50. The product measure $\rho^{\mathbb{Z}^d}$ obtained with $\rho_i \equiv \rho_0$ (some fixed distribution on Ω_0) is translation invariant. ◇

Example 6.51. The Gibbs measures of the Ising model, $\mu^+_{\beta,h}$ and $\mu^-_{\beta,h}$, are translation invariant. Namely, we saw in Theorem 3.17 that $\langle\cdot\rangle^+_{\beta,h}$ is invariant under any translation θ_j. Therefore, for each cylinder $C \in \mathscr{C}$, since $\mathbf{1}_C$ is local,

$$\theta_j\mu^+_{\beta,h}(C) = \mu^+_{\beta,h}(\theta_j^{-1}C) = \langle \mathbf{1}_C \circ \theta_j\rangle^+_{\beta,h} = \langle \mathbf{1}_C\rangle^+_{\beta,h} = \mu^+_{\beta,h}(C).$$

This implies that $\theta_j\mu^+_{\beta,h}$ and $\mu^+_{\beta,h}$ coincide on cylinders. Since the cylinders generate $\mathscr{F}$, Corollary B.37 implies $\theta_j\mu^+_{\beta,h} = \mu^+_{\beta,h}$. The same can be done with $\mu^-_{\beta,h}$. ◇

We will sometimes use the following notation:

$$\mathscr{M}_{1,\theta}(\Omega) \stackrel{\text{def}}{=} \{\mu \in \mathscr{M}_1(\Omega) : \mu \text{ is translation invariant}\}.$$

The study of translation-invariant measures is simplified thanks to the fact that *spatial averages of local observables,*

$$\lim_{n\to\infty} \frac{1}{|B(n)|} \sum_{j\in B(n)} \theta_j f \,,$$

exist almost surely (a.s.) and can be related to their expectation. To formulate this precisely, let $\mathscr{I}$ denote the **σ-algebra of translation-invariant events**:

$$\mathscr{I} \stackrel{\text{def}}{=} \{A \in \mathscr{F} : \theta_j A = A \,, \ \forall j \in \mathbb{Z}\} \,.$$

The following result is called the **Multidimensional Ergodic Theorem**. We state it without proof.[6]

Theorem 6.52. *Let $\mu \in \mathscr{M}_{1,\theta}(\Omega)$. Then, for any $f \in L^1(\mu)$,*

$$\frac{1}{|B(n)|} \sum_{j\in B(n)} \theta_j f \to \mu(f \mid \mathscr{I}) \quad \mu\text{-a.s. and in } L^1(\mu) \,. \tag{6.59}$$

Note that the limit (6.59) remains random in general. However, it becomes deterministic if one assumes that μ satisfies one further property.

Definition 6.53. $\mu \in \mathscr{M}_{1,\theta}(\Omega)$ is **ergodic** if each translation-invariant event A has probability $\mu(A) = 0$ or 1.

Theorem 6.54. *If $\mu \in \mathscr{M}_{1,\theta}(\Omega)$ is ergodic, then, for any $f \in L^1(\mu)$,*

$$\frac{1}{|B(n)|} \sum_{j\in B(n)} \theta_j f \to \mu(f) \quad \mu\text{-a.s. and in } L^1(\mu) \,.$$

Proof By Theorem 6.52, we only need to show that $\mu(f \mid \mathscr{I}) = \mu(f)$ almost surely. Notice that $g \stackrel{\text{def}}{=} \mu(f \mid \mathscr{I})$ is $\mathscr{I}$-measurable, and therefore $\{g \le \alpha\} \in \mathscr{I}$ for all $\alpha \in \mathbb{R}$, giving $\mu(g \le \alpha) \in \{0, 1\}$. Since $\alpha \mapsto \mu(g \le \alpha)$ is nondecreasing, there exists some $\alpha_* \in \mathbb{R}$ for which $\mu(g = \alpha_*) = 1$. But since $\mu(g) = \mu(f)$, we have $\alpha_* = \mu(f)$. □

6.7.1 Translation-Invariant Specifications

The action of a translation θ_j on a kernel π_Λ takes the form (see (6.58))

$$(\theta_j \pi)_\Lambda(A \mid \omega) \stackrel{\text{def}}{=} \pi_{\theta_j^{-1}\Lambda}(\theta_j^{-1} A \mid \theta_j^{-1}\omega) \,. \tag{6.60}$$

We say that $\pi = \{\pi_\Lambda\}_\Lambda \Subset \mathbb{Z}^d$ is **translation invariant** if $\theta_j \pi_\Lambda = \pi_\Lambda$ for all Λ and all $j \in \mathbb{Z}^d$.

Theorem 6.45 and its corollary also hold in this situation:

Exercise 6.21. Show that if π is translation invariant and $\mu \in \mathscr{G}(\pi)$, then $\theta_j \mu \in \mathscr{G}(\pi)$ for all $j \in \mathbb{Z}^d$. In particular, if $\mathscr{G}(\pi) = \{\mu\}$, then μ is translation invariant.

For example, if $\Phi = \{\Phi_B\}_{B \Subset \mathbb{Z}^d}$ is a translation-invariant (for all $B \Subset \mathbb{Z}^d$, $\Phi_{\theta_j B}(\omega) = \Phi_B(\theta_{-j}\omega)$) absolutely summable potential, then π^Φ is translation invariant, and $\theta_j\mu \in \mathscr{G}(\Phi)$ for each $\mu \in \mathscr{G}(\Phi)$.

We will sometimes use the following notation:

$$\mathscr{G}_\theta(\pi) \stackrel{\text{def}}{=} \big\{\mu \in \mathscr{G}(\pi) : \mu \text{ translation invariant}\big\}.$$

We leave it as an exercise to check that translation-invariant measures compatible with a translation-invariant specification always exist:

Exercise 6.22. Show that if π is translation invariant and such that $\mathscr{G}(\pi) \neq \varnothing$, then $\mathscr{G}_\theta(\pi) \neq \varnothing$. *Hint:* Take $\mu \in \mathscr{G}(\pi)$, and use $\mu_n \stackrel{\text{def}}{=} \frac{1}{|\mathrm{B}(n)|}\sum_{j \in \mathrm{B}(n)} \theta_j\mu$.

Let us stress, as we did in the preceding section in the case of internal transformations, that translation-invariant specifications do not necessarily yield translation-invariant measures:

Example 6.55. The specification associated with the Ising antiferromagnet defined in (3.76) is clearly translation invariant. Nevertheless, neither of the Gibbs measures μ_β^{even} and μ_β^{odd} constructed in Exercise 3.33 is translation invariant. ◇

6.8 Convexity and Extremal Gibbs Measures

We now investigate general properties of $\mathscr{G}(\pi)$, without assuming either symmetry or uniqueness, and derive fundamental properties of the measures $\mu \in \mathscr{G}(\pi)$.

Let $\nu_1, \nu_2 \in \mathscr{M}_1(\Omega)$, and $\lambda \in [0,1]$. Then the **convex combination** $\lambda\nu_1 + (1-\lambda)\nu_2$ is defined as follows: for $A \in \mathscr{F}$,

$$\big(\lambda\nu_1 + (1-\lambda)\nu_2\big)(A) \stackrel{\text{def}}{=} \lambda\nu_1(A) + (1-\lambda)\nu_2(A).$$

A set $\mathscr{M}' \subset \mathscr{M}_1(\Omega)$ is **convex** if it is stable under convex combination of its elements, that is, if $\nu_1, \nu_2 \in \mathscr{M}'$ and $\lambda \in (0,1)$ imply $\lambda\nu_1 + (1-\lambda)\nu_2 \in \mathscr{M}'$.

The following is a nice feature of the DLR approach, which the definition of Gibbs states in Chapter 3 does not enjoy in general. Let π be any specification.

Theorem 6.56. *$\mathscr{G}(\pi)$ is convex.*

Proof Let $\mu = \lambda\nu_1 + (1-\lambda)\nu_2$, with $\nu_1, \nu_2 \in \mathscr{G}(\pi)$. For all $\Lambda \Subset \mathbb{Z}^d$,

$$\mu\pi_\Lambda = \lambda\nu_1\pi_\Lambda + (1-\lambda)\nu_2\pi_\Lambda = \lambda\nu_1 + (1-\lambda)\nu_2 = \mu\,,$$

and so $\mu \in \mathscr{G}(\pi)$. □

Since $\mathscr{G}(\pi)$ is convex, it is natural to distinguish the measures that cannot be expressed as a nontrivial convex combination of other measures of $\mathscr{G}(\pi)$.

Definition 6.57. $\mu \in \mathscr{G}(\pi)$ is **extremal** if any decomposition of the form $\mu = \lambda\nu_1 + (1-\lambda)\nu_2$ (with $\lambda \in (0,1)$ and $\nu_1, \nu_2 \in \mathscr{G}(\pi)$) implies that $\mu = \nu_1 = \nu_2$. The set of extremal elements of $\mathscr{G}(\pi)$ is denoted by $\mathrm{ex}\,\mathscr{G}(\pi)$.

This in turn raises the following questions:

1. Is $\mathrm{ex}\,\mathscr{G}(\pi)$ nonempty?
2. Are there properties that distinguish the elements of $\mathrm{ex}\,\mathscr{G}(\pi)$ from the non-extremal ones?
3. Do extremal measures have special physical significance?

We will first answer the last two questions.

6.8.1 Properties of Extremal Gibbs Measures

We will see that extremal Gibbs measures are characterized by the fact that they possess *deterministic* macroscopic properties. These properties correspond to the following family of events, called the **tail-σ-algebra**:

$$\mathscr{T}_\infty \stackrel{\text{def}}{=} \bigcap_{\Lambda \Subset \mathbb{Z}^d} \mathscr{F}_{\Lambda^c}\,; \tag{6.61}$$

its elements are called **tail** (or **macroscopic**) **events**. Remembering that $\mathscr{F}_{\Lambda^c}$ is the σ-algebra of events that only depend on spins located outside Λ, we see that tail events are those *whose occurrence is not altered by local changes*: if $A \in \mathscr{T}_\infty$ and if ω and ω' coincide everywhere but on a finite set of vertices, then

$$\mathbf{1}_A(\omega) = \mathbf{1}_A(\omega')\,.$$

The σ-algebra $\mathscr{T}_\infty$ contains many important events. For example, particularly relevant in view of what we saw in Chapter 3, the event "the infinite-volume magnetization exists and is positive",

$$\Big\{\omega \in \Omega \,:\, \lim_{n\to\infty} \frac{1}{|\mathrm{B}(n)|} \sum_{j\in\mathrm{B}(n)} \omega_j \text{ exists and is positive}\Big\}$$

belongs to $\mathscr{T}_\infty$. Indeed, neither the existence nor the sign of the limit is altered if any finite number of spins are changed. The $\mathscr{T}_\infty$-measurable functions $f\colon \Omega \to \mathbb{R}$ are also called **macroscopic observables**, since they are not altered by local changes in a configuration. As the following exercise shows, their behavior contrasts sharply with that of local functions.

Exercise 6.23. Show that nonconstant $\mathscr{T}_\infty$-measurable functions are everywhere discontinuous.

We now present the main features that characterize the elements of $\mathrm{ex}\,\mathscr{G}(\pi)$.

Theorem 6.58. *Let π be a specification. Let $\mu \in \mathscr{G}(\pi)$. The following conditions are equivalent characterizations of extremality.*

1. *μ is extremal.*
2. *μ is **trivial on** $\mathscr{T}_\infty$: if $A \in \mathscr{T}_\infty$, then $\mu(A)$ is either* 1 *or* 0.
3. *All $\mathscr{T}_\infty$-measurable functions are μ-almost surely constant.*
4. *μ has **short-range correlations**: for all $A \in \mathscr{F}$ (or, equivalently, for all $A \in \mathscr{C}$),*

$$\lim_{\Lambda \uparrow \mathbb{Z}^d} \sup_{B \in \mathscr{F}_{\Lambda^c}} \left|\mu(A \cap B) - \mu(A)\mu(B)\right| = 0 . \qquad (6.62)$$

A few remarks need to be made:

- These characterizations all express the fact that, whenever a system is described by an extremal Gibbs measure, its macroscopic properties are deterministic: every macroscopic event occurs with probability 0 or 1. This is clearly a very desirable feature: as discussed at the beginning of Chapter 1, all observables associated with a given phase of a macroscopic system in thermodynamic equilibrium are determined once the thermodynamic parameters characterizing the macrostate (for example, (β, h) for an Ising ferromagnet) are fixed. Note however that, as mentioned there, the macrostate does not fully characterize the macroscopic state of the system when there is a first-order phase transition. The reason for that is made clear in the present context: all macroscopic observables are deterministic under each extremal measure, but the macrostate does not specify which of these measures is realized.
- The statement (6.62) implies that local events become asymptotically independent as the distance separating their support diverges. In fact, it even applies to nonlocal events, although the interpretation of the statement becomes more difficult.
- Notice also that condition 2 above provides a remarkable and far-reaching generalization of a famous result in probability theory: *Kolmogorov's* 0-1 *law*. Indeed, combined with Exercise 6.7, Theorem 6.58 implies triviality of the tail-σ-algebra associated with a collection of independent random variables indexed by $\mathbb{Z}^d$.

To prove Theorem 6.58, we first need two preliminary propositions. Since it will be convenient to specify the σ-algebra on which measures are defined, we temporarily write $\mathscr{M}_1(\Omega, \mathscr{F})$ instead of $\mathscr{M}_1(\Omega)$.

Let $\Lambda \Subset \mathbb{Z}^d$. We define the **restriction** $\mathsf{r}_\Lambda : \mathscr{M}_1(\Omega, \mathscr{F}) \to \mathscr{M}_1(\Omega, \mathscr{F}_{\Lambda^c})$ by

$$\mathsf{r}_\Lambda \mu(B) \stackrel{\text{def}}{=} \mu(B), \quad \forall B \in \mathscr{F}_{\Lambda^c} .$$

Observe that if $g\colon \Omega \to \mathbb{R}$ is $\mathscr{F}_{\Lambda^c}$-measurable, then $\mathsf{r}_\Lambda \mu(g) = \mu(g)$. Using a specification π, one can define for each $\Lambda \Subset \mathbb{Z}^d$ the **extension** $\mathsf{t}^\pi_\Lambda : \mathscr{M}_1(\Omega, \mathscr{F}_{\Lambda^c}) \to \mathscr{M}_1(\Omega, \mathscr{F})$ by

$$\mathsf{t}^\pi_\Lambda \nu(A) \stackrel{\text{def}}{=} \nu\pi_\Lambda(A), \quad \forall A \in \mathscr{F} .$$

Note that the composition of t^π_Λ with r_Λ is such that $\mathsf{t}^\pi_\Lambda \mathsf{r}_\Lambda : \mathscr{M}_1(\Omega, \mathscr{F}) \to \mathscr{M}_1(\Omega, \mathscr{F})$. We will prove the following new characterization of $\mathscr{G}(\pi)$:

Proposition 6.59. *$\mu \in \mathscr{G}(\pi)$ if and only if $\mu = \mathsf{t}^\pi_\Lambda \mathsf{r}_\Lambda \mu$ for all $\Lambda \Subset \mathbb{Z}^d$.*

This characterization[7] of $\mathscr{G}(\pi)$ can be interpreted as follows. Given a measure μ on $(\Omega, \mathscr{F})$, the restriction r_Λ results in a loss of information: from the measure $r_\Lambda \mu$, nothing can be said about what happens inside Λ. However, when $\mu \in \mathscr{G}(\pi)$, that lost information can be recovered using t^π_Λ: $t^\pi_\Lambda r_\Lambda \mu = \mu$. ◇

Proof of Proposition 6.59. Composing t^π_Λ with r_Λ gives, for all $A \in \mathscr{F}$,

$$\mathsf{t}^\pi_\Lambda \mathsf{r}_\Lambda \mu(A) = (\mathsf{r}_\Lambda \mu)\pi_\Lambda(A) = \int \pi_\Lambda(A \,|\, \omega)\mathsf{r}_\Lambda \mu(\mathrm{d}\omega) = \int \pi_\Lambda(A \,|\, \omega)\mu(\mathrm{d}\omega) = \mu\pi_\Lambda(A)\,.$$

In the third identity, we used the $\mathscr{F}_{\Lambda^c}$-measurability of $\pi_\Lambda(A \,|\, \cdot)$. □

Let $\hat{\mathscr{F}}$ be a sub-σ-algebra of $\mathscr{F}$ and let $\mu \in \mathscr{M}_1(\Omega, \hat{\mathscr{F}})$. For a nonnegative $\hat{\mathscr{F}}$-measurable function $f : \Omega \to \mathbb{R}$ that satisfies $\mu(f) = 1$, let $f\mu \in \mathscr{M}_1(\Omega, \hat{\mathscr{F}})$ denote the probability measure whose density with respect to μ is f:

$$f\mu(A) \overset{\text{def}}{=} \int_A f(\omega)\mu(\mathrm{d}\omega), \qquad \forall A \in \hat{\mathscr{F}}\,.$$

Observe that $f_1\mu = f_2\mu$ if and only if $f_1 = f_2$ μ-almost surely (Lemma B.42).

Lemma 6.60. *Let $\Lambda \Subset \mathbb{Z}^d$.*

1. *Let $\mu \in \mathscr{M}_1(\Omega, \mathscr{F})$ and let $f : \Omega \to \mathbb{R}_{\geq 0}$ be an $\mathscr{F}$-measurable function such that $\mu(f) = 1$. Then*
$$\mathsf{r}_\Lambda(f\mu) = \mu(f \,|\, \mathscr{F}_{\Lambda^c})\mathsf{r}_\Lambda \mu\,.$$
2. *Let $\nu \in \mathscr{M}_1(\Omega, \mathscr{F}_{\Lambda^c})$ and let $g : \Omega \to \mathbb{R}_{\geq 0}$ be an $\mathscr{F}_{\Lambda^c}$-measurable function such that $\nu(g) = 1$. Then*
$$\mathsf{t}^\pi_\Lambda(g\nu) = g \cdot \mathsf{t}^\pi_\Lambda \nu\,.$$

Proof For the first item, take $B \in \mathscr{F}_{\Lambda^c}$ and use the definition of conditional expectation:

$$\mathsf{r}_\Lambda(f\mu)(B) = \int_B f(\omega)\mu(\mathrm{d}\omega) = \int_B \mu(f \,|\, \mathscr{F}_{\Lambda^c})(\omega)\mu(\mathrm{d}\omega) = \int_B \mu(f \,|\, \mathscr{F}_{\Lambda^c})(\omega)\mathsf{r}_\Lambda \mu(\mathrm{d}\omega)\,.$$

For the second item, take $A \in \mathscr{F}$ and compute:

$$\mathsf{t}^\pi_\Lambda(g\nu)(A) = (g\nu)\pi_\Lambda(A) = g(\nu\pi_\Lambda)(A) = g \cdot \mathsf{t}^\pi_\Lambda \nu(A)\,. \tag{6.63}$$

□

Exercise 6.24. Justify the second identity in (6.63).

Proposition 6.61. *Let π be a specification.*

1. *Let $\mu \in \mathscr{G}(\pi)$. Let $f: \Omega \to \mathbb{R}_{\geq 0}$, $\mathscr{F}$-measurable, such that $\mu(f) = 1$. Then $f\mu \in \mathscr{G}(\pi)$ if and only if f is equal μ-almost everywhere to a $\mathscr{T}_\infty$-measurable function.*
2. *Let $\mu, \nu \in \mathscr{G}(\pi)$ be two probability measures that coincide on $\mathscr{T}_\infty$: $\mu(A) = \nu(A)$ for all $A \in \mathscr{T}_\infty$. Then $\mu = \nu$.*

We will need the following classical result, called the **Backward Martingale Convergence Theorem**: for all measurable $f: \Omega \to \mathbb{R}$, integrable with respect to μ,

$$\mu(f \mid \mathscr{F}_{\mathrm{B}(n)^{\mathrm{c}}}) \overset{n\to\infty}{\longrightarrow} \mu(f \mid \mathscr{T}_\infty), \ \mu\text{-a.s. and in } L^1(\mu). \tag{6.64}$$

See also Theorem B.52 in Appendix B.5.

Proof of Proposition 6.61. 1. If $f\mu \in \mathscr{G}(\pi)$, we use Lemma 6.60 and Proposition 6.59 to get, for all $\Lambda \Subset \mathbb{Z}^d$,

$$f\mu = \mathsf{t}^\pi_\Lambda \mathsf{r}_\Lambda(f\mu) = \mathsf{t}^\pi_\Lambda\{\mu(f \mid \mathscr{F}_{\Lambda^{\mathrm{c}}})\mathsf{r}_\Lambda\mu\} = \mu(f \mid \mathscr{F}_{\Lambda^{\mathrm{c}}}) \cdot \mathsf{t}^\pi_\Lambda\{\mathsf{r}_\Lambda\mu\} = \mu(f \mid \mathscr{F}_{\Lambda^{\mathrm{c}}}) \cdot \mu\,.$$

Therefore, again by Lemma B.42, this implies $f = \mu(f \mid \mathscr{F}_{\Lambda^{\mathrm{c}}})$ μ-almost surely. Since this holds in particular when $\Lambda = \mathrm{B}(n)$, and since $\mu(f \mid \mathscr{F}_{\mathrm{B}(n)^{\mathrm{c}}}) \to \mu(f \mid \mathscr{T}_\infty)$ almost surely as $n \to \infty$ (see (6.64)), we have shown that $f = \mu(f \mid \mathscr{T}_\infty)$ μ-almost surely. The latter is $\mathscr{T}_\infty$-measurable, which proves the claim. Inversely, if f coincides μ-almost surely with a $\mathscr{T}_\infty$-measurable function $\tilde f$, then $f\mu = \tilde f\mu$, and, since $\tilde f$ is $\mathscr{F}_{\Lambda^{\mathrm{c}}}$-measurable for all $\Lambda \Subset \mathbb{Z}^d$,

$$(f\mu)\pi_\Lambda(A) = (\tilde f\mu)\pi_\Lambda(A) = \tilde f(\mu\pi_\Lambda)(A) = \tilde f\mu(A) = f\mu(A)\,,$$

for all $A \in \mathscr{F}$ (we used Exercise 6.24 for the second identity), and so $f\mu \in \mathscr{G}(\pi)$.

2. Define $\lambda \overset{\text{def}}{=} \frac{1}{2}(\mu + \nu)$. Then, $\lambda \in \mathscr{G}(\pi)$ and both μ and ν are absolutely continuous with respect to λ. By the Radon–Nikodým Theorem (Theorem B.41), there exist $f, g \geq 0$, $\lambda(f) = \lambda(g) = 1$, such that $\mu = f\lambda$, $\nu = g\lambda$. For all $A \in \mathscr{T}_\infty$,

$$\int_A (f - g)\,\mathrm{d}\lambda = \mu(A) - \nu(A) = 0\,.$$

But, by item 1, there exist two $\mathscr{T}_\infty$-measurable functions $\tilde f$ and $\tilde g$, λ-almost surely equal to f, respectively g. Since $A = \{\tilde f > \tilde g\} \in \mathscr{T}_\infty$, we conclude that $\lambda(f > g) = \lambda(\tilde f > \tilde g) = 0$. In the same way, $\lambda(f < g) = 0$ and therefore $f = g$ λ-almost surely, which implies that $\mu = \nu$. □

Proof of Theorem 6.58. 1 ⇒ 2: Assume there exists $A \in \mathscr{T}_\infty$ such that $\alpha = \mu(A) \in (0, 1)$. By item 1 of Proposition 6.61, $\mu_1 \overset{\text{def}}{=} \frac{1}{\alpha}\mathbf{1}_A\mu$ and $\mu_2 \overset{\text{def}}{=} \frac{1}{1-\alpha}\mathbf{1}_{A^{\mathrm{c}}}\mu$ are both in $\mathscr{G}(\pi)$. But since $\mu = \alpha\mu_1 + (1 - \alpha)\mu_2$, μ cannot be extremal.

2 ⇒ 1: Let μ be trivial on $\mathscr{T}_\infty$, and assume that $\mu = \alpha\mu_1 + (1-\alpha)\mu_2$, with $\alpha \in (0,1)$ and $\mu_1, \mu_2 \in \mathscr{G}(\pi)$. Then μ_1 and μ_2 are absolutely continuous with respect to μ. Let now $A \in \mathscr{T}_\infty$. Then, since $\mu(A)$ can be either 0 or 1, $\mu_1(A)$ and $\mu_2(A)$ are either both 0 or both 1. By item 2 of Proposition 6.61, $\mu = \mu_1 = \mu_2$.

2 ⇒ 3: If f is $\mathscr{T}_\infty$-measurable, each $\{f \le c\} \in \mathscr{T}_\infty$ and thus $\mu(f \le c) \in \{0,1\}$ for all c. Setting $c_* = \inf\{c\colon \mu(f \le c) = 1\}$, we get $\mu(f = c_*) = 1$.

3 ⇒ 2: If $A \in \mathscr{T}_\infty$, then $\mathbf{1}_A$ is $\mathscr{T}_\infty$-measurable. Since it must be μ-almost surely constant, we necessarily have that $\mu(A) \in \{0,1\}$.

2 ⇒ 4: Let $A \in \mathscr{F}$, $\epsilon > 0$. Using (6.64) with $f = \mathbf{1}_A$, one can take n large enough so that

$$\|\mu(A \,|\, \mathscr{F}_{\mathrm{B}(n)^{\mathrm{c}}}) - \mu(A \,|\, \mathscr{T}_\infty)\|_1 \le \epsilon \,. \tag{6.65}$$

Since $\mu(A \,|\, \mathscr{T}_\infty)$ is $\mathscr{T}_\infty$-measurable, item 3 implies that it is μ-almost surely constant. This constant can only be $\mu(A)$, since $\mu(\mu(A \,|\, \mathscr{T}_\infty)) = \mu(A)$. Then, for all $B \in \mathscr{F}_{\mathrm{B}(n)^{\mathrm{c}}}$,

$$\begin{aligned}\big|\mu(A\cap B) - \mu(A)\mu(B)\big| &= \Big|\int_B \{\mathbf{1}_A - \mu(A)\}\,\mathrm{d}\mu\Big| \\ &= \Big|\int_B \{\mu(A \,|\, \mathscr{F}_{\mathrm{B}(n)^{\mathrm{c}}}) - \mu(A \,|\, \mathscr{T}_\infty)\}\,\mathrm{d}\mu\Big| \le \epsilon \,.\end{aligned}$$

4 ⇒ 2: Suppose that (6.62) holds for all $A \in \mathscr{C}$. Then, $\mu(A \cap B) = \mu(A)\mu(B)$ for all $A \in \mathscr{C}$ and all $B \in \mathscr{T}_\infty$ (since $B \in \mathscr{F}_{\Lambda^{\mathrm{c}}}$ for all $\Lambda \Subset \mathbb{Z}^d$). If this can be extended to

$$\mu(A\cap B) = \mu(A)\mu(B) \,, \qquad \forall A \in \mathscr{F}, B \in \mathscr{T}_\infty \,, \tag{6.66}$$

then, taking $A = B$ implies $\mu(B) = \mu(B \cap B) = \mu(B)^2$, which is only possible if $\mu(B) \in \{0,1\}$ for all $B \in \mathscr{T}_\infty$, that is, if μ is trivial on $\mathscr{T}_\infty$.

To prove (6.66), fix $B \in \mathscr{T}_\infty$ and define

$$\mathscr{D} \stackrel{\text{def}}{=} \big\{A \in \mathscr{F} \,:\, \mu(A\cap B) = \mu(A)\mu(B)\big\} \,.$$

If $A, A' \in \mathscr{D}$, with $A \subset A'$, then $\mu\big((A'\setminus A)\cap B\big) = \mu(A'\cap B) - \mu(A\cap B) = \mu(A'\setminus A)\mu(B)$, showing that $A'\setminus A \in \mathscr{D}$. Moreover, for any sequence $(A_n)_{n\ge 1} \subset \mathscr{D}$ such that $A_n \uparrow A$, we have that $\mu(A \cap B) = \lim_n \mu(A_n \cap B) = \lim_n \mu(A_n)\mu(B) = \mu(A)\mu(B)$, and so $A \in \mathscr{D}$. This implies that $\mathscr{D}$ is a **Dynkin system** (see Appendix B.5). Since $\mathscr{C} \subset \mathscr{D}$ by assumption, and since $\mathscr{C}$ is an algebra, we conclude that $\mathscr{D} = \sigma(\mathscr{C}) = \mathscr{F}$ (Theorem B.36), so (6.66) holds. □

In the following exercise, we consider a nonextremal measure for the Ising model, and we provide an example of events for which the property of short-range correlations does not hold.

Exercise 6.25. Consider the two-dimensional Ising model with $h = 0$ and $\beta > \beta_c(2)$. Take any $\lambda \in (0,1)$ and consider the (nonextremal) Gibbs measure

$$\mu = \lambda\mu^+_{\beta,0} + (1-\lambda)\mu^-_{\beta,0} \,.$$

Show that μ does not satisfy (6.62), by taking $A = \{\sigma_0 = 1\}$, $B_i = \{\sigma_i = 1\}$ and verifying that

$$\liminf_{\|i\|_1 \to \infty} \left| \mu(A \cap B_i) - \mu(A)\mu(B_i) \right| > 0 \,.$$

Hint: Use the symmetry between $\mu^+_{\beta,0}$ and $\mu^-_{\beta,0}$ and the FKG inequality.

To end this section, we mention that extremal measures of $\mathscr{G}(\pi)$ can be distinguished from each other by only considering tail events:

Lemma 6.62. *Distinct extremal measures $\mu, \nu \in \mathrm{ex}\mathscr{G}(\pi)$ are* ***singular****: there exists a tail event $A \in \mathscr{T}_\infty$ such that $\mu(A) = 0$ and $\nu(A) = 1$.*

Proof If $\mu, \nu \in \mathscr{G}(\pi)$ are distinct, then item 2 of Proposition 6.61 shows that there must exist $A \in \mathscr{T}_\infty$ such that $\mu(A) \neq \nu(A)$. But if μ and ν are extremal, they are trivial on $\mathscr{T}_\infty$ (Theorem 6.58, item 2), so either $\mu(A) = 0$ and $\nu(A) = 1$, or $\mu(A^c) = 0$ and $\nu(A^c) = 1$. □

6.8.2 Extremal Gibbs Measures and the Thermodynamic Limit

Since real macroscopic systems are always finite (albeit very large), the most physically relevant Gibbs measures are those that can be approximated by finite-volume Gibbs distributions, that is, those that can be obtained by a thermodynamic limit with some fixed boundary condition. It turns out that all *extremal* Gibbs measures enjoy this property:

Theorem 6.63. *Let $\mu \in \mathrm{ex}\mathscr{G}(\pi)$. Then, for μ-almost all ω,*

$$\pi_{\mathrm{B}(n)}(\cdot \,|\, \omega) \Rightarrow \mu \,.$$

Proof We need to prove that, for μ-almost all ω,

$$\pi_{\mathrm{B}(n)}(C \,|\, \omega) \overset{n\to\infty}{\longrightarrow} \mu(C) \,, \quad \forall C \in \mathscr{C} \,. \tag{6.67}$$

Let $C \in \mathscr{C}$. On the one hand (see (6.23)), there exists $\Omega_{n,C}$, $\mu(\Omega_{n,C}) = 1$, such that $\pi_{\mathrm{B}(n)}(C \,|\, \omega) = \mu(C \,|\, \mathscr{F}_{\mathrm{B}(n)^c})(\omega)$ for all $\omega \in \Omega_{n,C}$. On the other hand, the extremality of μ (see item 3 of Theorem 6.58) guarantees that there exists Ω_C, $\mu(\Omega_C) = 1$, such that $\mu(C) = \mu(C \,|\, \mathscr{T}_\infty)(\omega)$ for all $\omega \in \Omega_C$. Using (6.64) with $f = \mathbf{1}_C$, there also exists $\tilde{\Omega}_C$, $\mu(\tilde{\Omega}_C) = 1$, such that

$$\mu(C \,|\, \mathscr{F}_{\mathrm{B}(n)^c})(\omega) \to \mu(C \,|\, \mathscr{T}_\infty)(\omega), \quad \forall \omega \in \tilde{\Omega}_C.$$

Therefore, for all ω that belong to the countable intersection of all the sets Ω_C, $\tilde{\Omega}_C$ and $\Omega_{n,C}$, which has μ-measure 1, (6.67) holds. □

The above theorem shows yet another reason why extremal Gibbs measures are natural to consider: they can be prepared by taking limits of finite-volume systems.

However, we will see in Example 6.68 that the converse statement is not true: not all limits of finite-volume systems lead to extremal states.

A more basic question at this stage is whether all Gibbs measures can be obtained with the thermodynamic limit. The following example shows that this is not the case: $\mathscr{G}(\pi)$ can contain measures that do not appear in the approach of Chapter 3 relying on the thermodynamic limit.

Example 6.64.[8] Let us consider the three-dimensional Ising model, with $\beta > \beta_c(3)$ and $h = 0$, in the box $\mathrm{B}(n)$. We have seen in Section 3.10.7 that the sequence of finite-volume Gibbs distributions with Dobrushin boundary condition admits a converging subsequence, defining a Gibbs measure $\mu^{\mathrm{Dob}}_{\beta,0}$ satisfying, for any $\epsilon > 0$,

$$\langle \sigma_{(0,0,0)}\sigma_{(0,0,-1)}\rangle^{\mathrm{Dob}}_{\beta,0} \leq -1+\epsilon\,, \tag{6.68}$$

once β is large enough (see Corollary 3.60). Let us denote by $\mu^{\mathrm{Dob}}_{\beta,0}$ the corresponding Gibbs measure. Applying a global spin flip, we obtain another Gibbs measure, $\mu^{-\mathrm{Dob}}_{\beta,0} \stackrel{\mathrm{def}}{=} \tau_{\mathrm{f}}\mu^{\mathrm{Dob}}_{\beta,0}$, also satisfying (6.68). Since $\mathscr{G}(\beta,0)$ is convex, $\mu \stackrel{\mathrm{def}}{=} \frac{1}{2}\mu^{\mathrm{Dob}}_{\beta,0} + \frac{1}{2}\mu^{-\mathrm{Dob}}_{\beta,0} \in \mathscr{G}(\beta,0)$. We show that it cannot be obtained as a thermodynamic limit. Notice that $\langle\sigma_i\rangle_\mu = 0$ for all $i \in \mathbb{Z}^d$ and that one has, for any $\epsilon > 0$,

$$\langle \sigma_{(0,0,0)}\sigma_{(0,0,-1)}\rangle_{\mu} \leq -1+\epsilon\,, \tag{6.69}$$

once β is large enough. Suppose there exists a sequence $(\mu^{\eta_k}_{\mathrm{B}(n_k);\beta,0})_{k\geq 1}$ converging to μ. By the FKG inequality, for all $k \geq 1$,

$$\langle \sigma_{(0,0,0)}\sigma_{(0,0,-1)}\rangle^{\eta_k}_{\mathrm{B}(n_k);\beta,0} \geq \langle \sigma_{(0,0,0)}\rangle^{\eta_k}_{\mathrm{B}(n_k);\beta,0}\langle \sigma_{(0,0,-1)}\rangle^{\eta_k}_{\mathrm{B}(n_k);\beta,0}\,,$$

and thus

$$\langle \sigma_{(0,0,0)}\sigma_{(0,0,-1)}\rangle_{\mu} \geq \langle \sigma_{(0,0,0)}\rangle_{\mu}\langle \sigma_{(0,0,-1)}\rangle_{\mu} = 0\,.$$

This contradicts (6.69). ◇

So far, we have described general properties of extremal Gibbs measures. We still need to determine whether such measures exist in general, and what role they play in the description of $\mathscr{G}(\pi)$. Before pursuing the general description of the theory, we illustrate some of the ideas presented so far on our favorite example.

6.8.3 More on $\mu^+_{\beta,h}$, $\mu^-_{\beta,h}$ and $\mathscr{G}(\beta,h)$

In this section, using tools specific to the Ising model, we provide more information about $\mu^+_{\beta,h}$ and $\mu^-_{\beta,h}$.

Lemma 6.65. *$\mu^+_{\beta,h}$, $\mu^-_{\beta,h}$ are extremal.*

Proof We consider $\mu^+_{\beta,h}$. We start by showing that, for any $\nu \in \mathscr{G}(\beta,h)$,

$$\nu(f) \leq \mu^+_{\beta,h}(f) \quad \text{for every nondecreasing local function } f\,. \tag{6.70}$$

Remember that, for all $\Lambda \Subset \mathbb{Z}^d$ and all boundary conditions η, the FKG inequality implies that $\mu^{\eta}_{\Lambda;\beta,h}(f) \leq \mu^{+}_{\Lambda;\beta,h}(f)$ for every nondecreasing local function f (see Lemma 3.23). Therefore,

$$\nu(f) = \int \mu^{\eta}_{\Lambda;\beta,h}(f)\, \nu(\mathrm{d}\eta) \leq \mu^{+}_{\Lambda;\beta,h}(f).$$

Since $\lim_{\Lambda \uparrow \mathbb{Z}^d} \mu^{+}_{\Lambda;\beta,h}(f) = \mu^{+}_{\beta,h}(f)$, this establishes (6.70). Now assume that $\mu^{+}_{\beta,h}$ is not extremal:

$$\mu^{+}_{\beta,h} = \lambda \nu_1 + (1-\lambda)\nu_2,$$

where $\lambda \in (0,1)$ and $\nu_1, \nu_2 \in \mathscr{G}(\beta,h)$ are both distinct from $\mu^{+}_{\beta,h}$. We use (6.70) as follows. First, since $\nu_1 \neq \mu^{+}_{\beta,h}$, there must exist a local function f_* such that $\nu_1(f_*) \neq \mu^{+}_{\beta,h}(f_*)$. From Lemma 3.19, we can assume that f_* is nondecreasing. Therefore, (6.70) implies that $\nu_1(f_*) < \mu^{+}_{\beta,h}(f_*)$ and $\nu_2(f_*) \leq \mu^{+}_{\beta,h}(f_*)$. Consequently,

$$\mu^{+}_{\beta,h}(f_*) = \lambda \nu_1(f_*) + (1-\lambda)\nu_2(f_*) < \mu^{+}_{\beta,h}(f_*),$$

a contradiction. We conclude that $\mu^{+}_{\beta,h}$ is extremal. □

Since $\mu^{+}_{\beta,h}$ is extremal, it inherits all the properties described in Theorem 6.58. For example, property 4 of that theorem implies that the truncated 2-point function,

$$\langle \sigma_i ; \sigma_j \rangle^{+}_{\beta,h} \stackrel{\text{def}}{=} \langle \sigma_i \sigma_j \rangle^{+}_{\beta,h} - \langle \sigma_i \rangle^{+}_{\beta,h} \langle \sigma_j \rangle^{+}_{\beta,h},$$

tends to zero when $\|j-i\|_\infty \to \infty$. (Note that this claim was already established, by other means, in Exercise 3.15.) This can be used to obtain a Weak Law of Large Numbers:

Exercise 6.26. Consider

$$m_{\mathrm{B}(n)} \stackrel{\text{def}}{=} \frac{1}{|\mathrm{B}(n)|} \sum_{j \in \mathrm{B}(n)} \sigma_j.$$

Show that $m_{\mathrm{B}(n)} \to \mu^{+}_{\beta,h}(\sigma_0)$ in $\mu^{+}_{\beta,h}$-probability. That is, for all $\epsilon > 0$,

$$\mu^{+}_{\beta,h}\big(|m_{\mathrm{B}(n)} - \mu^{+}_{\beta,h}(\sigma_0)| \geq \epsilon\big) \to 0 \quad \text{when } n \to \infty.$$

Hint: Show that the variance of $m_{\mathrm{B}(n)}$ vanishes as $n \to \infty$.

One may wonder whether the convergence of $m_{\mathrm{B}(n)}$ to $\mu^{+}_{\beta,h}(\sigma_0)$ proved in the previous exercise also holds almost surely. Actually, we know that

$$m \stackrel{\text{def}}{=} \limsup_{n\to\infty} m_{\mathrm{B}(n)} \tag{6.71}$$

is almost surely constant, since it is a macroscopic observable. To show that the lim sup in (6.71) is a true limit, we use a further property of $\mu^{+}_{\beta,h}$.

Lemma 6.66. *$\mu^+_{\beta,h}$ and $\mu^-_{\beta,h}$ are ergodic.*

We start by proving the following general fact:

Lemma 6.67. *Let $\mu \in \mathscr{M}_1(\Omega, \mathscr{F})$ be invariant under translations. Then, for all $A \in \mathscr{I}$, there exists $B \in \mathscr{T}_\infty$ such that $\mu(A \triangle B) = 0$; in particular, $\mu(A) = \mu(B)$.*

Proof By Lemma B.34, there exists a sequence $(C_n)_{n\geq 1} \subset \mathscr{C}$ such that $\mu(A\triangle C_n) \leq 2^{-n}$. For each n, let $\Lambda(n) \Subset \mathbb{Z}^d$ be such that $C_n \in \mathscr{C}(\Lambda(n))$. By a property already used in Lemma 6.2, we can assume that $\Lambda(n) \uparrow \mathbb{Z}^d$. For each n, let $i_n \in \mathbb{Z}^d$ be such that $\Lambda(n) \cap \theta_{i_n}\Lambda(n) = \varnothing$. Let $C'_n = \theta_{i_n} C_n$. Since A and μ are invariant,

$$\mu(A\triangle C'_n) = \mu(\theta_{i_n}(\theta_{-i_n}A\triangle C_n)) = \mu(\theta_{i_n}(A\triangle C_n)) = \mu(A\triangle C_n) \leq 2^{-n}.$$

Since $C'_n \in \mathscr{F}_{\Lambda(n)^c}$, we have $B \stackrel{\text{def}}{=} \bigcap_n \bigcup_{m\geq n} C'_m \in \mathscr{T}_\infty$. Moreover,

$$\mu(A\triangle B) \leq \lim_{n\to\infty} \sum_{m\geq n} \mu(A\triangle C'_m) = 0.$$

□

Proof of Lemma 6.66. Since $\mu^+_{\beta,h}$ is invariant under translations, it follows from Lemma 6.67 that, for all $A \in \mathscr{I}$, there exists $B \in \mathscr{T}_\infty$ such that $\mu^+_{\beta,h}(A) = \mu^+_{\beta,h}(B)$. But $\mu^+_{\beta,h}$ is extremal, therefore $\mu^+_{\beta,h}(B) \in \{0, 1\}$. □

Since $\mu^+_{\beta,h}$ is ergodic, and since one can always write $\sigma_j = \sigma_0 \circ \theta_{-j}$, we deduce from Theorem 6.54 that the **infinite-volume magnetization**

$$m = \lim_{n\to\infty} \frac{1}{|\mathrm{B}(n)|} \sum_{j\in\mathrm{B}(n)} \sigma_j$$

exists $\mu^+_{\beta,h}$-almost surely, and equals $\mu^+_{\beta,h}(\sigma_0)$. A similar statement holds for $\mu^-_{\beta,h}$. Since $\mu^+_{\beta,0} \neq \mu^-_{\beta,0}$ when $\beta > \beta_c(d)$, we know from Lemma 6.62 that they are also singular. The events $\{m > 0\}$ and $\{m < 0\}$ thus provide examples of two tail events on which these measures differ.

Digression: On the Significance of Nonextremal Gibbs Measures

With the properties of extremal measures described in detail above, we can now understand better the significance of *nonextremal* Gibbs measures. We continue illustrating things on the Ising model, but this discussion applies to more general situations.

Let $\lambda \in (0, 1)$ and consider the convex combination

$$\mu \stackrel{\text{def}}{=} \lambda\mu^+_{\beta,0} + (1-\lambda)\mu^-_{\beta,0}.$$

Assume that $d \geq 2$ and $\beta > \beta_c(d)$, so that $\mu^+_{\beta,0} \neq \mu^-_{\beta,0}$. As explained below, a natural interpretation of the coefficient λ (resp. $1-\lambda$) is as the probability that a configuration sampled from μ is "typical" of $\mu^+_{\beta,0}$ (resp. $\mu^-_{\beta,0}$). The only minor difficulty is to give a reasonable meaning to the word "typical". One possible way to do that is to consider two tail-measurable events T^+ and T^- such that

$$\mu^+_{\beta,0}(T^+) = \mu^-_{\beta,0}(T^-) = 1, \quad \mu^+_{\beta,0}(T^-) = \mu^-_{\beta,0}(T^+) = 0 .$$

In other words, the event T^+ encodes macroscopic properties that are typically (that is, almost surely) verified under $\mu^+_{\beta,0}$, and similarly for T^-; moreover T^+ and T^- allow us to distinguish between these two measures. A configuration $\omega \in \Omega$ will then be said to be typical for $\mu^+_{\beta,0}$ (resp. $\mu^-_{\beta,0}$) if $\omega \in T^+$ (resp. T^-).

Since $\mu^+_{\beta,h}$ and $\mu^-_{\beta,h}$ are extremal and distinct, we know by Lemma 6.62 that events such as T^+ and T^- always exist. In the case of the Ising model, we can be more explicit. For example, since $\mu^+_{\beta,0}$ and $\mu^-_{\beta,0}$ are characterized by the probability they associate with cylinders, one can take

$$T^{\pm} = \bigcap_{C\in\mathscr{C}} \Big\{ \lim_{\Lambda\uparrow\mathbb{Z}^d} \frac{1}{|\Lambda|} \sum_{i\in\Lambda} \mathbf{1}_C \circ \theta_i \text{ exists and equals } \mu^{\pm}_{\beta,0}(C) \Big\} .$$

It is easy to verify that T^+ and T^- enjoy all the desired properties. First, Theorem 6.54 guarantees that $\mu^{\pm}_{\beta,0}(T^{\pm}) = 1$. Moreover, $T^+ \cap T^- = \varnothing$, since for example $\mu^+_{\beta,0}(\sigma_0) > 0 > \mu^-_{\beta,0}(\sigma_0)$.

Let us now check that if we sample a configuration according to μ, then it will be almost surely typical for either $\mu^+_{\beta,0}$ or $\mu^-_{\beta,0}$:

$$\mu(T^+ \cup T^-) \geq \lambda\, \mu^+_{\beta,0}(T^+) + (1-\lambda)\, \mu^-_{\beta,0}(T^-) = 1 .$$

Moreover, λ is the probability that the sampled configuration is typical for $\mu^+_{\beta,0}$:

$$\mu(T^+) = \lambda\, \mu^+_{\beta,0}(T^+) = \lambda \in (0,1) .$$

In the same way, $1-\lambda$ is the probability that the sampled configuration is typical of $\mu^-_{\beta,0}$. Let us ask the following question: *If the configuration sampled (under μ) was in T^+, what else can be said about its properties?* Since $\mu^+_{\beta,0}(T^+) = 1$ and $\mu^-_{\beta,0}(T^+) = 0$, we have, for all $B \in \mathscr{F}$,

$$\mu(B \cap T^+) = \lambda\, \mu^+_{\beta,0}(B \cap T^+) = \lambda\, \mu^+_{\beta,0}(B) = \mu(T^+)\, \mu^+_{\beta,0}(B) .$$

Therefore,

$$\mu(B \,|\, T^+) = \mu^+_{\beta,0}(B) .$$

In other words, *conditionally on the fact that one observes a configuration typical for $\mu^+_{\beta,0}$, the distribution is precisely given by $\mu^+_{\beta,0}$.*

For example, taking $T^+ = \{m > 0\}$, $T^- = \{m < 0\}$,

$$\mu(\cdot \,|\, m > 0) = \mu^+_{\beta,0} , \qquad \mu(\cdot \,|\, m < 0) = \mu^-_{\beta,0} .$$

This discussion shows that nonextremal Gibbs measures do not bring any new physics: everything that can be observed under such a measure is typical for one of the extremal Gibbs measures that appears in its decomposition. In this sense, the physically relevant elements of $\mathscr{G}(\pi)$ are the extremal ones.[9]

Digression: On the Simplex Structure for the Ising Model in $d = 2$
The nearest-neighbor Ising model on $\mathbb{Z}^2$ happens to be one of the very few models of equilibrium statistical mechanics for which the exact structure of $\mathscr{G}(\pi)$ is known. We make a few comments on this fact, whose full description is beyond the scope of this book, as it might help the reader to understand the following section on the extreme decomposition. This is also very closely related to the discussion in Section 3.10.8.

We continue with $h = 0$. The following can be proved for any $\beta > \beta_c(2)$:

1. $\mu^+_{\beta,0}$ *and* $\mu^-_{\beta,0}$ *are the* only *extremal Gibbs states*:
$$\text{ex}\,\mathscr{G}(\beta, 0) = \{\mu^-_{\beta,0}, \mu^+_{\beta,0}\}.$$
This follows from the discussion in Section 3.10.8.
2. *Any nonextremal Gibbs measure can be expressed in a unique manner as a convex combination of those two extremal elements*: if $\mu \in \mathscr{G}(\beta, 0)$, then there exists $\lambda \in [0, 1]$ such that
$$\forall B \in \mathscr{F}, \quad \mu(B) = \lambda\mu^+_{\beta,0}(B) + (1 - \lambda)\mu^-_{\beta,0}(B). \tag{6.72}$$
This representation induces in fact a one-to-one correspondence between measures in $\mathscr{G}(\beta)$ and the corresponding coefficient $\lambda \in [0, 1]$. Indeed, taking $B = \{\sigma_0 = 1\}$ in (6.72) shows that the coefficient λ associated with a measure $\mu \in \mathscr{G}(\beta)$ can be expressed as
$$\lambda = \frac{\mu(\sigma_0 = +1) - \mu^-_{\beta,0}(\sigma_0 = +1)}{\mu^+_{\beta,0}(\sigma_0 = +1) - \mu^-_{\beta,0}(\sigma_0 = +1)}.$$
In view of the discussion of the previous subsection, it is natural to interpret the pair $(\lambda, 1 - \lambda)$ as a probability distribution on $\text{ex}\,\mathscr{G}(\beta, 0)$.

All this can be compactly summarized by writing
$$\mathscr{G}(\beta, 0) = \left\{\lambda\mu^+_{\beta,0} + (1 - \lambda)\mu^-_{\beta,0}, \lambda \in [0, 1]\right\}. \tag{6.73}$$
This means that $\mathscr{G}(\beta, 0)$ is a **simplex**: it is a closed (Lemma 6.27), convex subset of $\mathscr{M}_1(\Omega)$, which is the convex hull of its extremal elements (that is, each of its elements can be written, in a unique way, as a convex combination of the extremal elements). Schematically,

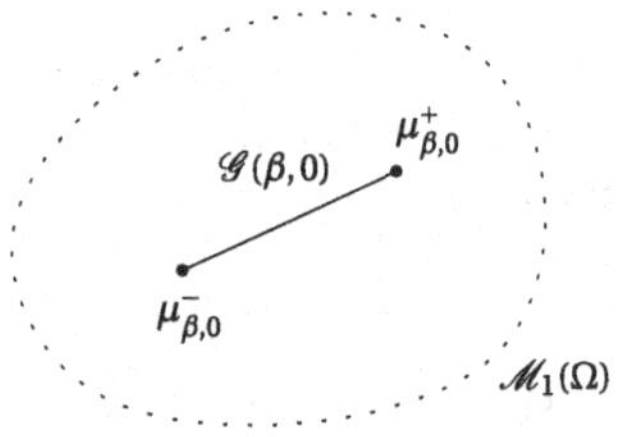

Accepting (6.73), we can clarify a point raised after Theorem 6.63: the thermodynamic limit does not always lead to an extremal state.

Example 6.68. Let $\mu^{\varnothing}_{\beta,0}$ be the Gibbs measure of the nearest-neighbor Ising model on $\mathbb{Z}^2$ prepared with the free boundary condition, which is constructed in Exercise 6.14. By (6.73), $\mu^{\varnothing}_{\beta,0}$ must be a convex combination of $\mu^{+}_{\beta,0}$ and $\mu^{-}_{\beta,0}$. But by symmetry, the only possibility is that $\mu^{\varnothing}_{\beta,0} = \frac{1}{2}\mu^{+}_{\beta,0} + \frac{1}{2}\mu^{-}_{\beta,0}$. Therefore, $\mu^{\varnothing}_{\beta,0}$ is not extremal as soon as $\mu^{+}_{\beta,0} \neq \mu^{-}_{\beta,0}$ (that is, when $\beta > \beta_c(2)$), although it was constructed using the thermodynamic limit. ◇

6.8.4 Extremal Decomposition

There are, unfortunately, very few nontrivial specifications π for which $\mathscr{G}(\pi)$ can be determined explicitly. However, as will be explained now, one can show in great generality that something similar to what we just saw in the case of the two-dimensional Ising model occurs: the set $\mathrm{ex}\,\mathscr{G}(\pi) \neq \varnothing$ and $\mathscr{G}(\pi)$ is always a simplex (although often an infinite-dimensional one).

Heuristics

Throughout the section, we assume that π is a specification for which $\mathscr{G}(\pi) \neq \varnothing$. (One can assume, for example, that π is quasilocal, but quasilocality itself is not necessary for the following results.) Our aim is to show that $\mathrm{ex}\,\mathscr{G}(\pi) \neq \varnothing$, and that any $\mu \in \mathscr{G}(\pi)$ can be expressed in a unique way as a convex combination of elements of $\mathrm{ex}\,\mathscr{G}(\pi)$. A priori, there can be uncountably many extremal Gibbs measures, so one can expect the combination to take the form of an integral:

$$\forall B \in \mathscr{F}, \quad \mu(B) = \int_{\mathrm{ex}\,\mathscr{G}(\pi)} \nu(B)\lambda_\mu(\mathrm{d}\nu). \tag{6.74}$$

Here, $\lambda_\mu(\cdot)$ is a probability distribution on $\mathrm{ex}\,\mathscr{G}(\pi)$ (the measurable structure on sets of probability measures will be introduced later) that plays the role of the coefficients $(\lambda, 1-\lambda)$ in (6.72); in particular,

$$\lambda_\mu(\mathrm{ex}\,\mathscr{G}(\pi)) = 1. \tag{6.75}$$

The main steps leading to (6.74) are as follows. To start, for each $B \in \mathscr{F}$, the definition of the conditional expectation allows us to write

$$\mu(B) = \int \mu(B \mid \mathscr{T}_\infty)(\omega)\mu(\mathrm{d}\omega).$$

The central ingredient will be to show that there exists a *regular version* of $\mu(\cdot \mid \mathscr{T}_\infty)$. This means that one can assign to each ω a probability measure $Q^\omega \in \mathscr{M}_1(\Omega)$ in such a way that

$$\mu(\cdot \mid \mathscr{T}_\infty)(\omega') = Q^{\omega'}(\cdot), \quad \text{for } \mu\text{-almost every } \omega'.$$

When such a family of measures Q^ω exists,

$$\mu(B) = \int Q^\omega(B)\mu(\mathrm{d}\omega), \tag{6.76}$$

which is a first step towards the decomposition of $\mu(B)$ we are after.

The idea behind the construction of $Q^{\cdot}$ given below can be illustrated as follows (although the true construction will be more involved). Consider the basic local property characterizing the measures of $\mathscr{G}(\pi)$, written in its integral form: for all $\Lambda \Subset \mathbb{Z}^d$,

$$\mu(B) = \int \pi_\Lambda(B\,|\,\omega)\mu(\mathrm{d}\omega)\,.$$

Then, taking $\Lambda \uparrow \mathbb{Z}^d$ formally in the previous display yields

$$\mu(B) = \int \underbrace{\lim_{\Lambda\uparrow\mathbb{Z}^d} \pi_\Lambda(B\,|\,\omega)}_{Q^\omega(B)}\,\mu(\mathrm{d}\omega)\,.$$

◇

Let us give two arguments in favor of the fact that, under the map $\omega \mapsto Q^\omega$, most of the configurations ω are mapped to a $Q^\omega \in \operatorname{ex}\mathscr{G}(\pi)$, in the sense that

$$\mu(Q^{\cdot} \in \operatorname{ex}\mathscr{G}(\pi)) = 1\,.$$

1. We have already seen (remember (6.64)) that $\mu(\cdot\,|\,\mathscr{T}_\infty)$ can be expressed as a limit:

   $$\mu(\cdot\,|\,\mathscr{F}_{\mathrm{B}(n)^{\mathrm{c}}}) \to \mu(\cdot\,|\,\mathscr{T}_\infty)\,.$$

   But since $\mu \in \mathscr{G}(\pi)$, we have $\mu(\cdot\,|\,\mathscr{F}_{\mathrm{B}(n)^{\mathrm{c}}})(\omega) = \pi_{\mathrm{B}(n)}(\cdot\,|\,\omega)$ for μ-almost all ω and for all n. We have also seen in Theorem 6.26 that the limits of sequences $\pi_{\mathrm{B}(n)}(\cdot\,|\,\omega)$, when they exist, belong to $\mathscr{G}(\pi)$. We therefore expect that

   $$Q^\omega(\cdot) \in \mathscr{G}(\pi)\,, \qquad \mu\text{-a.a. } \omega\,. \tag{6.77}$$

2. Moreover, if $A \in \mathscr{T}_\infty$, then $\mathbf{1}_A$ is $\mathscr{T}_\infty$-measurable and so $\mu(A\,|\,\mathscr{T}_\infty) = \mathbf{1}_A$ almost surely, which suggests that $Q^\omega(A) = \mathbf{1}_A(\omega)$; in other words, $Q^\omega(\cdot)$ should be trivial on $\mathscr{T}_\infty$, which by Theorem 6.58 means that

   $$Q^\omega \in \operatorname{ex}\mathscr{G}(\pi)\,, \qquad \mu\text{-a.a. } \omega\,.$$

The implementation of the above argument leads to a natural way of obtaining extremal elements: first take any $\mu \in \mathscr{G}(\pi)$, then condition it with respect to $\mathscr{T}_\infty$ and get (almost surely): $\mu(\cdot\,|\,\mathscr{T}_\infty) \in \operatorname{ex}\mathscr{G}(\pi)$. ◇

One should thus consider $\omega \mapsto Q^\omega$, roughly, as a mapping from Ω to $\operatorname{ex}\mathscr{G}(\pi)$:

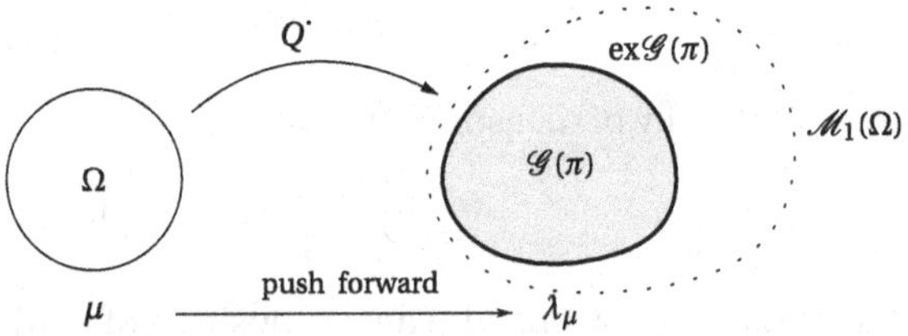

We would like to push μ forward onto $\mathscr{G}(\pi)$. Leaving aside the measurability issues, we proceed by letting, for $M \subset \mathscr{G}(\pi)$,

$$\lambda_\mu(M) \stackrel{\text{def}}{=} \mu(Q^{\cdot} \in M). \tag{6.78}$$

We then proceed as in elementary probability, and push the integration of μ over Ω onto an integration of λ_μ over $\mathrm{ex}\,\mathscr{G}(\pi)$. Namely,[1] for a function $\varphi\colon \mathscr{M}_1(\Omega) \to \mathbb{R}$,

$$\int_\Omega \varphi(Q^\omega)\mu(\mathrm{d}\omega) = \int_{\mathrm{ex}\,\mathscr{G}(\pi)} \varphi(\nu)\lambda_\mu(\mathrm{d}\nu). \tag{6.79}$$

If one defines, for all $B \in \mathscr{F}$, the **evaluation map** $e_B\colon \mathscr{M}_1(\Omega) \to [0,1]$ by

$$e_B(\nu) \stackrel{\text{def}}{=} \nu(B), \tag{6.80}$$

then (6.79) with $\varphi = e_B$ and (6.76) give (6.74).

Implementing the idea exposed in the two arguments given above is not trivial (albeit mostly technical); it will be rigorously established in Propositions 6.69 and 6.70 below.

Construction and Properties of the Kernel $Q^{\cdot}$

The family $\{Q^\omega\}_{\omega\in\Omega}$ is nothing but a regular conditional distribution for $\mu(\cdot \mid \mathscr{T}_\infty)$; it will be constructed using only the kernels of π. $Q^{\cdot}$ will be defined by a **probability kernel from $\mathscr{T}_\infty$ to $\mathscr{F}$**, which, similarly to the kernels introduced in Definition 6.9, is a mapping $\mathscr{F} \times \Omega \to [0,1]$, $(B,\omega) \mapsto Q^\omega(B)$ with the following properties:

- For each $\omega \in \Omega$, $B \mapsto Q^\omega(B)$ is a probability measure on $(\Omega, \mathscr{F})$.
- For each $B \in \mathscr{F}$, $\omega \mapsto Q^\omega(B)$ is $\mathscr{T}_\infty$-measurable.

Proposition 6.69. *There exists, for each $\omega \in \Omega$, a probability kernel Q^ω from $\mathscr{T}_\infty$ to $\mathscr{F}$ such that, for each $\mu \in \mathscr{G}(\pi)$,*

1. *For every bounded measurable $f\colon \Omega \to \mathbb{R}$,*

$$\mu(f \mid \mathscr{T}_\infty)(\cdot) = Q^{\cdot}(f), \quad \mu\text{-almost surely} \tag{6.81}$$

2. *$\{Q^{\cdot} \in \mathscr{G}(\pi)\} \in \mathscr{T}_\infty$ and $\mu(Q^{\cdot} \in \mathscr{G}(\pi)) = 1$.*

Proof ▶ *Construction of Q^ω.* Let $\pi = \{\pi_\Lambda\}_{\Lambda \Subset \mathbb{Z}^d}$ and let

$$\Omega_\pi \stackrel{\text{def}}{=} \bigcap_{C\in\mathscr{C}} \Bigl\{\omega \in \Omega\colon \lim_{n\to\infty} \pi_{\mathrm{B}(n)}(C \mid \omega) \text{ exists}\Bigr\}.$$

[1] What we are doing here is the exact analogue of the standard operation in probability theory. There, one defines the distribution of a random variable X, $\lambda_X(\cdot) \stackrel{\text{def}}{=} P(X \in \cdot)$, and uses it to express the expectation of functions of X as integrations over $\mathbb{R}$:

$$\int_\Omega g(X(\omega))P(\mathrm{d}\omega) = \int_{\mathbb{R}} g(x)\lambda_X(\mathrm{d}x).$$

Clearly, $\Omega_\pi \in \mathscr{T}_\infty$. When $\omega \in \Omega_\pi^c$, we define $Q^\omega \stackrel{\text{def}}{=} \mu_0$, where μ_0 is any fixed probability measure on Ω. When $\omega \in \Omega_\pi$, we define

$$Q^\omega(C) \stackrel{\text{def}}{=} \lim_{n\to\infty} \pi_{\mathrm{B}(n)}(C\,|\,\omega),$$

for each $C \in \mathscr{C}$. By construction, Q^ω is a probability measure on $\mathscr{C}$. By Theorem 6.96, it extends uniquely to $\mathscr{F}$. To prove $\mathscr{T}_\infty$-measurability, let

$$\mathscr{D} \stackrel{\text{def}}{=} \left\{B \in \mathscr{F} : \omega \mapsto Q^\omega(B) \text{ is } \mathscr{T}_\infty\text{-measurable}\right\}.$$

When $C \in \mathscr{C}$, we have, for all α, $\{Q^\cdot(C) \leq \alpha\} = \left(\{Q^\cdot(C) \leq \alpha\} \cap \Omega_\pi\right) \cup \left(\{Q^\cdot(C) \leq \alpha\} \cap \Omega_\pi^c\right) \in \mathscr{T}_\infty$. Therefore, $\mathscr{C} \subset \mathscr{D}$. We verify that $\mathscr{D}$ is a Dynkin class (see Appendix B.5): if $B, B' \in \mathscr{D}$, with $B \subset B'$, then $Q^\cdot(B' \setminus B) = Q^\cdot(B') - Q^\cdot(B)$ is $\mathscr{T}_\infty$-measurable, giving $B' \setminus B \in \mathscr{D}$. If $(B_n)_{n\geq 1} \subset \mathscr{D}$, $B_n \subset B_{n+1}$, $B_n \uparrow B$, then $Q^\cdot(B) = \lim_n Q^\cdot(B_n)$, and so $B \in \mathscr{D}$. Since $\mathscr{C}$ is stable under intersections, Theorem B.36 implies that $\mathscr{D} = \mathscr{F}$.

▶ *Relating Q^ω to $\mathscr{G}(\pi)$.* Let now $\mu \in \mathscr{G}(\pi)$. We have already seen in the proof of Theorem 6.63 that, on a set of μ-measure 1,

$$\pi_{\mathrm{B}(n)}(C\,|\,\cdot) = \mu(C\,|\,\mathscr{F}_{\mathrm{B}(n)^c})(\cdot) \xrightarrow{n\to\infty} \mu(C\,|\,\mathscr{T}_\infty)(\cdot) \quad \text{for all } C \in \mathscr{C}.$$

In particular, $\mu(\Omega_\pi) = 1$ and $\mu(C\,|\,\mathscr{T}_\infty) = Q^\cdot(C)$ μ-almost surely, for all $C \in \mathscr{C}$. Again, we can show that

$$\mathscr{D}' \stackrel{\text{def}}{=} \left\{B \in \mathscr{F} : \mu(B\,|\,\mathscr{T}_\infty) = Q^\cdot(B)\ \mu\text{-a.s.}\right\}$$

is a Dynkin class containing $\mathscr{C}$, giving $\mathscr{D}' = \mathscr{F}$. To show (6.81), one can assume that f is nonnegative, and take any sequence of simple functions $f_n \uparrow f$. Since each f_n is a finite sum of indicators and since $\mu(B\,|\,\mathscr{T}_\infty) = Q^\cdot(B)$ μ-a.s. for all $B \in \mathscr{F}$, it follows that $\mu(f_n\,|\,\mathscr{T}_\infty) = Q^\cdot(f_n)$ almost surely. The result follows by the Monotone Convergence Theorem.

To show that $\{Q^\cdot \in \mathscr{G}(\pi)\} \in \mathscr{T}_\infty$, we observe that

$$\begin{aligned}\{Q^\cdot \in \mathscr{G}(\pi)\} &= \bigcap_{\Lambda \Subset \mathbb{Z}^d} \bigcap_{A \in \mathscr{F}} \{Q^\cdot \pi_\Lambda(A) = Q^\cdot(A)\} \\ &= \bigcap_{\Lambda \Subset \mathbb{Z}^d} \bigcap_{C \in \mathscr{C}} \{Q^\cdot \pi_\Lambda(C) = Q^\cdot(C)\}. \end{aligned} \tag{6.82}$$

We used Lemma 6.22 in the second equality to obtain a countable intersection (over $C \in \mathscr{C}$). Since each $Q^\cdot(C)$ is $\mathscr{T}_\infty$-measurable, $Q^\cdot\pi_\Lambda(C)$ is also. Indeed, one can consider a sequence of simple functions $f_n \uparrow \pi_\Lambda(C\,|\,\cdot)$, giving $Q^\cdot\pi_\Lambda(C) = \lim_n Q^\cdot(f_n)$. Since each $Q^\cdot(f_n)$ is $\mathscr{T}_\infty$-measurable, its limit is also. This implies that each set $\{Q^\cdot\pi_\Lambda(C) = Q^\cdot(C)\} \in \mathscr{T}_\infty$ and, therefore, $\{Q^\cdot \in \mathscr{G}(\pi)\} \in \mathscr{T}_\infty$.

Now, if $\mu \in \mathscr{G}(\pi)$, we will show that $\mu(Q^\cdot\pi_\Lambda(C) = Q^\cdot(C)) = 1$ for all $C \in \mathscr{C}$, which with (6.82) implies $\mu(Q^\cdot \in \mathscr{G}(\pi)) = 1$, thus completing the proof of the proposition. Using (6.81), $\mathscr{F}_{\Lambda^c} \supset \mathscr{T}_\infty$ and the tower property of conditional expectation (the third to fifth inequalities below hold for μ-almost all ω),

$$\begin{aligned} Q^\omega \pi_\Lambda(C) &= Q^\omega(\pi_\Lambda(C\,|\,\cdot)) \\ &= Q^\omega(\mu(C\,|\,\mathscr{F}_{\Lambda^c})) \\ &= \mu(\mu(C\,|\,\mathscr{F}_{\Lambda^c})\,|\,\mathscr{T}_\infty)(\omega) \\ &= \mu(C\,|\,\mathscr{T}_\infty)(\omega) \\ &= Q^\omega(C)\,. \end{aligned}$$

□

Proposition 6.70. *If $\mu \in \mathscr{G}(\pi)$, then $\mu(Q^\cdot \in \mathrm{ex}\,\mathscr{G}(\pi)) = 1$.*

Note that this result has the following immediate, but crucial, consequence:

Corollary 6.71. *If $\mathscr{G}(\pi) \neq \varnothing$, then $\mathrm{ex}\,\mathscr{G}(\pi) \neq \varnothing$.*

The proof of Proposition 6.70 will rely partly on the characterization of $\mathscr{T}_\infty$ given in Theorem 6.58: extremal measures of $\mathscr{G}(\pi)$ are those that are trivial on $\mathscr{T}_\infty$. Furthermore, we have:

Exercise 6.27. Show that $\nu \in \mathscr{M}_1(\Omega)$ is trivial on $\mathscr{T}_\infty$ if and only if, for all $B \in \mathscr{F}$, $\nu(B\,|\,\mathscr{T}_\infty) = \nu(B)$ ν-almost surely. *Hint:* Half of the claim was already given in the proof of Theorem 6.58.

Proof of Proposition 6.70. Using Exercise 6.27,

$$\begin{aligned} \mathrm{ex}\,\mathscr{G}(\pi) &= \{\nu \in \mathscr{G}(\pi) : \nu \text{ is trivial on } \mathscr{T}_\infty\} \\ &= \{\nu \in \mathscr{G}(\pi) : \forall A \in \mathscr{F}, \nu(A\,|\,\mathscr{T}_\infty) = \nu(A), \nu\text{-a.s.}\} \\ &= \{\nu \in \mathscr{G}(\pi) : \forall C \in \mathscr{C}, \nu(C\,|\,\mathscr{T}_\infty) = \nu(C), \nu\text{-a.s.}\} \\ &= \{\nu \in \mathscr{G}(\pi) : \forall C \in \mathscr{C}, Q^\cdot(C) = \nu(C), \nu\text{-a.s.}\}\,. \end{aligned} \tag{6.83}$$

To prove the third identity, define $\mathscr{D}'' \stackrel{\text{def}}{=} \{A \in \mathscr{F} : \nu(A\,|\,\mathscr{T}_\infty) = \nu(A), \nu\text{-a.s.}\}$. Since $\mathscr{D}'' \supset \mathscr{C}$ and since $\mathscr{D}''$ is a Dynkin class (as can be verified easily), we have $\mathscr{D}'' = \mathscr{F}$. For all $C \in \mathscr{C}$, $\nu \in \mathscr{M}_1(\Omega)$, let $V_C(\nu)$ denote the variance of $Q^\cdot(C)$ under ν:

$$V_C(\nu) \stackrel{\text{def}}{=} E_\nu\left[(Q^\cdot(C) - E_\nu[Q^\cdot(C)])^2\right].$$

If $\nu \in \mathscr{G}(\pi)$, then $E_\nu[Q^\cdot(C)] = \nu(C)$ (because of (6.81)), and so

$$\mathrm{ex}\,\mathscr{G}(\pi) = \mathscr{G}(\pi) \cap \bigcap_{C \in \mathscr{C}} \{\nu \in \mathscr{M}_1(\Omega) : V_C(\nu) = 0\}.$$

Let $\mu \in \mathscr{G}(\pi)$. Since $\mu(Q^\cdot \in \mathscr{G}(\pi)) = 1$ (Proposition 6.69), we need to show that $\mu(V_C(Q^\cdot) = 0) = 1$ for each $C \in \mathscr{C}$. Since $V_C \geq 0$, it suffices to show that $\mu(V_C(Q^\cdot)) = 0$:

$$\mu(V_C(Q^{\cdot})) = \int \{E_{Q^\omega}[Q^{\cdot}(C)^2] - Q^\omega(C)^2\}\, \mu(\mathrm{d}\omega)$$
$$= \int \{E_\mu[Q^{\cdot}(C)^2 \mid \mathscr{T}_\infty](\omega) - Q^\omega(C)^2\}\, \mu(\mathrm{d}\omega) = 0\,,$$

where we used (6.81) for the second identity. □

Construction and Uniqueness of the Decomposition

Let $\mu \in \mathscr{G}(\pi)$. Since we are interested in having (6.79) valid for the evaluation maps e_B, we consider the smallest σ-algebra on $\mathscr{M}_1(\Omega)$ for which all the maps $\{e_B,\ B \in \mathscr{F}\}$ are measurable. Then, λ_μ in (6.78) defines a probability measure on this σ-algebra, and satisfies (6.75).

Theorem 6.72. *For all* $\mu \in \mathscr{G}(\pi)$,

$$\forall B \in \mathscr{F}, \quad \mu(B) = \int_{\mathrm{ex}\,\mathscr{G}(\pi)} \nu(B)\, \lambda_\mu(\mathrm{d}\nu)\,. \tag{6.84}$$

Moreover, λ_μ *is the unique measure on* $\mathscr{M}_1(\Omega)$ *for which such a representation holds.*

Proof The construction of (6.79) is standard. The definition $\lambda_\mu(M) \stackrel{\text{def}}{=} \mu(Q^{\cdot} \in M)$ can be expressed in terms of indicators:

$$\int_\Omega \mathbf{1}_M(Q^\omega)\, \mu(\mathrm{d}\omega) = \int_{\mathrm{ex}\,\mathscr{G}(\pi)} \mathbf{1}_M(\nu)\, \lambda_\mu(\mathrm{d}\nu).$$

An arbitrary bounded measurable function $\varphi\colon \mathscr{M}_1(\Omega) \to \mathbb{R}$ can be approximated by a sequence of finite linear combinations of indicator functions $\mathbf{1}_M$. Applying this with $\varphi = e_B$ yields (6.84).

Assume now that there exists another measure λ'_μ such that (6.84) holds with λ'_μ in place of λ_μ. Observe that any $\nu \in \mathrm{ex}\,\mathscr{G}(\pi)$ satisfies $\nu(Q^{\cdot} = \nu) = 1$ (see (6.83)). This implies that, for a measurable $M \subset \mathscr{M}_1(\Omega)$, $\nu(Q^{\cdot} \in M) = \mathbf{1}_M(\nu)$. Therefore,

$$\lambda'_\mu(M) = \int_{\mathrm{ex}\,\mathscr{G}(\pi)} \mathbf{1}_M(\nu)\, \lambda'_\mu(\mathrm{d}\nu)$$
$$= \int_{\mathrm{ex}\,\mathscr{G}(\pi)} \nu(Q^{\cdot} \in M)\, \lambda'_\mu(\mathrm{d}\nu) = \mu(Q^{\cdot} \in M) = \lambda_\mu(M)\,,$$

where we used (6.84) for the third identity. This shows that $\lambda_\mu = \lambda'_\mu$. □

The fact that any $\mu \in \mathscr{G}(\pi)$ can be decomposed over the extremal elements of $\mathscr{G}(\pi)$ is convenient when trying to establish uniqueness. Indeed, to show that $\mathscr{G}(\pi)$ is a singleton, by Theorem 6.72, it suffices to show that it contains a unique extremal element. Since the latter have distinguishing properties, proving that there is only one is often simpler. This is seen in the following proof of our result on uniqueness for one-dimensional systems, stated in Section 6.5.5.

Proof of Theorem 6.40. The proof consists in showing that $\mathscr{G}(\Phi)$ has a unique *extremal* measure. By Theorem 6.72, this implies that $\mathscr{G}(\Phi)$ is a singleton.

Let therefore $\mu, \nu \in \mathrm{ex}\mathscr{G}(\Phi)$. By Theorem 6.63, μ and ν can be constructed as thermodynamic limits. Let ω (resp. η) be such that $\pi^{\Phi}_{B(N)}(\cdot \mid \omega) \Rightarrow \mu$ (resp. $\pi^{\Phi}_{B(N)}(\cdot \mid \eta) \Rightarrow \nu$) as $N \to \infty$. By Lemma 6.43, we thus have, for all cylinders $C \in \mathscr{C}$,

$$\mu(C) = \lim_{N\to\infty} \pi^{\Phi}_{B(N)}(C \mid \omega) \geq e^{-2D} \lim_{N\to\infty} \pi^{\Phi}_{B(N)}(C \mid \eta) = e^{-2D}\nu(C).$$

It is easy to verify that $\mathscr{D} = \{A \in \mathscr{F} : \mu(A) \geq e^{-2D}\nu(A)\}$ is a monotone class. Since it contains $\mathscr{C}$, which is an algebra, it also coincides with $\mathscr{F}$, and so $\mu \geq e^{-2D}\nu$. In particular, ν is absolutely continuous with respect to μ. Since two distinct extremal measures are mutually singular (see Lemma 6.62), we conclude that $\mu = \nu$. □

6.9 The Variational Principle

The DLR formalism studied in the present chapter characterizes the Gibbs measures describing infinite systems through a collection of local conditions: $\mu\pi^{\Phi}_{\Lambda} = \mu$ for all $\Lambda \Subset \mathbb{Z}^d$. In this section, we present an alternative, variational characterization of translation-invariant Gibbs measures, which allows us to establish a relationship between the DLR formalism and the way equilibrium is described in thermostatics, as was presented in the introduction.

The idea behind the variational principle is of a different nature, and has a more thermodynamical flavor. It will only apply to *translation-invariant Gibbs measures*, and consists in defining an appropriate functional on the set of all translation-invariant probability measures, $\mathscr{W} : \mathscr{M}_{1,\theta}(\Omega) \to \mathbb{R}$, of the form

$$\mu \mapsto \mathscr{W}(\mu) = \text{Entropy}(\mu) - \beta \times \text{Energy}(\mu). \tag{6.85}$$

By analogy with (1.17) of Chapter 1, $-\frac{1}{\beta}\mathscr{W}(\mu)$ can be interpreted as the *free energy*. As seen at various places in Chapter 1, in particular in Section 1.3, equilibrium states are characterized as those that minimize the free energy; here, we will see that translation-invariant Gibbs measures are the minimizers of $-\frac{1}{\beta}\mathscr{W}(\cdot)$.

Remark 6.73. Since the relation between the material presented in this chapter and thermostatics is important from the physical point of view, we will again resort to the physicists' conventions regarding the inverse temperature and write, in particular, the potential as $\beta\Phi$. To ease notations, the temperature will usually not be explicitly indicated. ◇

6.9.1 Formulation in the Finite Case

To illustrate the content of the variational principle, let us consider the simplest case of a system living in a finite set $\Lambda \Subset \mathbb{Z}^d$. Let $\mathscr{M}_1(\Omega_\Lambda)$ denote the set of all probability distributions on Ω_Λ and let $\mathscr{H}_\Lambda : \Omega_\Lambda \to \mathbb{R}$ be a Hamiltonian. Define, for each $\mu_\Lambda \in \mathscr{M}_1(\Omega_\Lambda)$,

$$\mathscr{W}_\Lambda(\mu_\Lambda) \stackrel{\text{def}}{=} S_\Lambda(\mu_\Lambda) - \beta\langle\mathscr{H}_\Lambda\rangle_{\mu_\Lambda}, \tag{6.86}$$

where $S_\Lambda(\mu_\Lambda)$ is the Shannon entropy of μ_Λ, which has already been considered in Chapter 1. Here, we denote it by

$$S_\Lambda(\mu_\Lambda) \stackrel{\text{def}}{=} -\sum_{\omega_\Lambda\in\Omega_\Lambda} \mu_\Lambda(\omega_\Lambda)\log\mu_\Lambda(\omega_\Lambda), \tag{6.87}$$

and $\langle\mathscr{H}_\Lambda\rangle_{\mu_\Lambda}$ represents the **average energy** under μ_Λ. Our goal is to *maximize* $\mathscr{W}_\Lambda(\cdot)$ over all probability distributions $\mu_\Lambda \in \mathscr{M}_1(\Omega_\Lambda)$.

Notice that, at high temperature (small β), the dominant term is the entropy and $\mathscr{W}_\Lambda$ is maximal for the uniform distribution (remember Lemma 1.9). On the other hand, at low temperature (β large), the dominant term is the energy and $\mathscr{W}_\Lambda$ is maximal for distributions with a minimal energy. ◇

As we have already seen in Chapter 1, when μ_Λ is the Gibbs distribution associated with the Hamiltonian $\mathscr{H}_\Lambda$, $\mu_\Lambda^{\text{Gibbs}}(\omega_\Lambda) \stackrel{\text{def}}{=} \frac{e^{-\beta\mathscr{H}_\Lambda(\omega_\Lambda)}}{\mathbf{Z}_\Lambda}$, a simple computation shows that $\mathscr{W}(\mu_\Lambda^{\text{Gibbs}})$ coincides with the pressure of the system:

$$\mathscr{W}_\Lambda(\mu_\Lambda^{\text{Gibbs}}) = \log \mathbf{Z}_\Lambda\,.$$

Lemma 6.74 (Variational principle, finite version). *For all $\mu_\Lambda \in \mathscr{M}_1(\Omega_\Lambda)$,*

$$\mathscr{W}_\Lambda(\mu_\Lambda) \le \mathscr{W}_\Lambda(\mu_\Lambda^{\text{Gibbs}})\,.$$

Moreover, $\mu_\Lambda^{\text{Gibbs}}$ is the unique maximizer of $\mathscr{W}_\Lambda(\cdot)$.

Proof Since $\log(\cdot)$ is concave, Jensen's inequality gives, for all $\mu_\Lambda \in \mathscr{M}_1(\Omega_\Lambda)$,

$$\mathscr{W}_\Lambda(\mu_\Lambda) = \sum_{\omega_\Lambda\in\Omega_\Lambda}\mu_\Lambda(\omega_\Lambda)\log\frac{e^{-\beta\mathscr{H}_\Lambda(\omega_\Lambda)}}{\mu_\Lambda(\omega_\Lambda)} \le \log\sum_{\omega_\Lambda\in\Omega_\Lambda} e^{-\beta\mathscr{H}_\Lambda(\omega_\Lambda)} = \log\mathbf{Z}_\Lambda\,,$$

and equality holds if and only if $\frac{e^{-\beta\mathscr{H}_\Lambda(\omega_\Lambda)}}{\mu_\Lambda(\omega_\Lambda)}$ is constant, that is, if $\mu_\Lambda = \mu_\Lambda^{\text{Gibbs}}$. □

The variational principle can be expressed in a slightly different form, useful to understand what will be done later. Let us define the **relative entropy** of two distributions $\mu_\Lambda, \nu_\Lambda \in \mathscr{M}_1(\Omega_\Lambda)$ by

$$\mathsf{H}_\Lambda(\mu_\Lambda \mid \nu_\Lambda) \stackrel{\text{def}}{=} \begin{cases} \sum_{\omega_\Lambda\in\Omega_\Lambda}\mu_\Lambda(\omega_\Lambda)\log\frac{\mu_\Lambda(\omega_\Lambda)}{\nu_\Lambda(\omega_\Lambda)} & \text{if } \mu_\Lambda \ll \nu_\Lambda\,,\\ +\infty & \text{otherwise.}\end{cases} \tag{6.88}$$

(The interested reader can find a discussion of relative entropy and its basic properties in Appendix B.12.) First, H_Λ can be related to S_Λ by noting that, if λ_Λ denotes the uniform measure on Ω_Λ,

$$\mathsf{H}_\Lambda(\mu_\Lambda \mid \lambda_\Lambda) = \log|\Omega_\Lambda| - S_\Lambda(\mu_\Lambda). \tag{6.89}$$

Observe also that

$$\mathsf{H}_\Lambda(\mu_\Lambda \mid \mu_\Lambda^{\text{Gibbs}}) = \mathscr{W}_\Lambda(\mu_\Lambda^{\text{Gibbs}}) - \mathscr{W}_\Lambda(\mu_\Lambda), \tag{6.90}$$

so that the variational principle above can be reformulated as follows:

$$\mathsf{H}_\Lambda(\mu_\Lambda \mid \mu_\Lambda^{\text{Gibbs}}) \geq 0, \quad \textit{with equality if and only if} \quad \mu_\Lambda = \mu_\Lambda^{\text{Gibbs}}. \tag{6.91}$$

Exercise 6.28. Let $\mathscr{H}^{\text{per}}_{V_n;\beta,h}$ be the Hamiltonian of the Ising model in $V_n \stackrel{\text{def}}{=} \{0,\dots,n-1\}^d$ with periodic boundary condition. Show that, among all product probability measures $\mu_{V_n} = \bigotimes_{i\in V_n}\rho_i$ on $\{\pm1\}^{V_n}$ (where all ρ_i are equal), the unique measure maximizing

$$\mathscr{W}_{V_n}(\mu_{V_n}) \stackrel{\text{def}}{=} S_{V_n}(\mu_{V_n}) - \beta\langle\mathscr{H}^{\text{per}}_{V_n;\beta,h}\rangle_{\mu_{V_n}}$$

is the measure such that $\rho_i = \nu$ for all $i \in V_n$, where ν is the probability measure on $\{\pm1\}$ with mean m satisfying $m = \tanh(2d\beta m + h)$. In other words, the maximum is achieved by the product measure obtained through the "naive mean-field approach" of Section 2.5.1. In this sense, the latter is the best approximation of the original model among all product measures.

The rest of this section extends this point of view to infinite systems. To formulate the variational principle for infinite-volume Gibbs measures, we will need to introduce notions playing the role of the entropy and average energy for infinite systems. This will be done by considering the corresponding *densities*:

$$\lim_{\Lambda\uparrow\mathbb{Z}^d}\frac{1}{|\Lambda|}S_\Lambda(\mu_\Lambda), \qquad \lim_{\Lambda\uparrow\mathbb{Z}^d}\frac{1}{|\Lambda|}\langle\mathscr{H}_\Lambda\rangle_{\mu_\Lambda}.$$

The existence of these two limits will be established when μ_Λ is the marginal of a translation-invariant measure, in Propositions 6.75 and 6.78.

Remember that $\mathscr{G}(\beta\Phi)$ denotes the set of all probability measures compatible with the Gibbsian specification $\pi^{\beta\Phi}$, and $\mathscr{G}_\theta(\beta\Phi) \stackrel{\text{def}}{=} \mathscr{G}(\beta\Phi)\cap\mathscr{M}_{1,\theta}(\Omega)$.

In the following sections, we will define a functional $\mathscr{W}: \mathscr{M}_{1,\theta}(\Omega) \mapsto \mathbb{R}$, using the densities mentioned above, and characterize the Gibbs measures of $\mathscr{G}_\theta(\beta\Phi)$ as maximizers of $\mathscr{W}(\cdot)$. Notice that, since first-order phase transitions can occur on the infinite lattice, we do not expect uniqueness of the maximizer to hold in general.

6.9.2 Specific Entropy and Energy Density

Remember from the beginning of the chapter that the marginal of $\mu \in \mathscr{M}_1(\Omega)$ on $\Lambda \Subset \mathbb{Z}^d$ is $\mu|_\Lambda \stackrel{\text{def}}{=} \mu\circ\Pi_\Lambda^{-1} \in \mathscr{M}_1(\Omega_\Lambda)$. Since no confusion will be possible below, we will write μ_Λ instead of $\mu|_\Lambda$, to lighten the notations. One can then define the **Shannon entropy of μ in Λ** by

$$S_\Lambda(\mu) \stackrel{\text{def}}{=} S_\Lambda(\mu_\Lambda),$$

where $S_\Lambda(\mu_\Lambda)$ was defined in (6.87) (bearing in mind that μ_Λ is now the marginal of μ in Λ).

Proposition 6.75 (Existence of the specific entropy). *For all $\mu \in \mathcal{M}_{1,\theta}(\Omega)$,*

$$s(\mu) \stackrel{\text{def}}{=} \lim_{n\to\infty} \frac{1}{|\mathrm{B}(n)|} S_{\mathrm{B}(n)}(\mu) \tag{6.92}$$

exists and is called the ***specific entropy*** *of μ. Moreover,*

$$s(\mu) = \inf_{\Lambda\in\mathcal{R}} \frac{S_\Lambda(\mu)}{|\Lambda|}, \tag{6.93}$$

and $\mu \mapsto s(\mu)$ is ***affine****: for all $\mu, \nu \in \mathcal{M}_{1,\theta}(\Omega)$ and $\alpha \in (0,1)$,*

$$s(\alpha\mu + (1-\alpha)\nu) = \alpha s(\mu) + (1-\alpha)s(\nu).$$

Given $\mu, \nu \in \mathcal{M}_1(\Omega)$, let us use (6.88) to define the **relative entropy of μ with respect to ν (on Λ)**:

$$\mathsf{H}_\Lambda(\mu \,|\, \nu) \stackrel{\text{def}}{=} \mathsf{H}_\Lambda(\mu_\Lambda \,|\, \nu_\Lambda).$$

Lemma 6.76. *For all $\mu, \nu \in \mathcal{M}_1(\Omega)$ and all $\Lambda \Subset \mathbb{Z}^d$,*

1. *$\mathsf{H}_\Lambda(\mu \,|\, \nu) \geq 0$, with equality if and only if $\mu_\Lambda = \nu_\Lambda$,*
2. *$(\mu, \nu) \mapsto \mathsf{H}_\Lambda(\mu, \nu)$ is convex, and*
3. *if $\Delta \subset \Lambda$, then $\mathsf{H}_\Delta(\mu \,|\, \nu) \leq \mathsf{H}_\Lambda(\mu \,|\, \nu)$.*

Proof The first and second items are proved in Proposition B.66, so let us consider the third one. We can assume that $\mu_\Lambda \ll \nu_\Lambda$, otherwise the claim is trivial. Then,

$$\mathsf{H}_\Lambda(\mu \,|\, \nu) = \sum_{\omega_\Lambda} \phi\Big(\frac{\mu_\Lambda(\omega_\Lambda)}{\nu_\Lambda(\omega_\Lambda)}\Big)\nu_\Lambda(\omega_\Lambda),$$

where $\phi(x) \stackrel{\text{def}}{=} x\log x$ ($x \geq 0$) is convex. We then split ω_Λ into $\omega_\Lambda = \tau_\Delta\eta_{\Lambda\setminus\Delta}$, and consider the summation over $\eta_{\Lambda\setminus\Delta}$, for a fixed τ_Δ. Using Jensen's inequality for the distribution ν_Λ conditioned on τ_Δ, we get

$$\mathsf{H}_\Lambda(\mu \,|\, \nu) \geq \sum_{\tau_\Delta} \phi\Big(\frac{\mu_\Delta(\tau_\Delta)}{\nu_\Delta(\tau_\Delta)}\Big)\nu_\Delta(\tau_\Delta) = \mathsf{H}_\Delta(\mu \,|\, \nu).$$ □

Corollary 6.77. *$\mu \mapsto -S_\Lambda(\mu)$ is convex and, when $\Lambda, \Lambda' \Subset \mathbb{Z}^d$ are disjoint,*

$$S_{\Lambda\cup\Lambda'}(\mu) \leq S_\Lambda(\mu) + S_{\Lambda'}(\mu).$$

Proof The first claim follows from (6.89) and Lemma 6.76. A straightforward computation shows that, introducing $\nu \stackrel{\text{def}}{=} \mu_\Lambda \otimes \lambda_{\Lambda^c}$ with λ_{Λ^c} the uniform product measure on Ω_{Λ^c}, one gets

$$S_\Lambda(\mu) - S_{\Lambda\cup\Lambda'}(\mu) = \mathsf{H}_{\Lambda\cup\Lambda'}(\mu|\nu) - \log|\Omega_{\Lambda'}| \geq \mathsf{H}_{\Lambda'}(\mu|\nu) - \log|\Omega_{\Lambda'}| = -S_{\Lambda'}(\mu).$$

We used again Lemma 6.76 in the inequality. □

Proof of Proposition 6.75. By Corollary 6.77, the set function $a(\Lambda) \stackrel{\text{def}}{=} S_\Lambda(\mu)$ is both translation invariant and subadditive (see Section B.1.3): for all pairs of disjoint parallelepipeds Λ, Λ', $a(\Lambda \cup \Lambda') \le a(\Lambda) + a(\Lambda')$. The existence of the limit defining $s(\mu)$ is therefore guaranteed by Lemma B.6. By Corollary 6.77, $S_\Lambda(\cdot)$ is concave, which implies that $s(\cdot)$ is concave too. To verify that it is also convex, consider $\mu' \stackrel{\text{def}}{=} \alpha\mu + (1-\alpha)\nu$. Since $\log \mu'_\Lambda(\omega_\Lambda) \ge \log(\alpha\mu_\Lambda(\omega_\Lambda))$ and $\log \mu'_\Lambda(\omega_\Lambda) \ge \log((1-\alpha)\nu_\Lambda(\omega_\Lambda))$,

$$S_\Lambda(\mu') \le \alpha S_\Lambda(\mu) + (1-\alpha) S_\Lambda(\nu) - \alpha \log(1-\alpha) - (1-\alpha)\log(1-\alpha).$$

This implies that $s(\cdot)$ is also convex, proving the second claim of the proposition. □

Exercise 6.29. Show that $s(\cdot)$ is **upper semi-continuous**, that is,

$$\mu_k \Rightarrow \mu \quad \text{implies} \quad \limsup_{k\to\infty} s(\mu_k) \le s(\mu).$$

Let us now turn our attention to Gibbs measures and consider a potential $\Phi = \{\Phi_B\}_{B \Subset \mathbb{Z}^d}$. Until the end of the section, we will assume Φ to be absolutely summable and translation invariant, like the potentials considered in Section 6.7.1. Notice that translation invariance implies that Φ is in fact *uniformly* absolutely summable:

$$\sup_{i\in\mathbb{Z}^d} \sum_{\substack{B \Subset \mathbb{Z}^d:\\ B \ni i}} \|\Phi_B\|_\infty = \sum_{\substack{B \Subset \mathbb{Z}^d:\\ B \ni 0}} \|\Phi_B\|_\infty < \infty.$$

Proposition 6.78 (Existence of the average energy density). *For all $\mu \in \mathscr{M}_{1,\theta}(\Omega)$,*

$$\lim_{n\to\infty} \frac{1}{|\mathrm{B}(n)|} \big\langle \mathscr{H}_{\mathrm{B}(n);\Phi} \big\rangle_\mu = \langle u_\Phi \rangle_\mu, \tag{6.94}$$

where

$$u_\Phi \stackrel{\text{def}}{=} \sum_{\substack{B \Subset \mathbb{Z}^d:\\ B \ni 0}} \frac{1}{|B|} \Phi_B. \tag{6.95}$$

Proof First,

$$\Big| \mathscr{H}_{\mathrm{B}(n);\Phi} - \sum_{j\in \mathrm{B}(n)} \theta_j u_\Phi \Big| \le \sum_{j \in \mathrm{B}(n)} \sum_{\substack{B \ni j:\\ B \not\subset \mathrm{B}(n)}} \|\Phi_B\|_\infty \stackrel{\text{def}}{=} r_{\mathrm{B}(n);\Phi}. \tag{6.96}$$

The uniform absolute summability of Φ implies that $r_{\mathrm{B}(n);\Phi} = o(|\mathrm{B}(n)|)$ (see Exercise 6.30 below). By translation invariance, $\langle \theta_j u_\Phi \rangle_\mu = \langle u_\Phi \rangle_\mu$, which concludes the proof. □

Exercise 6.30. Show that, when Φ is absolutely summable and translation invariant,

$$\lim_{n\to\infty} \frac{r_{\mathrm{B}(n);\Phi}}{|\mathrm{B}(n)|} = 0\,. \tag{6.97}$$

The last object whose existence in the thermodynamic limit needs to be proved is the *pressure*.

Theorem 6.79. *When Φ is absolutely summable and translation invariant,*

$$\psi(\Phi) \stackrel{\text{def}}{=} \lim_{n\to\infty} \frac{1}{\beta|\mathrm{B}(n)|} \log \mathbf{Z}^{\eta}_{\mathrm{B}(n);\beta\Phi}$$

exists and does not depend on the boundary condition η; it is called the ***pressure****. Moreover, $\Phi \mapsto \psi(\Phi)$ is* ***convex*** *on the space of absolutely summable, translation-invariant potentials: if Φ^1, Φ^2 are two such potentials and $t \in (0,1)$, then*

$$\psi\big(t\Phi^1 + (1-t)\Phi^2\big) \le t\psi(\Phi^1) + (1-t)\psi(\Phi^2)\,.$$

We will actually see, as a byproduct of the proof, that the pressure equals

$$\beta\psi(\Phi) = s(\mu) - \beta\langle u_\Phi\rangle_\mu\,, \qquad \forall \mu \in \mathscr{G}_\theta(\beta\Phi)\,, \tag{6.98}$$

which is the analogue of the Euler relation (1.9).

We will show existence of the pressure by using the convergence proved above for the specific entropy and average energy. To start:

Lemma 6.80. *Let $\mu \in \mathscr{M}_{1,\theta}(\Omega)$ and $(\nu_n)_{n\ge1}$, $(\tilde\nu_n)_{n\ge1}$ be two arbitrary sequences in $\mathscr{M}_1(\Omega)$. If either of the sequences*

$$\Big(\frac{1}{|\mathrm{B}(n)|}\mathsf{H}_{\mathrm{B}(n)}(\mu \mid \nu_n\pi^{\beta\Phi}_{\mathrm{B}(n)})\Big)_{n\ge1}, \quad \Big(\frac{1}{|\mathrm{B}(n)|}\mathsf{H}_{\mathrm{B}(n)}(\mu \mid \tilde\nu_n\pi^{\beta\Phi}_{\mathrm{B}(n)})\Big)_{n\ge1} \tag{6.99}$$

has a limit as $n\to\infty$, then the other one does also, and the limits are equal.

Proof By the absolute summability of Φ, we have $\pi^{\beta\Phi}_{\mathrm{B}(n)}(\omega_{\mathrm{B}(n)} \mid \eta) > 0$ for all n, uniformly in $\omega_{\mathrm{B}(n)}$ and η. This guarantees that $\mu_{\mathrm{B}(n)} \ll \nu_n\pi^{\beta\Phi}_{\mathrm{B}(n)}$ and $\mu_{\mathrm{B}(n)} \ll \tilde\nu_n\pi^{\beta\Phi}_{\mathrm{B}(n)}$. Therefore,

$$\mathsf{H}_{\mathrm{B}(n)}(\mu \mid \nu_n\pi^{\beta\Phi}_{\mathrm{B}(n)}) - \mathsf{H}_{\mathrm{B}(n)}(\mu \mid \tilde\nu_n\pi^{\beta\Phi}_{\mathrm{B}(n)}) = \sum_{\omega_{\mathrm{B}(n)}} \mu_{\mathrm{B}(n)}(\omega_{\mathrm{B}(n)}) \log \frac{\tilde\nu_n\pi^{\beta\Phi}_{\mathrm{B}(n)}(\omega_{\mathrm{B}(n)})}{\nu_n\pi^{\beta\Phi}_{\mathrm{B}(n)}(\omega_{\mathrm{B}(n)})}\,.$$

Since

$$\sup_{\omega_{\mathrm{B}(n)},\eta,\tilde\eta} \big|\mathscr{H}_{\mathrm{B}(n);\Phi}(\omega_{\mathrm{B}(n)}\eta_{\mathrm{B}(n)^c}) - \mathscr{H}_{\mathrm{B}(n);\Phi}(\omega_{\mathrm{B}(n)}\tilde\eta_{\mathrm{B}(n)^c})\big| \le 2r_{\mathrm{B}(n);\Phi}\,, \tag{6.100}$$

where $r_{\mathrm{B}(n);\Phi}$ was defined in (6.96), we have,

$$e^{-4\beta r_{\mathrm{B}(n);\Phi}} \le \frac{\tilde{\nu}_n \pi^{\beta\Phi}_{\mathrm{B}(n)}(\omega_{\mathrm{B}(n)})}{\nu_n \pi^{\beta\Phi}_{\mathrm{B}(n)}(\omega_{\mathrm{B}(n)})} \le e^{4\beta r_{\mathrm{B}(n);\Phi}}.$$

By Exercise 6.30, $r_{\mathrm{B}(n);\Phi} = o(|\mathrm{B}(n)|)$, which proves the claim. □

Proof of Theorem 6.79. We use Lemma 6.80 with $\mu \in \mathscr{G}_\theta(\beta\Phi)$, $\nu_n = \mu$, and $\tilde{\nu}_n = \delta_\omega$, for some arbitrary fixed $\omega \in \Omega$. Then $\nu_n \pi^{\beta\Phi}_{\mathrm{B}(n)} = \mu$, so the first sequence in (6.99) is identically equal to zero. Therefore, the second sequence must converge to zero. By writing it explicitly, the second sequence becomes

$$\frac{1}{|\mathrm{B}(n)|} \mathsf{H}_{\mathrm{B}(n)}(\mu \mid \tilde{\nu}_n \pi^{\beta\Phi}_{\mathrm{B}(n)}) = \frac{1}{|\mathrm{B}(n)|} \log \mathbf{Z}^{\omega}_{\mathrm{B}(n);\beta\Phi} - \frac{1}{|\mathrm{B}(n)|} \big\{ S_{\mathrm{B}(n)}(\mu) - \beta \big\langle \mathscr{H}_{\mathrm{B}(n);\Phi}(\,\cdot\,\omega_{\mathrm{B}(n)^c}) \big\rangle_\mu \big\}. \qquad (6.101)$$

By Propositions 6.75 and 6.78, the second and third terms on the right-hand side are known to have limits, and the limit of the third one does not depend on ω, since, by (6.100),

$$\big\langle \mathscr{H}_{\mathrm{B}(n);\Phi}(\,\cdot\,\omega_{\mathrm{B}(n)^c}) \big\rangle_\mu = \big\langle \mathscr{H}_{\mathrm{B}(n);\Phi} \big\rangle_\mu + O(r_{\mathrm{B}(n);\Phi}).$$

Since the whole sequence on the right-hand side of (6.101) must converge to zero, this proves the existence of the pressure and justifies (6.98). As in the proof of Lemma 3.5, convexity follows from Hölder's inequality. □

6.9.3 Variational Principle for Gibbs Measures

Proposition 6.81. *Let Φ be absolutely convergent and translation invariant. Let $\nu \in \mathscr{G}_\theta(\beta\Phi)$. Then, for all $\mu \in \mathscr{M}_{1,\theta}(\Omega)$, the* ***Gibbs free energy***

$$\mathsf{h}(\mu \mid \Phi) \stackrel{\mathrm{def}}{=} \lim_{n\to\infty} \frac{1}{|\mathrm{B}(n)|} \mathsf{H}_{\mathrm{B}(n)}(\mu \mid \nu) \qquad (6.102)$$

exists and does not depend on ν (only on Φ). Moreover, $\mathsf{h}(\mu \mid \Phi)$ is nonnegative and satisfies

$$\mathsf{h}(\mu \mid \Phi) = \beta\psi(\Phi) - \{s(\mu) - \beta\langle u_\Phi\rangle_\mu\}. \qquad (6.103)$$

Proof If $\nu \in \mathscr{G}_\theta(\beta\Phi)$, then $\mathsf{H}_{\mathrm{B}(n)}(\mu \mid \nu) = \mathsf{H}_{\mathrm{B}(n)}(\mu \mid \nu\pi^{\beta\Phi}_{\mathrm{B}(n)})$. Therefore, to show the existence of the limit (6.102), we can use Lemma 6.80 with $\nu_n = \nu$. As earlier, we choose $\tilde{\nu}_n = \delta_\omega$, for which we know that the limit of the second sequence in (6.99) exists and equals $\beta\psi(\Phi) - \{s(\mu) - \beta\langle u_\Phi\rangle_\mu\}$, as seen after (6.101). □

We can now formulate the infinite-volume version of (6.91).

Theorem 6.82 (Variational principle). *Let $\mu \in \mathscr{M}_{1,\theta}(\Omega)$. Then*

$$\mu \in \mathscr{G}_\theta(\beta\Phi) \quad \textit{if and only if} \quad \mathsf{h}(\mu \mid \Phi) = 0. \qquad (6.104)$$

The variational principle stated above establishes the analogy between translation-invariant Gibbs measures and the basic principles of thermostatics, as announced at the beginning of the section.

We already know from (6.98) that $\mu \in \mathscr{G}_\theta(\beta\Phi)$ implies $\mathsf{h}(\mu \mid \Phi) = 0$. The proof of the converse statement is trickier; it will rely on the following lemma.

Lemma 6.83. *Let* $\mu, \nu \in \mathscr{M}_{1,\theta}(\Omega)$ *be such that*

$$\lim_{n\to\infty} \frac{1}{|\mathrm{B}(n)|} \mathsf{H}_{\mathrm{B}(n)}(\mu \mid \nu) = 0 . \tag{6.105}$$

Fix $\Delta \Subset \mathbb{Z}^d$*. Then, for all* $\delta > 0$ *and for all* k *for which* $\mathrm{B}(k) \supset \Delta$*, there exists some finite region* $\Lambda \supset \mathrm{B}(k)$ *such that*

$$0 \leq \mathsf{H}_\Lambda(\mu \mid \nu) - \mathsf{H}_{\Lambda\setminus\Delta}(\mu \mid \nu) \leq \delta . \tag{6.106}$$

We know from item 1 of Lemma 6.76 that, in a finite region, $\mathsf{H}_\Lambda(\mu \mid \nu) = 0$ *implies* $\mu_\Lambda = \nu_\Lambda$*. Although* (6.105) *does not necessarily imply that* $\mu = \nu$*,* (6.106) *will imply that* μ *and* ν *can be compared to each other on arbitrarily large regions* Λ *(see the proof of Theorem 6.82 below).* ◇

Proof We will repeatedly use the monotonicity of H_Λ in Λ, proved in Lemma 6.76. Fix $\delta > 0$ and k so that $\mathrm{B}(k) \supset \Delta$, and let n be such that

$$\frac{1}{|\mathrm{B}(n)|} \mathsf{H}_{\mathrm{B}(n)}(\mu \mid \nu) \leq \frac{\delta}{2|\mathrm{B}(k)|} .$$

If $m = \lfloor (2n+1)/(2k+1) \rfloor$, then at least m^d adjacent disjoint translates of $\mathrm{B}(k)$ can be arranged to fit in $\mathrm{B}(n)$; we denote them by $\mathrm{B}_1(k), \dots, \mathrm{B}_{m^d}(k)$. We assume for simplicity that $\mathrm{B}_1(k) = \mathrm{B}(k)$. For each $\ell \in \{2, \dots, m^d\}$, let i_ℓ be such that $\mathrm{B}_\ell(k) = i_\ell + \mathrm{B}(k)$. Define now $\Delta(\ell) \overset{\text{def}}{=} i_\ell + \Delta$ and $\Lambda(\ell) \overset{\text{def}}{=} \mathrm{B}_1(k) \cup \dots \cup \mathrm{B}_\ell(k)$. For convenience, let $\mathsf{H}_\varnothing(\mu \mid \nu) \overset{\text{def}}{=} 0$. Since $\Lambda(\ell) \setminus \mathrm{B}_\ell(k) \subset \Lambda(\ell) \setminus \Delta(\ell)$,

$$\begin{aligned}
\frac{1}{m^d} \sum_{\ell=1}^{m^d} \{\mathsf{H}_{\Lambda(\ell)}(\mu \mid \nu) - \mathsf{H}_{\Lambda(\ell)\setminus\Delta(\ell)}(\mu \mid \nu)\} &\leq \frac{1}{m^d} \sum_{\ell=1}^{m^d} \{\mathsf{H}_{\Lambda(\ell)}(\mu \mid \nu) - \mathsf{H}_{\Lambda(\ell)\setminus \mathrm{B}_\ell(k)}(\mu \mid \nu)\} \\
&= \frac{1}{m^d} \mathsf{H}_{\Lambda(m^d)}(\mu \mid \nu) \\
&\leq \frac{1}{m^d} \mathsf{H}_{\mathrm{B}(n)}(\mu \mid \nu) \\
&\leq \delta .
\end{aligned}$$

In the last line, we used the fact that $m^d \geq |\mathrm{B}(n)|/(2|\mathrm{B}(k)|)$. Since the first sum over ℓ corresponds to the arithmetic mean of a collection of nonnegative numbers, at least one of them must satisfy

$$\mathsf{H}_{\Lambda(\ell)}(\mu \mid \nu) - \mathsf{H}_{\Lambda(\ell)\setminus\Delta(\ell)}(\mu \mid \nu) \leq \delta .$$

One can thus take $\Lambda \stackrel{\text{def}}{=} \theta_{i_\ell}^{-1}\Lambda(\ell)$; translation invariance of μ and ν yields the desired result. □

Proof of Theorem 6.82. Let $\mu \in \mathcal{M}_{1,\theta}(\Omega)$ be such that $\mathsf{h}(\mu \mid \Phi) = 0$. We need to show that, for any $\Delta \Subset \mathbb{Z}^d$ and any local function f,

$$\mu\pi_\Delta^{\beta\Phi}(f) = \mu(f). \tag{6.107}$$

By Proposition 6.81, $\mathsf{h}(\mu \mid \Phi) = 0$ means that we can take any $\nu \in \mathcal{G}_\theta(\beta\Phi)$ and assume that

$$\lim_{n\to\infty} \frac{1}{|\mathrm{B}(n)|} \mathsf{H}_{\mathrm{B}(n)}(\mu \mid \nu) = 0.$$

In particular, $\mu_\Lambda \ll \nu_\Lambda$ for all $\Lambda \Subset \mathbb{Z}^d$; we will denote the corresponding Radon–Nikodým derivative by $\rho_\Lambda \stackrel{\text{def}}{=} \frac{d\mu_\Lambda}{d\nu_\Lambda}$. Observe that ρ_Λ is $\mathscr{F}_\Lambda$-measurable and that, for all $\mathscr{F}_\Lambda$-measurable functions g, $\mu(g) = \nu(\rho_\Lambda g)$.

Fix $\epsilon > 0$. To start, $\mu\pi_\Delta^{\beta\Phi}(f) = \mu(\pi_\Delta^{\beta\Phi} f)$ and, since $\pi_\Delta^{\beta\Phi} f$ is quasilocal and $\mathscr{F}_{\Delta^c}$-measurable, we can find some $\mathscr{F}_{\Delta^c}$-measurable local function g_* such that $\|\pi_\Delta^{\beta\Phi} f - g_*\|_\infty \leq \epsilon$. Let then k be large enough to ensure that $\mathrm{B}(k)$ contains Δ, as well as the supports of f and g_*. In this way, g_* is $\mathscr{F}_{\mathrm{B}(k)\setminus\Delta}$-measurable.

Let $\delta \stackrel{\text{def}}{=} \frac{r\epsilon}{2}$, and take $\Lambda \supset \mathrm{B}(k)$, as in Lemma 6.83. We write

$$\begin{aligned}\mu\pi_\Delta^{\beta\Phi}(f) - \mu(f) &= \mu(\pi_\Delta^{\beta\Phi} f - g_*) + \big(\mu(g_*) - \nu(\rho_{\Lambda\setminus\Delta} g_*)\big) \\ &\quad + \nu\big(\rho_{\Lambda\setminus\Delta}(g_* - \pi_\Delta^{\beta\Phi} f)\big) + \nu\big(\rho_{\Lambda\setminus\Delta}(\pi_\Delta^{\beta\Phi} f - f)\big) \\ &\quad + \nu\big((\rho_{\Lambda\setminus\Delta} - \rho_\Lambda) f\big) + \big(\nu(\rho_\Lambda f) - \mu(f)\big).\end{aligned}$$

We consider one by one the terms on the right-hand side of this last display. Since $\|\pi_\Delta^{\beta\Phi} f - g_*\|_\infty \leq \epsilon$, the first and third terms are bounded by ϵ. The second term is zero since g_* is $\mathscr{F}_{\Lambda\setminus\Delta}$-measurable, and the sixth term is zero since f is $\mathscr{F}_\Lambda$-measurable. Now the fourth term is zero too, since $\rho_{\Lambda\setminus\Delta}$ is $\mathscr{F}_{\Delta^c}$-measurable and since $\nu \in \mathcal{G}_\theta(\beta\Phi)$ implies that

$$\nu\big(\rho_{\Lambda\setminus\Delta}(\pi_\Delta^{\beta\Phi} f)\big) = \nu\big(\pi_\Delta^{\beta\Phi}(\rho_{\Lambda\setminus\Delta} f)\big) = \nu\pi_\Delta^{\beta\Phi}(\rho_{\Lambda\setminus\Delta} f) = \nu(\rho_{\Lambda\setminus\Delta} f).$$

Finally, consider the fifth term. First, notice that

$$\begin{aligned}\delta \geq \mathsf{H}_\Lambda(\mu \mid \nu) - \mathsf{H}_{\Lambda\setminus\Delta}(\mu \mid \nu) &= \mu\Big(\log \frac{\rho_\Lambda}{\rho_{\Lambda\setminus\Delta}}\Big) \\ &= \nu\Big(\rho_\Lambda \log \frac{\rho_\Lambda}{\rho_{\Lambda\setminus\Delta}}\Big) = \nu\big(\rho_{\Lambda\setminus\Delta}\phi(\rho_\Lambda/\rho_{\Lambda\setminus\Delta})\big),\end{aligned}$$

where $\phi(x) \stackrel{\text{def}}{=} 1 - x + x\log x$. It can be verified that there exists $r > 0$ such that

$$\phi(x) \geq r(|x-1| - \epsilon/2), \qquad \forall x \geq 0. \tag{6.108}$$

Using (6.108),

$$\nu\big(\rho_{\Lambda\setminus\Delta}\phi(\rho_\Lambda/\rho_{\Lambda\setminus\Delta})\big) \geq r\nu(|\rho_\Lambda - \rho_{\Lambda\setminus\Delta}|) - \frac{r\epsilon}{2}.$$

By definition of δ, this implies that $\nu(|\rho_\Lambda - \rho_{\Lambda\setminus\Delta}|) \le \epsilon$. The fifth term is therefore bounded by $\epsilon\|f\|_\infty$. Altogether, $|\mu\pi_\Lambda^{\beta\Phi}(f) - \mu(f)| \le (2 + \|f\|_\infty)\epsilon$, which proves (6.107) since ϵ was arbitrary. □

We have completed the program described at the beginning of the section:

Theorem 6.84. *Let Φ be an absolutely summable, translation-invariant potential. Define the affine functional $\mathscr{W} : \mathscr{M}_{1,\theta}(\Omega) \to \mathbb{R}$ by*

$$\mu \mapsto \mathscr{W}(\mu) \stackrel{\text{def}}{=} s(\mu) - \beta\langle u_\Phi\rangle_\mu .$$

Then the maximizers of $\mathscr{W}(\cdot)$ are the translation-invariant Gibbs measures compatible with $\pi^{\beta\Phi}$.

6.10 Continuous Spins

As we said at the beginning of the chapter, the DLR formalism can be developed for much more general single-spin spaces. In this section, we briefly discuss some of these extensions.

As long as the single-spin space is compact, most of the results stated and proved in the previous sections hold, although some definitions and some of the proofs given for $\Omega_0 = \{\pm 1\}$ need to be slightly adapted. In contrast, when the single-spin space is *not* compact, even the existence of Gibbs measures is not guaranteed (even for very reasonable interactions, as will be discussed in Chapter 8).

We discuss these issues briefly; for details on more general settings in which the DLR formalism can be developed, we refer the reader to [134].

6.10.1 General Definitions

Let Ω_0 be a separable metric space. Two guiding examples that the reader should keep in mind are $\mathbb{S}^1 = \{x \in \mathbb{R}^2 : \|x\|_2 = 1\}$ equipped with the Euclidean distance, or $\mathbb{R}$ with the usual distance $|\cdot|$.

The distance induces the Borel σ-algebra $\mathscr{B}_0$ on Ω_0, generated by the open sets. Let

$$\Omega_\Lambda \stackrel{\text{def}}{=} \Omega_0^\Lambda , \qquad \Omega \stackrel{\text{def}}{=} \Omega_0^{\mathbb{Z}^d} .$$

The measurable structure on Ω_Λ is the product σ-algebra

$$\mathscr{B}_\Lambda \stackrel{\text{def}}{=} \bigotimes_{i\in\Lambda} \mathscr{B}_0 ,$$

which is the smallest σ-algebra on Ω_Λ generated by the **rectangles**, that is, the sets of the form $\times_{i\in\Lambda} A_i$, $A_i \in \mathscr{B}_0$ for all $i \in \Lambda$. The projections $\Pi_\Lambda : \Omega \to \Omega_\Lambda$ are defined as before. For all $\Lambda \Subset \mathbb{Z}^d$,

$$\mathscr{C}(\Lambda) \stackrel{\text{def}}{=} \Pi_\Lambda^{-1}(\mathscr{B}_\Lambda)$$

denotes the σ-algebra of **cylinders with base in Λ** and, for $S \subset \mathbb{Z}^d$ (possibly infinite),

$$\mathscr{C}_S \stackrel{\text{def}}{=} \bigcup_{\Lambda \Subset \mathbb{Z}^d} \mathscr{C}(\Lambda)$$

is the **algebra of cylinders with base in** S, $\mathscr{F}_S \stackrel{\text{def}}{=} \sigma(\mathscr{C}_S)$. As we did earlier, we let $\mathscr{C} \stackrel{\text{def}}{=} \mathscr{C}_{\mathbb{Z}^d}$, $\mathscr{F} \stackrel{\text{def}}{=} \mathscr{F}_{\mathbb{Z}^d}$, and denote the set of probability measures on $(\Omega, \mathscr{F})$ by $\mathscr{M}_1(\Omega)$.

Rather than consider general specifications, we focus only on *Gibbsian* specifications. Let therefore $\Phi = \{\Phi_B\}_{B \Subset \mathbb{Z}^d}$ be a collection of maps $\Phi_B \colon \Omega \to \mathbb{R}$, where each Φ_B is $\mathscr{F}_B$-measurable, and absolutely summable, as in (6.25). We present two important examples.

- **The** $\mathsf{O}(N)$ **Model.** This model has single-spin space $\Omega_0 = \{x \in \mathbb{R}^N : \|x\|_2 = 1\}$ and its potential can be taken as

$$\Phi_B(\omega) = \begin{cases} -\beta\omega_i \cdot \omega_j & \text{if } B = \{i, j\},\ i \sim j, \\ 0 & \text{otherwise,} \end{cases} \tag{6.109}$$

  where $x \cdot y$ denotes the scalar product of $x, y \in \mathbb{R}^N$. The case $N = 2$ corresponds to the *XY* **model**, $N = 3$ corresponds to the **Heisenberg model**. These models, and their generalizations, will be discussed in Chapters 9 and 10.
- **The Gaussian Free Field.** Here, as already mentioned, $\Omega_0 = \mathbb{R}$ and

$$\Phi_B(\omega) = \begin{cases} \beta(\omega_i - \omega_j)^2 & \text{if } B = \{i, j\}, i \sim j, \\ \lambda\omega_i^2 & \text{if } B = \{i\}, \\ 0 & \text{otherwise.} \end{cases}$$

  This model will be discussed in Chapter 8, and some generalizations in Chapter 9.

Let $\mathscr{H}_{\Lambda;\Phi}$ denote the Hamiltonian associated with Φ, in a region $\Lambda \Subset \mathbb{Z}^d$, defined as in (6.24). For spins taking values in $\{\pm 1\}$, a Gibbsian specification associated with a Hamiltonian was defined pointwise in (6.30) through the numbers $\pi_\Lambda^\Phi(\tau_\Lambda \mid \omega)$. Due to the a-priori continuous nature of Ω_0 (as in the case $\Omega_0 = \mathbb{S}^1$), the finite-volume Gibbs distribution must be defined differently, since even configurations in a finite volume will usually have zero probability.

Assume therefore we are given a measure λ_0 on $(\Omega_0, \mathscr{B}_0)$, called the **reference measure**; λ_0 need not necessarily be a probability measure. In the case of $\mathbb{S}^1$ and $\mathbb{R}$, the most natural choice for λ_0 is the Lebesgue measure.[2] The product measure on $(\Omega_\Lambda, \mathscr{B}_\Lambda)$, usually denoted $\bigotimes_{i \in \Lambda} \lambda_0$ but which we will here abbreviate by λ_0^Λ, is defined by setting, for all rectangles $\times_{i \in \Lambda} A_i$,

$$\lambda_0^\Lambda(\times_{i \in \Lambda} A_i) \stackrel{\text{def}}{=} \prod_{i \in \Lambda} \lambda_0(A_i).$$

[2] In the case $\Omega_0 = \{-1, 1\}$, the (implicitly used) reference measure λ_0 was simply the counting measure $\lambda_0 = \delta_{-1} + \delta_1$.

We then define the **Gibbsian specification** $\pi^{\Phi} = \{\pi_{\Lambda}^{\Phi}\}_{\Lambda \Subset \mathbb{Z}^d}$ by setting, for all $A \in \mathscr{F}$ and all boundary conditions $\eta \in \Omega$ (compare with (6.30)),

$$\pi_{\Lambda}^{\Phi}(A \mid \eta) \stackrel{\text{def}}{=} \frac{1}{\mathbf{Z}_{\Lambda}^{\eta}} \int_{\Omega_\Lambda} \mathbf{1}_A(\omega_\Lambda \eta_{\Lambda^c})\, e^{-\mathscr{H}_{\Lambda;\Phi}(\omega_\Lambda \eta_{\Lambda^c})}\, \lambda_0^{\Lambda}(\mathrm{d}\omega_\Lambda), \tag{6.110}$$

where

$$\mathbf{Z}_{\Lambda;\Phi}^{\eta} \stackrel{\text{def}}{=} \int_{\Omega_\Lambda} e^{-\mathscr{H}_{\Lambda;\Phi}(\omega_\Lambda \eta_{\Lambda^c})}\, \lambda_0^{\Lambda}(\mathrm{d}\omega_\Lambda).$$

For convenience, we will sometimes use the following abbreviation:

$$\pi_{\Lambda}^{\Phi}(\mathrm{d}\omega \mid \eta) = \frac{e^{-\mathscr{H}_{\Lambda;\Phi}(\omega)}}{\mathbf{Z}_{\Lambda;\Phi}^{\eta}} \lambda_0^{\Lambda} \otimes \delta_\eta(\mathrm{d}\omega), \tag{6.111}$$

in which δ_η denotes the Dirac mass on Ω_{Λ^c} concentrated at η_{Λ^c}. The expectation of a function $f\colon \Omega \to \mathbb{R}$ with respect to $\pi_{\Lambda}^{\Phi}(\cdot \mid \eta)$ thus becomes, after integrating out over δ_η,

$$\pi_{\Lambda}^{\Phi} f(\eta) = \int_{\Omega_\Lambda} \frac{e^{-\mathscr{H}_{\Lambda;\Phi}(\omega_\Lambda \eta_{\Lambda^c})}}{\mathbf{Z}_{\Lambda;\Phi}^{\eta}} f(\omega_\Lambda \eta_{\Lambda^c}) \lambda_0^{\Lambda}(\mathrm{d}\omega_\Lambda).$$

All the notions introduced earlier, in particular the notion of consistency of kernels, of Gibbsian specification π^{Φ} and of the set of probability measures compatible with a specification π^{Φ}, $\mathscr{G}(\Phi)$, extend immediately to this more general setting.

The topological notions related to Ω have immediate generalizations. Let $\omega \in \Omega$. A sequence $(\omega^{(n)})_{n\geq 1} \subset \Omega$ **converges to** ω if, for all $j \in \mathbb{Z}^d$,

$$d(\omega_j^{(n)}, \omega_j) \to 0 \quad \text{as } n \to \infty.$$

The set $C(\Omega)$ of continuous functions is then defined as before. In the present general context, a sequence $\mu_n \in \mathscr{M}_1(\Omega)$ is said *to converge to* $\mu \in \mathscr{M}_1(\Omega)$, which we write $\mu_n \Rightarrow \mu$, if[2]

$$\mu_n(f) \to \mu(f) \quad \text{for all bounded local functions } f.$$

6.10.2 DLR Formalism for Compact Spin Space

When Ω_0 is compact, most of the important results of this chapter have immediate analogues. The starting point is that (as in the case $\Omega_0 = \{\pm 1\}$, see Proposition 6.20) the notions of convergence for configurations and measures make Ω and $\mathscr{M}_1(\Omega)$ sequentially compact. Proceeding exactly as in the proof of Theorem 6.26, the compactness of $\mathscr{M}_1(\Omega)$ and the Feller property allow us to show that there exists at least one Gibbs measure compatible with π^{Φ}: $\mathscr{G}(\Phi) \neq \varnothing$. Although some proofs need to be slightly adapted, all the main results presented on the structure of $\mathscr{G}(\Phi)$ when $\Omega_0 = \{\pm 1\}$ remain true when Ω_0 is a compact metric space. In particular, $\mathscr{G}(\Phi)$ is convex and its extremal elements enjoy the same properties as before:

Theorem 6.85. *(Compact spin space) Let Φ be an absolutely summable potential. Let $\mu \in \mathscr{G}(\Phi)$. The following conditions are equivalent characterizations of extremality.*

1. *μ is extremal.*
2. *μ is **trivial on** $\mathscr{T}_\infty$: if $A \in \mathscr{T}_\infty$, then $\mu(A)$ is either 1 or 0.*
3. *All $\mathscr{T}_\infty$-measurable functions are μ-almost surely constant.*
4. *μ has **short-range correlations:** for all $A \in \mathscr{F}$ (or, equivalently, for all $A \in \mathscr{C}$),*

$$\lim_{\Lambda \uparrow \mathbb{Z}^d} \sup_{B \in \mathscr{F}_{\Lambda^c}} \left|\mu(A \cap B) - \mu(A)\mu(B)\right| = 0\,. \tag{6.112}$$

Extremal elements can also be constructed using limits:

Theorem 6.86. *(Compact spin space) Let $\mu \in \mathrm{ex}\mathscr{G}(\Phi)$. Then, for μ-almost all ω,*

$$\pi^{\Phi}_{\mathrm{B}(n)}(\cdot \,|\, \omega) \Rightarrow \mu\,.$$

As in the case of finite single-spin space, to each $\mu \in \mathscr{G}(\Phi)$ corresponds a unique probability distribution λ_μ on $\mathscr{M}_1(\Omega)$, concentrated on the extremal measures of $\mathscr{G}(\Phi)$, leading to the following extremal decomposition:

Theorem 6.87. *(Compact spin space) For all $\mu \in \mathscr{G}(\Phi)$,*

$$\forall B \in \mathscr{F}, \quad \mu(B) = \int_{\mathrm{ex}\mathscr{G}(\pi)} \nu(B)\lambda_\mu(\mathrm{d}\nu)\,. \tag{6.113}$$

Moreover, λ_μ is the unique measure on $\mathscr{M}_1(\Omega)$ for which such a representation holds.

Finally, uniqueness results similar to Theorems 6.31 and 6.40 hold.

6.10.3 Symmetries

Let $(\mathsf{G}, \cdot)$ be a group acting on Ω_0 (this action can then be extended to Ω in the natural way as explained in Section 6.6). The notion of G-invariant potential is the same as before, but, in order to state the main result about symmetries, we need to assume that the reference measure is also invariant under G, that is, $\tau_{\mathsf{g}}\lambda_0 = \lambda_0$ for all $\mathsf{g} \in \mathsf{G}$.

To illustrate this, let us return to the two examples introduced above.

- For the $\mathsf{O}(N)$ models, the potential is invariant under the action of the orthogonal group $\mathsf{O}(N)$, which acts on Ω_0 via its representation as the set of all $N \times N$ orthogonal matrices. Observe that the reference measure is then $\mathsf{O}(N)$-invariant since the determinant of each such matrix is ± 1.
- For the Gaussian Free Field, the potential is invariant under the action of the group $(\mathbb{R}, +)$ (that is, under the addition of the same arbitrary real number

to all spins of the configuration). Since the reference measure is the Lebesgue measure, its invariance is clear.

One then gets:

Theorem 6.88. *Let* G *be an internal transformation group under which the reference measure is invariant. Let* π *be a* G*-invariant specification. Then,* $\mathscr{G}(\pi)$ *is preserved by* G*:*

$$\mu \in \mathscr{G}(\pi) \Rightarrow \tau_{\mathsf{g}}\mu \in \mathscr{G}(\pi), \quad \forall \mathsf{g} \in \mathsf{G}.$$

6.11 A Criterion for Nonuniqueness

The results on uniqueness and on the extremal decomposition mentioned earlier hold for a very general class of specifications. Unfortunately, it is much harder to establish *nonuniqueness* in a general setting and one usually has to resort to more model-specific methods. Two such approaches will be presented in Chapters 7 (the Pirogov–Sinai theory) and 10 (reflection positivity).

In the present section, we derive a criterion relating nonuniqueness to the nondifferentiability of a suitably defined pressure.[10] This provides a vast generalization of the corresponding discussion in Section 3.2.2. This criterion will be used in Chapter 10 to establish nonuniqueness in models with continuous spin. For simplicity, we will restrict our attention to translation-invariant potentials of finite range and assume that Ω_0 is compact.

Let Φ be a translation-invariant and finite-range potential, and g be any local function. For each $\Lambda \Subset \mathbb{Z}^d$, let $\Lambda(g) \stackrel{\text{def}}{=} \{i \in \mathbb{Z}^d : \text{supp}(g \circ \theta_i) \cap \Lambda \neq \varnothing\}$. Then define, for each $\omega \in \Omega$,

$$\psi^{\omega}_{\Lambda}(\lambda) \stackrel{\text{def}}{=} \frac{1}{|\Lambda(g)|} \log\Big\langle \exp\Big\{\lambda \sum_{j \in \Lambda(g)} g \circ \theta_j\Big\}\Big\rangle^{\omega}_{\Lambda;\Phi}, \tag{6.114}$$

where $\langle\cdot\rangle^{\omega}_{\Lambda;\Phi}$ denotes expectation with respect to the kernel $\pi^{\Phi}_{\Lambda}(\cdot \mid \omega)$.

Lemma 6.89. *For any sequence* $\Lambda_n \Uparrow \mathbb{Z}^d$ *and any sequence of boundary conditions* $(\omega_n)_{n\geq 1}$*, the limit*

$$\psi(\lambda) \stackrel{\text{def}}{=} \lim_{n\to\infty} \psi^{\omega_n}_{\Lambda_n}(\lambda) \tag{6.115}$$

exists and is independent of the choice of $(\Lambda_n)_{n\geq 1}$ *and* $(\omega_n)_{n\geq 1}$*. Moreover,* $\lambda \mapsto \psi(\lambda)$ *is convex.*

Although it could be adapted to the present situation, the existence of the pressure proved in Theorem 6.79 does not apply since, here, we do not assume that Ω_0 contains finitely many elements.

Proof Notice that $|\Lambda_n(g)|/|\Lambda_n| \to 0$. Using the fact that Φ has finite range and that g is local, one can repeat the same steps as in the proof of Theorem 3.6. Using the

Hölder inequality as we did in the proof of Lemma 3.5, we deduce that $\lambda \mapsto \psi^{\omega}_{\Lambda}(\lambda)$ is convex. □

Exercise 6.31. Complete the details of the proof of Lemma 6.89.

Remark 6.90. In (6.114), the expectation with respect to $\pi^{\Phi}_{\Lambda}(\cdot\,|\,\omega)$ can be substituted by the expectation with respect to any Gibbs measure $\mu \in \mathscr{G}(\Phi)$. Namely, let

$$\psi^{\mu}_{\Lambda}(\lambda) \stackrel{\text{def}}{=} \frac{1}{|\Lambda(g)|} \log \Big\langle \exp\Big\{\lambda \sum_{j\in\Lambda(g)} g \circ \theta_j\Big\}\Big\rangle_{\mu} . \tag{6.116}$$

Observe that, since $\langle f\rangle_{\mu} = \langle\langle f\rangle^{\cdot}_{\Lambda;\Phi}\rangle_{\mu}$, there exist ω', ω'' (depending on Λ, Φ, etc.) such that

$$\Big\langle \exp\Big\{\lambda \sum_{j\in\Lambda(g)} g \circ \theta_j\Big\}\Big\rangle^{\omega'}_{\Lambda;\Phi} \le \Big\langle \exp\Big\{\lambda \sum_{j\in\Lambda(g)} g \circ \theta_j\Big\}\Big\rangle_{\mu} \le \Big\langle \exp\Big\{\lambda \sum_{j\in\Lambda(g)} g \circ \theta_j\Big\}\Big\rangle^{\omega''}_{\Lambda;\Phi} .$$

(The existence of ω', ω'' follows from the fact that, for any local function f, the function $\omega \mapsto \langle f\rangle^{\omega}_{\Lambda;\Phi}$ is continuous and bounded and therefore attains its bounds.) Using Lemma 6.89, it follows that $\psi(\lambda) = \lim_{n\to\infty} \psi^{\mu}_{\Lambda_n}(\lambda)$. ◇

Remember that convexity guarantees that ψ possesses one-sided derivatives with respect to λ. As we did for the Ising model in Theorem 3.34 and Proposition 3.29, we can relate these derivatives to the expectation of g.

Proposition 6.91. *For all* $\mu \in \mathscr{G}_{\theta}(\Phi)$,

$$\frac{\partial\psi}{\partial\lambda^-}\Big|_{\lambda=0} \le \langle g\rangle_{\mu} \le \frac{\partial\psi}{\partial\lambda^+}\Big|_{\lambda=0} . \tag{6.117}$$

Moreover, there exist $\mu^+, \mu^- \in \mathscr{G}_{\theta}(\Phi)$, *such that*

$$\frac{\partial\psi}{\partial\lambda^+}\Big|_{\lambda=0} = \langle g\rangle_{\mu^+} , \qquad \frac{\partial\psi}{\partial\lambda^-}\Big|_{\lambda=0} = \langle g\rangle_{\mu^-} . \tag{6.118}$$

In practice, this result is used to obtain nonuniqueness (namely, the existence of distinct measures μ^+, μ^-) by finding a local function g for which ψ is not differentiable at $\lambda = 0$. In the Ising model, that function was $g = \sigma_0$.

Proof We first use the fact that $\psi = \lim_{n\to\infty} \psi^{\mu}_{\Lambda_n}(\lambda)$ (see Remark 6.90). Since ψ^{μ}_{Λ} is convex, it follows from (B.9) that, for all $\lambda > 0$,

$$\frac{\psi^{\mu}_{\Lambda}(\lambda) - \psi^{\mu}_{\Lambda}(0)}{\lambda} \ge \frac{\partial\psi^{\mu}_{\Lambda}}{\partial\lambda^+}\Big|_{\lambda=0} = \frac{1}{|\Lambda(g)|}\Big\langle \sum_{j\in\Lambda(g)} g\circ\theta_j\Big\rangle_{\mu} = \langle g\rangle_{\mu} ,$$

where the last identity is a consequence of the translation invariance of μ. Taking $\Lambda \Uparrow \mathbb{Z}^d$ followed by $\lambda \downarrow 0$, we get the upper bound in (6.117). The lower bound is obtained similarly.

Let us turn to the second claim. We now use the fact that $\psi = \lim_{n\to\infty} \psi^\omega_{\Lambda_n}$, for any ω. We *fix* $\lambda > 0$, and again use convexity: for all small $\epsilon > 0$,

$$\frac{\psi^\omega_\Lambda(\lambda) - \psi^\omega_\Lambda(\lambda - \epsilon)}{\epsilon} \le \frac{\partial \psi^\omega_\Lambda}{\partial \lambda^-}\Big|_\lambda = \frac{1}{|\Lambda(g)|} \frac{\big\langle (\sum_{i\in\Lambda(g)} g\circ\theta_i) e^{\lambda \sum_{i\in\Lambda(g)} g\circ\theta_i}\big\rangle^\omega_{\Lambda;\Phi}}{\big\langle e^{\lambda \sum_{i\in\Lambda(g)} g\circ\theta_i}\big\rangle^\omega_{\Lambda;\Phi}}$$

$$= \frac{1}{|\Lambda(g)|} \Big\langle \sum_{i\in\Lambda(g)} (g\circ\theta_i)\Big\rangle^\omega_{\Lambda;\Phi^\lambda}, \tag{6.119}$$

where the (translation-invariant) potential $\Phi^\lambda = \{\Phi^\lambda_B\}_{B\Subset\mathbb{Z}^d}$ is defined by

$$\Phi^\lambda_B \stackrel{\text{def}}{=} \begin{cases} \Phi_B + \lambda g\circ\theta_i & \text{if } B = \theta_i(\operatorname{supp} g), \\ \Phi_B & \text{otherwise.} \end{cases}$$

Now let $\mu^\lambda \in \mathscr{G}(\Phi^\lambda)$ be translation invariant (this is always possible by adapting the construction of Exercise 6.22). Integrating both sides of (6.119) with respect to μ^λ,

$$\Big\langle \frac{\psi^{\cdot}_\Lambda(\lambda) - \psi^{\cdot}_\Lambda(\lambda - \epsilon)}{\epsilon}\Big\rangle_{\mu^\lambda} \le \frac{1}{|\Lambda(g)|}\Big\langle \sum_{i\in\Lambda(g)} g\circ\theta_i\Big\rangle_{\mu^\lambda} = \langle g\rangle_{\mu^\lambda}.$$

Notice that $\|\psi^{\cdot}_\Lambda(\lambda)\|_\infty \le |\lambda|\|g\|_\infty < \infty$. Taking $\Lambda \Uparrow \mathbb{Z}^d$, followed by $\epsilon \downarrow 0$, we get

$$\langle g\rangle_{\mu^\lambda} \ge \frac{\partial\psi}{\partial\lambda^-}\Big|_\lambda \ge \frac{\partial\psi}{\partial\lambda^+}\Big|_{\lambda=0}, \qquad \forall \lambda > 0.$$

In the last step, we used item 3 of Theorem B.12. Consider now any sequence $(\lambda_k)_{k\ge1}$ decreasing to 0. By compactness (Theorem 6.24 applies here too), there exists a subsequence $(\lambda_{k_m})_{m\ge1}$ and a probability measure μ^+ such that $\mu^{\lambda_{k_m}} \Rightarrow \mu^+$ as $m \to \infty$. Clearly, μ^+ is also translation invariant and, by Exercise 6.32 below, $\mu^+ \in \mathscr{G}(\Phi)$. Since g is local, $\langle g\rangle_{\mu^{\lambda_k}} \to \langle g\rangle_{\mu^+}$. Applying (6.117) to μ^+, we conclude that $\langle g\rangle_{\mu^+} = \frac{\partial\psi}{\partial\lambda^+}\big|_{\lambda=0}$. □

Remark 6.92. It can be shown that the measures μ^+, μ^- in the second claim are in fact ergodic with respect to lattice translations. ◇

Exercise 6.32. Let $(\Phi^k)_{k\ge1}$, Φ be translation-invariant potentials of range at most r and such that $\|\Phi^k_B - \Phi_B\|_\infty \to 0$ when $k \to \infty$, for all $B \Subset \mathbb{Z}^d$. Let $\mu^k \in \mathscr{G}(\Phi^k)$ and μ be a probability measure such that $\mu^k \Rightarrow \mu$. Then $\mu \in \mathscr{G}(\Phi)$. *Hint:* Use a trick like (6.35).

Exercise 6.33. Consider the Ising model on $\mathbb{Z}^d$. Prove that $\beta \mapsto \psi^{\text{Ising}}(\beta, h)$ is differentiable whenever $|\mathscr{G}(\beta, h)| = 1$. *Hint:* To prove differentiability at β_0, combine Theorem 3.25 and Proposition 6.91 with $\lambda = \beta - \beta_0$ and $g = \frac{1}{2d}\sum_{i\sim0}\sigma_0\sigma_i$.

Remark 6.93. As a matter of fact, it can be proved[11] that the pressure of the Ising model on $\mathbb{Z}^d$ is differentiable with respect to β for *any* values of $\beta \geq 0$ and $h \in \mathbb{R}$, not only in the uniqueness regime. ◇

Exercise 6.34. Consider a one-dimensional model with a finite-range potential Φ, and let $\psi(\lambda)$ denote the pressure defined in (6.115), with $g \stackrel{\text{def}}{=} \sigma_0$. Use Proposition 6.91, combined with Theorem 6.40, to show that ψ is differentiable at $\lambda = 0$.

Remark 6.94. It can be shown that, in one-dimension, the pressure of a model with finite-range interactions is always *real-analytic* in its parameters.[12] ◇

6.12 Some Proofs

6.12.1 Proofs Related to the Construction of Probability Measures

The existence results of this chapter rely on the sequential compactness of Ω. This implies in particular the following property, actually equivalent to compactness:

Lemma 6.95. *Let $(C_n)_{n\geq 1} \subset \mathscr{C}$ be a decreasing ($C_{n+1} \subset C_n$) sequence of cylinders such that $\bigcap_n C_n = \varnothing$. Then $C_n = \varnothing$ for all large enough n.*

Proof Let each cylinder C_n be of the form $C_n = \Pi^{-1}_{\Lambda(n)}(A_n)$, where $\Lambda(n) \Subset \mathbb{Z}^d$ and $A_n \in \mathscr{P}(\Omega_{\Lambda(n)})$. With no loss of generality, we can assume that $\Lambda(n) \subset \Lambda(n+1)$ (remember the hint of Exercise 6.2). Assume that $C_n \neq \varnothing$ for all n, and let $\omega^{(n)} \in C_n$. Since $C_m \subset C_n$ for all $m > n$,

$$\Pi_{\Lambda(n)}(\omega^{(m)}) \in A_n \quad \text{for all } m > n\,.$$

By compactness of Ω, there exists ω^* and a subsequence $(\omega^{(n_k)})_{k\geq 1}$ such that $\omega^{(n_k)} \to \omega^*$. Of course,

$$\Pi_{\Lambda(n)}(\omega^*) \in A_n \quad \text{for all } n\,,$$

which implies $\omega^* \in C_n$ for all n. Therefore, $\bigcap_n C_n \neq \varnothing$. □

Theorem 6.96. *A finitely additive set function $\mu\colon \mathscr{C} \to \mathbb{R}_{\geq 0}$ with $\mu(\Omega) < \infty$ always has a unique extension to a measure on $\mathscr{F}$.*

Proof To use Carathéodory's Extension Theorem B.33, we must verify that if μ is finitely additive on $\mathscr{C}$, then it is also σ-additive on $\mathscr{C}$, in the sense that if $(C_n)_{n\geq 1} \subset \mathscr{C}$ is a sequence of pairwise disjoint cylinders such that $C = \bigcup_{n\geq 1} C_n \in \mathscr{C}$, then $\mu(C) = \sum_{n\geq 1} \mu(C_n)$. For this, it suffices to write $\bigcup_{n\geq 1} C_n = A_N \cup B_N$, where

$A_N = \bigcup_{n=1}^{N} C_n \in \mathscr{C}$, $B_N = \bigcup_{n>N} C_n \in \mathscr{C}$. Notice that, as $N \to \infty$, $A_N \uparrow C$ and $B_N \downarrow \varnothing$. Since $B_{N+1} \subset B_N$, Lemma 6.95 implies that there exists some N_0 such that $B_N = \varnothing$ when $N > N_0$. This also implies that $C_n = \varnothing$ for all $n > N_0$, and so $\mu(C) = \mu(A_{N_0}) = \sum_{n=1}^{N_0} \mu(C_n) = \sum_{n\geq 1} \mu(C_n)$. □

6.12.2 Proof of Theorem 6.5

We will prove Theorem 6.5 using the following classical result:

Theorem 6.97 (Riesz-Markov-Kakutani Representation Theorem on $\Omega = \{\pm 1\}^{\mathbb{Z}^d}$). *Let L: $C(\Omega) \to \mathbb{R}$ be a positive normalized linear functional, that is:*

1. *If $f \geq 0$, then $L(f) \geq 0$.*
2. *For all $f, g \in C(\Omega)$, $\alpha, \beta \in \mathbb{R}$, $L(\alpha f + \beta g) = \alpha L(f) + \beta L(g)$.*
3. *$L(1) = 1$.*

Then, there exists a unique measure $\mu \in \mathscr{M}_1(\Omega)$ such that

$$L(f) = \int f \mathrm{d}\mu\,, \quad \text{for all } f \in C(\Omega)\,.$$

This result holds in a much broader setting; its proof can be found in many textbooks. For the sake of concreteness, we give an elementary proof that makes use of the simple structure of $\Omega = \{-1, 1\}^{\mathbb{Z}^d}$.

Proof of Theorem 6.97. We use some of the notions developed in Section 6.4. Since $-\|f\|_\infty \leq f \leq \|f\|_\infty$, linearity and positivity of L yield $|L(f)| \leq \|f\|_\infty$. We have already seen that, for each cylinder $C \in \mathscr{C}$, $\mathbf{1}_C \in C(\Omega)$. Now let

$$\mu(C) \stackrel{\text{def}}{=} L(\mathbf{1}_C)\,.$$

Observe that $0 \leq \mu(C) \leq 1$, and that if $C_1, C_2 \in \mathscr{C}$ are disjoint, then $\mu(C_1 \cup C_2) = \mu(C_1) + \mu(C_2)$. By Theorem 6.96, μ extends uniquely to a measure on $(\Omega, \mathscr{F})$. To show that $\mu(f) = L(f)$ for all $f \in C(\Omega)$, let, for each n, f_n be a finite linear combination of the form $\sum_i a_i \mathbf{1}_{C_i}$, $C_i \in \mathscr{C}$, such that $\|f_n - f\|_\infty \to 0$. Then $\mu(f_n) = L(f_n)$ for all n, and therefore

$$|\mu(f) - L(f)| \leq |\mu(f) - \mu(f_n)| + |L(f_n) - L(f)| \leq 2\|f_n - f\|_\infty \to 0$$

as $n \to \infty$. □

We can now prove Theorem 6.5. Since a state $\langle\cdot\rangle$ is defined only on local functions, we must first extend it to continuous functions. Let $f \in C(\Omega)$ and let $(f_n)_{n\geq 1}$ be a sequence of local functions converging to f: $\|f_n - f\|_\infty \to 0$. Define $\langle f\rangle \stackrel{\text{def}}{=} \lim_n \langle f_n\rangle$. This definition does not depend on the choice of the sequence $(f_n)_{n\geq 1}$. Namely, if $(g_n)_{n\geq 1}$ is another such sequence, then $|\langle f_n\rangle - \langle g_n\rangle| \leq \|f_n - g_n\|_\infty \leq \|f_n - f\|_\infty + \|g_n - f\|_\infty \to 0$. The linear map $\langle\cdot\rangle$: $C(\Omega) \to \mathbb{R}$ then satisfies all the hypotheses of Theorem 6.97, which proves the result.

6.12.3 Proof of Theorem 6.6

We first define a probability measure on $\mathscr{C}$ and then extend it to $\mathscr{F}$ using Carathéodory's Extension Theorem (Theorem B.33). Consider a cylinder $C \in \mathscr{C}(\Lambda)$. Then C can be written in the form $C = \Pi_\Lambda^{-1}(A)$, where $A \in \mathscr{P}(\Omega_\Lambda)$. Now let

$$\mu(C) \stackrel{\text{def}}{=} \mu_\Lambda(A)\,.$$

The consistency condition (6.4) guarantees that this number is well defined. Namely, if C can also be written as $C = \pi_{\Lambda'}^{-1}(A')$, where $A' \in \mathscr{P}(\Omega_{\Lambda'})$, we must show that $\mu_\Lambda(A) = \mu_{\Lambda'}(A')$. But (remember Exercise 6.2), if Δ is large enough to contain both Λ and Λ', one can write $C = \pi_\Delta^{-1}(B)$, for some $B \in \mathscr{P}(\Omega_\Delta)$. But then $A = \Pi_\Lambda(\Pi_\Delta^{-1}(B))$, and so

$$\mu_\Lambda(A) = \mu_\Lambda\big(\Pi_\Lambda(\Pi_\Delta^{-1}(B))\big) = \mu_\Lambda(\Pi_\Lambda^\Delta(A)) = \mu_\Delta(B)\,.$$

The same with A' gives $\mu_\Lambda(A) = \mu_{\Lambda'}(A') = \mu_\Delta(B)$.

One then verifies that μ, defined as above, defines a probability measure on cylinders. For instance, if $C_1, C_2 \in \mathscr{C}$ are disjoint, then one can find some $\Delta \Subset \mathbb{Z}^d$ such that $C_1 = \Pi_\Delta^{-1}(A_1)$, $C_2 = \Pi_\Delta^{-1}(A_2)$, where $A_1, A_2 \in \mathscr{P}(\Omega_\Delta)$ are also disjoint. Then,

$$\begin{aligned}\mu(C_1 \cup C_2) = \mu(\Pi_\Delta^{-1}(A_1 \cup A_2)) &= \mu_\Delta(A_1 \cup A_2)\\ &= \mu_\Delta(A_1) + \mu_\Delta(A_2)\\ &= \mu(\Pi_\Delta^{-1}(A_1)) + \mu(\Pi_\Delta^{-1}(A_2)) = \mu(C_1) + \mu(C_2)\,.\end{aligned}$$

By Theorem 6.96, μ extends uniquely to a probability measure on $\mathscr{F}$, and (6.4) holds by construction.

Remark 6.98. Kolmogorov's Extension Theorem holds in more general settings, in particular for much more general single-spin spaces. ◇

6.12.4 Proof of Theorem 6.24

Let $\{C_1, C_2, \dots\}$ be an enumeration of all the cylinders of $\mathscr{C}$ (Exercise 6.2). First, we can extract from the sequence $(\mu_n(C_1))_{n\geq 1} \subset [0,1]$ a convergent subsequence $(\mu_{n_{1,j}}(C_1))_{j\geq 1}$ such that

$$\mu(C_1) \stackrel{\text{def}}{=} \lim_{j\to\infty} \mu_{n_{1,j}}(C_1) \quad \text{exists.}$$

Then, we extract from $(\mu_{n_{1,j}}(C_2))_{j\geq 1} \in [0,1]$ a convergent subsequence $(\mu_{n_{2,j}}(C_2))_{j\geq 1}$ such that

$$\mu(C_2) \stackrel{\text{def}}{=} \lim_{j\to\infty} \mu_{n_{2,j}}(C_2) \quad \text{exists.}$$

This process continues until we have, for each $k \geq 1$, a subsequence $(n_{k,j})_{j\geq 1}$ such that

$$\mu(C_k) \stackrel{\text{def}}{=} \lim_{j\to\infty} \mu_{n_{k,j}}(C_k)\,.$$

By considering the diagonal sequence $(n_{j,j})_{j\geq 1}$, we have that $\mu_{n_{j,j}}(C) \to \mu(C)$ for all $C \in \mathscr{C}$. Proceeding as in the proof of Theorem 6.97, using Lemma 6.95, we can verify that μ is a probability measure on $\mathscr{C}$ and use again Theorem 6.96 to extend it to a measure μ on $\mathscr{F}$. Obviously, $\mu_{n_{j,j}} \Rightarrow \mu$.

6.12.5 Proof of Proposition 6.39

To lighten the notations, we will omit $\beta\Phi$ most of the time. Let S_* denote the support of f. Assume that Λ is sufficiently large to contain S_*, and let $\Lambda' \stackrel{\text{def}}{=} \Lambda \setminus S_*$. Writing $\eta_\Lambda = \eta_{S_*}\eta_{\Lambda'}$, we have, by definition,

$$\begin{aligned}\pi_\Lambda f(\omega) &= \frac{1}{\mathbf{Z}_\Lambda^\omega}\sum_{\eta_\Lambda} f(\eta_{S_*})e^{-\beta\mathscr{H}_\Lambda(\eta_\Lambda\omega_{\Lambda^c})}\\ &= \frac{1}{\mathbf{Z}_\Lambda^\omega}\sum_{\eta_{S_*}} F_\Lambda^\omega(\eta_{S_*})\sum_{\eta_{\Lambda'}} e^{-\beta\mathscr{H}_{\Lambda'}(\eta_{\Lambda'}\eta_{S_*}\omega_{\Lambda^c})}\\ &= \sum_{\eta_{S_*}} F_\Lambda^\omega(\eta_{S_*})\frac{\mathbf{Z}_{\Lambda'}^{\omega'}}{\mathbf{Z}_\Lambda^\omega},\end{aligned} \tag{6.120}$$

where we have abbreviated $\eta_{S_*}\omega_{\Lambda^c}$ by ω', and defined

$$F_\Lambda^\omega(\eta_{S_*}) \stackrel{\text{def}}{=} f(\eta_{S_*})\exp\Big\{-\beta\sum_{\substack{B\cap S_*\neq\varnothing\\ B\cap\Lambda'=\varnothing}}\Phi_B(\eta_{S_*}\omega_{\Lambda^c})\Big\}.$$

First, observe that

$$\lim_{\Lambda\uparrow\mathbb{Z}^d} F_\Lambda^\omega(\eta_{S_*}) = f(\eta_{S_*})\exp\Big\{-\beta\sum_{B\subset S_*}\Phi_B(\eta_{S_*})\Big\}, \tag{6.121}$$

the latter expression being independent of Λ and ω. Indeed, by the absolute summability of the potential Φ,

$$\forall i,\ \lim_{r\to\infty}\sum_{\substack{B\ni i\\ \mathrm{diam}(B)>r}}\|\Phi_B\|_\infty = 0 \qquad \text{so that} \qquad \lim_{\Lambda\uparrow\mathbb{Z}^d}\sum_{\substack{B\cap S_*\neq\varnothing\\ B\cap\Lambda^c\neq\varnothing}}\|\Phi_B\|_\infty = 0\,.$$

$F_\Lambda^\omega(\eta_*)$ thus becomes independent of ω in the limit $\Lambda \uparrow \mathbb{Z}^d$. We will now prove that the same is true of the ratio appearing in (6.120). In order to do this, we will show that, when β is small, the ratio can be controlled using convergent cluster expansions, leading to crucial cancellations. We discuss explicitly only the case of $\mathbf{Z}_\Lambda^\omega$, the analysis being the same for $\mathbf{Z}_{\Lambda'}^{\omega'}$.

An application of the "+1 − 1 trick" (see Exercise 3.22) yields

$$e^{-\beta\mathscr{H}_\Lambda} = \prod_{B\cap\Lambda\neq\varnothing} e^{-\beta\Phi_B} = \sum_{\mathscr{B}}\prod_{B\in\mathscr{B}}\left(e^{-\beta\Phi_B}-1\right),$$

where the sum is over all finite collections $\mathscr{B}$ of finite sets B such that $B\cap\Lambda\neq\varnothing$. Of course, we can assume that the only sets B used are those for which $\Phi_B \not\equiv 0$ (this will be done implicitly from now on). We associate with each collection $\mathscr{B}$ a graph, as follows. With each $B \in \mathscr{B}$ is associated an abstract vertex x. We add an edge between two vertices x, x' if and only if they are associated with sets B, B' for which $B \cap B' \neq \varnothing$. The resulting graph is then decomposed into maximal connected components. To each such component, say with vertices $\{x_1, \dots, x_k\}$, corresponds a collection $\gamma = \{B_1, \dots, B_k\}$, called a **polymer**. The **support** of γ is defined by $\overline{\gamma} \stackrel{\text{def}}{=} B_1 \cup \cdots \cup B_k$. In $\mathbf{Z}_\Lambda^\omega$, one can interchange the summations over $\omega_\Lambda \in \Omega_\Lambda$ and $\mathscr{B}$ and obtain

$$\mathbf{Z}_\Lambda^\omega = 2^{|\Lambda|} \sum_\Gamma \prod_{\gamma \in \Gamma} \mathrm{w}(\gamma),$$

where the sum is over families Γ such that $\overline{\gamma} \cap \overline{\gamma}' = \varnothing$ whenever $\gamma = \{B_1, \dots, B_k\}$ and $\gamma' = \{B'_1, \dots, B'_{k'}\}$ are two distinct collections in Γ. The **weight** of γ is defined by

$$\mathrm{w}(\gamma) \stackrel{\text{def}}{=} 2^{-|\overline{\gamma} \cap \Lambda|} \sum_{\eta_{\overline{\gamma} \cap \Lambda}} \prod_{B \in \gamma} \left(e^{-\beta \Phi_B(\eta_{\overline{\gamma} \cap \Lambda} \omega_{\Lambda^c})} - 1 \right).$$

To avoid too heavy notation, we have not indicated the possible dependence of these weights on ω and Λ. Observe that, when $\overline{\gamma} \subset \Lambda$, $\mathrm{w}(\gamma)$ does not depend on ω. The following bound always holds:

$$|\mathrm{w}(\gamma)| \leq \prod_{B \in \gamma} \| e^{-\beta \Phi_B} - 1 \|_\infty . \tag{6.122}$$

We now show that, when β is small, the polymers and their weights satisfy condition (5.10), which guarantees convergence of the cluster expansion for $\log \mathbf{Z}_\Lambda^\omega$. We will use the function $a(\gamma) \stackrel{\text{def}}{=} |\overline{\gamma}|$.

Lemma 6.99. *Let $\epsilon \geq 0$. Assume that*

$$\alpha \stackrel{\text{def}}{=} \sup_{i \in \mathbb{Z}^d} \sum_{B \ni i} \| e^{-\beta \Phi_B} - 1 \|_\infty e^{(3+\epsilon)|B|} \leq 1 . \tag{6.123}$$

Then, for all γ_0, uniformly in ω and Λ,

$$\sum_{\gamma : \overline{\gamma} \cap \gamma_0 \neq \varnothing} |\mathrm{w}(\gamma)| e^{(1+\epsilon)|\overline{\gamma}|} \leq |\overline{\gamma}_0| . \tag{6.124}$$

Exercise 6.35. Show that, for any $\epsilon \geq 0$, any potential Φ satisfying (6.50) also satisfies (6.123) once β is small enough.

Proof of Lemma 6.99. Let $b(\gamma)$ denote the number of sets B_i contained in γ. Let, for all $n \geq 1$,

$$\xi(n) \stackrel{\text{def}}{=} \max_{i \in \Lambda} \sum_{\substack{\gamma : \overline{\gamma} \ni i \\ b(\gamma) \leq n}} |\mathrm{w}(\gamma)| e^{(1+\epsilon)|\overline{\gamma}|} . \tag{6.125}$$

We will show that, when (6.123) is satisfied,

$$\xi(n) \le \alpha\,, \qquad \forall n \ge 1, \tag{6.126}$$

which of course implies (6.124) after letting $n \to \infty$.

Let us first consider the case $n = 1$. In this case, γ contains a single set B and so $|\mathrm{w}(\gamma)| \le \|e^{-\beta\Phi_B} - 1\|_\infty$. This gives

$$\xi(1) \le \max_{i\in\Lambda} \sum_{B\ni i} \|e^{-\beta\Phi_B(\cdot)} - 1\|_\infty e^{(1+\epsilon)|B|} \le \alpha\,.$$

Let us then assume that (6.126) holds for n, and let us verify that it also holds for $n+1$. Since $\overline{\gamma} \ni i$, each polymer γ appearing in the sum for $n+1$ can be decomposed (not necessarily in a unique manner) as follows: $\gamma = \{B_0\} \cup \gamma^{(1)} \cup \cdots \cup \gamma^{(k)}$, where $B_0 \ni i$ and the $\gamma^{(j)}$s are polymers with disjoint support such that $b(\gamma^{(j)}) \le n$ and $\overline{\gamma}^{(j)} \cap B_0 \neq \varnothing$. Since the $\overline{\gamma}^{(j)}$s are disjoint,

$$|\mathrm{w}(\gamma)| \le 2^{|B_0|}\|e^{-\beta\Phi_{B_0}} - 1\|_\infty \prod_{j=1}^{k} |\mathrm{w}(\gamma^{(j)})|\,.$$

We have $|\overline{\gamma}| \le |B_0| + \sum_{j=1}^{k} |\overline{\gamma}^{(j)}|$ and therefore, for a fixed B_0, we can sum over the polymers $\gamma^{(j)}$ and use the induction hypothesis, obtaining a contribution bounded by

$$\sum_{k\ge 0}\frac{1}{k!} \sum_{\substack{\gamma^{(1)}:\\ \overline{\gamma}^{(1)}\cap B_0\neq\varnothing\\ b(\gamma^{(1)})\le n}} \cdots \sum_{\substack{\gamma^{(k)}:\\ \overline{\gamma}^{(k)}\cap B_0\neq\varnothing\\ b(\gamma^{(k)})\le n}} \prod_{j=1}^{k} |\mathrm{w}(\gamma^{(j)})| e^{(1+\epsilon)|\overline{\gamma}^{(j)}|} \le \sum_{k\ge 0}\frac{1}{k!}(|B_0|\xi(n))^k \le e^{\alpha|B_0|}\,,$$

and we are left with

$$\xi(n+1) \le \sum_{B_0\ni 0} 2^{|B_0|}\|e^{-\beta\Phi_{B_0}} - 1\|_\infty e^{(1+\epsilon)|B_0|} e^{\alpha|B_0|} \le \alpha\,.$$

In the last inequality, we used $\alpha \le 1$ and the definition of α. □

Proof of Proposition 6.39. Let $\epsilon > 0$ and let β_1 be such that (6.123) holds for all $\beta \le \beta_1$ (Exercise 6.35). We study the ratio in (6.120) by using convergent cluster expansions for its numerator and denominator. We use the terminology of Section 5.6. We denote by χ_Λ the set of clusters appearing in the expansion of $\log \mathbf{Z}_\Lambda^\omega$; the latter are made of polymers $\gamma = \{B_1, \dots, B_k\}$ for which $B_i \cap \Lambda \neq \varnothing$ for all i. The weight of $X \in \chi_\Lambda$ is denoted $\Psi_{\Lambda,\omega}(X)$ (see (5.20)); it is built using the weights $\mathrm{w}(\gamma)$, which can depend on ω if γ has a support that intersects Λ^c. Similarly, we denote by $\chi_{\Lambda'}$ the set of clusters appearing in the expansion of $\log \mathbf{Z}_{\Lambda'}^{\omega'}$; the latter are made of polymers $\gamma = \{B_1, \dots, B_k\}$ for which $B_i \cap \Lambda' \neq \varnothing$. The weight of $X \in \chi_{\Lambda'}$ is denoted $\Psi_{\Lambda',\omega'}(X)$. Let us denote the support of X by $\overline{X} \stackrel{\text{def}}{=} \bigcup_{\gamma\in X} \overline{\gamma}$. Taking $\beta \le \beta_1$ guarantees in particular that

$$\sum_{\gamma:\,\overline{\gamma}\cap\gamma_0\neq\varnothing} |\mathrm{w}(\gamma)| e^{|\overline{\gamma}|} \le |\overline{\gamma}_0|\,,$$

so we can expand that ratio using an absolutely convergent cluster expansion for each partition function:

$$\frac{\mathbf{Z}^{\omega'}_{\Lambda'}}{\mathbf{Z}^{\omega}_{\Lambda}} = 2^{-|S_*|}\frac{\exp\Bigl\{\sum_{X\in\chi_{\Lambda'}}\Psi_{\Lambda',\omega'}(X)\Bigr\}}{\exp\Bigl\{\sum_{X\in\chi_{\Lambda}}\Psi_{\Lambda,\omega}(X)\Bigr\}} = 2^{-|S_*|}\frac{\exp\Bigl\{\sum_{\substack{X\in\chi_{\Lambda'}\\ \overline{X}\cap S_*\neq\varnothing}}\Psi_{\Lambda',\omega'}(X)\Bigr\}}{\exp\Bigl\{\sum_{\substack{X\in\chi_{\Lambda}\\ \overline{X}\cap S_*\neq\varnothing}}\Psi_{\Lambda,\omega}(X)\Bigr\}}.$$

The second identity is due to the fact that each cluster $X \in \chi_{\Lambda'}$ in the numerator whose support does not intersect S_* also appears, with the same weight, in the denominator as a cluster $X \in \chi_\Lambda$. Their contributions thus cancel out. Among the remaining clusters, there are those that intersect Λ^c. These yield no contribution in the thermodynamic limit. Indeed, considering the denominator for example,

$$\sum_{\substack{X\in\chi_\Lambda:\\ \overline{X}\cap S_*\neq\varnothing\\ \overline{X}\cap\Lambda^c\neq\varnothing}} \bigl|\Psi_{\Lambda,\omega}(X)\bigr| \le |S_*| \max_{i\in S_*} \sum_{\substack{X:\,\overline{X}\ni i\\ \mathrm{diam}(\overline{X})\ge d(S_*,\Lambda^c)}} \bigl|\Psi_{\Lambda,\omega}(X)\bigr|, \tag{6.127}$$

and this last sum converges to zero when $\Lambda \uparrow \mathbb{Z}^d$. Indeed, we know (see (5.29)) that

$$\sum_{X:\,\overline{X}\ni i} \bigl|\Psi_{\Lambda,\omega}(X)\bigr| = \sum_{N\ge 1}\sum_{\substack{X:\,\overline{X}\ni i\\ \mathrm{diam}(\overline{X})=N}} \bigl|\Psi_{\Lambda,\omega}(X)\bigr|$$

is convergent. Therefore, the second sum above goes to zero when $N \to \infty$, allowing us to conclude that the contribution of the clusters intersecting Λ^c vanishes when $\Lambda \uparrow \mathbb{Z}^d$.

We are thus left with the clusters X that are strictly contained in Λ and intersect S_*. The weights of these do not depend on ω_{Λ^c} anymore (for that reason, the corresponding subscripts will be removed from their weights), but those that appear in the numerator have weights that still depend on η_{S_*} and their weights will be written, for simplicity, as $\Psi_{\eta_{S_*}}$. We get

$$\lim_{\Lambda\uparrow\mathbb{Z}^d} \frac{\mathbf{Z}^{\omega'}_{\Lambda'}}{\mathbf{Z}^{\omega}_{\Lambda}} = \exp\Bigl\{ \sum_{\substack{X:\,\overline{X}\not\subset S_*\\ \overline{X}\cap S_*\neq\varnothing}} \Psi_{\eta_{S_*}}(X) - \sum_{X:\,\overline{X}\cap S_*\neq\varnothing} \Psi(X)\Bigr\}.$$

Combined with (6.120) and (6.121), this completes the proof of the first claim.

Let us then see what more can be done when Φ has finite range: $r(\Phi) < \infty$. In this case, $F^{\omega}_{\Lambda}(\eta_{S_*})$ becomes *equal* to its limit as soon as Λ is large enough. Moreover, each cluster $X = \{\gamma_1, \dots, \gamma_n\}$ in the second sum of the right-hand side of (6.127) satisfies $\sum_{i=1}^{n} |\overline{\gamma}_i| \ge d(S_*, \Lambda^c)/r(\Phi)$. We can therefore write

$$\bigl|\Psi_{\Lambda,\omega}(X)\bigr| \le e^{-\epsilon d(S_*,\Lambda^c)/r(\Phi)}\bigl|\Psi^{\epsilon}_{\Lambda,\omega}(X)\bigr|,$$

where $\Psi^{\epsilon}_{\Lambda,\omega}(X)$ is defined as $\Psi_{\Lambda,\omega}(X)$, with $\mathrm{w}(\gamma)$ replaced by $\mathrm{w}(\gamma)e^{\epsilon|\overline{\gamma}|}$. Since this modified weight $\mathrm{w}(\gamma)e^{\epsilon|\overline{\gamma}|}$ also satisfies the condition ensuring the convergence of the cluster expansion (see (6.124)),

$$\sum_{\substack{X:\overline{X}\ni i\\ \mathrm{diam}(\overline{X})\geq d(S_*,\Lambda^c)}} \left|\Psi_{\Lambda,\omega}(X)\right| \leq e^{-\epsilon d(S_*,\Lambda^c)/r(\Phi)} \sum_{X:\overline{X}\ni i} \left|\Psi^{\epsilon}_{\Lambda,\omega}(X)\right|.$$

This last series is convergent as before. Gathering these bounds leads to (6.53). □

6.13 Bibliographical References

The notion of Gibbs measure was introduced independently by Dobrushin [88] and Lanford and Ruelle [204]. It has since then been firmly established as the proper probabilistic description of large classical systems of particles in equilibrium.

The standard reference to this subject is the well-known book by Georgii [134]. Although our aim is to be more introductory, large parts of the present chapter have benefited from [134], and the interested reader can consult that book for additional information and generalizations. We also strongly encourage the reader to have a look at its section on Bibliographical Notes, containing a large amount of information, presented in a very readable fashion.

Texts containing some introductory material on Gibbs measures include, for example, the books by Prum [282], Olivieri and Vares [258], Bovier [37] and Rassoul-Agha and Seppäläinen [283], the monograph by Preston [278], the lecture notes by Fernández [101] and by Le Ny [213]. The paper by van Enter, Fernández and Sokal [343] contains a very nice introduction, mostly without proofs, with a strong emphasis on the physical motivations behind the relevant mathematical concepts. The books by Israel [176] and Simon [308] also provide a general presentation of the subject, but their points of view are more functional-analytic than probabilistic.

Uniqueness. Dobrushin's Uniqueness Theorem, Theorem 6.31, was first proved in [88], but our presentation follows [108]; note that additional information can be extracted using the same strategy, such as exponential decay of correlations.

It can be shown that Dobrushin's condition of weak dependence cannot be improved in general [178, 309].

The one-dimensional uniqueness criterion given in Theorem 6.40 was originally proved in [49].

The approach in the proof of Theorem 6.38 is folklore.

Extremal decomposition. The integral decomposition (6.74) is usually derived from abstract functional-analytic arguments. Here, we follow the measure-theoretic approach of [134], itself based on an approach of Dynkin [97].

Variational principle. The exposition in Section 6.9 is inspired by [134, Chapter 15]. For a more general version of the variational principle, see Pfister's lecture notes [274]. Israel's book [176] develops the whole theory of Gibbs measures from the point of view of the variational principle and is a beautiful example of the kind of results that can be obtained within this framework.

6.14 Complements and Further Reading

6.14.1 The Equivalence of Ensembles

The variational principle allowed us to determine which translation-invariant infinite-volume measures are Gibbs measures. In this section, we explain, at a heuristic level, how the same approach might be used to prove a general version of the *equivalence of ensembles*, which we already mentioned in Chapter 1 and in Section 4.7.1.

For simplicity, we avoid the use of boundary conditions. Consider a finite region $\Lambda \Subset \mathbb{Z}^d$ (for example a box), and let Ω_Λ be as before. To stay simple, assume that the Hamiltonian is just a function $\mathscr{H}_\Lambda : \Omega_\Lambda \to \mathbb{R}$.

In Chapter 1, we introduced several probability distributions on Ω_Λ that were good candidates for the description of a system at equilibrium. The first was the microcanonical distribution $\nu^{\mathrm{Mic}}_{\Lambda;U}$, defined as the uniform distribution on the energy shell $\Omega_{\Lambda;U} \stackrel{\mathrm{def}}{=} \{\omega \in \Omega_\Lambda : \mathscr{H}_\Lambda(\omega) = U\}$. The second one was the canonical Gibbs distribution at inverse temperature β defined as $\mu_{\Lambda;\beta} \stackrel{\mathrm{def}}{=} e^{-\beta \mathscr{H}_\Lambda}/\mathbf{Z}_{\Lambda;\beta}$.

Obviously , these two distributions *differ* in finite volume. In view of the equivalence between these different descriptions in thermodynamics, one might however hope that these distributions yield similar predictions for large systems, or even become "identical" in the thermodynamic limit, at least when U and β are related in a suitable way. Properly stated, this is actually true and can be proved using the theory of large deviations.

In this section, we give a hint as to how this can be shown, but, since a full proof lies beyond the scope of this book, we will only motivate the result by a heuristic argument. The interested reader can find precise statements and detailed proofs in the papers of Lewis, Pfister and Sullivan [222, 223], or Georgii [133] and Deuschel, Stroock and Zessin [78]; a pedagogical account can be found in Pfister's lecture notes [274].

One way of trying to obtain the equivalence of $\nu^{\mathrm{Mic}}_{\Lambda;U}$ and $\mu_{\Lambda;\beta}$ in the thermodynamic limit is to proceed as in Proposition 6.81 and Theorem 6.82, and to find conditions under which

$$\frac{1}{|\Lambda|} \mathsf{H}_\Lambda(\nu^{\mathrm{Mic}}_{\Lambda;U} \,|\, \mu_{\Lambda;\beta}) \to 0\,, \qquad \text{when } \Lambda \uparrow \mathbb{Z}^d\,. \tag{6.128}$$

Although the setting is not the same as the one of Section 6.9.3 (in particular, the distributions under consideration are defined in finite volume and are thus not translation invariant), the variational principle at least makes it plausible that when this limit is zero, the thermodynamic limit of $\nu^{\mathrm{Mic}}_{\Lambda;U}$ is an infinite-volume Gibbs measure. (Let us emphasize that the proofs mentioned above do *not* proceed via (6.128); their approach however is similar in spirit.)

Remember that, for a finite system, a close relation between $\nu^{\mathrm{Mic}}_{\Lambda;U}$ and $\mu_{\Lambda;\beta}$ was established when it was shown, in Section 1.3, that if β is chosen properly as $\beta = \beta(U)$, then $\langle \mathscr{H}_\Lambda \rangle_{\mu_\beta} = U$ and $\mu_{\Lambda;\beta}$ has a maximal Shannon entropy among all distributions with this property. A new look can be given at this relation, in the light of the variational principle and the thermodynamic limit. Namely, observe that

$$\frac{1}{|\Lambda|}\mathsf{H}_\Lambda(\nu^{\mathrm{Mic}}_{\Lambda;U}\mid\mu_{\Lambda;\beta}) = -\frac{1}{|\Lambda|}S_\Lambda(\nu^{\mathrm{Mic}}_{\Lambda;U}) + \beta\Big\langle\frac{\mathscr{H}_\Lambda}{|\Lambda|}\Big\rangle^{\mathrm{Mic}}_{\Lambda;U} + \frac{1}{|\Lambda|}\log\mathbf{Z}_{\Lambda;\beta}$$
$$= -\frac{1}{|\Lambda|}\log|\Omega_{\Lambda;U}| + \beta\frac{U}{|\Lambda|} + \frac{1}{|\Lambda|}\log\mathbf{Z}_{\Lambda;\beta}\,.$$

In view of this expression, it is clear how (6.128) can be guaranteed. As was done for the variational principle in infinite volume, it is necessary to work with *densities*. So let us consider $\Lambda\uparrow\mathbb{Z}^d$, and assume that U also grows with the system, in such a way that $\frac{U}{|\Lambda|}\to u\in(h_{\min},h_{\max})$, where $h_{\min}\stackrel{\mathrm{def}}{=}\inf_\Lambda\inf_{\omega_\Lambda}\frac{\mathscr{H}_\Lambda(\omega_\Lambda)}{|\Lambda|}$, and $h_{\max}\stackrel{\mathrm{def}}{=}\sup_\Lambda\sup_{\omega_\Lambda}\frac{\mathscr{H}_\Lambda(\omega_\Lambda)}{|\Lambda|}$.

As we explained in (1.37),

$$\lim\frac{1}{V}\log\mathbf{Z}_{\Lambda;\beta} = -\inf_{\tilde u}\{\beta\tilde u - s_{\mathrm{Boltz}}(\tilde u)\}\,,$$

where s_{Boltz} is the Boltzmann entropy density

$$s_{\mathrm{Boltz}}(u)\stackrel{\mathrm{def}}{=}\lim\frac{1}{|\Lambda|}\log|\Omega_{\Lambda;U}|\,.$$

This shows that

$$\lim\frac{1}{|\Lambda|}\mathsf{H}_\Lambda(\nu^{\mathrm{Mic}}_{\Lambda;U}\mid\mu_{\Lambda;\beta}) = \beta u - s_{\mathrm{Boltz}}(u) - \inf_{\tilde u}\{\beta\tilde u - s_{\mathrm{Boltz}}(\tilde u)\}\,. \tag{6.129}$$

Now, the infimum above is realized for a particular value $\tilde u=\tilde u(\beta)$. If β is chosen in such a way that $\tilde u(\beta)=u$, we see that the right-hand side of (6.129) vanishes as desired. To see when this is possible, an analysis is required, along the same lines as what was done in Chapter 4 to prove the equivalence of the canonical and grand canonical ensembles at the level of thermodynamic potentials.

We thus conclude that *if equivalence of ensembles holds at the level of the thermodynamic potentials, then it should also hold at the level of measures.* As mentioned above, this conclusion can be made rigorous.

6.14.2 Pathologies of Transformations and Weaker Notions of Gibbsianness

The notion of Gibbs measure presented in this chapter, although efficient for the description of infinite systems in equilibrium, is not as robust as one might expect: the image of a Gibbs measure under natural transformations $T\colon\Omega\to\Omega$ can cease to be Gibbsian. An example of such a transformation has been mentioned in Section 3.10.11, when motivating the renormalization group.

Consider for example the two-dimensional Ising model at low temperature. Let $\mathscr{L}\stackrel{\mathrm{def}}{=}\{(i,0)\in\mathbb{Z}^2 : i\in\mathbb{Z}\}$ and consider the projection $\Pi_{\mathscr{L}}\colon\omega=(\omega_i)_{i\in\mathbb{Z}^2}\mapsto(\omega_j)_{j\in\mathscr{L}}$. The image of $\mu^+_{\beta,0}$ under $\Pi_{\mathscr{L}}$ is a measure ν^+_β on $\{\pm1\}^{\mathbb{Z}}$. It was shown by Schonmann [295] that ν^+_β *is not a Gibbs measure*: there exists no absolutely summable potential Φ so that ν_β^+ is compatible with the Gibbsian specification associated with Φ.

Before that, from a more general point of view, it had already been observed by Griffiths and Pearce [147], and Israel [177], that the same kind of phenomenon occurs when implementing rigorously certain renormalization group transformations. This is an important observation inasmuch as the renormalization group is often presented in the physics literature as a map defined on the space of all interactions (or Hamiltonians) (see the brief discussion in Section 3.10.11). What this shows is that such a map, which can always be defined on the set of probability measures, does not induce, in general, a map on the space of (physically reasonable) interactions.

More recently, there has been interest in whether the evolution of a Gibbs measure at temperature T under a stochastic dynamics corresponding to another temperature T' remains Gibbsian (which would again mean that one could follow the dynamics on the space of interactions). The observation is that the Gibbsian character can be quickly lost, depending on the values of T and T', see [341] for example.

A general discussion of this type of issues can be found in [343].

These so-called *pathologies* have led to the search for weaker notions of Gibbs measures, which would encompass the one presented in this chapter but would remain stable under transformations such as the one described above. This research was originally initiated by Dobrushin, and is nowadays known as *Dobrushin's restoration program.* A summary of the program can be found in the review of van Enter, Maes and Shlosman [342]. Other careful presentations of the subject are [101] and [213].

6.14.3 Gibbs Measures and the Thermodynamic Formalism

The ideas and techniques of equilibrium statistical mechanics have been useful in the theory of dynamical systems. For instance, Gibbs measures were introduced in ergodic theory by Sinai [311]. Moreover, the characterization of Gibbs measures via the variational principle of Section 6.9 is well suited for the definition of Gibbs measures in other settings. In symbolic dynamics, for instance, an invariant probability measure is said to be an *equilibrium measure* if it satisfies the variational principle. The monograph [40] by Bowen is considered as a pioneering contribution to this field. See also the books by Ruelle [291] or Keller [187], as well as Sarig's lecture notes [294].

7 Pirogov–Sinai Theory

As we have already discussed several times in previous chapters, a central task of equilibrium statistical physics is to characterize all possible macroscopic behaviors of the system under consideration, given the values of the relevant thermodynamic parameters. This includes, in particular, the determination of the *phase diagram* of the model. This can be tackled in at least two ways, as already seen in Chapter 3. In the first approach, one determines the set of all infinite-volume Gibbs measures as a function of the parameters of the model. In the second approach, one considers instead the associated pressure and studies its analytic properties as a function of its parameters; of particular interest is the determination of the set of values of the parameters at which the pressure fails to be differentiable.

Our goal in the present chapter is to introduce the reader to the *Pirogov–Sinai theory*, in which these two approaches can be implemented, at sufficiently low temperatures (or in other perturbative regimes), for a rather general class of models. This theory is one of the few frameworks in which first-order phase transitions can be established and phase diagrams constructed, under general assumptions.

To make the most out of this chapter, readers should preferably be familiar with the results derived for the Ising model in Chapter 3, as those provide useful intuition for the more complex problems addressed here. They should also be familiar with the cluster expansion technique described in Chapter 5, the latter being the basic tool we will use in our analysis. However, although it might help, a thorough understanding of the theory of Gibbs measures, as given in Chapter 6, is not required.

Conventions. We know from Corollary 6.41 that one-dimensional models with finite-range interactions do not exhibit phase transitions and thus possess a trivial phase diagram at all temperatures. We will therefore always assume, throughout the chapter, that $d \geq 2$.

It will once more be convenient to adopt the physicists' convention and let the inverse temperature β appear as a multiplicative constant in the Boltzmann weights and in the pressures. To lighten the notations, we will usually omit to mention β and the external fields, especially for partition functions.

7.1 Introduction

Most of Chapter 3 was devoted to the study of the phase diagram of the Ising model as a function of the inverse temperature β and magnetic field h. In particular, it was shown there that, at low temperature, the features that distinguish the regimes $h < 0, h = 0, h > 0$ are closely related to the *ground states* of the Ising Hamiltonian, that is, the configurations with lowest energy. These are given by η^- if $h < 0$, η^+ if $h > 0$ and both η^+ and η^- if $h = 0$ (we remind the reader that η^+ and η^- are the constant configurations $\eta_i^\pm = \pm 1$ for all $i \in \mathbb{Z}^d$). In dimension $d \geq 2$, the main features of the behavior of the model at low temperature can then be summarized as follows:

- When $h < 0$, resp. $h > 0$, there is a unique infinite-volume Gibbs measure: $\mathscr{G}(\beta, h) = \{\mu_{\beta,h}\}$. Moreover, the pressure $h \mapsto \psi_\beta(h)$ is differentiable (in fact, analytic) on these regions.
- At $h = 0$, a first-order phase transition occurs, characterized by the nondifferentiability of the pressure:
$$\left.\frac{\partial \psi_\beta}{\partial h^-}\right|_{h=0} \neq \left.\frac{\partial \psi_\beta}{\partial h^+}\right|_{h=0}.$$
When $h = 0$, the system becomes sensitive to the choice of boundary condition, in the sense that imposing $+$ or $-$ boundary conditions yields two distinct Gibbs measures in the thermodynamic limit,
$$\mu^+_{\beta,0} \neq \mu^-_{\beta,0}.$$
As seen when implementing Peierls' argument, at low temperature the typical configurations under each of these measures are described by small local deviations away from the ground state corresponding to the chosen boundary condition. Later, we will refer to this phenomenon as the *stability* of the two ground states (or of the two $+$ and $-$ boundary conditions) at the transition point.

These features can thus be summarized by the following picture:

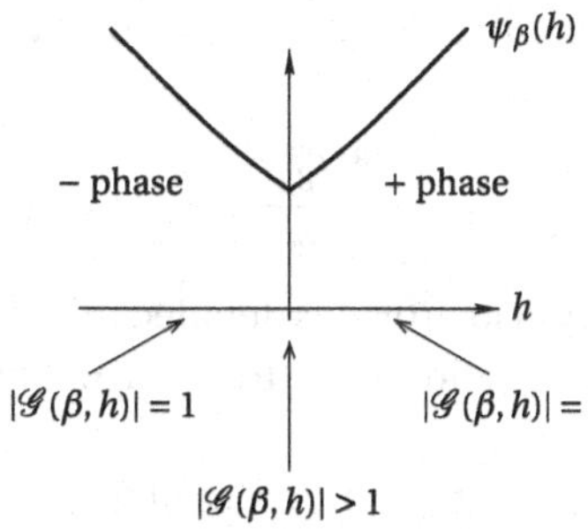

Figure 7.1. The phase diagram and pressure of the Ising model on $\mathbb{Z}^d$, $d \geq 2$, at low temperature.

We emphasize that the symmetry under the global spin flip enjoyed by the Ising model when $h = 0$ was a crucial simplifying feature when proving these results, and especially when implementing Peierls' argument (remember how spin flip symmetry was used on page 120).

In view of the above results, it is natural to wonder whether phase diagrams can be established rigorously for other models with more complicated interactions, in particular for models that do not enjoy any particular symmetry.

This is precisely the purpose of the Pirogov–Sinai theory (abbreviated PST below). Even though the theory applies in more general frameworks, we will only discuss models with finite single-spin space and finite-range interactions. Let us just mention two examples, the second of which will be the main subject of this chapter. (Other fields of application will be described in the Bibliographical References)

7.1.1 A Modified Ising Model

Consider, for example, the following modification of the formal Hamiltonian of the Ising model:

$$-\sum_{\{i,j\}\in\mathscr{E}_{\mathbb{Z}^d}} \omega_i\omega_j + \epsilon \sum_{\{i,j,k\}} \omega_i\omega_j\omega_k - h\sum_{i\in\mathbb{Z}^d} \omega_i\,, \tag{7.1}$$

where the second sum is over all triples $\{i,j,k\}$ having diameter bounded by 1, and ϵ is a small, fixed parameter.

When $\epsilon = 0$, this model coincides with the Ising model. But, as soon as $\epsilon \neq 0$, the Hamiltonian is no longer invariant under a global spin flip when $h = 0$, and there is no longer a reason for $h = 0$ to be the point of coexistence. Nevertheless, when $|\epsilon|$ is small, η^- and η^+ are the only possible ground states (see Exercise 7.6), and one might expect this model and the Ising model to have similar phase diagrams, except that the former's might not be symmetric in h when $\epsilon \neq 0$.

The above modification of the Ising model can be studied rigorously using the methods of PST. It can be proved that, once β is sufficiently large, there exists for all ϵ (not too large) a unique transition point $h_t = h_t(\beta,\epsilon)$ such that the pressure $h \mapsto \psi^{\mathrm{modif}}_{\beta,\epsilon}(h)$ is differentiable when $h < h_t$ and when $h > h_t$, but is not differentiable at h_t:

$$\frac{\partial \psi^{\mathrm{modif}}_{\beta,\epsilon}}{\partial h^-}\Big|_{h=h_t} \neq \frac{\partial \psi^{\mathrm{modif}}_{\beta,\epsilon}}{\partial h^+}\Big|_{h=h_t}.$$

In fact, the theory also provides detailed information on the behavior of h_t as a function of β and ϵ, and allows one to construct two distinct extremal Gibbs measures when $h = h_t$.

We will not discuss the properties of this model in detail here,[1] but after having read the chapter, the reader should be able to provide rigorous proofs of the above claims.

7.1.2 Models with Three or More Phases

The PST is however not restricted to models with only two equilibrium phases. In this chapter, in order to remain as concrete as possible, a large part of the discussion will be done for one particular model of interest: the **Blume–Capel model.**[2] In this model, spins take three values, $\omega_i \in \{+1, 0, -1\}$, and the formal Hamiltonian is defined by

$$\sum_{\{i,j\}\in\mathscr{E}_{\mathbb{Z}^d}} (\omega_i - \omega_j)^2 - h \sum_{i\in\mathbb{Z}^d} \omega_i - \lambda \sum_{i\in\mathbb{Z}^d} \omega_i^2 . \tag{7.2}$$

Depending on the values of λ and h, this Hamiltonian has three possible ground states, given by the constant configurations η^+, η^0 and η^- (ignoring, for the moment, possible boundary effects). The set of pairs $(\lambda, h) \in \mathbb{R}^2$ then splits into three regions $\mathscr{U}^+, \mathscr{U}^0, \mathscr{U}^-$ such that $\eta^\#$ is the unique ground state when (λ, h) belongs to the interior of $\mathscr{U}^\#$. The picture represented in Figure 7.2a illustrates this, and is called the **zero-temperature phase diagram.**

We will prove that, at low temperature, the phase diagram is a small deformation of the latter (in a sense that will be made precise later); see Figure 7.2b.

We will see in Theorem 7.36 that the pressure $(\lambda, h) \mapsto \psi_\beta(\lambda, h)$ is differentiable everywhere, except on the coexistence lines, across which its derivatives are discontinuous.

A qualitative plot of ψ_β can be found in Figure 7.3.

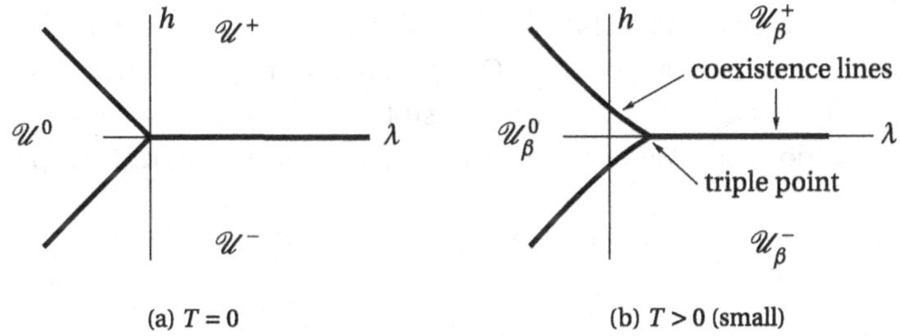

Figure 7.2. The Blume–Capel model at $T = 0$ ($\beta = \infty$) and small $T > 0$ ($\beta < \infty$, large). (a) At zero temperature, the phase diagram is just a partition of the (λ, h) plane into regions with different ground state(s): when $(\lambda, h) \in \mathscr{U}^\#$, $\eta^\#$ is a ground state. On the boundaries of these regions, several ground states coexist. In particular, there are three ground states when $(\lambda, h) = (0, 0)$. (b) At low temperature, the phase diagram is a small and smooth deformation of the zero-temperature one. When $(\lambda, h) \in \mathscr{U}_\beta^\#$, an extremal Gibbs measure $\mu^\#_{\beta;\lambda,h}$ can be constructed using the boundary condition #; typical configurations under this measure are described by small deviations from the ground state $\eta^\#$. There exists a triple point $(\lambda_t, 0)$, at which these three distinct extremal Gibbs measures coexist. From the triple point emanate three coexistence lines. On each of these, exactly two of these measures coexist. The rest of the diagram consists of uniqueness regions. The symmetry by a reflection across the λ-axis is due to the invariance of the Hamiltonian under the interchange of + and − spins. This phase diagram will be rigorously established in Section 7.4.

The analysis will also provide information on the structure of typical configurations, in Theorem 7.44.

Remark 7.1. The principles underlying the Pirogov–Sinai theory are rather general, robust and apply in many situations. Nevertheless, their current implementation requires perturbative techniques. As a consequence, this theory can provide precise information regarding the dependence of a model on its parameters only for regions of the parameter space that lie in a neighborhood of a regime that is already well understood. In this chapter, this will be the zero-temperature regime, and the results will thus only hold at sufficiently low temperatures (usually, *very* low temperatures). ◇

7.1.3 Overview of the Chapter

We will first introduce the general notion of *ground state* in Section 7.2 and describe the basic structure that a model with finite-range interactions should have in order to enter the framework of the Pirogov–Sinai theory. Ultimately, this will lead to a representation of its partition function as a polymer model in Section 7.3.

In a second step, we will study those polymer models at low temperature and construct the phase diagram in Section 7.4. For the sake of concreteness, as in the rest of the book, we will avoid adopting too general a point of view and implement this construction only for the Blume–Capel model. The reason for this choice is that this model is representative of the class of models to which this approach can be applied: Its analysis is sufficiently complicated to require the use of all the main ideas of PST, but simple enough to keep the discussion (and the notations) as elementary as possible. On the one hand, the absence of symmetry between the 0 and ± 1 spins makes it impossible to implement a "naive" Peierls' argument, as was done for the Ising model (remember how the ratio of partition functions was

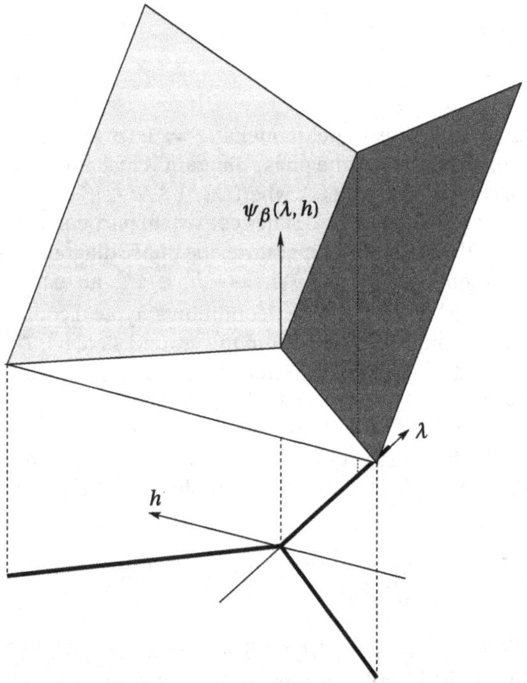

Figure 7.3. A qualitative plot of the pressure of the Blume–Capel model at low temperature.

bounded on page 120). On the other hand, since this model includes two external fields (h and λ), its phase diagram already has a nontrivial structure, containing coexistence lines and a triple point, as shown on Figure 7.2.

We are confident that, once they have carefully read the construction of the phase diagram of the Blume–Capel model, readers should be able to adapt the ideas to new situations.

7.1.4 Models with Finite-Range Translation-Invariant Interactions

The models to which PST applies are essentially those introduced in Section 6.3.2, with a potential Φ satisfying a set of extra conditions that will be described in the next section.

The distance on $\mathbb{Z}^d$ used throughout this chapter is the one associated with the norm $\|i\|_\infty \stackrel{\text{def}}{=} \max_{1\leq k\leq d} |i_k|$. The **diameter** of a set $B \subset \mathbb{Z}^d$, in particular, is defined by

$$\operatorname{diam}(B) \stackrel{\text{def}}{=} \sup\left\{d_\infty(i,j) \,:\, i,j \in B\right\},$$

where $d_\infty(i,j) \stackrel{\text{def}}{=} \|j-i\|_\infty$. We will use two notions of **boundary**: for $A \subset \mathbb{Z}^d$,

$$\partial^{\text{ex}} A \stackrel{\text{def}}{=} \left\{i \in A^{\text{c}} \,:\, d_\infty(i,A) \leq 1\right\}, \tag{7.3}$$

$$\partial^{\text{in}} A \stackrel{\text{def}}{=} \left\{i \in A \,:\, d_\infty(i,A^{\text{c}}) \leq 1\right\}. \tag{7.4}$$

As in Section 6.6, the **translation** by $i \in \mathbb{Z}^d$ will be denoted by θ_i and can act on configurations, events and measures.

We assume throughout that the single-spin space Ω_0 is *finite* and set, as usual, $\Omega_\Lambda \stackrel{\text{def}}{=} \Omega_0^\Lambda$ and $\Omega \stackrel{\text{def}}{=} \Omega_0^{\mathbb{Z}^d}$.

All the potentials $\Phi = \{\Phi_B\}_{B\Subset\mathbb{Z}^d}$ considered in this chapter will be of **finite range**,

$$r(\Phi) = \inf\left\{R > 0 \,:\, \Phi_B \equiv 0 \text{ for all } B \text{ with } \operatorname{diam}(B) > R\right\} < \infty,$$

and **invariant under translations**, meaning that

$$\Phi_{\theta_i B}(\theta_i\omega) = \Phi_B(\omega), \qquad \forall i \in \mathbb{Z}^d,\ \forall \omega \in \Omega.$$

The notations concerning Gibbs distributions associated with a potential Φ are those used in Section 6.3.2. For instance, the **Hamiltonian** in a region $\Lambda \Subset \mathbb{Z}^d$ is defined as usual by

$$\mathscr{H}_{\Lambda;\Phi}(\omega) \stackrel{\text{def}}{=} \sum_{\substack{B\Subset\mathbb{Z}^d:\\ B\cap\Lambda\neq\varnothing}} \Phi_B(\omega), \qquad \omega \in \Omega. \tag{7.5}$$

The partition function (denoted previously by $\mathbf{Z}^\eta_{\Lambda;\Phi}$, see (6.31)) will be denoted slightly differently, in order to emphasize its dependence on the set Λ, for reasons that will become clear later:

$$\mathbf{Z}_{\Phi}^{\eta}(\Lambda) \stackrel{\text{def}}{=} \sum_{\omega \in \Omega_{\Lambda}^{\eta}} \exp\bigl(-\beta \mathscr{H}_{\Lambda;\Phi}(\omega)\bigr). \tag{7.6}$$

We remind the reader that $\Omega_{\Lambda}^{\eta} \stackrel{\text{def}}{=} \{\omega \in \Omega : \omega_{\Lambda^c} = \eta_{\Lambda^c}\}$.

The **pressure** is obtained by considering the thermodynamic limit along a sequence $\Lambda \Uparrow \mathbb{Z}^d$:

$$\psi(\Phi) \stackrel{\text{def}}{=} \lim_{\Lambda \Uparrow \mathbb{Z}^d} \frac{1}{\beta|\Lambda|} \log \mathbf{Z}_{\Phi}^{\eta}(\Lambda). \tag{7.7}$$

In Theorem 6.79, we showed the existence of this limit along the sequence of boxes $\mathrm{B}(n)$, $n \to \infty$, for absolutely summable potentials. When the range is finite, existence can also be obtained by a simpler method.

Exercise 7.1. Adapting the proof of Theorem 3.6, show that $\psi(\Phi)$ exists, depends neither on η nor on the sequence $\Lambda \Uparrow \mathbb{Z}^d$ and is convex.

7.2 Ground States and Peierls' Condition

Loosely speaking, the main outcome of the Pirogov–Sinai theory is the determination of sufficient conditions to guarantee that typical configurations at (sufficiently low) positive temperatures are perturbations of those at zero temperature. As we already saw in the discussion of Section 1.4.3, typical configurations at zero temperature are those of minimal energy, that is, the ground states. Our first task is to find a suitable extension of this notion to infinite systems.

Remark 7.2. Note that what we call *ground states*, below, are in fact configurations (elements of Ω). This use of the word *state* should thus not be confused with that of earlier chapters, in which a state was a suitable linear functional acting on local functions. ◇

Since $\mathscr{H}_{\Lambda;\Phi}$ is usually not defined when $\Lambda = \mathbb{Z}^d$, defining ground states as the configurations minimizing the total energy (on $\mathbb{Z}^d$) raises the same difficulty we already encountered in Section 6.1. The resolution of this problem is based on the same observation we made there: the *difference* in energy between two configurations coinciding everywhere outside a finite set is always well defined. This leads to characterizing a ground state as a configuration whose energy cannot be lowered by changing its value at finitely many vertices. To make this idea precise, we start by introducing the following notion: two configurations $\omega, \tilde{\omega} \in \Omega$ are **equal at infinity** if they differ only at finitely many points, that is, if there exists a finite region $\Lambda \Subset \mathbb{Z}^d$ such that

$$\tilde{\omega}_{\Lambda^c} = \omega_{\Lambda^c}.$$

(As in Chapter 6, we use ω_{Λ^c} to denote the restriction of ω to Λ^c.) When $\tilde{\omega}$ and ω are equal at infinity, we write $\tilde{\omega} \stackrel{\infty}{=} \omega$; in such a case, $\tilde{\omega}$ can be considered as a *local perturbation of* ω (and vice versa). Then, the **relative Hamiltonian** is defined by

$$\mathscr{H}_\Phi(\tilde{\omega} \mid \omega) \stackrel{\text{def}}{=} \sum_{B \Subset \mathbb{Z}^d} \{\Phi_B(\tilde{\omega}) - \Phi_B(\omega)\}.$$

When $\tilde{\omega} \stackrel{\infty}{=} \omega$, the sum on the right-hand side is well defined, since it contains only finitely many nonzero terms (remember that $r(\Phi) < \infty$).

Definition 7.3. $\eta \in \Omega$ is called a **ground state** (for Φ) if

$$\mathscr{H}_\Phi(\omega \mid \eta) \geq 0 \qquad \text{for each} \quad \omega \stackrel{\infty}{=} \eta.$$

We denote the set of ground states for Φ by $g(\Phi)$.

Note that physically equivalent potentials (see Remark 6.17) yield the same relative Hamiltonian, and thus define the same set of ground states.

We will be interested mostly in *periodic* ground states. A configuration $\omega \in \Omega$ is **periodic** if there exist positive integers $l_1, \dots, l_d$ such that $\theta_{l_k \mathbf{e}_k}\omega = \omega$ for each $k = 1, \dots, d$ (remember that $\{\mathbf{e}_1, \dots, \mathbf{e}_d\}$ is the canonical basis of $\mathbb{R}^d$). The unique d-tuple $(l_1, \dots, l_d)$, in which each l_k is the smallest integer for which that property is satisfied, is called the **period** of ω. The set of periodic configurations is denoted by $\Omega^{\text{per}} \subset \Omega$ and the set of **periodic ground states** for Φ by $g^{\text{per}}(\Phi) \stackrel{\text{def}}{=} g(\Phi) \cap \Omega^{\text{per}}$.

We now provide a more global characterization of ground states. For $\omega \in \Omega^{\text{per}}$, the limit

$$e_\Phi(\omega) \stackrel{\text{def}}{=} \lim_{n \to \infty} \frac{1}{|B(n)|} \mathscr{H}_{B(n);\Phi}(\omega)$$

clearly exists; it is called the **energy density** of ω.

Lemma 7.4. *Let $\eta \in \Omega^{\text{per}}$. Then $\eta \in g^{\text{per}}(\Phi)$ if and only if its energy density is minimal:*

$$e_\Phi(\eta) = \underline{e}_\Phi \stackrel{\text{def}}{=} \inf_{\omega \in \Omega^{\text{per}}} e_\Phi(\omega).$$

Proof Let us introduce $\tilde{g}^{\text{per}}(\Phi) \stackrel{\text{def}}{=} \{\omega \in \Omega^{\text{per}} : e_\Phi(\omega) = \underline{e}_\Phi\}$.

We first assume that $\eta \in g^{\text{per}}(\Phi)$. For all $\omega \in \Omega^{\text{per}}$, we write

$$e_\Phi(\omega) = \lim_{n \to \infty} \frac{1}{|B(n)|} \{\mathscr{H}_{B(n);\Phi}(\omega) - \mathscr{H}_{B(n);\Phi}(\eta)\} + e_\Phi(\eta).$$

For all large n, define $\omega^{(n)} \stackrel{\text{def}}{=} \omega_{B(n)}\eta_{B(n)^c}$. Then $\omega^{(n)} \stackrel{\infty}{=} \eta$ and, since Φ has finite range,

$$\mathscr{H}_{B(n);\Phi}(\omega) - \mathscr{H}_{B(n);\Phi}(\eta) = \mathscr{H}_\Phi(\omega^{(n)} \mid \eta) + O(|\partial^{\text{ex}} B(n)|). \tag{7.8}$$

Since $\mathscr{H}_\Phi(\omega^{(n)} \mid \eta) \geq 0$ and $\lim_{n \to \infty} \frac{|\partial^{\text{ex}} B(n)|}{|B(n)|} = 0$, this proves that $e_\Phi(\omega) \geq e_\Phi(\eta)$. We conclude that $\eta \in \tilde{g}^{\text{per}}(\Phi)$.

Let us now assume that $\eta \in \tilde{g}^{\text{per}}(\Phi)$ and let ω be such that $\omega \stackrel{\infty}{=} \eta$. Since Φ has finite range, we can find k such that all the sets B that yield a nonzero contribution to $\mathscr{H}_\Phi(\omega \mid \eta)$ satisfy $B \subset B(k)$, and such that $\omega_{B(k)^c} = \eta_{B(k)^c}$. Let ω^{per} be the periodic configuration obtained by tiling $\mathbb{Z}^d$ with copies of $\omega_{B(k)}$ on all adjacent translates of $B(k)$. Proceeding as above, we write

$$\begin{aligned} e_\Phi(\omega^{\mathrm{per}}) &= \lim_{n\to\infty} \frac{1}{|B(n)|}\{\mathscr{H}_{B(n);\Phi}(\omega^{\mathrm{per}}) - \mathscr{H}_{B(n);\Phi}(\eta)\} + e_\Phi(\eta) \\ &= \frac{1}{|B(k)|}\{\mathscr{H}_{B(k);\Phi}(\omega^{\mathrm{per}}) - \mathscr{H}_{B(k);\Phi}(\eta)\} + e_\Phi(\eta) \\ &= \frac{1}{|B(k)|}\mathscr{H}_\Phi(\omega \mid \eta) + e_\Phi(\eta). \end{aligned}$$

Since $e_\Phi(\eta) \le e_\Phi(\omega^{\mathrm{per}})$, it follows that $\mathscr{H}_\Phi(\omega \mid \eta) \ge 0$. We conclude that $\eta \in g^{\mathrm{per}}(\Phi)$. □

Let us apply the above criterion to some examples. For reasons that will become clear later, we temporarily denote the potential by Φ^0 rather than Φ.

Example 7.5. Let us consider the **nearest-neighbor Ising model in the absence of a magnetic field.** Remember that, in this case, $\Omega_0 = \{\pm 1\}$ and, for all $B \Subset \mathbb{Z}^d$,

$$\Phi^0{}_B(\omega) \stackrel{\mathrm{def}}{=} \begin{cases} -\omega_i\omega_j & \text{if } B = \{i,j\},\ i \sim j, \\ 0 & \text{otherwise.} \end{cases} \tag{7.9}$$

(We remind the reader that, in this chapter, the inverse temperature is kept outside the Hamiltonian.) Consider the constant (and thus periodic) configurations η^+ and η^-. Then, for all $\omega \stackrel{\infty}{=} \eta^\pm$,

$$\mathscr{H}_{\Phi^0}(\omega \mid \eta^\pm) = \sum_{\{i,j\}\in\mathscr{E}_{\mathbb{Z}^d}} (1 - \omega_i\omega_j) \ge 0, \tag{7.10}$$

which shows that $\eta^+, \eta^- \in g^{\mathrm{per}}(\Phi^0)$. The associated energy densities can easily be computed explicitly: $e_{\Phi^0}(\eta^+) = e_{\Phi^0}(\eta^-) = -d$. Moreover, any periodic configuration $\omega \ne \eta^\pm$ satisfies $e_{\Phi^0}(\omega) > e_{\Phi^0}(\eta^\pm)$. Therefore, $\eta^\pm$ are the only periodic ground states:

$$g^{\mathrm{per}}(\Phi^0) = \{\eta^+, \eta^-\}.$$

There are, however, infinitely many other (nonperiodic) ground states (see Exercise 7.2). ◇

Exercise 7.2. Consider the Ising model on $\mathbb{Z}^2$ (still with no magnetic field). Fix $\mathbf{n} \in \mathbb{R}^2$ and define $\eta \in \Omega$ by $\eta_i = 1$ if and only if $\mathbf{n} \cdot i \ge 0$. Show that η and all its translates are ground states for the potential Φ^0 defined in (7.9).

Example 7.6. Let us now consider the **Blume-Capel model in the absence of external fields.** Remember that, in this model, $\Omega_0 = \{-1, 0, +1\}$ and

$$\Phi^0{}_B(\omega) \stackrel{\mathrm{def}}{=} \begin{cases} (\omega_i - \omega_j)^2 & \text{if } B = \{i,j\},\ i \sim j, \\ 0 & \text{otherwise.} \end{cases}$$

Let us consider again the constant configurations $\eta^+ \equiv +1$, $\eta^0 \equiv 0$ and $\eta^- \equiv -1$. Let $\# \in \{+,-,0\}$. Then, for all $\omega \stackrel{\infty}{=} \eta^\#$,

$$\mathscr{H}_{\Phi^0}(\omega \mid \eta^\#) = \sum_{\{i,j\}\in\mathscr{E}_{\mathbb{Z}^d}} (\omega_i - \omega_j)^2 \geq 0\,,$$

so that each $\eta^\#$ is a ground state. Since $\mathrm{e}_{\Phi^0}(\eta^+) = \mathrm{e}_{\Phi^0}(\eta^-) = \mathrm{e}_{\Phi^0}(\eta^0) = 0$ and any periodic, nonconstant configuration ω has $\mathrm{e}_{\Phi^0}(\omega) > 0$, we conclude that $g^{\mathrm{per}}(\Phi^0) = \{\eta^+, \eta^0, \eta^-\}$. ◇

Additional examples will be discussed in Section 7.2.2.

Exercise 7.3. Show that a model with a finite single-spin space and a finite-range potential always has at least one ground state.

7.2.1 Boundaries of a Configuration

From now on, we assume that the model under consideration has a *finite* number of periodic ground states:

$$g^{\mathrm{per}}(\Phi) = \{\eta^1, \dots, \eta^m\}\,.$$

We use the symbol $\# \in \{1, 2, \dots, m\}$ to denote an arbitrary index associated with the ground states of the model. Since our goal is to establish the existence of phase transitions, we assume that Φ has at least two periodic ground states: $m \geq 2$.

In view of what was proved for the Ising model, one might expect a typical configuration of an infinite system (with potential Φ) at low temperature to consist of large regions on each of which the configuration coincides with some ground state $\eta^\# \in g^{\mathrm{per}}(\Phi)$. Fix an integer $r > r(\Phi)$.

Definition 7.7. Let $\omega \in \Omega$. A vertex $i \in \mathbb{Z}^d$ is **#-correct (in ω)** if

$$\omega_j = \eta_j^\#, \qquad \forall j \in i + \mathrm{B}(r)\,.$$

The **boundary** of ω is defined by

$$\mathscr{B}(\omega) \stackrel{\text{def}}{=} \left\{ i \in \mathbb{Z}^d \;:\; i \text{ is not \#-correct in } \omega \text{ for any } \# \in \{1, \dots, m\} \right\}.$$

Before pursuing, let us make a specific choice for r. Let $(l_1^\#, \dots, l_d^\#)$ denote the period of $\eta^\#$. Until the end of this section, we use $r = r_*$, where

$$r_* \stackrel{\text{def}}{=} \text{least common multiple of } \left\{ l_k^\# : 1 \leq k \leq d,\ 1 \leq \# \leq m \right\} \text{ larger than or equal to } r(\Phi).$$

This choice implies that $\mathscr{B}(\omega) \cup \bigcup_\# \{\#\text{-correct vertices}\}$ forms a partition of $\mathbb{Z}^d$:

Lemma 7.8. *A vertex can be #-correct for at most one index #, and regions of #-correct and #′-correct vertices, #′ ≠ #, are separated by* $\mathscr{B}(\omega)$*, in the sense that, if* i *is #-correct and* i' *is #′-correct and if* $i_1 = i, i_2, \dots, i_{n-1}, i_n = i'$ *is a path such that* $d_\infty(i_k, i_{k+1}) \le 1$*, then there exists some* $1 < k < n$ *such that* $i_k \in \mathscr{B}(\omega)$*.*

Proof Observe first that our choice of r_* implies that any cube of sidelength r_* contains at least one period of each ground state. The first claim follows immediately, since $i + \mathrm{B}(r_*)$ contains such a cube. For the second claim, note that if i_k, i_{k+1} are two vertices at distance 1, then $\{i_k + \mathrm{B}(r_*)\} \cap \{i_{k+1} + \mathrm{B}(r_*)\}$ also contains a cube of sidelength r_* and thus i_k and i_{k+1} can be #-correct only for the same label #. □

Since the boundary of a configuration contains all vertices at which the energy is higher than in the ground states, it is natural to try to bound the relative Hamiltonian with respect to a ground state in terms of the size of the boundary.

Lemma 7.9. *Let* $\eta \in \mathrm{g}^{\mathrm{per}}(\Phi)$*. Then there exists a constant* $C > 0$ *(depending on* Φ*) such that, for any configuration* ω *such that* $\omega \stackrel{\infty}{=} \eta$*,*

$$\mathscr{H}_\Phi(\omega \mid \eta) \le C|\mathscr{B}(\omega)|\,. \tag{7.11}$$

Observe that

$$\mathscr{H}_{\Lambda;\Phi} = \sum_{\substack{B \Subset \mathbb{Z}^d:\\ B\cap\Lambda\neq\varnothing}} \Phi_B = \sum_{i\in\Lambda} \sum_{\substack{B \Subset \mathbb{Z}^d:\\ B\ni i}} \frac{1}{|B\cap\Lambda|}\Phi_B\,.$$

Introducing the functions (already used in (6.97))

$$u_{i;\Phi} \stackrel{\text{def}}{=} \sum_{\substack{B \Subset \mathbb{Z}^d:\\ B\ni i}} \frac{1}{|B|}\Phi_B\,, \qquad i \in \mathbb{Z}^d\,,$$

we have

$$\Big|\mathscr{H}_{\Lambda;\Phi} - \sum_{i\in\Lambda} u_{i;\Phi}\Big| \le c|\partial^{\mathrm{ex}}\Lambda|\,, \tag{7.12}$$

for some constant c that depends on Φ. We also have

$$\|u_{i;\Phi}\|_\infty \le \|\Phi\| \stackrel{\text{def}}{=} \sum_{\substack{B \Subset \mathbb{Z}^d:\\ B\ni 0}} \frac{1}{|B|}\|\Phi_B\|_\infty\,.$$

In the proof of the above lemma, but also in other arguments, it will be convenient to use the partition $\mathscr{P}$ of $\mathbb{Z}^d$ into adjacent cubic boxes of linear size r_*, of the form $b_k = kr_* + \{0,1,2,\dots,r_*-1\}^d$, where $k \in \mathbb{Z}^d$. Since each of these boxes contains an integer number of periods of each $\eta^\# \in \mathrm{g}^{\mathrm{per}}(\Phi)$, one has in particular, for all $k \in \mathbb{Z}^d$,

$$\frac{1}{|b_k|}\sum_{i\in b_k} u_{i;\Phi}(\eta^\#) = \mathrm{e}_\Phi(\eta^\#) = \underline{\mathrm{e}}_\Phi\,. \tag{7.13}$$

Proof of Lemma 7.9. Let $\omega \stackrel{\infty}{=} \eta$, and let $[\mathscr{B}](\omega)$ be the set of boxes $b \in \mathscr{P}$ whose intersection with $\mathscr{B}(\omega)$ is nonempty. Boxes b that are not part of $[\mathscr{B}](\omega)$ contain only correct vertices, and these are all correct for the same index # by Lemma 7.8. Then,

$$\begin{aligned}\mathscr{H}_\Phi(\omega \mid \eta) &= \sum_{i\in\mathbb{Z}^d}\{u_{i;\Phi}(\omega) - u_{i;\Phi}(\eta)\}\\ &= \sum_{b\in[\mathscr{B}](\omega)}\sum_{i\in b}\{u_{i;\Phi}(\omega) - u_{i;\Phi}(\eta)\} + \sum_{b\notin[\mathscr{B}](\omega)}\sum_{i\in b}\{u_{i;\Phi}(\omega) - u_{i;\Phi}(\eta)\}.\end{aligned}$$

The first double sum is upper-bounded by $2\|\Phi\| r_*^d |\mathscr{B}(\omega)|$. The second vanishes, since to each $b \notin [\mathscr{B}](\omega)$ corresponds some # such that $\omega_b = \eta_b^\#$, giving

$$\sum_{i\in b}\{u_{i;\Phi}(\omega) - u_{i;\Phi}(\eta)\} = |b|(\mathrm{e}_\Phi(\eta^\#) - \mathrm{e}_\Phi(\eta)) = 0. \qquad \square$$

For physical reasons, it is natural to expect that the energy of a configuration is proportional to the size of the boundary that separates regions with different periodic ground states, as happens in the Ising model. It is therefore natural to require that $\mathscr{H}_\Phi(\omega \mid \eta)$ should also grow proportionally to $|\mathscr{B}(\omega)|$. The notion that we will actually need is that of **thickened boundary**, defined by

$$\Gamma(\omega) \stackrel{\text{def}}{=} \bigcup\{i + \mathrm{B}(r_*) : i \in \mathscr{B}(\omega)\}. \tag{7.14}$$

The upper bound (7.11) implies of course that $\mathscr{H}_\Phi(\omega \mid \eta) \le C|\Gamma(\omega)|$ when $\omega \stackrel{\infty}{=} \eta$, but a corresponding *lower* bound does not hold in general. This turns out to be the main assumption of the Pirogov–Sinai theory:

Definition 7.10. Φ is said to **satisfy Peierls' condition** if

1. $g^{\mathrm{per}}(\Phi)$ is finite, and
2. there exists a constant $\rho > 0$ such that, for each $\eta \in g^{\mathrm{per}}(\Phi)$,

$$\mathscr{H}_\Phi(\omega \mid \eta) \ge \rho|\Gamma(\omega)|, \qquad \text{for all } \omega \stackrel{\infty}{=} \eta.$$

We call ρ **Peierls' constant.**

Peierls' condition can be violated even in simple models. An example will be given in Exercise 7.8.

Example 7.11. In Chapter 3, the *contours* of the **Ising model** on $\mathbb{Z}^2$ were defined as connected components of line segments (actually, edges of the dual lattice) separating + and − spins. Using the notations for contours adopted in Chapter 3, the relative Hamiltonian (7.10) can be expressed as

$$\mathscr{H}_{\Phi^0}(\omega \mid \eta^\pm) = 2\left|\{\{i,j\} \in \mathscr{E}_{\mathbb{Z}^2} : \omega_i \ne \omega_j\}\right| = 2\sum_{i=1}^n |\gamma_i|. \tag{7.15}$$

Since the ground states $\eta^\#$ are constant, we have $r_* = 1$. The difference between the corresponding set $\Gamma(\omega)$ and the contours γ_i is that $\Gamma(\omega)$ is a *thick* object, made of vertices of $\mathbb{Z}^2$ rather than edges of the dual lattice; see Figure 7.4.

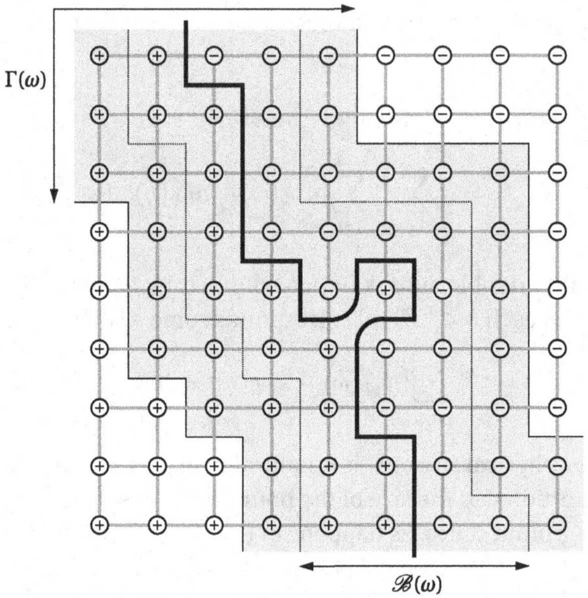

Figure 7.4. A portion of a configuration of the Ising model on $\mathbb{Z}^2$. The thick black line on the dual lattice, which separates + and − spins, is what was called a contour in Chapter 3. The set $\mathscr{B}(\omega)$ of vertices that are neither +- nor −-correct is delimited by the dotted line, and finally the shaded region represents the thickened boundary $\Gamma(\omega) \supset \mathscr{B}(\omega)$.

Note that, by construction, $\Gamma(\omega)$ is the union of translates of $\mathrm{B}(1)$ centered at vertices $i \in \mathbb{Z}^2$ located at Euclidean distance at most $\sqrt{2}/2$ from a contour. Since the total number of such vertices is at most twice the total length of the contours, we have

$$|\Gamma(\omega)| \leq 2|\mathrm{B}(1)| \sum_{i=1}^{n} |\gamma_i| .$$

We therefore see that Peierls' condition is satisfied, $\mathscr{H}_{\Phi^0}(\omega \mid \eta^{\pm}) \geq \rho|\Gamma(\omega)|$, with a Peierls' constant given by $\rho = |\mathrm{B}(1)|^{-1} = 1/9$. ◇

Example 7.12. For the **Blume-Capel model** on $\mathbb{Z}^d$, we also have $r_* = 1$. First, since $(\omega_i - \omega_j)^2 \geq 1$ when $\omega_i \neq \omega_j$, we get that

$$\mathscr{H}_{\Phi^0}(\omega \mid \eta^{\#}) \geq \left|\{\{i,j\} \in \mathscr{E}_{\mathbb{Z}^d} : \omega_i \neq \omega_j\}\right| , \tag{7.16}$$

for each ground state $\eta^{\#} \in \{\eta^+, \eta^0, \eta^-\}$. Then, since we clearly have

$$\Gamma(\omega) \subset \bigcup_{\substack{\{i,j\} \in \mathscr{E}_{\mathbb{Z}^d} \\ \omega_i \neq \omega_j}} (i + \mathrm{B}(1)) \cup (j + \mathrm{B}(1)) \subset \bigcup_{\substack{\{i,j\} \in \mathscr{E}_{\mathbb{Z}^d} \\ \omega_i \neq \omega_j}} (i + \mathrm{B}(2))$$

and thus

$$|\Gamma(\omega)| \leq \left|\{\{i,j\} \in \mathscr{E}_{\mathbb{Z}^d} : \omega_i \neq \omega_j\}\right| |\mathrm{B}(2)| ,$$

it follows that Peierls' condition also holds in this case, with $\rho = |\mathrm{B}(2)|^{-1} = 5^{-d}$. ◇

Exercise 7.4. 1. Show that

$$0 \le \psi(\Phi) - (-\underline{e}_\Phi) \le \beta^{-1} \log |\Omega_0| .$$

In particular, $\lim_{\beta\to\infty} \psi(\Phi) = -\underline{e}_\Phi$. *Hint:* Choose $\eta \in g^{\mathrm{per}}(\Phi)$, and start by writing $\mathscr{H}_{\Lambda;\Phi}(\omega) = \mathscr{H}_{\Lambda;\Phi}(\eta) + \{\mathscr{H}_{\Lambda;\Phi}(\omega) - \mathscr{H}_{\Lambda;\Phi}(\eta)\}$.

2. Assuming now that Φ satisfies Peierls' condition (with constant ρ), show that

$$0 \le \psi(\Phi) - (-\underline{e}_\Phi) \le |\Omega_0|\beta^{-1}e^{-\beta\rho} .$$

7.2.2 m-Potentials

Determining the set of ground states associated with a general potential Φ, as well as checking the validity of Peierls' condition, can be very difficult.[3] Ideally, one would like to do that by checking a finite set of *local* conditions.

Let us define, for each $B \Subset \mathbb{Z}^d$,

$$\phi_B \stackrel{\text{def}}{=} \min_\omega \Phi_B(\omega)$$

and set

$$g_m(\Phi) \stackrel{\text{def}}{=} \{\omega \in \Omega : \Phi_B(\omega) = \phi_B ,\ \forall B \Subset \mathbb{Z}^d\} .$$

If $g_m(\Phi) \neq \varnothing$, that is, if there exists at least one configuration that minimizes locally each Φ_B, then Φ is called an **m-potential**. We also let $g_m^{\mathrm{per}}(\Phi) \stackrel{\text{def}}{=} g_m(\Phi) \cap \Omega^{\mathrm{per}}$.

Lemma 7.13. *1.* $g_m(\Phi) \subset g(\Phi)$.
2. If $g_m^{\mathrm{per}}(\Phi) \neq \varnothing$*, then* $g_m^{\mathrm{per}}(\Phi) = g^{\mathrm{per}}(\Phi)$.
3. If $0 < |g_m(\Phi)| < \infty$*, then* $g_m(\Phi) = g_m^{\mathrm{per}}(\Phi) = g^{\mathrm{per}}(\Phi)$*, and* Φ *satisfies Peierls' condition.*

Proof The first claim is immediate and the second one follows from Lemma 7.4.

For the third claim, observe that $g_m(\Phi)$ is left invariant by any translation of the lattice, in the sense that $\omega \in g_m(\Phi)$ implies $\theta_i\omega \in g_m(\Phi)$ for all $i \in \mathbb{Z}^d$. Therefore, if $g_m(\Phi)$ is finite, all its elements must be periodic. Using the second claim yields $g_m(\Phi) = g_m^{\mathrm{per}}(\Phi) = g^{\mathrm{per}}(\Phi)$.

Let us now verify that Peierls' condition is satisfied when $0 < |g_m(\Phi)| < \infty$. We first claim that there exists $r \in (0, \infty)$ such that, for any configuration ω for which the vertex $i \in \mathbb{Z}^d$ is not correct, there exists $B \subset i + \mathrm{B}(r)$ such that $\Phi_B(\omega) \neq \phi_B$. Accepting this claim for the moment, the conclusion immediately follows: indeed, one can then set $\epsilon \stackrel{\text{def}}{=} \min\{\Phi_B(\omega) - \phi_B : \Phi_B(\omega) > \phi_B, B \subset i + \mathrm{B}(r), \omega \in \Omega \text{ incorrect at } i\} > 0$. Observe that, by translation invariance of Φ, ϵ does not depend on i. We can then write, for any $\eta \in g^{\mathrm{per}}(\Phi)$ and any $\omega \stackrel{\infty}{=} \eta$,

$$\mathscr{H}_\Phi(\omega \,|\, \eta) = \sum_{B \Subset \mathbb{Z}^d} \{\Phi_B(\omega) - \phi_B\} \ge \epsilon(2r+1)^{-d} |\mathscr{B}(\omega)| ,$$

which shows that Peierls' condition is satisfied.

We thus only need to establish the claim above. Let ω be some configuration such that i is incorrect. We claim that there exists $r' \in (0, \infty)$ such that, for any configuration ω' coinciding with ω on $i + \mathrm{B}(r_*)$, there exists $B \subset i + \mathrm{B}(r')$ such that $\Phi_B(\omega) \neq \phi_B$. (Note that this immediately implies the desired claim, since there are only finitely many possible configurations on $i+\mathrm{B}(r_*)$.) Let us assume the contrary: there exists a sequence of configurations $\omega^{(n)}$, all coinciding with ω on $i + \mathrm{B}(r_*)$, such that $\Phi_B(\omega^{(n)}) = \phi_B$ for all $B \subset i + \mathrm{B}(n)$. By sequential compactness of the set Ω (see Proposition 6.20), we can extract a subsequence converging to some configuration ω_*, still coinciding with ω on $i + \mathrm{B}(r_*)$ and such that $\Phi_B(\omega_*) = \phi_B$ for all $B \Subset \mathbb{Z}^d$. But this would mean that $\omega_* \in g_m(\Phi) = g^{\mathrm{per}}(\Phi)$, which would contradict the fact that i is incorrect. □

Example 7.14. For the Ising model, Φ^0 (defined in (7.9)) is an m-potential, since the only sets involved are the pairs of nearest neighbors, $B = \{i,j\} \in \mathscr{E}_{\mathbb{Z}^d}$, and the associated ${\Phi^0}_B(\omega) = -\omega_i\omega_j$ is minimized by taking either $\omega_i = \omega_j = +1$, or $\omega_i = \omega_j = -1$. Lemma 7.13 thus guarantees, as we already knew, that $g^{\mathrm{per}}(\Phi) = g_m(\Phi) = \{\eta^+, \eta^-\}$. ◇

Exercise 7.5. Study the periodic ground states of the **nearest-neighbor Ising antiferromagnet**, in which $\Omega_0 = \{\pm 1\}$ and, for $h \in \mathbb{R}$,

$$
{\Phi^0}_B(\omega) \stackrel{\text{def}}{=} \begin{cases} -h\omega_i & \text{if } B = \{i\}, \\ \omega_i\omega_j & \text{if } B = \{i,j\},\ i \sim j, \\ 0 & \text{otherwise.} \end{cases}
$$

Exercise 7.6. Consider the modification of the Ising model in (7.1), with ϵ sufficiently small, fixed. Study the ground states of that model, as a function of h. In particular: for which values of h are there two ground states?

The following exercise shows that it is sometimes possible to find an equivalent potential that is an m-potential when the original one is not. (However, this is *not* always possible.)

Exercise 7.7. Let $\mathscr{T}$ denote the set of all nearest-neighbor edges of $\mathbb{Z}^2$, to which are added all translates of the edge $\{0, i_c\}$, where $i_c = (1,1)$. Let $\Omega_0 \stackrel{\text{def}}{=} \{\pm 1\}$ and consider the potential $\Phi = \{\Phi_B\}$ of the **Ising antiferromagnet on the triangular lattice**, defined by

$$
\Phi_B(\omega) \stackrel{\text{def}}{=} \begin{cases} \omega_i\omega_j & \text{if } B = \{i,j\} \in \mathscr{T}, \\ 0 & \text{otherwise.} \end{cases}
$$

1. Check that Φ is not an m-potential.

2. Construct an m-potential $\widetilde{\Phi}$, physically equivalent to Φ. *Hint:* You can choose it such that $\widetilde{\Phi}_T > 0$ if and only if T is a triangle, $T = \{i,j,k\}$, with $\{i,j\}, \{j,k\}, \{k,i\} \in \mathscr{T}$.
3. Deduce that $\widetilde{\Phi}$ (and thus Φ) has an infinite number of periodic ground states.

Exercise 7.8. Consider the model on $\mathbb{Z}^2$ with $\Omega_0 \overset{\text{def}}{=} \{\pm 1\}$ and in which $\Phi^0{}_B \not\equiv 0$ only if $B = \{i,j,k,l\}$ is a square plaquette (see figure below). If ω coincides, on the plaquette B, with one of the following configurations,

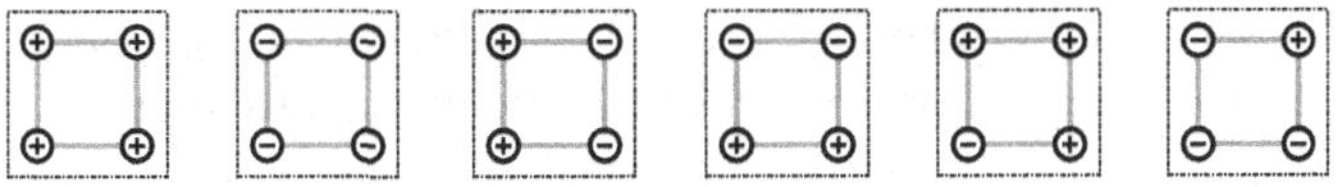

then $\Phi^0{}_B(\omega) = \alpha$. Otherwise, $\Phi^0{}_B(\omega) = \delta$, with $\delta > \alpha$. Find the periodic ground states of Φ^0, and give some examples of nonperiodic ground states. Then, show that Peierls' condition is not satisfied.

7.2.3 Lifting the Degeneracy

Consider a system with interactions Φ^0 and a finite set of periodic ground states $g^{\text{per}}(\Phi^0)$, at very low temperature. Using one of the ground states η as a boundary condition, one might wonder whether η is *stable* in the thermodynamic limit, in the sense that typical configurations under the corresponding infinite-volume Gibbs measure coincide with η with only sparse, local deviations.

In the Ising model on $\mathbb{Z}^d$, $d \geq 2$, this was the case for both η^+ and η^-. In more general situations, in particular in the absence of a symmetry relating all the elements of $g^{\text{per}}(\Phi^0)$, this issue is much more subtle.

To analyze this problem, we will first introduce a family of *external fields* which will be used to *lift* the degeneracy of the ground states, in the sense that, given any subset $g \subset g(\Phi^0)$, we can tune these external fields to obtain a potential whose set of periodic ground states is given by g. Eventually, these external fields will allow us to prepare the system in the desired Gibbs state and to drive the system from one phase to the other.

To lift the degeneracy, we *perturb* Φ^0 by considering a new potential Φ of the form

$$\Phi = \Phi^0 + W,$$

where $W = \{W_B\}_{B \Subset \mathbb{Z}^d}$ is the *perturbation potential.* We first verify that the perturbation, when small enough, does not lead to the appearance of new ground states.

Lemma 7.15. *If Φ^0 satisfies Peierls' condition with Peierls' constant $\rho > 0$ and if $\|W\| \leq \rho/4$, then $g^{\text{per}}(\Phi^0 + W) \subset g^{\text{per}}(\Phi^0)$.*

Proof Assume that $g^{\rm per}(\Phi^0) = \{\eta^1, \dots, \eta^m\}$. Let r_* and $\mathscr{P}$ be as before. Fix some $\omega \in \Omega^{\rm per}$, and let $[\Gamma](\omega)$ be the set of boxes $b \in \mathscr{P}$ whose intersection with $\Gamma(\omega)$ is nonempty. Again, by Lemma 7.8, boxes b not contained in $[\Gamma](\omega)$ contain only correct vertices, all of the same type. Let therefore $[\Pi_\#](\omega)$, $\# \in \{1, 2, \dots, m\}$, be the union of those boxes containing only #-correct vertices. Then, let

$$\pi_\#(\omega) \stackrel{\rm def}{=} \lim_{n\to\infty} \frac{|[\Pi_\#](\omega) \cap {\rm B}(n)|}{|{\rm B}(n)|}, \quad \gamma(\omega) \stackrel{\rm def}{=} \lim_{n\to\infty} \frac{|[\Gamma](\omega) \cap {\rm B}(n)|}{|{\rm B}(n)|}.$$

Observe that $\gamma(\omega) + \sum_{\#=1}^m \pi_\#(\omega) = 1$ and that $\omega \in g^{\rm per}(\Phi^0)$ if and only if $\gamma(\omega) = 0$. We will show below that, when $\|W\|$ is sufficiently small,

$$e_\Phi(\omega) - e_\Phi(\eta) \geq \sum_{\#=1}^m \pi_\#(\omega)[e_W(\eta^\#) - e_W(\eta)] + \tfrac{\rho}{2}\gamma(\omega), \tag{7.17}$$

for all $\eta \in g^{\rm per}(\Phi^0)$. Assuming this is true, let us take some $\eta \in g^{\rm per}(\Phi^0)$ for which $e_W(\eta) = \min_\# e_W(\eta^\#)$. If $\omega \in g^{\rm per}(\Phi)$, then in particular $e_\Phi(\omega) \leq e_\Phi(\eta)$ by Lemma 7.4. So (7.17) gives $\gamma(\omega) = 0$, that is, $\omega \in g^{\rm per}(\Phi^0)$.

To show (7.17), we start by writing

$$\begin{aligned}\mathscr{H}_{{\rm B}(n);\Phi}(\omega) - \mathscr{H}_{{\rm B}(n);\Phi}(\eta) = &\left\{\mathscr{H}_{{\rm B}(n);\Phi^0}(\omega) - \mathscr{H}_{{\rm B}(n);\Phi^0}(\eta)\right\} \\ &+ \left\{\mathscr{H}_{{\rm B}(n);W}(\omega) - \mathscr{H}_{{\rm B}(n);W}(\eta)\right\}.\end{aligned} \tag{7.18}$$

On the one hand, proceeding as in (7.8),

$$\begin{aligned}\mathscr{H}_{{\rm B}(n);\Phi^0}(\omega) - \mathscr{H}_{{\rm B}(n);\Phi^0}(\eta) &= \mathscr{H}_{\Phi^0}(\omega^{(n)}|\eta) + O(|\partial^{\rm ex}{\rm B}(n)|) \\ &\geq \rho|\Gamma(\omega^{(n)})| + O(|\partial^{\rm ex}{\rm B}(n)|) \\ &= \rho|\Gamma(\omega) \cap {\rm B}(n)| + O(|\partial^{\rm ex}{\rm B}(n)|).\end{aligned}$$

On the other hand, we can decompose:

$$\mathscr{H}_{{\rm B}(n);W}(\omega) = \sum_{i\in[\Gamma](\omega)\cap{\rm B}(n)} u_{i;W}(\omega) + \sum_{\#=1}^m \sum_{i\in[\Pi_\#](\omega)\cap{\rm B}(n)} u_{i;W}(\omega) + O(|\partial^{\rm ex}{\rm B}(n)|).$$

The first sum can be bounded by

$$\Big|\sum_{i\in[\Gamma](\omega)\cap{\rm B}(n)} u_{i;W}(\omega)\Big| \leq \big|[\Gamma](\omega) \cap {\rm B}(n)\big| \|W\|.$$

For the second one, using (7.13),

$$\begin{aligned}\sum_{i\in[\Pi_\#](\omega)\cap{\rm B}(n)} u_{i;W}(\omega) &= \sum_{i\in[\Pi_\#](\omega)\cap{\rm B}(n)} u_{i;W}(\eta^\#) \\ &= \big|[\Pi_\#](\omega) \cap {\rm B}(n)\big| e_W(\eta^\#) + O(|\partial^{\rm ex}{\rm B}(n)|).\end{aligned}$$

The other Hamiltonian is decomposed as follows:

$$\begin{aligned}\mathscr{H}_{\mathrm{B}(n);W}(\eta) &= |\mathrm{B}(n)|\mathrm{e}_W(\eta) + O(|\partial^{\mathrm{ex}}\mathrm{B}(n)|)\\ &\leq \|W\|\big|[\Gamma](\omega)\cap\mathrm{B}(n)\big| + \sum_{\#=1}^{m}\big|[\Pi_\#](\omega)\cap\mathrm{B}(n)\big|\mathrm{e}_W(\eta) + O(|\partial^{\mathrm{ex}}\mathrm{B}(n)|)\,.\end{aligned}$$

Inserting these estimates in (7.18), dividing by $|\mathrm{B}(n)|$, bounding $\|W\| \leq \rho/4$ and taking the limit $n \to \infty$ yields (7.17). □

The perturbation of Φ^0 will contain a certain number of parameters (which will play a role analogous to that of the magnetic field in the Ising model) that will allow us to *lift the degeneracy of the ground states of* Φ^0. This means that, if $|\mathrm{g}^{\mathrm{per}}(\Phi^0)| = m$, we need the perturbation W to contain $m-1$ parameters and it should be possible to tune the latter in order for $\mathrm{g}^{\mathrm{per}}(\Phi^0 + W)$ to be an arbitrary subset of $\mathrm{g}^{\mathrm{per}}(\Phi^0)$. This is best understood with some examples.

Example 7.16. The degeneracy of the potential Φ^0 of the **Ising model** can be lifted by introducing a magnetic field h and by considering the perturbation $W = \{W_B\}$ defined by

$$W_B(\omega) = \begin{cases} -h\omega_i & \text{if } B = \{i\}\,,\\ 0 & \text{otherwise.}\end{cases}$$

Lemma 7.15 guarantees that $\mathrm{g}^{\mathrm{per}}(\Phi) = \mathrm{g}^{\mathrm{per}}(\Phi^0 + W) \subset \{\eta^+, \eta^-\}$ when $\|W\| = |h|$ is sufficiently small. But, since the energy densities are given, for all h, by

$$\mathrm{e}_\Phi(\eta^\pm) = -d \mp h\,,$$

we get that $\mathrm{e}_\Phi(\eta^+) < \mathrm{e}_\Phi(\eta^-)$ when $h > 0$, and $\mathrm{e}_\Phi(\eta^+) > \mathrm{e}_\Phi(\eta^-)$ when $h < 0$. Therefore, we can describe $\mathrm{g}^{\mathrm{per}}(\Phi^0 + W)$ for all h (not only when $|h|$ is small):

$$\mathrm{g}^{\mathrm{per}}(\Phi^0 + W) = \begin{cases} \{\eta^+\} & \text{if } h > 0\,,\\ \{\eta^+, \eta^-\} & \text{if } h = 0\,,\\ \{\eta^-\} & \text{if } h < 0\,.\end{cases}$$

◇

Example 7.17. In the case of the **Blume–Capel model**, two parameters are necessary to lift the degeneracy. We denote these by h and λ, and consider the perturbation $W = \{W_B\}_{B\Subset\mathbb{Z}^d}$ defined by

$$W_B(\omega) \stackrel{\text{def}}{=} \begin{cases} -h\omega_i - \lambda\omega_i^2 & \text{if } B = \{i\}\,,\\ 0 & \text{otherwise.}\end{cases} \tag{7.19}$$

By Lemma (7.15), we know that $\mathrm{g}^{\mathrm{per}}(\Phi) \subset \{\eta^+, \eta^0, \eta^-\}$ when $\|W\| = |h| + |\lambda|$ is sufficiently small. The energy densities are given by

$$\mathrm{e}_\Phi(\eta^\pm) = \mp h - \lambda\,, \quad \mathrm{e}_\Phi(\eta^0) = 0\,, \tag{7.20}$$

and the periodic ground states are obtained by studying $\min_\# \mathrm{e}_\Phi(\eta^\#)$ as a function of (λ, h). Let us thus define the regions $\mathscr{U}^+, \mathscr{U}^0, \mathscr{U}^-$ by

$$\mathscr{U}^\# \stackrel{\text{def}}{=} \big\{(\lambda, h) : \mathrm{e}_\Phi(\eta^\#) = \min_{\#'} \mathrm{e}_\Phi(\eta^{\#'})\big\}\,. \tag{7.21}$$

The interior of these regions determines the values of (λ, h) for which there is a unique ground state. Except at $(0, 0)$, at which the three ground states coexist, two periodic ground states coexist on the boundaries of these regions, which are unions of lines,

$$\mathscr{L}^{\#\#'} \stackrel{\text{def}}{=} \mathscr{U}^{\#} \cap \mathscr{U}^{\#'} .$$

These are given explicitly by

$$\mathscr{L}^{+-} \stackrel{\text{def}}{=} \{(\lambda, h) : h = 0, \lambda \geq 0\} ,$$
$$\mathscr{L}^{-0} \stackrel{\text{def}}{=} \{(\lambda, h) : h = \lambda, \lambda \leq 0\} ,$$
$$\mathscr{L}^{+0} \stackrel{\text{def}}{=} \{(\lambda, h) : h = -\lambda, \lambda \leq 0\} .$$

Altogether, we recover the *zero-temperature* phase diagram already depicted on the left of Figure 7.2. ◇

In the following exercise, we see that it is always possible to lift the degeneracy.

Exercise 7.9. Suppose that $g^{\text{per}}(\Phi^0) = \{\eta^1, \dots, \eta^m\}$. Provide a collection of potentials $W^1, \dots, W^{m-1}$ such that, for all $I \subset \{1, \dots, m\}$, there exist $\lambda^1, \dots, \lambda^{m-1}$ such that $g^{\text{per}}(\Phi^0 + \sum_{i=1}^{m-1} \lambda_i W^i) = \{\eta^i, i \in I\}$.

7.2.4 A Glimpse of the Rest of This Chapter

Let us consider a model with potential $\Phi^{\underline{\lambda}} \stackrel{\text{def}}{=} \Phi^0 + \sum_{i=1}^{m-1} \lambda_i W^i$, where the W^i are potentials lifting the degeneracy of the periodic ground states $\eta^1, \dots, \eta^m$ of Φ^0, as explained in the previous section, and $\underline{\lambda} \stackrel{\text{def}}{=} (\lambda_i)_{1 \leq i \leq m-1} \in \mathbb{R}^{m-1}$. We can then construct the zero-temperature phase diagram, by specifying $g^{\text{per}}(\Phi^{\underline{\lambda}})$ for each value of the parameters $\underline{\lambda}$. This phase diagram thus consists of $(m-1)$-dimensional regions with a single periodic ground state, $(m-2)$-dimensional regions in which there are exactly two periodic ground states, etc.

Alternatively, notice that the energy density $\underline{\lambda} \mapsto \mathrm{e}_{\Phi^{\underline{\lambda}}}$ of the ground state is a piecewise linear function of $\underline{\lambda}$. The zero-temperature phase diagram characterizes the points $\underline{\lambda}$ at which $\mathrm{e}_{\Phi^{\underline{\lambda}}}$ fails to be differentiable.

Our goal in the rest of this chapter is to extend this construction to small positive temperatures. More precisely, we will prove that, in the limit $\beta \to \infty$, the set of values at which $\underline{\lambda} \mapsto \psi_\beta(\underline{\lambda})$ is nondifferentiable converges to the corresponding set at which $\mathrm{e}_{\Phi^{\underline{\lambda}}}$ fails to be differentiable.

This will be achieved by constructing C^1 functions, $\widehat{\psi}^1_\beta(\underline{\lambda}), \dots, \widehat{\psi}^m_\beta(\underline{\lambda})$, such that the following hold:

1. $\psi_\beta(\underline{\lambda}) = \max_i \widehat{\psi}^i_\beta(\underline{\lambda})$;
2. $\lim_{\beta \to \infty} \widehat{\psi}^i_\beta(\underline{\lambda}) = -\mathrm{e}_{\Phi^{\underline{\lambda}}}(\eta^i)$ for all $i \in \{1, \dots, m\}$;

3. $\lim_{\beta\to\infty} \frac{\partial \widehat{\psi}^i_\beta(\underline{\lambda})}{\partial \lambda_j} = -\frac{\partial e_{\Phi\underline{\lambda}}(\eta^i)}{\partial \lambda_j}$, for all $i \in \{1, \dots, m\}, j \in \{1, \dots, m-1\}$.

In addition, we will see that the only periodic extremal Gibbs measures at $\underline{\lambda}$ are precisely those obtained by taking the thermodynamic limit with boundary condition η^i for values of i such that $\psi_\beta(\underline{\lambda}) = \widehat{\psi}^i_\beta(\underline{\lambda})$.

Each $\widehat{\psi}^i_\beta$ is called a *truncated pressure*. It is obtained from the partition function with boundary condition η^i by adding the constraint that only "small" (in a sense to be made precise below) excitations are allowed. For certain values of the parameters $\underline{\lambda}$, the excitations turn out to be always small and the truncated pressure coincides with the usual pressure; for others, however, the constraint artificially stabilizes the boundary condition and yields a different, strictly smaller, truncated pressure.

7.2.5 From Finite-Range Interactions to Interactions of Range 1

In Section 7.4, we will initiate the low-temperature analysis of systems with a finite number of periodic ground states that satisfy Peierls' condition. This analysis will rely on the contour description of these systems, which we explain in detail in the next section.

It turns out that the contour description is considerably simplified if one assumes that the potential Φ under consideration has range 1. Fortunately, any model with a single-spin space Ω_0 and a potential Φ of range $r(\Phi) > 1$ can be mapped onto another model with a potential $\hat{\Phi}$ of range 1, at the cost of introducing a larger single-spin space $\hat{\Omega}_0$, such that the two models have the same pressure.

Earlier, the nuisance of having ground states with different periods was mitigated by considering the boxes $b_k \in \mathscr{P}$ with sidelength r_*, and this can be used further as follows. Assume that $\Phi = \Phi^0 + W$ has range $r(\Phi)$ and that Φ^0 has a finite set of periodic ground states. There are $N \stackrel{\text{def}}{=} |\Omega_0|^{|B(r_*)|}$ possible configurations inside each of the boxes b_k, and those configurations can be encoded into a new spin variable $\hat{\omega}_k$ taking values in $\hat{\Omega}_0 \stackrel{\text{def}}{=} \{1, 2, \dots, N\}$. Clearly, the set of configurations $\omega = (\omega_i)_{i\in\mathbb{Z}^d} \in \Omega$ is in one-to-one correspondence with the set of configurations $\hat{\omega} = (\hat{\omega}_k)_{k\in\mathbb{Z}^d} \in \hat{\Omega} \stackrel{\text{def}}{=} \hat{\Omega}_0^{\mathbb{Z}^d}$.

By the choice of r_*, it is clear that a spin $\hat{\omega}_k$ only interacts with spins $\hat{\omega}_{k'}$ at distance $d_\infty(k, k') \le 1$. Let us determine the corresponding potential. Denote by $\hat{B}$ a generic union of boxes b_k of diameter at most $2r_*$. For each set $B \Subset \mathbb{Z}^d$ contributing to the original Hamiltonian, let

$$N_B \stackrel{\text{def}}{=} \left|\{\hat{B} : \hat{B} \supset B\}\right|.$$

The terms of the formal Hamiltonian can be rearranged as follows:

$$\sum_B \Phi_B(\omega) = \sum_{\hat{B}} \Big\{ \sum_{B\subset\hat{B}} \frac{1}{N_B} \Phi_B(\omega) \Big\}.$$

We are led to defining the **rescaled potential** as

$$\hat{\Phi}_{\hat{B}}(\hat{\omega}) \stackrel{\text{def}}{=} \sum_{B \subset \hat{B}} \frac{1}{N_B} \Phi_B(\omega).$$

Clearly, all the information about the original model can be recovered from the rescaled model (with $\hat{\Omega}$ and $\hat{\Phi}$); in particular, they have the same pressure (up to a multiplicative constant).

By construction, the rescaled measure $\hat{\Phi}$ has range $r(\hat{\Phi}) = 1$ (as measured on the rescaled lattice $r_*\mathbb{Z}^d$). Of course, analyzing the set of ground states and the validity of Peierls' condition for Φ is equivalent to accomplishing these tasks for the rescaled model. Besides having interactions of range 1, this reformulation of the model presents the advantage that, now, the ground states correspond to *constant* configurations on $\hat{\Omega}$.

Exercise 7.10. Assume that the original potential Φ^0 satisfies Peierls' condition with constant $\rho > 0$. Show that Peierls' condition still holds for the rescaled model ($\omega \mapsto \hat{\omega}$, $\Phi \mapsto \hat{\Phi}$) and estimate the corresponding constant.

Since the above construction can always be implemented, we assume, from now on, that the model has been suitably formulated so as to have range 1 and a finite set of constant ground states. In this way, the analysis will become substantially simpler, without incurring any loss of generality.

7.2.6 Contours and Their Labels

Let us therefore consider a potential $\Phi = \Phi^0 + W$ with $r(\Phi^0) = 1$ and $r(W) \leq 1$ and such that the ground states of Φ^0,

$$g^{\text{per}}(\Phi^0) = \{\eta^1, \ldots, \eta^m\},$$

are all constant. We also assume that the parameters contained in W completely lift the degeneracy of the ground state. Since $r(\Phi) = 1$, the set $\Gamma(\omega)$ in (7.14) is defined using $r_* = 1$.

Since W does not introduce any new ground states (Lemma 7.15), we may expect, roughly, a typical configuration ω of the model associated with Φ to display, at low temperature, only small local deviations from one of the ground states $\eta^\#$. We thus start an analysis of the perturbed model in terms of the decomposition of $\Gamma(\omega)$ into *contours*, which separate regions on which the ground states $\{\eta^1, \ldots, \eta^m\}$ are seen. This is very similar to what was done when studying the low-temperature Ising model in Chapter 3.

Before continuing, let us define the notion of connectedness used in the rest of the chapter, based on the use of the distance $d_\infty(\cdot,\cdot)$: $A \subset \mathbb{Z}^d$ is **connected** if for each pair $j, j' \in A$ there exists a sequence $i_1 = j, i_2, \ldots, i_{n-1}, i_n = j'$ such that $i_k \in A$ for all $k = 1, \ldots, n$, and $d_\infty(i_k, i_{k+1}) = 1$. A connected component $A' \subset A$ is **maximal** if any set $B \neq A'$ such that $A' \subset B \subset A$ is necessarily disconnected.

When $\omega \stackrel{\infty}{=} \eta$ for some $\eta \in g^{\mathrm{per}}(\Phi)$, the set $\Gamma(\omega)$ is bounded and can be decomposed into maximal connected components:

$$\Gamma(\omega) = \{\overline{\gamma}_1, \dots, \overline{\gamma}_n\}.$$

For each component $\overline{\gamma} \in \Gamma(\omega)$, let $\omega_{\overline{\gamma}}$ denote the restriction of ω to $\overline{\gamma}$. The configuration $\omega_{\overline{\gamma}}$ should be considered as being part of the information contained in the component:

Definition 7.18. Each pair $\gamma \stackrel{\text{def}}{=} (\overline{\gamma}, \omega_{\overline{\gamma}})$ is called a **contour** of ω; $\overline{\gamma}$ is the **support** of γ.

The support of a contour γ splits $\mathbb{Z}^d$ into a finite number of maximal connected components (see Figure 7.5):

$$\overline{\gamma}^{\mathrm{c}} = \{A_0, A_1, \dots, A_k\}. \tag{7.22}$$

Exactly one of the components of $\overline{\gamma}^{\mathrm{c}}$ is unbounded; with no loss of generality we can assume it to be A_0. We call it the **exterior** of γ and denote it by $\mathrm{ext}\gamma$.

Let us say that a subset $A \subset \mathbb{Z}^d$ is **c-connected** if A^{c} is connected.

Exercise 7.11. Show that the subsets $A_0, \dots, A_k$ in the decomposition (7.22) are c-connected.

Remember the boundaries $\partial^{\mathrm{in}} A$ and $\partial^{\mathrm{ex}} A$ introduced in (7.3). Although the content of the following lemma might seem intuitively obvious to the reader (at least in low dimensions), we provide a proof in Appendix B.15.

Lemma 7.19. *Consider the decomposition* (7.22)*. For each $j = 0, 1, \dots, k$, $\partial^{\mathrm{ex}} A_j$ and $\partial^{\mathrm{in}} A_j$ are connected. Moreover, there exists $\# \in \{1, 2, \dots, m\}$, depending on j, such that $\omega_i = \eta_i^{\#}$ for all $i \in \partial^{\mathrm{ex}} A_j$. We call # the* ***label*** *of A_j, and denote it* $\mathrm{lab}(A_j)$.

Consider the decomposition (7.22) of some contour $\gamma = (\overline{\gamma}, \omega_{\overline{\gamma}})$. If the exterior has label $\mathrm{lab}(\mathrm{ext}\gamma) = \#$, we say that γ is of **type** #. The remaining components in (7.22), $A_1, \dots, A_k$, are all bounded and separated from $\mathrm{ext}\gamma$ by $\overline{\gamma}$. We group them according to their type. The **interior of type** # of γ is defined by

$$\mathrm{int}_{\#}\gamma \stackrel{\text{def}}{=} \bigcup_{\substack{i \in \{1,\dots,k\}:\\ \mathrm{lab}(A_i) = \#}} A_i .$$

We will also call $\mathrm{int}\gamma \stackrel{\text{def}}{=} \bigcup_{\#=1}^{m} \mathrm{int}_{\#}\gamma$ the **interior** of γ.

The collection of all possible contours of type # is denoted by $\mathscr{C}^{\#}$. Observe that, for each contour $\gamma = (\overline{\gamma}, \omega_{\overline{\gamma}}) \in \mathscr{C}^{\#}$, there exists a configuration that has γ as its unique contour. Namely, extend $\omega_{\overline{\gamma}} = (\omega_i)_{i \in \overline{\gamma}}$ to a configuration on the whole lattice by setting $\omega_i = \eta_i^{\#}$ for $i \in \mathrm{ext}\gamma$ and $\omega_i = \eta_i^{\#'}$ for each $i \in \mathrm{int}_{\#'}\gamma$. For notational convenience, we also denote this new configuration by $\omega_{\overline{\gamma}}$.

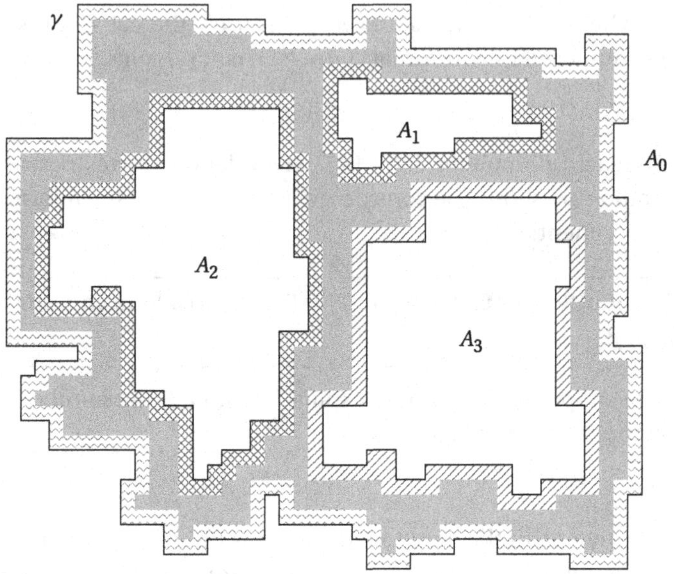

Figure 7.5. A contour γ for which $\overline{\gamma}^{\mathrm{c}} = \{A_0, A_1, A_2, A_3\}$. The labels, whose existence is guaranteed by Lemma 7.19, have been illustrated using different patterns. The components A_1 and A_2 have the same label. The label of $A_0 = \mathrm{ext}\gamma$ represents the type of γ. This picture shows how the labels induce a corresponding boundary condition on the interior components of γ (which we use later when defining $\mathbf{Z}^{\#}_{\Phi}(\Lambda)$ in (7.24)).

The type and labels associated with a contour will play an essential role in the next section.

7.3 Boundary Conditions and Contour Models

Notice that if γ, γ' are two contours in the same configuration, and if $\overline{\gamma} \subset \mathrm{int}\gamma'$, then $d_\infty(\overline{\gamma}, (\mathrm{int}\gamma')^{\mathrm{c}}) > 1$.

Let $\Lambda \Subset \mathbb{Z}^d$. *From now on, we always assume that* Λ *is* c-*connected.* To define contour models in Λ, it will be convenient to slightly modify the way in which boundary conditions are introduced: this will make it easier to consider the boundary condition induced by a contour on its interior.

Let $\eta^\# \in g^{\mathrm{per}}(\Phi^0)$ and $\omega \in \Omega_\Lambda^{\eta^\#}$. Since it is not guaranteed that $\Gamma(\omega) \subset \Lambda$, we define

$$\Omega_\Lambda^\# \stackrel{\text{def}}{=} \left\{\omega \in \Omega_\Lambda^{\eta^\#} : d_\infty(\Gamma(\omega), \Lambda^{\mathrm{c}}) > 1\right\}.$$

The additional restrictions imposed on the configurations in $\Omega_\Lambda^\#$ only affect vertices located near the boundary of Λ:

Lemma 7.20. *Let* $\omega \in \Omega_\Lambda^{\eta^\#}$. *Then,* $\omega \in \Omega_\Lambda^\#$ *if and only if* $\omega_i = \eta_i^\#$ *for every vertex* $i \in \Lambda$ *satisfying* $d_\infty(i, \Lambda^{\mathrm{c}}) \leq 3$.

Proof Let $\omega \in \Omega_\Lambda^{\eta^\#}$ and assume that there exists $i \in \Lambda$, $d_\infty(i, \Lambda^c) \le 3$, such that $\omega_i \neq \eta_i^\#$. This implies that there exists some $j \in i + \mathrm{B}(1)$ with $d_\infty(j, \Lambda^c) \le 2$ that is not #-correct, and so $j \in \mathscr{B}(\omega)$. Since $d_\infty(j + \mathrm{B}(1), \Lambda^c) \le 1$, this implies that $d_\infty(\Gamma(\omega), \Lambda^c) \le 1$ and, thus, $\omega \notin \Omega_\Lambda^\#$.

Conversely, suppose that $\omega_i = \eta_i^\#$ as soon as $d_\infty(i, \Lambda^c) \le 3$. Then, any vertex $i \in \Lambda$ with $d_\infty(i, \Lambda^c) \le 2$ is #-correct. Therefore, vertices that are not correct must satisfy $d_\infty(i, \Lambda^c) \ge 3$ and, thus, $d_\infty(i + \mathrm{B}(1), \Lambda^c) \ge 2$. This implies that $d_\infty(\Gamma(\omega), \Lambda^c) > 1$. □

In order to use contours and their weights for the description of finite systems, it will be convenient to introduce boundary conditions as above, using $\Omega_\Lambda^\#$ instead of $\Omega_\Lambda^{\eta^\#}$. We therefore consider the following Gibbs distributions: for all $\omega \in \Omega_\Lambda^\#$,

$$\mu_{\Lambda;\Phi}^\#(\omega) \stackrel{\text{def}}{=} \frac{e^{-\beta \mathscr{H}_{\Lambda;\Phi}(\omega)}}{\mathbf{Z}_\Phi^\#(\Lambda)}, \tag{7.23}$$

where

$$\mathbf{Z}_\Phi^\#(\Lambda) \stackrel{\text{def}}{=} \sum_{\omega \in \Omega_\Lambda^\#} e^{-\beta \mathscr{H}_{\Lambda;\Phi}(\omega)}. \tag{7.24}$$

It follows from Lemma 7.20 that (remember Exercise 7.1)

$$\lim_{\Lambda \Uparrow \mathbb{Z}^d} \frac{1}{\beta|\Lambda|} \log \mathbf{Z}_\Phi^\#(\Lambda) = \lim_{\Lambda \Uparrow \mathbb{Z}^d} \frac{1}{\beta|\Lambda|} \log \mathbf{Z}_\Phi^{\eta^\#}(\Lambda) = \psi(\Phi).$$

Let us say that Λ is **thin** if $d_\infty(i, \Lambda^c) \le 3$ for all $i \in \Lambda$. It follows from Lemma 7.20 that, whenever Λ is thin, $\Omega_\Lambda^\# = \{\eta^\#\}$ and, thus,

$$\mathbf{Z}_\Phi^\#(\Lambda) = e^{-\beta \mathscr{H}_{\Lambda;\Phi}(\eta^\#)}. \tag{7.25}$$

7.3.1 Extracting the Contribution from the Ground State

Let us fix a boundary condition $\# \in \{1, 2, \dots, m\}$ and relate the energy of each configuration $\omega \in \Omega_\Lambda^\#$ to the energy of $\eta^\#$:

$$\begin{aligned} \mathscr{H}_{\Lambda;\Phi}(\omega) &= \mathscr{H}_{\Lambda;\Phi}(\eta^\#) + \{\mathscr{H}_{\Lambda;\Phi}(\omega) - \mathscr{H}_{\Lambda;\Phi}(\eta^\#)\} \\ &= \mathscr{H}_{\Lambda;\Phi}(\eta^\#) + \mathscr{H}_\Phi(\omega \,|\, \eta^\#). \end{aligned} \tag{7.26}$$

One can thus write

$$\mathbf{Z}_\Phi^\#(\Lambda) = e^{-\beta \mathscr{H}_{\Lambda;\Phi}(\eta^\#)} \sum_{\omega \in \Omega_\Lambda^\#} e^{-\beta \mathscr{H}_\Phi(\omega \,|\, \eta^\#)} \stackrel{\text{def}}{=} e^{-\beta \mathscr{H}_{\Lambda;\Phi}(\eta^\#)} \Xi_\Phi^\#(\Lambda). \tag{7.27}$$

Notice that, since each ground state $\eta^\#$ is constant and Φ has range 1,

$$\mathscr{H}_{\Lambda;\Phi}(\eta^\#) = \mathrm{e}_\Phi(\eta^\#)|\Lambda|.$$

Our next goal is to express $\Xi_\Phi^\#(\Lambda)$ as the partition function of a polymer model having the same abstract structure as those of Section 5.2. To this end the contours introduced above will play the role of polymers. Remember, however, that the compatibility condition used in Section 5.2 was *pairwise*. Unfortunately, our contours have labels, and this yields a more complex compatibility condition.

To determine whether a given family of contours is compatible, that is, whether there exists a configuration yielding precisely this family of contours, we need to verify two conditions. The first one is that their supports are disjoint and sufficiently far apart, in a suitable sense; this can of course be expressed as a pairwise condition. However, we must also check that their labels match, *and this condition cannot be verified by only looking at pairs of contours. We illustrate this in Figure 7.6.* ◇

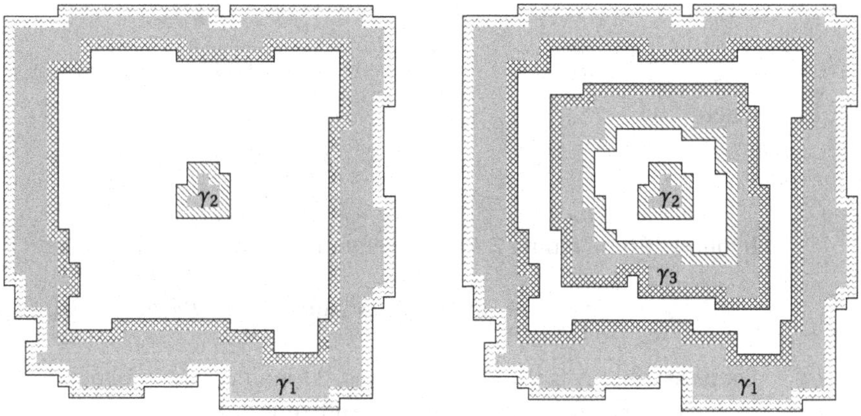

Figure 7.6. Two contours γ_1 and γ_2 (left). γ_2 is of type 1 and satisfies $\overline{\gamma}_2 \subset \text{int}_2\gamma_1$. These two contours can only be part of a configuration if there are other contours correcting the mismatch between the type of γ_2 and the label of the component of γ_1 it is located in. For example (right), there might be a third contour γ_3 of type 2 such that $\overline{\gamma}_3 \subset \text{int}_2\gamma_1$ and $\overline{\gamma}_2 \subset \text{int}_1\gamma_3$. This shows that the compatibility of a family of contours is a global property, which cannot be expressed pairwise.

To deal with this problem, we need to proceed with more care than in (7.27) and express $\Xi^{\#}_{\Phi}(\Lambda)$ as a polymer model in which the polymers are contours *all of the same type #*, for which the compatibility condition becomes purely geometrical, namely having supports that are far apart, in the following sense.

Definition 7.21. Two contours of the same type, γ_1 and γ_2, are said to be **compatible** if $d_\infty(\overline{\gamma}_1, \overline{\gamma}_2) > 1$.

By construction, all contours appearing in a same configuration ω are compatible. The important distinction that must be made among the contours of a configuration is the following:

Definition 7.22. Let $\omega \in \Omega^{\#}_{\Lambda}$. A contour $\gamma' \in \Gamma(\omega)$ is **external** if there exists no contour $\gamma \in \Gamma(\omega)$ such that $\overline{\gamma}' \subset \text{int}\gamma$.

We will group the configurations that contribute to $\mathbf{Z}^{\#}_{\Phi}(\Lambda)$ into families of configurations that have the same set of external contours. If $\Gamma(\omega) \neq \varnothing$, there exists at least one external contour, so let $\Gamma' \subset \Gamma(\omega)$ denote the collection of external contours of ω, which are pairwise compatible by construction. Let then

$$\text{ext} \stackrel{\text{def}}{=} \bigcap_{\gamma' \in \Gamma'} \text{ext}\gamma', \quad \Lambda^{\text{ext}} \stackrel{\text{def}}{=} \Lambda \cap \text{ext}.$$

The important property shared by the external contours of a configuration is that they all have the same type:

Lemma 7.23. *For all* $\omega \in \Omega^{\#}_{\Lambda}$, ext *is connected and* $\omega_i = \eta^{\#}_i$ *for each* $i \in$ ext. *As a consequence, all external contours of* $\Gamma(\omega)$ *are of type* #.

Proof Let $i', i'' \in$ ext and consider an arbitrary path $i' = i_1, \dots, i_n = i''$, $d_\infty(i_k, i_{k+1}) = 1$. If the path intersects the support of some external contour $\gamma' \in \Gamma(\omega)$, we define $k_- \stackrel{\text{def}}{=} \min\{k : i_k \in \overline{\gamma}'\} - 1$ and $k_+ \stackrel{\text{def}}{=} \max\{k : i_k \in \overline{\gamma}'\} + 1$. Clearly, $\{i_{k_-}, i_{k_+}\} \subset \partial^{\text{in}}\text{ext}\gamma'$. By Lemma 7.19, $\partial^{\text{in}}\text{ext}\gamma'$ is connected. One can therefore modify the path, between i_{k_-} and i_{k_+}, so that it is completely contained in $\partial^{\text{in}}\text{ext}\gamma'$. Since this can be done for each γ', we obtain in the end a path that is contained in each $\partial^{\text{in}}\text{ext}\gamma'$, hence in ext. This shows that ext is connected. The two other claims are immediate consequences. □

Remark 7.24. In this chapter, we always assume that the dimension is at least 2. Nevertheless, we invite the reader to stop and ponder over the peculiarities of the above-defined contours when $d = 1$. ◇

Now, since Λ is assumed to be c-connected, it can be partitioned into

$$\Lambda = \Lambda^{\text{ext}} \cup \bigcup_{\gamma' \in \Gamma'} \{\overline{\gamma}' \cup \bigcup_{\#'} \text{int}_{\#'}\gamma'\},$$

and we can then rearrange the sum over the sets $B \cap \Lambda \neq \varnothing$, in the Hamiltonian, to obtain:

$$\mathscr{H}_{\Lambda;\Phi}(\omega) = \mathscr{H}_{\Lambda^{\text{ext}};\Phi}(\omega) + \sum_{\gamma' \in \Gamma'} \Big\{ \sum_{B \subset \overline{\gamma}'} \Phi_B(\omega) + \sum_{\#'} \mathscr{H}_{\text{int}_{\#'}\gamma';\Phi}(\omega) \Big\}. \tag{7.28}$$

We have used the fact that the contours are thick, which implies that the components of their complement are at distances larger than the range of Φ (remember Lemma 7.8). Observe that, for each $B \subset \overline{\gamma}'$, $\Phi_B(\omega) = \Phi_B(\omega_{\overline{\gamma}'})$.

Let us characterize all configurations $\omega \in \Omega^{\#}_{\Lambda}$ that have the same set of external contours Γ':

1. Since Λ^{ext} does not contain any contours and since $\omega_i = \eta_i^\#$ for all $i \in \partial^{\mathrm{ex}}\Lambda^{\mathrm{ext}}$, we must have $\omega_i = \eta_i^\#$ for each $i \in \Lambda^{\mathrm{ext}}$. In particular, $\mathscr{H}_{\Lambda^{\mathrm{ext}};\Phi}(\omega) = \mathscr{H}_{\Lambda^{\mathrm{ext}};\Phi}(\eta^\#)$.
2. Each component of each $\mathrm{int}_{\#'}\gamma'$ has a boundary condition specified by the label of that component, namely $\#'$, and the contours of the configuration on that component must be at distances larger than 1 from γ'. The restrictions for the allowed configurations on $\mathrm{int}_{\#'}\gamma'$ therefore coincide exactly with those of $\Omega^{\#'}_{\mathrm{int}_{\#'}\gamma'}$.

Using (7.28), we thus get, after resumming over the allowed configurations on each component of $\mathrm{int}_{\#'}\overline{\gamma}'$:

$$\mathbf{Z}^\#_\Phi(\Lambda) = \sum_{\substack{\Gamma'\\ \text{compatible}\\ \text{external}}} e^{-\beta\mathscr{H}_{\Lambda^{\mathrm{ext}};\Phi}(\eta^\#)} \prod_{\gamma'\in\Gamma'} \Bigl\{\exp\Bigl(-\beta\sum_{B\subset\overline{\gamma}'}\Phi_B(\omega_{\overline{\gamma}'})\Bigr)\prod_{\#'}\mathbf{Z}^{\#'}_\Phi(\mathrm{int}_{\#'}\gamma')\Bigr\}. \tag{7.29}$$

Let us define, for each $\gamma \in \mathscr{C}^\#$, the **surface energy**

$$\|\gamma\| \stackrel{\text{def}}{=} \sum_{B\subset\overline{\gamma}}\bigl\{\Phi_B(\omega_{\overline{\gamma}}) - \Phi_B(\eta^\#)\bigr\}.$$

With this notation, (7.29) can be rewritten as

$$e^{\beta\mathscr{H}_{\Lambda;\Phi}(\eta^\#)}\mathbf{Z}^\#_\Phi(\Lambda) = \sum_{\substack{\Gamma'\\ \text{compatible}\\ \text{external}}} \prod_{\gamma'\in\Gamma'}\Bigl\{e^{-\beta\|\gamma'\|}\prod_{\#'} e^{\beta\mathscr{H}_{\mathrm{int}_{\#'}\gamma';\Phi}(\eta^\#)}\mathbf{Z}^{\#'}_\Phi(\mathrm{int}_{\#'}\gamma')\Bigr\}. \tag{7.30}$$

Our aim is then to go one step further and consider the external contours contained in each partition function $\mathbf{Z}^{\#'}_\Phi(\mathrm{int}_{\#'}\gamma')$ appearing on the right-hand side. Unfortunately, the external contours in $\mathbf{Z}^{\#'}_\Phi(\mathrm{int}_{\#'}\gamma')$ are of type $\#'$, and one needs to remove these from the analysis in order to avoid the global compatibility problem mentioned earlier.

In order to deal only with external contours of type #, we will use the following trick: [4] we multiply and divide the product over $\#'$, in (7.30), by the partition functions that involve only the #-boundary condition. That is, we write

$$\prod_{\#'}\mathbf{Z}^{\#'}_\Phi(\mathrm{int}_{\#'}\gamma') = \Bigl\{\prod_{\#'}\frac{\mathbf{Z}^{\#'}_\Phi(\mathrm{int}_{\#'}\gamma')}{\mathbf{Z}^{\#}_\Phi(\mathrm{int}_{\#'}\gamma')}\Bigr\}\prod_{\#'}\mathbf{Z}^{\#}_\Phi(\mathrm{int}_{\#'}\gamma'). \tag{7.31}$$

This introduces a nontrivial quotient that will be taken care of later, but it has the advantage of making the partition functions $\mathbf{Z}^\#_\Phi(\mathrm{int}_{\#'}\gamma')$ appear, which all share the same boundary condition #. This means that if one starts again summing over the external contours in $\mathbf{Z}^\#_\Phi(\mathrm{int}_{\#'}\gamma')$, *these will again be of type #*, as in the first step.

Let us express (7.30) using only the partition functions $\Xi^\#_\Phi(\cdot)$. Remembering that $e^{\beta\mathscr{H}_{\mathrm{int}_{\#'}\gamma';\Phi}(\eta^\#)}\mathbf{Z}^\#_\Phi(\mathrm{int}_{\#'}\gamma') \stackrel{\text{def}}{=} \Xi^\#_\Phi(\mathrm{int}_{\#'}\gamma')$, (7.30) becomes

$$\Xi^\#_\Phi(\Lambda) = \sum_{\substack{\Gamma'\\ \text{compatible}\\ \text{external}}} \prod_{\gamma'\in\Gamma'}\Bigl\{\mathrm{w}^\#(\gamma')\prod_{\#'}\Xi^\#_\Phi(\mathrm{int}_{\#'}\gamma')\Bigr\}, \tag{7.32}$$

where we introduced, for each $\gamma \in \mathscr{C}^{\#}$, the weight

$$\mathrm{w}^{\#}(\gamma) \stackrel{\text{def}}{=} e^{-\beta\|\gamma\|} \prod_{\#'} \frac{\mathbf{Z}_{\Phi}^{\#'}(\mathrm{int}_{\#'}\gamma)}{\mathbf{Z}_{\Phi}^{\#}(\mathrm{int}_{\#'}\gamma)} . \tag{7.33}$$

Looking at (7.32), it is clear that we can now repeat the procedure of fixing the external contours for each factor $\Xi_{\Phi}^{\#}(\mathrm{int}_{\#'}\gamma')$, these being all of type #. This process can be iterated and will automatically stop when one reaches contours whose interior is thin (remember the discussion on page 369), since the latter are too small to contain other contours. In this way, we end up with the following contour representation of the partition function:

$$\Xi_{\Phi}^{\#}(\Lambda) = \sum_{\substack{\Gamma \\ \text{compatible}}} \prod_{\gamma \in \Gamma} \mathrm{w}^{\#}(\gamma), \tag{7.34}$$

where the sum is over collections of contours of type #, in which the compatibility (in the sense of Definition 7.21) is purely geometrical, and can be encoded into

$$\delta(\gamma, \gamma') \stackrel{\text{def}}{=} \begin{cases} 1 & \text{if } d_{\infty}(\overline{\gamma}, \overline{\gamma}') > 1, \\ 0 & \text{otherwise.} \end{cases}$$

This pairwise hard-core interaction is similar to the one encountered in Section 5.7.1 (with a different distance). Notice, however, that the polymers considered here are more complex objects, which contain more information: their support and also the partial configuration $\omega_{\overline{\gamma}}$ (and the labels it induces).

Remark 7.25. It is important to emphasize that a compatible collection of contours contributing to (7.34) is an abstract collection, which *does not correspond*, in general, to the contours of any configuration $\omega \in \Omega_{\Lambda}^{\#}$. ◇

We have thus managed to express the partition function as a polymer model with a purely geometrical, pairwise compatibility condition. The price we had to pay for that was the introduction of the nontrivial weights $\mathrm{w}^{\#}(\cdot)$. The very nature of these suggests an *inductive* analysis. Namely, since $\mathrm{w}^{\#}(\gamma)$ can be written

$$\mathrm{w}^{\#}(\gamma) = e^{-\beta\|\gamma\|} \prod_{\#'} \frac{e^{-\beta \mathscr{H}_{\mathrm{int}_{\#'}\gamma;\Phi}(\eta^{\#'})}}{e^{-\beta \mathscr{H}_{\mathrm{int}_{\#'}\gamma;\Phi}(\eta^{\#})}} \frac{\Xi_{\Phi}^{\#'}(\mathrm{int}_{\#'}\gamma)}{\Xi_{\Phi}^{\#}(\mathrm{int}_{\#'}\gamma)}, \tag{7.35}$$

we see that $\mathrm{w}^{\#}(\gamma)$ depends on the weights of the smaller contours that appear in each $\Xi_{\Phi}^{\#'}(\mathrm{int}_{\#'}\gamma)$ (which are of type $\#' \neq \#$) and in $\Xi_{\Phi}^{\#}(\mathrm{int}_{\#'}\gamma)$.

7.3.2 Representing Probabilities Involving External Contours

Before going further, let us see how the contour models presented above can be used to represent probabilities involving external contours. Remember the definition of $\mu_{\Lambda;\Phi}^{\#}$ in (7.23).

Lemma 7.26. *Let Λ be* c*-connected and let $\{\gamma'_1, \dots, \gamma'_k\}$ be a collection of pairwise compatible contours of type # such that each γ'_i is contained in the exterior of the others, and $d_\infty(\gamma'_i, \Lambda^c) > 1$. Then*

$$\mu^{\#}_{\Lambda;\Phi}(\Gamma' \supset \{\gamma'_1, \dots, \gamma'_k\}) \leq \prod_{i=1}^{k} \mathrm{w}^{\#}(\gamma'_i). \tag{7.36}$$

Proof Follows the same steps that started with (7.29). □

Exercise 7.12. Complete the proof of Lemma 7.26.

7.4 Phase Diagram of the Blume–Capel Model

From now on, for the sake of concreteness, we will stick to the Blume–Capel model. As before, the three constant ground states are denoted by $\eta^{\#}$, with $\# \in \{+, 0, -\}$. We lift the degeneracy using W defined in (7.19). We continue to omit the dependence on β everywhere. We also drop Φ from the notations and only indicate the dependence on (λ, h) when it is really needed, that is, we write $\mathbf{Z}^{\#}(\Lambda)$ rather than $\mathbf{Z}^{\#}_{\lambda,h}(\Lambda)$, etc. We also write $\mathrm{e}^{\#}$ (or $\mathrm{e}^{\#}(\lambda, h)$ if necessary) instead of $\mathrm{e}_\Phi(\eta^{\#})$. With these conventions, (7.27) becomes

$$\mathbf{Z}^{\#}(\Lambda) = e^{-\beta \mathrm{e}^{\#}|\Lambda|}\Xi^{\#}(\Lambda).$$

We denote the pressure of the model, defined as in (7.7), by $\psi = \psi(\lambda, h)$.

7.4.1 Heuristics

We will construct the phase diagram by determining the *stable phases* of the system. Loosely speaking, this involves the determination, for each choice of boundary condition $\# \in \{+, 0, -\}$, of the set of pairs (λ, h) for which a Gibbs measure can be constructed using the thermodynamic limit with boundary condition #, whose typical configurations are small deviations from the ground state $\eta^{\#}$. Eventually, this will be done in Theorem 7.41.

But we first focus on the pressure, in particular the pressure in a finite volume with the boundary condition #:

$$\frac{1}{\beta|\Lambda|} \log \mathbf{Z}^{\#}(\Lambda) = -\mathrm{e}^{\#} + \frac{1}{\beta|\Lambda|} \log \Xi^{\#}(\Lambda).$$

In a regime where (λ, h) is such that $\mathrm{e}^{\#}$ is minimal among all $\mathrm{e}^{\#'}$, which happens when $(\lambda, h) \in \mathscr{U}^{\#}$, we expect typical configurations to be described by sparse local deviations away from $\eta^{\#}$; these configurations should also be the main contributions to $\mathbf{Z}^{\#}(\Lambda)$, in the sense that $-\mathrm{e}^{\#}$ should be the leading contribution to the pressure, the term $\frac{1}{\beta|\Lambda|}\log \Xi^{\#}(\Lambda)$ representing only *corrections* (for large values of β).

Making this argument rigorous requires having a control over $\log \Xi^{\#}(\Lambda)$; it will involve a detailed analysis of the weights $\mathrm{w}^{\#}(\cdot)$ and will eventually rely on a balance between the fields and an isoperimetric ratio related to the volume and support of the contours. Let us describe how these appear.

We know from Theorem 5.4 that $\log \Xi^{\#}(\Lambda)$ admits a convergent cluster expansion, in any finite region Λ, provided that one can find numbers $a(\gamma) \geq 0$ such that (the weights $\mathrm{w}^{\#}(\cdot)$ being real and nonnegative, there is no need for absolute values here and below)

$$\forall \gamma_* \in \mathscr{C}^{\#}, \quad \sum_{\gamma \in \mathscr{C}^{\#}} \mathrm{w}^{\#}(\gamma) e^{a(\gamma)} |\zeta(\gamma, \gamma_*)| \leq a(\gamma_*), \tag{7.37}$$

where $\zeta(\gamma, \gamma_*) \stackrel{\text{def}}{=} \delta(\gamma, \gamma_*) - 1$. As in Section 5.7.1, we observe that $\zeta(\gamma, \gamma_*) \neq 0$ if and only if $\overline{\gamma} \cap [\overline{\gamma}_*] \neq \varnothing$, where $[\overline{\gamma}_*] \stackrel{\text{def}}{=} \{j \in \mathbb{Z}^d : d_\infty(j, \overline{\gamma}_*) \leq 1\}$. This gives

$$\sum_{\gamma \in \mathscr{C}^{\#}} \mathrm{w}^{\#}(\gamma) e^{a(\gamma)} |\zeta(\gamma, \gamma_*)| \leq |[\overline{\gamma}_*]| \sup_{i \in \mathbb{Z}^d} \sum_{\gamma \in \mathscr{C}^{\#} : \overline{\gamma} \ni i} \mathrm{w}^{\#}(\gamma) e^{a(\gamma)} .$$

This shows that $a(\gamma) \stackrel{\text{def}}{=} |[\overline{\gamma}]|$ is a natural candidate. Since $|[\overline{\gamma}]| \leq 3^d |\overline{\gamma}|$, (7.37) is satisfied if

$$\sum_{\gamma \in \mathscr{C}^{\#} : \overline{\gamma} \ni 0} \mathrm{w}^{\#}(\gamma) e^{3^d |\overline{\gamma}|} \leq 1 . \tag{7.38}$$

Clearly, (7.38) can hold only if $\mathrm{w}^{\#}(\gamma)$ decreases exponentially fast with the size of the support of γ. We are thus naturally led to the following notion.

Definition 7.27. The weight $\mathrm{w}^{\#}(\gamma)$ is τ-**stable** if

$$\mathrm{w}^{\#}(\gamma) \leq e^{-\tau|\overline{\gamma}|} .$$

Below, in Lemma 7.30, we will show that (7.38) is indeed verified provided that all the weights $\mathrm{w}^{\#}(\gamma)$ are τ-stable (for a sufficiently large value of τ). Of course, this will be true only for certain values of (λ, h). For the moment, let us make a few comments about the difficulties encountered when trying to show that a weight is τ-stable.

Consider $\mathrm{w}^{\#}(\gamma)$, defined in (7.33). First, the **surface term**, $e^{-\beta\|\gamma\|}$, can always be bounded using Peierls' condition:

$$\|\gamma\| = \mathscr{H}_{\Phi^0}(\omega_{\overline{\gamma}} \mid \eta^{\#}) + \sum_{B \subset \overline{\gamma}} \{W_B(\omega_{\overline{\gamma}}) - W_B(\eta^{\#})\} \geq (\rho - 2\|W\|)|\overline{\gamma}| .$$

Remember from Example 7.12 that, for this model, Peierls' constant can be chosen to be $\rho = 5^{-d}$. Since $\|W\| \leq |h| + |\lambda|$, from now on we will always assume that $(\lambda, h) \in U$, where

$$U \stackrel{\text{def}}{=} \{(\lambda, h) \in \mathbb{R}^2 : |\lambda| \leq \rho/8,\ |h| \leq \rho/8\} , \tag{7.39}$$

which gives $\rho - 2\|W\| \geq \rho/2 \stackrel{\text{def}}{=} \rho_0$, yielding

$$e^{-\beta\|\gamma\|} \leq e^{-\beta\rho_0|\overline{\gamma}|}\,. \tag{7.40}$$

Let us then turn to the ratio of partition functions in (7.33).

The first observation is that this ratio is always *a boundary term: there exists a constant $c > 0$ (depending on Φ) such that*

$$e^{-c\beta|\partial^{\mathrm{ex}}\mathrm{int}_{\#'}\gamma|} \leq \frac{\mathbf{Z}^{\#'}(\mathrm{int}_{\#'}\gamma)}{\mathbf{Z}^{\#}(\mathrm{int}_{\#'}\gamma)} \leq e^{+c\beta|\partial^{\mathrm{ex}}\mathrm{int}_{\#'}\gamma|}\,.$$

(To check this, the reader can use the same type of arguments that were applied to prove, in Chapter 3, that the pressure of the Ising model does not depend on the boundary condition used.) Using (7.40), this gives $\mathrm{w}^{\#}(\gamma) \leq e^{-(\rho_0 - c)\beta|\overline{\gamma}|}$. Unfortunately, one can certainly not guarantee that $\rho_0 > c$. This naive argument shows that a more careful analysis is necessary to study those ratios, in order for the surface term to always be dominant. ◇

Let us then consider the weight $\mathrm{w}^{\#}(\gamma)$, but this time expressed as in (7.35). Since the ratios of polymer partition functions in that expression induce an intricate dependence of $\mathrm{w}^{\#}(\gamma)$ on (λ, h), let us ignore this ratio for a while and assume that

$$\prod_{\#'} \frac{\Xi^{\#'}(\mathrm{int}_{\#'}\gamma)}{\Xi^{\#}(\mathrm{int}_{\#'}\gamma)} = 1\,. \tag{7.41}$$

Of course, this is a serious over-simplification, since in general this ratio involves volume terms. (Note, however, that (7.41) indeed holds when each maximal component of each $\mathrm{int}_{\#'}\gamma$ is thin; remember (7.25).) Nevertheless, what remains of the weight after this simplification still contains volume terms, and the discussion below aims at showing how these will be handled.

Since $\mathscr{H}_{\mathrm{int}_{\#}\gamma;\Phi}(\eta^{\#}) = \mathrm{e}^{\#}|\mathrm{int}_{\#}\gamma|$, assuming (7.41) leaves us with

$$\mathrm{w}^{\#}(\gamma) = e^{-\beta\|\gamma\|}\prod_{\#'} e^{\beta(-\mathrm{e}^{\#'}+\mathrm{e}^{\#})|\mathrm{int}_{\#'}\gamma|} = e^{-\beta\|\gamma\|}\prod_{\#'} e^{\beta(\widehat{\psi}_0^{\#'}-\widehat{\psi}_0^{\#})|\mathrm{int}_{\#'}\gamma|}\,, \tag{7.42}$$

where we have introduced, for later convenience,

$$\widehat{\psi}_0^{\#} \stackrel{\text{def}}{=} -\mathrm{e}^{\#}\,. \tag{7.43}$$

Therefore, to guarantee that (7.42) decays exponentially fast with $|\overline{\gamma}|$, the key issue is to verify that the **volume term**, $\prod_{\#'} e^{\beta(\widehat{\psi}_0^{\#'}-\widehat{\psi}_0^{\#})|\mathrm{int}_{\#'}\gamma|}$, is not too large to destroy the exponential decay due to the surface term.

Of course, the simplest way to guarantee this is to assume that the exponents satisfy $\widehat{\psi}_0^{\#'} - \widehat{\psi}_0^{\#} \leq 0$ for each $\#'$, which occurs exactly when $(\lambda, h) \in \mathscr{U}^{\#}$ (see (7.21)), since we then have

$$\prod_{\#'} e^{\beta(\widehat{\psi}_0^{\#'}-\widehat{\psi}_0^{\#})|\mathrm{int}_{\#'}\gamma|} \leq 1\,.$$

This implies that the weight of $\gamma \in \mathscr{C}^{\#}$ is $\beta\rho_0$-stable uniformly on $\mathscr{U}^{\#}$:

$$\sup_{(\lambda,h)\in\mathscr{U}^{\#}} \mathrm{w}^{\#}(\gamma) \leq e^{-\beta\rho_0|\overline{\gamma}|}.$$

This bound is very natural, since Peierls' condition ensures that the creation of any contour represents a cost proportional to its support whenever $\eta^{\#}$ is a ground state for the pair (λ, h).

However, the construction of the phase diagram will require controlling the weights $\mathrm{w}^{\#}(\gamma)$ *in a neighborhood of the boundary of* $\mathscr{U}^{\#}$, that is, also for some values $(\lambda, h) \notin \mathscr{U}^{\#}$, for which $\widehat{\psi}_0^{\#'} - \widehat{\psi}_0^{\#} > 0$. In such a case, the volume term can be allowed to become large, but always less than the surface term. One can, for example, impose that

$$\prod_{\#'} e^{\beta(\widehat{\psi}_0^{\#'}-\widehat{\psi}_0^{\#})|\mathrm{int}_{\#'}\gamma|} \leq e^{\frac{1}{2}\beta\rho_0|\overline{\gamma}|}. \tag{7.44}$$

To guarantee this, one imposes restrictions on (λ, h) that depend on the geometrical properties of γ, namely on the ratios $\frac{|\overline{\gamma}|}{|\mathrm{int}_{\#}\gamma|}$. To make this dependence more explicit, we will use the following classical inequality, whose proof can be found in Section B.14 (see Corollary B.80).

Lemma 7.28. *Isoperimetric inequality, $d \geq 2$. For all $S \Subset \mathbb{Z}^d$,*

$$|\partial^{\mathrm{ex}} S| \geq |S|^{\frac{d-1}{d}}. \tag{7.45}$$

Although we will not use them later in this precise form, consider the sets

$$\mathscr{U}_{\gamma}^{\#} \stackrel{\mathrm{def}}{=} \{(\lambda, h) \in U : (\widehat{\psi}_0^{\#'} - \widehat{\psi}_0^{\#})|\mathrm{int}_{\#'}\gamma|^{1/d} \leq \tfrac{1}{2}\rho_0 \text{ for each } \#'\}. \tag{7.46}$$

Clearly, $\mathscr{U}_{\gamma}^{\#} \supset \mathscr{U}^{\#}$. Taking $(\lambda, h) \in \mathscr{U}_{\gamma}^{\#}$, we can use the isoperimetric inequality as follows:

$$\begin{aligned}\sum_{\#'}(\widehat{\psi}_0^{\#'} - \widehat{\psi}_0^{\#})|\mathrm{int}_{\#'}\gamma| &= \sum_{\#'}(\widehat{\psi}_0^{\#'} - \widehat{\psi}_0^{\#})|\mathrm{int}_{\#'}\gamma|^{1/d}|\mathrm{int}_{\#'}\gamma|^{(d-1)/d} \\ &\leq \tfrac{1}{2}\rho_0 \sum_{\#'} |\partial^{\mathrm{ex}}\mathrm{int}_{\#'}\gamma| \\ &= \tfrac{1}{2}\rho_0|\partial^{\mathrm{ex}}\mathrm{int}\gamma| \\ &\leq \tfrac{1}{2}\rho_0|\overline{\gamma}|.\end{aligned} \tag{7.47}$$

(We used the fact that the sets $\partial^{\mathrm{ex}}\mathrm{int}_{\#'}\gamma$ are pairwise disjoint subsets of $\overline{\gamma}$.) This implies (7.44) and yields $\frac{1}{2}\beta\rho_0$-stability uniformly on $\mathscr{U}_{\gamma}^{\#}$:

$$\sup_{(\lambda,h)\in\mathscr{U}_{\gamma}^{\#}} \mathrm{w}^{\#}(\gamma) \leq e^{-\frac{1}{2}\beta\rho_0|\overline{\gamma}|}.$$

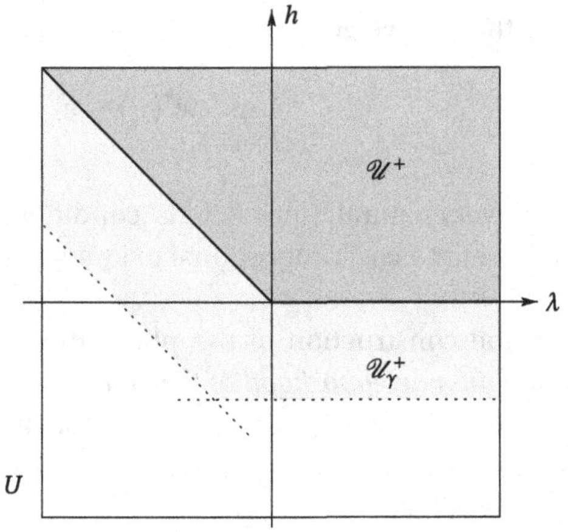

Figure 7.7. The weight of a contour $\gamma \in \mathscr{C}^+$ is $\frac{1}{2}\beta\rho_0$-stable in a region $\mathscr{U}_\gamma^+ \supset \mathscr{U}^+$, in particular in the neighborhood of the boundary of $\mathscr{U}^\#$. On the strip $\mathscr{U}_\gamma^+ \setminus \mathscr{U}^+$, which is small when the components $\text{int}_-\gamma$ and $\text{int}_0\gamma$ are large, (λ, h) has "the wrong sign" although $w^\#(\gamma)$ remains stable.

Let's be a little bit more explicit, for example in the case $\# = +$. Since $\widehat{\psi}_0^\pm = \pm h + \lambda$ and $\widehat{\psi}_0^0 = 0$, the set $\mathscr{U}_\gamma^+$ is given by

$$\mathscr{U}_\gamma^+ = \Big\{(\lambda, h) \in U : h \geq -\frac{\rho_0}{4|\text{int}_-\gamma|^{1/d}}\Big\} \cap \Big\{(\lambda, h) \in U : h \geq -\lambda - \frac{\rho_0}{2|\text{int}_0\gamma|^{1/d}}\Big\},$$

and is illustrated in Figure 7.7.

At this stage, the reader might benefit from having a look at the discussion in Section 3.10.10. ◇

The above discussion provides a sketch of the method that will be used later: *controlling the balance between volume and surface terms by combining the isoperimetric inequality with relevant thermodynamic quantities depending on* (λ, h). In our simplified discussion, which occurred only at the level of ground states, the thermodynamic quantities were represented by the differences $\widehat{\psi}_0^{\#'} - \widehat{\psi}_0^\#$. In the construction of the phase diagram, the inclusion of the ratios of partition functions neglected above will represent a technical nuisance and will be treated by a proof by induction, in which $\widehat{\psi}_0^{\#'} - \widehat{\psi}_0^\#$ will be replaced by $\widehat{\psi}_n^{\#'} - \widehat{\psi}_n^\#$. The induction index n will represent the size of the largest contour present in the system (in a sense to be made precise). Starting from the ground states ($n = 0$, no contours present in the system), we will progressively add contours of increasing size. At each step n, three pressures $\widehat{\psi}_n^\#$ will be introduced, constructed using contours of size smaller or equal to n. The weights of the newly added contours of volume n will be studied in detail; one will in particular determine the regions of parameters (λ, h) for which these weights are stable.

7.4.2 Polymer Models with τ-Stable Weights

During the induction argument below, we will use the cluster expansion to extract the surface and volume contributions to the polymer partition functions due to the quotients appearing in the weights $w^{\#}(\gamma)$. Before continuing, let us thus determine the conditions under which this procedure will be implemented and provide the main estimates that will be used throughout. Since cluster expansions will also be applied to auxiliary models that appear on the way, as well as to certain expressions involving derivatives with respect to λ and h, we first state a more general result, which will be applied in various situations.

Let $\mathscr{C}$ be a collection of contours, which we assume to be a subcollection of any of the families $\mathscr{C}^{\#}$, $\# \in \{+, 0, -\}$. For example, $\mathscr{C}$ can be the set of all contours of type $+$ whose interior has a size bounded by a constant. Assume that to each $\gamma \in \mathscr{C}$ corresponds a weight $w(\gamma) \geq 0$, possibly different from $w^{\#}(\gamma)$. We also assume that $\mathscr{C}$ and the weights $w(\cdot)$ are translation invariant, in the sense that if $\gamma \in \mathscr{C}$ and if γ' is any translate of γ, then $\gamma' \in \mathscr{C}$ and $w(\gamma') = w(\gamma)$. We will denote the size of the support of the smallest contour of $\mathscr{C}$ by

$$\ell_0 \overset{\text{def}}{=} \min\left\{|\overline{\gamma}| \,:\, \gamma \in \mathscr{C}\right\}.$$

For all $\Lambda \Subset \mathbb{Z}^d$, define

$$\Xi(\Lambda) \overset{\text{def}}{=} \sum_{\substack{\Gamma \subset \mathscr{C} \\ \text{compatible}}} \prod_{\gamma \in \Gamma} w(\gamma), \tag{7.48}$$

where the sum is over all families of pairwise compatible (in the sense of Definition 7.21) families Γ, such that $\overline{\gamma} \subset \Lambda$ and $d_\infty(\overline{\gamma}, \Lambda^c) > 1$ for all $\gamma \in \Gamma$.

Exercise 7.13. Show that the following limit exists:

$$g \overset{\text{def}}{=} \lim_{k \to \infty} \frac{1}{|B(k)|} \log \Xi(B(k)). \tag{7.49}$$

Hint: Use a subadditivity argument.

The cluster expansion for $\log \Xi(\Lambda)$, when it converges, is given by

$$\log \Xi(\Lambda) = \sum_X \Psi(X), \tag{7.50}$$

where the sum is over clusters X made of contours $\gamma \in \mathscr{C}$ such that $\overline{\gamma} \subset \Lambda$ and $d_\infty(\overline{\gamma}, \Lambda^c) > 1$, and $\Psi(X)$ is defined as in (5.20):

$$\Psi(X) \overset{\text{def}}{=} \alpha(X) \prod_{\gamma \in X} w(\gamma). \tag{7.51}$$

Remember that a contour can appear more than once in a cluster (in which case its weight appears more than once in the previous product), and that $\alpha(X)$ is a purely combinatorial factor.

Everywhere below, we will use the function

$$\eta(\tau,\ell) \stackrel{\text{def}}{=} 2e^{-\tau\ell/3}\,.$$

Theorem 7.29. *Assume that, for all $\gamma \in \mathscr{C}$, the weight $\mathrm{w}(\gamma)$ is C^1 in a parameter $s \in (a,b)$, and that, uniformly on (a,b),*

$$\mathrm{w}(\gamma) \le e^{-\tau|\overline{\gamma}|}\,, \qquad \Big|\frac{d\mathrm{w}(\gamma)}{ds}\Big| \le D|\overline{\gamma}|^{d/(d-1)}e^{-\tau|\overline{\gamma}|}\,, \tag{7.52}$$

where $D \ge 1$ is a constant. There exists $\tau_1 = \tau_1(D,d) < \infty$ such that the following holds. If $\tau > \tau_1$, then g defined in (7.49) is given by the following absolutely convergent series,

$$g = \sum_{X:\,\overline{X}\ni 0} \frac{1}{|\overline{X}|}\Psi(X)\,, \tag{7.53}$$

where the sum is over clusters X made of contours $\gamma \in \mathscr{C}$ and $\overline{X} \stackrel{\text{def}}{=} \bigcup_{\gamma\in X}\overline{\gamma}$. Moreover,

$$|g| \le \eta(\tau,\ell_0) \le 1\,,$$

and, for all $\Lambda \Subset \mathbb{Z}^d$, g provides the volume contribution to $\log \Xi(\Lambda)$, in the sense that

$$\Xi(\Lambda) = \exp\big(g|\Lambda| + \Delta\big)\,, \tag{7.54}$$

where Δ is a boundary term:

$$|\Delta| \le \eta(\tau,\ell_0)|\partial^{\text{in}}\Lambda|\,.$$

Finally, g is also C^1 in $s \in (a,b)$; its derivative equals

$$\frac{dg}{ds} = \sum_{X:\,\overline{X}\ni 0} \frac{1}{|\overline{X}|}\frac{d\Psi(X)}{ds} \tag{7.55}$$

and

$$\Big|\frac{dg}{ds}\Big| \le D\eta(\tau,\ell_0)\,.$$

We can express (7.54) in the following manner

$$\frac{1}{|\Lambda|}\log \Xi(\Lambda) = g + O\Big(\frac{|\partial^{\text{in}}\Lambda|}{|\Lambda|}\Big)\,.$$

◇

We have already seen in (7.38) that, when $\mathrm{w}(\gamma) \le e^{-\tau|\overline{\gamma}|}$, a sufficient condition for the convergence of the cluster expansion is that

$$\sum_{\gamma\in\mathscr{C}:\,\overline{\gamma}\ni 0} e^{-\tau|\overline{\gamma}|}e^{3^d|\overline{\gamma}|} \le 1\,. \tag{7.56}$$

We will actually choose τ so large that a stronger condition is satisfied, which will be needed in the proofs of Theorem 7.29 and Lemma 7.31.

Lemma 7.30. *There exists $\tau_0 < \infty$ such that, when $\tau > \tau_0$,*

$$\sum_{\gamma\in\mathscr{C}:\,\overline{\gamma}\ni 0} |\overline{\gamma}|^{d/(d-1)} e^{-(\tau/2-1)|\overline{\gamma}|} e^{3^d|\overline{\gamma}|} \le \eta(\tau,\ell_0) \le 1\,, \tag{7.57}$$

uniformly in the collection $\mathscr{C}$.

Proof First,

$$\sum_{\substack{\gamma\in\mathscr{C}:\\ \overline{\gamma}\ni 0}} |\overline{\gamma}|^{d/(d-1)} e^{-(\tau/2-1)|\overline{\gamma}|} e^{3^d|\overline{\gamma}|} \le \sum_{k\ge\ell_0} k^{d/(d-1)} e^{-(\tau/2-1-3^d)k}\, \#\{\gamma\in\mathscr{C} : \overline{\gamma}\ni 0,\ |\overline{\gamma}| = k\}\,.$$

Once the support $\overline{\gamma}$ is fixed, the number of possible configurations $\omega_{\overline{\gamma}}$ is bounded above by $|\Omega_0|^{|\overline{\gamma}|} = 3^{|\overline{\gamma}|}$. Therefore, proceeding as in Exercise 5.3, we can show that there exists a constant $c > 0$ (depending on the dimension, different from the one of Exercise 5.3) such that

$$\#\{\gamma\in\mathscr{C} : \overline{\gamma}\ni 0,\ |\overline{\gamma}| = k\} \le e^{ck}\,.$$

We assume that τ is so large that

$$\tau' \stackrel{\text{def}}{=} \tau/2 - 1 - 3^d - d/(d-1) - c \ge \tau/3 \tag{7.58}$$

and $e^{-\tau'} < 1/2$. Then, since $k^{d/(d-1)} < e^{dk/(d-1)}$,

$$\sum_{\gamma\in\mathscr{C}:\,\overline{\gamma}\ni 0} |\overline{\gamma}|^{d/(d-1)} e^{-(\tau/2-1)|\overline{\gamma}|} e^{3^d|\overline{\gamma}|} \le \sum_{k\ge\ell_0} e^{-\tau' k} = \frac{e^{-\tau'\ell_0}}{1-e^{-\tau'}} \le 2e^{-\tau'\ell_0} \le \eta(\tau,\ell_0)\,.$$

For each $\# \in \{+,0,-\}$, let $\tau^\#$ be the smallest constant τ satisfying the above requirements when using $\mathscr{C} = \mathscr{C}^\#$. We can then take $\tau_0 \stackrel{\text{def}}{=} \max_\# \tau^\#$. □

Everywhere below, τ_0 will refer to the number that appeared in Lemma 7.30.

Proof of Theorem 7.29. Denote by τ_1 the smallest $\tau > \tau_0$ such that $D\eta(\tau,\ell_0) \le 1$. Since (7.57) implies (7.56), Theorem 5.4 guarantees that the series on the right-hand side of (7.50) converges absolutely. In fact, proceeding as in (5.29), using the fact that $a(\gamma) \stackrel{\text{def}}{=} |[\overline{\gamma}]| \le 3^d|\overline{\gamma}|$,

$$\sum_{X:\,\overline{X}\ni 0} |\Psi(X)| \le \sum_{\gamma_1\in\mathscr{C}:\,\overline{\gamma}_1\ni 0} w(\gamma_1) e^{3^d|\overline{\gamma}_1|} \le \eta(\tau,\ell_0)\,. \tag{7.59}$$

This yields the convergence of the series for g, as well as the upper bound $|g| \le \eta(\tau,\ell_0)$. The same computations as those preceding Remark 5.9 and translation invariance give (7.54). The boundary term Δ is bounded in the same way. Since g is defined by an absolutely convergent series, we can rearrange its terms as $g = \sum_n f_n$, where

$$f_n \stackrel{\text{def}}{=} \sum_{k\geq 1} \sum_{\substack{X=\{\gamma_1,\dots,\gamma_k\}:\\ \overline{X}\ni 0\\ \sum|\gamma_i|=n}} \frac{1}{|\overline{X}|}\Psi(X).$$

Since only finitely many terms contribute to f_n and since each $\Psi(X)$ is C^1, f_n is also C^1. Moreover, for a cluster $X = \{\gamma_1, \dots, \gamma_k\}$, (7.52) gives

$$\begin{aligned}\Big|\frac{\mathrm{d}\Psi(X)}{\mathrm{d}s}\Big| &= \Big|\alpha(X)\sum_{j=1}^{k}\frac{\mathrm{dw}(\gamma_j)}{\mathrm{d}s}\prod_{\substack{i=1\\(i\neq j)}}^{k}\mathrm{w}(\gamma_i)\Big| \\ &\leq |\alpha(X)|\prod_{j=1}^{k}\{D|\overline{\gamma}_j|^{d/(d-1)}e^{-(\tau-1)|\overline{\gamma}_j|}\} = |\overline{\Psi}(X)|,\end{aligned}$$

where we used $\sum_{j=1}^{k} 1 \leq \prod_{j=1}^{k} e^{|\overline{\gamma}_j|}$; $\overline{\Psi}(X)$ is defined as $\Psi(X)$ in (7.51), but with $\mathrm{w}(\gamma)$ replaced by

$$\overline{\mathrm{w}}(\gamma) \stackrel{\text{def}}{=} D|\overline{\gamma}|^{d/(d-1)}e^{-(\tau-1)|\overline{\gamma}|}.$$

Again by Lemma 7.30, the analogue of (7.56) with $\overline{\mathrm{w}}(\gamma)$ replaced by $\mathrm{w}(\gamma)$ is satisfied. Therefore,

$$\begin{aligned}\sum_n\Big|\frac{\mathrm{d}f_n}{\mathrm{d}s}\Big| &\leq \sum_{n\geq 1}\sum_{k\geq 1}\sum_{\substack{X=\{\gamma_1,\dots,\gamma_k\}:\\ \overline{X}\ni 0\\ \sum|\gamma_i|=n}}\frac{1}{|\overline{X}|}|\overline{\Psi}(X)| \\ &\leq \sum_{X:\,\overline{X}\ni 0}|\overline{\Psi}(X)| \leq \sum_{\gamma_1\in\mathscr{C}:\,\overline{\gamma}_1\ni 0}\overline{\mathrm{w}}(\gamma_1)e^{3^d|\overline{\gamma}_1|} \leq D\eta(\tau,\ell_0).\end{aligned}$$

Theorem B.7 thus guarantees that g is also C^1, and that its derivative is given by $\frac{\mathrm{d}g}{\mathrm{d}s} = \sum_n \frac{\mathrm{d}f_n}{\mathrm{d}s}$, which proves (7.55). □

We will also need bounds on the sums of the weights of clusters that contain at least one contour with a large support.

Lemma 7.31. *Assume that* $\mathrm{w}(\gamma) \leq e^{-\tau|\overline{\gamma}|}$ *for each* $\gamma \in \mathscr{C}$ *and some* $\tau > \tau_0$. *Then, for all* $L \geq \ell_0$,

$$\sum_{\substack{X:\,\overline{X}\ni 0\\ |\overline{X}|\geq L}} |\Psi(X)| \leq e^{-\frac{1}{2}\tau L}. \tag{7.60}$$

Proof Proceeding as we did at the end of Section 5.7.4,

$$1 = e^{-\tau|\overline{X}|/2}e^{\tau|\overline{X}|/2} \leq e^{-\tau|\overline{X}|/2}\prod_{\gamma\in X}e^{\tau|\overline{\gamma}|/2},$$

which can be inserted into

$$\sum_{\substack{X:\,\overline{X}\ni 0\\ |\overline{X}|\geq L}} |\Psi(X)| \leq e^{-\tau L/2} \sum_{X:\,\overline{X}\ni 0} |\Psi(X)| \prod_{\gamma\in X} e^{\frac{\tau}{2}|\overline{\gamma}|}$$

$$= e^{-\tau L/2} \sum_{X:\,\overline{X}\ni 0} |\overline{\Psi}(X)| \leq e^{-\tau L/2}\eta(\tau,\ell_0) \leq e^{-\tau L/2}.$$

In the equality, we defined $\overline{\Psi}(X)$ as $\Psi(X)$ in (7.51), but with $\mathrm{w}(\gamma)$ replaced by $\overline{\mathrm{w}}(\gamma) \stackrel{\text{def}}{=} e^{-\frac{\tau}{2}|\overline{\gamma}|}$. We then again used Lemma 7.30. □

7.4.3 Truncated Weights and Pressures, Upper Bounds on Partition Functions

As explained above, we will construct the phase diagram by progressively adding contours on top of the ground states $\eta^{\#}$. To this end, we need to order contours according to their size.

Definition 7.32. A contour $\gamma \in \mathscr{C}^{\#}$ is **of class** n if $|\mathrm{int}\gamma| = n$. The **collection of contours of type # and class** n is denoted by $\mathscr{C}_n^{\#}$.

Clearly, a component of the interior of a contour $\gamma \in \mathscr{C}_n^{\#}$ can contain only contours of class strictly smaller than n.

We know that the weight of a contour with a large interior might not be stable for all values of the fields (λ, h). It will, however, turn out to be very useful to control the weights on the whole region $(\lambda, h) \in U$. To deal with the problem of unstable phases, we will *truncate* the weight of a contour as soon as its volume term becomes too large.

Contours of class zero contain no volume term, so they need not be truncated. Contours of large class do, however, contain volume terms and we will suppress the latter as soon as they become too important. We therefore fix some choice of **cutoff function** $\chi : \mathbb{R} \to [0,1]$, satisfying the following properties: (i)$\chi(s) = 1$ if $s \leq \rho_0/4$, (ii) $\chi(s) = 0$ if $s \geq \rho_0/2$, (iii) χ is C^1. Such a cutoff satisfies $\|\chi'\| \stackrel{\text{def}}{=} \sup_s |\chi'(s)| < \infty$.

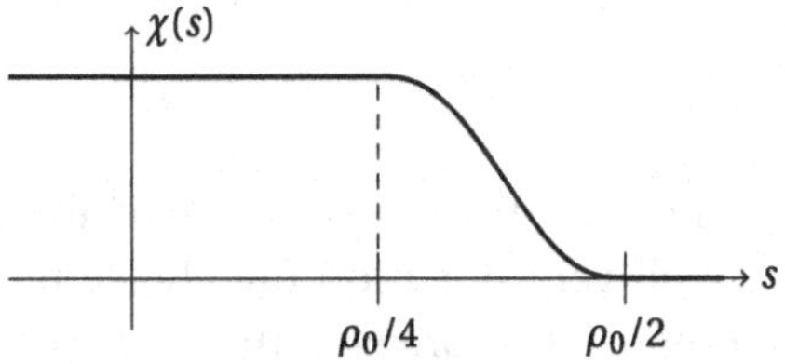

We start by defining the truncated quantities associated with $n = 0$. First, the **truncated pressures** (which we have already encountered) are defined by

$$\widehat{\psi}_0^{\#} \stackrel{\text{def}}{=} -\mathrm{e}^{\#}.$$

We define

$$\forall \gamma \in \mathscr{C}_0^{\#}, \quad \widehat{\mathrm{w}}^{\#}(\gamma) \stackrel{\text{def}}{=} \mathrm{w}^{\#}(\gamma) = e^{-\beta\|\gamma\|}.$$

Everywhere below, we assume that $(\lambda, h) \in U$ so that we can use the bound (7.40).

Assume now that the truncated weights $\widehat{\mathrm{w}}^\#(\cdot)$ have been defined for all contours of class $\leq n$. For a c-connected $\Lambda \Subset \mathbb{Z}^d$, let $\widehat{\Xi}^\#_n(\Lambda)$ denote the polymer model defined as in (7.34), but where the collections contain only contours of class $\leq n$, and with $\mathrm{w}^\#(\gamma)$ replaced by $\widehat{\mathrm{w}}^\#(\gamma)$. Then, set

$$\widehat{\mathbf{Z}}^\#_n(\Lambda) \stackrel{\text{def}}{=} e^{-\beta \mathrm{e}^\# |\Lambda|}\, \widehat{\Xi}^\#_n(\Lambda)\,.$$

For each $\# \in \{+, 0, -\}$, we use Exercise 7.13 to define the **truncated pressure** by

$$\begin{aligned}\widehat{\psi}^\#_n = \widehat{\psi}^\#_n(\lambda, h) &\stackrel{\text{def}}{=} \lim_{k\to\infty} \frac{1}{\beta|\mathrm{B}(k)|} \log \widehat{\mathbf{Z}}^\#_n(\mathrm{B}(k)) \\ &= -\mathrm{e}^\# + \lim_{k\to\infty} \frac{1}{\beta|\mathrm{B}(k)|} \log \widehat{\Xi}^\#_n(\mathrm{B}(k))\,.\end{aligned}$$

Notice that, since $\widehat{\Xi}^\#_n(\mathrm{B}(k)) \geq 1$, we have

$$\widehat{\psi}^\#_n \geq -\mathrm{e}^\#\,. \tag{7.61}$$

Definition 7.33. The **truncated weight** of $\gamma \in \mathscr{C}^\#_{n+1}$ is defined by

$$\widehat{\mathrm{w}}^\#(\gamma) \stackrel{\text{def}}{=} e^{-\beta\|\gamma\|} \prod_{\#'} \Big\{ \chi\big((\widehat{\psi}^{\#'}_n - \widehat{\psi}^\#_n)|\mathrm{int}_{\#'}\gamma|^{1/d}\big) \frac{\mathbf{Z}^{\#'}(\mathrm{int}_{\#'}\gamma)}{\mathbf{Z}^\#(\mathrm{int}_{\#'}\gamma)} \Big\}\,.$$

Intuitively, the goal of the previous definition is to eliminate all contours that could lead to instability. The reason we do not simply use a hard constraint of the form $\mathbf{1}_{\{(\widehat{\psi}^{\#'}_n - \widehat{\psi}^\#_n)|\mathrm{int}_{\#'}\gamma|^{1/d} \leq \rho_0/2\}}$, rather than a soft cutoff, is that the smoothness of the latter will allow us to obtain useful information on the regularity of the pressure and of the phase diagram. ◇

Notice that, since $0 \leq \chi \leq 1$, we have

$$\widehat{\mathrm{w}}^\#(\gamma) \leq \mathrm{w}^\#(\gamma)\,, \qquad \forall \gamma \in \mathscr{C}^\#\,. \tag{7.62}$$

Actually, unlike the true pressure ψ of the model, the truncated pressures *do* in fact depend very much on the choice of the boundary condition, that is, $\widehat{\psi}^\#_n \neq \widehat{\psi}^{\#'}_n$ in general. Moreover, the truncated weights and pressures depend on the specific choice of the cutoff function. Of course, as we will see later in Remark 7.37, this has no impact on the final construction of the phase diagram (but has an influence on what information on the latter can be extracted from our construction).

Other useful quantities will be important in the sequel. The first is

$$\widehat{\psi}_n \stackrel{\text{def}}{=} \max_{\#} \widehat{\psi}^\#_n\,.$$

Then, the following will be handy to relate the original weights to their truncated versions:

$$a_n^\# \stackrel{\text{def}}{=} \max_{\#'}\{\widehat{\psi}_n^{\#'} - \widehat{\psi}_n^\#\} = \widehat{\psi}_n - \widehat{\psi}_n^\#.$$

By definition, $a_n^\# \geq 0$ and, for all $\gamma \in \mathscr{C}_{n+1}^\#$,

$$a_n^\# |\mathrm{int}\gamma|^{1/d} \leq \rho_0/4 \quad \Longrightarrow \quad \widehat{\mathrm{w}}^\#(\gamma) = \mathrm{w}^\#(\gamma). \tag{7.63}$$

The following proposition is the main technical result of this chapter. Remember that τ_1 was defined in Theorem 7.29.

Proposition 7.34. *Let*

$$\tau \stackrel{\text{def}}{=} \tfrac{1}{2}\beta\rho_0 - 6. \tag{7.64}$$

There exists $0 < \beta_0 < \infty$ *such that the following holds. If* $\beta > \beta_0$*, then* $\tau > \tau_1$ *and there exists an increasing sequence* $c_n \uparrow c_\infty < \infty$ *such that, for all* # *and all* $n \geq 0$*, the following statements hold.*

1. (Bounds on the truncated weights.) *For all* $k \leq n$*, the weight of each* $\gamma \in \mathscr{C}_k^\#$ *is* τ*-stable uniformly on* U*:*

$$\widehat{\mathrm{w}}^\#(\gamma) \leq e^{-\tau|\overline{\gamma}|}, \tag{7.65}$$

and

$$a_n^\# |\mathrm{int}\gamma|^{1/d} \leq \rho_0/8 \quad \textit{implies} \quad \widehat{\mathrm{w}}^\#(\gamma) = \mathrm{w}^\#(\gamma). \tag{7.66}$$

Moreover, $\lambda \mapsto \widehat{\mathrm{w}}^\#(\gamma)$ *and* $h \mapsto \widehat{\mathrm{w}}^\#(\gamma)$ *are* C^1 *and, uniformly on* U*,*

$$\left|\frac{\partial \widehat{\mathrm{w}}^\#(\gamma)}{\partial \lambda}\right| \leq D|\overline{\gamma}|^{d/(d-1)} e^{-\tau|\overline{\gamma}|}, \qquad \left|\frac{\partial \widehat{\mathrm{w}}^\#(\gamma)}{\partial h}\right| \leq D|\overline{\gamma}|^{d/(d-1)} e^{-\tau|\overline{\gamma}|}, \tag{7.67}$$

where $D \stackrel{\text{def}}{=} 4(\beta + \|\chi'\|)$.

2. (Bounds on the partition functions.) *Assume that* $\Lambda \Subset \mathbb{Z}^d$ *is c-connected and* $|\Lambda| \leq n$*. Then*

$$\mathbf{Z}^\#(\Lambda) \leq e^{\beta\widehat{\psi}_n|\Lambda| + c_n|\partial^{\mathrm{ex}}\Lambda|}, \tag{7.68}$$

$$\left|\frac{\partial \mathbf{Z}^\#(\Lambda)}{\partial \lambda}\right| \leq \beta|\Lambda| e^{\beta\widehat{\psi}_n|\Lambda| + c_n|\partial^{\mathrm{ex}}\Lambda|}, \tag{7.69}$$

$$\left|\frac{\partial \mathbf{Z}^\#(\Lambda)}{\partial h}\right| \leq \beta|\Lambda| e^{\beta\widehat{\psi}_n|\Lambda| + c_n|\partial^{\mathrm{ex}}\Lambda|}, \tag{7.70}$$

uniformly in $(\lambda, h) \in U$*.*

Notice that the way the proposition is formulated allows one to obtain asymptotic bounds also in the limit $n \to \infty$.

Fixing the Constants. Before turning to the proof, we fix the relevant constants. Theorem 7.29 will be used repeatedly. Remember that $\eta(\tau, \ell_0) \stackrel{\text{def}}{=} 2e^{-\tau\ell_0/3}$, where

$\ell_0 \geq |\mathrm{B}(1)|$ is the size of the smallest support of a contour. We assume that β_0 satisfies $\beta_0 \geq 1$ and that it is large enough to ensure that, for all $\beta > \beta_0$, we have both $\tau > \tau_1$ and

$$D3^d\,\eta(\tau, \ell_0) \leq 1,$$

where $D \stackrel{\text{def}}{=} 4(\beta + \|\chi'\|)$. This will, in particular, always allow us to control the boundary terms that appear when using the cluster expansion in a region Λ:

$$|\Delta| \leq \eta(\tau, \ell_0)|\partial^{\text{in}}\Lambda| \leq \eta(\tau, \ell_0)3^d|\partial^{\text{ex}}\Lambda| \leq |\partial^{\text{ex}}\Lambda|. \tag{7.71}$$

We will also assume that β_0 is large enough to guarantee that

$$\forall k \geq 1, \qquad 2\beta^{-1}k^{1/d}e^{-\tau k^{(d-1)/d}/2} \leq \rho_0/8. \tag{7.72}$$

Let $c_0 \stackrel{\text{def}}{=} 2$ and

$$c_{n+1} \stackrel{\text{def}}{=} c_n + (n+1)^{1/d}e^{-\tau n^{(d-1)/d}/2}.$$

We then have $2 \leq c_n \uparrow c_\infty \stackrel{\text{def}}{=} 2 + \sum_{n\geq 1}(n+1)^{1/d}e^{-\tau n^{(d-1)/d}/2}$ and we can thus assume that β_0 is so large that $c_\infty \leq 3$. Finally, we assume β_0 is such that

$$\forall a > 0, \qquad \beta^{-1}\exp\bigl(-\max\{(\rho_0/4a)^{d-1}, \ell_0\}\tau/2\bigr) \leq \frac{a}{2}. \tag{7.73}$$

Proof of Proposition 7.34. Let us first prove the proposition in the case $n = 0$. When $\gamma \in \mathscr{C}_0^{\#}$, (7.66) is always true, $\widehat{\mathrm{w}}^{\#}(\gamma) = e^{-\beta\|\gamma\|}$, and (see (7.40)) $e^{-\beta\|\gamma\|} \leq e^{-\beta\rho_0|\overline{\gamma}|} < e^{-\tau|\overline{\gamma}|}$ when $(\lambda, h) \in U$. Since

$$\left|\frac{\partial\|\gamma\|}{\partial\lambda}\right| = \left|\sum_{i\in\overline{\gamma}}\{(\eta_i^{\#})^2 - (\omega_{\overline{\gamma}})_i^2\}\right| \leq 2|\overline{\gamma}|, \tag{7.74}$$

we have

$$\left|\frac{\partial\widehat{\mathrm{w}}^{\#}(\gamma)}{\partial\lambda}\right| \leq 2\beta|\overline{\gamma}|e^{-\beta\|\gamma\|} < D|\overline{\gamma}|^{d/(d-1)}e^{-\tau|\overline{\gamma}|}.$$

The bound on the derivative with respect to h is obtained in exactly the same way. Finally, (7.68)-(7.70) are trivial when $|\Lambda| = 0$, since the corresponding partition functions are all equal to 1.

We now assume that the claims of the proposition have been proved up to n, and prove that they also hold for $n+1$.

▶ *Controlling the truncated pressures* $\widehat{\psi}_n^{\#}$. Since the contours appearing in $\widehat{\Xi}_n^{\#}(\mathrm{B}(k))$ are all of class at most n and since their weights are τ-stable on U by the induction hypothesis, we can use Theorem 7.29 to express $\widehat{\psi}_n^{\#} = -\mathrm{e}^{\#} + \hat{g}_n^{\#}$, with $\hat{g}_n^{\#}$ given by the absolutely convergent series (notice that now there appears a division by β)

$$\hat{g}_n^{\#} = \sum_{\substack{X\in\mathcal{X}_n^{\#}:\\ \overline{X}\ni 0}} \frac{1}{\beta|\overline{X}|}\widehat{\Psi}^{\#}(X), \tag{7.75}$$

where $\chi_n^\#$ is the collection of all clusters made of contours of type # and class at most n, and where $\widehat{\Psi}^\#$ are defined as in (7.51), but with the weights $\widehat{w}^\#$. Moreover, $\hat{g}_n^\#$ is C^1 in λ and h on U, and

$$\left|\frac{\partial \hat{g}_n^\#}{\partial \lambda}\right| \le D\beta^{-1}\eta(\tau,\ell_0) \le 1\,. \tag{7.76}$$

The same upper bound holds for the derivative with respect to h.

▶ *Studying the truncated weights of contours of class* $n+1$. We first prove that (7.65) holds when $\gamma \in \mathscr{C}_{n+1}^\#$. Observe that $\widehat{w}^\#(\gamma) = 0$ whenever there exists #′ such that $(\widehat{\psi}_n^{\#'} - \widehat{\psi}_n^\#)|\mathrm{int}_{\#'}\gamma|^{1/d} > \frac{1}{2}\rho_0$. So we can assume that

$$(\widehat{\psi}_n^{\#'} - \widehat{\psi}_n^\#)|\mathrm{int}_{\#'}\gamma|^{1/d} \le \tfrac{1}{2}\rho_0 \qquad \text{for all } \#'. \tag{7.77}$$

Since $|\mathrm{int}\gamma| = n+1$, all contours contributing to the partition functions appearing in $\widehat{w}^\#(\gamma)$ are of type at most n. We can thus apply the induction hypothesis to deduce that, for any #′,

$$\mathbf{Z}^{\#'}(\mathrm{int}_{\#'}\gamma) \le e^{\beta\widehat{\psi}_n|\mathrm{int}_{\#'}\gamma| + c_n|\partial^{\mathrm{ex}}\mathrm{int}_{\#'}\gamma|} \le e^{\beta\widehat{\psi}_n|\mathrm{int}_{\#'}\gamma| + 3|\partial^{\mathrm{ex}}\mathrm{int}_{\#'}\gamma|}\,.$$

(Remember that $c_n \uparrow c_\infty \le 3$.) The truncated weight of each contour contributing to $\widehat{\mathbf{Z}}_n^\#(\mathrm{int}_{\#'}\gamma)$ is τ-stable by the induction hypothesis. Therefore, after using (7.62), (7.54) and (7.71),

$$\mathbf{Z}^\#(\mathrm{int}_{\#'}\gamma) \ge \widehat{\mathbf{Z}}_n^\#(\mathrm{int}_{\#'}\gamma) = e^{-\beta e^\#|\Lambda|}\widehat{\Xi}^\#(\Lambda) = e^{\beta\widehat{\psi}_n^\#|\mathrm{int}_{\#'}\gamma| + \Delta} \ge e^{\beta\widehat{\psi}_n^\#|\mathrm{int}_{\#'}\gamma| - |\partial^{\mathrm{ex}}\mathrm{int}_{\#'}\gamma|}\,. \tag{7.78}$$

Combining these two bounds, using the isoperimetric inequality as in (7.47), we obtain

$$\frac{\mathbf{Z}^{\#'}(\mathrm{int}_{\#'}\gamma)}{\mathbf{Z}^\#(\mathrm{int}_{\#'}\gamma)} \le e^{\beta(\widehat{\psi}_n - \widehat{\psi}_n^\#)|\mathrm{int}_{\#'}\gamma| + 4|\partial^{\mathrm{ex}}\mathrm{int}_{\#'}\gamma|} \le e^{(\frac{1}{2}\beta\rho_0 + 4)|\partial^{\mathrm{ex}}\mathrm{int}_{\#'}\gamma|}\,. \tag{7.79}$$

Bounding the cutoff function by 1, using (7.40) and $\sum_{\#'}|\partial^{\mathrm{ex}}\mathrm{int}_{\#'}\gamma| \le |\overline{\gamma}|$, we conclude that (7.65) indeed holds for γ:

$$\widehat{w}^\#(\gamma) \le e^{-\frac{1}{2}\beta\rho_0|\overline{\gamma}| + 4|\overline{\gamma}|} \le e^{-\tau|\overline{\gamma}|}\,. \tag{7.80}$$

Let us turn to (7.67). The derivative with respect to λ equals

$$\frac{\partial\widehat{w}^\#(\gamma)}{\partial\lambda} = -\beta\frac{\partial\|\gamma\|}{\partial\lambda}\widehat{w}^\#(\gamma) + e^{-\beta\|\gamma\|}\sum_{\#'}\frac{\partial}{\partial\lambda}\Big\{\chi(\cdot)\frac{\mathbf{Z}^{\#'}(\mathrm{int}_{\#'}\gamma)}{\mathbf{Z}^\#(\mathrm{int}_{\#'}\gamma)}\Big\}\prod_{\#''\neq\#'}\Big\{\chi(\cdot)\frac{\mathbf{Z}^{\#''}(\mathrm{int}_{\#''}\gamma)}{\mathbf{Z}^\#(\mathrm{int}_{\#''}\gamma)}\Big\}\,. \tag{7.81}$$

The only term appearing in (7.81) that we have not yet estimated is

$$\Big|\frac{\partial}{\partial\lambda}\Big\{\chi\big((\widehat{\psi}_n^{\#'} - \widehat{\psi}_n^\#)|\mathrm{int}_{\#'}\gamma|^{1/d}\big)\frac{\mathbf{Z}^{\#'}(\mathrm{int}_{\#'}\gamma)}{\mathbf{Z}^\#(\mathrm{int}_{\#'}\gamma)}\Big\}\Big|\,.$$

Using the chain rule together with (7.76), we see that the latter is bounded above by

$$4|\mathrm{int}_{\#'}\gamma|^{1/d}\|\chi'\|\frac{\mathbf{Z}^{\#'}(\mathrm{int}_{\#'}\gamma)}{\mathbf{Z}^\#(\mathrm{int}_{\#'}\gamma)} + \Big|\frac{\partial}{\partial\lambda}\frac{\mathbf{Z}^{\#'}(\mathrm{int}_{\#'}\gamma)}{\mathbf{Z}^\#(\mathrm{int}_{\#'}\gamma)}\Big|\,.$$

(In the second term, the cutoff was bounded by 1.) Equation (7.79) already leads to a bound on the first term, so we only have to consider the second one. Of course,

$$\left|\frac{\partial}{\partial\lambda}\frac{\mathbf{Z}^{\#'}(\mathrm{int}_{\#'}\gamma)}{\mathbf{Z}^{\#}(\mathrm{int}_{\#'}\gamma)}\right| \le \frac{|\partial\mathbf{Z}^{\#'}(\mathrm{int}_{\#'}\gamma)/\partial\lambda|}{\mathbf{Z}^{\#}(\mathrm{int}_{\#'}\gamma)} + \frac{\mathbf{Z}^{\#'}(\mathrm{int}_{\#'}\gamma)}{\mathbf{Z}^{\#}(\mathrm{int}_{\#'}\gamma)}\frac{|\partial\mathbf{Z}^{\#}(\mathrm{int}_{\#'}\gamma)/\partial\lambda|}{\mathbf{Z}^{\#}(\mathrm{int}_{\#'}\gamma)}. \tag{7.82}$$

Observe that

$$\begin{aligned}\left|\frac{\partial\mathbf{Z}^{\#}(\mathrm{int}_{\#'}\gamma)}{\partial\lambda}\right| &= \left|\sum_{\omega\in\Omega_{\mathrm{int}_{\#'}\gamma}} e^{-\beta\mathscr{H}_{\mathrm{int}_{\#'}\gamma;\Phi}(\omega)}\frac{\partial(-\beta\mathscr{H}_{\mathrm{int}_{\#'}\gamma;\Phi}(\omega))}{\partial\lambda}\right| \\ &\le \beta|\mathrm{int}_{\#'}\gamma|\,\mathbf{Z}^{\#}(\mathrm{int}_{\#'}\gamma).\end{aligned} \tag{7.83}$$

This bounds the last ratio in (7.82). The remaining ratios can be estimated using the induction hypothesis (7.69), and (7.79), yielding

$$\left|\frac{\partial}{\partial\lambda}\frac{\mathbf{Z}^{\#'}(\mathrm{int}_{\#'}\gamma)}{\mathbf{Z}^{\#}(\mathrm{int}_{\#'}\gamma)}\right| \le 2\beta|\mathrm{int}_{\#'}\gamma|e^{(\frac{1}{2}\beta\rho_0+4)|\partial^{\mathrm{ex}}\mathrm{int}_{\#'}\gamma|}. \tag{7.84}$$

Using (7.40), (7.74), (7.79), (7.80) and (7.84) in (7.81) leads to

$$\left|\frac{\partial\widehat{\mathrm{w}}^{\#}(\gamma)}{\partial\lambda}\right| \le \{2\beta|\overline{\gamma}| + (2\beta+4\|\chi'\|)|\mathrm{int}\gamma|\}e^{-\tau|\overline{\gamma}|};$$

(7.67) then follows after an application of the isoperimetric inequality. Once again, the derivative with respect to h is treated in the same way.

► *Estimating the differences* $|\widehat{\psi}^{\#}_{n+1} - \widehat{\psi}^{\#}_{k}|$, $k \le n$. Notice that $|\widehat{\psi}^{\#}_{n+1} - \widehat{\psi}^{\#}_{k}| = |\hat{g}^{\#}_{n+1} - \hat{g}^{\#}_{k}|$. By what was done above, all truncated weights of contours of class $\le n+1$ are τ-stable. We can therefore consider the expansions for each of the functions $\hat{g}^{\#}_{k}$, $k \le n+1$, as in (7.75). Then, observe that the clusters $X \in \chi^{\#}_{n+1}\setminus\chi^{\#}_{k}$ that contribute to $\hat{g}^{\#}_{n+1} - \hat{g}^{\#}_{k}$ contain at least one contour γ_* with $|\mathrm{int}\gamma_*| > k$. By the isoperimetric inequality, $|\overline{\gamma}_*| \ge k^{(d-1)/d}$. By Lemma 7.31,

$$\sum_{\substack{X:\,\overline{X}\ni 0\\ |\overline{X}|\ge k^{(d-1)/d}}} |\widehat{\Psi}^{\#}(X)| \le e^{-\frac{1}{2}\tau k^{(d-1)/d}}.$$

As a consequence,

$$|\widehat{\psi}^{\#}_{n+1} - \widehat{\psi}^{\#}_{k}| \le \beta^{-1}e^{-\frac{1}{2}\tau k^{(d-1)/d}}, \qquad |\widehat{\psi}_{n+1} - \widehat{\psi}_{k}| \le \beta^{-1}e^{-\frac{1}{2}\tau k^{(d-1)/d}}. \tag{7.85}$$

► *Showing that n, in* (7.66), *can be replaced by* $n+1$. For $\gamma \in \mathscr{C}^{\#}_{n+1}$, (7.66) holds by (7.63). Let us fix $\gamma \in \mathscr{C}^{\#}_{k}$, $k \le n$, and write

$$\begin{aligned}a^{\#}_{k}|\mathrm{int}\gamma|^{1/d} &= a^{\#}_{n+1}|\mathrm{int}\gamma|^{1/d} + (a^{\#}_{k} - a^{\#}_{n+1})|\mathrm{int}\gamma|^{1/d} \\ &\le a^{\#}_{n+1}|\mathrm{int}\gamma|^{1/d} + 2\beta^{-1}k^{1/d}e^{-\tau k^{(d-1)/d}/2} \\ &\le a^{\#}_{n+1}|\mathrm{int}\gamma|^{1/d} + \rho_0/8.\end{aligned}$$

(We used (7.72) in the last inequality.) Therefore, $a^{\#}_{n+1}|\mathrm{int}\gamma|^{1/d} \le \rho_0/8$ implies $a^{\#}_{k}|\mathrm{int}\gamma|^{1/d} \le \rho_0/4$. According to how the cutoffs were defined, this implies $\widehat{\mathrm{w}}^{\#}(\gamma) = \mathrm{w}^{\#}(\gamma)$.

We now move on to the most delicate part of the proof:

► *Showing that* (7.68) *holds if* $|\Lambda| = n+1$. Let $\Lambda \Subset \mathbb{Z}^d$ be an arbitrary c-connected set satisfying $|\Lambda| = n+1$, fix $(\lambda, h) \in U$ and consider $\mathbf{Z}^{\#}(\Lambda) = e^{-\mathrm{e}^{\#}|\Lambda|}\Xi^{\#}(\Lambda)$. Let $\gamma \in \mathscr{C}^{\#}$ be any contour appearing in the contour representation of $\Xi^{\#}(\Lambda)$ (therefore necessarily of class at most n). If

$$a^{\#}_{n}|\mathrm{int}\gamma|^{1/d} \le \rho_0/4\,,$$

we say that γ is **stable**; otherwise, we say that γ is **unstable**. By definition, a stable contour satisfies $\mathrm{w}^{\#}(\gamma) = \widehat{\mathrm{w}}^{\#}(\gamma)$.

Whether a contour is stable or not depends on the point $(\lambda, h) \in U$ we are considering. Note that when $a^{\#}_{n} = 0$, all contours appearing in $\mathbf{Z}^{\#}(\Lambda)$ are stable; in such a case, we can use Theorem 7.29 to conclude that

$$\mathbf{Z}^{\#}(\Lambda) = \widehat{\mathbf{Z}}^{\#}(\Lambda) = e^{-\mathrm{e}^{\#}|\Lambda|}\Xi^{\#}(\Lambda) = e^{\beta\widehat{\psi}_n|\Lambda|+\Delta} \le e^{\beta\widehat{\psi}_n|\Lambda|+|\partial^{\mathrm{ex}}\Lambda|} \le e^{\beta\widehat{\psi}_n|\Lambda|+c_n|\partial^{\mathrm{ex}}\Lambda|}\,.$$

We can thus assume that $a^{\#}_{n} > 0$. Note that the possible presence of unstable contours now prevents us from using the representation $\mathbf{Z}^{\#}(\Lambda) = e^{-\beta\mathrm{e}^{\#}}\Xi^{\#}(\Lambda)$ to analyze $\mathbf{Z}^{\#}(\Lambda)$.

Let us fix the set of external *unstable* contours. Once these are fixed, we can resum over the configurations on their exterior Λ^{ext}, with the restriction of allowing only stable contours. Observe that being stable is hereditary: if γ is stable, then any contour contained in its interior is also stable, so that we are guaranteed that none of these contours will surround one of the fixed unstable contours.

Proceeding similarly to what we did in (7.29), this first step gives

$$\mathbf{Z}^{\#}(\Lambda) = \sum_{\substack{\Gamma' \\ \text{compatible} \\ \text{external} \\ \text{unstable}}} \mathbf{Z}^{\#}_{\text{stable}}(\Lambda^{\mathrm{ext}}) \prod_{\gamma'\in\Gamma'} \Big\{\exp\big(-\beta\sum_{B\subset\overline{\gamma}'}\Phi_B(\omega_{\overline{\gamma}'})\big)\prod_{\#'}\mathbf{Z}^{\#'}(\mathrm{int}_{\#'}\gamma')\Big\}\,,$$

where $\mathbf{Z}^{\#}_{\text{stable}}(\Lambda^{\mathrm{ext}})$ denotes the partition function restricted to configurations in which all contours are stable. Since $\mathrm{w}^{\#}(\gamma) = \widehat{\mathrm{w}}^{\#}(\gamma)$ when γ is stable and since the truncated weights are τ-stable, we can use a convergent cluster expansion to study $\mathbf{Z}^{\#}_{\text{stable}}(\Lambda^{\mathrm{ext}})$:

$$\mathbf{Z}^{\#}_{\text{stable}}(\Lambda^{\mathrm{ext}}) = e^{-\beta\mathrm{e}^{\#}|\Lambda^{\mathrm{ext}}|}\Xi^{\#}_{\text{stable}}(\Lambda^{\mathrm{ext}}) \le e^{-\beta(\mathrm{e}^{\#}-\hat{g}^{\#}_{n,\text{stable}})|\Lambda^{\mathrm{ext}}|}e^{|\partial^{\mathrm{ex}}\Lambda^{\mathrm{ext}}|}\,.$$

For the partition functions in the interior of unstable contours, we apply the induction hypothesis (7.68):

$$\prod_{\#'}\mathbf{Z}^{\#'}(\mathrm{int}_{\#'}\gamma') \le \prod_{\#'} e^{\beta\widehat{\psi}_n|\mathrm{int}_{\#'}\gamma'|+c_n|\partial^{\mathrm{ex}}\mathrm{int}_{\#'}\gamma'|} \le e^{\beta\widehat{\psi}_n|\mathrm{int}\gamma'|}e^{3|\overline{\gamma}'|}\,.$$

Using $|\partial^{\mathrm{ex}}\Lambda^{\mathrm{ext}}| \le |\partial^{\mathrm{ex}}\Lambda| + \sum_{\gamma'\in\Gamma'}|\overline{\gamma}'|$ and extracting $e^{\beta\widehat{\psi}_n|\Lambda|}$ from the sum, we obtain

$$\mathbf{Z}^{\#}(\Lambda) \leq e^{\beta\widehat{\psi}_n|\Lambda|}e^{|\partial^{\mathrm{ex}}\Lambda|} \times \sum_{\substack{\Gamma' \\ \text{compatible} \\ \text{external} \\ \text{unstable}}} e^{-\beta(\widehat{\psi}_n + \mathrm{e}^{\#} - \hat{g}^{\#}_{n,\mathrm{stable}})|\Lambda^{\mathrm{ext}}|} \prod_{\gamma' \in \Gamma'} \exp\Big(-\beta \sum_{B \subset \overline{\gamma}'} \Phi_B(\omega_{\overline{\gamma}'})\Big) e^{(4-\beta\widehat{\psi}_n)|\overline{\gamma}'|}.$$

Observe now that $\widehat{\psi}_n|\overline{\gamma}| \geq \widehat{\psi}^{\#}_n|\overline{\gamma}| \geq -\mathrm{e}^{\#}|\overline{\gamma}| = -\sum_{B\subset\overline{\gamma}} \Phi_B(\eta^{\#})$ (indeed, remember (7.61), and observe that all pairwise interactions, in a ground state, are zero). Defining $\widehat{\psi}^{\#}_{n,\mathrm{stable}} \stackrel{\mathrm{def}}{=} -\mathrm{e}^{\#} + \hat{g}^{\#}_{n,\mathrm{stable}}$,

$$\mathbf{Z}^{\#}(\Lambda) \leq e^{\beta\widehat{\psi}_n|\Lambda|}e^{|\partial^{\mathrm{ex}}\Lambda|} \sum_{\substack{\Gamma' \\ \text{compatible} \\ \text{external} \\ \text{unstable}}} e^{-\beta(\widehat{\psi}_n - \widehat{\psi}^{\#}_{n,\mathrm{stable}})|\Lambda^{\mathrm{ext}}|} \prod_{\gamma'\in\Gamma'} e^{-\beta\|\gamma\|}e^{4|\overline{\gamma}'|}. \tag{7.86}$$

We will show that this last sum is bounded by $e^{|\partial^{\mathrm{ex}}\Lambda|}$, which will allow us to conclude that, indeed,

$$\mathbf{Z}^{\#}(\Lambda) \leq e^{\beta\widehat{\psi}_n|\Lambda|}e^{2|\partial^{\mathrm{ex}}\Lambda|} \leq e^{\beta\widehat{\psi}_n|\Lambda| + c_n|\partial^{\mathrm{ex}}\Lambda|}. \tag{7.87}$$

In order to do that, we will prove that $\widehat{\psi}_n - \widehat{\psi}^{\#}_{n,\mathrm{stable}}$ is positive and sufficiently large to strongly penalize families Γ' for which $|\Lambda^{\mathrm{ext}}|$ is large. First, let us write

$$\widehat{\psi}_n - \widehat{\psi}^{\#}_{n,\mathrm{stable}} = a^{\#}_n + (\hat{g}^{\#}_n - \hat{g}^{\#}_{n,\mathrm{stable}}).$$

The clusters that contribute to $\hat{g}^{\#}_n - \hat{g}^{\#}_{n,\mathrm{stable}}$ necessarily contain at least one unstable contour. Therefore, since an unstable contour γ satisfies

$$|\overline{\gamma}| \geq |\mathrm{int}\gamma|^{(d-1)/d} \geq \Big(\frac{\rho_0}{4a^{\#}_n}\Big)^{d-1},$$

we have

$$|\hat{g}^{\#}_n - \hat{g}^{\#}_{n,\mathrm{stable}}| \leq \beta^{-1}\exp\big(-\max\{(\rho_0/4a^{\#}_n)^{d-1}, \ell_0\}\tau/2\big) \leq \tfrac{1}{2}a^{\#}_n,$$

where we used (7.73). We conclude that $\widehat{\psi}_n - \widehat{\psi}^{\#}_{n,\mathrm{stable}} \geq a^{\#}_n/2$. Then, let us define new weights as follows: for each $\gamma \in \mathscr{C}^{\#}$, set

$$\mathrm{w}^{\#}_{*}(\gamma) \stackrel{\mathrm{def}}{=} \begin{cases} e^{-(\beta\rho_0 - 5)|\overline{\gamma}|} & \text{if } \gamma \text{ is unstable,} \\ 0 & \text{otherwise.} \end{cases}$$

We denote by $\Xi^{\#}_{*}(\cdot)$ the associated polymer partition function, and let

$$\hat{g}^{\#}_{*} \stackrel{\mathrm{def}}{=} \lim_{n\to\infty} \frac{1}{\beta|\mathrm{B}(n)|}\log \Xi^{\#}_{*}(\mathrm{B}(n)).$$

Since $\beta\rho_0 - 5 \geq \tau$, $\hat{g}^{\#}_{*}$ can be controlled by a convergent cluster expansion. Once again, the clusters that contribute to $\hat{g}^{\#}_{*}$ contain only unstable contours, and therefore, using again (7.73),

$$|\hat{g}^{\#}_{*}| \leq \beta^{-1}\exp\big(-\max\{(\rho_0/4a^{\#}_n)^{d-1}, \ell_0\}\tau/2\big) \leq \tfrac{1}{2}a^{\#}_n. \tag{7.88}$$

One can thus guarantee that

$$\widehat{\psi}_n - \widehat{\psi}^{\#}_{\text{stable}} \geq \hat{g}^{\#}_* \,. \tag{7.89}$$

We can now use this to show that the sum in (7.86) is bounded above by

$$\begin{aligned}
\sum_{\substack{\Gamma' \\ \text{compatible} \\ \text{external} \\ \text{unstable}}} e^{-\beta\hat{g}^{\#}_*|\Lambda^{\text{ext}}|} \prod_{\gamma'\in\Gamma'} e^{-(\beta\rho_0-4)|\overline{\gamma}'|} &\leq e^{-\beta\hat{g}^{\#}_*|\Lambda|} \sum_{\substack{\Gamma' \\ \text{compatible} \\ \text{external} \\ \text{unstable}}} \prod_{\gamma'\in\Gamma'} e^{-(\beta\rho_0-5)|\overline{\gamma}'|} e^{\beta\hat{g}^{\#}_*|\text{int}\gamma'|} \\
&\leq e^{-\beta\hat{g}^{\#}_*|\Lambda|} \sum_{\substack{\Gamma' \\ \text{compatible} \\ \text{external} \\ \text{unstable}}} \prod_{\gamma'\in\Gamma'} e^{-(\beta\rho_0-6)|\overline{\gamma}'|} \Xi^{\#}_*(\text{int}\gamma') \\
&= e^{-\beta\hat{g}^{\#}_*|\Lambda|} \Xi^{\#}_*(\Lambda) \\
&\leq e^{|\partial^{\text{ex}}\Lambda|} \,.
\end{aligned}$$

In the first inequality, we used $|\hat{g}^{\#}_*| \leq 1$, which follows from the first inequality in (7.88); in the second, we again used Theorem 7.29: $\Xi^{\#}_*(\text{int}\gamma) \geq e^{\beta\hat{g}^{\#}_*|\text{int}\gamma|-|\overline{\gamma}|}$. This proves the earlier claim.

▶ *Showing that* (7.69) *and* (7.70) *hold for* $|\Lambda| = n+1$. Proceeding as in (7.83), we see that

$$\left|\frac{\partial \mathbf{Z}^{\#}(\Lambda)}{\partial \lambda}\right| \leq \beta|\Lambda|\mathbf{Z}^{\#}(\Lambda)\,, \tag{7.90}$$

and, therefore, (7.69) follows from (7.87). The same argument yields (7.70).

▶ *Showing that n, in the right-hand side of* (7.68)-(7.70)*, can be replaced by* $n+1$. Using (7.85) and the isoperimetric inequality we get, for all $|\Lambda| \leq n+1$,

$$\begin{aligned}
\beta\widehat{\psi}_n|\Lambda| + c_n|\partial^{\text{ex}}\Lambda| &= \beta\widehat{\psi}_{n+1}|\Lambda| + c_n|\partial^{\text{ex}}\Lambda| + \beta(\widehat{\psi}_n - \widehat{\psi}_{n+1})|\Lambda| \\
&\leq \beta\widehat{\psi}_{n+1}|\Lambda| + (c_n + (n+1)^{1/d} e^{-\frac{1}{2}\tau n^{(d-1)/d}})|\partial^{\text{ex}}\Lambda| \\
&= \beta\widehat{\psi}_{n+1}|\Lambda| + c_{n+1}|\partial^{\text{ex}}\Lambda| \,.
\end{aligned}$$

□

7.4.4 Construction of the Phase Diagram

Let us now exploit the consequences of Proposition 7.34. We assume throughout this section that $\beta > \beta_0$. Since it appears at several places, we define

$$\epsilon = \epsilon(\beta) \stackrel{\text{def}}{=} D3^d \eta(\tau, \ell_0)\,.$$

When needed, β can be taken larger to make ϵ smaller. We will assume, for instance, that

$$\epsilon < \rho/32\,.$$

Proposition 7.34 dealt with the truncated pressures $\widehat{\psi}^{\#}_n$, to which only contours with an interior of size at most n contributed. Let us first see how the limit $n \to \infty$ restores the full model.

It follows from (7.85) that $\hat{g}^{\#}_n$ is a Cauchy sequence, which guarantees the existence of

$$\hat{g}^{\#} \stackrel{\text{def}}{=} \lim_{n\to\infty} \hat{g}_n^{\#}.$$

Moreover, $\hat{g}^{\#}$ can be expressed as the convergent series

$$\hat{g}^{\#} = \sum_{\substack{X\in\chi^{\#}:\\ \overline{X}\ni 0}} \frac{1}{\beta|\overline{X}|}\widehat{\Psi}^{\#}(X), \tag{7.91}$$

where $\chi^{\#}$ is the collection of all clusters made of contours of type # using the weights $\widehat{\mathrm{w}}^{\#}$. Namely, the difference between this series and the one in (7.75) is an infinite sum over clusters X such that (i) their support contains 0 and (ii) they contain at least one contour of class larger than n. By Lemma 7.31,

$$|\hat{g}^{\#} - \hat{g}_n^{\#}| \le \beta^{-1} e^{-\frac{1}{2}\tau n^{(d-1)/d}}. \tag{7.92}$$

We can thus also define

$$\widehat{\psi}^{\#} \stackrel{\text{def}}{=} \lim_{n\to\infty} \widehat{\psi}_n^{\#}.$$

The series in (7.91) can be bounded as usual: $|\hat{g}^{\#}| \le \epsilon$. This shows that $\widehat{\psi}^{\#}$ is a small perturbation of minus the energy density of the ground state $\eta^{\#}$:

$$\left|\widehat{\psi}^{\#} - (-\mathrm{e}^{\#})\right| \le \epsilon.$$

In order to compare the original weights and their truncated versions, we define

$$a^{\#} \stackrel{\text{def}}{=} \lim_{n\to\infty} a_n^{\#} = \widehat{\psi} - \widehat{\psi}^{\#},$$

where

$$\widehat{\psi} \stackrel{\text{def}}{=} \max_{\#} \widehat{\psi}^{\#}.$$

Letting $n \to \infty$ in (7.66) implies that

$$\forall \gamma \in \mathscr{C}^{\#}, \qquad a^{\#}|\mathrm{int}\gamma|^{1/d} \le \rho_0/8 \quad \text{implies} \quad \widehat{\mathrm{w}}^{\#}(\gamma) = \mathrm{w}^{\#}(\gamma). \tag{7.93}$$

Using this, we can define regions of parameters on which *all contours of type # will coincide with their truncated versions* (we indicate the dependence on β, to distinguish these sets from those defined in (7.21), which were associated with the energy densities of the ground states):

$$\mathscr{U}_\beta^{\#} \stackrel{\text{def}}{=} \{(\lambda,h)\in U : a^{\#}(\lambda,h) = 0\} = \{(\lambda,h)\in U : \widehat{\psi}^{\#}(\lambda,h) = \max_{\#'} \widehat{\psi}^{\#'}(\lambda,h)\}.$$

Observe that at a given point $(\lambda, h) \in U$, there is always at least one # for which $a^{\#}(\lambda, h) = 0$. This means that the regions $\mathscr{U}^{\#}$ cover U. By (7.93), we obtain the following:

Theorem 7.35. *There exists $0 < \beta_0 < \infty$ such that the following holds for all $\beta > \beta_0$:*

$$\forall \gamma \in \mathscr{C}^{\#}, \qquad (\lambda,h) \in \mathscr{U}_\beta^{\#} \quad \textit{implies} \quad \widehat{\mathrm{w}}^{\#}(\gamma) = \mathrm{w}^{\#}(\gamma).$$

In particular, when $(\lambda, h) \in \mathscr{U}_\beta^\#$, the true pressure of the model equals

$$\psi(\lambda,h) = \lim_{n\to\infty} \frac{1}{\beta|\mathrm{B}(n)|} \log \mathbf{Z}^\#(\mathrm{B}(n)) = \lim_{n\to\infty} \frac{1}{\beta|\mathrm{B}(n)|} \log \widehat{\mathbf{Z}}^\#(\mathrm{B}(n)) = \widehat{\psi}^\#(\lambda,h).$$

In other words,

$$\psi(\lambda,h) = \begin{cases} \widehat{\psi}^+(\lambda,h) & \text{if } (\lambda,h) \in \mathscr{U}_\beta^+, \\ \widehat{\psi}^0(\lambda,h) & \text{if } (\lambda,h) \in \mathscr{U}_\beta^0, \\ \widehat{\psi}^-(\lambda,h) & \text{if } (\lambda,h) \in \mathscr{U}_\beta^-. \end{cases} \tag{7.94}$$

In particular, we can extract properties of the (true) pressure by studying the truncated pressures and determining the regions $\mathscr{U}_\beta^\#$.

Up to now, even though we restricted our discussion to the Blume–Capel model for pedagogical reasons, the specific properties of this model were not used in any important way. In order to obtain more precise information about this model however, it will be useful to exploit these properties from now on. For example, the $+ \leftrightarrow -$ symmetry provides us immediately with useful information about the truncated pressures.

Exercise 7.14. Check that the $+ \leftrightarrow -$ symmetry implies $\widehat{\psi}^+(\lambda,-h) = \widehat{\psi}^-(\lambda,h)$.

Let us write the truncated pressures more explicitly, using the expressions for the ground state energy densities $\mathrm{e}^\#$ given in (7.20):

$$\widehat{\psi}^\pm(\lambda,h) = \pm h + \lambda + \hat{g}^\pm(\lambda,h), \qquad \widehat{\psi}^0(\lambda,h) = \hat{g}^0(\lambda,h).$$

Since the weights $\widehat{\mathrm{w}}^\#(\gamma)$ are C^1, Theorem 7.29 guarantees again that $\hat{g}^\#$ is C^1 on U and that, uniformly on U,

$$\Big|\frac{\partial \hat{g}^\#}{\partial \lambda}\Big| \le \epsilon, \qquad \Big|\frac{\partial \hat{g}^\#}{\partial h}\Big| \le \epsilon. \tag{7.95}$$

Therefore,

$$\Big|\frac{\partial \widehat{\psi}^\pm}{\partial h} \mp 1\Big| \le \epsilon, \qquad \Big|\frac{\partial \widehat{\psi}^\pm}{\partial \lambda} - 1\Big| \le \epsilon, \qquad \Big|\frac{\partial \widehat{\psi}^0}{\partial h}\Big| \le \epsilon, \qquad \Big|\frac{\partial \widehat{\psi}^0}{\partial \lambda}\Big| \le \epsilon.$$

The regions $\mathscr{U}_\beta^\#$. Let us start with $\mathscr{U}_\beta^+$, which can be written as

$$\mathscr{U}_\beta^+ = \{(\lambda,h) \in U : \widehat{\psi}^+ \ge \widehat{\psi}^-\} \cap \{(\lambda,h) \in U : \widehat{\psi}^+ \ge \widehat{\psi}^0\}.$$

Since the truncated pressures are continuous, the boundary of the set $\{\widehat{\psi}^+ \ge \widehat{\psi}^-\}$ is given by $\{\widehat{\psi}^+ = \widehat{\psi}^-\}$. By the $+ \leftrightarrow -$ symmetry (Exercise 7.14),

$$\widehat{\psi}^+(\lambda,0) = \widehat{\psi}^-(\lambda,0), \quad \forall \lambda.$$

Since $\frac{\partial \widehat{\psi}^+}{\partial h} > \frac{\partial \widehat{\psi}^-}{\partial h}$, uniformly on U, this shows that $\{\widehat{\psi}^+ = \widehat{\psi}^-\} = \{h = 0\}$ and that $\{\widehat{\psi}^+ \geq \widehat{\psi}^-\} = \{h \geq 0\}$. In the same way, the boundary of $\{\widehat{\psi}^+ \geq \widehat{\psi}^0\}$ equals $\{\widehat{\psi}^+ = \widehat{\psi}^0\}$. Since there is no symmetry between $+$ and 0, we will fix λ and search for the value of h such that $\widehat{\psi}^+(\lambda, h) = \widehat{\psi}^0(\lambda, h)$, which can also be written as

$$G(\lambda, h) \stackrel{\text{def}}{=} h + \lambda + \hat{g}^+(\lambda, h) - \hat{g}^0(\lambda, h) = 0\,. \tag{7.96}$$

To guarantee that (7.96) has solutions in U, we restrict our attention to a slightly smaller region. Remember that U is defined by $|\lambda|, |h| < \rho/8$. Let $\delta > 0$ be any number satisfying $2\epsilon < \delta < \rho/16$, and set

$$\check{U} \stackrel{\text{def}}{=} \big\{(\lambda, h) \,:\, |\lambda| < \rho/8 - \delta, |h| < \rho/8\big\}\,.$$

Then take $\lambda \in (-\rho/8 + \delta, \rho/8 - \delta)$ and define $h_\pm \stackrel{\text{def}}{=} -\lambda \pm \delta$. We have $(\lambda, h_\pm) \in U$ and $G(\lambda, h_-) < 0 < G(\lambda, h_+)$. Since $\frac{\partial G}{\partial h} > 0$ uniformly on U, this implies that there exists a unique $h = h(\lambda) \in (h_-, h_+)$, such that $G(\lambda, h) = 0$. The Implicit Function Theorem guarantees that $\lambda \mapsto h(\lambda)$ is actually C^1, and differentiating $G(\lambda, h(\lambda)) = 0$ with respect to λ leads to $|h'(\lambda) + 1| \leq 2\epsilon$.

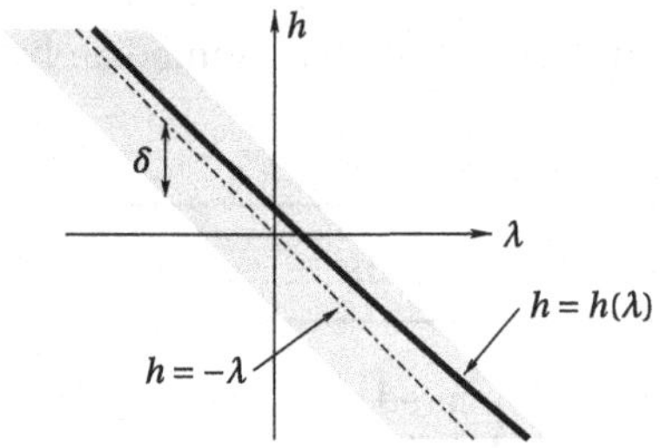

Figure 7.8. The construction of $\{\widehat{\psi}^+ = \widehat{\psi}^0\}$, which can be parametrized by a smooth map $\lambda \mapsto h(\lambda)$, whose graph lies in the strip of width 2δ around $h = -\lambda$.

One then has $\mathscr{U}_\beta^+ = \Big\{(\lambda, h) \in \check{U} \,:\, h \geq \max\{h(\lambda), 0\}\Big\}$. By symmetry,

$$\mathscr{U}_\beta^- = \big\{(\lambda, h) \in \check{U} \,:\, (\lambda, -h) \in \mathscr{U}_\beta^+\big\}\,,$$

and $\{\widehat{\psi}^- = \widehat{\psi}^0\}$ can be parametrized by $\lambda \mapsto -h(\lambda)$. Finally, $\mathscr{U}_\beta^0$ is the closure of $\check{U} \setminus (\mathscr{U}_\beta^+ \cup \mathscr{U}_\beta^-)$. The regions $\mathscr{U}_\beta^\#$ are separated by **coexistence lines,**

$$\mathscr{L}_\beta^{\#\#'} \stackrel{\text{def}}{=} \mathscr{U}_\beta^\# \cap \mathscr{U}_\beta^{\#'}\,, \qquad \# \neq \#'\,.$$

7.4.5 Results for the Pressure

We summarize the results obtained so far about the pressure in the following theorem (see also the qualitative picture in Figure 7.3).

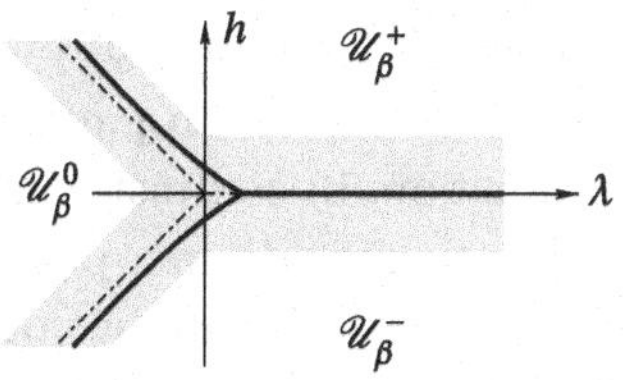

Figure 7.9. The phase diagram of the Blume-Capel model, which lies in a neighborhood of size δ of the zero-temperature diagram (dashed lines). Actually, by the $+/-$ symmetry of the model, we know that the line separating $\mathscr{U}_\beta^+$ and $\mathscr{U}_\beta^-$ lies *exactly* on the line $\{h = 0\}$.

Theorem 7.36 (The pressure of the Blume-Capel model at low temperature). *Let β_0 be as in Proposition 7.34. For all $\beta > \beta_0$,*

$$\psi(\lambda, h) = \max_{\#} \widehat{\psi}^{\#}(\lambda, h) = \begin{cases} \widehat{\psi}^{+}(\lambda, h) & \text{if } (\lambda, h) \in \mathscr{U}_\beta^+, \\ \widehat{\psi}^{0}(\lambda, h) & \text{if } (\lambda, h) \in \mathscr{U}_\beta^0, \\ \widehat{\psi}^{-}(\lambda, h) & \text{if } (\lambda, h) \in \mathscr{U}_\beta^-. \end{cases} \tag{7.97}$$

As a consequence:

1. *The pressure $\psi(\lambda, h)$ is C^1 in λ and h, everywhere in the interior of each region $\mathscr{U}_\beta^\#$, $\# \in \{+, 0, -\}$.*
2. ***First-order phase transitions** occur across each of the coexistence lines, in the sense that*

$$\frac{\partial \psi}{\partial \lambda^+} > \frac{\partial \psi}{\partial \lambda^-} \qquad \text{at each } (\lambda, h) \in \mathscr{L}_\beta^{\pm 0},$$

and, for all $\# \neq \#'$,

$$\frac{\partial \psi}{\partial h^+} > \frac{\partial \psi}{\partial h^-} \qquad \text{at each } (\lambda, h) \in \mathscr{L}_\beta^{\#\#'}.$$

Remark 7.37. Remember that the construction of the truncated pressures depends on the choice of the cutoff $\chi(\cdot)$ used in the definition of the truncated weights. This choice of course only affects the truncated pressures. It has, however, an impact on what we could extract from the analysis above. Namely, the assumption that the cutoff was C^1 yields, ultimately, the corresponding regularity of the pressure in the interior of the regions $\mathscr{U}_\beta^\#$, as well as the regularity of the boundary of these regions. Choosing a cutoff with higher regularity would yield a corresponding enhancement of these properties (but would require control of higher-order derivatives of the truncated pressures in Proposition 7.34). ◇

Remark 7.38. The fact that the pressure coincides with the maximal truncated pressure (see (7.97)) means that the truncated pressures provide natural continuations of the pressure through the coexistence lines. A similar conclusion has been drawn for the Curie–Weiss model (see Figure 3.19 on page 169). Nevertheless, the continuations of $h \mapsto \psi_\beta^{\mathrm{CW}}(h)$ through the transition point were *analytic*, while those obtained here are only C^1. In particular, there are infinitely many ways to make such a continuation, in contrast with the analytic case. For example, each choice of a cutoff function yields an a-priori different C^1 continuation. Nevertheless, *analytic* continuations through the coexistence lines do not exist, in general, in the framework of PST. This will be discussed in Section 7.6.6. ◇

Remark 7.39. In this chapter, we started with the Blume–Capel model at parameters $h = \lambda = 0$ and considered perturbations around this point, constructing the phase diagram in its vicinity. In the same way, we could have started with $\lambda = \lambda_0$ and $h = h_0$, and constructed the phase diagram around this point. This allows, in particular, the description of coexistence lines outside the domain U. ◇

7.4.6 Gibbs Measures at Low Temperature

So far, the phase diagram has been constructed by studying partition functions and truncated pressures. In this section, we consider the consequences of the previous study at low temperature, from a probabilistic point of view.

Let us take β large, fix $(\lambda, h) \in \check{U}$, and denote by $\mathscr{G}(\beta, \lambda, h)$ the set of infinite-volume Gibbs measures associated with the Blume–Capel model. As in Chapter 6, the latter are defined as the probability measures compatible with the specification associated with the potential $\beta\Phi = \beta(\Phi^0 + W)$ (remember (7.19)). Let

$$\Upsilon(\beta, \lambda, h) \stackrel{\text{def}}{=} \{\# \in \{+, 0, -\} : \mathscr{U}_\beta^\# \ni (\lambda, h)\}$$

denote the set of stable periodic ground states at β, λ, h. We will show that, *for each* $\# \in \Upsilon(\beta, \lambda, h)$, *a Gibbs measure* $\mu_{\beta,\lambda,h}^\# \in \mathscr{G}(\beta, \lambda, h)$ *can be prepared using the boundary condition* $\eta^\#$, *under which typical configurations are described by small local perturbations away from* $\eta^\#$. Moreover, these measures are extremal and ergodic. As will be explained in Section 7.6.1, this construction yields the *complete* phase diagram: any other translation-invariant Gibbs measure can be written as a convex combination of the measures $\mu_{\beta,\lambda,h}^+$, $\mu_{\beta,\lambda,h}^0$ and $\mu_{\beta,\lambda,h}^-$.

Remark 7.40. Notice that, using the boundary condition # and proceeding as we did in the proof of Theorem 6.26, we can extract from any sequence $\Lambda_n \uparrow \mathbb{Z}^d$ a subsequence $(\Lambda_{n_k})_{k\geq 1}$ such that the limit

$$\lim_{k\to\infty} \mu_{\Lambda_{n_k};\beta,\lambda,h}^{\eta^\#}(f)$$

exists for every local function f, thus defining a Gibbs measure. A priori, this measure depends on the subsequence $(\Lambda_{n_k})_{k\geq 1}$. ◇

Below, we show that, when the temperature is sufficiently low and $\# \in \Upsilon(\beta, \lambda, h)$, the thermodynamic limit used to construct this measure does not depend on the sequence $(\Lambda_n)_{n\geq 1}$ and can be controlled in a much more precise way.

In order to use the earlier results that rely on the contour representations of configurations, we will use the Gibbs distributions $\mu^{\#}_{\Lambda;\beta,\lambda,h}$ for the Blume–Capel model, defined as $\mu^{\#}_{\Lambda;\Phi}$ in Section 7.3, rather than those of Chapter 6.

Theorem 7.41. *There exists β_0 such that, for all $\beta \geq \beta_0$ and all $(\lambda, h) \in \check{U}$, the following holds.*

1. *Let $\# \in \Upsilon(\beta, \lambda, h)$. For all sequences of c-connected sets $\Lambda_n \uparrow \mathbb{Z}^d$ and every local function f, the following limit exists,*

$$\mu^{\#}_{\beta,\lambda,h}(f) \stackrel{\text{def}}{=} \lim_{n\to\infty} \mu^{\#}_{\Lambda_n;\beta,\lambda,h}(f), \tag{7.98}$$

and defines a Gibbs measure $\mu^{\#}_{\beta,\lambda,h} \in \mathscr{G}(\beta, \lambda, h)$.

2. *The measures $\mu^{\#}_{\beta,\lambda,h}$, $\# \in \Upsilon(\beta, \lambda, h)$ are translation invariant, extremal and ergodic. Moreover, they are distinct, since*

$$\mu^{\#}_{\beta,\lambda,h}(\sigma_0 = \#) \geq 1 - \delta(\beta), \tag{7.99}$$

where $\delta(\beta) > 0$ tends to zero as $\beta \to \infty$.

In particular, two (resp. three) distinct Gibbs measures can be constructed for each pair (λ, h) living on a coexistence line (resp. at the triple point). (Note that a similar statement could also be derived from Theorem 6.91 and the nondifferentiability of the pressure.) Geometric properties of typical configurations under these measures will be described in Theorem 7.44.

Proof of Theorem 7.41. Fix (β, λ, h) and let $\# \in \Upsilon(\beta, \lambda, h)$. To lighten the notations, we omit β, λ and h everywhere in the indices.

► *Proof of* (7.98). Let us fix some local function f. We will first show that, for all $n \geq 1$,

$$\left|\mu^{\#}_{\Lambda}(f) - \mu^{\#}_{\Delta}(f)\right| \leq c\|f\|_{\infty} n^d e^{-\tau' n}, \tag{7.100}$$

whenever $\Lambda, \Delta \Subset \mathbb{Z}^d$ are c-connected and both contain $\mathrm{B}(2n)$. This implies that $(\mu^{\#}_{\Lambda_n}(f))_{n\geq 1}$ is a Cauchy sequence, which proves the existence of the limit in (7.98). We have already seen that this is sufficient to define the measure $\mu^{\#}$; we leave it as an exercise to verify that $\mu^{\#} \in \mathscr{G}(\beta, \lambda, h)$ (adapt the proof of Theorem 6.26). Observe also that (7.100) shows that the limit does not depend on the chosen sequence $(\Lambda_n)_{n\geq 1}$.

We prove (7.100) using a *coupling* of $\mu^{\#}_{\Lambda}$ and $\mu^{\#}_{\Delta}$. Let

$$\Omega_\Lambda \times \Omega_\Delta \stackrel{\text{def}}{=} \left\{(\omega, \omega') : \omega \in \Omega_\Lambda, \omega' \in \Omega_\Delta\right\}.$$

Let n be large enough to ensure that $\mathrm{B}(n)$ contains the support of f, and assume Λ and Δ are both c-connected and large enough to contain $\mathrm{B}(2n)$. On $\Omega_\Lambda \times \Omega_\Delta$, define

$$\mathbb{P}^{\#}_{\Lambda,\Delta} \stackrel{\text{def}}{=} \mu^{\#}_{\Lambda} \otimes \mu^{\#}_{\Delta}.$$

We call $D \subset \mathbb{Z}^d$ **#-surrounding** if (i) D is connected, (ii) $B(n) \subset D \subset B(2n)$, (iii) $\omega_i = \omega'_i = \#$ for all $i \in \partial^{ex} D$.

Let us consider the event $C_{n,\#,\#} \subset \Omega_\Lambda \times \Omega_\Delta$ defined as follows: $(\omega, \omega') \in C_{n,\#,\#}$ if and only if there exists at least one #-surrounding set. Observe that if D_1 and D_2 are #-surrounding, then $D_1 \cup D_2$ is also #-surrounding. When $C_{n,\#,\#}$ occurs, we therefore denote the *largest* (with respect to inclusion) #-surrounding set by $D_\#$. We denote by $[D_\#]$ the event that $C_{n;\#,\#}$ occurs and $D_\#$ is the largest #-surrounding set. Letting $F(\omega, \omega') \stackrel{\text{def}}{=} f(\omega) - f(\omega')$,

$$|\mu^\#_\Lambda(f) - \mu^\#_\Delta(f)| = |\mathbb{E}^\#_{\Lambda,\Delta}[F]| \le |\mathbb{E}^\#_{\Lambda,\Delta}[F \mathbf{1}_{C_{n,\#,\#}}]| + 2\|f\|_\infty \mathbb{P}^\#_{\Lambda,\Delta}({C_{n,\#,\#}}^{\mathrm{c}}).$$

For the first term in the right-hand side, we can sum over the possible $D_\#$ (for simplicity, we denote the realization of each such set also by $D_\#$). Note that, by construction, it is sufficient to look at the configuration outside $D_\#$ to determine whether the event $[D_\#]$ occurs; in other words, $[D_\#] \in \mathscr{F}_{D^{\mathrm{c}}_\#}$. Therefore,

$$\mathbb{E}^\#_{\Lambda,\Delta}[F \mathbf{1}_{[D_\#]}] = \mathbb{E}^\#_{\Lambda,\Delta}\big[\mathbb{E}^\#_{\Lambda,\Delta}[F \mid \mathscr{F}_{D^{\mathrm{c}}_\#}] \mathbf{1}_{[D_\#]}\big] = 0,$$

since $\mathbb{P}^\#_{\Lambda,\Delta}\big((\omega_{D_\#}, \omega'_{D_\#}) \mid \mathscr{F}_{D^{\mathrm{c}}_\#}\big) = \mu^\#_{D_\#}(\omega_{D_\#})\mu^\#_{D_\#}(\omega'_{D_\#})$ on $[D_\#]$. This implies that

$$\mathbb{E}^\#_{\Lambda,\Delta}[F \mathbf{1}_{C_{n,\#,\#}}] = \sum_{D_\#} \mathbb{E}^\#_{\Lambda,\Delta}[F \mathbf{1}_{[D_\#]}] = 0.$$

We now show that, at low temperature, #-surrounding sets exist with probability close to 1.

Lemma 7.42. *When β is sufficiently large, there exists $\tau' = \tau'(\beta) > 0$ such that*

$$\mathbb{P}^\#_{\Lambda,\Delta}({C_{n,\#,\#}}^{\mathrm{c}}) \le 2|B(n)|e^{-\tau' n}. \tag{7.101}$$

Proof Let $\mathscr{P}_n$ denote the family of all self-avoiding paths $(i_0, i_1, \dots, i_k)$ inside $B(2n) \setminus B(n)$, with $i_0 \in \partial^{ex} B(n)$, $i_k \in \partial^{in} B(2n)$, $i_j \sim i_{j+1}$. We say that i is **(#, #)-correct** if it is #-correct both in ω and in ω'. We claim that

$${C_{n,\#,\#}}^{\mathrm{c}} \subset \big\{\exists (i_0, \dots, i_k) \in \mathscr{P}_n \text{ such that each } i_j \text{ is not } (\#,\#)\text{-correct}\big\}. \tag{7.102}$$

Indeed, assume that each $\pi = (i_0, \dots, i_k) \in \mathscr{P}_n$ is such that i_j is (#, #)-correct for at least one index $j \in \{0, 1, \dots, k\}$, and let $j(\pi)$ denote the smallest such index. Consider the truncated path $\widetilde{\pi} \stackrel{\text{def}}{=} (i_0, i_1, \dots, i_{j(\pi)})$. Then, clearly, $D \stackrel{\text{def}}{=} B(n) \cup \bigcup_{\pi \in \mathscr{P}_n} \widetilde{\pi}$ is #-surrounding.

Now, if each vertex of a path $\pi \in \mathscr{P}_n$ is not (#, #)-correct, then there must exist two collections $\Gamma' = \{\gamma'_1, \dots, \gamma'_l\} \subset \Lambda$ and $\Gamma'' = \{\gamma''_1, \dots, \gamma''_m\} \subset \Delta$ of *external* contours of type # such that

$$\pi \subset \bigcup_{k=1}^{l} \overline{\operatorname{int}\gamma'_k} \cup \bigcup_{k=1}^{m} \overline{\operatorname{int}\gamma''_k},$$

where we introduced the notation $\overline{\operatorname{int}\gamma} \stackrel{\text{def}}{=} \overline{\gamma} \cup \operatorname{int}\gamma$.

- From the collection $\Gamma' \cup \Gamma''$, we can always extract an ordered subcollection $Y = (\gamma_1, \dots, \gamma_k) \subset \Gamma' \cup \Gamma''$ enjoying the following properties: (i) $\overline{\text{int}\gamma_1} \cap \text{B}(n) \neq \varnothing$; (ii) either all contours $\gamma_i \in Y$ whose index i is odd belong to Γ' and those whose index i is even belong to Γ'', or vice versa; (iii) Y is a *chain* in the sense that each $\gamma_i \in Y$ is compatible neither with γ_{i-1} nor with γ_{i+1}, but is compatible with all other γ_js; (iv) if $\overline{Y} \stackrel{\text{def}}{=} \bigcup_{i=1}^k \overline{\gamma}_i$ then $|\overline{Y}| \geq n$ (since π has diameter at least n).
- By construction, Y is made of contours belonging to subcollections $\Gamma_1 \subset \Gamma'$ and $\Gamma_2 \subset \Gamma''$. Using Lemma 7.26 and the fact that all contours of type # are τ-stable when $(\lambda, h) \in \mathscr{U}_\beta^\#$, we get

$$\mu_\Lambda^\#\big(\text{each } \gamma' \in \Gamma_1 \text{ is external}\big) \leq \prod_{\gamma' \in \Gamma_1} \mathrm{w}^\#(\gamma') \leq \prod_{\gamma' \in \Gamma_1} e^{-\tau|\overline{\gamma}'|}.$$

A similar bound holds for $\mu_\Delta^\#\big(\text{each } \gamma'' \in \Gamma_2 \text{ is external}\big)$.

We can gather these into the following bound:

$$\begin{aligned}\mathbb{P}_{\Lambda,\Delta}^\#(C_{n,\#,\#}{}^{\mathrm{c}}) &\leq 2\sum_{k\geq 1} \sum_{\substack{Y=(\gamma_1,\cdots,\gamma_k):\\ \overline{\text{int}\gamma_1}\cap \text{B}(n)\neq\varnothing \\ |\overline{Y}|\geq n}} \prod_{i=1}^k e^{-\tau|\overline{\gamma}_i|} \\ &\leq 2e^{-\tau' n} \sum_{k\geq 1} \sum_{\substack{Y=(\gamma_1,\cdots,\gamma_k):\\ \overline{\text{int}\gamma_1}\cap \text{B}(n)\neq\varnothing}} \prod_{i=1}^k e^{-\tau'|\overline{\gamma}_i|},\end{aligned} \tag{7.103}$$

where $\tau' \stackrel{\text{def}}{=} \tau/2$. We sum over $Y = (\gamma_1, \dots, \gamma_k)$, starting with γ_k:

$$\sum_{\gamma_k:\, \gamma_k \not\sim \gamma_{k-1}} e^{-\tau'|\overline{\gamma}_k|} \leq 3^d |\overline{\gamma}_{k-1}| \sum_{\gamma_k \overline{\gamma}_k \ni 0} e^{-\tau'|\overline{\gamma}_k|} \leq e^{3^d|\overline{\gamma}_{k-1}|} \sum_{\gamma_k:\, \overline{\gamma}_k \ni 0} e^{-\tau'|\overline{\gamma}_k|}.$$

Then, for $j = k-1, k-2, \dots, 2$,

$$\sum_{\gamma_j:\, \gamma_j \not\sim \gamma_{j-1}} e^{-(\tau'-3^d)|\overline{\gamma}_j|} \leq e^{3^d|\overline{\gamma}_{j-1}|} \sum_{\gamma_j:\, \overline{\gamma}_j \ni 0} e^{-(\tau'-3^d)|\overline{\gamma}_j|}.$$

In the end, we are left with $j = 1$:

$$\sum_{\gamma_1:\, \overline{\text{int}\gamma_1} \cap \text{B}(n) \neq \varnothing} e^{-(\tau'-3^d)|\overline{\gamma}_j|} \leq |\text{B}(n)| \sum_{\gamma_1:\, \overline{\text{int}\gamma_1} \ni 0} e^{-(\tau'-3^d)|\overline{\gamma}_j|}.$$

This last sum can be bounded as in Lemma 7.30 and shown to be smaller than some $\eta_1 = \eta_1(\tau', \ell_0) < 1/2$ if τ is large enough. In particular, $\sum_{k\geq 1} \eta_1^k < 1$ and the conclusion follows. □

We have proved (7.100).

► *Proof of translation invariance.* That $\mu^\#$ is translation invariant can be shown exactly as in the proof of Theorem 3.17 (p. 108): for any translation θ_i and any local function f (remember Figure 3.8),

$$|\mu^{\#}_{\Lambda_n}(f) - \mu^{\#}_{\Lambda_n}(f \circ \theta_i)| = |\mu^{\#}_{\Lambda_n}(f) - \mu^{\#}_{\theta_i\Lambda_n}(f)|,$$

and, by (7.100), the right-hand side converges to zero.

▶ *Proof of extremality.* We will use the characterization of extremality given in item 4 of Theorem 6.58. Let $A \in \mathscr{C}$ be any cylinder, and n be so large that $A \in \mathscr{F}_{\mathrm{B}(n)}$. Let also $B \in \mathscr{F}_{\Lambda^c}$, where Λ is large enough to contain $\mathrm{B}(4n)$. Consider the event $C_{2n,\#} \subset \Omega$ that there exists a largest connected set $D_{\#}$ such that $\mathrm{B}(2n) \subset D_{\#} \subset \mathrm{B}(4n)$ and $\omega_i = \#$ for each $i \in \partial^{\mathrm{ex}} D_{\#}$. Using a decomposition similar to the one used earlier,

$$\mu^{\#}(A \cap B) = \sum_{D_{\#}} \mu^{\#}(A \cap B \cap [D_{\#}]) + \mu^{\#}(A \cap B \cap C_{2n,\#}{}^{\mathrm{c}}).$$

Since $C_{2n,\#}$ is local, $\mu^{\#}(C_{2n,\#}{}^{\mathrm{c}}) = \lim_{N\to\infty} \mu^{\#}_{\Lambda_N}(C_{2n,\#}{}^{\mathrm{c}})$. Using Lemma 7.42, for all $N \geq 4n$,

$$\mu^{\#}_{\Lambda_N}(C_{2n,\#}{}^{\mathrm{c}}) \leq \mathbb{P}^{\#}_{\Lambda_N,\Lambda_N}(C_{2n,\#,\#}{}^{\mathrm{c}}) \leq 2|\mathrm{B}(2n)|e^{-2\tau' n}. \tag{7.104}$$

Therefore, $\mu^{\#}(A \cap B \cap C_{2n,\#}{}^{\mathrm{c}}) \leq 2|\mathrm{B}(2n)|e^{-2\tau' n}$. Now, for a fixed $D_{\#}$, $\mathscr{F}_{\Lambda^c} \subset \mathscr{F}_{D_{\#}^{\mathrm{c}}}$, and so

$$\mu^{\#}(A \cap B \cap [D_{\#}]) = \mu^{\#}\big(\mu^{\#}(A \cap B \,|\, \mathscr{F}_{D_{\#}^{\mathrm{c}}})\mathbf{1}_{[D_{\#}]}\big) = \mu^{\#}\big(\mu^{\#}(A \,|\, \mathscr{F}_{D_{\#}^{\mathrm{c}}})\mathbf{1}_{B\cap[D_{\#}]}\big).$$

Since $\mu^{\#} \in \mathscr{G}(\beta, \lambda, h)$, we have, on $[D_{\#}]$, $\mu^{\#}(A \,|\, \mathscr{F}_{D_{\#}^{\mathrm{c}}}) = \mu^{\eta^{\#}}_{D_{\#}}(A)$ almost surely. Adapting (7.106) below gives

$$\mu^{\eta^{\#}}_{D_{\#}}(A) = \mu^{\#}(A) + O(n^d e^{-\tau' n}).$$

Altogether, we get

$$\mu^{\#}(A \cap B) = \mu^{\#}(A)\mu^{\#}(B) + O(n^d e^{-\tau' n}). \tag{7.105}$$

This implies that $\mu^{\#}$ is extremal.

▶ *Proof of ergodicity.* Ergodicity follows from extremality and translation invariance, exactly as in the proof of Lemma 6.66.

▶ *Proof of* (7.99). We reformulate Peierls' argument. We fix some $\Lambda \Subset \mathbb{Z}^d$ and observe that, in any configuration $\omega \in \Omega^{\#}_{\Lambda}$ such that $\omega_0 \neq \#$, there exists an external contour $\gamma' \subset \Lambda$ such that $\overline{\mathrm{int}\gamma'} \ni 0$. We can then use Lemma 7.26 again and the stability of the weight of contours of type # when $(\lambda, h) \in \mathscr{U}^{\#}_{\beta}$ to obtain, uniformly in Λ, a bound involving the same sum as before:

$$\mu^{\#}_{\Lambda}(\sigma_0 \neq \#) \leq \sum_{\gamma' :\, \overline{\mathrm{int}\gamma'} \ni 0} e^{-\tau|\overline{\gamma'}|}.$$

This sum can be made arbitrarily small when β (and hence τ) is large enough. This concludes the proof of Theorem 7.41. □

In the above proof, we have actually established more, namely **exponential relaxation** and **exponential mixing** at low temperature.

Corollary 7.43. *Under the same hypotheses as Theorem 7.41, if* $\# \in \Upsilon(\beta,\lambda,h)$, *there exists* $c<\infty$ *such that, for any function* f *having its support inside* $\mathrm{B}(n)$ *and for all* c*-connected* $\Lambda \supset \mathrm{B}(2n)$,

$$|\mu^{\#}_{\Lambda;\beta,\lambda,h}(f) - \mu^{\#}_{\beta,\lambda,h}(f)| \le c\|f\|_\infty n^d e^{-\tau' n}. \tag{7.106}$$

Moreover, for any $\mathscr{F}_{\mathrm{B}(4n)^c}$*-measurable function* g,

$$|\mu^{\#}_{\beta,\lambda,h}(fg) - \mu^{\#}_{\beta,\lambda,h}(f)\mu^{\#}(g)_{\beta,\lambda,h}| = c\|f\|_\infty \|g\|_\infty n^d e^{-\tau' n}.$$

Proof The first claim follows by taking $\Delta \uparrow \mathbb{Z}^d$ in (7.100). The second follows from the same argument that led to (7.105). □

A Characterization of "the Sea of # with Small Islands". The bound (7.99) suggests that a typical configuration under $\mu^{\#}_{\beta,\lambda,h}$ displays only small local deviations from the ground state $\eta^{\#}$. Here, we will provide a more global characterization, by giving a description of configurations on the whole lattice, which holds almost surely.

For instance, (7.104) implies that, for all $\# \in \Upsilon(\beta,\lambda,h)$,

$$\sum_n \mu^{\#}_{\beta,\lambda,h}(C_{n,\#}{}^{\mathrm{c}}) < \infty .$$

Therefore, the Borel-Cantelli Lemma implies that, $\mu^{\#}_{\beta,\lambda,h}$-almost surely, all but a finite number of the events $C_{n,\#}$ occur simultaneously. This means that the origin is always surrounded by an infinite number of #-surrounding sets, of arbitrarily large sizes. But it does not yet rule out the presence of #′-surrounding sets, for other labels #′.

To remedy this problem, let $N > n$ and consider $E_{N,n,\#} \stackrel{\mathrm{def}}{=} F_{N,n,\#} \cap C_{n,\#}$, where $C_{n,\#}$ was defined earlier and $F_{N,n,\#}$ is the event that there exists a self-avoiding path $\pi = (i_0, i_1, \dots, i_k) \subset \mathrm{B}(N)\setminus\mathrm{B}(n)$, with $i_0 \in \partial^{\mathrm{ex}}\mathrm{B}(n)$, $i_k \in \partial^{\mathrm{in}}\mathrm{B}(N)$, $i_j \sim i_{j+1}$, such that $\omega_{i_j} = \#$ for all j. On the event

$$E_{n,\#} \stackrel{\mathrm{def}}{=} \bigcap_{N>n} E_{N,n,\#} ,$$

there exists a #-surrounding $\mathrm{B}(n) \subset D_\# \subset \mathrm{B}(2n)$, and there exists an infinite self-avoiding path π (connecting $D_\#$ to $+\infty$) of vertices i with $\omega_i = \#$ (see Figure 7.10).

Theorem 7.44. *Let* $\# \in \Upsilon(\beta,\lambda,h)$. *Then,*

$$\mu^{\#}_{\beta,\lambda,h}(\exists M < \infty \text{ such that } E_{M,\#} \text{ occurs}) = 1 .$$

Proof Let us study $F_{N,n,\#}$ under $\mu^{\#}_{\Lambda;\beta,\lambda,h}$. Observe that

$$F_{N,n,\#}{}^{\mathrm{c}} \subset \{\text{there exists an external contour } \gamma' \subset \Lambda \text{ such that } \overline{\mathrm{int}\gamma'} \supset \mathrm{B}(n)\} .$$

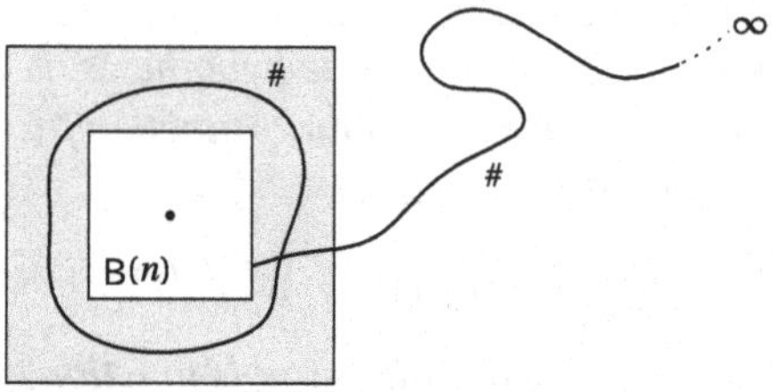

Figure 7.10. Almost surely under $\mu^{\#}_{\beta,\lambda,h}$, the origin (as well as every other vertex of the lattice) is surrounded by a circuit (in $d=2$; otherwise, a closed surface) of #-spins, and this circuit is itself connected to $+\infty$ by a path of #-spins.

Now, if $\overline{\text{int}\gamma'} \supset \mathrm{B}(n)$, then $|\overline{\gamma}'| \geq |\partial^{\text{ex}}\mathrm{B}(n)| \geq n^{d-1}$. Therefore, we can proceed as earlier to obtain, for all c-connected $\Lambda \supset \mathrm{B}(N)$,

$$\mu^{\#}_{\Lambda;\beta,\lambda,h}({F_{N,n,\#}}^{\mathrm{c}}) \leq \sum_{\substack{\gamma':\,\overline{\text{int}\gamma'}\cap\mathrm{B}(n)\neq\varnothing\\ |\overline{\gamma}'|\geq n^{d-1}}} e^{-\tau|\overline{\gamma}'|}$$

$$\leq |\mathrm{B}(n)| \sum_{\substack{\gamma':\,\overline{\text{int}\gamma'}\ni 0\neq\varnothing\\ |\overline{\gamma}'|\geq n^{d-1}}} e^{-\tau|\overline{\gamma}'|} \leq |\mathrm{B}(n)|e^{-\tau'' n^{d-1}},$$

uniformly in N and Λ, for some $\tau''>0$ depending on β. Since $E_{N,n,\#}$ is decreasing in N,

$$\mu^{\#}_{\beta,\lambda,h}(E_{n,\#}) = \lim_{N\to\infty} \mu^{\#}_{\beta,\lambda,h}(E_{N,n,\#}) \geq 1-\epsilon_n\,,$$

where $\epsilon_n \overset{\text{def}}{=} |\mathrm{B}(n)|(2e^{-\tau' n} + e^{-\tau'' n^{d-1}})$. Since ϵ_n is summable, we can again apply the Borel–Cantelli Lemma to conclude that all but a finite number of events $E_{n,\#}$ occur $\mu^{\#}_{\beta,\lambda,h}$-almost surely. □

Combining the previous result with translation invariance, we summarize the almost-sure properties of typical configurations in a theorem:

Theorem 7.45. *Let β be large enough and $(\lambda,h)\in\mathscr{U}^{\#}_{\beta}$. For all $\#\in\Upsilon(\beta,\lambda,h)$, under $\mu^{\#}_{\beta,\lambda,h}$, a typical configuration consists in a* ***sea of # (the ground state $\eta^{\#}$) with local bounded deformations****, in the following sense: every vertex $i\in\mathbb{Z}^d$ is either connected to $+\infty$ by a self-avoiding path along which all spins are #, or there exists a finite external contour γ such that $\overline{\text{int}\gamma}\ni i$.*

The study of the largest contours in a box can be done as for the Ising model:

Exercise 7.15. Let β be large, as above. Fix $\#\in\Upsilon(\beta,\lambda,h)$, and consider the Blume–Capel model in $\mathrm{B}(n)$. Adapting the method of Exercise 3.18, show that under $\mu^{\#}_{\mathrm{B}(n);\beta,\lambda,h}$ the largest contours in $\mathrm{B}(n)$ have a support of size of order $\log n$.

7.5 Bibliographical References

Although it is sometimes unfairly referred to as a "generalization of Peierls' argument", the Pirogov–Sinai theory (PST) actually uses several important concepts of equilibrium statistical mechanics and introduces important new ideas. The original method introduced by Pirogov and Sinai (English translations of their original papers can be found in [313]) was based on the use of contours and of a Hamiltonian satisfying Peierls' condition, but the phase diagram was constructed by using an abstract approach involving a fixed-point argument. Later, Zahradník [354] contributed substantially to the theory by introducing fundamental new ideas. In particular, he introduced the notion of truncated pressure, which eventually superseded the original fixed-point argument and became the core of the current understanding of the theory. For that reason, it would be more correct to call it the *Pirogov–Sinai–Zahradník theory.*

Pedagogical texts on PST include the paper [34] by Borgs and Imbrie, and the lecture notes by Fernández [104]. The review paper of Slawny [316], although based on the fixed-point method of Pirogov and Sinai, is clear and applies the theory to various models.

Our Section 7.2, devoted to ground states, is inspired by the presentation in Chapter 2 of Sinai's book [312]. Other notions of ground states exist in the literature (see [85]). The method to determine the set of periodic ground states based on m-potentials, as presented in Section 7.2.2, is due to [164]. An interesting account of the main ideas of PST as well as a description of more general notions of ground state can be found in [343].

The model considered in Example 7.8 is due to Pechersky in [265]. Prior to that counter-example, it had been conjectured [312] that a finite-range potential with a finite number of periodic ground states would always satisfy Peierls' condition.

Our analysis of the phase diagram is based on the ideas introduced by Zahradník and followers. In particular, the C^1-truncation used when defining the truncated weights is a simpler version of the C^k-truncations used by Borgs and Kotecký in [35].

Although we only implemented the details in the case of the Blume–Capel model, the methods used are general and can be applied in many other situations. As an exercise, the interested reader can use them to provide a full description of the low-temperature phase diagram of the modified Ising model described early in Section 7.1.1.

7.6 Complements and Further Reading

7.6.1 Completeness of the Phase Diagram

One of the main results in this chapter was the construction of low-temperature translation-invariant extremal Gibbs measures for the Blume–Capel model using stable periodic (actually constant) ground states as boundary conditions: $\mu^{\#}_{\beta,\lambda,h}$, $\# \in \Upsilon(\beta,\lambda,h)$.

At this stage, it is very natural to wonder whether there are other Gibbs measures, in addition to those constructed here. By the general theory of Chapter 6, the set of infinite-volume Gibbs measures is a simplex, so we can restrict our discussion to extremal measures. The following remarkable result shows that the Gibbs measures we constructed exhaust the set of translation-invariant Gibbs measures: at sufficiently low temperature, any translation-invariant measure in $\mathscr{G}_\theta(\beta,\lambda,h)$ can be represented as a convex combination of $\mu^{\#}_{\beta,\lambda,h}$, $\# \in \Upsilon(\beta,\lambda,h)$.

Theorem 7.46. *There exists β_0 such that, for all $\beta \geq \beta_0$, the following holds. For all $(\lambda,h) \in \check{U}$ and all $\mu \in \mathscr{G}_\theta(\beta,\lambda,h)$, there exist coefficients $(\alpha_\#)_{\#\in\Upsilon(\beta,\lambda,h)} \subset [0,1]$ such that*

$$\mu = \sum_{\#\in\Upsilon(\beta,\lambda,h)} \alpha_\# \mu^{\#}_{\beta,\lambda,h}.$$

In particular, in the regions where only one of the boundary conditions $+,-,0$ is stable (the interior of the regions $\mathscr{U}^{\#}_{\beta}$ on Figure 7.2), there is a *unique* translation-invariant Gibbs measure.

The same statement holds for the general class of models to which the theory applies; see the original paper of Zahradník [354], where a proof can be found. This kind of statement is usually referred to as the **completeness of the phase diagram.**

We therefore see that, while PST is limited to perturbative regimes (here, very low temperature), it provides in such regimes a complete description of the set of all translation-invariant Gibbs measures. Of course, there are, in general, other non-translation-invariant Gibbs measures, such as Dobrushin states in the Ising model in dimensions $d \geq 3$ (see the discussion in Section 3.10.7). Extensions of PST dealing with such states have been developed; see for instance [162].

7.6.2 Generalizations

Pirogov–Sinai theory has been extended in various directions, for instance to systems with continuous spins [48, 87], quasiperiodic interactions [199] or long-range interactions [262, 263].

One important application of PST was in the seminal work of Lebowitz, Mazel and Presutti [214], in which a first-order phase transition is proved for a model of particles in the continuum with Kać interactions (of finite range), similar to those we considered in Section 4.10. At the core of their technique lies a nontrivial definition of *contour* associated with configurations of point particles in the continuum, and the use of the main ideas of PST.

7.6.3 Large-β Asymptotics of the Phase Diagram

The analysis of this chapter showed that the low-temperature phase diagram of the Blume-Capel model is a small perturbation of the corresponding one at zero temperature. A more delicate analysis is required if one wants to derive more *quantitative* information on this diagram as a function of β.

For instance, $(\lambda, h) = (0, 0)$ is the triple point of the phase diagram at zero temperature, at which η^+, η^0 and η^- are ground states. The following question is natural: in which direction does the triple point move when the inverse temperature is finite? In other words: which are the stable phases at $(0, 0)$ when $\beta < \infty$?

In principle, this question can be answered by determining which truncated pressure $\widehat{\psi}^\#(0, 0) = \hat{g}^\#(0, 0)$ (remember that $e^\#(0, 0) = 0$), $\# \in \{+, 0, -\}$, is maximal. But, each $\hat{g}^\#(0, 0)$ is a series made of products of τ-stable weights, where τ can be made large when β is large. Computing the first terms of these expansions should allow one to determine which one dominates. Unfortunately, the structure of the truncated weights makes extracting such information difficult. We briefly describe an alternative approach, informally, providing references for the interested reader.

Let us describe the contributions from the smallest perturbations of the ground state $\eta^\#$, which provide the main contributions to $\hat{g}^\#$. We do this at a heuristic level.

Consider first the case $\# = +$. Among the configurations that coincide with η^+ everywhere except on a finite set, the configurations with lowest energy are those that have a single 0 spin. The energy associated with such an *excitation* is $2d$. It therefore seems plausible that the leading term in the expansion of $\hat{g}^+(0, 0)$ is due to the cluster made of exactly one such excitation, that is,

$$\beta\widehat{\psi}^+(0, 0) = e^{-2d\beta} + \cdots ,$$

where the dots stand for higher-order terms in $e^{-\beta}$.

Now, among the configurations that coincide with η^0 everywhere except on a finite set, the configurations with lowest energy are those that have either a single $+$ spin or a single $-$ spin. Both these excitations have an energy equal to $2d$, as before, which leads to

$$\beta\widehat{\psi}^0(0, 0) = 2e^{-2d\beta} + \cdots .$$

Provided the higher-order terms yield significantly smaller contributions, this gives, at large β,

$$\widehat{\psi}^0(0, 0) - \widehat{\psi}^+(0, 0) = \frac{1}{\beta} e^{-2d\beta} + \cdots \geq \frac{1}{2\beta} e^{-2d\beta} > 0,$$

implying that only the 0 phase is stable at $(0, 0)$. In other words, the triple point shifts to the *right* at positive temperatures, as depicted in Figure 7.9.

In fact, this argument can be repeated when (λ, h) lies in the neighborhood of $(0, 0)$, allowing one to construct the coexistence line $\mathscr{L}_\beta^{0\pm}$. Namely, fix β large and take λ, h such that $\beta\lambda \ll 1$ and $\beta h \ll 1$. Arguing as above, considering the excitations of smallest energy, we get (remember that $e^+(\lambda, h) = \lambda + h$)

$$\beta\widehat{\psi}^0(\lambda, h) = e^{-2d\beta+\beta\lambda+\beta h} + e^{-2d\beta+\beta\lambda-\beta h} + \cdots ,$$

$$\beta\widehat{\psi}^+(\lambda, h) = \beta\lambda + \beta h + e^{-2d\beta-\beta\lambda-\beta h} + \cdots .$$

Therefore, for very large β, the smooth map describing $\mathscr{L}_\beta^{0+}$ (see Figure 7.8) can be obtained, in a first approximation, by equating these two expressions, yielding

$$\lambda \mapsto h(\lambda) = \beta^{-1}e^{-2d\beta} - \lambda(1 - 4e^{-2d\beta}) + O(\lambda^2).$$

In particular, the position of the triple point $(\lambda_*(\beta), 0)$ is obtained by solving $h(\lambda) = 0$, which yields

$$\lambda_*(\beta) = \frac{e^{-2d\beta}}{\beta}\big(1 + O(e^{-2d\beta})\big).$$

These computations are purely formal, but their conclusions can be made rigorous. The idea is to replace the notion of ground state by the notion of **restricted ensemble**. In the context described above, the restricted ensemble $\mathscr{R}^0$ is defined as the set of all configurations ω such that $\omega_i \neq 0 \implies \omega_j = 0$ for all $j \sim i$. That is, they correspond to the ground state η^0 on top of which only the smallest possible excitations are allowed. Starting from a general configuration, one then erases all such smallest energy excitations and constructs contours for the resulting configuration. Of course, this is more delicate than before, since, in contrast to the ground state η^0, the restricted phase $\mathscr{R}^0$ has a nontrivial pressure that, in particular, depends on the volume. Nevertheless, the analysis can be done along similar lines. We refer to the lecture notes [50] by Bricmont and Slawny for a pedagogical introduction to this problem (and a proof that the triple point of the Blume–Capel model is indeed shifted to the right) and to their paper [51] for a more detailed account.

7.6.4 Other Regimes

Generically, the methods of PST can be used to study models whose partition function can be written as a system of contours, with some equivalent of Peierls' condition. The perturbation parameter need not be the temperature, and the parameter driving the transition need not be related to some external field as we saw in the Blume–Capel model. Consider, for example, the Potts model with spins $\omega_i \in \{0, 1, 2, \dots, q-1\}$, at inverse temperature β, and denote its pressure by $\psi_q(\beta)$. It turns out that, when q is large, q^{-1} can be used as a perturbation parameter to study $\beta \mapsto \psi_q(\beta)$. Using the methods of PST, it was shown in [48, 203, 233] that a first-order phase transition in β occurs when q is large enough: there exists $\beta_c = \beta_c(q)$ such that the pressure is differentiable when $\beta < \beta_c$ and $\beta > \beta_c$, and that it is nondifferentiable at β_c. (This result was first proved, using reflection positivity, in [197]. In two dimensions, the simplest proof, relying on a variation of Peierls' argument, can be found in [93].)

In addition to perturbations of a finite collection of ground states, PST can also be applied successfully to analyze perturbations of other well-understood regimes, usually involving constraints of a certain type. This appears, for instance, in the study of Kać potentials which applies in the neighborhood of the mean-field regine [214], or in the use of restricted phases as in [48, 50, 113].

7.6.5 Finite-Size Scaling

In [35], Borgs and Kotecký used the ideas of PST to initiate a theory of *finite-size scaling*, that is, a thorough analysis of the *rate* at which certain thermodynamic quantities (the magnetization, for example) converge to their asymptotic values in the thermodynamic limit. In addition to its obvious theoretical interest, such an analysis also plays an essential role when extrapolating to infinite systems the information obtained from the observation of the relatively small systems that can be analyzed using numerical simulations.

7.6.6 Complex Parameters, Lee–Yang Zeros and Singularities

With minor changes, most of the material presented in this chapter can be extended to include complex fields; see [34], for example.

As an interesting application, it has been shown in [20] and [21] that the Lee-Yang Theorem, applied to the Ising model in Section 3.7.3, can be extended to other models, allowing one to determine the locus of the zeros of their partition function. Of course, since they rely on the main results of PST, these results hold only in a perturbative regime.

Furthermore, the techniques of PST can be used to obtain finer analytic properties of the pressure. Consider, for instance, the Ising model in a complex magnetic field $h \in \mathbb{C}$. For simplicity, let us consider the contours defined in Section 5.7.4. The magnetic field leads one to introduce contours of two types: $+$ and $-$. Then, using the trick (7.31), one is led to two types of weights: $\mathrm{w}^+(\gamma)$ and $\mathrm{w}^-(\gamma)$. When $h = 0$, these coincide with those defined in (5.41), but otherwise they contain ratios of partition functions. The weight of a contour of type $+$, for example, takes the form

$$\mathrm{w}^+(\gamma) = e^{-2\beta|\gamma|}\frac{\mathbf{Z}^-(\mathrm{int}_-\gamma)}{\mathbf{Z}^+(\mathrm{int}_-\gamma)} = e^{-2\beta|\gamma|}\frac{e^{-\beta h|\mathrm{int}_-\gamma|}\Xi^-(\mathrm{int}_-\gamma)}{e^{+\beta h|\mathrm{int}_-\gamma|}\Xi^+(\mathrm{int}_-\gamma)}.$$

The analysis can then be done following the main induction used earlier for the Blume-Capel model. When the field is real, the symmetry between $+$ and $-$ implies that all weights are stable at $h = 0$. For general complex values of h, the symmetry implies that all weights are stable on the imaginary axis $\{\mathfrak{Re}\, h = 0\}$. The weights are then shown to be well defined and analytic in regions of the complex plane analogous to the stability regions defined earlier; if γ is of type $+$, its weight is analytic in

$$\mathscr{U}_\gamma^+ \stackrel{\mathrm{def}}{=} \Big\{\mathfrak{Re}\, h > -\frac{\theta}{|\mathrm{int}_-\gamma|^{1/d}}\Big\},$$

for a suitable constant θ. Implementing this analysis allows one to obtain a more quantitative version of the Lee-Yang Theorem (restricted to low temperature).

Then, in a second step, Isakov's analysis [174] provides an estimate of high-order derivatives of the pressure (see Sections 3.10.9 and 4.12.3), showing that the function $h \mapsto \psi_\beta(h)$ cannot be analytically continued through $h = 0$, along either of the real paths $h \downarrow 0$, $h \uparrow 0$. This analysis was generalized by Friedli and Pfister [114] to all two-phase models to which PST applies, which implies in particular that the pressure of the Blume–Capel model has no analytic continuation accross the lines of coexistence of its phase diagram, at least for low temperatures, away from the triple point.

8 The Gaussian Free Field on $\mathbb{Z}^d$

The model studied in this chapter, the *Gaussian Free Field* (GFF), is the only one we will consider whose single-spin space, $\mathbb{R}$, is *noncompact.* Its sets of configurations in finite and infinite volume are therefore, respectively,

$$\Omega_\Lambda \stackrel{\text{def}}{=} \mathbb{R}^\Lambda \qquad \text{and} \qquad \Omega \stackrel{\text{def}}{=} \mathbb{R}^{\mathbb{Z}^d}.$$

Although most of the general structure of the DLR formalism developed in Chapter 6 applies, the *existence* of infinite-volume Gibbs measures is no longer guaranteed under the most general hypotheses, and requires more care.

One possible physical interpretation of this model is as follows. In $d = 1$, the spin at vertex $i \in \Lambda$, $\omega_i \in \mathbb{R}$, can be interpreted as the *height of a random line* above the x-axis:

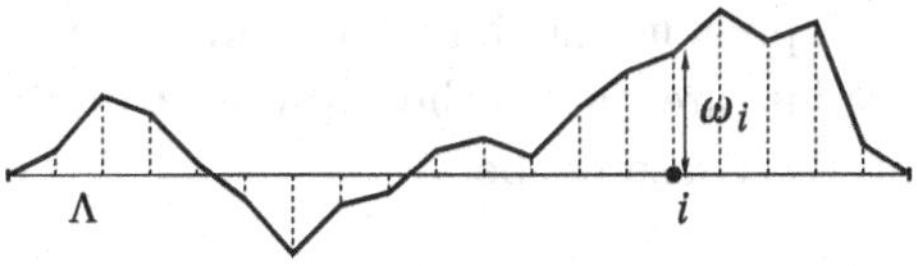

Figure 8.1. A configuration of the Gaussian Free Field in a one-dimensional box Λ, with boundary condition $\eta \equiv 0$.

The behavior of the model in large volumes is therefore intimately related to the *fluctuations* of the line away from the x-axis. Similarly, in $d = 2$, ω_i can be interpreted as the *height of a surface* above the (x, y)-plane:

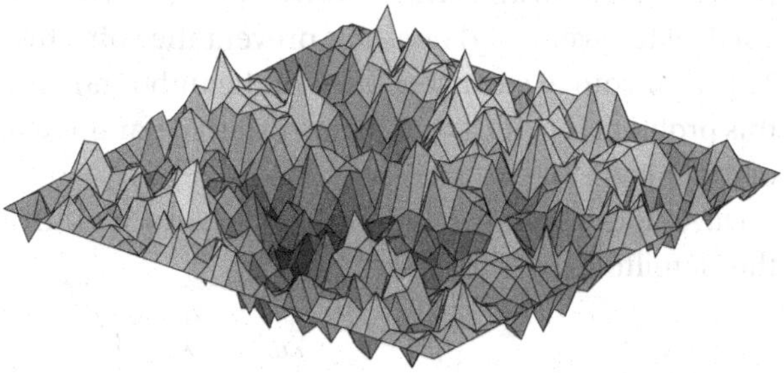

Figure 8.2. A configuration of the Gaussian Free Field in $d = 2$, in a 30×30 box with boundary condition $\eta \equiv 0$, which can be interpreted as a random surface.

The techniques we will use to study the GFF will be very different from those used in the previous chapters. In particular, *Gaussian vectors* and *random walks* will play a central role in the analysis of the model. The basic results required on these two topics are collected in Appendices B.9 and B.13.

8.1 Definition of the Model

We consider a configuration $\omega \in \Omega$ of the GFF, in which a variable $\omega_i \in \mathbb{R}$ is associated with each vertex $i \in \mathbb{Z}^d$; as usual, we refer to ω_i as the **spin** at i. We define the interactions between the spins located inside a region $\Lambda \Subset \mathbb{Z}^d$, and between these spins and those located outside Λ. We motivate the definition of the Hamiltonian of the GFF by a few natural assumptions.

1. We first assume that only spins located at nearest-neighbor vertices of $\mathbb{Z}^d$ interact.
2. Our second requirement is that the interaction favors agreement of neighboring spins. This is achieved by assuming that the contribution to the energy due to two neighboring spins ω_i and ω_j is given by

$$\beta V(\omega_i - \omega_j), \tag{8.1}$$

   for some $V: \mathbb{R} \to \mathbb{R}_{\geq 0}$, which is assumed to be even, $V(-x) = V(x)$. Models with this type of interaction, depending only on the difference between neighboring spins, are often called **gradient models**. In the case of the GFF, the function V is chosen to be

$$V(x) \stackrel{\text{def}}{=} x^2 .$$

   An interaction of the type (8.1) has the following property: the interaction between two neighboring spins, ω_i and ω_j, does not change if the spins are shifted by the same value a: $\omega_i \mapsto \omega_i + a$, $\omega_j \mapsto \omega_j + a$. As will be explained later in Section 9.3, this invariance is at the origin of the mechanism that prevents the existence of infinite-volume Gibbs measures in low dimensions. The point is that local agreement between neighboring spins (that is, having $|\omega_j - \omega_i|$ small whenever $i \sim j$) does not prevent the spins from taking very large values. This is of course a consequence of the unboundedness of $\mathbb{R}$. One way to avoid this problem is to introduce some external parameter that penalizes large values of the spins.
3. To favor localization of the spin ω_i near zero, we introduce an additional term to the Hamiltonian, of the form

$$\lambda \omega_i^2 , \quad \lambda \geq 0 .$$

   This guarantees that when $\lambda > 0$ large values of $|\omega_i|$ represent large energies, and are therefore penalized.

We are thus led to consider a formal Hamiltonian of the following form:

$$\beta \sum_{\{i,j\}\in\mathscr{E}_{\mathbb{Z}^d}} (\omega_i - \omega_j)^2 + \lambda \sum_{i\in\mathbb{Z}^d} \omega_i^2 .$$

For convenience, we will replace β and λ by coefficients better suited to the manipulations that will come later.

Definition 8.1. The **Hamiltonian of the GFF in $\Lambda \Subset \mathbb{Z}^d$** is defined by

$$\mathscr{H}_{\Lambda;\beta,m}(\omega) \stackrel{\text{def}}{=} \frac{\beta}{4d} \sum_{\{i,j\}\in\mathscr{E}_\Lambda^{\text{b}}} (\omega_i - \omega_j)^2 + \frac{m^2}{2} \sum_{i\in\Lambda} \omega_i^2 , \quad \omega\in\Omega, \tag{8.2}$$

where $\beta \geq 0$ is the inverse temperature and $m \geq 0$ is the **mass.**[1] The model is **massive** when $m > 0$, **massless** if $m = 0$.

Once we have a Hamiltonian, finite-volume Gibbs measures are defined in the usual way. The measurable structures on Ω_Λ and Ω were defined in Section 6.10; we use the Borel sets $\mathscr{B}_\Lambda$ on Ω_Λ, and the σ-algebra $\mathscr{F}$ generated by cylinders on Ω. Since the spins are real-valued, a natural reference measure for the spin at site i is the Lebesgue measure, which we simply denote $\mathrm{d}\omega_i$. We remind the reader that $\omega_\Lambda\eta_{\Lambda^c} \in \Omega$ is the configuration that agrees with ω_Λ on Λ, and with η on Λ^c.

So, given $\Lambda \Subset \mathbb{Z}^d$ and $\eta \in \Omega$, the Gibbs distribution for the GFF in Λ with boundary condition η, at inverse temperature $\beta \geq 0$ and mass $m \geq 0$, is the probability measure $\mu^\eta_{\Lambda;\beta,m}$ on $(\Omega, \mathscr{F})$ defined by

$$\forall A \in \mathscr{F}, \quad \mu^\eta_{\Lambda;\beta,m}(A) = \int \frac{e^{-\mathscr{H}_{\Lambda;\beta,m}(\omega_\Lambda\eta_{\Lambda^c})}}{\mathbf{Z}^\eta_{\Lambda;\beta,m}} \mathbf{1}_A(\omega_\Lambda\eta_{\Lambda^c}) \prod_{i\in\Lambda} \mathrm{d}\omega_i . \tag{8.3}$$

The partition function is of course

$$\mathbf{Z}^\eta_{\Lambda;\beta,m} \stackrel{\text{def}}{=} \int e^{-\mathscr{H}_{\Lambda;\beta,m}(\omega_\Lambda\eta_{\Lambda^c})} \prod_{i\in\Lambda} \mathrm{d}\omega_i .$$

Exercise 8.1. Show that $\mathbf{Z}^\eta_{\Lambda;\beta,m}$ is well defined, for all $\eta \in \Omega$, $\beta > 0$, $m \geq 0$.

Remark 8.2. In the previous chapters, we also considered other types of boundary conditions, namely free and periodic. As shown in the next exercise, this cannot be done for the massless GFF. Sometimes (in particular when using reflection positivity, see Chapter 10), it is nevertheless necessary to use periodic boundary conditions. In such situations, a common way of dealing with this problem is to take first the thermodynamic limit with a positive mass and then send the mass to zero: $\lim_{m\downarrow 0} \lim_{n\to\infty} \mu^{\text{per}}_{V_n;m}$, remember Definition 3.2. ◇

[1] The terminology "mass" is inherited from quantum field theory, where the corresponding quadratic terms in the Lagrangian indeed give rise to the mass of the associated particles.

Exercise 8.2. Check that, for all nonempty $\Lambda \Subset \mathbb{Z}^d$ and all $\beta > 0$,

$$\mathbf{Z}^{\varnothing}_{\Lambda;\beta,0} = \mathbf{Z}^{\mathrm{per}}_{\Lambda;\beta,0} = \infty .$$

In particular, it is not possible to define the massless GFF with free or periodic boundary conditions.

Before continuing, observe that the scaling properties of the Gibbs measure imply that one of the parameters, β or m, plays an irrelevant role when studying the GFF. Indeed, the change of variables $\omega'_i \stackrel{\text{def}}{=} \beta^{1/2}\omega_i$, $i \in \Lambda$, leads to

$$\mathbf{Z}^{\eta}_{\Lambda;\beta,m} = \beta^{-|\Lambda|/2}\mathbf{Z}^{\eta'}_{\Lambda;1,m'} ,$$

where $m' \stackrel{\text{def}}{=} \beta^{-1/2}m$ and $\eta' \stackrel{\text{def}}{=} \beta^{1/2}\eta$, and, similarly,

$$\mu^{\eta}_{\Lambda;\beta,m}(A) = \mu^{\eta'}_{\Lambda;1,m'}(\beta^{1/2}A), \quad \forall A \in \mathscr{F} .$$

This shows that there is no loss of generality in assuming that $\beta = 1$, which we will do from now on; of course, we will then also omit β from the notation.

The next step is to define infinite-volume Gibbs measures. We will do so by using the approach described in detail in Chapter 6. Readers not comfortable with this material can skip to the next subsection. We emphasize that, although we will from time to time resort to this abstract setting in the following, most of our estimates actually pertain to finite-volume Gibbs measures, and therefore do not require this level of abstraction.

We proceed as in Section 6.10. First, the specification $\pi = \{\pi^m_\Lambda\}_{\Lambda \Subset \mathbb{Z}^d}$ of the GFF is defined by the kernels

$$\pi^m_\Lambda(\cdot \mid \eta) \stackrel{\text{def}}{=} \mu^{\eta}_{\Lambda;m}(\cdot) .$$

Then, one defines the set of Gibbs measures compatible with π, by

$$\mathscr{G}(m) \stackrel{\text{def}}{=} \{\mu \in \mathscr{M}_1(\Omega) \,:\, \mu\pi^m_\Lambda = \mu \text{ for all } \Lambda \Subset \mathbb{Z}^d\} .$$

We remind the reader (see Section 6.3.1) of the following equivalent characterization: $\mu \in \mathscr{G}(m)$ if and only if, for all $\Lambda \Subset \mathbb{Z}^d$ and all $A \in \mathscr{F}$,

$$\mu(A \mid \mathscr{F}_{\Lambda^c})(\omega) = \pi^m_\Lambda(A \mid \omega) \quad \text{for } \mu\text{-almost all } \omega. \tag{8.4}$$

Usually, a Gibbs measure in $\mathscr{G}(m)$ will be denoted μ_m, or μ^η_m when constructed via a limiting procedure using a boundary condition η. Expectation of a function f with respect to μ^η_m will be denoted $\mu^\eta_m(f)$ or $E^\eta_m[f]$.

8.1.1 Overview

The techniques used to study the GFF are very different from those used in previous chapters. Let us first introduce the random variables $\varphi_i \colon \Omega \to \mathbb{R}$, defined by

$$\varphi_i(\omega) \stackrel{\text{def}}{=} \omega_i , \quad i \in \mathbb{Z}^d .$$

Similarly to what was done in Chapter 3, we will consider first the distribution of $\varphi_\Lambda = (\varphi_i)_{i\in\Lambda}$ in a finite region $\Lambda \subset \mathrm{B}(n) \Subset \mathbb{Z}^d$, under $\mu^\eta_{\mathrm{B}(n);m}(\cdot)$. We will then determine under which conditions the random vector φ_Λ possesses a limiting distribution when $n \to \infty$. The first step will be to observe that, under $\mu^\eta_{\mathrm{B}(n);m}$, $\varphi_{\mathrm{B}(n)}$ is actually distributed as a *Gaussian vector*. This will give us access to various tools from the theory of Gaussian processes, in particular when studying the thermodynamic limit. Namely, as explained in Appendix B.9, the limit of a Gaussian vector, when it exists, is also Gaussian. This will lead to the construction, in the limit $n \to \infty$, of a *Gaussian field* $\varphi = (\varphi_i)_{i\in\mathbb{Z}^d}$. The distribution of this field, denoted μ^η_m, will be shown to be a Gibbs measure in $\mathscr{G}(m)$. But μ^η_m is entirely determined by its mean $E^\eta_m[\varphi_i]$ and by its covariance matrix, which measures the correlations between the variables φ_i:

$$\mathrm{Cov}^\eta_m(\varphi_i, \varphi_j) \stackrel{\text{def}}{=} E^\eta_m\big[(\varphi_i - E^\eta_m[\varphi_i])(\varphi_j - E^\eta_m[\varphi_j])\big] .$$

It turns out that the mean and covariance matrix will take on a particularly nice form, with a probabilistic interpretation in terms of the *symmetric simple random walk on* $\mathbb{Z}^d$. This will make it possible to compute explicitly various quantities of interest. More precise statements will be given later, but the behavior established for the Gaussian Free Field will be roughly the following:

- *Massless case ($m = 0$), low dimensions:* In dimensions $d = 1$ and 2, the random variables φ_i, when considered in a large box $\mathrm{B}(n) = \{-n, \dots, n\}^d$ with an arbitrary fixed boundary condition, present large fluctuations, unbounded as $n \to \infty$. For example, the variance of the spin located at the center of the box is of order

  $$\mathrm{Var}^\eta_{\mathrm{B}(n);0}(\varphi_0) \approx \begin{cases} n & \text{when } d = 1 , \\ \log n & \text{when } d = 2 . \end{cases}$$

  In such a situation, the field is said to **delocalize**. As we will see, delocalization implies that *there are no infinite-volume Gibbs measures* in this case: $\mathscr{G}(0) = \varnothing$.
- *Massless case ($m = 0$), high dimensions:* In $d \geq 3$, the presence of a larger number of neighbors renders the field sufficiently more rigid to remain localized, in the sense that it has fluctuations of bounded variance. In particular, there exist (infinitely many extremal) infinite-volume Gibbs measures in this case. We will also show that the correlations under these measures are nonnegative and decay slowly with the distance:

  $$\mathrm{Cov}^\eta_0(\varphi_i, \varphi_j) \approx \|j - i\|_2^{-(d-2)} .$$

  In particular, the susceptibility is infinite:

  $$\sum_{j\in\mathbb{Z}^d} \mathrm{Cov}^\eta_0(\varphi_i, \varphi_j) = +\infty .$$

- *Massive case ($m > 0$), all dimensions:* The presence of a mass term in the Hamiltonian prevents the delocalization observed in dimensions 1 and 2 in the massless case. However, we will show that, even in this case, there are infinitely many infinite-volume Gibbs measures. As we will see, the presence of a mass term also makes the correlations decay exponentially fast: there exists $c_+ = c_+(m) > 0, c_- = c_-(m) < \infty, C_+ = C_+(m) < \infty$ and $C_- = C_-(m) > 0$ such that

$$C_- e^{-c_- \|j-i\|_2} \leq \mathrm{Cov}_m^\eta(\varphi_i, \varphi_j) \leq C_+ e^{-c_+ \|j-i\|_2} \qquad \forall i,j \in \mathbb{Z}^d .$$

Moreover, $c_\pm(m) = O(m)$ as $m \downarrow 0$. This shows that the correlation length of the model is of the order of the inverse of the mass, m^{-1}, when the mass is small.

As seen from this short description, the GFF has no *uniqueness regime* (except in the trivial case $\beta = 0, m > 0$).

8.2 Parenthesis: Gaussian Vectors and Fields

Before continuing, we recall a few generalities about Gaussian vectors, which in our case will be a family $(\varphi_i)_{i\in\Lambda}$ of random variables, indexed by the vertices of a finite region $\Lambda \Subset \mathbb{Z}^d$. A more detailed account of Gaussian vectors can be found in Appendix B.9.

8.2.1 Gaussian Vectors

Let $\varphi_\Lambda = (\varphi_i)_{i\in\Lambda} \in \Omega_\Lambda$ be a random vector, defined on some probability space. We do not yet assume that the distribution of this vector is Gibbsian. We consider the following scalar product on Ω_Λ: for $t_\Lambda = (t_i)_{i\in\Lambda}, \varphi_\Lambda = (\varphi_i)_{i\in\Lambda}$,

$$t_\Lambda \cdot \varphi_\Lambda \stackrel{\text{def}}{=} \sum_{i\in\Lambda} t_i \varphi_i .$$

Definition 8.3. The random vector φ_Λ is **Gaussian** if, for all fixed t_Λ, $t_\Lambda \cdot \varphi_\Lambda$ is a Gaussian variable (possibly degenerate, that is, with zero variance).

The distribution of a Gaussian variable X is determined entirely by its mean and variance, and its characteristic function is given by

$$E[e^{\mathrm{i}tX}] = \exp\big(\mathrm{i}\, tE[X] - \tfrac{1}{2}t^2 \,\mathrm{Var}(X)\big) .$$

Let us assume that $\varphi_\Lambda = (\varphi_i)_{i\in\Lambda}$ is Gaussian, and let us denote its distribution by μ_Λ. Expectation (resp. variance, covariance) with respect to μ_Λ will be denoted E_Λ (resp. Var_Λ, Cov_Λ). The mean and variance of $t_\Lambda \cdot \varphi_\Lambda$ depend on t_Λ as follows:

$$E_\Lambda[t_\Lambda \cdot \varphi_\Lambda] = \sum_{i\in\Lambda} t_i E_\Lambda[\varphi_i] = t_\Lambda \cdot a_\Lambda , \tag{8.5}$$

where $a_\Lambda = (a_i)_{i\in\Lambda}$, $a_i \stackrel{\text{def}}{=} E_\Lambda[\varphi_i]$, is the **average** (or **mean**) of φ_Λ. Moreover,

$$\mathrm{Var}_\Lambda(t_\Lambda \cdot \varphi_\Lambda) = E_\Lambda\big[(t_\Lambda \cdot \varphi_\Lambda - E_\Lambda[t_\Lambda \cdot \varphi_\Lambda])^2\big] = \sum_{i,j\in\Lambda} \Sigma_\Lambda(i,j) t_i t_j = t_\Lambda \cdot \Sigma_\Lambda t_\Lambda , \tag{8.6}$$

where $\Sigma_\Lambda = (\Sigma_\Lambda(i,j))_{i,j\in\Lambda}$ is the **covariance matrix** of φ_Λ, defined by

$$\Sigma_\Lambda(i,j) \stackrel{\text{def}}{=} \mathrm{Cov}_\Lambda(\varphi_i, \varphi_j). \tag{8.7}$$

Therefore, for each t_Λ, the characteristic function of $t_\Lambda \cdot \varphi_\Lambda$ is given by

$$E_\Lambda[e^{\mathrm{i}t_\Lambda\cdot\varphi_\Lambda}] = \exp\bigl(\mathrm{i}t_\Lambda \cdot a_\Lambda - \tfrac{1}{2}t_\Lambda \cdot \Sigma_\Lambda t_\Lambda\bigr), \tag{8.8}$$

and the moment generating function by

$$E_\Lambda[e^{t_\Lambda\cdot\varphi_\Lambda}] = \exp\bigl(t_\Lambda \cdot a_\Lambda + \tfrac{1}{2}t_\Lambda \cdot \Sigma_\Lambda t_\Lambda\bigr). \tag{8.9}$$

The distribution of a Gaussian vector φ_Λ is thus entirely determined by the pair $(a_\Lambda, \Sigma_\Lambda)$; it is traditionally denoted by $\mathscr{N}(a_\Lambda, \Sigma_\Lambda)$. We say that φ_Λ is **centered** if $a_\Lambda \equiv 0$.

Clearly, Σ_Λ is symmetric: $\Sigma_\Lambda(i,j) = \Sigma_\Lambda(j,i)$. Moreover, since $\mathrm{Var}_\Lambda(t_\Lambda \cdot \varphi_\Lambda) \geq 0$, we see from (8.6) that Σ_Λ is nonnegative definite. In fact, to any $a_\Lambda \in \Omega_\Lambda$ and any symmetric nonnegative definite matrix Σ_Λ corresponds a (possibly degenerate) Gaussian vector φ_Λ having a_Λ as mean and Σ_Λ as covariance matrix. Moreover, when Σ_Λ is *positive* definite, the density with respect to the Lebesgue measure takes the following well-known form:

Theorem 8.4. *Assume that $\varphi_\Lambda \sim \mathscr{N}(a_\Lambda, \Sigma_\Lambda)$, with a covariance matrix Σ_Λ which is positive definite (and, therefore, invertible). Then, the distribution of φ_Λ is absolutely continuous with respect to the Lebesgue measure $\mathrm{d}x_\Lambda$ (on Ω_Λ), with a density given by*

$$\frac{1}{(2\pi)^{|\Lambda|/2}\sqrt{|\det \Sigma_\Lambda|}} \exp\bigl(-\tfrac{1}{2}(x_\Lambda - a_\Lambda)\cdot \Sigma_\Lambda^{-1}(x_\Lambda - a_\Lambda)\bigr), \quad x_\Lambda \in \Omega_\Lambda. \tag{8.10}$$

Conversely, if φ_Λ is a random vector whose distribution is absolutely continuous with respect to the Lebesgue measure on Ω_Λ with a density of the form (8.10)*, then φ_Λ is Gaussian, with mean a_Λ and covariance matrix Σ_Λ.*

We emphasize that, once a vector is Gaussian, $\varphi_\Lambda \sim \mathscr{N}(a_\Lambda, \Sigma_\Lambda)$, various quantities of interest have immediate expressions in terms of the mean and covariance matrix. For example, to study the random variable φ_{i_0} (which is, of course, Gaussian) at some given vertex $i_0 \in \Lambda$, one can consider the vector $t_\Lambda = (\delta_{i_0 j})_{j\in\Lambda}$, write φ_{i_0} as $\varphi_{i_0} = t_\Lambda \cdot \varphi_\Lambda$ and conclude that the mean and variance of φ_{i_0} are given by

$$E_\Lambda[\varphi_{i_0}] = a_{i_0}, \qquad \mathrm{Var}_\Lambda(\varphi_{i_0}) = \Sigma_\Lambda(i_0, i_0).$$

Although it will not be used later, the following exercise shows that correlation functions of Gaussian vectors enjoy a remarkable factorization property, known as *Wick's formula* or *Isserlis' Theorem.*

Exercise 8.3. Let φ_Λ be a centered Gaussian vector. Show that, for any $n \in \mathbb{N}$:

1. the **$2n+1$-point correlation functions** all vanish: for any collection of (not necessarily distinct) vertices $i_1, \ldots, i_{2n+1} \in \Lambda$,

$$E_\Lambda[\varphi_{i_1} \cdots \varphi_{i_{2n+1}}] = 0; \tag{8.11}$$

2. the **$2n$-point correlation function** can always be expressed in terms of the 2-point correlation functions: for any collection of (not necessarily distinct) vertices $i_1, \dots, i_{2n} \in \Lambda$,

$$E_\Lambda[\varphi_{i_1} \cdots \varphi_{i_{2n}}] = \sum_{\mathscr{P}} \prod_{\{\ell,\ell'\} \in \mathscr{P}} E_\Lambda[\varphi_{i_\ell} \varphi_{i_{\ell'}}], \tag{8.12}$$

where the sum is over all **pairings** $\mathscr{P}$ of $\{1, \dots, 2n\}$, that is, all families of n pairs $\{\ell, \ell'\} \subset \{1, \dots, 2n\}$ whose union equals $\{1, \dots, 2n\}$. *Hint:* Use (8.9). Expanding the exponential in the right-hand side of that expression, determine the coefficient of $t^{i_1} \cdots t^{i_m}$.

In fact, this factorization property characterizes Gaussian vectors.

Exercise 8.4. Consider a random vector φ_Λ satisfying (8.11) and (8.12). Show that φ_Λ is centered Gaussian.

8.2.2 Gaussian Fields and the Thermodynamic Limit

Gaussian fields are infinite collections of random variables whose local behavior is Gaussian:

Definition 8.5. An infinite collection of random variables $\varphi = (\varphi_i)_{i \in \mathbb{Z}^d}$ is a **Gaussian field** if, for each $\Lambda \Subset \mathbb{Z}^d$, the restriction $\varphi_\Lambda = (\varphi_i)_{i \in \Lambda}$ is Gaussian. The distribution of a Gaussian field is called a **Gaussian measure.**

Consider now the sequence of boxes $\mathrm{B}(n)$, $n \geq 0$, and assume that, for each n, a Gaussian vector $\varphi_{\mathrm{B}(n)}$ is given, $\varphi_{\mathrm{B}(n)} \sim \mathscr{N}(a_{\mathrm{B}(n)}, \Sigma_{\mathrm{B}(n)})$, whose distribution we denote by $\mu_{\mathrm{B}(n)}$. A meaning can be given to the thermodynamic limit as follows. We fix $\Lambda \Subset \mathbb{Z}^d$. If n is large, then $\mathrm{B}(n) \supset \Lambda$. Notice that the distribution of $\varphi_\Lambda = (\varphi_i)_{i \in \Lambda}$, seen as a collection of variables indexed by vertices of $\mathrm{B}(n)$, can be computed by taking a vector $t_{\mathrm{B}(n)} = (t_i)_{i \in \mathrm{B}(n)}$ for which $t_i = 0$ for all $i \in \mathrm{B}(n) \setminus \Lambda$. In this way,

$$E_{\mathrm{B}(n)}[e^{\mathrm{i}t_\Lambda \cdot \varphi_\Lambda}] = E_{\mathrm{B}(n)}[e^{\mathrm{i}t_{\mathrm{B}(n)} \cdot \varphi_{\mathrm{B}(n)}}] = e^{\mathrm{i}t_{\mathrm{B}(n)} \cdot a_{\mathrm{B}(n)} - \frac{1}{2} t_{\mathrm{B}(n)} \cdot \Sigma_{\mathrm{B}(n)} t_{\mathrm{B}(n)}}.$$

Remembering that only a fixed number of components of $t_{\mathrm{B}(n)}$ are nonzero, we see that the limiting distribution of φ_Λ can be controlled if $a_{\mathrm{B}(n)}$ and $\Sigma_{\mathrm{B}(n)}$ have limits as $n \to \infty$.

Theorem 8.6. *Let, for all n, $\varphi_{\mathrm{B}(n)} = (\varphi_i)_{i \in \mathrm{B}(n)}$ be a Gaussian vector, $\varphi_{\mathrm{B}(n)} \sim \mathscr{N}(a_{\mathrm{B}(n)}, \Sigma_{\mathrm{B}(n)})$. Assume that, for all $i, j \in \mathbb{Z}^d$, the limits*

$$a_i \overset{\text{def}}{=} \lim_{n\to\infty} (a_{\mathrm{B}(n)})_i \quad \text{and} \quad \Sigma(i,j) \overset{\text{def}}{=} \lim_{n\to\infty} \Sigma_{\mathrm{B}(n)}(i,j)$$

exist and are finite. Then the following holds.

1. *For all $\Lambda \Subset \mathbb{Z}^d$, the distribution of $\varphi_\Lambda = (\varphi_i)_{i\in\Lambda}$ converges, when $n \to \infty$, to that of a Gaussian vector $\mathscr{N}(a_\Lambda, \Sigma_\Lambda)$, with mean and covariance given by the* ***restrictions***

$$a_\Lambda \stackrel{\text{def}}{=} (a_i)_{i\in\Lambda} \quad \text{and} \quad \Sigma_\Lambda \stackrel{\text{def}}{=} (\Sigma(i,j))_{i,j\in\Lambda}\,.$$

2. *There exists a Gaussian field $\widetilde{\varphi}$ whose restriction $\widetilde{\varphi}_\Lambda$ to each $\Lambda \Subset \mathbb{Z}^d$ is a Gaussian vector with distribution $\mathscr{N}(a_\Lambda, \Sigma_\Lambda)$.*

Proof The first claim is a consequence of Proposition B.56. For the second one, fix any $\Lambda \Subset \mathbb{Z}^d$, and let μ_Λ denote the limiting distribution of φ_Λ. By construction, the collection $\{\mu_\Lambda\}_{\Lambda\Subset\mathbb{Z}^d}$ is consistent in the sense of Kolmogorov's Extension Theorem (Theorem 6.6 and Remark 6.98). This guarantees the existence of a probability measure μ on $(\Omega, \mathscr{F})$ whose marginal on each $\Lambda \Subset \mathbb{Z}^d$ is exactly μ_Λ. Under μ, the random variables $\widetilde{\varphi}_i(\omega) \stackrel{\text{def}}{=} \omega_i$ then form a Gaussian field such that, for each Λ, $\widetilde{\varphi}_\Lambda = (\widetilde{\varphi}_i)_{i\in\Lambda}$ has distribution μ_Λ. □

Consider now the GFF in Λ, defined by the measure $\mu^\eta_{\Lambda;m}$ in (8.3). Although the latter is a probability measure on $(\Omega, \mathscr{F})$, it acts in a trivial way on the spins outside Λ (for each $j \notin \Lambda$, $\varphi_j = \eta_j$ almost surely). We will therefore, without loss of generality, consider it as a distribution on $(\Omega_\Lambda, \mathscr{F}_\Lambda)$.

By definition, $\mu^\eta_{\Lambda;m}$ is absolutely continuous with respect to the Lebesgue measure on Ω_Λ. We will show that it can be put in the form (8.10), which will prove that $(\varphi_i)_{i\in\Lambda}$ is a nondegenerate Gaussian vector. We thus need to reformulate the Hamiltonian $\mathscr{H}_{\Lambda;m}$ in such a way that it takes the form of the exponent that appears in the density (8.10). We do this following a step-by-step procedure that will take us on a detour.

8.3 Harmonic Functions and the Discrete Green Identities

Given a collection $f = (f_i)_{i\in\mathbb{Z}^d}$ of real numbers, we define, for each pair $\{i,j\} \in \mathscr{E}_{\mathbb{Z}^d}$, the **discrete gradient**

$$(\nabla f)_{ij} \stackrel{\text{def}}{=} f_j - f_i\,,$$

and, for all $i \in \mathbb{Z}^d$, the **discrete Laplacian**

$$(\Delta f)_i \stackrel{\text{def}}{=} \sum_{j:\, j\sim i} (\nabla f)_{ij}\,. \tag{8.13}$$

Lemma 8.7 (Discrete Green identities). *Let $\Lambda \Subset \mathbb{Z}^d$. Then, for all collections of real numbers $f = (f_i)_{i\in\mathbb{Z}^d}$, $g = (g_i)_{i\in\mathbb{Z}^d}$,*

$$\sum_{\{i,j\}\in\mathscr{E}^{\mathrm{b}}_\Lambda} (\nabla f)_{ij}(\nabla g)_{ij} = -\sum_{i\in\Lambda} g_i(\Delta f)_i + \sum_{\substack{i\in\Lambda\\ j\in\Lambda^{\mathrm{c}}, j\sim i}} g_j(\nabla f)_{ij}\,, \tag{8.14}$$

$$\sum_{i\in\Lambda}\{f_i(\Delta g)_i - g_i(\Delta f)_i\} = \sum_{\substack{i\in\Lambda\\ j\in\Lambda^c, j\sim i}} \{f_j(\nabla g)_{ij} - g_j(\nabla f)_{ij}\}. \tag{8.15}$$

Remark 8.8. The continuous analogues of (8.14) and (8.15) are the classical *Green identities*, which, on a smooth domain $U \subset \mathbb{R}^n$, provide a higher-dimensional version of the classical integration by parts formula. That is, for all smooth functions f and g,

$$\int_U \nabla f \cdot \nabla g \, \mathrm{d}V = -\int_U g\Delta f \, \mathrm{d}V + \oint_{\partial U} g(\nabla f \cdot n)\, \mathrm{d}S,$$

$$\int_U \{f\Delta g - g\Delta f\}\, \mathrm{d}V = \oint_{\partial U} \{f\nabla g \cdot n - g\nabla f \cdot n\}\, \mathrm{d}S,$$

where n is the outward normal unit vector and $\mathrm{d}V$ and $\mathrm{d}S$ denote, respectively, the volume and surface elements. ◇

Proof of Lemma 8.7. Using the symmetry between i and j (in all the sums below, j is always assumed to be a nearest neighbor of i):

$$\begin{aligned}\sum_{\{i,j\}\subset\Lambda} (\nabla f)_{ij}(\nabla g)_{ij} &= \sum_{\{i,j\}\subset\Lambda} g_j(f_j - f_i) - \sum_{\{i,j\}\subset\Lambda} g_i(f_j - f_i)\\ &= -\sum_{i\in\Lambda} g_i \sum_{j\in\Lambda}(f_j - f_i)\\ &= -\sum_{i\in\Lambda} g_i(\Delta f)_i + \sum_{i\in\Lambda} g_i \sum_{j\in\Lambda^c}(f_j - f_i).\end{aligned}$$

Therefore,

$$\begin{aligned}\sum_{\{i,j\}\in\mathscr{E}^{\mathrm{b}}_\Lambda} (\nabla f)_{ij}(\nabla g)_{ij} &= \sum_{\{i,j\}\subset\Lambda} (\nabla f)_{ij}(\nabla g)_{ij} + \sum_{i\in\Lambda, j\in\Lambda^c} (\nabla f)_{ij}(\nabla g)_{ij}\\ &= -\sum_{i\in\Lambda} g_i(\Delta f)_i + \sum_{i\in\Lambda, j\in\Lambda^c} g_j(f_j - f_i).\end{aligned}$$

The second identity (8.15) is obtained using the first one twice, interchanging the roles of f and g. □

We can write the action of the Laplacian on $f = (f_i)_{i\in\mathbb{Z}^d}$ as:

$$(\Delta f)_i = \sum_{j\in\mathbb{Z}^d} \Delta_{ij} f_j, \quad i \in \mathbb{Z}^d,$$

where the matrix elements $(\Delta_{ij})_{i,j\in\mathbb{Z}^d}$ are defined by

$$\Delta_{ij} = \begin{cases} -2d & \text{if } i = j,\\ 1 & \text{if } i \sim j,\\ 0 & \text{otherwise.}\end{cases} \tag{8.16}$$

To obtain a representation of $\mathscr{H}_{\Lambda;m}$ in terms of the scalar product in Λ, we introduce the **restriction** of Δ to Λ, defined by $\Delta_\Lambda \stackrel{\text{def}}{=} (\Delta_{ij})_{i,j\in\Lambda}$.

Remark 8.9. Let $i \in \Lambda$. In what follows, it will be important to distinguish between $(\Delta f)_i$, defined in (8.13) and which may depend on some of the variables f_j located outside Λ, and $(\Delta_\Lambda f)_i$, which is a shorthand notation for $\sum_{j\in\Lambda} \Delta_{ij} f_j$ (and thus involves only variables f_j inside Λ). In particular, we will use the notation

$$f \cdot \Delta_\Lambda g \stackrel{\text{def}}{=} \sum_{i,j\in\Lambda} \Delta_{ij} f_i g_j \,,$$

which clearly satisfies

$$f \cdot \Delta_\Lambda g = (\Delta_\Lambda f) \cdot g \,. \tag{8.17}$$

◇

From now on, we assume that f coincides with η outside Λ and denote by B_Λ any **boundary term**, that is, any quantity (possibly changing from place to place) depending only on the values η_j, $j \in \Lambda^c$.

Let us see how the quadratic term in the Hamiltonian will be handled. Applying (8.14) with $f = g$ and rearranging terms, we get

$$\begin{aligned}\sum_{\{i,j\}\in\mathscr{E}^b_\Lambda} (f_j - f_i)^2 &= \sum_{\{i,j\}\in\mathscr{E}^b_\Lambda} (\nabla f)^2_{ij} \\ &= -f \cdot \Delta_\Lambda f - 2\sum_{i\in\Lambda}\sum_{j\in\Lambda^c, j\sim i} f_i f_j + B_\Lambda \,. \end{aligned}\tag{8.18}$$

One can then introduce $u = (u_i)_{i\in\mathbb{Z}^d}$, to be determined later, depending on η and Λ, and playing the role of the mean of f. Our aim is to rewrite (8.18) in the form $-(f-u)\cdot\Delta_\Lambda(f-u)$, up to boundary terms. We can, in particular, include in B_Λ any expression that depends only on the values of u. We have, using (8.17),

$$\begin{aligned}(f-u)\cdot\Delta_\Lambda(f-u) &= f\cdot\Delta_\Lambda f - 2f\cdot\Delta_\Lambda u + u\cdot\Delta_\Lambda u \\ &= f\cdot\Delta_\Lambda f - 2\sum_{i\in\Lambda} f_i(\Delta u)_i + 2\sum_{i\in\Lambda}\sum_{j\in\Lambda^c, j\sim i} f_i u_j + B_\Lambda \,.\end{aligned}$$

Comparing with (8.18), we deduce that

$$\begin{aligned}\sum_{\{i,j\}\in\mathscr{E}^b_\Lambda} (f_j - f_i)^2 &= -(f-u)\cdot\Delta_\Lambda(f-u) \\ &\quad - 2\sum_{i\in\Lambda} f_i \underbrace{(\Delta u)_i}_{\text{(i)}} + 2\sum_{i\in\Lambda}\sum_{j\in\Lambda^c, j\sim i} f_i \underbrace{(u_j - f_j)}_{\text{(ii)}} + B_\Lambda \,.\end{aligned}\tag{8.19}$$

A look at the second line in this last display indicates exactly the restrictions one should impose on u in order for $-(f-u)\cdot\Delta_\Lambda(f-u)$ to be the one and only contribution to the Hamiltonian (up to boundary terms). To cancel the nontrivial terms that depend on the values of f inside Λ, we need to ensure that: (i) u is **harmonic in Λ**:

$$(\Delta u)_i = 0, \quad \forall i \in \Lambda \,.$$

(ii) u coincides with f (hence with η) outside Λ. We have thus proved:

Lemma 8.10. *Let f coincide with η outside Λ. Assume that u* ***solves the Dirichlet problem in Λ with boundary condition η****:*

$$\begin{cases} u \text{ harmonic in } \Lambda\,, \\ u_j = \eta_j \text{ for all } j \in \Lambda^c\,. \end{cases} \tag{8.20}$$

Then,

$$\sum_{\{i,j\}\in\mathscr{E}_\Lambda^b} (f_j - f_i)^2 = -(f-u)\cdot\Delta_\Lambda(f-u) + B_\Lambda\,. \tag{8.21}$$

Existence of a solution to the Dirichlet problem will be proved in Lemma 8.15. Uniqueness can be verified easily:

Lemma 8.11. (8.20) *has at most one solution.*

Proof We first consider the boundary condition $\eta \equiv 0$, and show that $u \equiv 0$ is the unique solution. Namely, assume v is any solution, and let $i_* \in \Lambda$ be such that $|v_{i_*}| = \max_{j\in\Lambda}|v_j|$. With no loss of generality, assume $v_{i_*} \geq 0$. Since $(\Delta v)_{i_*} = 0$ implies $v_{i_*} = \frac{1}{2d}\sum_{j\sim i_*} v_j$, and $v_j \leq v_{i_*}$ for all $j \sim i_*$, we conclude that $v_j = v_{i_*}$ for all $j \sim i_*$. Repeating this procedure until the boundary of Λ is reached, we deduce that v must be constant, and this constant can only be 0. Now let u and v be two solutions of (8.20). Then, $h = u - v$ is a solution to the Dirichlet problem in Λ with boundary condition $\eta' \equiv 0$. By the previous argument, $h \equiv 0$ and thus $u = v$. □

Exercise 8.5. Show that, when $d = 1$, the solution of the Dirichlet problem on an interval $\Lambda = \{a, \dots, b\}$ is of the form $u_i = ai + c$, for some $a, c \in \mathbb{R}$ determined by the boundary condition.

8.4 The Massless Case

Let us consider the massless Hamiltonian $\mathscr{H}_{\Lambda;0}$, expressed in terms of the variables $\varphi = (\varphi_i)_{i\in\mathbb{Z}^d}$, which are assumed to satisfy $\varphi_i = \eta_i$ for all $i \notin \Lambda$. We apply Lemma 8.10 with $f = \varphi$, assuming for the moment that one can find a solution u to the Dirichlet problem (in Λ, with boundary condition η). Since it does not alter the Gibbs distribution, the constant B_Λ in (8.21) can always be subtracted from the Hamiltonian. We get

$$\mathscr{H}_{\Lambda;0} = \tfrac{1}{2}(\varphi - u)\cdot(-\tfrac{1}{2d}\Delta_\Lambda)(\varphi - u)\,. \tag{8.22}$$

Our next tasks are, first, to invert the matrix $-\frac{1}{2d}\Delta_\Lambda$, in order to obtain an explicit expression for the covariance matrix, and, second, to find an explicit expression for the solution u to the Dirichlet problem.

8.4.1 Random Walk Representation

We need to determine whether there exists some positive-definite covariance matrix Σ_Λ such that $-\frac{1}{2d}\Delta_\Lambda = \Sigma_\Lambda^{-1}$. Observe first that

$$-\tfrac{1}{2d}\Delta_\Lambda = I_\Lambda - P_\Lambda\,,$$

where $I_\Lambda = (\delta_{ij})_{i,j\in\Lambda}$ is the identity matrix and $P_\Lambda = (P(i,j))_{i,j\in\Lambda}$ is the matrix with elements

$$P(i,j) \stackrel{\text{def}}{=} \begin{cases} \frac{1}{2d} & \text{if } j \sim i\,, \\ 0 & \text{otherwise.} \end{cases}$$

The numbers $(P(i,j))_{i,j\in\mathbb{Z}^d}$ are the **transition probabilities of the symmetric simple random walk $X = (X_k)_{k\geq 0}$ on $\mathbb{Z}^d$**, which at each time step jumps to any one of its $2d$ nearest neighbors with probability $\frac{1}{2d}$:

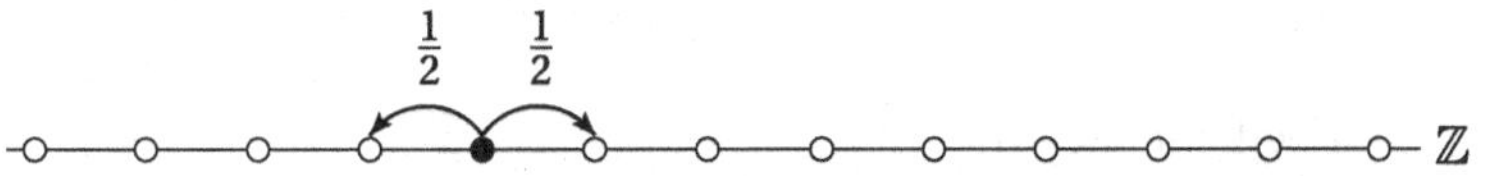

Figure 8.3. The one-dimensional symmetric simple random walk.

We denote by $\mathbb{P}_i$ the distribution of the walk starting at $i \in \mathbb{Z}^d$. That is, we have $\mathbb{P}_i(X_0 = i) = 1$ and, for $n \geq 0$,

$$\mathbb{P}_i(X_{n+1} = k \,|\, X_n = j) = P(j,k), \quad \forall j,k \in \mathbb{Z}^d.$$

(Information on the simple random walk on $\mathbb{Z}^d$ can be found in Appendix B.13.)

We will need to know that the walk almost surely exits a finite region in a finite time:

Lemma 8.12. *For $\Lambda \Subset \mathbb{Z}^d$, let $\tau_{\Lambda^c} \stackrel{\text{def}}{=} \inf\{k \geq 0 : X_k \in \Lambda^c\}$ be the **first exit time from Λ**. Then $\mathbb{P}_i(\tau_{\Lambda^c} < \infty) = 1$. More precisely, there exists $c = c(\Lambda) > 0$ such that, for all $i \in \Lambda$,*

$$\mathbb{P}_i(\tau_{\Lambda^c} > n) \leq e^{-cn}\,. \tag{8.23}$$

Proof If we let $R = \sup_{l\in\Lambda}\inf_{k\in\Lambda^c}\|k-l\|_1$, then, starting from i, one can find a nearest-neighbor path of length at most R that exits Λ. This means that, during any time interval of length R, there is a probability at least $(2d)^{-R}$ that the random walk exits Λ (just force it to follow the path). In particular,

$$\mathbb{P}_i(\tau_{\Lambda^c} > n) \leq (1-(2d)^{-R})^{[n/R]}\,. \qquad \square$$

The next lemma shows that the matrix $I_\Lambda - P_\Lambda$ is invertible, and provides a probabilistic interpretation for its inverse:

Lemma 8.13. *The* $|\Lambda| \times |\Lambda|$ *matrix* $I_\Lambda - P_\Lambda$ *is invertible. Moreover, its inverse* $G_\Lambda \stackrel{\text{def}}{=} (I_\Lambda - P_\Lambda)^{-1}$ *is given by* $G_\Lambda = (G_\Lambda(i,j))_{i,j\in\Lambda}$*, the* ***Green function in* Λ *of the simple random walk on* $\mathbb{Z}^d$***, defined by*

$$G_\Lambda(i,j) \stackrel{\text{def}}{=} \mathbb{E}_i\Big[\sum_{n=0}^{\tau_{\Lambda^c}-1} \mathbf{1}_{\{X_n=j\}}\Big]. \tag{8.24}$$

The Green function $G_\Lambda(i,j)$ represents the average number of visits at j made by a walk started at i, before it leaves Λ.

Proof To start, observe that (below, P_Λ^n denotes the nth power of P_Λ)

$$(I_\Lambda - P_\Lambda)(I_\Lambda + P_\Lambda + P_\Lambda^2 + \cdots + P_\Lambda^n) = I_\Lambda - P_\Lambda^{n+1}. \tag{8.25}$$

We claim that there exists $c = c(\Lambda)$ such that, for all $i,j \in \Lambda$ and all $n \geq 1$,

$$P_\Lambda^n(i,j) \leq e^{-cn}. \tag{8.26}$$

Indeed, for each $n \geq 1$,

$$P_\Lambda^n(i,j) = \sum_{i_1,\dots,i_{n-1}\in\Lambda} P(i,i_1)P(i_1,i_2)\cdots P(i_{n-1},j) = \mathbb{P}_i(X_n = j, \tau_{\Lambda^c} > n).$$

Since $\mathbb{P}_i(X_n = j, \tau_{\Lambda^c} > n) \leq \mathbb{P}_i(\tau_{\Lambda^c} > n)$, (8.26) follows from (8.23). This implies that the matrix $G_\Lambda = (G_\Lambda)_{i,j\in\Lambda}$, defined by

$$G_\Lambda(i,j) = (I_\Lambda + P_\Lambda + P_\Lambda^2 + \cdots)(i,j) = \sum_{n\geq 0} \mathbb{P}_i(X_n = j, \tau_{\Lambda^c} > n), \tag{8.27}$$

is well defined and, by (8.25), that it satisfies $(I_\Lambda - P_\Lambda)G_\Lambda = I_\Lambda$. Of course, by symmetry, we also have $G_\Lambda(I_\Lambda - P_\Lambda) = I_\Lambda$. The conclusion follows, since the right-hand side of (8.27) can be rewritten in the form given in (8.24). □

Remark 8.14. The key ingredient in the above proof that $I_\Lambda - P_\Lambda$ is invertible is the fact that P_Λ is **substochastic**: $\sum_{j\in\Lambda} P(i,j) < 1$ for those vertices i that lie along the inner boundary of Λ. This property was crucial in establishing (8.26). ◇

Let us now prove the existence of a solution to the Dirichlet problem (uniqueness was shown in Lemma 8.11), also expressed in terms of the simple random walk. Let $X_{\tau_{\Lambda^c}}$ denote the position of the walk at the time of first exit from Λ.

Lemma 8.15. *The solution to the Dirichlet problem* (8.20) *is given by the function* $u = (u_i)_{i\in\mathbb{Z}^d}$ *defined by*

$$u_i \stackrel{\text{def}}{=} \mathbb{E}_i[\eta_{X_{\tau_{\Lambda^c}}}], \quad \forall i \in \mathbb{Z}^d. \tag{8.28}$$

Proof When $j \in \Lambda^c$, $\mathbb{P}_j(\tau_{\Lambda^c} = 0) = 1$ and, thus, $u_j = \mathbb{E}_j[\eta_{X_0}] = \eta_j$. When $i \in \Lambda$, by conditioning on the first step of the walk,

$$\begin{aligned} u_i = \mathbb{E}_i[\eta_{X_{\tau_{\Lambda^c}}}] &= \sum_{j\sim i} \mathbb{E}_i[\eta_{X_{\tau_{\Lambda^c}}}, X_1 = j] \\ &= \sum_{j\sim i} \mathbb{P}_i(X_1 = j)\, \mathbb{E}_i[\eta_{X_{\tau_{\Lambda^c}}} \,|\, X_1 = j] \\ &= \sum_{j\sim i} \tfrac{1}{2d} \mathbb{E}_j[\eta_{X_{\tau_{\Lambda^c}}}] = \tfrac{1}{2d} \sum_{j\sim i} u_j, \end{aligned}$$

which implies $(\Delta u)_i = 0$. □

Remark 8.16. Observe that if $\eta = (\eta_i)_{i\in\mathbb{Z}^d}$ is itself harmonic, then $u \equiv \eta$ *is* the solution to the Dirichlet problem in Λ with boundary condition η. ◇

Theorem 8.17. *Under* $\mu^\eta_{\Lambda;0}$, $\varphi_\Lambda = (\varphi_i)_{i\in\Lambda}$ *is Gaussian, with mean* $u_\Lambda = (u_i)_{i\in\Lambda}$ *defined in* (8.28), *and positive-definite covariance matrix* $G_\Lambda = (G_\Lambda(i,j))_{i,j\in\Lambda}$ *given by the Green function* (8.24).

Proof The claim follows from the representation (8.22) of the Hamiltonian, the expression (8.28) for the solution to the Dirichlet problem, the expression (8.24) for the inverse of $I_\Lambda - P_\Lambda$ and Theorem 8.4. □

The reader should note the remarkable fact that the distribution of φ_Λ under $\mu^\eta_{\Lambda;0}$ depends on the boundary condition η only through its mean: the covariance matrix is only sensitive to the choice of Λ.

Example 8.18. Consider the GFF in dimension $d = 1$. Let η be any boundary condition, fixed outside an interval $\Lambda = \{a, \dots, b\} \Subset \mathbb{Z}$. As we saw in Exercise 8.5, the solution to the Dirichlet problem is the affine interpolation between (a, η_a) and (b, η_b). A typical configuration under $\mu^\eta_{\Lambda;0}$ should therefore be thought of as describing fluctuations around this line (however, these fluctuations can be large on the microscopic scale, as will be seen below):

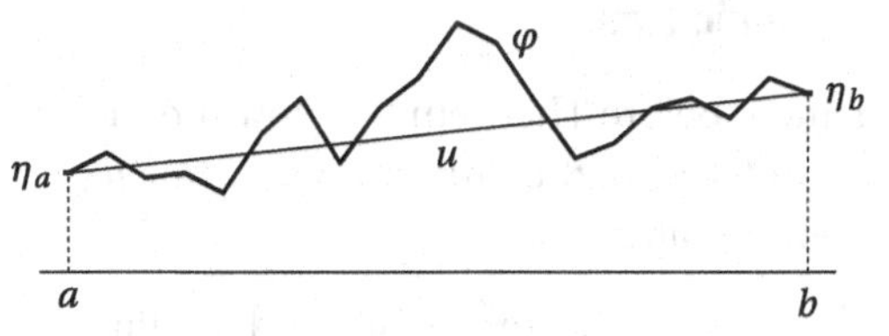

Figure 8.4. A configuration of the one-dimensional GFF under $\mu^\eta_{\Lambda;0}$, whose mean u_Λ is the harmonic function given by the linear interpolation between the values of η on the boundary.

◇

Before we start our analysis of the thermodynamic limit, let us exploit the representation derived in Theorem 8.17 in order to study the fluctuations of $\varphi_{B(n)}$ in a large box $B(n)$.

For the sake of concreteness, consider the spin at the origin, φ_0. The latter is a Gaussian random variable with variance given by

$$\mathrm{Var}^{\eta}_{B(n);0}(\varphi_0) = G_{B(n)}(0,0).$$

Notice first that the time $\tau_{B(n)^c}$ it takes for the random walk, starting at 0, to leave the box $B(n)$ is increasing in n and is always larger than n. Therefore, by monotone convergence,

$$\lim_{n\to\infty} G_{B(n)}(0,0) = \mathbb{E}_0\Big[\sum_{k\geq 0} \mathbf{1}_{\{X_k=0\}}\Big] \tag{8.29}$$

is just the expected number of visits of the walk at the origin. In particular, the variance of φ_0 diverges in the limit $n \to \infty$, whenever the symmetric simple random walk is recurrent, that is, in dimensions 1 and 2 (see Appendix B.13.4). When this happens, the field is said to **delocalize**. A closer analysis of the Green function (see Theorem B.76) yields the following more precise information:

$$G_{B(n)}(0,0) \approx \begin{cases} n & \text{if } d = 1, \\ \log n & \text{if } d = 2. \end{cases} \tag{8.30}$$

In contrast, in dimensions $d \geq 3$, the variance remains bounded, and so the field remains localized close to its mean value even in the limit $n \to \infty$.

In the next section, we will relate these properties to the problem of existence of infinite-volume Gibbs measures for the massless GFF.

Exercise 8.6. Consider the one-dimensional GFF in $B(n)$ with 0 boundary condition (see Figure 8.1). Interpreting the values of the field, $\varphi_{-n}, \ldots, \varphi_n$, as the successive positions of a random walk on $\mathbb{R}$ with Gaussian increments, starting at $\varphi_{-n-1} = 0$ and conditioned on $\{\varphi_{n+1} = 0\}$, prove directly (that is, without using the random walk representation of $G(0,0)$) that $\varphi_0 \sim \mathcal{N}(0, n+1)$.

8.4.2 Thermodynamic Limit

We explained, before Theorem 8.6, how the thermodynamic limit $n \to \infty$ can be expressed in terms of the limits of the means $u_{B(n)}$ and of the covariance matrices $G_{B(n)}$, when the latter exist:

$$\lim_{n\to\infty} \mathbb{E}_i[\eta_{X_{\tau_{B(n)^c}}}], \quad \lim_{n\to\infty} G_{B(n)}(i,j), \tag{8.31}$$

for all fixed pairs $i, j \in \mathbb{Z}^d$.

Low Dimensions. We have seen that, when $d = 1$ or $d = 2$, $\lim_{n\to\infty} G_{B(n)}(0,0) = \infty$. This has the following consequence:

Theorem 8.19. *When $d = 1$ or $d = 2$, the massless Gaussian Free Field has no infinite-volume Gibbs measures: $\mathscr{G}(0) = \varnothing$.*

Proof Assume there exists a probability measure $\mu \in \mathscr{G}(0)$. Since $\mu = \mu\pi^0_{\mathrm{B}(n)}$ for all n, we have

$$\mu(\varphi_0 \in [a,b]) = \mu\pi^0_{\mathrm{B}(n)}(\varphi_0 \in [a,b]) = \int \mu^\eta_{\mathrm{B}(n);0}(\varphi_0 \in [a,b])\,\mu(\mathrm{d}\eta),$$

for any interval $[a,b] \subset \mathbb{R}$. But, uniformly in η,

$$\begin{aligned}\mu^\eta_{\mathrm{B}(n);0}(\varphi_0 \in [a,b]) &= \frac{1}{\sqrt{2\pi G_{\mathrm{B}(n)}(0,0)}}\int_a^b \exp\Big\{-\frac{\big(x-\mu^\eta_{\mathrm{B}(n);0}(\varphi_0)\big)^2}{2G_{\mathrm{B}(n)}(0,0)}\Big\}\,\mathrm{d}x \\ &\le \frac{b-a}{\sqrt{2\pi G_{\mathrm{B}(n)}(0,0)}}\,. \end{aligned} \tag{8.32}$$

In dimensions 1 and 2, the right-hand side tends to 0 as $n \to \infty$. We conclude that $\mu(\varphi_0 \in [a,b]) = 0$, for all $a < b$, and thus $\mu(\varphi_0 \in \mathbb{R}) = 0$, which contradicts the assumption that μ is a probability measure. □

Remark 8.20. The lack of infinite-volume Gibbs measures for the massless GFF $\varphi = (\varphi_i)_{i\in\mathbb{Z}^d}$ in dimensions 1 and 2, as seen above, is due to the fact that the fluctuations of each spin φ_i become unbounded when $\mathrm{B}(n) \uparrow \mathbb{Z}^d$. This is not incompatible, nevertheless, with the fact that some *random* translation of the field does have a well-defined thermodynamic limit. Namely, define the random variables $\widetilde{\varphi} = (\widetilde{\varphi}_i)_{i\in\mathbb{Z}^d}$ by

$$\widetilde{\varphi}_i \stackrel{\text{def}}{=} \varphi_i - \varphi_0\,. \tag{8.33}$$

Then, as shown in the next exercise, these random variables have a well-defined thermodynamic limit, even when $d = 1, 2$. ◇

Exercise 8.7. Consider the variables $\widetilde{\varphi}_i$ defined in (8.33). Show that under $\mu^0_{\Lambda;0}$ (0 boundary condition), $(\widetilde{\varphi}_i)_{i\in\Lambda}$ is Gaussian, centered, with covariance matrix given by

$$\widetilde{G}_\Lambda(i,j) \stackrel{\text{def}}{=} G_\Lambda(i,j) - G_\Lambda(i,0) - G_\Lambda(0,j) + G_\Lambda(0,0)\,. \tag{8.34}$$

It can be shown that the matrix elements $\widetilde{G}_\Lambda(i,j)$ in (8.34) have a finite limit when $\Lambda \uparrow \mathbb{Z}^d$; see the comments in Section 8.7.2.

Higher Dimensions. When $d \ge 3$, transience of the symmetric simple random walk implies that

$$G(i,j) \stackrel{\text{def}}{=} \lim_{n\to\infty} G_{\mathrm{B}(n)}(i,j) = \mathbb{E}_i\Big(\sum_{k\ge 0} \mathbf{1}_{\{X_k=j\}}\Big) \tag{8.35}$$

is finite. This will allow us to construct infinite-volume Gibbs measures. We say that $\eta = (\eta_i)_{i\in\mathbb{Z}^d}$ is **harmonic (in $\mathbb{Z}^d$)** if

$$(\Delta\eta)_i = 0 \quad \forall i \in \mathbb{Z}^d .$$

Theorem 8.21. *In dimensions $d \geq 3$, the massless Gaussian Free Field possesses infinitely many infinite-volume Gibbs measures: $|\mathscr{G}(0)| = \infty$. More precisely, given any harmonic function η on $\mathbb{Z}^d$, there exists a Gaussian Gibbs measure μ_0^η with mean η and covariance matrix $G = (G(i,j))_{i,j\in\mathbb{Z}^d}$ given in (8.35).*

Remark 8.22. It can be shown that the Gibbs measures μ_0^η of Theorem 8.21 are precisely the extremal elements of $\mathscr{G}(0)$: $\mathrm{ex}\,\mathscr{G}(0) = \{\mu_0^\eta : \eta \text{ harmonic}\}$. ◇

Clearly, there exist infinitely many harmonic functions. For example, any constant function is harmonic, or, more generally, any function of the form

$$\eta_i \stackrel{\text{def}}{=} \alpha_1 i_1 + \cdots + \alpha_d i_d + c, \qquad \forall i = (i_1, \ldots, i_d) \in \mathbb{Z}^d, \tag{8.36}$$

with $\alpha = (\alpha_1, \ldots, \alpha_d) \in \mathbb{R}^d$. But, in $d \geq 2$, the variety of harmonic functions is much larger:

Exercise 8.8. Show that all harmonic functions $u\colon \mathbb{Z}^d \to \mathbb{R}$ can be obtained by fixing arbitrary values of u_i at all vertices i belonging to the strip

$$\{i = (i_1, i_2, \ldots, i_d) \in \mathbb{Z}^d : i_d \in \{0,1\}\}$$

and extending u to the whole of $\mathbb{Z}^d$ using $\Delta u = 0$.

An analysis of the Green function $G(i,j)$ as $\|j - i\|_1 \to \infty$ (see Theorem B.76) yields the following information on the asymptotic behavior of the covariance.

Proposition 8.23. *Assume that $d \geq 3$ and $m = 0$. Then, the infinite-volume Gibbs measures μ_0^η of Theorem 8.21 satisfy, as $\|i - j\|_2 \to \infty$,*

$$\mathrm{Cov}_0^\eta(\varphi_i, \varphi_j) = \frac{r(d)}{\|j - i\|_2^{d-2}}(1 + o(1)) \tag{8.37}$$

for some constant $r(d) > 0$.

Proof of Theorem 8.21. Fix some harmonic function η. The restriction of η to any finite box $\mathrm{B}(n)$ obviously solves the Dirichlet problem on $\mathrm{B}(n)$ (with boundary condition η). Since the limits $\lim_{n\to\infty} G_{\mathrm{B}(n)}(i,j)$ exist when $d \geq 3$, we can use Theorem 8.6 to construct a Gaussian field $\varphi = (\varphi_i)_{i\in\mathbb{Z}^d}$ whose restriction to any finite region Λ is a Gaussian vector $(\varphi_i)_{i\in\Lambda}$ with mean $(\eta_i)_{i\in\Lambda}$ and covariance matrix $G_\Lambda = (G(i,j))_{i,j\in\Lambda}$. If we let μ_0^η denote the distribution of φ, then $\mu_0^\eta(\varphi_i) = \eta_i$ and

$$\mathrm{Cov}_0^\eta(\varphi_i, \varphi_j) = G(i,j). \tag{8.38}$$

It remains to show that $\mu_0^\eta \in \mathscr{G}(0)$. This could be done following the same steps used to prove Theorem 6.26. For pedagogical reasons, we will give a different proof relying on the Gaussian properties of φ.

We use the criterion (8.4), and show that, for all $\Lambda \Subset \mathbb{Z}^d$ and all $A \in \mathscr{F}$,

$$\mu_0^\eta(A \mid \mathscr{F}_{\Lambda^c})(\omega) = \mu_{\Lambda;0}^\omega(A) \quad \text{for } \mu_0^\eta\text{-almost all } \omega.$$

For that, we will verify that the field $\varphi = (\varphi_i)_{i \in \mathbb{Z}^d}$, when conditioned on $\mathscr{F}_{\Lambda^c}$, remains Gaussian (Lemma 8.24 below), and that, for all t_Λ,

$$E_0^\eta\big[e^{\mathrm{i}t_\Lambda \cdot \varphi_\Lambda} \mid \mathscr{F}_{\Lambda^c}\big](\omega) = e^{\mathrm{i}t_\Lambda \cdot a_\Lambda(\omega) - \frac{1}{2} t_\Lambda \cdot G_\Lambda t_\Lambda}, \tag{8.39}$$

where $a_i(\omega) = \mathbb{E}_i[\omega_{X_{\tau_{\Lambda^c}}}]$ is the solution of the Dirichlet problem in Λ with boundary condition ω.

Lemma 8.24. *Let φ be the Gaussian field constructed above. Let, for all $i \in \Lambda$,*

$$a_i(\omega) \stackrel{\text{def}}{=} E_0^\eta[\varphi_i \mid \mathscr{F}_{\Lambda^c}](\omega).$$

Then, μ-almost surely, $a_i(\omega) = \mathbb{E}_i[\omega_{X_{\tau_{\Lambda^c}}}]$. In particular, each $a_i(\omega)$ is a finite linear combination of the variables ω_j and $(a_i)_{i \in \mathbb{Z}^d}$ is a Gaussian field.

Proof When $i \in \Lambda$, we use the characterization of the conditional expectation given in Lemma B.50: up to equivalence, $E_0^\eta[\varphi_i|\mathscr{F}_{\Lambda^c}]$ is the unique $\mathscr{F}_{\Lambda^c}$-measurable random variable ψ for which

$$E_0^\eta\big[(\varphi_i - \psi)\varphi_j\big] = 0 \quad \text{for all } j \in \Lambda^c. \tag{8.40}$$

We verify that this condition is indeed satisfied when $\psi(\omega) = \mathbb{E}_i[\omega_{X_{\tau_{\Lambda^c}}}]$. By (8.38),

$$\begin{aligned} E_0^\eta\big[(\varphi_i - \mathbb{E}_i[\varphi_{X_{\tau_{\Lambda^c}}}])\varphi_j\big] &= E_0^\eta[\varphi_i\varphi_j] - E_0^\eta\big[\mathbb{E}_i[\varphi_{X_{\tau_{\Lambda^c}}}]\varphi_j\big] \\ &= G(i,j) + \eta_i\eta_j - E_0^\eta\big[\mathbb{E}_i[\varphi_{X_{\tau_{\Lambda^c}}}]\varphi_j\big]. \end{aligned}$$

Using again (8.38),

$$\begin{aligned} E_0^\eta\big[\mathbb{E}_i[\varphi_{X_{\tau_{\Lambda^c}}}]\varphi_j\big] &= \sum_{k \in \partial^{\mathrm{ex}}\Lambda} E_0^\eta[\varphi_k\varphi_j]\,\mathbb{P}_i[X_{\tau_{\Lambda^c}} = k] \\ &= \mathbb{E}_i\big[E_0^\eta[\varphi_{X_{\tau_{\Lambda^c}}}\varphi_j]\big] \\ &= \mathbb{E}_i\big[G(X_{\tau_{\Lambda^c}}, j)\big] + \mathbb{E}_i\big[E_0^\eta[\varphi_{X_{\tau_{\Lambda^c}}}]E_0^\eta[\varphi_j]\big]. \end{aligned} \tag{8.41}$$

On the one hand, since $i \in \Lambda$ and $j \in \Lambda^c$, any trajectory of the random walk that contributes to $G(i,j)$ must intersect $\partial^{\mathrm{ex}}\Lambda$ at least once, so the strong Markov property gives

$$G(i,j) = \sum_{k \in \partial^{\mathrm{ex}}\Lambda} \mathbb{P}_i(X_{\tau_{\Lambda^c}} = k)G(k,j) = \mathbb{E}_i[G(X_{\tau_{\Lambda^c}}, j)].$$

On the other hand, since φ has mean η and since η is the solution of the Dirichlet problem in Λ with boundary condition η, we have

$$\mathbb{E}_i\big[E_0^\eta[\varphi_{X_{\tau_{\Lambda^c}}}]E_0^\eta[\varphi_j]\big] = \mathbb{E}_i[\eta_{X_{\tau_{\Lambda^c}}}\eta_j] = \mathbb{E}_i[\eta_{X_{\tau_{\Lambda^c}}}]\eta_j = \eta_i\eta_j\,.$$

This shows that $a_i(\omega) = \mathbb{E}_i[\omega_{X_{\tau_{\Lambda^c}}}]$. In particular, it is a linear combination of the ω_js:

$$a_i(\omega) = \sum_{k\in\partial^{\text{ex}}\Lambda} \omega_k \mathbb{P}_i(X_{\tau_{\Lambda^c}} = k)\,,$$

which implies that $(a_i)_{i\in\mathbb{Z}^d}$ is also a Gaussian field. □

Corollary 8.25. *Under μ_0^η, the random vector $(\varphi_i - a_i)_{i\in\Lambda}$ is independent of $\mathscr{F}_{\Lambda^c}$.*

Proof We know that the variables $\varphi_i - a_i$, $i \in \Lambda$, and φ_j, $j \in \Lambda^c$, form a Gaussian field. Therefore, a classical result (Proposition (B.58)) implies that $(\varphi_i - a_i)_{i\in\Lambda}$, which is centered, is independent of $\mathscr{F}_{\Lambda^c}$ if and only if each pair $\varphi_i - a_i$ ($i \in \Lambda$) and φ_j ($j \in \Lambda^c$) is uncorrelated. But this follows from (8.40). □

Let $a_\Lambda = (a_i)_{i\in\Lambda}$. By Corollary 8.25 and since a_Λ is $\mathscr{F}_{\Lambda^c}$-measurable,

$$E_0^\eta\big[e^{it_\Lambda\cdot\varphi_\Lambda} \mid \mathscr{F}_{\Lambda^c}\big] = e^{it_\Lambda\cdot a_\Lambda}E_0^\eta\big[e^{it_\Lambda\cdot(\varphi_\Lambda - a_\Lambda)} \mid \mathscr{F}_{\Lambda^c}\big] = e^{it_\Lambda\cdot a_\Lambda}E_0^\eta\big[e^{it_\Lambda\cdot(\varphi_\Lambda - a_\Lambda)}\big]\,.$$

We know that the variables $\varphi_i - a_i$, $i \in \Lambda$, form a Gaussian vector under μ_0^η. Since it is centered, we need only compute its covariance. For $i, j \in \Lambda$, write

$$(\varphi_i - a_i)(\varphi_j - a_j) = \varphi_i\varphi_j - (\varphi_i - a_i)a_j - (\varphi_j - a_j)a_i - a_ia_j\,.$$

Using Corollary 8.25 again, we see that $E_0^\eta\big[(\varphi_i - a_i)a_j\big] = 0$ and $E_0^\eta\big((\varphi_j - a_j)a_i\big) = 0$ (since a_i and a_j are $\mathscr{F}_{\Lambda^c}$-measurable). Therefore,

$$\begin{aligned}\mathrm{Cov}_0^\eta\big((\varphi_i - a_i), (\varphi_j - a_j)\big) &= E_0^\eta[\varphi_i\varphi_j] - E_0^\eta(a_ia_j)\\ &= G(i,j) + \eta_i\eta_j - E_0^\eta[a_ia_j]\,.\end{aligned}$$

Proceeding as in (8.41),

$$E_0^\eta[a_ia_j] = \mathbb{E}_{i,j}\big[G(X_{\tau_{\Lambda^c}}, X'_{\tau_{\Lambda^c}})\big] + \mathbb{E}_{i,j}\big[E_0^\eta[\varphi_{X_{\tau_{\Lambda^c}}}]E_0^\eta[\varphi_{X'_{\tau'_{\Lambda^c}}}]\big]\,, \tag{8.42}$$

where X and X' are two independent symmetric simple random walks, starting respectively at i and j, $\mathbb{P}_{i,j}$ denotes their joint distribution, and τ'_{Λ^c} is the first exit time of X' from Λ. As was done earlier,

$$\mathbb{E}_{i,j}\big[E_0^\eta[\varphi_{X_{\tau_{\Lambda^c}}}]E_0^\eta[\varphi_{X'_{\tau'_{\Lambda^c}}}]\big] = \mathbb{E}_{i,j}[\eta_{X_{\tau_{\Lambda^c}}}\eta_{X'_{\tau'_{\Lambda^c}}}] = \mathbb{E}_i[\eta_{X_{\tau_{\Lambda^c}}}]\mathbb{E}_j[\eta_{X_{\tau_{\Lambda^c}}}] = \eta_i\eta_j\,.$$

Let us then define the modified Green function

$$K_\Lambda(i,j) \stackrel{\text{def}}{=} \mathbb{E}_i\Big[\sum_{n\geq\tau_{\Lambda^c}} \mathbf{1}_{\{X_n=j\}}\Big] = G(i,j) - G_\Lambda(i,j)\,.$$

Observe that $K_\Lambda(i,j) = K_\Lambda(j,i)$, since G and G_Λ are both symmetric; moreover, $K_\Lambda(i,j) = G(i,j)$ if $i \in \Lambda^c$. We can thus write

$$\begin{aligned}
\mathbb{E}_{i,j}\big[G(X_{\tau_{\Lambda^c}}, X'_{\tau'_{\Lambda^c}})\big] &= \sum_{k,l\in\partial^{\mathrm{ext}}\Lambda} \mathbb{P}_i(X_{\tau_{\Lambda^c}} = k)\mathbb{P}_j(X_{\tau_{\Lambda^c}} = l)G(k,l) \\
&= \sum_{l\in\partial^{\mathrm{ext}}\Lambda} \mathbb{P}_j(X_{\tau_{\Lambda^c}} = l)K_\Lambda(i,l) \\
&= \sum_{l\in\partial^{\mathrm{ext}}\Lambda} \mathbb{P}_j(X_{\tau_{\Lambda^c}} = l)K_\Lambda(l,i) \\
&= \sum_{l\in\partial^{\mathrm{ext}}\Lambda} \mathbb{P}_j(X_{\tau_{\Lambda^c}} = l)G(l,i) \\
&= K_\Lambda(j,i) = G(i,j) - G_\Lambda(i,j)\,.
\end{aligned}$$

We have thus shown that $\mathrm{Cov}_0^\eta\big((\varphi_i - a_i), (\varphi_j - a_j)\big) = G_\Lambda(i,j)$, which implies that

$$E_0^\eta\big[e^{it_\Lambda\cdot\varphi_\Lambda} \mid \mathscr{F}_{\Lambda^c}\big] = e^{it_\Lambda\cdot a_\Lambda} e^{-\frac{1}{2}t_\Lambda\cdot G_\Lambda t_\Lambda}\,.$$

This shows that, under $\mu_0^\eta(\cdot \mid \mathscr{F}_{\Lambda^c})$, φ_Λ is Gaussian with distribution given by $\mu^\eta_{\Lambda;0}(\cdot)$. We have thereby proved (8.39) and Theorem 8.21. □

The proof given above that the limiting Gaussian field belongs to $\mathscr{G}(0)$ only depends on having a convergent expression for the Green function of the associated random walk; it will be used again in the massive case.

8.5 The Massive Case

A similar analysis, based on a Gaussian description of the finite-volume Gibbs distribution, holds in the massive case $m > 0$. Nevertheless, the presence of a mass term in the Hamiltonian leads to a change in the probabilistic interpretation, which eventually leads to a completely different behavior.

Consider the Hamiltonian $\mathscr{H}_{\Lambda;m}$, which contains the term $\frac{m^2}{2}\sum_{i\in\Lambda}\varphi_i^2$. To express $\mathscr{H}_{\Lambda;m}$ as a scalar product involving the inverse of a covariance matrix, we use (8.19), but this time including the mass term. After rearrangement, this yields

$$\begin{aligned}
\mathscr{H}_{\Lambda;m} = {}& \tfrac{1}{2}(\varphi - u)\cdot\big(-\tfrac{1}{2d}\Delta_\Lambda + m^2\big)(\varphi - u) \\
&+ \sum_{i\in\Lambda}\varphi_i \underbrace{\big((-\tfrac{1}{2d}\Delta + m^2)u\big)_i}_{\text{(i)}} + \tfrac{1}{2d}\sum_{i\in\Lambda}\sum_{j\in\Lambda^c, j\sim i}\varphi_i\underbrace{(u_j - \varphi_j)}_{\text{(ii)}} + B_\Lambda\,.
\end{aligned} \tag{8.43}$$

As before, we choose u so as to cancel the extra terms on the second line. The mean $u = (u_i)_{i\in\mathbb{Z}^d}$ we are after must solve a modified Dirichlet problem. Let us say that u is **m-harmonic on Λ (resp. $\mathbb{Z}^d$)** if

$$\big((-\tfrac{1}{2d}\Delta + m^2)u\big)_i = 0\,, \quad \forall i \in \Lambda \text{ (resp. } i \in \mathbb{Z}^d)\,.$$

We say that u **solves the massive Dirichlet problem in Λ** if

$$\begin{cases} u \text{ is } m\text{-harmonic on } \Lambda\,, \\ u_j = \eta_j \text{ for all } j \in \Lambda^c\,. \end{cases} \tag{8.44}$$

We will give a probabilistic solution of the massive Dirichlet problem, by again representing the scalar product in the Hamiltonian, using a different random walk.

Exercise 8.9. Verify that the solution to (8.44) (whose existence will be proved below) is unique.

When $d = 1$, one can determine all m-harmonic functions explicitly:

Exercise 8.10. Show that all m-harmonic functions on $\mathbb{Z}$ are of the following type: $u_k = Ae^{\alpha k} + Be^{-\alpha k}$, where $\alpha \overset{\text{def}}{=} \log(1 + m^2 + \sqrt{2m^2 + m^4})$.

8.5.1 Random Walk Representation

Consider a random walker on $\mathbb{Z}^d$ which, as before, only jumps to nearest neighbors but which, at each step, has a probability $\frac{m^2}{1+m^2} > 0$ of *dying*. That is, assume that, before taking each new step, the walker flips a coin with probability $P(\text{head}) = \frac{1}{1+m^2}$, $P(\text{tail}) = \frac{m^2}{1+m^2}$. If the outcome is a head, the walker survives and jumps to a nearest neighbor on $\mathbb{Z}^d$ uniformly, with probability $\frac{1}{2d}$. If the outcome is a tail, the walker dies (and remains dead for all subsequent times).

This process can be defined by considering

$$\mathbb{Z}^d_\star \overset{\text{def}}{=} \mathbb{Z}^d \cup \{\star\}\,,$$

where $\star \notin \mathbb{Z}^d$ is a new vertex which we call the **graveyard**. We define the following transition probabilities on $\mathbb{Z}^d_\star$:

$$P_m(i,j) \overset{\text{def}}{=} \begin{cases} \frac{1}{1+m^2}\frac{1}{2d} & \text{if } i,j \in \mathbb{Z}^d,\, i \sim j, \\ 1 - \frac{1}{1+m^2} & \text{if } i \in \mathbb{Z}^d \text{ and } j = \star, \\ 1 & \text{if } i = j = \star, \\ 0 & \text{otherwise.} \end{cases}$$

Let $Z = (Z_k)_{k \geq 0}$ denote the **killed random walk**, that is, the Markov chain on $\mathbb{Z}^d_\star$ associated with the transition matrix P_m.

We denote by $\mathbb{P}^m_i$ the distribution of the process Z starting at $i \in \mathbb{Z}^d$. By definition, $\star$ is an absorbing state for Z:

$$\forall n \geq 0\,, \quad \mathbb{P}^m_i(Z_{n+1} = \star \,|\, Z_n = \star) = 1\,.$$

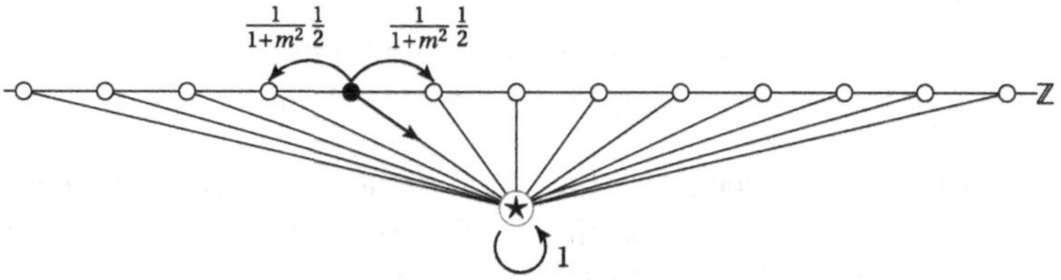

Figure 8.5. The one-dimensional symmetric simple random walk that has a probability $\frac{m^2}{1+m^2}$ of dying at each step.

Moreover, when $m > 0$, at all time $n \geq 0$ (up to which the walk has not yet entered the graveyard), the walk has a positive probability of dying:

$$\forall k \in \mathbb{Z}^d, \quad \mathbb{P}_i^m(Z_{n+1} = \star \mid Z_n = k) = \tfrac{m^2}{1+m^2} > 0\,.$$

Let

$$\tau_\star \stackrel{\text{def}}{=} \inf\{n \geq 0 : Z_n = \star\}$$

be the time at which the walker dies. Since $\mathbb{P}_i^m(\tau_\star > n) = (1 + m^2)^{-n}$, $\tau_\star$ is $\mathbb{P}_i^m$-almost surely finite.

Notice that, when $m = 0$, Z reduces to the symmetric simple walk considered in the previous section. The processes X and Z are in fact related by

$$\mathbb{P}_i^m(Z_n = j) = \mathbb{P}_i^m(\tau_\star > n)\mathbb{P}_i(X_n = j) = (1 + m^2)^{-n}\mathbb{P}_i(X_n = j)\,, \tag{8.45}$$

for all $i, j \in \mathbb{Z}^d$.

The process Z underlies the probabilistic representation of the mean and covariance of the finite-volume Gibbs distribution of the massive GFF:

Theorem 8.26. *Let $\Lambda \Subset \mathbb{Z}^d$, $d \geq 1$, and η be any boundary condition. Define $\eta_\star \stackrel{\text{def}}{=} 0$. Then, under $\mu^\eta_{\Lambda;m}$, $\varphi_\Lambda = (\varphi_i)_{i\in\Lambda}$ is Gaussian, with mean $u^m_\Lambda = (u^m_i)_{i\in\Lambda}$ given by*

$$u_i^m \stackrel{\text{def}}{=} \mathbb{E}_i^m[\eta_{Z_{\tau_{\Lambda^c}}}]\,, \quad \forall i \in \Lambda\,, \tag{8.46}$$

where $\tau_{\Lambda^c} \stackrel{\text{def}}{=} \inf\{k \geq 0 : Z_k \notin \Lambda\}$, and covariance matrix $G_{m;\Lambda} = (G_{m;\Lambda}(i,j))_{i,j\in\Lambda}$ given by

$$G_{m;\Lambda}(i,j) = \frac{1}{1+m^2}\mathbb{E}_i^m\Big[\sum_{n=0}^{\tau_{\Lambda^c}-1} \mathbf{1}_{\{Z_n=j\}}\Big]\,. \tag{8.47}$$

Exercise 8.11. Returning to the original Hamiltonian (8.2) with β not necessarily equal to 1, check that the mean and covariance matrix are given by

$$u_i^{\beta,m} = \beta^{-1/2}\, u_i^{m\beta^{-1/2}}\,, \qquad G_{\beta,m;\Lambda}(i,j) = \beta^{-1}\, G_{m\beta^{-1/2};\Lambda}(i,j)\,.$$

Observe that there are now two ways for the process Z to reach the exit time τ_{Λ^c}: either by stepping on a vertex $j \notin \Lambda$ or by dying.

Proof We proceed as in the massless case. First, it is easy to verify that u_i^m defined in (8.46) provides a solution to the massive Dirichlet problem (8.44). Then, we use (8.43), in which only the term involving $-\frac{1}{2d}\Delta_\Lambda + m^2$ remains. By introducing the restriction $P_{m;\Lambda} = (P_m(i,j))_{i,j\in\Lambda}$, we write

$$-\tfrac{1}{2d}\Delta_\Lambda + m^2 = (1+m^2)I_\Lambda - P_\Lambda = (1+m^2)\{I_\Lambda - P_{m;\Lambda}\}.$$

Since $P^k_{m;\Lambda}(i,j) = \mathbb{P}_i^m(Z_k = j, \tau_{\Lambda^c} > k)$ and, by (8.45),

$$\mathbb{P}_i^m(Z_k = j) \le (1+m^2)^{-k}, \tag{8.48}$$

we conclude, as before, that the matrix $I_\Lambda - P_{m;\Lambda}$ is invertible and that its inverse is given by the convergent series

$$G_{m;\Lambda} = \frac{1}{1+m^2}\big(I_\Lambda + P_{m;\Lambda} + P^2_{m;\Lambda} + \cdots\big).$$

Clearly, the entries of $G_{m;\Lambda}$ are exactly those given in (8.47). □

Example 8.27. Consider the one-dimensional massive GFF in $\{-n,\dots,n\}$, with a boundary condition η. Using Exercises 8.9 and 8.10, we can easily check that the solution to the Dirichlet problem with boundary condition η is given by $u_k^m = Ae^{\alpha k} + Be^{-\alpha k}$, where $\alpha = \log(1 + m^2 + \sqrt{2m^2 + m^4})$ and

$$A = \frac{\eta_{n+1}e^{\alpha(n+1)} - \eta_{-n-1}e^{-\alpha(n+1)}}{e^{2\alpha(n+1)} - e^{-2\alpha(n+1)}}, \qquad B = \frac{\eta_{-n-1}e^{\alpha(n+1)} - \eta_{n+1}e^{-\alpha(n+1)}}{e^{2\alpha(n+1)} - e^{-2\alpha(n+1)}}. \quad \diamond$$

Exercise 8.12. Let $(p(i))_{i\in\mathbb{Z}^d}$ be nonnegative real numbers such that $\sum_i p(i) = 1$. Consider the generalization of the GFF in which the Hamiltonian is given by

$$\frac{\beta}{2}\sum_{\substack{\{i,j\}\subset\mathbb{Z}^d\\ \{i,j\}\cap\Lambda\neq\varnothing}} p(j-i)(\omega_i-\omega_j)^2 + \frac{m^2}{2}\sum_{i\in\Lambda}\omega_i^2, \qquad \omega\in\Omega.$$

Show that the random walk representation derived above extends to this more general situation, provided that one replaces the simple random walk on $\mathbb{Z}^d$ by the random walk on $\mathbb{Z}^d$ with transition probabilities $p(\cdot)$.

8.5.2 Thermodynamic Limit

We can easily show that the massive GFF always has at least one infinite-volume Gibbs measure, in any dimension $d \ge 1$. Namely, the boundary condition $\eta \equiv 0$

is m-harmonic, so $u \equiv 0$ is the solution of the corresponding massive Dirichlet problem (8.44). Moreover,

$$G_m(i,j) \stackrel{\text{def}}{=} \lim_{n\to\infty} G_{m;\mathrm{B}(n)}(i,j) = \frac{1}{1+m^2}\sum_{n\geq 0}\mathbb{P}_i^m(Z_n=j)\,. \tag{8.49}$$

In view of (8.48), this series always converges when $m > 0$. By Theorem 8.6, this yields the existence of the Gaussian field with mean zero and covariance matrix G_m, whose distribution we denote by μ_m^0. As in the proof of Theorem 8.21, one then shows that its distribution μ_m^0 belongs to $\mathscr{G}(m)$. Of course, the same argument can be used starting with any m-harmonic function η on $\mathbb{Z}^d$; observe that Exercise 8.8 can be extended readily to the massive case, providing a description of all m-harmonic functions. We have therefore proved the following result:

Theorem 8.28. *In any dimension $d \geq 1$, the massive Gaussian Free Field possesses infinitely many infinite-volume Gibbs measures: $|\mathscr{G}(m)| = \infty$. More precisely, given any m-harmonic function η on $\mathbb{Z}^d$, there exists a Gaussian Gibbs measure μ_m^η with mean η and covariance matrix $G_m = (G_m(i,j))_{i,j\in\mathbb{Z}^d}$ given in* (8.49).

Remark 8.29. As in the massless case, it can be shown that m-harmonic functions parametrize extremal Gibbs measures: $\mathrm{ex}\,\mathscr{G}(m) = \{\mu_m^\eta : \eta \text{ is } m\text{-harmonic}\}$. ◇

In contrast to the massless case in dimension $d \geq 3$, in which $G(0,i)$ decreases algebraically when $\|i\|_2 \to \infty$, we will now see that the decay in the massive case is always exponential. Let us thus define the rate

$$\xi_m(i) \stackrel{\text{def}}{=} \lim_{\ell\to\infty} -\frac{1}{\ell}\log G_m(0,\ell i).$$

Proposition 8.30. *Let $d \geq 1$. For any $i \in \mathbb{Z}^d$, $\xi_m(i)$ exists and*

$$G_m(0,i) \leq G_m(0,0)e^{-\xi_m(i)}\,.$$

Moreover,

$$\log(1+m^2) \leq \frac{\xi_m(i)}{\|i\|_1} \leq \log(2d) + \log(1+m^2)\,. \tag{8.50}$$

Proof Let, for all $j \in \mathbb{Z}^d$, $\tau_j \stackrel{\text{def}}{=} \min\{n \geq 0 : Z_n = j\}$. Observe that

$$G_m(0,\ell i) = \mathbb{P}_0^m(\tau_{\ell i} < \tau_\star)G_m(\ell i,\ell i)\,.$$

Therefore, since $G_m(\ell i,\ell i) = G_m(0,0) < \infty$ for any $m > 0$,

$$\lim_{\ell\to\infty} -\frac{1}{\ell}\log G_m(0,\ell i) = \lim_{\ell\to\infty} -\frac{1}{\ell}\log\mathbb{P}_0^m(\tau_{\ell i} < \tau_\star)\,.$$

Now, for all $\ell_1, \ell_2 \in \mathbb{N}$, it follows from the strong Markov property that

$$\mathbb{P}_0^m(\tau_{(\ell_1+\ell_2)i} < \tau_\star) \geq \mathbb{P}_0^m(\tau_{\ell_1 i} < \tau_{(\ell_1+\ell_2)i} < \tau_\star) = \mathbb{P}_0^m(\tau_{\ell_1 i} < \tau_\star)\mathbb{P}_0^m(\tau_{\ell_2 i} < \tau_\star).$$

This implies that the sequence $\left(-\log \mathbb{P}_0^m(\tau_{\ell i} < \tau_\star)\right)_{\ell \geq 1}$ is subadditive; Lemma B.5 then guarantees the existence of $\xi_m(i)$, and provides the desired upper bound on $G_m(0,i)$, after taking $\ell = 1$.

Let us now turn to the bounds on $\xi_m(i)/\|i\|_1$. For the lower bound, we use (8.48):

$$\begin{aligned}(1+m^2)G_m(i,j) &= \sum_{n\geq 0} \mathbb{P}_i^m(Z_n = j)\\ &\leq \sum_{n \geq \|j-i\|_1} (1+m^2)^{-n} \leq \frac{1+m^2}{m^2}(1+m^2)^{-\|j-i\|_1}.\end{aligned}$$

For the upper bound, we can use

$$(1+m^2)G_m(i,j) \geq \mathbb{P}_i^m(\tau_j < \tau_\star) \geq (2d(1+m^2))^{-\|j-i\|_1},$$

where the second inequality is obtained by fixing an arbitrary shortest path from i to j and then forcing the walk to follow it. □

Using m-harmonic functions as a boundary condition allows one to construct infinitely many distinct Gibbs measures. It turns out, however, that if we only consider boundary conditions growing not too fast, then the corresponding Gaussian field is unique:

Theorem 8.31. *Let $d \geq 1$. For any boundary condition η satisfying*

$$\limsup_{k\to\infty} \max_{i:\, \|i\|_1 = k} \frac{\log|\eta_i|}{k} < \log(1+m^2), \tag{8.51}$$

the Gaussian Gibbs measure μ_m^η constructed in Theorem 8.28 is the same as the one obtained with the boundary condition $\eta \equiv 0$: $\mu_m^\eta = \mu_m^0$.

Since each m-harmonic function leads to a distinct infinite-volume Gibbs measure, Theorem 8.31 shows that the only m-harmonic function with subexponential growth is $\eta \equiv 0$. This is in sharp contrast with the massless case, for which distinct Gibbs measures can be constructed using boundary conditions of the form (8.36), in which η_i diverges linearly in $\|i\|_1$.

Proof of Theorem 8.31. It suffices to prove that $\lim_{n\to\infty} \mathbb{E}_i^m[\eta_{Z_{\tau_{B(n)^c}}}] = 0$ whenever η satisfies (8.51). Let $\epsilon > 0$ be such that $e^\epsilon/(1+m^2) < 1$ and n be large enough to ensure that $|\eta_i| \leq e^{\epsilon n}$ for all $i \in \partial^{\mathrm{ext}}B(n)$. Then,

$$\left|\mathbb{E}_i^m[\eta_{Z_{\tau_{B(n)^c}}}]\right| \leq e^{\epsilon n}\mathbb{P}_i^m(\tau_{B(n)^c} > d_1(i, B(n)^c)) \leq e^{\epsilon n}(1+m^2)^{-n+\|i\|_1},$$

which tends to 0 as $n \to \infty$. □

8.5.3 Limit $m \downarrow 0$

We have seen that, when $d = 1$ or $d = 2$, the large fluctuations of the field prevent the existence of any infinite-volume Gibbs measure for the massless GFF. It is thus natural to study how these large fluctuations build up as $m \downarrow 0$. One way to quantify the change in behavior as $m \downarrow 0$ (in dimensions 1 and 2) is to consider how fast the variance $\mathrm{Var}_m(\varphi_0)$ diverges and how the rate of exponential decay of the covariance $\mathrm{Cov}_m(\varphi_i, \varphi_j)$ decays to zero.

Divergence of the Variance in $d = 1, 2$

We first study the variance of the field in the limit $m \downarrow 0$, when $d = 1$ or 2.

Proposition 8.32. *Let φ be any massive Gaussian Free Field on $\mathbb{Z}^d$. Then, as $m \downarrow 0$,*

$$\mathrm{Var}_m(\varphi_0) \simeq \begin{cases} \frac{1}{\sqrt{2}m} & \text{in } d = 1\,, \\ \frac{2}{\pi}|\log m| & \text{in } d = 2\,. \end{cases} \tag{8.52}$$

Proof Let $e^{\lambda} = 1 + m^2$, and remember that

$$\mathrm{Var}_m(\varphi_0) = G_m(0,0) = (1+m^2)^{-1} \sum_{n \geq 0} e^{-\lambda n} \mathbb{P}_0(X_n = 0)\,.$$

We first consider the case $d = 1$. From the Local Limit Theorem (Theorem B.70), for all $\epsilon > 0$, there exists K_0 such that

$$\frac{1-\epsilon}{\sqrt{\pi k}} \leq \mathbb{P}_0(X_{2k} = 0) \leq \frac{1+\epsilon}{\sqrt{\pi k}}\,, \qquad \forall k \geq K_0\,. \tag{8.53}$$

This leads to the lower bound

$$\sum_{n \geq 0} e^{-\lambda n} \mathbb{P}_0(X_n = 0) \geq \frac{1-\epsilon}{\sqrt{\pi}} \sum_{k \geq K_0} \frac{e^{-2\lambda k}}{\sqrt{k}} \geq \frac{1-\epsilon}{\sqrt{\pi}} \int_{K_0}^{\infty} \frac{e^{-2\lambda x}}{\sqrt{x}}\, \mathrm{d}x = \frac{1-\epsilon}{\sqrt{2\lambda}} (1 - O(\sqrt{\lambda}))\,,$$

where we used the change of variable $2\lambda x \equiv y^2/2$. For the upper bound, we bound the first K_0 terms of the series by 1, and obtain

$$\sum_{k \geq 0} e^{-2\lambda k} \mathbb{P}_0(X_{2k} = 0) \leq K_0 + 1 + \frac{1+\epsilon}{\sqrt{\pi}} \sum_{k=K_0+1}^{\infty} \frac{e^{-2\lambda k}}{\sqrt{k}} \leq K_0 + 1 + \frac{1+\epsilon}{\sqrt{\pi}} \int_{K_0}^{\infty} \frac{e^{-2\lambda x}}{\sqrt{x}}\, \mathrm{d}x\,.$$

The case $d = 2$ is similar and is left as an exercise; the main difference is that the integral obtained cannot be computed explicitly. □

Rate of Decay for Small Masses

Proposition 8.30 shows that, as $m \to \infty$,

$$G_m(i,j) = e^{-2 \log m\,(1+o(1)) \|j-i\|_1}\,, \qquad \forall i \neq j \in \mathbb{Z}^d\,.$$

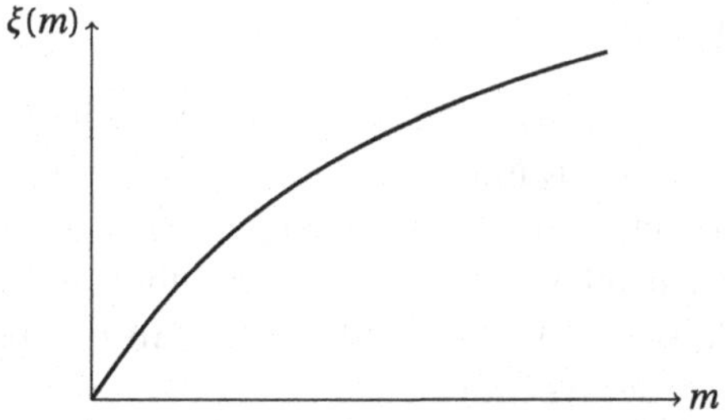

Figure 8.6. The rate of exponential decay ξ_m of the massive Green function in dimension 1.

It turns out that the rate of decay for small values of m has a very different behavior. We first consider the one-dimensional case, in which an exact computation can be made, valid for all $m > 0$:

Theorem 8.33. *Let $d = 1$ and $m > 0$. For all $i, j \in \mathbb{Z}^d$,*

$$G_m(i,j) = A_m \exp\bigl(-\xi_m |j - i|\bigr), \tag{8.54}$$

where $A_m, \xi_m > 0$ are given in (8.55). *In particular,* $\lim_{m\downarrow 0} \frac{\xi_m}{m} = \sqrt{2}$.

Proof Since $G_m(i,j) = G_m(0, j-i)$, it suffices to consider $i = 0$. Let $\lambda > 0$ be such that $e^\lambda = 1 + m^2$ and use (8.45) to write

$$(1+m^2)G_m(0,j) = \sum_{n\geq 0} e^{-\lambda n}\mathbb{P}_0(X_n = j) = \mathbb{E}_0\Bigl[\sum_{n\geq 0} e^{-\lambda n}\mathbf{1}_{\{X_n=j\}}\Bigr].$$

We then use a Fourier representation for the indicator: for all $j \in \mathbb{Z}$,

$$\mathbf{1}_{\{X_n=j\}} = \frac{1}{2\pi}\int_{-\pi}^{\pi} e^{\mathrm{i}k(X_n - j)}\,\mathrm{d}k.$$

The position of the symmetric simple random walk after n steps, X_n, can be expressed as a sum of independent identically distributed increments: $X_n = \xi_1 + \cdots + \xi_n$, with $\mathbb{P}_0(\xi_1 = \pm 1) = \frac{1}{2}$. Let $\phi(k) \stackrel{\text{def}}{=} \mathbb{E}_0[e^{\mathrm{i}k\xi_1}] = \cos(k)$ denote the characteristic function of the increment. Since the increments are independent, $E[e^{\mathrm{i}kX_n}] = \phi(k)^n$. Since $\lambda > 0$, we can interchange the sum and the integral and get

$$\begin{aligned}(1+m^2)G_m(0,j) &= \frac{1}{2\pi}\int_{-\pi}^{\pi} e^{-\mathrm{i}kj}\sum_{n\geq 0}(e^{-\lambda}\phi(k))^n\,\mathrm{d}k\\ &= \frac{1}{2\pi}\int_{-\pi}^{\pi}\frac{e^{-\mathrm{i}kj}}{1-e^{-\lambda}\phi(k)}\,\mathrm{d}k.\end{aligned}$$

We will study the behavior of this last integral using the residue theorem. To start, we look for the singularities of $z \mapsto \frac{e^{-\mathrm{i}zj}}{1-e^{-\lambda}\cos z}$ in the complex plane. Solving $\cos z = e^\lambda$, we find $z_\pm = \mathrm{i}t_\pm$, with $t_\pm = t_\pm(\lambda) = -\log(e^\lambda \mp \sqrt{e^{2\lambda}-1})$. Observe that $t_-(\lambda) <$

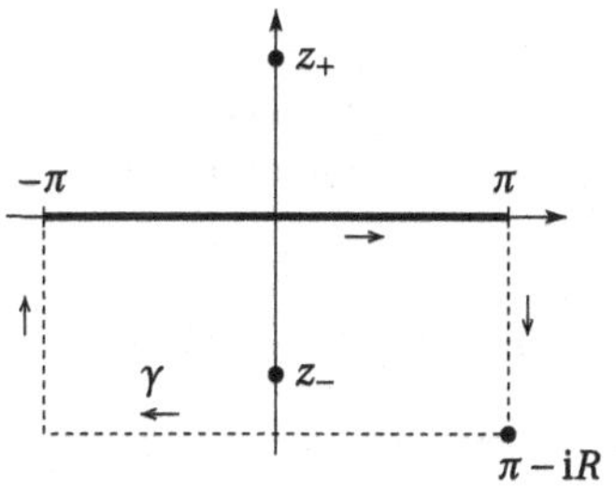

$0 < t_+(\lambda)$. Let γ denote the closed clockwise-oriented path in $\mathbb{C}$ depicted on the figure:

We decompose

$$\oint_\gamma = \int_{-\pi}^{\pi} + \int_{\pi}^{\pi-\mathrm{i}R} + \int_{\pi-\mathrm{i}R}^{-\pi-\mathrm{i}R} + \int_{-\pi-\mathrm{i}R}^{-\pi} .$$

Uniformly for all z on the path of integration from $\pi - \mathrm{i}R$ to $-\pi - \mathrm{i}R$, when R is large enough, $|1 - e^{-\lambda}\cos(z)| \geq e^{R-\lambda}/3$. Therefore, as $R \to \infty$,

$$\Big| \int_{\pi-\mathrm{i}R}^{-\pi-\mathrm{i}R} \frac{e^{-\mathrm{i}zj}}{1 - e^{-\lambda}\phi(z)} \, \mathrm{d}z \Big| \to 0 .$$

On the other hand, since the integrand is periodic, the integrals $\int_{\pi}^{\pi-\mathrm{i}R}$ and $\int_{-\pi-\mathrm{i}R}^{-\pi}$ cancel each other. By the residue theorem (since the path is oriented clockwise),

$$\begin{aligned} -\oint_\gamma \frac{e^{-\mathrm{i}zj}}{1 - e^{-\lambda}\cos(z)} \, \mathrm{d}z &= 2\pi\mathrm{i} \, \mathrm{Res}\Big(\frac{e^{-\mathrm{i}zj}}{1 - e^{-\lambda}\cos(z)}; z_- \Big) \\ &= 2\pi\mathrm{i} \lim_{z \to z_-} (z - z_-) \frac{e^{-\mathrm{i}zj}}{1 - e^{-\lambda}\cos(z)} . \end{aligned}$$

This yields

$$G_m(0, j) = \frac{e^{t_-(\lambda)j}}{\sinh |t_-(\lambda)|} \equiv A_m e^{-\xi_m j} , \tag{8.55}$$

with $\xi_m = \log(1 + m^2 + \sqrt{2m^2 + m^4})$. □

The previous result shows in particular that the rate of decay $\xi_m(i)/\|i\|_2$ behaves linearly in m as $m \downarrow 0$. We now extend this to all dimensions, using a more probabilistic approach, which has the additional benefit of shedding more light on the underlying mechanism.

Theorem 8.34. *There exist* $m_0 > 0$ *and constants* $0 < \alpha \leq \delta$ *such that, for all* $0 < m < m_0$ *and all* $i \in \mathbb{Z}^d$,

$$\alpha m \|i\|_2 \leq \xi_m(i) \leq \delta m \|i\|_2 . \tag{8.56}$$

Let us explain why this behavior should be expected. Let $j \in \mathbb{Z}^d$ (with $\|j\|_2$ large) and let τ_j (resp. $\tau_\star$) be the time at which the walk first reaches j (resp. dies). As we observed earlier,

$$G_m(0,j) = \mathbb{P}_0^m(\tau_j < \tau_\star)G_m(j,j) = \mathbb{P}_0^m(\tau_j < \tau_\star)G_m(0,0)\,. \tag{8.57}$$

On the one hand, it is unlikely that the walker survives for a time much longer than $1/m^2$. Indeed, for all $r > 0$ for which r/m^2 is an integer,

$$\mathbb{P}_0^m(\tau_\star > r/m^2) = (1+m^2)^{-r/m^2} \le e^{-r/2}\,, \tag{8.58}$$

for all sufficiently small m. On the other hand, in a time at most r/m^2, the walker typically cannot get to a distance further than r/m:

$$\begin{aligned}\mathbb{P}_0^m(\|Z_{r/m^2}\|_2 \ge r/m) &\le \mathbb{P}_0(\|X_{r/m^2}\|_2 \ge r/m)\\ &\le \frac{\mathbb{E}_0[\|X_{r/m^2}\|_2^2]}{r^2/m^2} = \frac{r/m^2}{r^2/m^2} = \frac{1}{r}\,.\end{aligned} \tag{8.59}$$

However, in order for a random walk started at 0 to reach j, such an event has to occur at least $\|j\|_2/(r/m)$ times. Therefore, the probability that the random walk reaches j should decay exponentially with $\|j\|_2/(r/m) = (m/r)\|j\|_2$. The proof below makes this argument precise. ◇

Proof *Lower bound.* Let $r \ge 8$ be such that r/m^2 is a positive integer and $m/r < 1$. Set $M \overset{\text{def}}{=} \lfloor \frac{m}{r}\|j\|_2 \rfloor$. Let us introduce the following sequence of random times: $T_0 \overset{\text{def}}{=} 0$ and, for $k > 0$,

$$T_k \overset{\text{def}}{=} \inf\{n > T_{k-1} \,:\, \|Z_n - Z_{T_{k-1}}\|_2 \ge r/m\}\,.$$

Note that, by definition, $T_M \le \tau_j$. Applying the strong Markov property at times $T_1, T_2, \dots, T_{M-1}$,

$$\mathbb{P}_0^m(\tau_j < \tau_\star) \le \prod_{k=0}^{M-1} \mathbb{P}_0^m(T_1 < \tau_\star) = \mathbb{P}_0^m(T_1 < \tau_\star)^M\,.$$

Following the heuristics described before the proof, we use the decomposition

$$\begin{aligned}\mathbb{P}_0^m(T_1 < \tau_\star) &= \mathbb{P}_0^m(T_1 < \tau_\star, T_1 \le r/m^2) + \mathbb{P}_0^m(T_1 < \tau_\star, T_1 > r/m^2)\\ &\le \mathbb{P}_0^m(T_1 \le r/m^2) + \mathbb{P}_0^m(\tau_* > r/m^2)\,.\end{aligned}$$

Now, on the one hand, it follows from (8.58) that $\mathbb{P}_0^m(\tau_* > r/m^2) \le e^{-r/2}$, which is smaller than $\frac{1}{4}$ by our choice of r. On the other hand,

$$\begin{aligned}\mathbb{P}_0^m(\|Z_{r/m^2}\|_2 \ge r/m) &\ge \mathbb{P}_0^m(\|Z_{r/m^2}\|_2 \ge r/m \mid T_1 \le r/m^2)\,\mathbb{P}_0^m(T_1 \le r/m^2)\\ &\ge \tfrac{1}{2}\mathbb{P}_0^m(T_1 \le r/m^2)\,,\end{aligned}$$

since, by symmetry,

$$\mathbb{P}_0^m\big(\|Z_\ell\|_2 \geq r/m \;\big|\; \|Z_k\|_2 \geq r/m\big) \geq \frac{1}{2},$$

for all $\ell \geq k$. Therefore, it follows from (8.59) that

$$\mathbb{P}_0^m(T_1 \leq r/m^2) \leq 2\mathbb{P}_0^m\big(\|Z_{r/m^2}\|_2 \geq r/m\big) \leq \frac{2}{r} \leq \frac{1}{4},$$

again by our choice of r. We conclude that

$$G_m(0,j) \leq 2G_m(0,0)\, e^{-(\log 2/r)m\|j\|_2}.$$

Upper bound. In (8.57), we write

$$\mathbb{P}_0^m(\tau_j < \tau_*) \geq \mathbb{P}_0(X_{[\|j\|_2/m]} = j)\, \mathbb{P}_0^m(\tau_* > \|j\|_2/m),$$

where we assume $[\|j\|_2/m]$ to be either $\lfloor\|j\|_2/m\rfloor$ or $\lfloor\|j\|_2/m\rfloor + 1$ in such a way that $\{Z_{[\|j\|_2/m]} = j\} \neq \varnothing$. The first factor in the right-hand side can then be estimated using the Local Limit Theorem, Theorem B.70. Namely, provided that m is sufficiently small, Theorem B.70 implies the existence of constants c_1, c_2 such that

$$\mathbb{P}_0(X_{[\|j\|_2/m]} = j) \geq \frac{e^{-c_1 m\|j\|_2}}{c_1(\|j\|_2/m)^{d/2}},$$

for all $j \in \mathbb{Z}^d$ with $\|j\|_2 > c_2$. Since

$$\mathbb{P}_0^m(\tau_* > \|j\|_2/m) = (1+m^2)^{-\lfloor\|j\|_2/m\rfloor} \geq e^{-m\|j\|_2},$$

the conclusion follows easily. □

8.6 Bibliographical References

The study of the Gaussian Free Field (often also called the *harmonic crystal* in the literature) was initiated in the 1970s. More details can be found in Chapter 13 of Georgii's book [134], in particular, proofs of the facts mentioned in Remarks 8.22 and 8.29, as well as an extensive bibliography. Some parts of Section 8.4.2 were inspired by the lecture notes of Spitzer [320].

8.7 Complements and Further Reading

8.7.1 Random Walk Representations

The random walk representation presented in this chapter (Theorems 8.17 and 8.26 and Exercise 8.12) can be extended in (at least) two directions.

In the first generalization, one replaces φ_i^2 in the mass term by a more general smooth function $U_i(\varphi_i)$ with a sufficiently fast growth. Building on earlier work by Symanzik [324], a generalization of the random walk representation to this context was first derived by Brydges, Fröhlich and Spencer in [56], which is still a nice place to learn about this material. Another source we recommend is the book [102] by Fernández, Fröhlich and Sokal, which also contains several important applications of this representation. In fact, as explained in these references, the spins φ_i themselves can be allowed to take values in $\mathbb{R}^\nu$, $\nu \geq 1$. Considering suitable sequences of functions $U_i^{(n)}$, this makes it possible to obtain random walk representations for the types of continuous spin models discussed in Chapters 9 and 10.

In the second generalization, it is the quadratic interaction $(\varphi_i - \varphi_j)^2$ that is replaced by a more general function $V(\varphi_i - \varphi_j)$ of the gradients. (Models of this type will be briefly considered in Section 9.3.) In this case, a generalization of the random walk representation was obtained by Helffer and Sjöstrand [158]. A good account can be found in Section 2 of the article [76] by Deuschel, Giacomin and Ioffe.

8.7.2 Gradient Gibbs States

As discussed in this chapter, the massless GFF delocalizes in dimensions 1 and 2, which leads to the absence of any Gibbs measure in the thermodynamic limit in those two cases.

Nevertheless, we have seen in Exercise 8.7 that, under $\mu^0_{\Lambda;0}$, the random vector $(\widetilde{\varphi}_i)_{i\in\Lambda}$, where $\widetilde{\varphi}_i \stackrel{\text{def}}{=} \varphi_i - \varphi_0$, is centered Gaussian with covariance matrix given by

$$\widetilde{G}_\Lambda(i,j) \stackrel{\text{def}}{=} G_\Lambda(i,j) - G_\Lambda(i,0) - G_\Lambda(0,j) + G_\Lambda(0,0).$$

It can be shown [209] that, as $\Lambda \uparrow \mathbb{Z}^d$, the limit of this quantity is given by the convergent series

$$\widetilde{G}(i,j) \stackrel{\text{def}}{=} \sum_{n\geq 0} \mathbb{P}_i(X_n = j, \tau_0 > n), \tag{8.60}$$

where $\tau_0 \stackrel{\text{def}}{=} \min\{n \geq 0 : X_n = 0\}$. In particular, the limiting Gaussian field is always well defined.

More generally, the joint distribution of the *gradients* $\varphi_i - \varphi_j$, $\{i,j\} \in \mathscr{E}_{\mathbb{Z}^d}$, remains well defined in all dimensions. It is thus possible to define Gibbs measures for this collection of random variables, instead of the original random variables φ_i, $i \in \mathbb{Z}^d$. This approach was pursued in a systematic way by Funaki and Spohn in [124], where the reader can find much more information. Other good source are Funaki's lecture notes [125] and Sheffield's thesis [302].

8.7.3 Effective Interface Models

As mentioned in the text, the GFF on $\mathbb{Z}^d$, as well as the more general class of gradient models, are often used as caricatures of the interfaces in more realistic lattice systems, such as the three-dimensional Ising model. Such caricatures are known as *effective interface models*. They are much simpler to analyze than the objects they approximate and their analysis yields valuable insights into the properties of the objects. In particular, they are used to study the effect of various external potentials or constraints on interfaces. More information on these topics can be found in the review article [46] by Bricmont, El Mellouki and Fröhlich, and in the lecture notes by Giacomin [136], Funaki [125] and Velenik [347]. In addition, the reader would probably also enjoy the older, but classical, review paper [107] by Fisher, although it only covers one-dimensional effective interface models.

8.7.4 Continuum Gaussian Free Field

In this chapter, we only considered the GFF on the lattice $\mathbb{Z}^d$. It turns out that it is possible to define an analogous model on $\mathbb{R}^d$. The latter object plays a crucial role in the analysis of the scaling limit of critical systems in two dimensions. Good introductions to this topic can be found in the review [303] by Sheffield and the lecture notes [348] by Werner.

8.7.5 A Link to Discrete Spin Systems

We saw in Section 2.5.2 how the Hubbard–Stratonovich transformation can be used to compute the pressure of the Curie–Weiss model. Let us use the same idea and explain how discrete spin systems can sometimes be expressed in terms of the GFF. The approach is very general but, for simplicity, we only consider an Ising ferromagnet with periodic boundary conditions. That is, we work on the torus $\mathbb{T}_n$, whose set of vertices is denoted V_n, as in Chapter 3.

Let us thus consider the Ising ferromagnet on $\mathbb{T}_n$, with the following Hamiltonian:

$$\mathscr{H}_{V_n;\mathbf{J},h} \stackrel{\text{def}}{=} -\tfrac{1}{2}\beta \sum_{i,j\in V_n} J_{ij}\sigma_i\sigma_j - h\sum_{i\in V_n}\sigma_i .$$

We will see that an interesting link can be made between the partition function $\mathbf{Z}^{\text{per}}_{V_n;\beta,\mathbf{J},h}$ and the GFF, provided that the coupling constants $\mathbf{J} = (J_{ij})_{i,j\in V_n}$ are well chosen.

The starting point is the following generalization of (2.20): for any positive-definite matrix $\Sigma = (\Sigma(i,j))_{1\le i,j\le N}$ and any vector $x = (x_1,\dots,x_N)\in\mathbb{R}^N$,

$$\exp\Bigl[\tfrac{1}{2}\sum_{i,j=1}^{N}\Sigma(i,j)x_ix_j\Bigr] = \bigl((2\pi)^N\det\Sigma\bigr)^{-1/2} \times \int_{-\infty}^{\infty}\mathrm{d}y_1\cdots\int_{-\infty}^{\infty}\mathrm{d}y_N \exp\Bigl[-\tfrac{1}{2}\sum_{i,j=1}^{N}\Sigma^{-1}(i,j)y_iy_j\Bigr]\exp\Bigl[\sum_{i=1}^{N}x_iy_i\Bigr]. \tag{8.61}$$

(See Exercise B.22.) We will apply this identity to the quadratic part of the Boltzmann weight of the ferromagnet introduced above, with $x_i \stackrel{\text{def}}{=} \sqrt{\beta}\sigma_i$, $\Sigma(i,j) \stackrel{\text{def}}{=} J_{ij}$. To establish a correspondence with the GFF, we choose $\mathbf{J}$ so that the inverse Σ^{-1} can be related to the GFF. Let us therefore take

$$J_{ij} \stackrel{\text{def}}{=} G_{m;\mathbb{T}_n}(i,j),$$

where $G_{m;\mathbb{T}_n}(i,j)$ denotes the massive Green function of the symmetric simple random walk $(X_n)_{n\geq 0}$ on $\mathbb{T}_n$, given by

$$G_{m;\mathbb{T}_n}(i,j) \stackrel{\text{def}}{=} \sum_{n\geq 0}(1+m^2)^{-n-1}\mathbb{P}_i(X_n=j)\,. \tag{8.62}$$

A straightforward adaptation of the proof of Theorem 8.26 shows that

$$(G_{m;\mathbb{T}_n})^{-1} = -\tfrac{1}{2d}\Delta + m^2\,,$$

where $\Delta = (\Delta_{ij})_{i,j\in\mathbb{T}_n}$ denotes the discrete Laplacian on $\mathbb{T}_n$, defined as in (8.16).

Notice that, even though the coupling constants J_{ij} defined above depend on n and involve long-range interactions, they converge as $n \to \infty$ and decay exponentially fast in $\|j-i\|_1$, uniformly in n, as can be seen from (8.62).

With this choice of coupling constants, (8.61) can be written as

$$\exp\Big[\tfrac{1}{2}\beta\sum_{i,j\in V_n} J_{ij}\sigma_i\sigma_j\Big] = \big((2\pi)^{|V_n|}\det G_{m;\mathbb{T}_n}\big)^{-1/2}$$
$$\times\int \exp\Big[-\tfrac{1}{2}\sum_{i,j\in V_n} y_i\Big(-\tfrac{1}{2d}\Delta_{ij}+m^2\Big)y_j\Big]\exp\Big[\beta^{1/2}\sum_{i\in V_n} y_i\sigma_i\Big]\prod_{i\in V_n}\mathrm{d}y_i\,,$$

where each y_i, $i\in V_n$, is integrated over $\mathbb{R}$. Since we recognize the Boltzmann weight of the massive centered GFF on $\mathbb{T}_n$, we get

$$\exp\Big[\tfrac{1}{2}\beta\sum_{i,j\in V_n} J_{ij}\sigma_i\sigma_j\Big] = \Big\langle \exp\Big[\beta^{1/2}\sum_{i\in V_n}\varphi_i\sigma_i\Big]\Big\rangle^{\mathrm{GFF}}_{V_n;\beta,m}\,.$$

We can now perform the summation over configurations in the partition function of the ferromagnet, which yields

$$\mathbf{Z}^{\mathrm{per}}_{V_n;\beta,\mathbf{J},h} = 2^{|V_n|}\Big\langle\prod_{i\in V_n}\cosh(\beta^{1/2}\varphi_i+h)\Big\rangle^{\mathrm{GFF,per}}_{V_n;m}\,.$$

Note that the numerator on the right-hand side corresponds to a massless GFF with an additional term $\sum_{i\in V_n} W(\varphi_i)$ in the Hamiltonian, where $W(\cdot)$ is an **external potential** defined by (see Figure 8.7)

$$W(x) \stackrel{\text{def}}{=} \tfrac{m^2}{2}x^2 - \log\cosh(\beta^{1/2}x+h)\,. \tag{8.63}$$

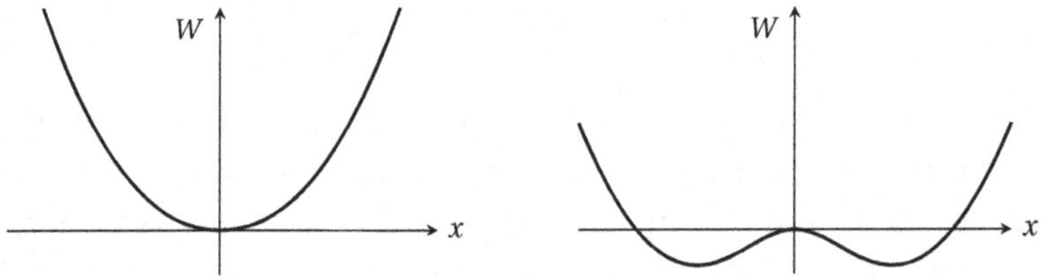

Figure 8.7. The external potential W with $m=1$ and $h=0$. Left: $\beta=0.5$. Right: $\beta=2$.

More generally, the same argument leads to a similar representation for any correlation function:

$$\langle \sigma_A \rangle^{\mathrm{per}}_{V_n;\beta,h} = \Big\langle \prod_{i \in A} \tanh\big(\beta^{1/2}\varphi_i + h\big) \Big\rangle^{\mathrm{GFF,per}}_{V_n;W},$$

where the measure is that of the massless GFF in the external potential W.

In a sense, the above transformation (sometimes called the **sine-Gordon** transformation) allows us to replace the discrete ± 1 spins of the Ising model by the continuous (and unbounded) spins of a Gaussian Free Field. A trace of the two values can still be seen in the resulting double-well potential (8.63) to which this field is submitted when β is sufficiently large; see Figure 8.7. Even though we will not make use of this in the present book, this continuous setting turns out to be very convenient when rigorously implementing the renormalization group approach. We refer to [57] for more information.

9 Models with Continuous Symmetry

In Chapter 3, we analyzed the phase transition occurring in the Ising model. We saw, in particular, that the change of behavior observed (when $h = 0$ and $d \geq 2$) as the inverse temperature β crosses the critical value $\beta_c = \beta_c(d)$ was associated with the spontaneous breaking of a discrete symmetry: when $\beta < \beta_c$, there is a unique infinite-volume Gibbs measure, invariant under a global spin flip (that is, interchange of all $+$ and $-$ spins); on the contrary, when $\beta > \beta_c$, uniqueness fails, and we proved the existence of two distinct infinite-volume Gibbs measures $\mu^{+}_{\beta,0}$ and $\mu^{-}_{\beta,0}$, which are not invariant under a global spin flip (since $\langle \sigma_0 \rangle^{+}_{\beta,0} > 0 > \langle \sigma_0 \rangle^{-}_{\beta,0}$).

Our goal in the present chapter is to analyze the effect of the existence of a *continuous* symmetry (that is, corresponding to a Lie group) on phase transitions. We will see that, in one- and two-dimensional models, a global continuous symmetry is in general never spontaneously broken. In this sense, continuous symmetries are more robust.

9.1 O(*N*)-Symmetric Models

The systems we consider in this chapter are models for which the spins are N-dimensional unit vectors, living at the vertices of $\mathbb{Z}^d$.

Let us thus fix some $N \in \mathbb{N}$, and define the single-spin space

$$\Omega_0 \stackrel{\text{def}}{=} \{v \in \mathbb{R}^N : \|v\|_2 = 1\} \equiv \mathbb{S}^{N-1}.$$

Correspondingly, the set of configurations in a finite set $\Lambda \Subset \mathbb{Z}^d$ (resp. in $\mathbb{Z}^d$) is given by

$$\Omega_\Lambda \stackrel{\text{def}}{=} \Omega_0^\Lambda \qquad (\text{resp. } \Omega = \Omega_0^{\mathbb{Z}^d}).$$

with each vertex $i \in \mathbb{Z}^d$, we associate the random variable $\mathbf{S_i} = (S_i^1, S_i^2, \dots, S_i^N)$ defined by

$$\mathbf{S}_i(\omega) \stackrel{\text{def}}{=} \omega_i,$$

which we call, as usual, the **spin** at i. We assume that spins interact only with their nearest neighbors and, most importantly, that *the interaction is invariant under simultaneous rotations of all the spins.* We can therefore assume that the interaction between two spins located at nearest-neighbor vertices i and j contributes an amount to the total energy that is a function of their scalar product $\mathbf{S}_i \cdot \mathbf{S}_j$.

Definition 9.1. Let $W\colon [-1,1] \to \mathbb{R}$. The **Hamiltonian of an $\mathsf{O}(N)$-symmetric model** in $\Lambda \Subset \mathbb{Z}^d$ is defined by

$$\mathscr{H}_{\Lambda;\beta} \stackrel{\text{def}}{=} \beta \sum_{\{i,j\} \in \mathscr{E}^{\mathrm{b}}_\Lambda} W(\mathbf{S}_i \cdot \mathbf{S}_j). \tag{9.1}$$

A particularly important class of models is given by the **$\mathsf{O}(N)$ models**, for which $W(x) = -x$:

$$\mathscr{H}_{\Lambda;\beta} = -\beta \sum_{\{i,j\} \in \mathscr{E}^{\mathrm{b}}_\Lambda} \mathbf{S}_i \cdot \mathbf{S}_j. \tag{9.2}$$

With this choice, different values of N then lead to different models, some of which have their own names. When $N = 1$, $\Omega_0 = \{\pm \mathbf{e}_1\}$ can be identified with $\{\pm 1\}$, so that the $O(1)$ model reduces to the Ising model. The case $N = 2$ corresponds to the *XY* **model**, and $N = 3$ corresponds to the **(classical) Heisenberg model.**

Given the Hamiltonian (9.1), we can define finite-volume Gibbs distributions and Gibbs measures in the usual way. We use the measurable structures on Ω_Λ and Ω, denoted respectively $\mathscr{F}_\Lambda$ and $\mathscr{F}$, introduced in Section 6.10. The reference measure for the spin at vertex i is the Lebesgue measure on $\mathbb{S}^{N-1}$, denoted simply $\mathrm{d}\omega_i$.

Given $\Lambda \Subset \mathbb{Z}^d$ and $\eta \in \Omega$, the Gibbs distribution of the $\mathsf{O}(N)$-symmetric models in Λ with boundary condition η is the probability measure $\mu^\eta_{\Lambda;\beta}$ on $(\Omega, \mathscr{F})$ defined by

$$\forall A \in \mathscr{F}, \quad \mu^\eta_{\Lambda;\beta}(A) \stackrel{\text{def}}{=} \int_{\Omega_\Lambda} \frac{e^{-\mathscr{H}_{\Lambda;\beta}(\omega_\Lambda \eta_{\Lambda^c})}}{\mathbf{Z}^\eta_{\Lambda;\beta}} \mathbb{1}_A(\omega_\Lambda \eta_{\Lambda^c}) \prod_{i \in \Lambda} \mathrm{d}\omega_i,$$

where the partition function is given by

$$\mathbf{Z}^\eta_{\Lambda;\beta} \stackrel{\text{def}}{=} \int_{\Omega_\Lambda} e^{-\mathscr{H}_{\Lambda;\beta}(\omega_\Lambda \eta_{\Lambda^c})} \prod_{i \in \Lambda} \mathrm{d}\omega_i.$$

As in Chapter 6, we can consider the specification associated with the kernels $(A, \eta) \mapsto \pi_\Lambda(A \mid \eta) \stackrel{\text{def}}{=} \mu^\eta_{\Lambda;\beta}(A)$, $\Lambda \Subset \mathbb{Z}^d$, and then denote by $\mathscr{G}(N)$ the set of associated infinite-volume Gibbs measures. (To lighten the notation, we do not indicate the dependence of $\mathscr{G}(N)$ on the choice of W and β.) Notice that Ω_0, and hence Ω, are compact, and so the results of Section 6.10.2 guarantee that the model has at least one infinite-volume Gibbs measure: $\mathscr{G}(N) \neq \varnothing$.

Even though our results below are stated in terms of infinite-volume Gibbs measures, the estimates in the proofs are actually valid for large finite systems. Therefore, readers not comfortable with the DLR formalism of Chapter 6 should be able to understand most of the content of this chapter.

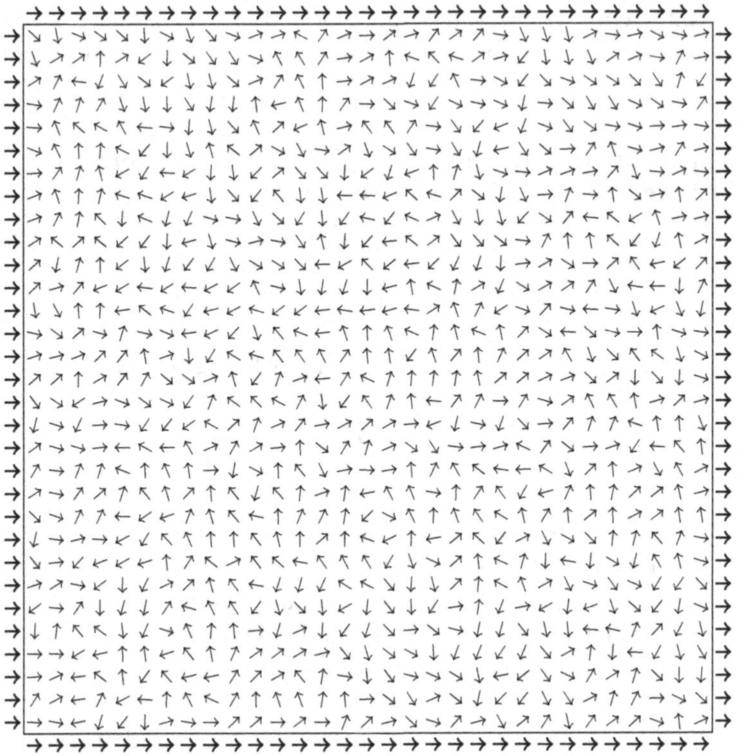

Figure 9.1. A configuration of the two-dimensional *XY* model with $\mathbf{e}_1$ boundary conditions, at high temperature: $\beta = 0.7$.

9.1.1 Overview

Inspired by what was done for the Ising model, one of our goals in this chapter will be to determine whether suitable boundary conditions can lead to **orientational long-range order**, that is, whether spins align macroscopically along a preferred direction, giving rise to a nonzero spontaneous magnetization. For the sake of concreteness, one can think of those Gibbs measures obtained by fixing a boundary condition η and taking the thermodynamic limit:

$$\mu^{\eta}_{\mathrm{B}(n);\beta} \Rightarrow \mu .$$

- *Dimensions* 1 *and* 2*;* $N \geq 2$. We will see that, when $d = 1$ or $d = 2$, under any measure $\mu \in \mathscr{G}(N)$, the distribution $\mu(\mathbf{S}_i \in \cdot)$ of each individual spin $\mathbf{S}_i$ is *uniform* on $\mathbb{S}^{N-1}$; in particular,

$$\langle \mathbf{S}_i \rangle_{\mu} = \mathbf{0} .$$

  Therefore, even in dimension 2 at very low temperature, orientational order *does not* occur in $\mathrm{O}(N)$-symmetric models. This is due, as will be seen, to the existence of order-destroying excitations of arbitrarily low energy (Proposition 9.7 below).

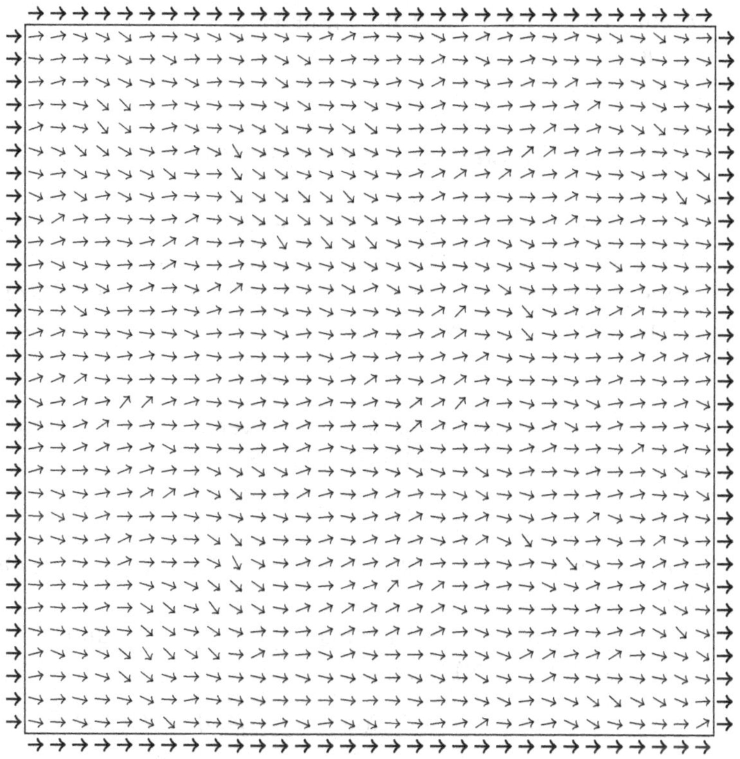

Figure 9.2. A configuration of the two-dimensional *XY* model with $\mathbf{e}_1$ boundary conditions, at low temperature: $\beta = 5$. In spite of the apparent ordering of the spins, we will prove that, in large enough systems, there is no orientational long-range order at any temperature.

The above will actually be a consequence of a more general result, the celebrated *Mermin–Wagner Theorem* (Theorem 9.2).

- *Dimension* 2; $N = 2$. Even though there is no orientational long-range order when $d = 2$, we will see (but not prove) that there is **quasi-long-range order** at low temperatures: the 2-point correlation function decays only algebraically with the distance:

$$\langle \mathbf{S}_i \cdot \mathbf{S}_j \rangle_\mu \approx \|j - i\|_2^{-C/\beta},$$

for some $C > 0$. This is in sharp contrast with $d = 1$ (for all $\beta \geq 0$), or with $d \geq 2$ at sufficiently high temperatures, where the 2-point correlation functions decay exponentially.

- *Dimensions* $d \geq 3$; $N \geq 2$. It turns out that spontaneous breaking of the continuous symmetry does indeed occur at low enough temperatures in the O(N) models, in dimensions $d \geq 3$, as will be discussed in Remark 9.5, and proved later in Chapter 10 (Theorem 10.25).

Additional information, including some outstanding open problems, can be found in the complements.

9.2 Absence of Continuous Symmetry Breaking

Symmetries in the study of Gibbs measures were described in Section 6.6. Let $R \in \mathsf{SO}(N)$ be any rotation on $\mathbb{S}^{N-1}$; R can of course be represented as an $N \times N$ orthogonal matrix of determinant 1. We can use R to define a global rotation r on a configuration $\omega \in \Omega$ by

$$(\mathsf{r}\omega)_i \stackrel{\text{def}}{=} R\,\omega_i\,, \qquad \forall i \in \mathbb{Z}^d\,.$$

A global rotation can also be defined on events $A \in \mathscr{F}$, by letting $\mathsf{r}A \stackrel{\text{def}}{=} \{\mathsf{r}\omega : \omega \in A\}$, as well as on functions and probability measures:

$$\mathsf{r}f(\omega) \stackrel{\text{def}}{=} f(\mathsf{r}^{-1}\omega)\,, \qquad \mathsf{r}(\mu)(A) \stackrel{\text{def}}{=} \mu(\mathsf{r}^{-1}A)\,.$$

We will often write $\mathsf{r} \in \mathsf{SO}(N)$, meaning that r is a global rotation associated with some element of $\mathsf{SO}(N)$.

By construction, the Hamiltonian (9.1) is invariant under global rotations of the spins: for all $\mathsf{r} \in \mathsf{SO}(N)$,

$$\mathscr{H}_{\Lambda;\beta}(\mathsf{r}\omega) = \mathscr{H}_{\Lambda;\beta}(\omega), \qquad \forall \omega \in \Omega\,. \tag{9.3}$$

Therefore, as a consequence of Theorem 6.45, $\mathscr{G}(N)$ is invariant under r: if $\mu \in \mathscr{G}(N)$, then $\mathsf{r}(\mu) \in \mathscr{G}(N)$. What Theorem 6.45 does not say is whether $\mathsf{r}(\mu)$ coincides with μ. A remarkable fact is that this is necessarily the case when $d = 1, 2$.

Theorem 9.2 (Mermin–Wagner Theorem). *Assume that $N \geq 2$, and that W is twice continuously differentiable. Then, when $d = 1$ or 2, all infinite-volume Gibbs measures are invariant under the action of $\mathsf{SO}(N)$: for all $\mu \in \mathscr{G}(N)$,*

$$\mathsf{r}(\mu) = \mu, \qquad \forall \mathsf{r} \in \mathsf{SO}(N)\,.$$

Of course, the claim is wrong when $N = 1$, since in this case the global spin flip symmetry can be broken at low temperature in $d = 2$. Let us make a few important comments.

Theorem 9.2 implies that, in an infinite system whose equilibrium properties are described by a Gibbs measure $\mu \in \mathscr{G}(N)$, the distribution of each individual spin $\mathbf{S}_i$ is uniform on $\mathbb{S}^{N-1}$. Namely, let $I \subset \mathbb{S}^{N-1}$; for any $\mathsf{r} \in \mathsf{SO}(N)$,

$$\mu(\mathbf{S}_i \in I) = \mathsf{r}(\mu)(\mathbf{S}_i \in I) = \mu(\mathbf{S}_i \in \mathsf{r}^{-1}(I))\,.$$

As a consequence, spontaneous magnetization (that is, some global orientation observed at the macroscopic level) cannot be observed in low-dimensional systems with continuous symmetries, even at very low temperature:

$$\langle \mathbf{S}_0 \rangle_\mu = \mathbf{0} \qquad (d = 1, 2)\,. \tag{9.4}$$

The above is in sharp contrast with the symmetry breaking observed in the two-dimensional Ising model at low temperature. There, when $h = 0$, the Hamiltonian was invariant under the discrete global spin flip, $\tau_{\text{g.s.f.}}$, but $\tau_{\text{g.s.f.}}(\mu^+_{\beta,0}) = \mu^-_{\beta,0} \neq \mu^+_{\beta,0}$ when $\beta > \beta_c(d)$.

Although this was not stated explicitly above, the $\mathsf{SO}(N)$-invariance of the infinite-volume Gibbs measures also implies absence of orientational long-range order. Namely, let $k \in \mathbb{Z}^d$ be fixed, far from the origin. Let n be large, but small enough to have $k \in \mathrm{B}(n)^c$ (for example: $n = \|k\|_\infty - 1$). If $\mu \in \mathscr{G}(N)$, then the DLR compatibility conditions $\mu = \mu\pi_\Lambda$, $\forall \Lambda \Subset \mathbb{Z}^d$, imply that

$$\langle \mathbf{S}_0 \cdot \mathbf{S}_k \rangle_\mu = \int \langle \mathbf{S}_0 \cdot \mathbf{S}_k \rangle^\eta_{\mathrm{B}(n);\beta}\, \mu(\mathrm{d}\eta) = \int \langle \mathbf{S}_0 \rangle^\eta_{\mathrm{B}(n);\beta} \cdot \mathbf{S}_k(\eta)\, \mu(\mathrm{d}\eta). \tag{9.5}$$

We will actually obtain a quantitative version of (9.4) in Proposition 9.7, a consequence of which will be that $\lim_{n\to\infty} \langle \mathbf{S}_0 \rangle^\eta_{\mathrm{B}(n);\beta} = \mathbf{0}$, uniformly in the boundary condition η (see Exercise 9.5). Therefore, by dominated convergence,

$$\langle \mathbf{S}_0 \cdot \mathbf{S}_k \rangle_\mu \to 0 \quad \text{when } \|k\|_\infty \to \infty .$$

Exercise 9.1. Prove that this also implies that

$$\lim_{n\to\infty} \langle \|m_{\mathrm{B}(n)}\|_2^2 \rangle_\mu = 0 ,$$

where $m_{\mathrm{B}(n)} \overset{\text{def}}{=} |\mathrm{B}(n)|^{-1} \sum_{i\in\mathrm{B}(n)} \mathbf{S}_i$ is the **magnetization density** in $\mathrm{B}(n)$.

Remark 9.3. As explained above, Theorem 9.2 implies the absence of spontaneous magnetization and long-range order. Nevertheless, this theorem does not imply that there is a unique infinite-volume Gibbs measure.[1] ◇

Remark 9.4. It is interesting to see what happens if one considers perturbations of the above models in which the continuous symmetry is explicitly broken. As an example, consider the **anisotropic** *XY* **model**, which has the same single-spin space as the *XY* model, but a more general Hamiltonian

$$\mathscr{H}_{\Lambda;\beta,\alpha} = -\beta \sum_{\{i,j\}\in\mathscr{E}^b_\Lambda} \{S^1_i S^1_j + \alpha S^2_i S^2_j\}$$

depending on an **anisotropy parameter** $\alpha \in [0,1]$. Observe that this Hamiltonian is $\mathsf{SO}(2)$-invariant only when $\alpha = 1$, in which case one recovers the usual *XY* model.

It turns out that there is always orientational long-range order at sufficiently low temperatures when $\alpha \in [0,1)$ and $d \geq 2$ (in $d = 1$, uniqueness always holds thanks to a suitable generalization of Theorem 6.40). Indeed, using reflection positivity, we will prove in Theorem 10.2 that, for any $\alpha \in [0,1)$ and all β sufficiently large, there exist at least two infinite-volume Gibbs measures μ^+ and μ^- such that

$$\langle \mathbf{S}_0 \cdot \mathbf{e}_1 \rangle_{\mu^+} > 0 > \langle \mathbf{S}_0 \cdot \mathbf{e}_1 \rangle_{\mu^-} .$$

This shows that having continuous spins is not sufficient to prevent orientational long-range order in low dimensions: the presence of a continuous symmetry is essential. ◇

Remark 9.5. Theorem 9.2 is restricted to dimensions 1 and 2. Let us briefly mention what happens in higher dimensions, restricting the discussion to the *XY* model: as soon as $d \geq 3$, for all β sufficiently large, there exist a number $m(\beta) > 0$ and a family of extremal infinite-volume Gibbs measures $(\mu_\beta^\psi)_{-\pi<\psi\leq\pi}$ such that

$$\langle \mathbf{S}_0 \rangle_\beta^\psi = m(\beta)\,(\cos\psi, \sin\psi)\,.$$

The proof of this claim (actually, for all values of N) will be given in Chapter 10.

Additional information on the role of the dimension, as well as on the corresponding results for $O(N)$ models with more general (not necessarily nearest-neighbor) interactions will be provided in Section 9.6.2. ◇

Remark 9.6. Both the proof of Theorem 9.2 and the heuristic argument below rely in a seemingly crucial way on the smoothness of the interaction W. The reader might thus wonder whether the latter is a necessary condition. It turns out that Theorem 9.2 can be extended to all piecewise continuous interactions W; see Section 9.6.2. ◇

9.2.1 Heuristic Argument

Before turning to the proof of Theorem 9.2, let us emphasize a crucial difference between continuous and discrete spin systems. For this heuristic discussion, W can be any twice continuously differentiable function, but we only consider the case $N = 2$ and, mostly, $d = 2$.

Let us therefore consider a two-dimensional $\mathrm{O}(2)$-symmetric model in the box $\mathrm{B}(n)$, with boundary condition $\eta_i = \mathbf{e}_1 = (1, 0)$ for all $i \in \mathrm{B}(n)^c$. If we assume that W is decreasing on $[-1, 1]$, then the ground state (that is, the configuration with the lowest energy) is the one that agrees everywhere with the boundary condition. We denote it by $\omega^{\mathbf{e}_1}$: $\omega_i^{\mathbf{e}_1} = \mathbf{e}_1$ for all i. We would like to determine the energetic cost of flipping the spin in the middle of the box. More precisely: among all configurations ω that agree with the boundary condition outside $\mathrm{B}(n)$ but in which the spin at the origin is flipped, $\omega_0 = -\mathbf{e}_1$, which one minimizes the Hamiltonian, and what is the corresponding value of the energy?

Remember that for the two-dimensional Ising model ($\mathrm{O}(N)$ with $N = 1$), the energetic cost required to flip the spin at the center of the box, with $+$ boundary condition, is at least 8β (since the shortest Peierls contour surrounding the origin has length 4), uniformly in the size of the box. Due to the presence of a continuous symmetry, the situation is radically different for the two-dimensional $\mathrm{O}(2)$-symmetric model: by slowly rotating the spins between the boundary and the center of the box, the spin at the origin can be flipped at an arbitrarily low cost (see Figure 9.3).

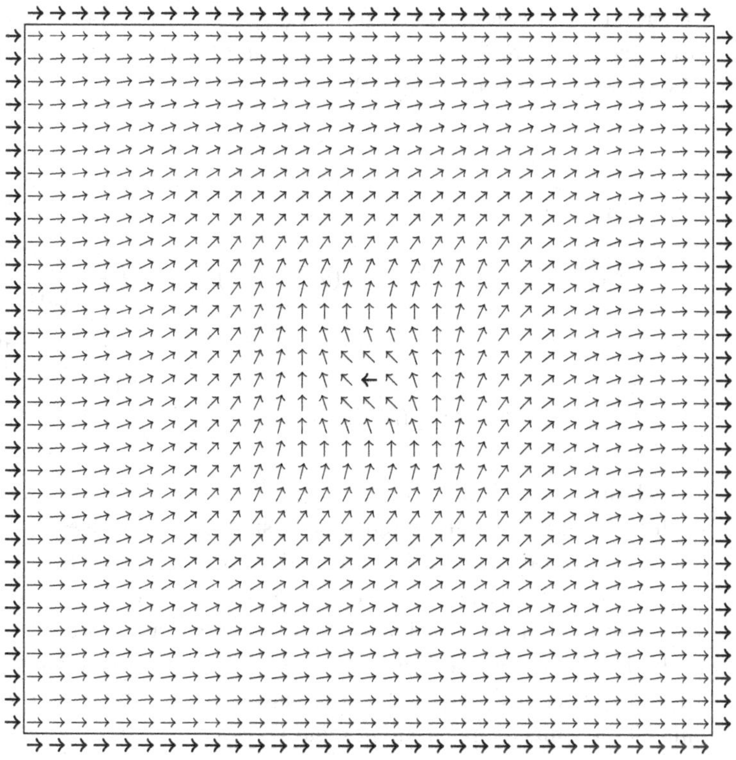

Figure 9.3. The spin wave ω^{SW} (see (9.6)), which flips the spin at the center of $\mathrm{B}(n)$, at a cost that can be made arbitrarily small by taking n large enough.

To understand this quantitatively, let us describe each configuration by the family $(\vartheta_i)_{i\in\mathbb{Z}^2}$, where $\vartheta_i \in (-\pi,\pi]$ is the angle such that $\mathbf{S}_i = (\cos\vartheta_i, \sin\vartheta_i)$. Let us also write $V(\theta) = W(\cos(\theta))$, so that

$$\mathscr{H}_{\mathrm{B}(n);\beta} = \beta \sum_{\{i,j\}\in\mathscr{E}^{\mathrm{b}}_{\mathrm{B}(n)}} W(\mathbf{S}_i\cdot\mathbf{S}_j) = \beta \sum_{\{i,j\}\in\mathscr{E}^{\mathrm{b}}_{\mathrm{B}(n)}} V(\vartheta_j - \vartheta_i)\,.$$

Let us consider the configuration $\omega^{\mathrm{SW}}_i = (\cos\theta^{\mathrm{SW}}_i, \sin\theta^{\mathrm{SW}}_i)$, where

$$\theta^{\mathrm{SW}}_i \stackrel{\text{def}}{=} \Bigl(1 - \frac{\log(1+\|i\|_\infty)}{\log(1+n)}\Bigr)\pi\,, \qquad i \in \mathrm{B}(n)\,, \tag{9.6}$$

and $\theta^{\mathrm{SW}}_i = 0$ for $i \notin \mathrm{B}(n)$ (see Figure 9.3). Clearly, the only nonzero contributions to $\mathscr{H}_{\mathrm{B}(n);\beta}(\omega^{\mathrm{SW}})$ are those due to pairs of neighboring vertices i and j such that $\|i\|_\infty = \|j\|_\infty - 1$. For each such pair,

$$\theta^{\mathrm{SW}}_i - \theta^{\mathrm{SW}}_j = \pi\frac{\log\Bigl(1+\frac{1}{\|j\|_\infty}\Bigr)}{\log(1+n)} \le \frac{\pi}{\log(1+n)}\frac{1}{\|j\|_\infty}\,.$$

Therefore, if n is large, each term $V(\theta_i^{\rm SW} - \theta_j^{\rm SW})$ can be estimated using a Taylor expansion of V at $\theta = 0$. Moreover, since V is twice continuously differentiable, there exists a constant C such that

$$\sup_{\theta \in (-\pi,\pi]} V''(\theta) \leq C, \tag{9.7}$$

and we have, since $V'(0) = 0$,

$$V(\theta_i^{\rm SW} - \theta_j^{\rm SW}) \leq V(0) + \tfrac{1}{2}C(\theta_i^{\rm SW} - \theta_j^{\rm SW})^2 \leq V(0) + \frac{C\pi^2}{2\,(\log(1+n))^2}\frac{1}{\|j\|_\infty^2}.$$

Summing over the contributing pairs of neighboring vertices i and j,

$$0 \leq \mathscr{H}_{\mathrm{B}(n)}(\omega^{\rm SW}) - \mathscr{H}_{\mathrm{B}(n)}(\omega^{\mathbf{e}_1}) \leq \frac{C\beta\pi^2}{2\,(\log(1+n))^2}\sum_{r=1}^{n+1} 4(2r-1)\frac{1}{r^2} \leq \frac{8C\beta\pi^2}{\log(1+n)},$$

which indeed tends to 0 when $n \to \infty$.

It is the existence of configurations such as $\omega^{\rm SW}$, representing collective excitations of arbitrarily low energy, called **spin waves**, which renders impossible the application of a naive Peierls-type argument. We will see that spin waves are the key ingredient in the proof of the Mermin–Wagner Theorem, given in Section 9.2.2.

In the above argument, we only flipped the spin located at the center of the box. It is easy to check that similar spin waves can also be constructed if one wants to flip all the spins in an extended region.

Exercise 9.2. ($d = 2$) Adapt the previous computation to show that the lowest energy required to flip all the spins in a smaller box $\mathrm{B}(\ell) \subset \mathrm{B}(n)$ goes to zero when $n \to \infty$.

Let us now briefly discuss what happens in dimensions $d \neq 2$. First, in the next exercise, the reader is encouraged to check that one can also construct a spin wave as above in dimension 1 (actually, one can take a much simpler one in that case).

Exercise 9.3. Construct a suitable spin wave for the one-dimensional $\mathrm{O}(2)$-symmetric model.

In higher dimensions, $d \geq 3$, it is not possible to repeat the argument given above. In the following exercise (see also Lemma 9.8), the reader is asked to show that the second-order term in the Taylor expansion remains bounded away from zero as $n \to \infty$.

Exercise 9.4. Let $f\colon \mathbb{Z}_{\geq 0} \to \mathbb{R}$ be such that $f(r) = 1$ if $0 \leq r \leq \ell$ and $f(r) = 0$ if $r \geq n$. Show that

$$\sum_{\{i,j\} \in \mathscr{E}^{\rm b}_{\mathrm{B}(n)}} \big(f(\|i\|_\infty) - f(\|j\|_\infty)\big)^2 \geq \Big\{\sum_{k \geq \ell} k^{-(d-1)}\Big\}^{-1}.$$

Conclude that, in contrast to the case $d = 2$ (Exercise 9.2), the minimal energy required to flip all spins in the box $\mathrm{B}(\ell)$ does not tend to zero when $n \to \infty$ when $d \geq 3$ (it is not even bounded in ℓ). *Hint:* To derive the inequality, use the Cauchy-Schwarz inequality for functions $g\colon \{0, \dots, n\} \to \mathbb{R}$.

9.2.2 Proof of the Mermin–Wagner Theorem for $N = 2$

We first give a proof of the result in the case $N = 2$, and then use it to address the general case in Section 9.2.3. We write $\mathbf{S}_i = (\cos\vartheta_i, \sin\vartheta_i)$ and set $V(\theta) \stackrel{\text{def}}{=} W(\cos(\theta))$, as in the heuristic argument above.

Let $\mu \in \mathscr{G}(2)$ and let $\mathsf{r}_\psi \in \mathrm{SO}(2)$ denote the rotation of angle $\psi \in (-\pi, \pi]$. To show that $\mathsf{r}_\psi(\mu) = \mu$, we will show that $\langle f\rangle_\mu = \langle \mathsf{r}_\psi f\rangle_\mu$ for each local bounded measurable function f. But, by the DLR compatibility conditions, we can write, for any $\Lambda \Subset \mathbb{Z}^d$,

$$|\langle f\rangle_\mu - \langle \mathsf{r}_\psi f\rangle_\mu| = \Big|\int \big\{\langle f\rangle^\eta_{\Lambda;\beta} - \langle \mathsf{r}_\psi f\rangle^\eta_{\Lambda;\beta}\big\}\,\mu(\mathrm{d}\eta)\Big| \leq \int \big|\langle f\rangle^\eta_{\Lambda;\beta} - \langle \mathsf{r}_\psi f\rangle^\eta_{\Lambda;\beta}\big|\,\mu(\mathrm{d}\eta)\,. \tag{9.8}$$

We study the differences $|\langle f\rangle^\eta_{\Lambda;\beta} - \langle \mathsf{r}_\psi f\rangle^\eta_{\Lambda;\beta}|$ quantitatively in the following proposition. In view of (9.8), Theorem 9.2 is a direct consequence of the following proposition:

Proposition 9.7. *Assume that $d = 1$ or $d = 2$ and fix $N = 2$. Under the hypotheses of Theorem 9.2, there exist constants c_1, c_2 such that, for any boundary condition $\eta \in \Omega$, any inverse temperature $\beta < \infty$, any angle $\psi \in (-\pi, \pi]$ and any $\ell \in \mathbb{Z}_{\geq 0}$,*

$$\big|\langle f\rangle^\eta_{\mathrm{B}(n);\beta} - \langle \mathsf{r}_\psi f\rangle^\eta_{\mathrm{B}(n);\beta}\big| \leq \beta^{1/2}|\psi|\,\|f\|_\infty \times \begin{cases} \frac{c_1}{\sqrt{n-\ell}} & \text{if } d = 1,\\ \frac{c_2\sqrt{\ell}}{\sqrt{\log(n-\ell)}} & \text{if } d = 2, \end{cases} \tag{9.9}$$

for all $n > \ell$ and all bounded functions f such that $\mathrm{supp}(f) \subset \mathrm{B}(\ell)$.

Below, we will use the notation $T_n(1) \stackrel{\text{def}}{=} \sqrt{n}$, $T_n(2) \stackrel{\text{def}}{=} \sqrt{\log n}$ when we want to treat the cases $d = 1$ and $d = 2$ simultaneously.

Exercise 9.5. Deduce from (9.9) that, under $\mu^\eta_{\mathrm{B}(n);\beta}$, the distribution of ϑ_0 converges to the uniform distribution on $(-\pi, \pi]$. In particular, for any η,

$$\lim_{n\to\infty} \big\|\langle \mathbf{S}_0\rangle^\eta_{\mathrm{B}(n);\beta}\big\|_2 = 0\,.$$

Most of the proof of the bounds (9.9) does not depend on the shape of the system considered. So, let us first consider an arbitrary connected Λ, which will later be taken to be the box $\mathrm{B}(n)$. Our starting point is to express

$$\langle \mathsf{r}_\psi f\rangle^\eta_{\Lambda;\beta} = (\mathbf{Z}^\eta_{\Lambda;\beta})^{-1} \int f(\mathsf{r}_{-\psi}\omega) e^{-\mathscr{H}_{\Lambda;\beta}(\omega_\Lambda \eta_{\Lambda^c})} \prod_{i\in\Lambda} \mathrm{d}\omega_i$$

as the expectation of f under a modified distribution.

We let Λ and ℓ be large enough so that $\Lambda \supset \mathrm{B}(\ell) \supset \mathrm{supp}(f)$:

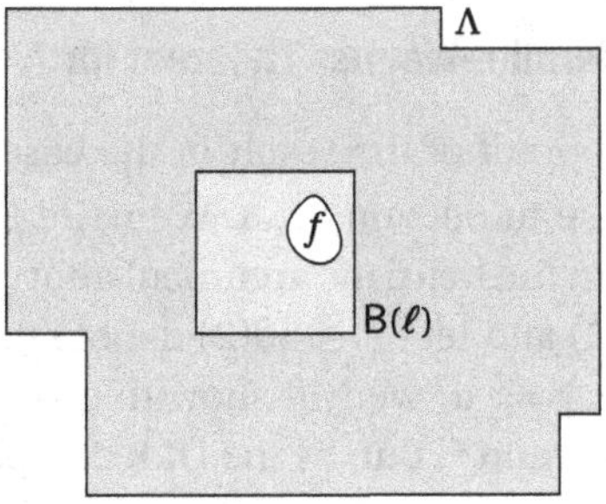

Let $\Psi\colon \mathbb{Z}^d \to (-\pi, \pi]$ satisfy $\Psi_i = \psi$ for all $i \in \mathrm{B}(\ell)$, and $\Psi_i = 0$ for all $i \notin \Lambda$. An explicit choice for Ψ will be made later. Let $\mathsf{t}_\Psi : \Omega \to \Omega$ denote the transformation under which

$$\vartheta_i(\mathsf{t}_\Psi \omega) = \vartheta_i(\omega) + \Psi_i\,, \quad \forall \omega \in \Omega\,.$$

That is, t_Ψ acts as the identity on spins located outside Λ and as the rotation r_ψ on spins located inside $\mathrm{B}(\ell)$. Observe that $\mathsf{t}_{-\Psi} = \mathsf{t}_\Psi^{-1}$. Now, since $\mathsf{t}_{-\Psi}\omega$ and $\mathsf{r}_{-\psi}\omega$ coincide on $\mathrm{supp}(f) \subset \mathrm{B}(\ell)$,

$$\begin{aligned} \int f(\mathsf{r}_{-\psi}\omega) e^{-\mathscr{H}_{\Lambda;\beta}(\omega_\Lambda \eta_{\Lambda^c})} \prod_{i\in\Lambda} \mathrm{d}\omega_i &= \int f(\mathsf{t}_{-\Psi}\omega) e^{-\mathscr{H}_{\Lambda;\beta}(\omega_\Lambda \eta_{\Lambda^c})} \prod_{i\in\Lambda} \mathrm{d}\omega_i \\ &= \int f(\omega) e^{-\mathscr{H}_{\Lambda;\beta}(\mathsf{t}_\Psi(\omega_\Lambda \eta_{\Lambda^c}))} \prod_{i\in\Lambda} \mathrm{d}\omega_i\,. \end{aligned}$$

In the second equality, we used the fact that the mapping $\omega_\Lambda \mapsto (\mathsf{t}_{-\Psi}\omega)_\Lambda$ has a Jacobian equal to 1. Let $\langle\cdot\rangle^{\eta,\Psi}_{\Lambda;\beta}$ denote the expectation under the probability measure

$$\mu^{\eta;\Psi}_{\Lambda;\beta}(A) \stackrel{\text{def}}{=} (\mathbf{Z}^{\eta;\Psi}_{\Lambda;\beta})^{-1} \int_{\Omega_\Lambda} e^{-\mathscr{H}_{\Lambda;\beta}(\mathsf{t}_\Psi(\omega_\Lambda \eta_{\Lambda^c}))} \mathbf{1}_A(\omega_\Lambda \eta_{\Lambda^c}) \prod_{i\in\Lambda} \mathrm{d}\omega_i\,, \quad A \in \mathscr{F}\,.$$

Observe that, for the same reasons as above (the Jacobian being equal to 1 and the boundary condition being preserved by t_Ψ), the partition function is actually left unchanged:

$$\mathbf{Z}^{\eta;\Psi}_{\Lambda;\beta} = \mathbf{Z}^\eta_{\Lambda;\beta}\,. \tag{9.10}$$

We can then write $\langle \mathsf{r}_\psi f\rangle^\eta_{\Lambda;\beta} = \langle f\rangle^{\eta,\Psi}_{\Lambda;\beta}$, and therefore

$$\bigl|\langle f\rangle^\eta_{\Lambda;\beta} - \langle \mathsf{r}_\psi f\rangle^\eta_{\Lambda;\beta}\bigr| = \bigl|\langle f\rangle^\eta_{\Lambda;\beta} - \langle f\rangle^{\eta;\Psi}_{\Lambda;\beta}\bigr|\,,$$

which reduces the problem to comparing the expectation of f under the measures $\mu^\eta_{\Lambda;\beta}$ and $\mu^{\eta;\Psi}_{\Lambda;\beta}$.

The measures $\mu^{\eta}_{\Lambda;\beta}$ and $\mu^{\eta;\Psi}_{\Lambda;\beta}$ only differ by the "addition" of the spin wave Ψ. However, we saw in Section 9.2.1 that the latter can be chosen such that its energetic cost is arbitrarily small. One would thus expect such excitations to proliferate in the system, and thus the two Gibbs distributions to be "very close" to each other. ◇

One convenient way of measuring the "closeness" of two measures μ, ν is the **relative entropy**:

$$\mathsf{h}(\mu \mid \nu) \overset{\text{def}}{=} \begin{cases} \left\langle \frac{\mathrm{d}\mu}{\mathrm{d}\nu} \log \frac{\mathrm{d}\mu}{\mathrm{d}\nu} \right\rangle_\nu & \text{if } \mu \ll \nu, \\ \infty & \text{otherwise,} \end{cases}$$

where $\frac{\mathrm{d}\mu}{\mathrm{d}\nu}$ is the Radon–Nikodym derivative of μ with respect to ν. The relevant properties of the relative entropy can be found in Appendix B.12. Particularly well suited to our needs, **Pinsker's inequality**, see Lemma B.67, states that, for any measurable function f with $\|f\|_\infty \le 1$,

$$|\langle f\rangle_\mu - \langle f\rangle_\nu| \le \sqrt{2\mathsf{h}(\mu \mid \nu)}\,.$$

In our case, thanks to (9.10),

$$\frac{\mathrm{d}\mu^{\eta}_{\Lambda;\beta}}{\mathrm{d}\mu^{\eta,\Psi}_{\Lambda;\beta}}(\omega) = e^{\mathscr{H}_{\Lambda;\beta}(\mathsf{t}_\Psi\omega) - \mathscr{H}_{\Lambda;\beta}(\omega)}.$$

Using Pinsker's inequality,

$$\begin{aligned} |\langle f\rangle^{\eta}_{\Lambda;\beta} - \langle f\rangle^{\eta,\Psi}_{\Lambda;\beta}| &\le \|f\|_\infty \sqrt{2\mathsf{h}(\mu^{\eta}_{\Lambda;\beta} \mid \mu^{\eta;\Psi}_{\Lambda;\beta})} \\ &= \|f\|_\infty \sqrt{2\big\langle \mathscr{H}_{\Lambda;\beta} \circ \mathsf{t}_\Psi - \mathscr{H}_{\Lambda;\beta}\big\rangle^{\eta}_{\Lambda;\beta}}\,. \end{aligned} \tag{9.11}$$

A second-order Taylor expansion yields, using again (9.7),

$$\begin{aligned} \big\langle \mathscr{H}_{\Lambda;\beta} \circ \mathsf{t}_\Psi - \mathscr{H}_{\Lambda;\beta}\big\rangle^{\eta}_{\Lambda;\beta} &= \beta \sum_{\{i,j\} \in \mathscr{E}^{\mathrm{b}}_\Lambda} \Big\langle V\big(\vartheta_j - \vartheta_i + \Psi_j - \Psi_i\big) - V\big(\vartheta_j - \vartheta_i\big)\Big\rangle^{\eta}_{\Lambda;\beta} \\ &\le \beta \sum_{\{i,j\} \in \mathscr{E}^{\mathrm{b}}_\Lambda} \Big\{ \big\langle V'\big(\vartheta_j - \vartheta_i\big)\big\rangle^{\eta}_{\Lambda;\beta} (\Psi_j - \Psi_i) + \tfrac{C}{2}(\Psi_j - \Psi_i)^2 \Big\}\,. \end{aligned}$$

Note the parallel with the heuristic discussion of Section 9.2.1. There, however, the first-order terms trivially vanished. We need an alternative way to see that the same occurs here, since the contribution of these terms would be too large to prove our claim. In order to get rid of them, we use the following trick: since the relative entropy is always nonnegative (Lemma B.65), we can write

$$\mathsf{h}(\mu^{\eta}_{\Lambda;\beta} \mid \mu^{\eta;\Psi}_{\Lambda;\beta}) \le \mathsf{h}(\mu^{\eta}_{\Lambda;\beta} \mid \mu^{\eta;\Psi}_{\Lambda;\beta}) + \mathsf{h}(\mu^{\eta}_{\Lambda;\beta} \mid \mu^{\eta;-\Psi}_{\Lambda;\beta}). \tag{9.12}$$

The second term on the right-hand side of the latter expression can be treated as above, and gives rise to the same first-order terms *but with the opposite sign.* These thus cancel, and we are left with (remember the notation $(\nabla\Psi)_{ij} \overset{\text{def}}{=} \Psi_j - \Psi_i$)

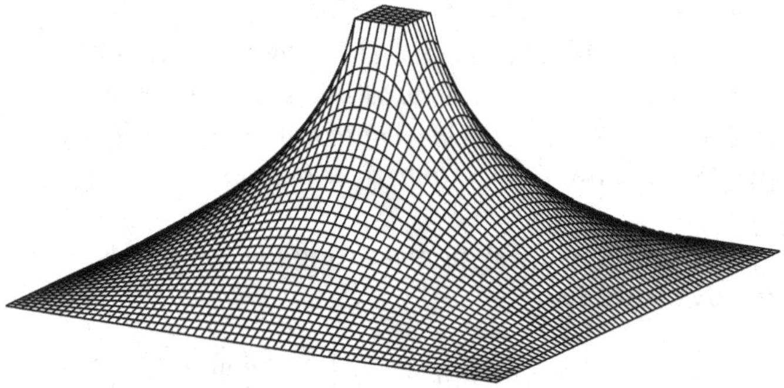

Figure 9.4. The minimizer of the Dirichlet energy given by (9.14). In this picture, $\Lambda = \mathrm{B}(30)$ and $\ell = 3$.

$$\mathsf{h}(\mu^{\eta}_{\Lambda;\beta} \,|\, \mu^{\eta;\Psi}_{\Lambda;\beta}) \leq C\beta \sum_{\{i,j\}\in\mathscr{E}^{\mathrm{b}}_{\Lambda}} (\nabla\Psi)^2_{ij} \,. \tag{9.13}$$

We will now choose the values Ψ_i, $i \in \Lambda \setminus \mathrm{B}(\ell)$. One possible choice is to take $\Lambda = \mathrm{B}(n)$, $n > \ell$, and to take for Ψ the spin wave introduced in Exercise 9.2. Nevertheless, since it is instructive, we will provide a more detailed study of the problem of minimizing the above sum under constraints. This will also shed some light on the role played by the dimension d.

Define the **Dirichlet energy** (in $\Lambda \setminus \mathrm{B}(\ell)$) of a function $\Psi \colon \mathbb{Z}^d \to \mathbb{R}$ by

$$\mathscr{E}(\Psi) \stackrel{\text{def}}{=} \tfrac{1}{2} \sum_{\{i,j\}\in\mathscr{E}^{\mathrm{b}}_{\Lambda\setminus\mathrm{B}(\ell)}} (\nabla\Psi)^2_{ij} \,.$$

We will determine the minimizer of $\mathscr{E}(\Psi)$ among all functions Ψ such that $\Psi \equiv \psi$ on $\mathrm{B}(\ell)$ and $\Psi \equiv 0$ on Λ^{c}. As we will see, in dimensions 1 and 2, the minimum value of that minimizer tends to zero when $\Lambda \uparrow \mathbb{Z}^d$.

Lemma 9.8. *The Dirichlet energy possesses a unique minimizer among all functions $u \colon \mathbb{Z}^d \to \mathbb{R}$ satisfying $u_i = 0$ for all $i \notin \Lambda$, and $u_i = 1$ for all $i \in \mathrm{B}(\ell)$. This minimizer is given by (see Figure 9.4)*

$$u^*_i \stackrel{\text{def}}{=} \mathbb{P}_i\big(X \text{ enters } \mathrm{B}(\ell) \text{ before exiting } \Lambda\big)\,, \tag{9.14}$$

where $X = (X_k)_{k\geq 0}$ is the symmetric simple random walk on $\mathbb{Z}^d$ and $\mathbb{P}_i(X_0 = i) = 1$. Moreover,

$$\mathscr{E}(u^*) = d \sum_{j\in\partial^{\mathrm{int}}\mathrm{B}(\ell)} \mathbb{P}_j\big(X \text{ exits } \Lambda \text{ before returning to } \mathrm{B}(\ell)\big)\,. \tag{9.15}$$

Proof of Lemma 9.8. Let us first characterize the critical points of $\mathscr{E}$. Namely, assume u is a critical point of $\mathscr{E}$, satisfying the constraints. Then we must have

$$\frac{\mathrm{d}}{\mathrm{d}s}\mathscr{E}(u + s\delta)\Big|_{s=0} = 0\,, \tag{9.16}$$

for all perturbations $\delta\colon \mathbb{Z}^d \to \mathbb{R}$ such that $\delta_i = 0$ for all $i \notin \Lambda \setminus \mathrm{B}(\ell)$. However, a simple computation yields

$$\frac{\mathrm{d}}{\mathrm{d}s}\mathscr{E}(u+s\delta)\Big|_{s=0} = \sum_{\{i,j\}\in\mathscr{E}^{\mathrm{b}}_{\Lambda\setminus\mathrm{B}(\ell)}} (\nabla u)_{ij}(\nabla\delta)_{ij}\,,$$

and, by the discrete Green identity (8.14),

$$\sum_{\{i,j\}\in\mathscr{E}^{\mathrm{b}}_{\Lambda\setminus\mathrm{B}(\ell)}} (\nabla u)_{ij}(\nabla\delta)_{ij} = -\sum_{i\in\Lambda\setminus\mathrm{B}(\ell)} \delta_i(\Delta u)_i + \sum_{\substack{i\in\Lambda\setminus\mathrm{B}(\ell)\\ j\notin\Lambda\setminus\mathrm{B}(\ell), j\sim i}} \delta_j(\nabla u)_{ij}\,. \tag{9.17}$$

The second sum in the right-hand side vanishes since $\delta_j = 0$ outside $\Lambda \setminus \mathrm{B}(\ell)$. In order for the first sum to be equal to zero for all δ, Δu must vanish everywhere on $\Lambda \setminus \mathrm{B}(\ell)$. This shows that the minimizer we are after is harmonic on $\Lambda \setminus \mathrm{B}(\ell)$. We know from Lemma 8.15 that the solution to the Dirichlet problem on $\Lambda \setminus \mathrm{B}(\ell)$ with boundary condition η is unique and given by

$$u_i^* \stackrel{\text{def}}{=} \mathbb{E}_i[\eta_{X_{\tau_{(\Lambda\setminus\mathrm{B}(\ell))^{\mathrm{c}}}}}]\,, \tag{9.18}$$

where $\mathbb{P}_i$ is the law of the simple random walk on $\mathbb{Z}^d$ with initial condition $X_0 = i$. Using our boundary condition ($\eta_i = 1$ on $\mathrm{B}(\ell)$, $\eta_i = 0$ on Λ^{c}), we easily write u_i^* as in (9.14).

We still have to check that u^* is actually a minimizer of the Dirichlet energy. But this follows from (9.17), since, for all δ as above, the latter implies that $\mathscr{E}(u^*+\delta) = \mathscr{E}(u^*) + \mathscr{E}(\delta) \geq \mathscr{E}(u^*)$.

Finally, using (9.17) with $u = \delta = u^*$,

$$\begin{aligned}
\mathscr{E}(u^*) &= \tfrac{1}{2}\sum_{i\in\Lambda\setminus\mathrm{B}(\ell)}\sum_{j\in\mathrm{B}(\ell), j\sim i} (u_j^* - u_i^*)\\
&= \tfrac{1}{2}\sum_{j\in\mathrm{B}(\ell)}\sum_{\substack{i\in\partial^{\mathrm{ext}}\mathrm{B}(\ell)\\ i\sim j}} \mathbb{P}_i\big(X \text{ exits } \Lambda \text{ before hitting } \mathrm{B}(\ell)\big)\\
&= \tfrac{1}{2}\sum_{j\in\mathrm{B}(\ell)}\sum_{i\sim j} \mathbb{P}_i\big(X \text{ exits } \Lambda \text{ before hitting } \mathrm{B}(\ell)\big)\\
&= d\sum_{j\in\partial^{\mathrm{int}}\mathrm{B}(\ell)} \mathbb{P}_j\big(X \text{ exits } \Lambda \text{ before returning to } \mathrm{B}(\ell)\big)\,,
\end{aligned}$$

where we used the Markov property for the fourth equality. □

We can now complete the proof of the Mermin–Wagner Theorem for $N = 2$:

Proof of Proposition 9.7. Take $\Lambda = \mathrm{B}(n)$. Let u^* be the minimizer (9.18) and set $\Psi = \psi u^*$. Observe that this choice of Ψ has all the required properties and that $\mathscr{E}(\Psi) = \psi^2\mathscr{E}(u^*)$. Using (9.11) and (9.13), we thus have

$$\big|\langle f\rangle^{\eta}_{\mathrm{B}(n);\beta} - \langle \mathrm{r}_{\psi} f\rangle^{\eta}_{\mathrm{B}(n);\beta}\big| \leq \|f\|_\infty\sqrt{4C\beta\psi^2\mathscr{E}(u^*)}\,.$$

Since

$$\begin{aligned}\mathbb{P}_j\bigl(X \text{ exits } \mathrm{B}(n) \text{ before returning to } \mathrm{B}(\ell)\bigr) \\ &\leq \mathbb{P}_j\bigl(X \text{ exits } \mathrm{B}(n-\ell)+j \text{ before returning to } j\bigr) \\ &= \mathbb{P}_0\bigl(X \text{ exits } \mathrm{B}(n-\ell) \text{ before returning to } 0\bigr),\end{aligned}$$

we finally get

$$\begin{aligned}\bigl|\langle f\rangle^{\eta}_{\mathrm{B}(n);\beta} - \langle \mathsf{r}_{\psi} f\rangle^{\eta}_{\mathrm{B}(n);\beta}\bigr| \leq & \|f\|_\infty \sqrt{4Cd\beta\psi^2\,|\partial^{\mathrm{int}}\mathrm{B}(\ell)|} \\ & \times \mathbb{P}_0\bigl(X \text{ exits } \mathrm{B}(n-\ell) \text{ before returning to } 0\bigr)^{1/2}.\end{aligned}$$

In dimensions $d = 1$ and $d = 2$, recurrence of the symmetric simple random walk implies that the latter probability goes to zero as $n \to \infty$. The rate at which this occurs is given in Theorem B.74. □

9.2.3 Proof of the Mermin–Wagner Theorem for $N \geq 3$

To prove Theorem 9.2 when $N \geq 3$, we essentially reduce the problem to the case $N = 2$. The main observation is that, given an arbitrary rotation $R \in \mathrm{SO}(N)$, there exists an orthonormal basis, an integer $n \leq N/2$ and n numbers $\psi_i \in (-\pi, \pi]$, such that R can be represented as a block diagonal matrix of the following form:[2]

$$\begin{pmatrix} M(\psi_1) & & & & \\ & M(\psi_2) & & & \\ & & \ddots & & \\ & & & M(\psi_n) & \\ & & & & I_{N-2n} \end{pmatrix},$$

where I_{N-2n} is the identity matrix of dimension $N - 2n$, and the matrix $M(\psi)$ is given by

$$M(\psi) = \begin{pmatrix} \cos\psi & -\sin\psi \\ \sin\psi & \cos\psi \end{pmatrix}.$$

In particular, R is the composition of n two-dimensional rotations. Therefore, it suffices to prove that any infinite-volume Gibbs measure μ is invariant under such a rotation. This can be achieved almost exactly as was done in the case $N = 2$, as we briefly explain now.

In view of the above, we can assume without loss of generality that R has the following block diagonal matrix representation:

$$\begin{pmatrix} M(\psi) & 0 \\ 0 & I_{N-2} \end{pmatrix},$$

for some $-\pi < \psi \leq \pi$. Let r be the global rotation associated with R. Since r only affects nontrivially the first two components S_i^1 and S_i^2 of the spins $\mathbf{S}_i$, we introduce the random variables r_i and ϑ_i, $i \in \mathbb{Z}^d$, such that

$$S_i^1 = r_i \cos\vartheta_i, \qquad S_i^2 = r_i \sin\vartheta_i.$$

(Notice that $r_i > 0$ almost surely, so that ϑ_i is almost surely well defined.)

As in the case $N = 2$, we consider an application $\Psi\colon \mathbb{Z}^d \to (-\pi,\pi]$ such that $\Psi_i = \psi$ for all $i \in \mathrm{B}(\ell)$, and $\Psi_i = 0$ for all $i \notin \mathrm{B}(n)$, and let $\mathtt{t}_\Psi : \Omega \to \Omega$ be the transformation such that $\vartheta_i(\mathtt{t}_\Psi\omega) = \vartheta_i(\omega) + \Psi_i$ for all configurations $\omega \in \Omega$.

From this point on, the proof is identical to the one given in Section 9.2.2. The only thing to check is that the relative entropy estimate still works in the same way. But $W(\mathbf{S}_i \cdot \mathbf{S}_j)$ is actually a function of $\vartheta_j - \vartheta_i$, r_i, r_j and the components S_i^l, S_j^l with $l \geq 3$. Since all these quantities except the first one remain constant under the action of $\mathtt{t}_\Psi$, and since the first one becomes $\vartheta_j - \vartheta_i + \Psi_j - \Psi_i$, the conclusion follows exactly as before. □

9.3 Digression on Gradient Models

Before turning to the study of correlations in $\mathrm{O}(N)$-symmetric models, we take advantage of the technique developed in the previous section to take a new look at the gradient models of Chapter 8.

We proved in Theorem 8.19 that the massless GFF possesses no infinite-volume Gibbs measures in dimensions 1 and 2, a consequence of the divergence of the variance of the field in the thermodynamic limit. Our aim here is to explain how this divergence can actually be seen as resulting from the presence of a continuous symmetry at the level of the Hamiltonian. As a by-product, this will allow us to extend the proof of nonexistence of infinite-volume Gibbs measures in dimensions 1 and 2 to a rather large class of models. So, only for this section, we switch to the models and notation of Chapter 8. In particular, spins take their values in $\mathbb{R}$, and Ω now represents $\mathbb{R}^{\mathbb{Z}^d}$.

Remember that gradient models have Hamiltonians of the form

$$\mathscr{H}_\Lambda = \sum_{\{i,j\}\in\mathscr{E}_\Lambda^{\mathrm{b}}} V(\varphi_j - \varphi_i),$$

where the inverse temperature has been included in $V\colon \mathbb{R} \to \mathbb{R}_{\geq 0}$. One must assume that V increases fast enough at infinity to make the finite-volume Gibbs measure well defined (that is, to make the partition function finite). The massless GFF corresponds to taking $V(x) = \frac{1}{4d}x^2$.

Let $t \in \mathbb{R}$, and consider the transformation $\mathtt{v}_t\colon \Omega \to \Omega$ defined by

$$(\mathtt{v}_t\omega)_i \stackrel{\text{def}}{=} \omega_i - t\,.$$

Since the interaction, by definition, depends only on the gradients $\omega_j - \omega_i$,

$$\mathscr{H}_\Lambda(\mathtt{v}_t\omega) = \mathscr{H}_\Lambda(\omega), \quad \forall t \in \mathbb{R}.$$

Of course, the setting here differs from the one we studied earlier in this chapter, in particular because the transformation group now is noncompact (it is actually isomorphic to $(\mathbb{R}, +)$). Let us assume for a moment that an analogue of the

Mermin–Wagner Theorem still applies in the present setting. Suppose also that μ is an infinite-volume Gibbs measure. We would then conclude that the distribution of φ_0 under μ should be uniform over $\mathbb{R}$, but then it would not be a probability distribution. This contradiction would show that such an infinite-volume Gibbs measure μ cannot exist! ◇

We will now show how the above can be turned into a rigorous argument. In fact, we will obtain (rather good) lower bounds on fluctuations for finite-volume Gibbs distributions:

Theorem 9.9. *($d = 1, 2$) Consider the gradient model introduced above, with $V\colon \mathbb{R} \to \mathbb{R}_{\geq 0}$ even, twice differentiable, satisfying $V(0) = 0$ and $\sup_{x\in\mathbb{R}} V''(x) < C < \infty$. Then there exists a constant $c > 0$ such that, for any boundary condition η, the following holds: for all $K > 0$, when n is large enough,*

$$\mu^{\eta}_{\mathrm{B}(n)}\big(|\varphi_0| > K\big) \geq \begin{cases} \frac{1}{c}\exp\{-c\,CK^2/n\} & \text{if } d = 1,\\ \frac{1}{c}\exp\{-c\,CK^2/\log n\} & \text{if } d = 2.\end{cases}$$

Exercise 9.6. Using Theorem 9.9, show that there exist no Gibbs measures for such gradient models in $d = 1, 2$. *Hint:* Argue as in the proof of Theorem 8.19.

Proof of Theorem 9.9. We assume that $\mu^{\eta}_{\mathrm{B}(n)}(\varphi_0 > 0) \geq \frac{1}{2}$ (if this fails, then consider the boundary condition $-\eta$). Let $T_n(d)$ be defined as in Exercise 9.5. To study $\mu^{\eta}_{\mathrm{B}(n)}(|\varphi_0| > K)$, we change $\mu^{\eta}_{\mathrm{B}(n)}$ into a new measure under which the event is likely to occur. Namely, let $\Psi\colon \mathbb{Z}^d \to \mathbb{R}$ be such that $\Psi_0 = K$, and $\Psi_i = 0$ when $i \notin \mathrm{B}(n)$. Then let $\mathsf{v}_\Psi : \Omega \to \Omega$ be defined by $(\mathsf{v}_\Psi\omega)_i \stackrel{\text{def}}{=} \omega_i - \Psi_i$. Let us consider the following deformed probability measure:

$$\mu^{\eta;\Psi}_{\mathrm{B}(n)}(A) \stackrel{\text{def}}{=} \mu^{\eta}_{\mathrm{B}(n)}(\mathsf{v}_\Psi A), \qquad \forall A \in \mathscr{F}.$$

Under this new measure, the event we are considering has probability

$$\mu^{\eta;\Psi}_{\mathrm{B}(n)}\big(|\varphi_0| > K\big) \geq \mu^{\eta;\Psi}_{\mathrm{B}(n)}\big(\varphi_0 > K\big) = \mu^{\eta}_{\mathrm{B}(n)}\big(\varphi_0 > 0\big) \geq \tfrac{1}{2}. \tag{9.19}$$

The probability of the same event under the original measure can be estimated by using the relative entropy inequality of Lemma B.68. This allows us to compare the probability of an event A under two different (nonsingular) probability measures μ, ν:

$$\mu(A) \geq \nu(A)\exp\left(-\frac{\mathsf{h}(\nu\,|\,\mu) + e^{-1}}{\nu(A)}\right).$$

Together with (9.19), this gives, in our case,

$$\mu^{\eta}_{\mathrm{B}(n)}\big(|\varphi_0| > K\big) \geq \tfrac{1}{2}\exp\big\{-2\big(\mathsf{h}(\mu^{\eta;\Psi}_{\mathrm{B}(n)}\,\big|\,\mu^{\eta}_{\mathrm{B}(n)}) + e^{-1}\big)\big\}.$$

To conclude, we must now choose Ψ so as to bound $\mathsf{h}(\mu^{\eta;\Psi}_{\mathrm{B}(n)} \mid \mu^{\eta}_{\mathrm{B}(n)})$ uniformly in n. Proceeding as in the proof of Theorem 9.7, we obtain

$$\mathsf{h}(\mu^{\eta;\Psi}_{\mathrm{B}(n)} \mid \mu^{\eta}_{\mathrm{B}(n)}) \leq C \sum_{\{i,j\}\in\mathscr{E}^{\mathrm{b}}_{\mathrm{B}(n)}} (\nabla\Psi)^2_{ij} = 2C\mathscr{E}(\Psi).$$

Lemma 9.8 thus implies that the choice of Ψ that minimizes $\mathscr{E}$ is

$$\Psi_i = K\,\mathbb{P}_i(X \text{ hits } 0 \text{ before exiting } \mathrm{B}(n)).$$

Moreover, for this choice of Ψ, it follows from (9.15) that

$$\begin{aligned}\mathscr{E}(\Psi) &= \tfrac{1}{2}K^2 \sum_{i\sim 0} \mathbb{P}_i(\mathrm{X} \text{ exits } \mathrm{B}(n) \text{ before hitting } 0) \\ &= dK^2\mathbb{P}_0(\mathrm{X} \text{ exits } \mathrm{B}(n) \text{ before returning to } 0).\end{aligned}$$

Since the latter probability is of order $T_n(d)^{-2}$ (Theorem B.74), this concludes the proof. □

9.4 Decay of Correlations

We have already seen the following consequence of Theorem 9.2: for any $\mu \in \mathscr{G}(2)$, there is no orientational long-range order in dimensions 1 and 2:

$$\langle \mathbf{S}_i \cdot \mathbf{S}_j \rangle_\mu \to 0, \quad \|j-i\|_2 \to \infty.$$

The estimates in the proof of Proposition 9.7 can be used to provide some information on the speed at which these correlations decay to zero. Namely, using (9.5) and Exercise 9.5 with $n = \|j-i\|_\infty - 1$, one obtains the upper bound

$$\left|\langle \mathbf{S}_i \cdot \mathbf{S}_j \rangle_\mu\right| \leq \begin{cases} \frac{C}{\sqrt{\|j-i\|_\infty}} & \text{in } d = 1, \\ \frac{C}{\sqrt{\log \|j-i\|_\infty}} & \text{in } d = 2. \end{cases}$$

Unfortunately, these bounds are far from being optimal. In the next sections, we discuss various improvements.

9.4.1 One-Dimensional Models

For one-dimensional models, it can actually be proved that the 2-point function decays exponentially fast in $\|j-i\|_\infty$ for all $\beta < \infty$. In this section, we will prove this result for $\mathsf{O}(N)$ models.

There are several ways of obtaining this result; we will proceed by comparison with the Ising model, for which this issue has already been considered in Chapter 3. The main result we will use is the following simple inequality between 2-point functions in the $\mathsf{O}(N)$ and the Ising models.

Theorem 9.10. *For any $d \geq 1$, any $N \geq 1$, any $\beta \geq 0$, any Gibbs measure μ of the $\mathsf{O}(N)$ model at inverse temperature β on $\mathbb{Z}^d$,*

$$|\langle \mathbf{S}_0 \cdot \mathbf{S}_i \rangle_\mu| \leq N \langle \sigma_0 \sigma_i \rangle^{+,\text{Ising}}_{\beta,0} ,$$

where the expectation on the right-hand side is with respect to the Gibbs measure $\mu^+_{\beta,0}$ of the Ising model on $\mathbb{Z}^d$ at inverse temperature β and $h = 0$.

Proof Let n be large enough to ensure that $\{0, i\} \subset \mathrm{B}(n)$. By the DLR compatibility conditions,

$$\langle \mathbf{S}_0 \cdot \mathbf{S}_i \rangle_\mu = \big\langle \langle \mathbf{S}_0 \cdot \mathbf{S}_i \rangle^{\cdot}_{\mathrm{B}(n)} \big\rangle_\mu = \sum_{\ell=1}^{N} \big\langle \langle S^\ell_0 S^\ell_i \rangle^{\cdot}_{\mathrm{B}(n)} \big\rangle_\mu .$$

It is thus sufficient to prove that

$$|\langle S^1_0 S^1_i \rangle^\eta_{\mathrm{B}(n)}| \leq \langle \sigma_0 \sigma_i \rangle^{+,\text{Ising}}_{\beta,0} ,$$

for any boundary condition η.

Let $\sigma_j \stackrel{\text{def}}{=} S^1_j / |S^1_j| \in \{\pm 1\}$ (of course, $S^1_j \neq 0$, for all $j \in \mathrm{B}(n)$, almost surely). Since

$$S^1_j = |S^1_j| \sigma_j = \Big\{ 1 - \sum_{\ell=2}^{N} (S^\ell_j)^2 \Big\}^{1/2} \sigma_j ,$$

conditionally on the values of $S^2_j, \dots, S^N_j$, all the randomness in S^1_j is contained in the sign σ_j. Introducing the σ-algebra $\mathscr{F}^{\neq 1}_{\mathrm{B}(n)} \stackrel{\text{def}}{=} \sigma\big\{ S^\ell_j : j \in \mathrm{B}(n), \ell \neq 1 \big\}$, we can thus write

$$\langle S^1_0 S^1_i \rangle^\eta_{\mathrm{B}(n)} = \big\langle \langle S^1_0 S^1_i \mid \mathscr{F}^{\neq 1}_{\mathrm{B}(n)} \rangle^\eta_{\mathrm{B}(n)} \big\rangle = \big\langle |S^1_0|\, |S^1_i| \big\langle \sigma_0 \sigma_i \mid \mathscr{F}^{\neq 1}_{\mathrm{B}(n)} \big\rangle^\eta_{\mathrm{B}(n)} \big\rangle .$$

Observe now that the joint distribution of the random variables $(\sigma_j)_{j \in \mathrm{B}(n)}$ is given by an inhomogeneous Ising model in $\mathrm{B}(n)$, with Hamiltonian

$$\mathscr{H}_{\mathrm{B}(n);\mathbf{J}} \stackrel{\text{def}}{=} - \sum_{\{u,v\} \in \mathscr{E}^{\mathrm{b}}_{\mathrm{B}(n)}} J_{uv} \sigma_u \sigma_v ,$$

where the coupling constants are given by $J_{uv} \stackrel{\text{def}}{=} \beta |S^1_u|\, |S^1_v|$ and the boundary condition by $\bar\eta = (\eta^1_j / |\eta^1_j|)_{j \in \mathbb{Z}^d}$. Since $0 \leq J_{uv} \leq \beta$, it follows from Exercise 3.31 that, almost surely,

$$\big\langle \sigma_0 \sigma_i \,\big|\, \mathscr{F}^{\neq 1}_{\mathrm{B}(n)} \big\rangle^\eta_{\mathrm{B}(n)} \leq \langle \sigma_0 \sigma_i \rangle^{\bar\eta,\text{Ising}}_{\mathrm{B}(n);\mathbf{J}} \leq \langle \sigma_0 \sigma_i \rangle^{+,\text{Ising}}_{\mathrm{B}(n);\beta,0} .$$

For the lower bound, set $\bar J_{0j} = -J_{0j}$ for all $j \sim 0$ and $\bar J_{uv} = J_{uv}$ for all other pairs and use $\langle \sigma_0 \sigma_i \rangle^{\bar\eta,\text{Ising}}_{\mathrm{B}(n);\mathbf{J}} = -\langle \sigma_0 \sigma_i \rangle^{\bar\eta,\text{Ising}}_{\mathrm{B}(n);\bar{\mathbf{J}}} \geq -\langle \sigma_0 \sigma_i \rangle^{+,\text{Ising}}_{\mathrm{B}(n);\beta,0}$ as before. □

Applying this lemma in dimension 1, we immediately deduce the desired estimate from Exercise 3.25.

Corollary 9.11. *Let μ be the unique Gibbs measure of the $\mathsf{O}(N)$ model on $\mathbb{Z}$. Then, for any $0 \leq \beta < \infty$,*

$$|\langle \mathbf{S}_0 \cdot \mathbf{S}_i \rangle_\mu| \leq N (\tanh \beta)^{|i|} .$$

Alternatively, one can explicitly compute the 2-point function, by integrating one spin at a time.

Exercise 9.7. Consider the one-dimensional *XY* model at inverse temperature β. Let μ be its unique Gibbs measure. Compute the pressure and the correlation function $\langle \mathbf{S}_0 \cdot \mathbf{S}_i \rangle_\mu$ in terms of the **modified Bessel functions of the first kind**:

$$I_n(x) \stackrel{\text{def}}{=} \frac{1}{\pi} \int_0^\pi e^{x \cos t} \cos(nt) \mathrm{d}t \,.$$

Hint: Use free boundary conditions.

9.4.2 Two-Dimensional Models

We now investigate whether it is also possible to improve the estimate in dimension 2. To keep the matter as simple as possible, we only consider the *XY* model, although similar arguments apply for a much larger class of two-dimensional models, as described in Section 9.6.2.

Heuristic Argument

Let us start with some heuristic considerations, which lead to a conjecture on the rate at which $\langle \mathbf{S}_i \cdot \mathbf{S}_j \rangle_\mu$ should decrease to 0 at low temperature.

As before, we write the spin at i as $\mathbf{S}_i = (\cos \vartheta_i, \sin \vartheta_i)$. We are interested in the asymptotic behavior of

$$\langle \mathbf{S}_i \cdot \mathbf{S}_j \rangle_{\Lambda;\beta}^{\mathbf{e}_1} = \langle \cos(\vartheta_j - \vartheta_i) \rangle_{\Lambda;\beta}^{\mathbf{e}_1} = \langle e^{\mathrm{i}(\vartheta_j - \vartheta_i)} \rangle_{\Lambda;\beta}^{\mathbf{e}_1} ,$$

where the expectation is taken with boundary condition $\eta \equiv \mathbf{e}_1$; the last identity relies on the symmetry, which makes the imaginary part vanish.

At very low temperatures, most neighboring spins are typically nearly aligned, $|\vartheta_i - \vartheta_j| \ll 1$. In this regime, it makes sense to approximate the interaction term in the Hamiltonian using a Taylor expansion to second order:

$$-\beta \sum_{\{i,j\} \in \mathscr{E}_\Lambda^{\mathrm{b}}} \mathbf{S}_i \cdot \mathbf{S}_j = -\beta \sum_{\{i,j\} \in \mathscr{E}_\Lambda^{\mathrm{b}}} \cos(\vartheta_j - \vartheta_i) \cong -\beta |\mathscr{E}_\Lambda^{\mathrm{b}}| + \tfrac{1}{2}\beta \sum_{\{i,j\} \in \mathscr{E}_\Lambda^{\mathrm{b}}} (\vartheta_j - \vartheta_i)^2 .$$

We may also assume that, when β is very large, the behavior of the field is not much affected by replacing the angles ϑ_i, which take their values in $(-\pi, \pi]$, by variables φ_i taking values in $\mathbb{R}$, especially since we are interested in the expectation value of the 2π-periodic function $e^{\mathrm{i}(\varphi_j - \varphi_i)}$. This discussion leads us to conclude that the very-low-temperature properties of the *XY* model should be closely approximated by those of the GFF at inverse temperature 4β.

In particular, if we temporarily denote the expectations of the *XY* model with boundary condition $\eta_i \equiv \mathbf{e}_1$ by $\langle \cdot \rangle_{\Lambda;\beta}^{XY}$, and the expectation of the corresponding GFF with boundary condition $\eta_i \equiv 0$ by $\langle \cdot \rangle_{\Lambda,4\beta}^{GFF}$, we conclude that

$$\langle \mathbf{S}_i \cdot \mathbf{S}_j \rangle_{\Lambda;\beta}^{XY} = \langle e^{\mathrm{i}(\vartheta_j - \vartheta_i)} \rangle_{\Lambda;\beta}^{XY} \cong \langle e^{\mathrm{i}(\varphi_j - \varphi_i)} \rangle_{\Lambda;4\beta}^{GFF} .$$

Now, since $(\varphi_i)_{i\in\Lambda}$ is Gaussian, (8.8) gives

$$\langle e^{\mathrm{i}(\varphi_j-\varphi_i)}\rangle^{GFF}_{\Lambda;4\beta} = e^{-\frac{1}{8\beta}(G_\Lambda(i,i)+G_\Lambda(j,j)-2G_\Lambda(i,j))}, \tag{9.20}$$

where $G_\Lambda(i,j)$ is the Green function of the simple random walk in Λ (see Section 8.4.1). We will see at the end of the section that, as $\|j-i\|_2 \to \infty$,

$$\tfrac{1}{2}\lim_{\Lambda\uparrow\mathbb{Z}^2}(G_\Lambda(i,i)+G_\Lambda(j,j)-2G_\Lambda(i,j)) \simeq \frac{2}{\pi}\log\|j-i\|_2\,,$$

which leads to the following conjectural behavior for correlations at low temperatures:

$$\langle \mathbf{S}_i\cdot\mathbf{S}_j\rangle^{XY}_{\Lambda;\beta} \cong e^{-\frac{1}{2\pi\beta}\log\|j-i\|_2} = \|j-i\|_2^{-1/(2\pi\beta)}\,. \tag{9.21}$$

Algebraic Decay at Low Temperature

The following theorem provides, for large β, an essentially optimal upper bound of the type (9.21). The lower bound will be discussed (but not proved) in Section 9.6.1.

Theorem 9.12. *Let μ be an infinite-volume Gibbs measure associated with the two-dimensional XY model at inverse temperature β. For all $\epsilon > 0$, there exists $\beta_0(\epsilon) < \infty$ such that, for all $\beta > \beta_0(\epsilon)$ and all $i \neq j \in \mathbb{Z}^2$,*

$$|\langle \mathbf{S}_i\cdot\mathbf{S}_j\rangle_\mu| \leq \|j-i\|_2^{-(1-\epsilon)/(2\pi\beta)}\,.$$

Before turning to the proof, let us try to motivate the approach that will be used, which might otherwise seem rather uncanny. To do this, let us return to (9.20). To actually compute the expectation $\langle e^{\mathrm{i}(\varphi_j-\varphi_i)}\rangle^{GFF}_{\Lambda;4\beta}$, one should remember that $\varphi_j-\varphi_i$ has a normal distribution $\mathcal{N}(0,\sigma^2)$, with (see (8.6))

$$\sigma^2 = \frac{1}{4\beta}(G_\Lambda(i,i)+G_\Lambda(j,j)-2G_\Lambda(i,j))\,.$$

Its characteristic function can be computed by first completing the square:

$$\langle e^{\mathrm{i}(\varphi_j-\varphi_i)}\rangle^{GFF}_{\Lambda;4\beta} = \frac{1}{\sqrt{2\pi\sigma^2}}\int_{\mathbb{R}} e^{\mathrm{i}x-\frac{1}{2\sigma^2}x^2}\,\mathrm{d}x = \frac{1}{\sqrt{2\pi\sigma^2}}e^{-\frac{\sigma^2}{2}}\int_{\mathbb{R}} e^{-\frac{1}{2\sigma^2}(x-\mathrm{i}\sigma^2)^2}\,\mathrm{d}x\,.$$

Once the leading term $e^{-\frac{\sigma^2}{2}}$ is extracted, the remaining integral can be computed by translating the path of integration from $\mathbb{R}$ to $\mathbb{R}+\mathrm{i}\sigma^2$, an operation vindicated, through Cauchy's Integral Theorem, by the analyticity and rapid decay at infinity of the integrand:

$$\int_{\mathbb{R}} e^{-\frac{1}{2\sigma^2}(x-\mathrm{i}\sigma^2)^2}\,\mathrm{d}x = \int_{\mathbb{R}+\mathrm{i}\sigma^2} e^{-\frac{1}{2\sigma^2}(z-\mathrm{i}\sigma^2)^2}\,\mathrm{d}z = \int_{\mathbb{R}} e^{-\frac{1}{2\sigma^2}x^2}\,\mathrm{d}x = \sqrt{2\pi\sigma^2}\,.$$

The proof below follows a similar scheme, but applied directly to the *XY* spins instead of the GFF approximation.

Proof of Theorem 9.12. Without loss of generality, we consider $i = 0, j = k$. Similarly to what was done in (9.5), we first rely on the DLR property: for all n such that $\mathrm{B}(n) \ni k$,

$$\langle \mathbf{S}_0 \cdot \mathbf{S}_k \rangle_\mu = \int \langle \mathbf{S}_0 \cdot \mathbf{S}_k \rangle^\eta_{\mathrm{B}(n);\beta} \, \mu(\mathrm{d}\eta). \tag{9.22}$$

We will estimate the expectation on the right-hand side, uniformly in the boundary condition η. Observe first that

$$|\langle \mathbf{S}_0 \cdot \mathbf{S}_k \rangle^\eta_{\mathrm{B}(n);\beta}| = |\langle \cos(\vartheta_k - \vartheta_0) \rangle^\eta_{\mathrm{B}(n);\beta}| \le \left| \langle e^{\mathrm{i}(\vartheta_k - \vartheta_0)} \rangle^\eta_{\mathrm{B}(n);\beta} \right|,$$

since $|z| \ge |\mathfrak{Re}\, z| \; \forall z \in \mathbb{C}$. We will write the expectation $\langle \cdot \rangle^\eta_{\mathrm{B}(n);\beta}$ using explicit integrals over the angle variables $\vartheta_i \in (-\pi, \pi]$, $i \in \mathrm{B}(n)$. As a shorthand, we use the notation

$$\int_{-\pi}^{\pi} \cdots \int_{-\pi}^{\pi} \prod_{i \in \mathrm{B}(n)} \mathrm{d}\theta_i \equiv \int \mathrm{d}\theta_{\mathrm{B}(n)}.$$

Therefore,

$$\langle e^{\mathrm{i}(\vartheta_k - \vartheta_0)} \rangle^\eta_{\mathrm{B}(n);\beta} = \frac{1}{\mathbf{Z}^\eta_{\mathrm{B}(n);\beta}} \int \mathrm{d}\theta_{\mathrm{B}(n)} \exp\Big\{ \mathrm{i}(\theta_k - \theta_0) + \beta \sum_{\{i,j\} \in \mathscr{E}^{\mathrm{b}}_{\mathrm{B}(n)}} \cos(\theta_i - \theta_j) \Big\},$$

where we have set $\theta_i = \vartheta_i(\eta)$ for each $i \notin \mathrm{B}(n)$. Following the approach sketched before the proof, we add an imaginary part to the variables θ_j, $j \in \mathrm{B}(n)$. Since the integrand is clearly analytic, we can easily deform the integration path associated with the variable θ_j away from the real axis: we shift the integration interval from $[-\pi, \pi]$ to $[-\pi, \pi] + \mathrm{i}r_j$, where r_j will be chosen later (also as a function of β and n):

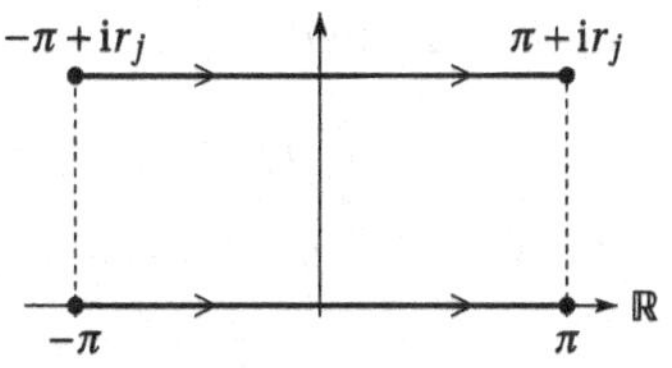

Figure 9.5. Shifting the integration path of θ_j. The shift depends on the vertex j, on n and on β.

Notice that the periodicity of the integrand guarantees that the contributions coming from the two segments connecting these two intervals cancel each other. We extend the r_is to a function $r: \mathbb{Z}^2 \to \mathbb{R}$, with $r_i = 0$ for all $i \notin \mathrm{B}(n)$. Observe now that

$$|e^{\mathrm{i}(\theta_k + \mathrm{i}r_k - \theta_0 - \mathrm{i}r_0)}| = e^{-(r_k - r_0)},$$
$$|e^{\cos(\theta_i + \mathrm{i}r_i - \theta_j - \mathrm{i}r_j)}| = e^{\cosh(r_i - r_j)\cos(\theta_i - \theta_j)}.$$

We thus have, letting ω_θ denote the spin configuration associated with the angles θ_i,

$$
\begin{aligned}
|\langle \mathbf{S}_0 \cdot \mathbf{S}_k \rangle^\eta_{\mathrm{B}(n);\beta}| &\le \frac{e^{-(r_k - r_0)}}{\mathbf{Z}^\eta_{\mathrm{B}(n);\beta}} \int \mathrm{d}\theta_{\mathrm{B}(n)} \exp\Big\{\beta \sum_{\{i,j\} \in \mathscr{E}^{\mathrm{b}}_{\mathrm{B}(n)}} \cosh(r_i - r_j) \cos(\theta_i - \theta_j)\Big\} \\
&= e^{-(r_k - r_0)} \int \mathrm{d}\theta_{\mathrm{B}(n)} \exp\Big\{\beta \sum_{\{i,j\} \in \mathscr{E}^{\mathrm{b}}_{\mathrm{B}(n)}} (\cosh(r_i - r_j) - 1) \cos(\theta_i - \theta_j)\Big\} \frac{e^{-\mathscr{H}_{\mathrm{B}(n);\beta}(\omega_\theta)}}{\mathbf{Z}^\eta_{\mathrm{B}(n);\beta}} \\
&= e^{-(r_k - r_0)} \Big\langle \exp\Big\{\beta \sum_{\{i,j\} \in \mathscr{E}^{\mathrm{b}}_{\mathrm{B}(n)}} (\cosh(r_i - r_j) - 1) \cos(\vartheta_i - \vartheta_j)\Big\}\Big\rangle^\eta_{\mathrm{B}(n);\beta} \\
&\le e^{-(r_k - r_0)} \exp\Big\{\beta \sum_{\{i,j\} \in \mathscr{E}^{\mathrm{b}}_{\mathrm{B}(n)}} (\cosh(r_i - r_j) - 1)\Big\}. \qquad (9.23)
\end{aligned}
$$

In the last inequality, we used the fact that $\cosh(r_i - r_j) \ge 1$ and $\cos(\vartheta_i - \vartheta_j) \le 1$. Assume that r can be chosen in such a way that

$$
|r_i - r_j| \le C/\beta, \quad \forall \{i,j\} \in \mathscr{E}^{\mathrm{b}}_{\mathrm{B}(n)}, \qquad (9.24)
$$

for some constant C. This allows us to replace the cosh term by a simpler quadratic term: given $\epsilon > 0$, we can assume that β_0 is large enough to ensure that $\beta \ge \beta_0$ implies $\cosh(r_i - r_j) - 1 \le \frac{1}{2}(1+\epsilon)(r_i - r_j)^2$ for all $\{i,j\} \in \mathscr{E}^{\mathrm{b}}_{\mathrm{B}(n)}$. In particular, we can write

$$
\sum_{\{i,j\} \in \mathscr{E}^{\mathrm{b}}_{\mathrm{B}(n)}} (\cosh(r_i - r_j) - 1) \le \tfrac{1}{2}(1+\epsilon) \sum_{\{i,j\} \in \mathscr{E}^{\mathrm{b}}_{\mathrm{B}(n)}} (r_i - r_j)^2 = (1+\epsilon)\mathscr{E}(r), \qquad (9.25)
$$

where $\mathscr{E}(\cdot)$ is the Dirichlet energy functional defined on maps $r \colon \mathbb{Z}^2 \to \mathbb{R}$ that vanish outside $\mathrm{B}(n)$. We thus have

$$
|\langle \mathbf{S}_0 \cdot \mathbf{S}_k \rangle^\eta_{\mathrm{B}(n);\beta}| \le \exp\{-\mathscr{D}(r)\}, \qquad (9.26)
$$

with $\mathscr{D}(\cdot)$ the functional defined by

$$
\mathscr{D}(r) \stackrel{\text{def}}{=} r_k - r_0 - \beta' \mathscr{E}(r),
$$

where we have set $\beta' \stackrel{\text{def}}{=} (1+\epsilon)\beta$. We now search for a maximizer of $\mathscr{D}$.

Lemma 9.13. *For a fixed $0 \ne k \in \mathrm{B}(n)$, the functional $\mathscr{D}$ possesses a unique maximizer r^* among the functions r that satisfy $r_i = 0$ for all $i \notin \mathrm{B}(n)$. That maximizer is the unique such function that satisfies*

$$
(\Delta r)_i = (\mathbb{1}_{\{i=0\}} - \mathbb{1}_{\{i=k\}})/\beta', \quad i \in \mathrm{B}(n). \qquad (9.27)
$$

It can be expressed explicitly as (see Figure 9.6)

$$
r^*_i = (G_{\mathrm{B}(n)}(i,k) - G_{\mathrm{B}(n)}(i,0))/(4\beta'), \quad i \in \mathrm{B}(n), \qquad (9.28)
$$

where $G_{\mathrm{B}(n)}(\cdot,\cdot)$ is the Green function of the symmetric simple random walk in $\mathrm{B}(n)$.

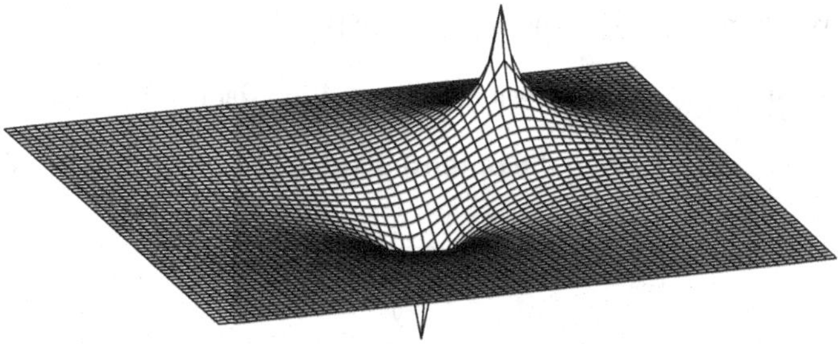

Figure 9.6. The maximizer (9.28) of the functional $\mathscr{D}$. In this picture, $\Lambda = \mathrm{B}(30)$ and $k = (12, 12)$.

Proof As in the proof of Lemma 9.8, we start by observing that a critical point r of $\mathscr{D}$ must be such that $\frac{\mathrm{d}}{\mathrm{d}s}\mathscr{D}(r+s\delta)\big|_{s=0} = 0$ for all perturbations $\delta\colon \mathbb{Z}^d \to \mathbb{R}$ that vanish outside $\mathrm{B}(n)$. But, as a straightforward computation shows,

$$\frac{\mathrm{d}}{\mathrm{d}s}\mathscr{D}(r+s\delta)\Big|_{s=0} = \delta_k - \delta_0 + \beta' \sum_{i\in\mathrm{B}(n)} \delta_i(\Delta r)_i$$
$$= \delta_k\big(1+\beta'(\Delta r)_k\big) - \delta_0\big(1-\beta'(\Delta r)_0\big) + \beta' \sum_{i\in\mathrm{B}(n)\setminus\{0,k\}} \delta_i(\Delta r)_i\,.$$

Since this sum of three terms must vanish for all δ, we see that r must satisfy (9.27). Since $r_i = 0$ outside $\mathrm{B}(n)$, we have $(\Delta r)_i = (\Delta_{\mathrm{B}(n)} r)_i$ (remember Remark 8.9), so that (9.27) can be written

$$(\Delta_{\mathrm{B}(n)} r)_i = (\mathbb{1}_{\{i=0\}} - \mathbb{1}_{\{i=k\}})/\beta'\,.$$

In Lemma 8.13, we saw that $G_{\mathrm{B}(n)}$ is precisely the inverse of $-\frac{1}{4}\Delta_{\mathrm{B}(n)}$. Therefore, multiplying by $G_{\mathrm{B}(n)}(j,i)$ on both sides of the previous display and summing over i gives (9.28). To prove that r^* actually maximizes $\mathscr{D}(\cdot)$, let δ be such that $\delta_i = 0$ outside $\mathrm{B}(n)$. Proceeding as we have already done several times before,

$$\mathscr{D}(r^*+\delta) = \mathscr{D}(r^*) - \beta'\mathscr{E}(\delta) + \delta_k - \delta_0 - \beta' \sum_{\{i,j\}\in\mathscr{E}^{\mathrm{b}}_{\mathrm{B}(n)}} (\nabla\delta)_{ij}(\nabla r^*)_{ij}$$
$$= \mathscr{D}(r^*) - \beta'\mathscr{E}(\delta) + \delta_k - \delta_0 + \beta' \sum_{i\in\mathrm{B}(n)} \delta_i \underbrace{(\Delta r^*)_i}_{\text{use (9.27)}}$$
$$= \mathscr{D}(r^*) - \beta'\mathscr{E}(\delta)\,.$$

Since $\mathscr{E}(\delta) \geq 0$, we conclude that $\mathscr{D}(r^*+\delta) \leq \mathscr{D}(r^*)$. □

It follows from Theorem B.77 that there exists C such that $|G_{\mathrm{B}(n)}(i,v) - G_{\mathrm{B}(n)}(j,v)| \leq 2C$, uniformly in n, in $v \in \mathrm{B}(n)$ and in $\{i,j\} \in \mathscr{E}^{\mathrm{b}}_{\mathrm{B}(n)}$. In particular, $|r^*_i - r^*_j| \leq C/\beta$, meaning that (9.24) is satisfied. We now use (9.26) with r^*. First, one easily verifies that

$$\mathscr{D}(r^*) = \tfrac{1}{2}(r^*_k - r^*_0) = \tfrac{1}{8\beta'}\big\{\big(G_{\mathrm{B}(n)}(k,k) - G_{\mathrm{B}(n)}(k,0)\big) + \big(G_{\mathrm{B}(n)}(0,0) - G_{\mathrm{B}(n)}(0,k)\big)\big\}\,.$$

We then let $n \to \infty$; using Exercise B.24,

$$\lim_{n\to\infty}(G_{B(n)}(k,k) - G_{B(n)}(k,0)) = \lim_{n\to\infty}(G_{B(n)}(0,0) - G_{B(n)}(0,k)) = a(k), \quad (9.29)$$

where $a(k)$ is called the **potential kernel** of the symmetric simple random walk on $\mathbb{Z}^2$, defined by

$$a(k) \stackrel{\text{def}}{=} \sum_{m\geq 0}\{\mathbb{P}_0(X_m = 0) - \mathbb{P}_k(X_m = 0)\}.$$

Therefore,

$$|\langle \mathbf{S}_0 \cdot \mathbf{S}_k\rangle_\mu| \leq e^{-a(k)/4\beta'}.$$

The conclusion now follows since, by Theorem B.76,

$$a(k) = \frac{2}{\pi}\log\|k\|_2 + O(1) \quad \text{as } \|k\|_2 \to \infty. \qquad \square$$

9.5 Bibliographical References

The problems treated in this chapter have their origins in celebrated work by Mermin and Wagner [241, 242]. This triggered a long series of subsequent investigations, leading to stronger claims under weaker assumptions: [32, 83, 126, 169, 179, 193, 237, 240, 243, 252, 273, 317] to mention just a few.

Mermin–Wagner Theorem. The proof of Theorem 9.2 follows the approach of Pfister [273], with some improvements from [169]. Being really classical material, alternative presentations of this material can be found in many books, such as [134, 282, 308, 312].

Effective Interface Models in $d = 1$ and 2. That the same type of arguments can be used to prove the absence of any Gibbs state in models with unbounded spins, as in our Theorem 9.9, was first realized by Dobrushin and Shlosman [84] and by Fröhlich and Pfister [119], although they did not derive quantitative lower bounds. Generalizations of Theorem 9.9 to more general potentials can be found in [169, 246]. Let us mention that results of this type can also be derived by a very different method. Namely, relying on an inequality derived by Brascamp and Lieb in [42], it is possible to compare, under suitable assumptions, the variance of an effective interface model with that of the GFF; this alternative approach is described in [41].

Comparison with the Ising Model. Our proof of Theorem 9.10 is original, as far as we know. However, a similar claim can be found in [250], with a proof based on correlation inequalities. Arguments similar to those used in the proof of Theorem 9.10 have already been used, for example, in [69].

Algebraic Decay of Correlation in Two Dimensions. The proof of Theorem 9.12 is originally due to McBryan and Spencer [237]; the argument presented in the chapter is directly based on their work. Again, alternative presentations can be found in many places, such as [140, 308].

9.6 Complements and Further Reading

9.6.1 Berezinskiĭ–Kosterlitz–Thouless Phase Transition

Theorem 9.12 provides an algebraically decaying upper bound on the 2-point function of the two-dimensional *XY* model at low temperatures, which improves substantially on the bound that can be extracted from Proposition 9.7. Nevertheless, one might wonder whether this bound could be further improved, an issue that we briefly discuss now.

Consider again the two-dimensional *XY* model. Theorem 9.2 shows that all Gibbs measures are SO(2)-invariant, but, as already mentioned, it does not imply uniqueness.[1] It is however possible to prove, using suitable correlation inequalities, that absence of spontaneous magnetization for the *XY* model entails the existence of a unique *translation-invariant* infinite-volume Gibbs measure [47]. Consequently, the Mermin–Wagner Theorem implies at least that there is a unique translation-invariant infinite-volume Gibbs measure at all $\beta \geq 0$ for the two-dimensional *XY* model (moreover, this Gibbs measure is extremal). It is in fact expected that uniqueness holds for this model, but this has not yet been proved.

The following remarkable result proves that a phase transition of a more subtle kind nevertheless occurs in this model.

Theorem 9.14. *Consider the unique translation-invariant Gibbs measure of the two-dimensional XY model. There exist* $0 < \beta_1 < \beta_2 < \infty$ *such that*

- *for all* $\beta < \beta_1$*, there exist* $C(\beta)$ *and* $m(\beta) > 0$ *such that*
$$|\langle \mathbf{S}_0 \cdot \mathbf{S}_k \rangle_\beta| \leq C(\beta) \exp(-m(\beta)\|k\|_2),$$
for all $k \in \mathbb{Z}^2$;
- *for all* $\beta > \beta_2$*, there exist* $c(\beta) > 0$ *and* $D > 0$ *such that*
$$\langle \mathbf{S}_0 \cdot \mathbf{S}_k \rangle_\beta \geq c(\beta)\, \|k\|_2^{-D/\beta},$$
for all $k \in \mathbb{Z}^2$.

Proof The first claim follows immediately from Theorem 9.10 and Exercise 3.24. The proof of the second part, which is due to Fröhlich and Spencer [122], is however quite involved and goes beyond the scope of this book. □

Note that, combined with the McBryan–Spencer [237] upper bound of Theorem 9.12, this shows that the 2-point function of the two-dimensional *XY* model really decays algebraically at low temperature.

A sharp transition is expected between the two regimes (exponential vs. algebraic decay) described in Theorem 9.14, at a value β_{BKT} of the inverse temperature. This so-called **Berezinskiĭ–Kosterlitz–Thouless phase transition**, named after the physicists who studied this problem in the early 1970s [19, 195], exhibits several remarkable properties, including the fact that the pressure remains infinitely differentiable (but not analytic) at the transition. One should point out, however, that the analytic properties at this phase transition are not universal, and other $\mathsf{O}(2)$-symmetric models display very different behavior, such as a first-order phase transition [344].

To conclude this discussion, let us mention an outstanding open problem in this area. As explained in Section 9.6.2, the proof of Theorem 9.12 can be adapted to obtain similar upper bounds for a general class of $\mathsf{O}(N)$-symmetric models, and in particular for all $\mathsf{O}(N)$ models with $N > 2$. However, it is conjectured that this upper bound is very poor when $N > 2$, Namely, it is expected that the 2-point function then *decays exponentially at all temperatures*. Interestingly, it is the fact that $\mathsf{SO}(2)$ is abelian, while the groups $\mathsf{SO}(N)$, $N \geq 3$, are not, that is deemed to be responsible for the difference in behavior [307].

9.6.2 Generalizations

For pedagogical reasons, we have restricted our discussion to the simplest setup. The results presented here can, however, be extended in various directions. We briefly describe one possible such framework and provide some relevant references.

We assume that the spins $\mathbf{S}_i$ take values in some topological space $\mathscr{S}$, on which a compact, connected Lie group G acts continuously (we simply denote the action of $\mathsf{g} \in \mathsf{G}$ on $x \in \mathscr{S}$ by $\mathsf{g}x$). We replace the Hamiltonian in (9.1) by

$$\sum_{\substack{\{i,j\}\subset\mathbb{Z}^d:\\ \{i,j\}\cap\Lambda\neq\varnothing}} J_{j-i}\tilde{W}(S_i, S_j),$$

where $(J_i)_{i\in\mathbb{Z}^d}$ is a collection of real numbers such that $\sum_{i\in\mathbb{Z}^d}|J_i| = 1$ and $\tilde{W} : \mathscr{S} \times \mathscr{S} \to \mathbb{R}$ is continuous and G-invariant, in the sense that $\tilde{W}(\mathsf{g}x, \mathsf{g}y) = \tilde{W}(x, y)$ for all $x, y \in \mathscr{S}$ and all $\mathsf{g} \in \mathsf{G}$.

Theorem 9.2 (and the more quantitative Proposition 9.7) can then be extended to this more general setup, under the assumption that the random walk on $\mathbb{Z}^d$, which jumps from i to j with probability $|J_{j-i}|$, is recurrent. This result was proved by Ioffe, Shlosman and Velenik [169], building on earlier works by Dobrushin and Shlosman [83] and Pfister [273]. We emphasize that the recurrence assumption cannot be improved in general, as there are examples of models for which spontaneous symmetry breaking at low temperatures occurs as soon as the corresponding random walk is transient [32, 201]; see also [134, Theorem 20.15].

Using a similar approach and building on the earlier works of McBryan and Spencer [237] and of Messager, Miracle-Solé and Ruiz [243], Theorem 9.12 has been extended by Gagnebin and Velenik [126] to $\mathrm{O}(N)$-symmetric models with a Hamiltonian as above, provided that $|J_i| \leq J\|i\|_1^{-\alpha}$ for some $J < \infty$ and $\alpha > 4$.

10 Reflection Positivity

In this chapter, we study models whose Gibbs distribution possesses a remarkable property: *reflection positivity*. Two consequences of this property, the *chessboard estimate* and the *infrared bound*, will be described in a general setting.

Before that, in order to motivate this approach, we describe the two main applications that will be discussed in this chapter. Of course, there are many other such applications, reflection positivity playing a crucial role in a large number of proofs in this field.

10.1 Motivation: Some New Results for O(*N*)-Type Models

We remind the reader that in O(N) models, which were discussed in Section 9.1, the spins take their values in the N-dimensional sphere ($N \geq 2$),

$$\Omega_0 \stackrel{\text{def}}{=} \mathbb{S}^{N-1} \subset \mathbb{R}^N ,$$

and have the formal Hamiltonian

$$-\beta \sum_{\{i,j\} \in \mathscr{E}_{\mathbb{Z}^d}} \mathbf{S}_i \cdot \mathbf{S}_j ,$$

where $\mathbf{S}_i(\omega) \stackrel{\text{def}}{=} \omega_i$ denotes the spin at $i \in \mathbb{Z}^d$ and the symbol $\cdot$ denotes the scalar product in $\mathbb{R}^N$. We denote by $\mathscr{G}(\beta)$ the set of Gibbs measures for this model at inverse temperature β. In Chapter 9, we proved that, on $\mathbb{Z}^2$, the invariance of Φ under a global rotation of the spins leads to the absence of orientational long-range order at any positive temperature. In particular, we showed that the distribution of the spin at the origin is uniform on $\mathbb{S}^{N-1}$: for all $\mu \in \mathscr{G}(\beta)$, $\langle \mathbf{S}_0 \rangle_\mu = \mathbf{0}$.

In contrast, in Section 10.5.2, we will prove that, in larger dimensions, the global symmetry under rotations is spontaneously broken at low temperature.

Theorem 10.1. *Assume that $N \geq 2$ and $d \geq 3$. There exists $0 < \beta_0 < \infty$ and $m^* = m^*(\beta) > 0$ such that, whenever $\beta > \beta_0$, for each direction $\mathbf{e} \in \mathbb{S}^{N-1}$, there exists $\mu^{\mathbf{e}} \in \mathscr{G}(\beta)$ such that*

$$\langle \mathbf{S}_0 \rangle_{\mu^{\mathbf{e}}} = m^* \mathbf{e} .$$

In our second application, in Section 10.4.3, we will consider the anisotropic *XY* model on $\mathbb{Z}^2$ (although the argument applies as well to higher values of d and N). This model was introduced in Remark 9.4; its spins take values in $\mathbb{S}^1$ and the formal Hamiltonian is given by

$$-\beta \sum_{\{i,j\}\in\mathscr{E}_{\mathbb{Z}^d}} \{S_i^1 S_j^1 + \alpha S_i^2 S_j^2\},$$

where $\alpha \in [0,1]$ is the **anisotropy parameter** and we have written $\mathbf{S}_i = (S_i^1, S_i^2)$. We denote the set of Gibbs measures at inverse temperature β and anisotropy α by $\mathscr{G}(\beta,\alpha)$.

When $\alpha = 1$, this model reduces to the *XY* model, and we have seen in Section 9.2 that there is no spontaneous magnetization in this case. The next theorem shows that, in the presence of an arbitrary weak anisotropy, there are Gibbs measures displaying spontaneous magnetization at low temperature.

Theorem 10.2. *Assume that $N = 2$ and $d = 2$. For any $0 \le \alpha < 1$, there exists $\beta_0 = \beta_0(\alpha)$ such that, for all $\beta > \beta_0$, there exist $\mu^+, \mu^- \in \mathscr{G}(\beta,\alpha)$ such that*

$$\langle \mathbf{S}_0 \cdot \mathbf{e}_1 \rangle_{\mu^+} > 0 > \langle \mathbf{S}_0 \cdot \mathbf{e}_1 \rangle_{\mu^-}.$$

Remark 10.3. Whenever $\alpha \in [0,1)$, the system possesses exactly two configurations with minimal energy: those in which the spins are either all equal to $+\mathbf{e}_1$ or all equal to $-\mathbf{e}_1$ (see Exercise 10.5). This makes it reasonable to implement a suitable version of Peierls' argument. Note, however, that the continuous nature of the spins does not allow us to apply directly the results of Pirogov–Sinai theory developed in Chapter 7, although extensions covering such situations exist. [1] ◇

10.2 Models Defined on the Torus

Positivity under reflections is naturally formulated for measures that are invariant under reflections through planes perpendicular to some coordinate axis of $\mathbb{Z}^d$. Since most of the finite systems considered previously in the book are only left invariant by a few, if any, such reflections, it turns out to be much more convenient, in this chapter, to consider finite-volume Gibbs measures with *periodic boundary conditions.*

Let us therefore denote by $\mathbb{T}_L$ the ***d*-dimensional torus of linear size** $L > 0$, which is obtained by identifying the opposite sides of the box $\{0,1,\dots,L\}^d$ (remember the one- and two-dimensional tori depicted in Figure 3.1). Equivalently, we can set $\mathbb{T}_L \overset{\text{def}}{=} (\mathbb{Z}/L\mathbb{Z})^d$. Note that, to lighten the notation, we will only indicate the dimensionality of the torus explicitly when it might not be clear from the context.

We will transfer various notions from $\mathbb{Z}^d$ to the torus. For example, we will continue using the translation by i, denoted θ_i. We denote by $\mathscr{E}_L$ the set of all edges between nearest-neighbor vertices of $\mathbb{T}_L$. (The models that fit the framework of this chapter are not restricted to nearest-neighbor interactions, but we introduce this set for later convenience.)

As always, the single-spin space is denoted Ω_0 and the set of spin configurations on the torus is

$$\Omega_L \stackrel{\text{def}}{=} \Omega_0^{\mathbb{T}_L} = \{\omega = (\omega_i)_{i\in\mathbb{T}_L} : \omega_i \in \Omega_0\} .$$

Even though the models to which we later apply the theory will have either $\Omega_0 = \mathbb{S}^{N-1}$ or $\Omega_0 = \mathbb{R}^N$, the theory has no such limitations. In fact, one of the arguments used below will require allowing Ω_0 to be far more general. Let us thus assume that Ω_0 is some topological space, on which one can define the usual Borel σ-algebra $\mathscr{B}_0$, generated by the open sets. (These notions are introduced in Section 6.10.1 and Appendix B.5.) The product σ-algebra of events on Ω_L is denoted simply $\mathscr{F}_L = \bigotimes_{i\in\mathbb{T}_L} \mathscr{B}_0$. The set of measures on $(\Omega_L, \mathscr{F}_L)$ is denoted $\mathscr{M}(\Omega_L, \mathscr{F}_L)$.

Remark 10.4. In the following, we will always assume L to be even. Moreover, since *all* the models considered in this chapter will be defined on $\mathbb{T}_L$, we will substantially lighten the notation by using everywhere a subscript L instead of $\mathbb{T}_L$. For example, a Gibbs distribution on Ω_L will be denoted μ_L instead of $\mu_{\mathbb{T}_L}$. ◇

10.3 Reflections

We consider transformations on the torus,

$$\Theta : \mathbb{T}_L \to \mathbb{T}_L ,$$

associated with *reflections* through planes that split the torus in two. (This Θ is not to be mistaken for the translation θ_i.) Before moving on to the precise definitions, the reader is invited to take a look at Figures 10.1 and 10.2, where the meaning of these reflections is made clear.

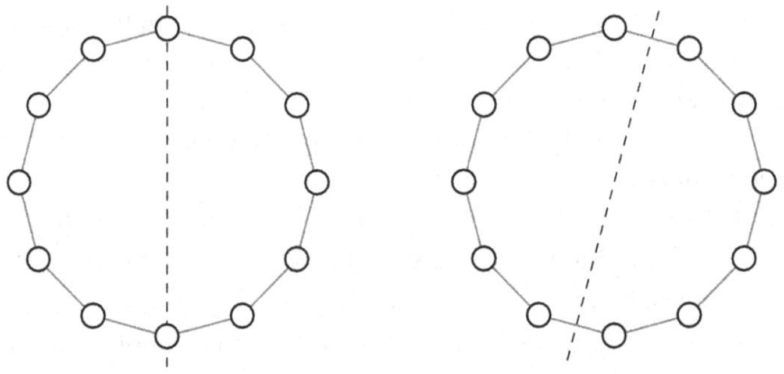

Figure 10.1. The one-dimensional torus $\mathbb{T}_{12}$, with a reflection through vertices (left) and through edges (right).

► *Reflection through vertices.* Let $k \in \{1, \dots, d\}$ denote one among the d possible directions parallel to the coordinate axes and $n \in \{0, \dots, \frac{1}{2}L - 1\}$. The **reflection through vertices** $\Theta\colon \mathbb{T}_L \to \mathbb{T}_L$ (associated with k and n), which maps $i = (i_1, \dots, i_d)$ to $\Theta(i) = (\Theta(i)_1, \dots, \Theta(i)_d)$, is defined by

$$\Theta(i)_\ell \stackrel{\text{def}}{=} \begin{cases} (2n - i_k) \mod L & \text{if } \ell = k, \\ i_\ell & \text{if } \ell \neq k. \end{cases} \tag{10.1}$$

Θ is a reflection of the torus through a plane Π orthogonal to the direction $\mathbf{e}_k$. The intersection between the plane and the torus is given by

$$\Pi \cap \mathbb{T}_L = \{i \in \mathbb{T}_L : i_k = n \text{ or } i_k = n + L/2\} .$$

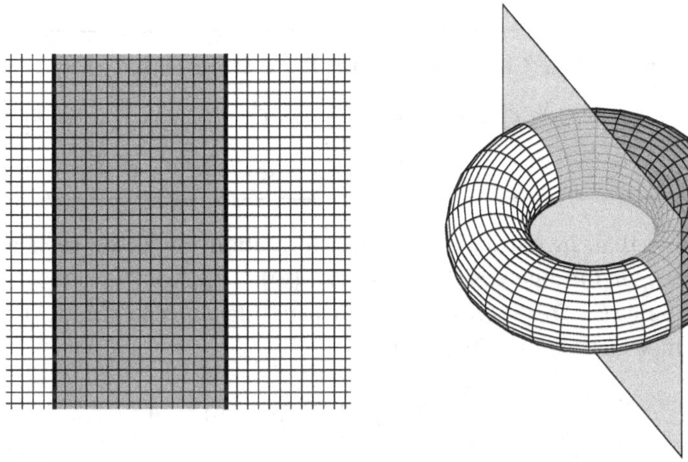

Figure 10.2. In $d = 2$, a reflection through the vertices of $\mathbb{T}_{32}$. The sets $\mathbb{T}_{32,+}$ and $\mathbb{T}_{32,-}$ are drawn in white and gray, respectively, and the intersection of the reflection plane and the torus is represented by the two thick lines. Left: Planar representation. Right: Spatial representation.

This also leads to a natural decomposition of the torus into two overlapping halves: $\mathbb{T}_L = \mathbb{T}_{L,+} \cup \mathbb{T}_{L,-}$, where

$$\mathbb{T}_{L,+} = \mathbb{T}_{L,+}(\Theta) \stackrel{\text{def}}{=} \{i \in \mathbb{T}_L : n \le i_k \le n + L/2\},$$
$$\mathbb{T}_{L,-} = \mathbb{T}_{L,-}(\Theta) \stackrel{\text{def}}{=} \{i \in \mathbb{T}_L : 0 \le i_k \le n \text{ or } n + L/2 \le i_k \le L - 1\}.$$

► *Reflection through edges.* The **reflection through edges** $\Theta\colon \mathbb{T}_L \to \mathbb{T}_L$ (associated with k and n) is defined exactly as in (10.1), but with $n \in \{\frac{1}{2}, \frac{3}{2}, \dots, \frac{L-1}{2}\}$. Now, Θ should be seen as a reflection of the torus through a plane Π with $\Pi \cap \mathbb{T}_L = \varnothing$, so that the corresponding decomposition of the torus, $\mathbb{T}_L = \mathbb{T}_{L,+} \cup \mathbb{T}_{L,-}$, is into two disjoint halves.

By definition, each transformation Θ is an involution: $\Theta^{-1} = \Theta$. Below, it will always be clear from the context whether the Θ under consideration is a reflection through vertices or edges.

10.3.1 Reflection Positive Measures

A reflection Θ can be made to act on spin configurations, $\Theta\colon \Omega_L \to \Omega_L$, by setting

$$(\Theta(\omega))_i \stackrel{\text{def}}{=} \omega_{\Theta(i)}, \quad \forall i \in \mathbb{T}_L .$$

Similarly, its action on functions $f\colon \Omega_L \to \mathbb{R}$ is defined by

$$\Theta(f)(\omega) \stackrel{\text{def}}{=} f(\Theta^{-1}(\omega)) .$$

We denote by $\mathfrak{A}_+(\Theta)$, respectively $\mathfrak{A}_-(\Theta)$, the algebra of all bounded measurable functions f on Ω_L with support inside $\mathbb{T}_{L,+}(\Theta)$, respectively $\mathbb{T}_{L,-}(\Theta)$. The following properties will be used constantly in the following.

Exercise 10.1. Check that, for any $f, g \in \mathfrak{A}_+(\Theta)$ and $\lambda \in \mathbb{R}$,

$$\Theta^2(f) = f, \quad \Theta(\lambda f) = \lambda\Theta(f), \quad \Theta(f+g) = \Theta(f) + \Theta(g),$$
$$\Theta(fg) = \Theta(f)\Theta(g), \quad \Theta(e^f) = e^{\Theta(f)} .$$

Note that, in particular, the transformation Θ can be seen as an isomorphism between the algebras $\mathfrak{A}_+(\Theta)$ and $\mathfrak{A}_-(\Theta)$.

Since $\Theta\colon \Omega_L \to \Omega_L$ is clearly measurable, one can also define the action of Θ on a measure $\mu \in \mathscr{M}(\Omega_L, \mathscr{F}_L)$ by

$$\Theta(\mu)(A) \stackrel{\text{def}}{=} \mu(\Theta^{-1}A), \quad A \in \mathscr{F}_L .$$

Of course, this implies that, for every bounded measurable function f (remember that $\langle f \rangle_\mu \stackrel{\text{def}}{=} \int f \, d\mu$),

$$\langle f \rangle_{\Theta(\mu)} = \langle \Theta(f) \rangle_\mu .$$

Definition 10.5. Let Θ be a reflection. A measure $\mu \in \mathscr{M}(\Omega_L, \mathscr{F}_L)$ is **reflection positive with respect to Θ** if

1. $\langle f\,\Theta(g)\rangle_\mu = \langle g\,\Theta(f)\rangle_\mu$, for all $f, g \in \mathfrak{A}_+(\Theta)$;
2. $\langle f\,\Theta(f)\rangle_\mu \geq 0$, for all $f \in \mathfrak{A}_+(\Theta)$.

The set of measures that are reflection positive with respect to Θ is denoted by $\mathcal{M}_{\mathrm{RP}(\Theta)}$.

In other words, μ is reflection positive if and only if the bilinear form $(f, g) \mapsto \langle f\,\Theta(g)\rangle_\mu$ on $\mathfrak{A}_+(\Theta)$ is symmetric and positive semi-definite. This immediately implies the validity of a Cauchy–Schwarz-type inequality, which will be the basis of the properties to be derived later:

Lemma 10.6. *Let $\mu \in \mathcal{M}_{\mathrm{RP}(\Theta)}$. Then, for all $f, g \in \mathfrak{A}_+(\Theta)$,*

$$\langle f\Theta(g)\rangle_\mu^2 \leq \langle f\Theta(f)\rangle_\mu \langle g\Theta(g)\rangle_\mu .$$

Proof Let $\mu \in \mathcal{M}_{\mathrm{RP}(\Theta)}$. We have, for all $\lambda \in \mathbb{R}$,

$$0 \leq \langle (\lambda f + g)\, \Theta(\lambda f + g)\rangle_\mu = \langle f\, \Theta(f)\rangle_\mu\, \lambda^2 + 2\langle f\, \Theta(g)\rangle_\mu\, \lambda + \langle g\, \Theta(g)\rangle_\mu .$$

This implies that the latter quadratic polynomial in λ has at most one root and, therefore, the associated discriminant cannot be positive. The claim follows. □

As seen in the following exercise, the first condition in Definition 10.5 is equivalent to saying that μ is invariant under Θ; it is thus both natural and rather mild. It will always be trivially satisfied in the cases considered later.

Exercise 10.2. Show that the first condition in Definition 10.5 holds if and only if

$$\Theta(\mu) = \mu . \tag{10.2}$$

10.3.2 Examples of Reflection Positive Measures

As a starting point, we consider product measures. Let ρ be a measure on $(\Omega_0, \mathscr{F}_0)$, which we will refer to as the **reference** measure, and let

$$\mu_0 \stackrel{\text{def}}{=} \bigotimes_{i \in \mathbb{T}_L} \rho . \tag{10.3}$$

Lemma 10.7. *μ_0 is reflection positive with respect to all reflections Θ.*

Proof Notice that

$$\Theta(\mu_0) = \mu_0 \tag{10.4}$$

for each reflection Θ. Indeed, the measure of any rectangle $\times_{k=1}^{|\mathbb{T}_L|} B_k$ ($B_k \in \mathscr{F}_0$) is the same under μ_0 or $\Theta(\mu_0)$, since μ_0 is invariant under any relabeling of the vertices of $\mathbb{T}_L$. By Exercise 10.2, this implies that μ_0 satisfies the first condition of Definition 10.5; let us check the second one.

We first consider a reflection Θ through edges. Let $f \in \mathfrak{A}_+(\Theta)$. Since $\mathbb{T}_{L,+}(\Theta) \cap \mathbb{T}_{L,-}(\Theta) = \varnothing$, f and $\Theta(f)$ have disjoint supports and (10.4) yields

$$\langle f\, \Theta(f)\rangle_{\mu_0} = \langle f\rangle_{\mu_0} \langle \Theta(f)\rangle_{\mu_0} = (\langle f\rangle_{\mu_0})^2 \geq 0 ,$$

thus showing that $\mu_0 \in \mathcal{M}_{\mathrm{RP}(\Theta)}$.

Let us now assume that Θ is a reflection through vertices and, again, let us take $f \in \mathfrak{A}_+(\Theta)$. In this case, the supports of f and $\Theta(f)$ may intersect. Therefore let P be the set of all vertices of $\mathbb{T}_L$ belonging to the reflection plane and remember that

$\mathscr{F}_P$ denotes the σ-algebra generated by the spins attached to vertices in P. We then have

$$\langle f\Theta(f)\rangle_{\mu_0} = \big\langle \mu_0(f\,\Theta(f)\,|\,\mathscr{F}_P)\big\rangle_{\mu_0} = \big\langle \mu_0(f\,|\,\mathscr{F}_P)\mu_0(\Theta(f)\,|\,\mathscr{F}_P)\big\rangle_{\mu_0}$$
$$= \big\langle \mu_0(f\,|\,\mathscr{F}_P)^2\big\rangle_{\mu_0} \geq 0\,,$$

and reflection positivity follows again. (In the second equality, we used the fact that $\mu_0(\cdot|\mathscr{F}_P)$ is again a product measure.) □

From now on, we let ρ denote some reference measure on $(\Omega_0, \mathscr{F}_0)$, which we assume to be compactly supported, with $\rho(\Omega_0) < \infty$. We define μ_0 as in (10.3). We can then define the Gibbs distribution on $(\Omega_L, \mathscr{F}_L)$, associated with a Hamiltonian $\mathscr{H}_L \colon \Omega_L \to \mathbb{R}$, by

$$\forall A \in \mathscr{F}_L\,, \quad \mu_L(A) \stackrel{\text{def}}{=} \int_{\Omega_L} \frac{e^{-\mathscr{H}_L(\omega)}}{\mathbf{Z}_L} 1_A(\omega)\,\mu_0(\mathrm{d}\omega)\,, \tag{10.5}$$

where

$$\mathbf{Z}_L = \int_{\Omega_L} e^{-\mathscr{H}_L(\omega)}\mu_0(\mathrm{d}\omega) = \langle e^{-\mathscr{H}_L}\rangle_{\mu_0}\,.$$

(Of course, for this definition to make sense, we must have $\mathbf{Z}_L < \infty$. This will always be the case below.)

Lemma 10.8. *Let μ_L be as above. Let Θ be a reflection on $\mathbb{T}_L$ and assume that the Hamiltonian can be written as*

$$-\mathscr{H}_L = A + \Theta(A) + \sum_\alpha C_\alpha \Theta(C_\alpha)\,, \tag{10.6}$$

for some functions $A, C_\alpha \in \mathfrak{A}_+(\Theta)$. Then $\mu_L \in \mathcal{M}_{\mathrm{RP}(\Theta)}$.

Proof Using a Taylor expansion for the factor $\exp\big(\sum_\alpha C_\alpha \Theta(C_\alpha)\big)$,

$$\langle f\,\Theta(g)\rangle_{\mu_L} = \frac{1}{\mathbf{Z}_L}\big\langle f\,\Theta(g)\,e^{A+\Theta(A)+\sum_\alpha C_\alpha\Theta(C_\alpha)}\big\rangle_{\mu_0}$$
$$= \frac{1}{\mathbf{Z}_L}\sum_{n\geq 0}\frac{1}{n!}\sum_{\alpha_1,\dots,\alpha_n}\big\langle f\,e^A C_{\alpha_1}\cdots C_{\alpha_n}\,\Theta(g\,e^A C_{\alpha_1}\cdots C_{\alpha_n})\big\rangle_{\mu_0}\,.$$

The result now follows from Lemma 10.7, since $\mu_0 \in \mathcal{M}_{\mathrm{RP}(\Theta)}$. □

As usual, the Hamiltonian can be constructed from a potential $\Phi = \{\Phi_B\}$:

$$\mathscr{H}_L \stackrel{\text{def}}{=} \sum_{B\subset\mathbb{T}_L} \Phi_B\,,$$

where Φ_B is a measurable function with support in B. To ensure that $\mathscr{H}_L$ can be put in the form (10.6), some symmetry assumptions will be made about the functions Φ_B.

Example 10.9. Consider a translation-invariant potential $\{\Phi_B\}_{B\subset\mathbb{T}_L}$ involving interactions only between pairs (that is, satisfying $\Phi_B = 0$ whenever $|B| \neq 2$) and such that $\Phi_{\{i,j\}} = 0$ whenever $\|j-i\|_\infty > 1$. Assume Θ is a reflection through the *vertices* of $\mathbb{T}_L$ satisfying

$$\Phi_{\{i,j\}}(\omega) = \Phi_{\{\Theta(i),\Theta(j)\}}(\Theta(\omega)), \quad \forall\omega, \tag{10.7}$$

for all $\{i,j\} \subset \mathbb{T}_L$. This holds, for example, if $\Phi_{\{i,j\}}$ depends only on the distance $\|j-i\|_1$.

Let us show that $\mu_L \in \mathcal{M}_{\mathrm{RP}(\Theta)}$. Namely, let again P denote the set of vertices of $\mathbb{T}_L$ lying in the reflection plane of Θ. Notice that, since the only pairs $e=\{i,j\}$ to be considered involve points with $\|j-i\|_\infty \le 1$, the Hamiltonian can be written as

$$\mathscr{H}_L = \sum_{e\subset\mathbb{T}_L} \Phi_e = \sum_{e\subset P} \Phi_e + \sum_{\substack{e\subset\mathbb{T}_{L,+}\\ e\not\subset P}} \Phi_e + \sum_{\substack{e\subset\mathbb{T}_{L,-}\\ e\not\subset P}} \Phi_e .$$

Each pair $e=\{i,j\} \subset \mathbb{T}_{L,-}$ can be paired with its reflection $\Theta(e) = \{\Theta(i),\Theta(j)\} \subset \mathbb{T}_{L,+}$. Therefore, a change of variables yields, using (10.7),

$$\sum_{\substack{e\subset\mathbb{T}_{L,-}\\ e\not\subset P}} \Phi_e(\omega) = \sum_{\substack{e\subset\mathbb{T}_{L,+}\\ e\not\subset P}} \Phi_{\Theta(e)}(\omega) = \sum_{\substack{e\subset\mathbb{T}_{L,+}\\ e\not\subset P}} \Phi_e(\Theta(\omega)) .$$

This means that $-\mathscr{H}_L = A + \Theta(A)$, with $A \in \mathfrak{A}_+(\Theta)$ given by

$$A \stackrel{\text{def}}{=} -\tfrac{1}{2}\sum_{e\subset P} \Phi_e - \sum_{\substack{e\subset\mathbb{T}_{L,+}\\ e\not\subset P}} \Phi_e .$$

Lemma 10.8 now implies that $\mu_L \in \mathcal{M}_{\mathrm{RP}(\Theta)}$. ◇

Example 10.10. Let $\Omega_0 = \mathbb{R}^\nu$ and ρ be compactly supported. We assume that, for each $1 \le m \le \nu$ and each $1 \le k \le d$, J_k^m is a fixed nonnegative number. We consider a Hamiltonian of the form

$$\mathscr{H}_L \stackrel{\text{def}}{=} -\sum_{\{i,j\}\in\mathscr{E}_L} \sum_{m=1}^{\nu} J_{i,j}^m S_i^m S_j^m , \tag{10.8}$$

where $J_{i,j}^m = J_k^m$ when i and j differ in their kth component and S_i^m is the mth component of $\mathbf{S}_i$. This Hamiltonian actually covers all the applications we are going to consider in this chapter.

Let Θ be a reflection through edges of the torus. Proceeding similarly to what we did in Example 10.9, it is easy to check that

$$-\mathscr{H}_L = A + \Theta(A) + \sum_{m=1}^{\nu} \sum_{\substack{i\in\mathbb{T}_{L,+}:\\ \{i,\Theta(i)\}\in\mathscr{E}_L}} C_i^m \,\Theta(C_i^m),$$

where the functions $A, C_i^m \in \mathfrak{A}_+(\Theta)$ are given by

$$A \stackrel{\text{def}}{=} \sum_{m=1}^{\nu} \sum_{\substack{\{i,j\}\in\mathscr{E}_L: \\ i,j\in\mathbb{T}_{L,+}}} J_{i,j}^m S_i^m S_j^m \quad \text{and} \quad C_i^m \stackrel{\text{def}}{=} \sqrt{J^m(i,\Theta(i))}\, S_i^m .$$

Lemma 10.8 implies again that $\mu_L \in \mathcal{M}_{\mathrm{RP}(\Theta)}$. ◇

Exercise 10.3. Give an example of a translation-invariant measure $\mu \in \mathscr{M}(\Omega_L, \mathscr{F}_L)$ which is not reflection positive.

10.4 The Chessboard Estimate

In this section, we establish a first major consequence of reflection positivity, the chessboard estimate, and provide two applications.

10.4.1 Proof of the Estimate

To simplify the exposition, we will focus on the case of reflections through edges; however, both the statement and the proof can be adapted straightforwardly to the case of reflections through vertices.

Let $B < L$ be two positive integers such that $2B$ divides L and let us define $\Lambda_B \stackrel{\text{def}}{=} \{0, \dots, B-1\}^d \subset \mathbb{T}_L$. We decompose the torus into a disjoint union of translates of Λ_B, called **blocks**. These can be indexed by $t \in \mathbb{T}_{L/B}$:

$$\mathbb{T}_L = \bigcup_{t\in\mathbb{T}_{L/B}} (\Lambda_B + Bt) .$$

A function f with support inside Λ_B is said to be **Λ_B-local**. Given a Λ_B-local function f and $t \in \mathbb{T}_{L/B}$, we define a $(\Lambda_B + Bt)$-local function $f^{[t]}$ by successive reflections: Let $t_0 = 0, t_1, \dots, t_k = t$ be a self-avoiding nearest-neighbor path in $\mathbb{T}_{L/B}$ and let Θ_i be the reflection through the plane going through the edges connecting $\Lambda_B + t_{i-1}B$ and $\Lambda_B + t_iB$; we set

$$f^{[t]} \stackrel{\text{def}}{=} \Theta_k \circ \Theta_{k-1} \circ \cdots \circ \Theta_1(f) .$$

A glance at Figure 10.3 shows that the definition of $f^{[t]}$ does not depend on the chosen path (observe that this relies on L/B being even).

Let us say that $\mu \in \mathscr{M}(\Omega_L, \mathscr{F}_L)$ is **B-periodic** if it is invariant under translations by B along any coordinate axis: $\mu = \mu \circ \theta_{B\mathbf{e}_k}$ for all $k \in \{1, \dots, d\}$.

Theorem 10.11 (Chessboard estimate). *Let $\mu \in \mathscr{M}(\Omega_L, \mathscr{F}_L)$ be B-periodic and such that $\mu \in \mathcal{M}_{\mathrm{RP}(\Theta)}$ for all reflections Θ between neighboring blocks (that is,*

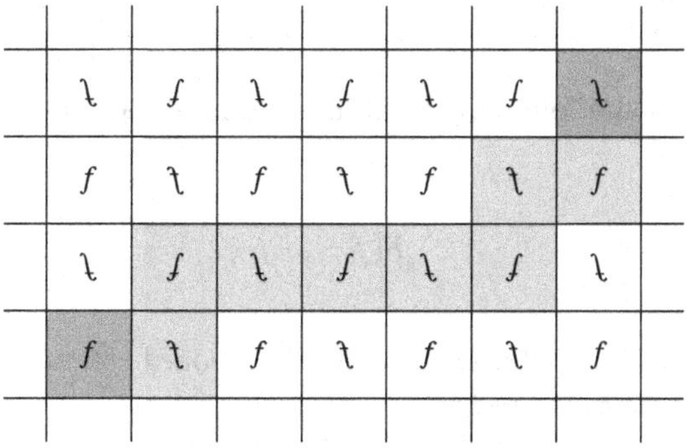

Figure 10.3. A graphical evocation of the definition of $f^{[t]}$ obtained by applying reflections to the original function f (located in the bottom left block Λ_B), until reaching the block indexed by t (top right). The definition of $f^{[t]}$ is independent of the chosen path (shaded cells).

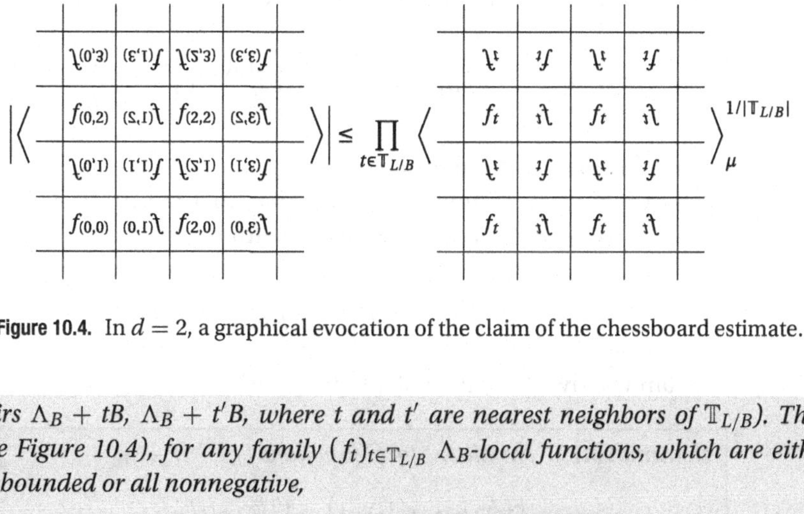

Figure 10.4. In $d = 2$, a graphical evocation of the claim of the chessboard estimate.

pairs $\Lambda_B + tB$, $\Lambda_B + t'B$, where t and t' are nearest neighbors of $\mathbb{T}_{L/B}$). Then (see Figure 10.4), for any family $(f_t)_{t\in\mathbb{T}_{L/B}}$ Λ_B-local functions, which are either all bounded or all nonnegative,

$$\Big|\Big\langle \prod_{t\in\mathbb{T}_{L/B}} f_t^{[t]}\Big\rangle_\mu\Big| \le \prod_{t\in\mathbb{T}_{L/B}} \Big[\Big\langle \prod_{s\in\mathbb{T}_{L/B}} f_t^{[s]}\Big\rangle_\mu\Big]^{1/|\mathbb{T}_{L/B}|}. \tag{10.9}$$

Proof of Theorem 10.11. We can assume that the functions f_t, $t \in \mathbb{T}_{L/B}$, are bounded. Indeed, if they are unbounded (but nonnegative), we can apply the result to the bounded functions $f_t \wedge K$ ($K \in \mathbb{N}$) and use monotone convergence to take the limit $K \uparrow \infty$.

The proof is done by induction on the dimension.

The Case $d = 1$: In the one-dimensional case, the boxes are simply intervals indexed by $t \in \{0, 1, \ldots, 2N - 1\}$, where $N = L/(2B) \in \mathbb{N}$. Observe that, in this case, given a Λ_B-local function f, each function $f^{[t]}$ coincides either with a translate of f or with a translate of the Λ_B-local function defined by

$$\bar{f}(\omega_0, \omega_1, \dots, \omega_{B-1}, \omega_B, \dots, \omega_{L-1}) \stackrel{\text{def}}{=} f(\omega_{B-1}, \omega_{B-2}, \dots, \omega_0, \omega_{L-1}, \dots, \omega_B).$$

Consider the following multilinear functional on the $2N$-tuples of Λ_B-local functions:

$$F(f_0, \dots, f_{2N-1}) \stackrel{\text{def}}{=} \Big\langle \prod_{t=0}^{2N-1} f_t^{[t]} \Big\rangle_\mu .$$

Reformulated in terms of F, the chessboard estimate (10.9) that we want to establish can be expressed as

$$\big|F(f_0, \dots, f_{2N-1})\big| \le \prod_{t=0}^{2N-1} F(f_t, \dots, f_t)^{1/2N} . \tag{10.10}$$

Observe that each of the expectations on the right-hand side of (10.10) is nonnegative. Indeed, we can write

$$F(f_t, \dots, f_t) = \Big\langle \big(\prod_{t=0}^{N-1} f_t^{[t]}\big)\, \Theta\big(\prod_{t=0}^{N-1} f_t^{[t]}\big) \Big\rangle_\mu ,$$

where Θ is the reflection through the edge between the blocks $N-1$ and N (and $2N-1$ and 0). This implies, in particular, that (10.10) trivially holds whenever $F(f_0, \dots, f_{2N-1}) = 0$. We will thus assume from now on that $(f_0, \dots, f_{2N-1})$ is fixed and that

$$F(f_0, \dots, f_{2N-1}) \neq 0 . \tag{10.11}$$

We start with two fundamental properties of F.

Lemma 10.12. *For all Λ_B-local functions $f_0, \dots, f_{2N-1}$,*

$$F(f_0, f_1, \dots, f_{2N-1}) = F(\bar{f}_{2N-1}, \bar{f}_0, \bar{f}_1, \dots, \bar{f}_{2N-2}) \tag{10.12}$$

and

$$F(f_0, \dots, f_{N-1}, f_N, \dots, f_{2N-1})^2 \le F(f_0, \dots, f_{N-1}, f_{N-1}, \dots, f_0) F(f_{2N-1}, \dots, f_N, f_N, \dots, f_{2N-1}). \tag{10.13}$$

Exercise 10.4. Show that, in general, $F(f_0, \dots, f_{2N-1}) \neq F(\bar{f}_0, \dots, \bar{f}_{2N-1})$.

Proof The first identity is a simple consequence of the B-periodicity of μ and of the definition of F. To prove the second one, let again Θ denote the reflection through the edge between the blocks $N-1$ and N (and $2N-1$ and 0). Observe that, for each $N \le t \le 2N-1$, $f_t^{[t]} = \Theta(f_{2N-1-t'}^{[t']})$, where $t' = 2N-1-t \in \{0, \dots, N-1\}$.

Therefore, by Lemma 10.6,

$$F(f_0,\dots,f_{N-1},f_N,\dots,f_{2N-1})^2 = \Big\langle \big(\prod_{t=0}^{N-1} f_t^{[t]}\big)\,\Theta\big(\prod_{t=0}^{N-1} f_{2N-1-t}^{[t]}\big)\Big\rangle_\mu^2$$
$$\le F(f_0,\dots,f_{N-1},f_{N-1},\dots,f_0)F(f_{2N-1},\dots,f_N,f_N,\dots,f_{2N-1})\,. \qquad \square$$

When $2N$ is a power of 2, repeated use of the above lemma leads directly to (10.10). *For simplicity, assume first that $2N = 4$. In this case,* (10.13) *yields*

$$F(f_0,f_1,f_2,f_3)^4 \le F(f_0,f_1,f_1,f_0)^2 F(f_3,f_2,f_2,f_3)^2\,.$$

Now, by (10.12), $F(f_0,f_1,f_1,f_0) = F(\bar{f}_0,\bar{f}_0,\bar{f}_1,\bar{f}_1)$, $F(f_3,f_2,f_2,f_3) = F(\bar{f}_2,\bar{f}_2,\bar{f}_3,\bar{f}_3)$. *But, using again* (10.13) *twice,*

$$F(\bar{f}_0,\bar{f}_0,\bar{f}_1,\bar{f}_1)^2 F(\bar{f}_2,\bar{f}_2,\bar{f}_3,\bar{f}_3)^2 \le \prod_{t=0}^{3} F(\bar{f}_t,\bar{f}_t,\bar{f}_t,\bar{f}_t)\,.$$

This implies (10.10), *since* $F(\bar{f}_t,\bar{f}_t,\bar{f}_t,\bar{f}_t) = F(f_t,f_t,f_t,f_t)$ *by* (10.12). *Clearly, if $2N = 2^M$, the same argument can be used repeatedly. The proof of* (10.10) *for general values of N relies on a variant of this argument, as we explain below.* ◇

Let us consider the auxiliary functional

$$G(f_0,\dots,f_{2N-1}) \stackrel{\text{def}}{=} \frac{|F(f_0,\dots,f_{2N-1})|}{\prod_{t=0}^{2N-1} F(f_t,\dots,f_t)^{1/2N}}\,,$$

which is well defined thanks to the following property.

Lemma 10.13. *For each $t \in \{0,1,\dots,2N-1\}$, $F(f_t,\dots,f_t) > 0$.*

Proof For the sake of readability, we treat explicitly only the case $2N = 6$. The extension to general values of $2N$ is straightforward, as explained below. Let $K_N \stackrel{\text{def}}{=} (\max_t \|f_t\|_\infty)^{2N}$. Applying (10.13), we get

$$|F(f_0,f_1,f_2,f_3,f_4,f_5)| \le F(f_0,f_1,f_2,f_2,f_1,f_0)^{1/2} F(f_5,f_4,f_3,f_3,f_4,f_5)^{1/2}$$
$$\le K_3^{1/2}\, F(f_0,f_1,f_2,f_2,f_1,f_0)^{1/2}\,.$$

We now apply (10.12) in order to push the two copies of f_0 in the first two slots:

$$K_3^{1/2}\, F(f_0,f_1,f_2,f_2,f_1,f_0)^{1/2} = K_3^{1/2}\, F(\bar{f}_0,\bar{f}_0,\bar{f}_1,\bar{f}_2,\bar{f}_2,\bar{f}_1)^{1/2}\,.$$

Using (10.13) once more, we obtain

$$K_3^{1/2}\, F(\bar{f}_0,\bar{f}_0,\bar{f}_1,\bar{f}_2,\bar{f}_2,\bar{f}_1)^{1/2} \le K_3^{3/4}\, F(\bar{f}_0,\bar{f}_0,\bar{f}_1,\bar{f}_1,\bar{f}_0,\bar{f}_0)^{1/4}\,.$$

Again, (10.12) allows us to push the four copies of $\bar{f}_0$ in the first four slots:

$$K_3^{3/4}\, F(\bar{f}_0,\bar{f}_0,\bar{f}_1,\bar{f}_1,\bar{f}_0,\bar{f}_0)^{1/4} = K_3^{3/4}\, F(\bar{f}_0,\bar{f}_0,\bar{f}_0,\bar{f}_0,\bar{f}_1,\bar{f}_1)^{1/4}\,.$$

Applying (10.13) one last time yields

$$K_3^{3/4} F(\bar{f}_0,\bar{f}_0,\bar{f}_0,\bar{f}_0,\bar{f}_1,\bar{f}_1)^{1/4} \leq K_3^{7/8} F(\bar{f}_0,\bar{f}_0,\bar{f}_0,\bar{f}_0,\bar{f}_0,\bar{f}_0)^{1/8}$$

and thus, since $F(f_0,f_0,f_0,f_0,f_0,f_0) = F(\bar{f}_0,\bar{f}_0,\bar{f}_0,\bar{f}_0,\bar{f}_0,\bar{f}_0)$ by (10.12),

$$F(f_0,f_0,f_0,f_0,f_0,f_0) \geq K_3^{-7}|F(f_0,f_1,f_2,f_3,f_4,f_5)|^8 > 0\,.$$

(We used our assumption (10.11).) General values of $2N$ are treated in exactly the same way, applying (10.12) and (10.13) alternatively until all $2N$ slots of F are filled by copies of f_0. Since the number of such copies doubles at each stage, the number of required iterations is given by the smallest integer M such that $2^M \geq 2N$, which yields

$$F(f_0,\dots,f_0) \geq K_N^{1-2^M}|F(f_0,\dots,f_{2N-1})|^{2^M} > 0\,.$$

The same argument applies to other values of t using (10.12). □

By construction, G verifies the same properties as those satisfied by F in (10.12) and (10.13). Moreover, $G(f_t,\dots,f_t) = 1$ for all t. In terms of G, we will obtain (10.10) by showing that $G(f_0,\dots,f_{2N-1}) \leq 1$, which is equivalent to saying that G reaches its maximum value on $2N$-tuples of functions (from the collection $\{f_0,f_1,\dots,f_{2N-1}\}$) which are composed of a single function f_t.

Let $(g_0,\dots,g_{2N-1})$ be such that

(i) $g_i \in \{f_0,\dots,f_{2N-1}\}$ for each $i \in \{0,\dots,2N-1\}$;
(ii) $(g_0,\dots,g_{2N-1})$ maximizes G;
(iii) $(g_0,\dots,g_{2N-1})$ is minimal, in the sense that it contains the longest contiguous substring of the form $f_i,\dots,f_i$ for some $i \in \{0,\dots,2N-1\}$. Here g_{2N-1} and g_0 are considered contiguous (because of property (10.12)).

Let k be the length of the substring in (iii). Thanks to (10.12), we can assume that the latter occurs at the beginning of the string $(g_0,\dots,g_{2N-1})$, that is, that $g_0 = g_1 = \dots = g_{k-1} = f_i$ (or $\bar{f}_i$, with bars on each of the $2N$ entries). We will now check that $k = 2N$, which will conclude the proof of the one-dimensional case.

Suppose that $k < 2N$. We have

$$\begin{aligned} G(g_0,\dots,g_{2N-1})^2 &\leq G(g_0,\dots,g_{N-1},g_{N-1},\dots,g_0)G(g_{2N-1},\dots,g_N,g_N,\dots,g_{2N-1}) \\ &\leq G(g_0,\dots,g_{N-1},g_{N-1},\dots,g_0)G(g_0,\dots,g_{2N-1})\,, \end{aligned}$$

since $(g_0,\dots,g_{2N-1})$ maximizes G. Therefore ($G(g_0,\dots,g_{2N-1}) > 0$ by (10.11)),

$$G(g_0,\dots,g_{2N-1}) \leq G(g_0,\dots,g_{N-1},g_{N-1},\dots,g_0)\,,$$

which means that $(g_0,\dots,g_{N-1},g_{N-1},\dots,g_0)$ is also a maximizer of G. But this is impossible, since the string $(g_0,\dots,g_{N-1},g_{N-1},\dots,g_0)$ possesses a substring $f_i,\dots,f_i$ of length $\min\{2N,2k\} > k$, which would violate our minimality assumption (iii).

The Case $d \geq 2$: We now assume that the chessboard estimate (10.9) has been established for all dimensions $d' \in \{1, \ldots, d\}$ and show that it also holds in dimension $d+1$. Although this induction step is rather straightforward, it involves a few subtleties which we discuss after the proof, in Remark 10.14.

We temporarily denote the d-dimensional torus by $\mathbb{T}_L^d$ and consider $\mathbb{T}_L^{d+1}$ as L adjacent copies of $\mathbb{T}_L^d$:

$$\mathbb{T}_L^{d+1} = \mathbb{T}_L^1 \times \mathbb{T}_L^d .$$

We can thus write $u \in \mathbb{T}_{L/B}^{d+1}$ as $u = (i,t)$, with $i \in \mathbb{T}_{L/B}^1$ and $t \in \mathbb{T}_{L/B}^d$, and use the shorthand notation $f^{[u]} = f^{[(i,t)]} \equiv f^{[i,t]}$. Therefore, applying (10.9) with $d' = 1$,

$$\begin{aligned}
\Big|\Big\langle \prod_{u\in\mathbb{T}_{L/B}^{d+1}} f_u^{[u]} \Big\rangle_\mu\Big| &= \Big|\Big\langle \prod_{i\in\mathbb{T}_{L/B}^1} \Big\{ \prod_{t\in\mathbb{T}_{L/B}^d} f_{(i,t)}^{[t]} \Big\}^{[i]} \Big\rangle_\mu\Big| \\
&\leq \prod_{i\in\mathbb{T}_{L/B}^1} \Big[\Big\langle \prod_{j\in\mathbb{T}_{L/B}^1} \Big\{ \prod_{t\in\mathbb{T}_{L/B}^d} f_{(i,t)}^{[t]} \Big\}^{[j]} \Big\rangle_\mu\Big]^{1/|\mathbb{T}_{L/B}^1|} && (10.14)\\
&= \prod_{i\in\mathbb{T}_{L/B}^1} \Big[\Big\langle \prod_{t\in\mathbb{T}_{L/B}^d} \Big\{ \prod_{j\in\mathbb{T}_{L/B}^1} f_{(i,t)}^{[j]} \Big\}^{[t]} \Big\rangle_\mu\Big]^{1/|\mathbb{T}_{L/B}^1|} . && (10.15)
\end{aligned}$$

The expectation on the right-hand side can be bounded using (10.9) once more, this time with $d' = d$: for each $i \in \mathbb{T}_{L/B}^1$,

$$\begin{aligned}
\Big\langle \prod_{t\in\mathbb{T}_{L/B}^d} \Big\{ \prod_{j\in\mathbb{T}_{L/B}^1} f_{(i,t)}^{[j]} \Big\}^{[t]} \Big\rangle_\mu &\leq \prod_{t\in\mathbb{T}_{L/B}^d} \Big[\Big\langle \prod_{s\in\mathbb{T}_{L/B}^d} \Big\{ \prod_{j\in\mathbb{T}_{L/B}^1} f_{(i,t)}^{[j]} \Big\}^{[s]} \Big\rangle_\mu\Big]^{1/|\mathbb{T}_{L/B}^d|} && (10.16)\\
&= \prod_{t\in\mathbb{T}_{L/B}^d} \Big[\Big\langle \prod_{v\in\mathbb{T}_{L/B}^{d+1}} f_{(i,t)}^{[v]} \Big\rangle_\mu\Big]^{1/|\mathbb{T}_{L/B}^d|} .
\end{aligned}$$

Inserting the latter bound into (10.15),

$$\begin{aligned}
\Big|\Big\langle \prod_{u\in\mathbb{T}_{L/B}^d} f_u^{[u]} \Big\rangle_\mu\Big| &\leq \prod_{i\in\mathbb{T}_{L/B}^1} \prod_{t\in\mathbb{T}_{L/B}^d} \Big[\Big\langle \prod_{v\in\mathbb{T}_{L/B}^{d+1}} f_{(i,t)}^{[v]} \Big\rangle_\mu\Big]^{1/|\mathbb{T}_{L/B}^{d+1}|} \\
&= \prod_{u\in\mathbb{T}_{L/B}^{d+1}} \Big[\Big\langle \prod_{v\in\mathbb{T}_{L/B}^{d+1}} f_u^{[v]} \Big\rangle_\mu\Big]^{1/|\mathbb{T}_{L/B}^{d+1}|} .
\end{aligned}$$

This completes the proof of Theorem 10.11. □

Remark 10.14. Let us make a comment about what was done in the last part of the proof. The verification of certain claims made below is left as an exercise to the reader.

With $\mathbb{T}_L^{d+1} = \mathbb{T}_L^1 \times \mathbb{T}_L^d$, we are naturally led to identify Ω_L, the set of configurations on $\mathbb{T}_L^{d+1}$, with the set of configurations on $\mathbb{T}_L^1$ defined by

$$\widetilde{\Omega}_L \stackrel{\text{def}}{=} \big\{\widetilde{\omega} = (\widetilde{\omega}_i)_{i\in\mathbb{T}_L^1} : \widetilde{\omega}_i \in \widetilde{\Omega}_0\big\} ,$$

where we introduced the new single-spin space

$$\widetilde{\Omega}_0 \stackrel{\text{def}}{=} \bigtimes_{j \in \mathbb{T}_L^d} \Omega_0 .$$

Let us denote this identification by $\phi\colon \Omega_L \to \widetilde{\Omega}_L$. Each $f\colon \Omega_L \to \mathbb{R}$ can be identified with $\widetilde{f}\colon \widetilde{\Omega}_L \to \mathbb{R}$, by $\widetilde{f}(\widetilde{\omega}) \stackrel{\text{def}}{=} f(\phi^{-1}(\widetilde{\omega}))$. The single-spin space $\widetilde{\Omega}_0$ can of course be equipped with its natural σ-algebra of Borel sets, leading to the product σ-algebra $\widetilde{\mathscr{F}}_L$ on $\widetilde{\Omega}_L$. The measure μ on $(\Omega_L, \mathscr{F}_L)$ can be identified with the measure $\widetilde{\mu}$ on $(\widetilde{\Omega}_L, \widetilde{\mathscr{F}}_L)$ defined by $\widetilde{\mu} \stackrel{\text{def}}{=} \mu \circ \phi^{-1}$. We then have

$$\langle f \rangle_\mu = \langle \widetilde{f} \rangle_{\widetilde{\mu}} ,$$

for every bounded measurable function f and, clearly, $\widetilde{\mu}$ is reflection positive with respect to all reflections of $\mathbb{T}_L^1$. This is what guarantees that the one-dimensional chessboard estimate can be used to prove (10.14). A similar argument justifies the second use of the chessboard estimate in (10.16). ◇

Remark 10.15. We will actually make use of a version of Theorem 10.11 in which the cubic block Λ_B (and its translates) is replaced by a rectangular box $\times_{i=1}^d \{0, \dots, B_i\}$ (and its translates) such that $2B_i$ divides L for all i. Of course, the conditions of periodicity and reflection positivity have to be correspondingly modified, but the proof applies essentially verbatim. ◇

10.4.2 Application: The Ising Model in a Large Magnetic Field

In this section, we show a use of the chessboard estimate in the simplest possible setting. A more involved application is described in the following sections.

We have studied the Ising model in a large magnetic field in Section 5.7.1. In particular, we obtained in (5.34) a convergent cluster expansion for the pressure of the model, in terms of $z = e^{-2\beta h}$. When $h > 0$, $\frac{\partial \psi_\beta}{\partial h} = \langle \sigma_0 \rangle_{\beta,h}^+$ and, therefore,

$$\mu_{\beta,h}^+(\sigma_0 = -1) = \tfrac{1}{2}\bigl(1 - \tfrac{\partial \psi_\beta}{\partial h}\bigr) .$$

The expansion (5.34) thus implies that $\mu_{\beta,h}^+(\sigma_0 = -1) = e^{-2h-4d\beta} + O(e^{-4h})$ for $h > 0$ large enough. Here, we show how a simple application of the chessboard estimate leads to an upper bound for this probability (on the torus) valid for all $h, \beta \geq 0$.

For convenience, we write the Hamiltonian of the d-dimensional Ising model on $\mathbb{T}_L$ as

$$\mathscr{H}_{L;\beta,h}(\omega) \stackrel{\text{def}}{=} -\beta \sum_{\{i,j\} \in \mathscr{E}_L} (\omega_i \omega_j - 1) - h \sum_{i \in \mathbb{T}_L} \omega_i . \tag{10.17}$$

Let $\mu_{L;\beta,h}$ be the corresponding Gibbs distribution.

Proposition 10.16. *For all $h \geq 0$, uniformly in L (even) and $\beta \geq 0$,*

$$\mu_{L;\beta,h}(\sigma_0 = -1) \leq e^{-2h}\,. \tag{10.18}$$

Proof The first observation is that $\mathscr{H}_{L;\beta,h}$ can be put in the form (10.8) (up to an irrelevant constant), from which we conclude that $\mu_{L;\beta,h}$ is reflection positive with respect to all reflections through edges.

Using 1×1 blocks (which we naturally identify with the vertices of $\mathbb{T}_L$) and setting $f_0 \stackrel{\text{def}}{=} \mathbf{1}_{\{\sigma_0=-1\}}$ and $f_t \stackrel{\text{def}}{=} 1$ for all $t \in \mathbb{T}_L \setminus \{0\}$, the chessboard estimate yields

$$\langle \mathbf{1}_{\{\sigma_0=-1\}} \rangle_{L;\beta,h} \leq \Big\langle \prod_{s\in\mathbb{T}_L} \mathbf{1}_{\{\sigma_s=-1\}} \Big\rangle_{L;\beta,h}^{1/|\mathbb{T}_L|}. \tag{10.19}$$

(Just observe that all the factors corresponding to $t \neq 0$ in the product in (10.9) are equal to 1.) This can be rewritten as

$$\mu_{L;\beta,h}(\sigma_0 = -1) \leq \mu_{L;\beta,h}(\eta^-)^{1/|\mathbb{T}_L|} = \Big\{ \frac{e^{-\mathscr{H}_{L;\beta,h}(\eta^-)}}{\mathbf{Z}_{L;\beta,h}} \Big\}^{1/|\mathbb{T}_L|},$$

where $\eta_j^- = -1$ for all $j \in \mathbb{T}_L$. On the one hand, $\mathscr{H}_{L;\beta,h}(\eta^-) = h|\mathbb{T}_L|$. On the other hand, we obtain a lower bound on the partition function by keeping only the configuration $\eta^+ \equiv 1$: $\mathbf{Z}_{L;\beta,h} \geq e^{-\mathscr{H}_{L;\beta,h}(\eta^+)} = e^{+h|\mathbb{T}_L|}$. This proves (10.18). □

In probabilistic terms, (10.19) *shows how the chessboard estimate allows us to bound the probability of a* local *event, namely* $\{\sigma_0 = -1\}$, *by the probability of the same event, but "spread out throughout the system":* $\bigcap_{s\in\mathbb{T}_L}\{\sigma_s = -1\}$. *This* global *event is much easier to estimate.* ◇

10.4.3 Application: The Two-Dimensional Anisotropic *XY* Model

We now consider the two-dimensional anisotropic *XY* model, in which the spins take values in $\Omega_0 = \mathbb{S}^1$ and whose Hamiltonian on $\mathbb{T}_L$ is defined by

$$\mathscr{H}_{L;\beta,\alpha} \stackrel{\text{def}}{=} -\beta \sum_{\{i,j\}\in\mathscr{E}_L} \big\{ S_i^1 S_j^1 + \alpha S_i^2 S_j^2 \big\}, \tag{10.20}$$

where $0 \leq \alpha \leq 1$ is the **anisotropy parameter** and we have written $\mathbf{S}_i = (S_i^1, S_i^2)$ for the spin at i. We denote by $\mu_{L;\beta,\alpha}$ the corresponding Gibbs distribution on Ω_L (see (10.5)), with a reference measure ρ on Ω_0 given by the normalized Lebesgue measure (that is, such that $\rho(\Omega_0) = 1$).

To quantify global ordering, we will again use the magnetization density:

$$\mathbf{m}_L \stackrel{\text{def}}{=} \frac{1}{|\mathbb{T}_L|} \sum_{i\in\mathbb{T}_L} \mathbf{S}_i\,,$$

which now takes values in the unit disk $\{u \in \mathbb{R}^2 : \|u\|_2 \le 1\}$. By translation invariance and symmetry,

$$\langle \mathbf{m}_L \rangle_{L;\beta,\alpha} = \langle \mathbf{S}_0 \rangle_{L;\beta,\alpha} = \mathbf{0}.$$

Nevertheless, we will see that the distribution of $\mathbf{m}_L$ is far from uniform at low temperatures when $\alpha < 1$. This, in turn, will lead to the proof of Theorem 10.2.

First, as the following exercise shows, when $\alpha < 1$, this model possesses exactly two ground states: one in which all spins take the value $\mathbf{e}_1$ and one in which this value is $-\mathbf{e}_1$.

Exercise 10.5. Let $\mathbf{S}_i = (S_i^1, S_i^2)$ and $\mathbf{S}_j = (S_j^1, S_j^2)$ be two unit vectors in $\mathbb{R}^2$. Show that, when $0 \le \alpha < 1$, the function

$$f(\mathbf{S}_i, \mathbf{S}_j) \stackrel{\text{def}}{=} -S_i^1 S_j^1 - \alpha S_i^2 S_j^2, \text{ when } 0 \le \alpha < 1,$$

is minimal when either $\mathbf{S}_i = \mathbf{S}_j = \mathbf{e}_1$ or $\mathbf{S}_i = \mathbf{S}_j = -\mathbf{e}_1$.

In view of this, it is reasonable to expect that, at sufficiently low temperature, typical configurations should be given by local perturbations of these two ground states, even in the thermodynamic limit. This is confirmed by the following result.

Theorem 10.17. *For each $0 \le \alpha < 1$ and each $\epsilon > 0$, there exists $\beta_0 = \beta_0(\alpha, \epsilon)$ such that, for all $\beta > \beta_0$,*

$$\left\langle (\mathbf{m}_L \cdot \mathbf{e}_1)^2 \right\rangle_{L;\beta,\alpha} \ge 1 - \epsilon, \qquad \textit{and therefore} \qquad \left\langle (\mathbf{m}_L \cdot \mathbf{e}_2)^2 \right\rangle_{L;\beta,\alpha} \le \epsilon,$$

uniformly in L (multiple of 4).

This result is in sharp contrast with the case $\alpha = 1$ (see Exercise 9.1); it will be a consequence of the orientational long-range order that occurs at low enough temperatures. Observe that

$$\left\langle (\mathbf{m}_L \cdot \mathbf{e}_1)^2 \right\rangle_{L;\beta,\alpha} = \frac{1}{|\mathbb{T}_L|^2} \sum_{i,j \in \mathbb{T}_L} \langle S_i^1 S_j^1 \rangle_{L;\beta,\alpha}.$$

We will use reflection positivity to prove the following result, of which Theorem 10.17 is a direct consequence.

Proposition 10.18. *For each $0 \le \alpha < 1$ and each $\epsilon > 0$, there exists $\beta_0 = \beta_0(\alpha, \epsilon)$ such that, for all $\beta > \beta_0$,*

$$\langle S_i^1 S_j^1 \rangle_{L;\beta,\alpha} \ge 1 - \epsilon, \qquad \forall i, j \in \mathbb{T}_L, \tag{10.21}$$

uniformly in L (multiple of 4).

Proof First, we easily check that $\mathscr{H}_{L;\beta,\alpha}$ can be put in the form (10.8), from which we conclude that $\mu_{L;\beta,\alpha}$ is reflection positive with respect to all reflections through edges of the torus.

Second, since $\mu_{L;\beta,\alpha}$ is invariant under all translations of the torus, we only need to prove that

$$\langle S_0^1 S_j^1 \rangle_{L;\beta,\alpha} \geq 1 - \epsilon(\beta)\,, \tag{10.22}$$

uniformly in L and in $j \in \mathbb{T}_L^2$, with $\epsilon(\beta) \to 0$ when $\beta \to \infty$.

Now, in view of the discussion before Theorem 10.17, we expect that $|S_i^1|$ should be close to 1 for most spins in the torus and that the sign of S_i^1 should be the same at most vertices. To quantify this, let us fix some $\delta \in (0,1)$. If (i) $|S_0^1| \geq \delta$, (ii) $|S_j^1| \geq \delta$ and (iii) $S_0^1 S_j^1 > 0$, then $S_0^1 S_j^1 \geq \delta^2$. Therefore, we can write

$$\langle S_0^1 S_j^1 \rangle_{L;\beta,\alpha} \geq \delta^2 - \mu_{L;\beta,\alpha}(|S_0^1| < \delta) - \mu_{L;\beta,\alpha}(|S_j^1| < \delta) - \mu_{L;\beta,\alpha}(S_0^1 S_j^1 \leq 0)\,. \tag{10.23}$$

Since $\mu_{L;\beta,\alpha}(|S_j^1| < \delta) = \mu_{L;\beta,\alpha}(|S_0^1| < \delta)$ by translation invariance, the claim of Proposition 10.18 follows immediately from Lemmas 10.19 and 10.20 below: choose $\delta^2 = 1 - \frac{1}{4}\epsilon$ and let β be sufficiently large to ensure that the last three terms in (10.23) are smaller than $\epsilon/4$. □

Lemma 10.19. *For any* $0 \leq \alpha < 1$, $0 < \delta < 1$ *and* $\epsilon > 0$, *there exists* $\beta_0' = \beta_0'(\epsilon,\alpha,\delta)$ *such that, for all* $\beta > \beta_0'$,

$$\mu_{L;\beta,\alpha}(|S_0^1| < \delta) \leq \epsilon\,,$$

uniformly in L (even).

Lemma 10.20. *For any* $0 \leq \alpha < 1$, $0 < \delta < 1$ *and* $\epsilon > 0$, *there exists* $\beta_0'' = \beta_0''(\epsilon,\alpha)$ *such that, for all* $\beta > \beta_0''$,

$$\mu_{L;\beta,\alpha}(S_0^1 S_j^1 \leq 0) \leq \epsilon\,,$$

uniformly in $j \in \mathbb{T}_L$ *and L (multiple of 4).*

Proof of Lemma 10.19. We proceed as in the proof of Proposition 10.16. Applying the chessboard estimate, Theorem 10.11, with $d = 2$, $B = 1$, $f_0 \stackrel{\text{def}}{=} \mathbf{1}_{\{|S_0^1| < \delta\}}$ and $f_t \stackrel{\text{def}}{=} 1$ for $t \in \mathbb{T}_L \setminus \{0\}$, we obtain

$$\mu_{L;\beta,\alpha}(|S_0^1| < \delta) \leq \mu_{L;\beta,\alpha}(|S_i^1| < \delta,\ \forall i \in \mathbb{T}_L)^{1/|\mathbb{T}_L|}\,. \tag{10.24}$$

We write

$$\mu_{L;\beta,\alpha}(|S_i^1| < \delta,\ \forall i \in \mathbb{T}_L) = \frac{\big\langle e^{-\mathscr{H}_{L;\beta,\alpha}} \mathbf{1}_{\{|S_i^1| < \delta,\ \forall i \in \mathbb{T}_L\}} \big\rangle_{\mu_0}}{\langle e^{-\mathscr{H}_{L;\beta,\alpha}} \rangle_{\mu_0}}\,, \tag{10.25}$$

where we remind the reader that $\mu_0(\mathrm{d}\omega) = \bigotimes_{i \in \mathbb{T}_L} \rho(\mathrm{d}\omega_i)$, with ρ the uniform probability measure on $\mathbb{S}^1$.

We first bound the numerator in (10.25) from above. When $|S_i^1| < \delta$ for all $i \in \mathbb{T}_L$, a simple computation shows that

$$\mathscr{H}_{L;\beta,\alpha} \geq -\beta \sum_{\{i,j\}\in\mathscr{E}_L} \left(\delta^2 + \alpha(1-\delta^2)\right) = -2\beta\left(\delta^2 + \alpha(1-\delta^2)\right) |\mathbb{T}_L| .$$

(We used the fact that $|\mathscr{E}_L| = 2|\mathbb{T}_L|$ in $d = 2$.) Consequently, since μ_0 is normalized by assumption,

$$\left\langle e^{-\mathscr{H}_{L;\beta,\alpha}} \mathbf{1}_{\{|S_i^1|<\delta,\ \forall i\in\mathbb{T}_L\}} \right\rangle_{\mu_0} \leq e^{2\beta(\delta^2+\alpha(1-\delta^2))\,|\mathbb{T}_L|} . \tag{10.26}$$

To obtain a lower bound on the denominator, let $0 < \tilde{\delta} < 1$ and write

$$\begin{aligned} \langle e^{-\mathscr{H}_{L;\beta,\alpha}} \rangle_{\mu_0} &\geq \left\langle e^{-\mathscr{H}_{L;\beta,\alpha}} \mathbf{1}_{\{S_i^1 \geq \tilde{\delta},\ \forall i\in\mathbb{T}_L\}} \right\rangle_{\mu_0} \\ &= \left\langle e^{-\mathscr{H}_{L;\beta,\alpha}} \ \middle|\ S_i^1 \geq \tilde{\delta},\ \forall i \in \mathbb{T}_L \right\rangle_{\mu_0} \mu_0(S_i^1 \geq \tilde{\delta},\ \forall i \in \mathbb{T}_L) \\ &= \left\langle e^{-\mathscr{H}_{L;\beta,\alpha}} \right\rangle_{\tilde{\mu}_0} \mu_0(S_i^1 \geq \tilde{\delta},\ \forall i \in \mathbb{T}_L), \end{aligned} \tag{10.27}$$

where we have introduced the probability measure $\tilde{\mu}_0(\cdot) \stackrel{\text{def}}{=} \mu(\cdot \mid S_i^1 \geq \tilde{\delta},\ \forall i \in \mathbb{T}_L)$. On the one hand, observe that $\langle S_i^2 \rangle_{\tilde{\mu}_0} = 0$, by symmetry, and $\langle S_i^1 \rangle_{\tilde{\mu}_0} \geq \tilde{\delta}$. Therefore,

$$\langle \mathscr{H}_{L;\beta,\alpha} \rangle_{\tilde{\mu}_0} = -\beta \sum_{\{i,j\}\in\mathscr{E}_L} \langle S_i^1 \rangle_{\tilde{\mu}_0} \langle S_j^1 \rangle_{\tilde{\mu}_0} \leq -2\beta\tilde{\delta}^2\, |\mathbb{T}_L| .$$

So, an application of Jensen's inequality yields

$$\langle e^{-\mathscr{H}_{L;\beta,\alpha}} \rangle_{\tilde{\mu}_0} \geq e^{-\langle \mathscr{H}_{L;\beta,\alpha} \rangle_{\tilde{\mu}_0}} \geq e^{2\beta\tilde{\delta}^2\,|\mathbb{T}_L|} . \tag{10.28}$$

On the other hand,

$$\mu_0(S_i^1 \geq \tilde{\delta},\ \forall i \in \mathbb{T}_L) = \left(\tfrac{1}{\pi} \arccos(\tilde{\delta})\right)^{|\mathbb{T}_L|} = e^{-b(\tilde{\delta})|\mathbb{T}_L|} ,$$

where $b(\tilde{\delta}) \stackrel{\text{def}}{=} -\log(\frac{1}{\pi} \arccos(\tilde{\delta})) > 0$. Inserting this and (10.28) into (10.27) yields

$$\langle e^{-\mathscr{H}_{L;\beta,\alpha}} \rangle_{\mu_0} \geq \exp\left\{ \left(2\beta\tilde{\delta}^2 - b(\tilde{\delta})\right) |\mathbb{T}_L| \right\} . \tag{10.29}$$

Let us then choose $\tilde{\delta}$ such that $\tilde{\delta}^2 = \frac{1}{2}(1 + \delta^2 + \alpha(1-\delta^2)) \in (0,1)$. By (10.26) and (10.29),

$$\begin{aligned} \mu_{L;\beta,\alpha}(|S_i^1| < \delta,\ \forall i \in \mathbb{T}_L) &\leq \exp\left[-\beta\{(1-\delta^2)(1-\alpha) - b(\tilde{\delta})/\beta\}|\mathbb{T}_L|\right] \\ &\leq \exp\left[-\tfrac{1}{2}(1-\delta^2)(1-\alpha)\beta|\mathbb{T}_L|\right], \end{aligned}$$

for all $\beta \geq \beta_1(\alpha,\delta) \stackrel{\text{def}}{=} b(\tilde{\delta})/((1-\delta)^2(1-\alpha))$. By (10.24), this ensures that, for any $\alpha, \delta < 1$ and any $\beta \geq \beta_1(\alpha,\delta)$,

$$\mu_{L;\beta,\alpha}(|S_0^1| < \delta) \leq \exp\left[-\tfrac{1}{2}(1-\delta^2)(1-\alpha)\beta\right] .$$

The right-hand side can be made as small as desired by taking β large enough. □

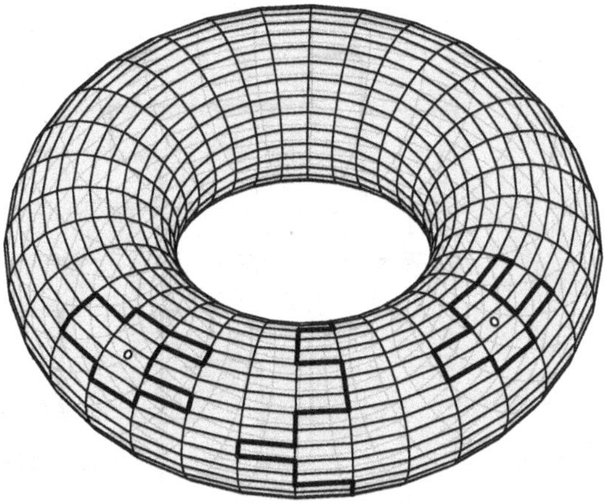

Figure 10.5. The three types of contours on a torus, separating the vertices 0 and j (indicated by the two circles). The mesh corresponds to the dual lattice here, with the vertices in the middle of the faces.

Proof of Lemma 10.20 This proof relies on a variant of Peierls' argument, as shown in Section 3.7.2. We assume that the reader is familiar with this material.

Let, for each $i \in \mathbb{T}_L$, $I_i^+ \stackrel{\text{def}}{=} \mathbf{1}_{\{S_i^1 \geq 0\}}$ and $I_i^- \stackrel{\text{def}}{=} \mathbf{1}_{\{S_i^1 \leq 0\}}$. We have, by symmetry,

$$\mu_{L;\beta,\alpha}(S_0^1 S_j^1 \leq 0) = 2\langle I_0^+ I_j^- \rangle_{L;\beta,\alpha} .$$

Since, almost surely, $I_i^+ + I_i^- = 1$ for all $i \in \mathbb{T}_L$ (namely, $\{S_i^1 = 0\}$ has measure zero under the reference measure and therefore also under $\mu_{L;\beta,\alpha}$), we can write

$$\langle I_0^+ I_j^- \rangle_{L;\beta,\alpha} = \Big\langle I_0^+ I_j^- \prod_{i \in \mathbb{T}_L \setminus \{0,j\}} (I_i^+ + I_i^-) \Big\rangle_{L;\beta,\alpha} = \sum_{\substack{\eta \in \{-1,1\}^{\mathbb{T}_L} \\ \eta_0 = 1, \eta_j = -1}} \Big\langle \prod_{i \in \mathbb{T}_L} I_i^{\eta_i} \Big\rangle_{L;\beta,\alpha} .$$

With each configuration η appearing in the sum, we associate the corresponding set of contours $\Gamma(\eta)$, exactly as in Section 3.7.2 (including the deformation rule). Note, however, that it would not be possible to reconstruct a configuration ω only from the geometry of its contours: the contours only determine the configuration up to a global spin flip. In order to avoid this problem, we consider contours that are not purely geometrical objects, but also include information of the values of the spins on both "sides". When u, v denote neighbors separated by γ, we will use the convention that $\eta_u = +1$ and $\eta_v = -1$.

The configurations η appearing in the sum above are such that $\eta_0 \neq \eta_j$. Therefore, there exists (at least) one contour γ separating 0 and j, in the sense that it satisfies one of the three following conditions (see Figure 10.5): (i) γ surrounds 0 but not j, (ii) γ surrounds j but not 0, (iii) γ is winding around the torus (of course, in this case, there must be at least one other such contour). We can thus write

$$\big\langle I_0^+ I_j^- \big\rangle_{L;\beta,\alpha} \leq \sum_{\gamma} \sum_{\eta:\, \Gamma(\eta) \ni \gamma} \Big\langle \prod_{i \in \mathbb{T}_L} I_i^{\eta_i} \Big\rangle_{L;\beta,\alpha} , \tag{10.30}$$

where the first sum is taken over all contours separating 0 and j.

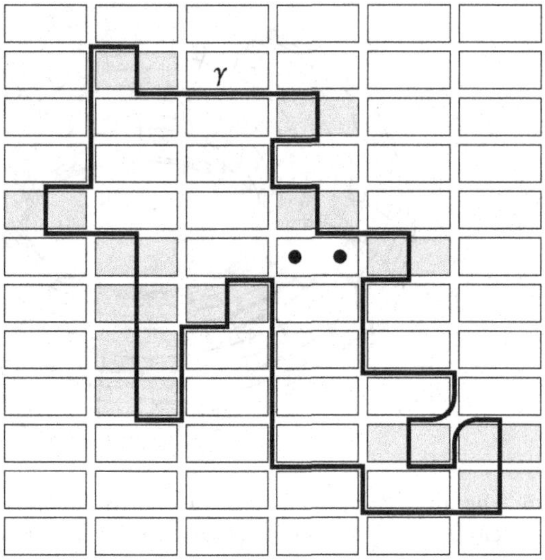

Figure 10.6. The partition of $\mathbb{T}_L$ (here, $L = 12$) into 2×1 blocks, using translates of $\{(0,0),(1,0)\}$ (represented by the two dots). For a contour γ, the set $\mathscr{E}_\gamma^{\mathrm{h},0}$ (whose corresponding blocks are highlighted) represents the blocks of that partition which are crossed by γ in their centre.

Given a contour γ, we denote by $\mathscr{E}_\gamma$ the set of all edges of $\mathscr{E}_L$ that are crossed by γ. Notice that

$$\Big\langle \prod_{\{u,v\}\in\mathscr{E}_\gamma} I_u^+ I_v^- \Big\rangle_{L;\beta,\alpha} = \Big\langle \prod_{\{u,v\}\in\mathscr{E}_\gamma} I_u^+ I_v^- \prod_{k\in\mathbb{T}_L} (I_k^+ + I_k^-) \Big\rangle_{L;\beta,\alpha}$$
$$\geq \sum_{\eta:\,\Gamma(\eta)\ni\gamma} \Big\langle \prod_{i\in\mathbb{T}_L} I_i^{\eta_i} \Big\rangle_{L;\beta,\alpha}. \tag{10.31}$$

The last inequality is due to the fact that forcing $\eta_u \neq \eta_v$ for each $\{u,v\} \in \mathscr{E}_\gamma$ is not sufficient to guarantee that $\gamma \in \Gamma(\eta)$ (remember, in particular, the deformation rule used in the definition of contours). Putting all this together,

$$\mu_{L;\beta,\alpha}(S_0^1 S_j^1 \leq 0) \leq 2\sum_\gamma \Big\langle \prod_{\{u,v\}\in\mathscr{E}_\gamma} I_u^+ I_v^- \Big\rangle_{L;\beta,\alpha}. \tag{10.32}$$

The chessboard estimate will be used to show that the presence of a contour is strongly suppressed when $\alpha < 1$ and β is taken sufficiently large:

$$\Big\langle \prod_{\{u,v\}\in\mathscr{E}_\gamma} I_u^+ I_v^- \Big\rangle_{L;\beta,\alpha} \leq e^{-c(\alpha)\beta|\gamma|}, \tag{10.33}$$

where $c(\alpha) \stackrel{\text{def}}{=} (1-\alpha)/16 > 0$.

In order to use the chessboard estimate, we consider four distinct partitions of the torus into blocks. Consider first the partition of $\mathbb{T}_L$ into blocks of sizes 2×1, translates of $\{(0,0),(1,0)\}$ by all vectors of the form $2m\mathbf{e}_1 + n\mathbf{e}_2$ (m, n are integers)

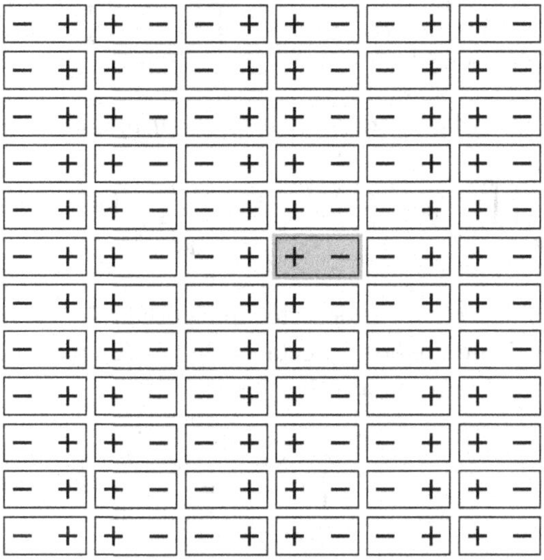

Figure 10.7. The configuration $\eta^{\mathrm{h},0}$ (on the same torus as in Figure 10.6). The shaded block, containing the origin, is the support of ϕ^{+-}, which is then spread throughout the torus by successive reflections through the edges separating the blocks.

(see Figure 10.6). This partition can be identified with the set $\mathscr{E}_L^{\mathrm{h},0} \subset \mathscr{E}_L$ of horizontal nearest-neighbor edges with both endpoints in the same block of the partition. We will use $\{u, v\} \in \mathscr{E}_L^{\mathrm{h},0}$ to index the $|\mathbb{T}_L|/2$ blocks of this partition.

Similarly, one defines the partition made of 2×1 blocks that are translates of $\{(1, 0), (2, 0)\}$; the corresponding set of horizontal edges is written $\mathscr{E}_L^{\mathrm{h},1} \subset \mathscr{E}_L$.

Finally, we define two partitions made of 1×2 blocks, which are, respectively, translates of the block $\{(0, 0), (0, 1)\}$ or of the block $\{(0, 1), (0, 2)\}$. The corresponding sets of vertical edges are denoted $\mathscr{E}_L^{\mathrm{v},0}$ and $\mathscr{E}_L^{\mathrm{v},1}$, respectively.

This leads us to split the edges crossing γ into four families, according to the element of the partition to which they belong: $\mathscr{E}_\gamma = \mathscr{E}_\gamma^{\mathrm{h},0} \cup \mathscr{E}_\gamma^{\mathrm{h},1} \cup \mathscr{E}_\gamma^{\mathrm{v},0} \cup \mathscr{E}_\gamma^{\mathrm{v},1}$. Applying twice the Cauchy–Schwarz inequality,

$$\Big\langle \prod_{\{u,v\}\in\mathscr{E}_\gamma} I_u^+ I_v^- \Big\rangle_{L;\beta,\alpha} \leq \prod_{\substack{a\in\{\mathrm{h},\mathrm{v}\}\\ \#\in\{0,1\}}} \Big(\Big\langle \prod_{\{u,v\}\in\mathscr{E}_\gamma^{a,\#}} I_u^+ I_v^- \Big\rangle_{L;\beta,\alpha} \Big)^{1/4}. \tag{10.34}$$

The four factors on the right-hand side can be treated in the same way. To be specific, we consider the factor with $a = \mathrm{h}$ and $\# = 0$. Notice that, for each $\{u, v\} \in \mathscr{E}_\gamma^{\mathrm{h},0}$, the function $I_u^+ I_v^-$ can be obtained by successive reflections through edges (between the blocks of $\mathscr{E}_L^{\mathrm{h},0}$) of one of the two following $\{(0, 0), (1, 0)\}$-local functions: $\phi^{+-} \overset{\mathrm{def}}{=} I_{(0,0)}^+ I_{(1,0)}^-$ and $\phi^{-+} \overset{\mathrm{def}}{=} I_{(0,0)}^- I_{(1,0)}^+$; we denote the corresponding function $f_{\{u,v\}}$. For each $\{u, v\} \in \mathscr{E}_L^{\mathrm{h},0} \setminus \mathscr{E}_\gamma^{\mathrm{h},0}$, we take $f_{\{u,v\}} \overset{\mathrm{def}}{=} 1$. The chessboard estimate (which we use for nonsquare blocks here, see Remark 10.15) yields

$$\Big\langle \prod_{\{u,v\}\in\mathscr{E}_\gamma^{h,0}} I_u^+ I_v^- \Big\rangle_{L;\beta,\alpha} \leq \prod_{\{u,v\}\in\mathscr{E}_\gamma^{h,0}} \Big\langle \prod_{\{u',v'\}\in\mathscr{E}_\gamma^{h,0}} f_{\{u,v\}}^{[\{u',v'\}]} \Big\rangle_{L;\beta,\alpha}^{1/(|\mathbb{T}_L|/2)}. \tag{10.35}$$

Now, by translation invariance, for all $\{u,v\} \in \mathscr{E}_\gamma^{h,0}$,

$$\Big\langle \prod_{\{u',v'\}\in\mathscr{E}_\gamma^{h,0}} f_{\{u,v\}}^{[\{u',v'\}]} \Big\rangle_{L;\beta,\alpha} = \Big\langle \prod_{\{u',v'\}\in\mathscr{E}_\gamma^{h,0}} (\phi^{+-})^{[\{u',v'\}]} \Big\rangle_{L;\beta,\alpha} = \Big\langle \prod_{i\in\mathbb{T}_L} I_i^{\eta_i^{h,0}} \Big\rangle_{L;\beta,\alpha}, \tag{10.36}$$

where $\eta^{h,0} \in \{\pm 1\}^{\mathbb{T}_L}$ is depicted in Figure 10.7.

To evaluate the last expectation in (10.36), we proceed similarly to what we did in the proof of Lemma 10.19:

$$\Big\langle \prod_{i\in\mathbb{T}_L} I_i^{\eta_i^{h,0}} \Big\rangle_{\mu_0} = \frac{\big\langle e^{-\mathscr{H}_{L;\beta,\alpha}} \prod_{i\in\mathbb{T}_L} I_i^{\eta_i^{h,0}} \big\rangle_{\mu_0}}{\big\langle e^{-\mathscr{H}_{L;\beta,\alpha}} \big\rangle_{\mu_0}}.$$

Let us bound from below the energy of any configuration for which $\prod_{i\in\mathbb{T}_L} I_i^{\eta_i^{h,0}} = 1$. Each edge between two vertices at which $\eta^{h,0}$ takes the same value contributes at least $-\beta$ to the energy, and those edges account for $\frac{3}{4}$ of all the edges of the torus. However, it is easy to check that, for spins located at the endpoints of the remaining edges, the minimal energy is obtained when both their first components vanish; this yields a minimal contribution of $-\beta\alpha$. We conclude that the energy of the relevant configurations is always at least $-2(\frac{3}{4} + \frac{1}{4}\alpha)\beta|\mathbb{T}_L|$ and, therefore,

$$\Big\langle e^{-\mathscr{H}_{L;\beta,\alpha}} \prod_{i\in\mathbb{T}_L} I_i^{\eta_i^{h,0}} \Big\rangle_{\mu_0} \leq \exp\big\{\tfrac{1}{2}\beta(3+\alpha)|\mathbb{T}_L|\big\}.$$

Combining this with the lower bound (10.29), we obtain, choosing $\tilde{\delta}^2 = (7+\alpha)/8$,

$$\begin{aligned}
\Big\langle \prod_{i\in\mathbb{T}_L} I_i^{\eta_i^{h,0}} \Big\rangle_{L;\beta,\alpha} &\leq \exp\Big\{ -2\beta\big(\tilde{\delta}^2 - \frac{b(\tilde{\delta})}{2\beta} - \tfrac{1}{4}(3+\alpha)\big)|\mathbb{T}_L|\Big\} \\
&= \exp\Big\{ -\tfrac{1}{4}\beta\big(1-\alpha-\frac{4b(\tilde{\delta})}{\beta}\big)|\mathbb{T}_L|\Big\} \\
&\leq \exp\big\{-\tfrac{1}{8}(1-\alpha)\beta\,|\mathbb{T}_L|\big\},
\end{aligned}$$

for all $\beta \geq 8b(\tilde{\delta})/(1-\alpha)$. Inserting this into (10.35), we obtain

$$\Big\langle \prod_{\{u,v\}\in\mathscr{E}_\gamma^{h,0}} I_u^+ I_v^- \Big\rangle_{L;\beta,\alpha} \leq \exp\big\{-\tfrac{1}{4}(1-\alpha)\beta\,|\mathscr{E}_\gamma^{h,0}|\big\}.$$

Doing this for the other three partitions, and using (10.34) and the fact that $|\mathscr{E}_\gamma^{h,0}| + |\mathscr{E}_\gamma^{h,1}| + |\mathscr{E}_\gamma^{v,0}| + |\mathscr{E}_\gamma^{v,1}| = |\gamma|$, (10.33) follows. Using this estimate, (10.32) becomes

$$\mu_{L;\beta,\alpha}(S_0^1 S_j^1 \leq 0) \leq 2\sum_\gamma e^{-c(\alpha)\beta|\gamma|}.$$

We can now conclude the proof following the energy–entropy argument used when implementing Peierls' argument in Section 3.7.2. There is only one minor difference: in the sum over γ, there are also contours that wind around the torus, a situation we did not have to consider in Chapter 3. However, since such contours have length at least L,

$$\sum_{\gamma,\,\text{winding}} e^{-c(\alpha)\beta|\gamma|} = \sum_{k\geq L} e^{-c(\alpha)\beta k} \#\{\gamma : \text{winding}, |\gamma| = k\} \leq \sum_{k\geq L} e^{-c(\alpha)\beta k}(L^2 8^k).$$

Taking β sufficiently large, this last sum is bounded uniformly in L and can be made as small as desired. The conclusion thus follows exactly as in Chapter 3. □

Remark 10.21. As the reader can check, the arguments above apply more generally. In particular, they extend readily to the anisotropic $\mathsf{O}(N)$ model, in which the spins $\mathbf{S}_i = (S_i^1, \ldots, S_i^N) \in \mathbb{S}^{N-1}$ and the Hamiltonian is given by

$$\mathscr{H}_{L;\beta,\alpha} = -\beta \sum_{\{i,j\}\in\mathscr{E}_L} \{S_i^1 S_j^1 + \alpha(S_i^2 S_j^2 + \cdots + S_i^N S_j^N)\}.$$

Also, the extension to $d \geq 3$ is rather straightforward (only the implementation of Peierls' argument is affected and can be dealt with as in Exercise 3.20). ◇

We can finally conclude the proof of Theorem 10.2. We will rely on the main result of Section 6.11.

Proof of Theorem 10.2 Let $m_L^1 \stackrel{\text{def}}{=} \mathbf{m}_L \cdot \mathbf{e}_1$ and

$$\begin{aligned}\psi(h) &\stackrel{\text{def}}{=} \lim_{L\to\infty} \frac{1}{|\mathbb{T}_L|} \log \Big\langle \exp\Big\{h \sum_{j\in\mathbb{T}_L} S_j^1\Big\}\Big\rangle_{L;\beta,\alpha} \\ &= \lim_{L\to\infty} \frac{1}{|\mathbb{T}_L|} \log \big\langle e^{hm_L^1|\mathbb{T}_L|}\big\rangle_{L;\beta,\alpha}.\end{aligned}$$

Existence of this limit and its convexity in h follow from Lemma 6.89 (used with $g \stackrel{\text{def}}{=} S_0^1$ and with periodic boundary conditions, for which that result also holds). We have seen that m_L^1 remains bounded away from zero with high probability, uniformly in L, when β is large. We are going to show that this implies that ψ is not differentiable at $h = 0$.

Let $0 \leq \alpha < 1$, $\epsilon > 0$ and $\beta > \beta_0(\alpha, \epsilon)$, where $\beta_0(\alpha, \epsilon)$ was introduced in Theorem 10.17. Also let $0 < \delta < 1$ be such that $\delta^2 < 1 - \epsilon$. To start, observe that, for any $h \geq 0$,

$$\big\langle e^{hm_L^1|\mathbb{T}_L|}\big\rangle_{L;\beta,\alpha} \geq \big\langle e^{hm_L^1|\mathbb{T}_L|}\mathbf{1}_{\{m_L^1\geq\delta\}}\big\rangle_{L;\beta,\alpha} \geq e^{\delta h|\mathbb{T}_L|}\mu_{L;\beta,\alpha}(m_L^1 \geq \delta). \tag{10.37}$$

But Theorem 10.17 implies that, uniformly in L (multiple of 4),

$$1 - \epsilon \leq \big\langle (m_L^1)^2\big\rangle_{L;\beta,\alpha} \leq \delta^2 + \mu_{L;\beta,\alpha}\big((m_L^1)^2 \geq \delta^2\big).$$

Therefore, again uniformly in L (multiple of 4),

$$\mu_{L;\beta,\alpha}(m_L^1 \geq \delta) = \tfrac{1}{2}\mu_{L;\beta,\alpha}\big((m_L^1)^2 \geq \delta^2\big) \geq \tfrac{1}{2}(1 - \epsilon - \delta^2) > 0.$$

Inserting this estimate in (10.37), we conclude that $(\psi(h) - \psi(0))/h = \psi(h)/h \geq \delta$, for all $h > 0$. Letting $h \downarrow 0$ yields

$$\frac{\partial \psi}{\partial h^+}\Big|_{h=0} \geq \delta > 0\,.$$

Since $\psi(-h) = \psi(h)$, this implies that ψ is not differentiable at $h = 0$. Proposition 6.91 then guarantees the existence of two Gibbs measures $\mu^+ \neq \mu^-$ such that

$$\langle S_0^1 \rangle_{\mu^+} = \frac{\partial \psi}{\partial h^+}\Big|_{h=0} > 0 > \frac{\partial \psi}{\partial h^-}\Big|_{h=0} = \langle S_0^1 \rangle_{\mu^-}\,,$$

thereby completing the proof of Theorem 10.2. □

10.5 The Infrared Bound

We now turn to the second major consequence of reflection positivity, the *infrared bound*, which provides one of the few known approaches to proving spontaneous breaking of a continuous symmetry. In this section we go back to the d-dimensional torus $\mathbb{T}_L$, $d \geq 1$.

In order to motivate the infrared bound, we start by showing how it appears as the central tool to prove Theorem 10.1, in Section 10.5.2. The proof of the infrared bound and of the related Gaussian domination is provided in Section 10.5.3.

10.5.1 Models to Be Considered

The infrared bound holds for a wide class of models, but still requires a little more structure than just reflection positivity. Namely, we will assume that the spins are ν-dimensional vectors,

$$\Omega_0 \stackrel{\text{def}}{=} \mathbb{R}^\nu\,,$$

and that the Hamiltonian is given by

$$\mathscr{H}_{L;\beta} \stackrel{\text{def}}{=} \beta \sum_{\{i,j\} \in \mathscr{E}_L} \|\mathbf{S}_i - \mathbf{S}_j\|_2^2\,, \tag{10.38}$$

with $\beta \geq 0$. As before, we assume that the reference measure ρ on Ω_0 (equipped with the Borel subsets of $\mathbb{R}^\nu$) is supported on a compact subset of $\mathbb{R}^\nu$ and write $\mu_0 \stackrel{\text{def}}{=} \bigotimes_{i \in \mathbb{T}_L} \rho$. The Gibbs distribution $\mu_{L;\beta}$ on $(\Omega_L, \mathscr{F}_L)$, associated with $\mathscr{H}_{L;\beta}$, is then defined exactly as in (10.5).

The choice of the reference measure ρ leads to various interesting models encountered in previous chapters.

Example 10.22. Choose $\nu = N$ and let ρ be the Lebesgue measure on the sphere $\mathbb{S}^{N-1} \subset \mathbb{R}^N$. Since $\|\mathbf{S}_i\|_2 = 1$ for all $i \in \mathbb{T}_L$, almost surely, the Hamiltonian can be rewritten as

$$\mathscr{H}_{L;\beta} = 2\beta|\mathscr{E}_L| - 2\beta \sum_{\{i,j\}\in\mathscr{E}_L} \mathbf{S}_i \cdot \mathbf{S}_j .$$

We recognize (up to an irrelevant constant $2\beta|\mathscr{E}_L|$) the Hamiltonian of the **O(N) model.** ◇

Example 10.23. Choose $\nu = q - 1$ and let ρ be the uniform distribution concentrated on the vertices of the regular ν-simplex (see Figure 10.8). The vertices of this simplex lie on $\mathbb{S}^{\nu-1}$, and the scalar product of any two vectors from the origin to two distinct vertices of the simplex is always the same. Note that this is just the **q-state Potts model** in disguise. Indeed, a configuration can almost surely be identified with a configuration $\omega' \in \{1,\dots,q\}^{\mathbb{T}_L}$, where $1,\dots,q$ is a numbering of the vertices of the simplex. Then, up to an irrelevant constant, we see that the Hamiltonian becomes $-\beta_\nu \sum_{\{i,j\}\in\mathscr{E}_L} \delta_{\omega'_i,\omega'_j}$, for some $\beta_\nu \geq 0$ proportional to β. ◇

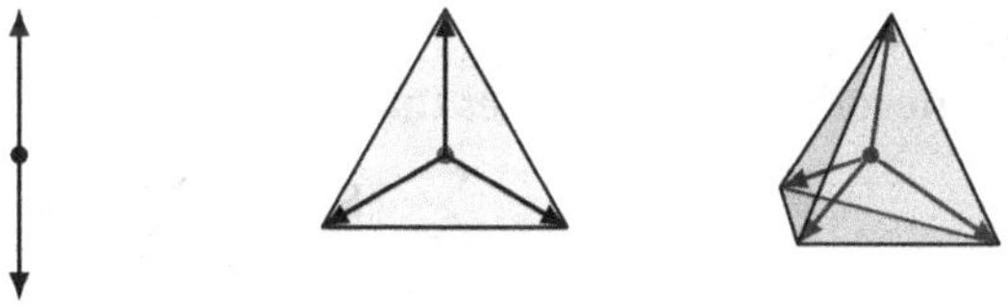

Figure 10.8. The simplex representation for the 2-, 3- and 4-state Potts model.

10.5.2 Application: Orientational Long-Range Order in the O(N) Model

In order to motivate the infrared bound, we start with one of its major applications: the proof that, when $d \geq 3$, there is orientational long-range order at low temperatures for models with continuous spins of the type described above.

We have seen, when proving the Mermin–Wagner Theorem in Chapter 9, that the absence of orientational long-range order in the two-dimensional O(N) model was due to the fact that *spin waves*, by which we meant spin configurations varying slowly over macroscopic regions, were created in the system, at arbitrarily low cost (remember Figure 9.3). If we want to establish orientational long-range order, we have to exclude the existence of such excitations. In order to do that, it is very convenient to consider the *Fourier representation* of the variables $(\mathbf{S}_j)_{j\in\mathbb{T}_L}$.

Consider the **reciprocal torus**, defined by

$$\mathbb{T}_L^\star \stackrel{\text{def}}{=} \left\{ \frac{2\pi}{L}(n_1,\dots,n_d) \,:\, 0 \leq n_i < L \right\} .$$

Note that $|\mathbb{T}_L^\star| = |\mathbb{T}_L|$. The **Fourier transform** of $(\mathbf{S}_j)_{j\in\mathbb{T}_L}$ is $(\widehat{\mathbf{S}}_p)_{p\in\mathbb{T}_L^\star}$, defined by

$$\widehat{\mathbf{S}}_p \stackrel{\text{def}}{=} \frac{1}{|\mathbb{T}_L|^{1/2}} \sum_{j\in\mathbb{T}_L} e^{\mathrm{i}p\cdot j} \mathbf{S}_j , \qquad p \in \mathbb{T}_L^\star .$$

Let us recall two important properties:

1. The original variables can be reconstructed from their Fourier transform, by the inversion formula:

$$\mathbf{S}_j = \frac{1}{|\mathbb{T}_L^\star|^{1/2}} \sum_{p \in \mathbb{T}_L^\star} e^{-\mathrm{i}p\cdot j} \widehat{\mathbf{S}}_p \,, \qquad j \in \mathbb{T}_L \,.$$

   Each index p is called a **mode** and corresponds to an oscillatory term $e^{-\mathrm{i}p\cdot j}$. This sum should be interpreted as the contributions of the different Fourier modes to the field variable $\mathbf{S}_j$. The importance of mode p is measured by $\|\widehat{\mathbf{S}}_p\|_2$. On the one hand, modes with small values of p describe slow variations of $\mathbf{S}_j$, meaning variations detectable only on macroscopic regions, at the scale of the torus. In particular, the mode $p = 0$ corresponds to the nonoscillating ("infinite wavelength") component of $\mathbf{S}_j$ and is proportional to the magnetization of the system (see below). On the other hand, modes with large p represent rapid oscillations present in $\mathbf{S}_j$.
2. **Plancherel's Theorem** states that

$$\sum_{p \in \mathbb{T}_L^\star} \|\widehat{\mathbf{S}}_p\|_2^2 = \sum_{j \in \mathbb{T}_L} \|\mathbf{S}_j\|_2^2 \,. \tag{10.39}$$

Exercise 10.6. Prove the above two properties.

As mentioned above, the magnetization density $\mathbf{m}_L = \frac{1}{|\mathbb{T}_L|}\sum_{i \in \mathbb{T}_L} \mathbf{S}_i$ is simply related to the $p = 0$ mode by

$$\mathbf{m}_L = \frac{1}{|\mathbb{T}_L|^{1/2}} \widehat{\mathbf{S}}_0 \,.$$

Therefore, the importance of the $p = 0$ mode characterizes the presence or absence of orientational long-range order in the system. For example, we have seen in Exercise 9.1 that the contribution of the $p = 0$ mode becomes negligible in the thermodynamic limit for the two-dimensional *XY* model; this was due to the appearance of spin waves. Therefore, to prove that orientational long-range order *does* occur, one must show that the $p = 0$ mode has a nonzero contribution even in the thermodynamic limit.

In order to do this, we add a new restriction to the class of models we consider. Namely, we assume in the rest of this section that the reference measure ρ is such that, almost surely,

$$\|\mathbf{S}_j\|_2 = 1, \quad \forall j \in \mathbb{T}_L \,.$$

This is of course the case in the $\mathsf{O}(N)$ and Potts models. With this assumption, (10.39) implies that $\sum_{p \in \mathbb{T}_L^\star} \|\widehat{\mathbf{S}}_p\|_2^2 = |\mathbb{T}_L|$, which yields

$$\|\widehat{\mathbf{S}}_0\|_2^2 = |\mathbb{T}_L| - \sum_{\substack{p \in \mathbb{T}_L^\star \\ p \neq 0}} \|\widehat{\mathbf{S}}_p\|_2^2 \,.$$

Moreover, by translation invariance,

$$\langle \|\hat{\mathbf{S}}_p\|_2^2 \rangle_{L;\beta} = \frac{1}{|\mathbb{T}_L|} \sum_{i,j \in \mathbb{T}_L} e^{\mathrm{i}p\cdot(j-i)} \langle \mathbf{S}_i \cdot \mathbf{S}_j \rangle_{L;\beta} = \sum_{j \in \mathbb{T}_L} e^{\mathrm{i}p\cdot j} \langle \mathbf{S}_0 \cdot \mathbf{S}_j \rangle_{L;\beta} .$$

Gathering these identities, we conclude that

$$\langle \|\mathbf{m}_L\|_2^2 \rangle_{L;\beta} = \frac{1}{|\mathbb{T}_L|} \langle \|\widehat{\mathbf{S}}_0\|_2^2 \rangle_{L;\beta} = 1 - \Big\{ \frac{1}{|\mathbb{T}_L|} \sum_{\substack{p \in \mathbb{T}_L^\star \\ p \neq 0}} \sum_{j \in \mathbb{T}_L} e^{\mathrm{i}p\cdot j} \langle \mathbf{S}_0 \cdot \mathbf{S}_j \rangle_{L;\beta} \Big\} . \tag{10.40}$$

To obtain a lower bound on $\langle \|\mathbf{m}_L\|_2^2 \rangle_{L;\beta}$, we thus need to find an upper bound on the double sum appearing on the right-hand side of the previous display. It is precisely at this stage that the infrared bound becomes crucial; its proof will be provided in Section 10.5.3.

> **Theorem 10.24** (Infrared bound). *Let $\mu_{L;\beta}$ be the Gibbs distribution associated with a Hamiltonian of the form* (10.38). *Then, for any $p \in \mathbb{T}_L^\star \setminus \{0\}$,*
>
> $$\sum_{j \in \mathbb{T}_L} e^{\mathrm{i}p\cdot j} \langle \mathbf{S}_0 \cdot \mathbf{S}_j \rangle_{L;\beta} \leq \frac{\nu}{4\beta d} \Big\{ 1 - \frac{1}{2d} \sum_{j \sim 0} \cos(p \cdot j) \Big\}^{-1} .$$

Using the infrared bound in (10.40), we get

$$\langle \|\mathbf{m}_L\|_2^2 \rangle_{L;\beta} \geq 1 - \frac{\nu}{4\beta d} \frac{1}{|\mathbb{T}_L|} \sum_{\substack{p \in \mathbb{T}_L^\star \\ p \neq 0}} \Big\{ 1 - \frac{1}{2d} \sum_{j \sim 0} \cos(p \cdot j) \Big\}^{-1} .$$

The reader can recognize a Riemann sum on the right-hand side, which implies that

$$\begin{aligned} \beta_0 &\stackrel{\text{def}}{=} \frac{\nu}{4d} \lim_{L \to \infty} \frac{1}{|\mathbb{T}_L|} \sum_{\substack{p \in \mathbb{T}_L^\star \\ p \neq 0}} \Big\{ 1 - \frac{1}{2d} \sum_{j \sim 0} \cos(p \cdot j) \Big\}^{-1} \\ &= \frac{\nu}{4d} \int_{[-\pi,\pi]^d} \Big\{ 1 - \frac{1}{2d} \sum_{j \sim 0} \cos(p \cdot j) \Big\}^{-1} \mathrm{d}p . \end{aligned} \tag{10.41}$$

Notice that this integral is improper, precisely because of the singularity of the integrand at $p = 0$. Therefore,

$$\liminf_{L \to \infty} \langle \|\mathbf{m}_L\|_2^2 \rangle_{L;\beta} \geq 1 - \frac{\beta_0}{\beta} .$$

This proves that there is orientational long-range order for all $\beta > \beta_0$. The only remaining task is to make sure that β_0 is indeed finite.

It turns out (Theorem B.72) that β_0 is finite if and only if the symmetric simple random walk on $\mathbb{Z}^d$ is transient. As shown in Corollary B.73 (by directly studying the integral above), this occurs if and only if $d \geq 3$.

We have thus proved the following result.

Theorem 10.25. *Assume that $d \geq 3$. Let $\mu_{L;\beta}$ be defined with respect to a reference measure μ_0 under which $\|\mathbf{S}_i\|_2 = 1$, almost surely, for all $i \in \mathbb{T}_L$. Let*

$$\beta_0 \stackrel{\text{def}}{=} \frac{\nu}{4d} \int_{[-\pi,\pi]^d} \Big\{1 - \frac{1}{2d} \sum_{j\sim 0} \cos(p \cdot j)\Big\}^{-1} \mathrm{d}p.$$

Then $\beta_0 < \infty$ and, for any $\beta > \beta_0$,

$$\liminf_{L\to\infty} \langle \|\mathbf{m}_L\|_2^2 \rangle_{L;\beta} \geq 1 - \frac{\beta_0}{\beta}. \tag{10.42}$$

Remark 10.26. Theorem 10.25 implies the existence of orientational long-range order for the q-state Potts model on $\mathbb{Z}^d$, $d \geq 3$. The latter, however, displays orientational long-range order also in dimension 2, even though this cannot be inferred from the infrared bound. (This can be proved, for example, via a Peierls argument as in Section 3.7.2 or using the chessboard estimate, using a variant of the proof of Lemma 10.19.) The crucial difference, of course, is that the symmetry group is discrete in this case. ◇

With the help of Theorem 10.25, we can now prove the existence of a continuum of distinct Gibbs states in such models, as stated in Theorem 10.1.

Proof of Theorem 10.1. First, fix some unit vector $\mathbf{e} \in \mathbb{S}^{N-1}$. For simplicity and with no loss of generality, we can take $\mathbf{e} = \mathbf{e}_1$ (indeed, $\mu_{L;\beta}$ is invariant under any global rotation of the spins). Then, define

$$\psi(h) \stackrel{\text{def}}{=} \lim_{L\to\infty} \frac{1}{|\mathbb{T}_L|} \log \Big\langle \exp\Big\{h \sum_{j\in\mathbb{T}_L} \mathbf{S}_j \cdot \mathbf{e}_1\Big\}\Big\rangle_{L;\beta}.$$

We again use Theorem 10.25 to show that ψ is not differentiable at $h = 0$, following the pattern used to prove Theorem 10.2, and conclude using Proposition 6.91. The only difference here is when showing that $\mu_{L;\beta}(\mathbf{m}_L \cdot \mathbf{e}_1 \geq \delta)$ is bounded away from zero. To use the lower bound we have on $\langle \|\mathbf{m}_L\|_2 \rangle_{L;\beta}$, we can use the following comparisons:

$$\mu_{L;\beta}(\mathbf{m}_L \cdot \mathbf{e}_1 \geq \delta) \geq \tfrac{1}{2d} \mu_{L;\beta}(\|\mathbf{m}_L\|_\infty \geq \delta) \geq \tfrac{1}{2d} \mu_{L;\beta}(\|\mathbf{m}_L\|_2 \geq \delta\sqrt{d}).$$

When δ is sufficiently small, a lower bound on the latter, uniform in L, can be obtained as before. □

10.5.3 Gaussian Domination and the Infrared Bound

The infrared bound relies on the following proposition, whose proof will use reflection positivity. Let $h = (h_i)_{i\in\mathbb{T}_L} \in (\mathbb{R}^\nu)^{\mathbb{T}_L}$ and

$$\mathbf{Z}_{L;\beta}(h) \stackrel{\text{def}}{=} \Big\langle \exp\Big\{-\beta \sum_{\{i,j\}\in\mathscr{E}_L} \|\mathbf{S}_i - \mathbf{S}_j + h_i - h_j\|_2^2\Big\}\Big\rangle_{\mu_0}.$$

Notice that $\mathbf{Z}_{L;\beta}(h)$ is well defined (since we are assuming ρ to have compact support) and that $\mathbf{Z}_{L;\beta}(0)$ coincides with the partition function $\mathbf{Z}_{L;\beta}$ associated with $\mathscr{H}_{L;\beta}$.

Proposition 10.27 (Gaussian domination). *For all* $h = (h_i)_{i\in\mathbb{T}_L}$,

$$\mathbf{Z}_{L;\beta}(h) \le \mathbf{Z}_{L;\beta}(0)\,. \tag{10.43}$$

Proposition 10.27 will be a consequence of the following lemma, which is a version of the Cauchy-Schwarz-type inequality of Lemma 10.6.

Lemma 10.28. *Let* $\mu \in \mathcal{M}_{\mathrm{RP}(\Theta)}$ *and* $A, B, C_\alpha, D_\alpha \in \mathfrak{A}_+(\Theta)$. *Then*

$$\big\{\big\langle e^{A+\Theta(B)+\sum_\alpha C_\alpha\Theta(D_\alpha)}\big\rangle_\mu\big\}^2 \le \big\langle e^{A+\Theta(A)+\sum_\alpha C_\alpha\Theta(C_\alpha)}\big\rangle_\mu \big\langle e^{B+\Theta(B)+\sum_\alpha D_\alpha\Theta(D_\alpha)}\big\rangle_\mu\,.$$

Proof Expanding the exponential,

$$\big\langle e^{A+\Theta(B)+\sum_\alpha C_\alpha\Theta(D_\alpha)}\big\rangle_\mu = \sum_{n\ge 0}\frac{1}{n!}\sum_{\alpha_1,\dots,\alpha_n}\big\langle e^A C_{\alpha_1}\cdots C_{\alpha_n}\Theta(e^B D_{\alpha_1}\cdots D_{\alpha_n})\big\rangle_\mu\,. \tag{10.44}$$

By Lemma 10.6,

$$\begin{aligned}\big\langle e^A C_{\alpha_1}\cdots C_{\alpha_n}\Theta(e^B D_{\alpha_1}\cdots D_{\alpha_n})\big\rangle_\mu &\le \big\langle e^A C_{\alpha_1}\cdots C_{\alpha_n}\Theta(e^A C_{\alpha_1}\cdots C_{\alpha_n})\big\rangle_\mu^{1/2}\\ &\quad\times \big\langle e^B D_{\alpha_1}\cdots D_{\alpha_n}\Theta(e^B D_{\alpha_1}\cdots D_{\alpha_n})\big\rangle_\mu^{1/2}.\end{aligned}$$

The classical Cauchy-Schwarz inequality, $\sum_k |a_k b_k| \le (\sum_k a_k^2)^{1/2}(\sum_k b_k^2)^{1/2}$, then yields

$$\begin{aligned}&\sum_{\alpha_1,\dots,\alpha_n}\langle e^A C_{\alpha_1}\cdots C_{\alpha_n}\Theta(e^B D_{\alpha_1}\cdots D_{\alpha_n})\rangle_\mu\\ &\quad\le \Big[\sum_{\alpha_1,\dots,\alpha_n}\big\langle e^A C_{\alpha_1}\cdots C_{\alpha_n}\Theta(e^A C_{\alpha_1}\cdots C_{\alpha_n})\big\rangle_\mu\Big]^{1/2}\\ &\quad\quad\times\Big[\sum_{\alpha_1,\dots,\alpha_n}\big[\big\langle e^B D_{\alpha_1}\cdots D_{\alpha_n}\Theta(e^B D_{\alpha_1}\cdots D_{\alpha_n})\big\rangle_\mu\Big]^{1/2}.\end{aligned}$$

Inserting this in (10.44), again using the Cauchy-Schwarz inequality (this time, to the sum over n) and resumming the series, we get

$$\begin{aligned}\big\langle e^{A+\Theta(B)+\sum_\alpha C_\alpha\Theta(D_\alpha)}\big\rangle_\mu &\le \Big[\sum_{n\ge 0}\frac{1}{n!}\sum_{\alpha_1,\dots,\alpha_n}\big\langle e^A C_{\alpha_1}\cdots C_{\alpha_n}\Theta(e^A C_{\alpha_1}\cdots C_{\alpha_n})\big\rangle_\mu\Big]^{1/2}\\ &\quad\times\Big[\sum_{n\ge 0}\frac{1}{n!}\sum_{\alpha_1,\dots,\alpha_n}\big\langle e^B D_{\alpha_1}\cdots D_{\alpha_n}\Theta(e^B D_{\alpha_1}\cdots D_{\alpha_n})\big\rangle_\mu\Big]^{1/2}\\ &= \Big[\big\langle e^{A+\Theta(A)+\sum_\alpha C_\alpha\Theta(C_\alpha)}\big\rangle_\mu\big\langle e^{B+\Theta(B)+\sum_\alpha D_\alpha\Theta(D_\alpha)}\big\rangle_\mu\Big]^{1/2}.\end{aligned}$$

□

Proof of Proposition 10.27. First, notice that $\mathbf{Z}_{L;\beta}(h) = \mathbf{Z}_{L;\beta}(h')$ whenever there exists $c \in \mathbb{R}$ such that $h_i - h'_i = c$ for all $i \in \mathbb{T}_L$. There is thus no loss of generality in assuming that $h_0 = 0$, which we do from now on. Next, observe that $\mathbf{Z}_{L;\beta}(h)$ tends to 0 as any $\|h_i\|_2 \to \infty$, $i \neq 0$. In particular, there exists C such that $\sum_i \|h_i\|_2^2 \leq C$ for all h that maximize $\mathbf{Z}_{L;\beta}(h)$. Among the latter, let us denote by $h^\star$ a maximizer that minimizes the quantity $N(h) \stackrel{\text{def}}{=} \#\big\{\{i,j\} \in \mathscr{E}_L : h_i \neq h_j\big\}$. We claim that $N(h^\star) = 0$. Since $h^\star_0 = 0$, this will then imply that $h^\star_i = 0$ for all $i \in \mathbb{T}_L$, which will conclude the proof.

Let us therefore suppose to the contrary that $N(h^\star) > 0$. In that case, we can find $\{i,j\} \in \mathscr{E}_L$ such that $h^\star_i \neq h^\star_j$. Let Π be the reflection plane going through the middle of the edge $\{i,j\}$ and let Θ denote the reflection through Π. Below, we use $\{i',j'\}$ to denote the edges that cross Π, with $i' \in \mathbb{T}_{L,+}$ and $j' = \Theta(i') \in \mathbb{T}_{L,-}$. Since $\|\omega_{i'} - \omega_{j'} + h_{i'} - h_{j'}\|_2^2 = \|\omega_{i'} + h_{i'}\|_2^2 + \|\omega_{j'} + h_{j'}\|_2^2 - 2(\omega_{i'} + h_{i'}) \cdot (\omega_{j'} + h_{j'})$, we can write

$$-\beta \sum_{\{i,j\} \in \mathscr{E}_L} \|\omega_i - \omega_j + h_i - h_j\|_2^2 = A + \Theta(B) + \sum_{i'} C_{i'} \cdot \Theta(D_{i'}),$$

where $A, B, C_i, D_i \in \mathfrak{A}_+(\Theta)$, and

$$A \stackrel{\text{def}}{=} -\beta \sum_{\substack{\{i,j\} \in \mathscr{E}_L: \\ i,j \in \mathbb{T}_{L,+}(\Theta)}} \|\omega_i - \omega_j + h_i - h_j\|_2^2 - \beta \sum_{i'} \|\omega_{i'} + h_{i'}\|_2^2,$$

$$\Theta(B) \stackrel{\text{def}}{=} -\beta \sum_{\substack{\{i,j\} \in \mathscr{E}_L: \\ i,j \in \mathbb{T}_{L,-}(\Theta)}} \|\omega_i - \omega_j + h_i - h_j\|_2^2 - \beta \sum_{j'} \|\omega_{j'} + h_{j'}\|_2^2,$$

$$C_{i'} \stackrel{\text{def}}{=} \sqrt{2\beta}(\omega_{i'} + h_{i'}), \qquad \Theta(D_{i'}) \stackrel{\text{def}}{=} \sqrt{2\beta}(\omega_{j'} + h_{j'}).$$

(Remember that Θ acts on ω, not on h; this implies that, in general, $A \neq B$ and $C_{i'} \neq D_{i'}$.) One can thus use Lemma 10.28 to obtain

$$\mathbf{Z}_{L;\beta}(h^\star)^2 \leq \mathbf{Z}_{L;\beta}(h^{\star,+})\mathbf{Z}_{L;\beta}(h^{\star,-}),$$

where

$$h^{\star,+}_i = \begin{cases} h^\star_i & \forall i \in \mathbb{T}_{L,+}(\Theta), \\ h^\star_{\Theta(i)} & \forall i \in \mathbb{T}_{L,-}(\Theta), \end{cases} \qquad h^{\star,-}_i = \begin{cases} h^\star_i & \forall i \in \mathbb{T}_{L,-}(\Theta), \\ h^\star_{\Theta(i)} & \forall i \in \mathbb{T}_{L,+}(\Theta). \end{cases}$$

Our choice of Θ guarantees that $\min\{N(h^{\star,+}), N(h^{\star,-})\} < N(h^\star)$. To be specific, let us assume that $N(h^{\star,+}) < N(h^\star)$. Then, since $h^\star$ is a maximizer,

$$\mathbf{Z}_{L;\beta}(h^\star)^2 \leq \mathbf{Z}_{L;\beta}(h^{\star,+})\mathbf{Z}_{L;\beta}(h^{\star,-}) \leq \mathbf{Z}_{L;\beta}(h^{\star,+})\mathbf{Z}_{L;\beta}(h^\star),$$

that is, $\mathbf{Z}_{L;\beta}(h^{\star,+}) \geq \mathbf{Z}_{L;\beta}(h^\star)$. This implies that $h^{\star,+}$ is also a maximizer that satisfies $N(h^{\star,+}) < N(h^\star)$. This contradicts our choice of $h^\star$, and therefore implies that $N(h^\star) = 0$. □

The following exercise provides some motivation for the terminology "Gaussian domination". One can define the discrete Laplacian of $h = (h_i)_{i\in\mathbb{T}_L}$, Δh, by

$$(\Delta h)_i \stackrel{\text{def}}{=} \sum_{j\sim i}(h_j - h_i), \qquad i \in \mathbb{Z}^d .$$

The discrete Green identities of Lemma 8.7 can also be used here; they take slightly simpler forms due to the absence of boundary terms on the torus.

Exercise 10.7. Show that (10.43) can be rewritten as

$$\Big\langle \exp\Big\{2\beta \sum_{i\in\mathbb{T}_L}(\Delta h)_i \cdot (\mathbf{S}_i - \mathbf{S}_0)\Big\}\Big\rangle_{L;\beta} \le \exp\Big\{-\beta \sum_{i\in\mathbb{T}_L}(\Delta h)_i \cdot h_i\Big\} . \tag{10.45}$$

Let $\nu_{L;\beta}$ be the Gibbs distribution corresponding to the prior measure given by the Lebesgue measure: $\rho(d\omega_i) = d\omega_i$. Show that

$$\Big\langle \exp\Big\{2\beta \sum_{i\in\mathbb{T}_L}(\Delta h)_i \cdot (\mathbf{S}_i - \mathbf{S}_0)\Big\}\Big\rangle_{\nu_{L;\beta}} = \exp\Big\{-\beta \sum_{i\in\mathbb{T}_L}(\Delta h)_i \cdot h_i\Big\},$$

so that the bound (10.45) is saturated by the Gaussian measure $\nu_{L;\beta}$.

We can now turn to the proof of the infrared bound.

Proof of Theorem 10.24. We know from Proposition 10.27 that $\mathbf{Z}_{L;\beta}(h)$ is maximal at $h \equiv 0$. Consequently, at fixed h,

$$\frac{\partial}{\partial\lambda}\mathbf{Z}_{L;\beta}(\lambda h)\big|_{\lambda=0} = 0 \quad \text{and} \quad \frac{\partial^2}{\partial\lambda^2}\mathbf{Z}_{L;\beta}(\lambda h)\big|_{\lambda=0} \le 0 . \tag{10.46}$$

The first claim in (10.46) does not provide any nontrivial information, but the second one is instrumental in the proof. Elementary computations show that

$$\begin{aligned}
&\frac{\partial^2}{\partial\lambda^2}\mathbf{Z}_{L;\beta}(\lambda h)\big|_{\lambda=0} \\
&\quad = 4\beta^2\Big\langle \Big|\sum_{\{i,j\}\in\mathscr{E}_L}(\mathbf{S}_i - \mathbf{S}_j)\cdot(h_i - h_j)\Big|^2 \exp\Big\{-\beta\sum_{\{i,j\}\in\mathscr{E}_L}\|\mathbf{S}_i - \mathbf{S}_j\|_2^2\Big\}\Big\rangle_{\mu_0} \\
&\qquad - 2\beta\sum_{\{i,j\}\in\mathscr{E}_L}\|h_i - h_j\|_2^2 \Big\langle \exp\Big\{-\beta\sum_{\{i,j\}\in\mathscr{E}_L}\|\mathbf{S}_i - \mathbf{S}_j\|_2^2\Big\}\Big\rangle_{\mu_0} .
\end{aligned}$$

The inequality in (10.46) is thus equivalent to

$$\Big\langle \Big|\sum_{\{i,j\}\in\mathscr{E}_L}(\mathbf{S}_i - \mathbf{S}_j)\cdot(h_i - h_j)\Big|^2\Big\rangle_{L;\beta} \le \frac{1}{2\beta}\sum_{\{i,j\}\in\mathscr{E}_L}\|h_i - h_j\|_2^2 . \tag{10.47}$$

The latter holds for any $h \in (\mathbb{R}^\nu)^{\mathbb{T}_L}$, but it is easily seen that it also extends to any $h \in (\mathbb{C}^\nu)^{\mathbb{T}_L}$ (just treat the real and imaginary parts separately). Let us fix $p \in \mathbb{T}_L^\star\setminus\{0\}$, $\ell \in \{1,\dots,\nu\}$, and make the following specific choice:

$$\forall j \in \mathbb{T}_L, \qquad \alpha_j \stackrel{\text{def}}{=} e^{\mathrm{i}\, p\cdot j}, \qquad h_j \stackrel{\text{def}}{=} \alpha_j \mathbf{e}_\ell .$$

The Green identity (8.14) yields ($\bar{\alpha}$ denoting the complex conjugate of α)

$$\sum_{\{i,j\}\in\mathscr{E}_L} \|h_i - h_j\|_2^2 = \sum_{\{i,j\}\in\mathscr{E}_L} (\nabla\bar{\alpha})_{ij}(\nabla\alpha)_{ij} = \sum_{i\in\mathbb{T}_L} \bar{\alpha}_i(-\Delta\alpha)_i$$
$$= 2d|\mathbb{T}_L|\Big\{1 - \frac{1}{2d}\sum_{j\sim 0}\cos(p\cdot j)\Big\},$$

since, for any $i \in \mathbb{T}_L$,

$$(-\Delta\alpha)_i = \sum_{j\sim i}(\alpha_i - \alpha_j) = e^{\mathrm{i}p\cdot i}\sum_{j\sim i}(1 - e^{\mathrm{i}p\cdot(j-i)}) = 2de^{\mathrm{i}p\cdot i}\Big\{1 - \frac{1}{2d}\sum_{j\sim 0}\cos(p\cdot j)\Big\}.$$

Similarly, denoting by $S_i^\ell \stackrel{\text{def}}{=} \mathbf{S}_i \cdot \mathbf{e}_\ell$ the ℓth component of $\mathbf{S}_i$,

$$\sum_{\{i,j\}\in\mathscr{E}_L} (\mathbf{S}_i - \mathbf{S}_j)\cdot(h_i - h_j) = \sum_{\{i,j\}\in\mathscr{E}_L} (\nabla S^\ell)_{ij}(\nabla\alpha)_{ij} = \sum_{i\in\mathbb{T}_L} S_i^\ell(-\Delta\alpha)_i$$
$$= 2d\Big\{1 - \frac{1}{2d}\sum_{j\sim 0}\cos(p\cdot j)\Big\}\sum_{i\in\mathbb{T}_L} S_i^\ell e^{\mathrm{i}p\cdot i},$$

and therefore

$$\Big\langle\Big|\sum_{\{i,j\}\in\mathscr{E}_L} (\mathbf{S}_i - \mathbf{S}_j)\cdot(h_i - h_j)\Big|^2\Big\rangle_{L;\beta} = 4d^2\Big|1 - \frac{1}{2d}\sum_{j\sim 0}\cos(p\cdot j)\Big|^2\Big\langle\Big|\sum_{i\in\mathbb{T}_L} S_i^\ell e^{\mathrm{i}p\cdot i}\Big|^2\Big\rangle_{L;\beta}.$$

The inequality (10.47) thus implies that

$$\Big\langle\Big|\sum_{i\in\mathbb{T}_L} S_j^\ell e^{\mathrm{i}p\cdot i}\Big|^2\Big\rangle_{L;\beta} \le \frac{|\mathbb{T}_L|}{4d\beta}\Big\{1 - \frac{1}{2d}\sum_{j\sim 0}\cos(p\cdot j)\Big\}^{-1}.$$

Since, by translation invariance of $\mu_{L;\beta}$,

$$\Big\langle\Big|\sum_{i\in\mathbb{T}_L} S_j^\ell e^{\mathrm{i}p\cdot i}\Big|^2\Big\rangle_{L;\beta} = \sum_{i,j\in\mathbb{T}_L} e^{\mathrm{i}p\cdot(j-i)}\langle S_i^\ell S_j^\ell\rangle_{L;\beta} = |\mathbb{T}_L|\sum_{j\in\mathbb{T}_L} e^{\mathrm{i}p\cdot j}\langle S_0^\ell S_j^\ell\rangle_{L;\beta},$$

the conclusion follows by summing over $\ell \in \{1,\dots,\nu\}$ to recover the inner product. □

10.6 Bibliographical Remarks

There exist several nice reviews on reflection positivity, which can serve as complements to what is discussed in this chapter and provide additional examples of applications. These include the reviews by Shlosman [305] and Biskup [22] and the books by Sinai [312, Chapter 3], Prum [282, Chapter 7] and Georgii [134, Part IV]. The present chapter was largely inspired by the presentation in [22].

Reflection Positivity. Reflection positivity was first introduced in the context of constructive quantum field theory, where it plays a fundamental role. Its use in equilibrium statistical mechanics started in the late 1970s, see [98, 115, 116, 118, 157].

Infrared Bound and Long-Range Order in O(N) Models. The infrared bound, Theorem 10.24, was first proved by Fröhlich, Simon and Spencer in [118]. In this paper, among other applications, they use this bound to establish existence of spontaneous magnetization in O(N) models on $\mathbb{Z}^d$, $d \geq 3$, at low temperature (Theorem 10.1 in this chapter).

Chessboard Estimate and the Anisotropic *XY* Model. The chessboard estimate, in the form stated in Theorem 10.11, was first proved by Fröhlich and Lieb [117]. There were however earlier versions of it, see [134, Notes on Chapter 17]. The application to the anisotropic *XY* model, Theorem 10.2, was first established using other methods in [226] and [202]. The first proof relying on the chessboard estimate appeared in [117] and served as the basis for Section 10.2.

Appendix A **Notes**

Chapter 1

[1] (p. 4) The property described in (1.1) is usually referred to as *additivity* rather than extensivity. Extensivity of the energy is often valid and equivalent to additivity in the thermodynamic limit, at least for systems with finite-range interactions, as considered usually in this book. For systems with long-range interactions, extensivity does not always hold.

[2] (p. 20) This terminology was introduced by Gibbs [137], but the statistical ensembles were first introduced by Boltzmann under a different name (*ergode* for the microcanonical ensemble and *holode* for the canonical).

[3] (p. 21) We adopt here the following point of view explained by Jaynes in [181]:

> This problem of specification of probabilities in cases where little or no information is available, is as old as the theory of probability. Laplace's "Principle of Insufficient Reason" was an attempt to supply a criterion of choice, in which one said that two events are to be assigned equal probabilities if there is no reason to think otherwise.

Of course, some readers might not consider such a point of view to be fully satisfactory. In particular, one might dislike the interpretation of a probability distribution as a description of a state of knowledge, rather than as a quantity intrinsic to the system. After all, *there is* a more fundamental theory and it would be satisfactory to *derive* this probability distribution from the latter. Many attempts have been done, but no fully satisfactory derivation has been obtained. We will not discuss such issues further here, but refer the interested reader to the extensive literature on this topic; see for example [130].

[4] (p. 22) Historically, the entropy of a probability density had already been introduced by Gibbs in [137].

[5] (p. 39) In our brief description of a ferromagnet and its basic properties, we are neglecting many physically important aspects of the corresponding phenomena. Our goal is not to provide a faithful account, but rather to provide the uninitiated reader with an idea of what ferromagnetic and paramagnetic behaviors correspond to. We refer readers who would prefer a more thorough description to any of the many books on condensed matter physics, such as [1, 14, 65, 356].

[6] (p. 41). Let us briefly recall a famous anecdote originally reported by Uhlenbeck (see [260]). In November 1937, during the van der Waals Centenary Conference, a morning-long debate took place about the following question: does the partition function contain the information necessary to describe a sharp phase transition? As the debate turned out to be inconclusive, Kramers, who was the chairman, put the question to a vote, the result of which was nearly a tie (the "yes" winning by a small margin).

[7] (p. 41) The importance of Peierls' contribution was not immediately recognized. Rather, it was the groundbreaking mathematical analysis by Lars Onsager in 1944 that convinced the physics community. In particular, Onsager's formula for the pressure of the two-dimensional Ising model in the thermodynamic limit showed explicitly the existence of a singularity of this function. Moreover, and perhaps even more importantly, it showed that the behavior at the transition was completely different from what all of the former approximation schemes were predicting. The ensuing necessity of developing more refined approximation methods triggered the development of the modern theory of critical phenomena, in which the Ising model played a central role.

[8] (p. 41). In this book, we provide only brief and very qualitative physical motivations and background information for the Ising model. Much more can be found in many statistical physics textbooks aimed at physicists, such as [165, 264, 298, 299, 331]; see also the (old) review by Fisher [105]. An interesting and detailed description of the major role played by this model in the development of statistical mechanics in the twentieth century is given in the series of papers [255, 256, 257], while a shorter one can be found in [55].

[9] (p. 48). The determination of the explicit expression for the spontaneous magnetization of the two-dimensional Ising model given in (1.51) is due to Onsager and Kaufman and was announced by Onsager in 1949. However, they did not publish their result since they still had to work out "how to fill out the holes in the mathematics and show the epsilons and the deltas and all of that" [159]. The first published proof appeared in 1952 and is due to Yang [352]. See [15, 16] for more information.

[10] (p. 52) As an example, let us cite this passage from Peierls' famous paper [266]:

> In the meantime it was shown by Heisenberg that the forces leading to ferromagnetism are due to electron exchange. Therefore the energy function is of a more complicated nature than was assumed by Ising; it depends not only on the arrangement of the elementary magnets, but also on the speed with which they exchange their places. *The Ising model is therefore now only of mathematical interest* [emphasis added].

[11] (p. 53) In the words of Fisher [105]:

> [I]t is appropriate to ask what the main aim of theory should be. This is sometimes held (implicitly or explicitly) to be the calculation of the observable properties of a system from first principles using the full microscopic quantum-mechanical description of the constituent electrons, protons and neutrons. Such a calculation, however, even if feasible for a many-particle system which undergoes a phase transition need not and, in all probability, would not increase one's understanding of the observed behaviour of the system. Rather, the aim of the theory of a complex phenomenon should be to elucidate which general features

of the Hamiltonian of the system lead to the most characteristic and typical observed properties. Initially one should aim at a broad qualitative understanding, successively refining one's quantitative grasp of the problem when it becomes clear that the main features have been found.

Chapter 3

[1] (p. 91). It is known that subadditivity is not sufficient to prove convergence along arbitrary sequences $\Lambda_n \Uparrow \mathbb{Z}^d$ and that it has to be replaced by *strong subadditivity*, see [148]. Subadditivity is however sufficient to prove convergence *in the sense of Fisher*, that is, for sequences $\Lambda_n \uparrow \mathbb{Z}^d$ such that, for all $n \geq 1$, there exists a cube K_n such that $\Lambda_n \subset K_n$ and $\sup_n |K_n|/|\Lambda_n| < \infty$.

[2] (p. 110). The statement of Theorem 3.25 does not indicate what happens at the critical point $(\beta, h) = (\beta_c(d), 0)$. In that case, one can prove that, in all dimensions $d \geq 2$, uniqueness holds. In dimension 2, this can be proved in many ways; see, for example, [350]. In dimension $d \geq 4$, the proof is due to Aizenman and Fernández [7]. The case of dimension 3 was treated recently by Aizenman, Duminil-Copin and Sidoravicius [8]. Both are based on the random-current representation, a geometric representation of the Ising model which we briefly present in Section 3.10.6.

[3] (pp. 114 and 126). Much is known about the decay of correlations in the Ising model. In two dimensions, explicit computations show that $\langle \sigma_0 \sigma_x \rangle_{\beta,0}$ decays exponentially in $\|x\|_2$ for all $\beta \neq \beta_c(2)$ (with a rate that can be determined) and that $\langle \sigma_0 \sigma_x \rangle_{\beta_c(2),0} \sim \|x\|_2^{-1/4}$; see, for instance, [239, 261].

In any dimension, Aizenman, Barsky and Fernández have proved that there is exponential decay of the 2-point function $\langle \sigma_0 \sigma_x \rangle_{\beta,0}$ for all $\beta < \beta_c(d)$ [5]. In the same regime, it is actually possible to prove [60] that the 2-point function has *Ornstein–Zernike* behavior: $\langle \sigma_0 \sigma_x \rangle_{\beta,0} \simeq \Psi_\beta(x/\|x\|_2)\|x\|_2^{-(d-1)/2} e^{-\xi_\beta(x/\|x\|_2)\|x\|_2}$, as $\|x\|_2 \to \infty$, where Ψ_β and ξ_β are positive, analytic functions.

Sakai proved [292] that $\langle \sigma_0 \sigma_x \rangle_{\beta_c(d),0} \simeq c_d \|x\|_2^{2-d}$, for some constant c_d, in large enough dimensions d.

In the remaining cases, the 2-point function remains uniformly bounded away from 0 (by the FKG inequality), and the relevant quantity is the truncated 2-point function $\langle \sigma_i ; \sigma_j \rangle^+_{\beta,h} \stackrel{\text{def}}{=} \langle \sigma_i \sigma_j \rangle^+_{\beta,h} - \langle \sigma_i \rangle^+_{\beta,h} \langle \sigma_j \rangle^+_{\beta,h}$.

It is known that $\langle \sigma_i ; \sigma_j \rangle^+_{\beta,0}$ decays exponentially for all $\beta > \beta_c(d)$ in dimension 2 [67]. A proof for sufficiently low temperatures is given in Section 5.7.4.

Finally, $\langle \sigma_i ; \sigma_j \rangle^+_{\beta,h}$ decays exponentially for all $h \neq 0$ [95]; a simple geometric proof relying on the random-current representation can be found in [172].

[4] (p. 128). Even this statement should be qualified, since procedures allowing an experimental observation of the effect of complex values of physical parameters have recently been proposed and implemented. We refer the interested readers to [268] for more information.

[5] (p. 161). These two conjectures are supported by proofs of a similar behavior for a simplified model of the interface, known as the SOS (or Solid-On-Solid) model.

For this model, the analogue of the first claim above is already highly nontrivial, and was proved in [122], while the second claim was proved in [52].

[6] (p. 161). In sufficiently large dimensions (conjecturally: for all dimensions $d \geq 3$), there exists $\beta_p(d) > \beta_c(d)$ such that the $-$ spins percolate under $\mu^+_{\beta,0}$ for all $\beta \in (\beta_c(d), \beta_p(d))$ [6]. As a consequence, Peierls contours, and in particular the interface as defined in Section 3.10.7, are not very relevant anymore. One should then consider analogous objects defined on a coarser scale. For example, one might partition $\mathbb{Z}^d$ into blocks of $R \times R$ spins, with $R \uparrow \infty$ as $\beta \downarrow \beta_c(d)$. A block would then be said to be of type $+$ if the corresponding portion of configuration is "typical of the $+$ phase", of type $-$ if it is "typical of the $-$ phase", and of type 0 otherwise. Provided one defines these notions in a suitable way, then $+$ and $-$ blocks are necessarily separated by 0 blocks, and one can define contours as connected components of 0 blocks. If R diverges fast enough as $\beta \downarrow \beta_c(d)$, then this notion of contours makes sense for all $\beta > \beta_c(d)$. We refer to [276] for an explicit example of such a construction.

[7] (p. 170) It can be shown, nevertheless, that the series (3.98) provides an **asymptotic expansion** for ψ_β at $h = 0$:

$$\left|\psi_\beta(h) - \sum_{k=0}^{n} a_k h^k\right| = o(h^n), \quad \forall n \geq 1.$$

Chapter 4

[1] (p. 178) These were made in van der Waals' thesis [339].

[2] (p. 196) In Section 6.14.1, we give a sketch of one way by which equivalence can be approached for systems with interactions. We refer to the papers of Lanford [205] and of Lewis, Pfister and Sullivan [222, 223] for a much more complete and general treatment.

Chapter 6

[1] (p. 269). This statement should be qualified. Indeed, there are very specific cases in which such an approach allows one to construct infinite-volume Gibbs measures. The main example concerns models on trees (instead of lattices such as $\mathbb{Z}^d$, $d \geq 2$). In such a case, the absence of loops in the graph makes it possible to compute explicitly the marginal of the field in a finite subset. Roughly speaking, it yields explicit (finite) sets of equations, each of whose solutions correspond to one possible compatible family of marginals. In this way, it is possible to have multiple infinite-volume measures, even though one is still relying on Kolmogorov's Extension Theorem. A general reference for Gibbs measures on trees is [288]; see also [134, Chapter 12].

[2] (p. 282). The equivalence in Lemma 6.21 does not always hold if the single-spin space is not finite. When working on more general spaces, it turns out that quasilocal, rather than continuous, functions are the natural objects to consider. Note that the equivalences stated in Exercise 6.12 also fail to hold in general.

[3] (p. 286). In fact, the class of specifications that can be constructed in this way is very general. A specification π is **non-null** if, for all $\Lambda \Subset \mathbb{Z}^d$ and $\omega \in \Omega$, $\pi(\eta_\Lambda \mid \omega) > 0$ for all $\eta_\Lambda \in \Omega_\Lambda$. (An alternative terminology, often used in the context of percolation models, is that the specification π has **finite energy**.) It can then be shown [134, Section 2.3] that, if π is quasilocal and non-null, then there exists an absolutely summable potential Φ such that $\pi^\Phi = \pi$. This result is known as the *Kozlov–Sullivan Theorem*.

[4] (p. 286) This counter-example was taken from [134].

[5] (p. 296) The fact that a phase transition occurs in this model was proved by Dyson [99] when $-1 < \epsilon < 0$ and by Fröhlich and Spencer [123] when $\epsilon = 0$.

[6] (p. 300). A proof can be found, for example, in [134, Section 14.A].

[7] (p. 304) The use of the operations r_Λ and t^π_Λ is taken from [278].

[8] (p. 308). The argument in Example 6.64 is due to Miyamoto and first appeared in his book [249] (in Japanese). The argument was rediscovered independently by Coquille [72].

[9] (p. 311). This statement should be slightly nuanced. For concreteness, let us consider the two-dimensional Ising model with $\beta > \beta_c(2)$ and $h = 0$. On the one hand, when the free boundary condition is chosen, typical configurations show the box $B(n)$ to be entirely filled with either the $+$ phase or the $-$ phase, both occurring with equal probability:

On the other hand, when the Dobrushin boundary condition is applied (see the discussion in Section 3.10.7), typical configurations display coexistence of both $+$ and $-$ phases, separated by an interface:

Nevertheless, letting $n \to \infty$, both these sequences of finite-volume Gibbs distributions converge to the same Gibbs measure $\frac{1}{2}\mu^+_{\beta,0} + \frac{1}{2}\mu^-_{\beta,0}$. So, even though all the physics in the latter measure is already present in the two extremal measures $\mu^+_{\beta,0}$ and $\mu^-_{\beta,0}$, the physical mechanism leading to this particular nonextremal

Gibbs state are very different. In this sense, there can be hidden physics behind the coefficients of the extremal decomposition.

[10] (p. 332) The nonuniqueness criterion presented in Section 6.11 was inspired by [22].

[11] (p. 335) In dimension 2, this follows, for example, from the explicit expression (3.14) for the pressure at $h = 0$. For general dimensions, the claim follows from the fact that continuity of the magnetization implies differentiability of the pressure with respect to β [218] and the results on continuity of the magnetization [7, 8, 27, 352].

[12] (p. 335) This result can be found in [289, Theorem 5.6.2].

Chapter 7

[1] (p. 348) A detailed analysis of this model can be found in [35].

[2] (p. 348) This model was first studied by Blume [25] and Capel [61].

[3] (p. 359) More specifically, the problem of determining the ground states of a lattice model can be shown to fall, in general, in the class of NP-hard problems. For a discussion of this notion in the context of statistical mechanics, we recommend the book [245].

[4] (p. 372) This trick is known as the "Minlos–Sinai trick", and seems to have appeared first in [248].

Chapter 9

[1] (pp. 449 and 469). The first example of a two-dimensional O(N)-symmetric model with several infinite-volume Gibbs measures at low temperature was provided by Shlosman [304]. His model has $N = 2$ and formal Hamiltonian

$$-\beta \sum_{\substack{i,j\in\mathbb{Z}^2\\ \|j-i\|_2=\sqrt{2}}} \cos(\vartheta_i - \vartheta_j) + \beta J \sum_{\substack{i,j\in\mathbb{Z}^2\\ \|j-i\|_2=1}} \cos(2(\vartheta_i - \vartheta_j)),$$

where J is nonnegative, and we have written $\mathbf{S}_i = (\cos\vartheta_i, \sin\vartheta_i)$ for the spin at $i \in \mathbb{Z}^2$. The crucial feature of this model is that, in addition to the SO(2)-invariance of the Hamiltonian, the latter is also preserved under the simultaneous transformation

$$\vartheta_i \mapsto \vartheta_i, \qquad \vartheta_j \mapsto \vartheta_j + \pi,$$

for all $i \in \mathbb{Z}^2_{\text{even}} \stackrel{\text{def}}{=} \{i = (i_1, i_2) \in \mathbb{Z}^2 : i_1 + i_2 \text{ is even}\}$ and $j \in \mathbb{Z}^2_{\text{odd}} \stackrel{\text{def}}{=} \mathbb{Z}^2 \setminus \mathbb{Z}^2_{\text{even}}$. It is this discrete symmetry that is spontaneously broken at low temperatures, yielding two Gibbs measures under which, in typical configurations, either most nearest-neighbor spins differ by approximately $\pi/2$, or most differ by approximately $-\pi/2$; see Figure A.1.

[2] (p. 458), A proof can be found, for example, in [111].

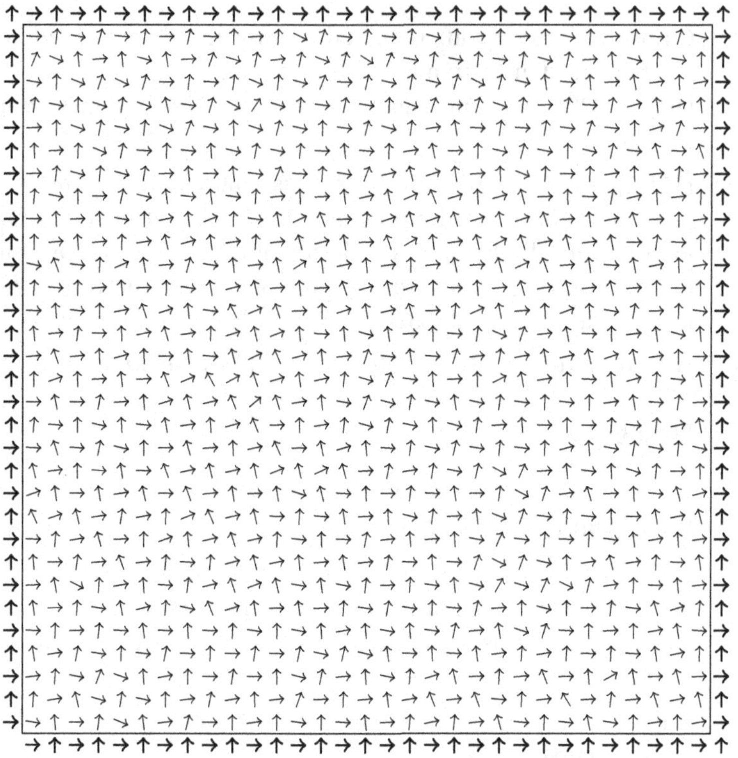

Figure A.1. A typical low-temperature configuration of the model in Note 1 of Chapter 9.

Chapter 10

[1] (p. 473) Extensions of Pirogov–Sinai theory covering some models with continuous spins can be found, for example, in [87], [168] and [355].

Appendix B Mathematical Appendices

In this appendix, the reader can find a number of basic definitions and results concerning some of the mathematical tools that are used throughout the book. Given their wide range, it is not possible to discuss these tools in a self-contained manner in this appendix. Nevertheless, we believe that gathering a coherent set of notions and notations could be useful to the reader.

Although most of the proofs can be found in the literature (we provide references for most of them), often in a much more general form, we have occasionally provided explicit elementary derivations tailored for the particular use made in the book. The results are not always stated in their most general form, in order to avoid introducing too many concepts and notations.

Since the elementary notions borrowed from topology are used only in the case of metric spaces and are always presented and developed from scratch, they are not explained in a systematic way.

B.1 Real Analysis

B.1.1 Elementary Inequalities

Lemma B.1 (Comparing arithmetic and geometric means). *For any collection $x_1, \dots, x_n$ of nonnegative real numbers,*

$$\frac{1}{n}\sum_{i=1}^{n} x_i \geq \Big\{\prod_{i=1}^{n} x_i\Big\}^{1/n}, \tag{B.1}$$

with equality if and only if $x_1 = x_2 = \cdots = x_n$.

Lemma B.2 (Hölder's inequality, finite form). *For all $(x_1, \dots, x_n), (y_1, \dots, y_n) \in \mathbb{R}^n$ and all $p, q > 1$ such that $\frac{1}{p} + \frac{1}{q} = 1$,*

$$\sum_{k=1}^{n} |x_k y_k| \leq \Big(\sum_{k=1}^{n} |x_k|^p\Big)^{1/p} \Big(\sum_{k=1}^{n} |x_k|^q\Big)^{1/q}.$$

Lemma B.3 (Stirling's formula). *For all $n \geq 1$,*

$$e^{\frac{1}{12n+1}}\sqrt{2\pi n}n^n e^{-n} \leq n! \leq e^{\frac{1}{12n}}\sqrt{2\pi n}n^n e^{-n}. \tag{B.2}$$

A proof of this version of Stirling's formula can be found in [285].

B.1.2 Double Sequences

We say that a double sequence $(a_{m,n})_{m,n\geq 1}$ is **nondecreasing** if

$$m \leq m', n \leq n' \implies a_{m,n} \leq a_{m',n'}$$

and **nonincreasing** if $(-a_{m,n})_{m,n\geq 1}$ is nondecreasing. It is **bounded above** (resp. **below**) if there exists $C < \infty$ such that $a_{m,n} \leq C$ (resp. $a_{m,n} \geq -C$), for all $m, n \geq 1$.

Lemma B.4. *Let $(a_{m,n})_{m,n\geq 1}$ be a nondecreasing double sequence bounded above. Then,*

$$\lim_{m\to\infty}\lim_{n\to\infty} a_{m,n} = \lim_{n\to\infty}\lim_{m\to\infty} a_{m,n} = \lim_{m,n\to\infty} a_{m,n} = \sup\big\{a_{m,n} : m, n \geq 1\big\}. \tag{B.3}$$

Proof $(a_{m,n})$ being bounded, $s \stackrel{\text{def}}{=} \sup_{m,n} a_{m,n}$ is finite. Let $\epsilon > 0$, and take m_0, n_0 such that $a_{m_0,n_0} \geq s - \epsilon$. $(a_{m,n})$ being nondecreasing, we deduce that

$$s \geq a_{m,n} \geq s - \epsilon, \qquad \forall m \geq m_0, n \geq n_0.$$

Consequently, $\lim_{m,n\to\infty} a_{m,n} = s$. For all fixed $m \geq 1$, the sequence $(a_{m,n})_{n\geq 1}$ is nondecreasing and bounded, and thus converges to some limit s_m. For a fixed $\epsilon > 0$, let m_1, n_1 be such that

$$|a_{m,n} - s| \leq \tfrac{\epsilon}{2}, \qquad \forall m \geq m_1, n \geq n_1.$$

For fixed m, we can also find $n_2(m)$ such that

$$|a_{m,n} - s_m| \leq \tfrac{\epsilon}{2}, \qquad \forall n \geq n_2(m).$$

Consequently,

$$|s_m - s| \leq \epsilon, \qquad \forall m \geq m_1,$$

which implies that $\lim_{m\to\infty} s_m = s$. We have thus proved (B.3). □

B.1.3 Subadditive Sequences

A sequence $(a_n)_{n\geq 1} \subset \mathbb{R}$ is called **subadditive** if

$$a_{n+m} \leq a_n + a_m, \qquad \forall m, n.$$

Lemma B.5. *If (a_n) is subadditive, then*

$$\lim_{n\to\infty} \frac{a_n}{n} = \inf_n \frac{a_n}{n}.$$

Proof Let $\alpha \stackrel{\text{def}}{=} \inf_n \frac{a_n}{n}$, and fix $\alpha' > \alpha$. Let ℓ be such that $\frac{a_\ell}{\ell} \leq \alpha'$. For all n, there exists k and $0 \leq j < \ell$ such that $n = k\ell + j$. We can then use the definition of α, and k times the subadditivity of (a_n) to write

$$\alpha n \leq a_n = a_{k\ell+j} \leq k a_\ell + a_j.$$

Dividing by n,

$$\alpha \leq \liminf_{n\to\infty} \frac{a_n}{n} \leq \limsup_{n\to\infty} \frac{a_n}{n} \leq \frac{a_\ell}{\ell} \leq \alpha'.$$

The desired result follows by letting $\alpha' \downarrow \alpha$. □

On the lattice $\mathbb{Z}^d$, a similar property holds. Let us denote by $\mathscr{R}$ the set of all **parallelepipeds** of $\mathbb{Z}^d$, that is, sets of the form $\Lambda = [a_1, b_1] \times [a_2, b_2] \times \cdots \times [a_d, b_d] \cap \mathbb{Z}^d$. A set function $a\colon \mathscr{R} \to \mathbb{R}$ is **subadditive** if $R_1, R_2 \in \mathscr{R}$, $R_1 \cup R_2 \in \mathscr{R}$ implies

$$a(R_1 \cup R_2) \leq a(R_1) + a(R_2).$$

Let, as usual, $\mathrm{B}(n) = \{-n, \ldots, n\}^d$.

Lemma B.6. *Let $a\colon \mathscr{R} \to \mathbb{R}$ be subadditive and such that $a(\Lambda + i) = a(\Lambda)$ for all $\Lambda \in \mathscr{R}$ and all $i \in \mathbb{Z}^d$. Then*

$$\lim_{n\to\infty} \frac{a(\mathrm{B}(n))}{|\mathrm{B}(n)|} = \inf_{\Lambda\in\mathscr{R}} \frac{a(\Lambda)}{|\Lambda|}.$$

The proof is a d-dimensional adaptation of the one given above for sequences $(a_n)_{n\geq 1}$; we leave it as an exercise (a proof can be found in [134]).

B.1.4 Functions Defined by Series

Theorem B.7. *Let $I \subset \mathbb{R}$ be an open interval. For each $k \geq 1$, let $\phi_k \colon I \to \mathbb{R}$ be C^1. Assume that there exists a summable sequence $(\epsilon_k)_{k\geq 1} \subset \mathbb{R}_{\geq 0}$ such that $\sup_{x\in I} |\phi_k(x)| \leq \epsilon_k$, $\sup_{x\in I} |\phi_k'(x)| \leq \epsilon_k$. Then $f(x) \stackrel{\text{def}}{=} \sum_{k\geq 1} \phi_k(x)$ is well defined and C^1 on I. Moreover, $f'(x) = \sum_{k\geq 1} \phi_k'(x)$.*

Proof Since $\sum_k \epsilon_k < \infty$, $\sum_k \phi_k(x)$ is an absolutely convergent series for all $x \in I$, defining a function $f \colon I \to \mathbb{R}$. Then, fix $x \in I$ and take some small $h > 0$:

$$\frac{f(x+h) - f(x)}{h} = \sum_k \frac{\phi_k(x+h) - \phi_k(x)}{h}.$$

By the Mean Value Theorem, there exists $\tilde{x} \in [x, x+h]$ such that $|\frac{\phi_k(x+h)-\phi_k(x)}{h}| = |\phi_k'(\tilde{x})| \leq \epsilon_k$. Using Exercise B.15, we can therefore interchange $h \downarrow 0$ with $\sum_k$. The same argument with $h \uparrow 0$ then gives $f'(x) = \sum_k \phi_k'(x)$. A similar argument guarantees that f is C^1. □

B.2 Convex Functions

In this section, we gather a few elementary results about convex functions of one real variable. Rockafellar's book [287] is a standard reference on the subject; another nice and accessible reference is the book [286] by Roberts and Varberg.

We will use I to denote an (not necessarily bounded) open interval in $\mathbb{R}$, that is, $I = (a, b)$ with $-\infty \leq a < b \leq +\infty$.

Definition B.8. A function $f \colon I \to \mathbb{R}$ is **convex** if

$$f(\alpha x + (1-\alpha)y) \leq \alpha f(x) + (1-\alpha) f(y), \qquad \forall x, y \in I, \forall \alpha \in [0,1]. \tag{B.4}$$

When the inequality is strict for all $x \neq y$ and all $\alpha \in (0,1)$, f is **strictly convex.** If $-f$ is (strictly) convex, then f is said to be **(strictly) concave.**

For the cases considered in the book, f always has a continuous extension to the boundary of I (when I is finite). Sometimes, we need to extend the domain of f from a finite I to the whole of $\mathbb{R}$; in such cases, one can do that by setting $f(x) \stackrel{\text{def}}{=} +\infty$ for all $x \notin I$. The definition of convexity can then be extended, allowing f to take infinite values in (B.4).

The following exercise is elementary, but emphasizes a property of convex functions that will be used repeatedly in the following; it is illustrated in Figure B.1.

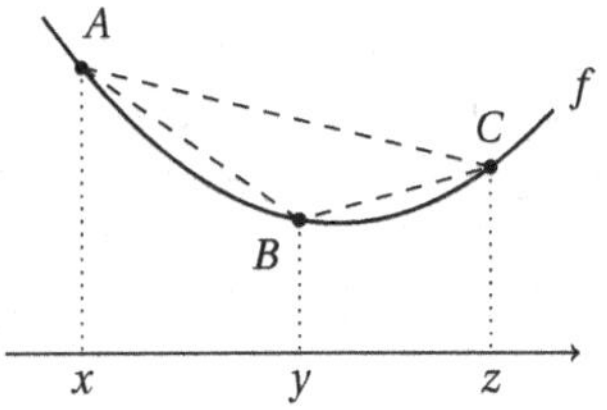

Figure B.1. The geometrical meaning of (B.6): for any triple of points on the graph of a convex function, slope(AB) $\leq$ slope(AC) $\leq$ slope(BC).

Exercise B.1. Show that $f: I \to \mathbb{R}$ is convex if and only if, for any $x < y < z$ in I,

$$f(y) \leq \frac{z-y}{z-x} f(x) + \frac{y-x}{z-x} f(z). \tag{B.5}$$

From this, deduce that if f is finite, then, for any $x < y < z$ in I,

$$\frac{f(y)-f(x)}{y-x} \leq \frac{f(z)-f(x)}{z-x} \leq \frac{f(z)-f(y)}{z-y}. \tag{B.6}$$

Exercise B.2. Show that $f: I \to \mathbb{R}$ is convex if and only if, for all $\alpha_1, \dots, \alpha_n \in [0,1]$ such that $\alpha_1 + \cdots + \alpha_n = 1$ and all $x_1, \dots, x_n \in I$,

$$f\Big(\sum_{k=1}^{n} \alpha_k x_k\Big) \leq \sum_{k=1}^{n} \alpha_k f(x_k).$$

An important property is that limits of convex functions are convex:

Exercise B.3. Show that if $(f_n)_{n\geq 1}$ is a sequence of convex functions from I to $\mathbb{R}$, then $x \mapsto \limsup_{n\to\infty} f_n(x)$ is convex. In particular, if $f(x) \stackrel{\text{def}}{=} \lim_{n\to\infty} f_n(x)$ exists (in $\mathbb{R} \cup \{+\infty\}$) for all $x \in I$, then it is also convex.

B.2.1 Convexity vs. Continuity

Proposition B.9. *Let $f: I \to \mathbb{R}$ be convex. Then f is locally Lipschitz: for all compact $K \subset I$, there exists $C_K < \infty$ such that $|f(x) - f(y)| \leq C_K |x-y|$ for all $x, y \in K$. In particular, f is continuous.*

Proof Let $K \subset I$ be compact, and let $\epsilon > 0$ be small enough to ensure that $K_\epsilon \stackrel{\text{def}}{=} \{z : d(z,K) \leq \epsilon\} \subset I$. Let also $M \stackrel{\text{def}}{=} \sup_{z\in K_\epsilon} f(z)$, $m \stackrel{\text{def}}{=} \inf_{z\in K_\epsilon} f(z)$. Observe that both m and M are finite. (Otherwise, there would exist an interior point $x_* \in I$ and a sequence $x_n \to x_*, f(x_n) \uparrow +\infty$. Then, for all pairs $z < x_* < z'$ one would get, for all sufficiently large n, $f(x_n) > \max\{f(z), f(z')\}$, a contradiction with the convexity of f.) Let $x, y \in K$, and set $z \stackrel{\text{def}}{=} y + \epsilon\frac{y-x}{|y-x|} \in K_\epsilon$. Then $y = (1-\lambda)x + \lambda z$ with $\lambda = \frac{|y-x|}{\epsilon+|y-x|}$, and therefore $f(y) \leq (1-\lambda)f(x) + \lambda f(z)$, which gives after rearrangement

$$f(y) - f(x) \leq \lambda(f(z) - f(x)) \leq \lambda(M - m) \leq \frac{M-m}{\epsilon}|y-x|. \qquad \square$$

Lemma B.10. *Let $(f_n)_{n\geq 1}$ be a sequence of convex functions on I converging pointwise to $f: I \to \mathbb{R}$. Then $f_n \to f$ uniformly on all compacts $K \subset I$.*

Proof Fix some compact $K \subset I$, and let $a' < a < b < b'$ in I such that $[a, b] \supset K$. It follows from (B.6) that, for all n and all distinct $x, y \in [a, b]$,

$$\frac{f_n(a) - f_n(a')}{a - a'} \leq \frac{f_n(y) - f_n(x)}{y - x} \leq \frac{f_n(b') - f_n(b)}{b' - b}.$$

By pointwise convergence, the leftmost and rightmost ratios converge to finite values as $n \to \infty$. Therefore, there exists C, independent of n, such that

$$|f_n(y) - f_n(x)| \leq C|y - x|, \qquad \forall x, y \in [a, b].$$

Letting $n \to \infty$ in the last display shows that the same is also true for the limiting function f.

Fix $\epsilon > 0$. Let $N \in \mathbb{N}$ and define $\delta = (b-a)/N$ and $x_k = a + k\delta$, $k = 0, \ldots, N$. Pointwise convergence implies that there exists n_0 such that, for all $n \geq n_0$,

$$|f_n(x_k) - f(x_k)| < \tfrac{1}{3}\epsilon, \qquad \forall k \in \{0, \ldots, N\}.$$

Let $z \in [a, b]$ and let $k \in \{0, \ldots, N\}$ be such that $|x_k - z| < \delta$. Then, for all $n \geq n_0$,

$$|f_n(z) - f(z)| \leq \underbrace{|f_n(z) - f_n(x_k)|}_{\leq C\delta} + \underbrace{|f_n(x_k) - f(x_k)|}_{\leq \epsilon/3} + \underbrace{|f(x_k) - f(z)|}_{\leq C\delta} \leq \epsilon,$$

provided we choose N such that $C\delta \leq \epsilon/3$. $\square$

A function $f: I \to \mathbb{R}$ is said to be **midpoint-convex** if

$$f\Big(\frac{x+y}{2}\Big) \leq \frac{f(x) + f(y)}{2}, \qquad \forall x, y \in I. \tag{B.7}$$

Clearly, a convex function is also midpoint-convex.

Lemma B.11. *If f is midpoint-convex and continuous, then it is convex.*

Proof Using the continuity of f, it suffices to show that (B.4) holds in the case where $\alpha \in \mathscr{D} \stackrel{\text{def}}{=} \bigcup_{m\geq 1} \mathscr{D}_m$, with $\mathscr{D}_m \stackrel{\text{def}}{=} \{\frac{k}{2^m} : 0 \leq k < 2^m\}$. Observe first that (B.7) means that (B.4) holds for all $x, y \in I$ and for $\alpha \in \mathscr{D}_1$.

We can now proceed by induction. Assume that (B.4) holds for all $\alpha \in \mathscr{D}_m$. Let $z = \alpha x + (1-\alpha)y$, with $\alpha \in \mathscr{D}_{m+1} \setminus \mathscr{D}_m$; with no loss of generality, we can assume that $\alpha > 1/2$. Let $z' \stackrel{\text{def}}{=} 2z - x = \alpha' x + (1-\alpha')y$, where $\alpha' \stackrel{\text{def}}{=} 2\alpha - 1 \in \mathscr{D}_m$. Applying (B.7) and the induction assumption, we get

$$f(z) = f(\tfrac{1}{2}x + \tfrac{1}{2}z') \leq \tfrac{1}{2}f(x) + \tfrac{1}{2}f(z')$$
$$\leq \tfrac{1}{2}f(x) + \tfrac{1}{2}\{\alpha' f(x) + (1-\alpha')f(y)\} = \alpha f(x) + (1-\alpha)f(y),$$

so (B.4) also holds for all $\alpha \in \mathscr{D}_{m+1}$. □

B.2.2 Convexity vs. Differentiability

The **one-sided derivatives** of a function f at a point x are defined by

$$\partial^+ f(x) = \frac{\partial f}{\partial x^+} \stackrel{\text{def}}{=} \lim_{z \downarrow x} \frac{f(z) - f(x)}{z - x},$$
$$\partial^- f(x) = \frac{\partial f}{\partial x^-} \stackrel{\text{def}}{=} \lim_{z \uparrow x} \frac{f(z) - f(x)}{z - x}.$$

These quantities are always well defined for a convex function, and enjoy several useful properties:

Theorem B.12. *Let $f: I \to \mathbb{R}$ be convex. The following properties hold:*

1. *$\partial^+ f(x)$ and $\partial^- f(x)$ exist at all points $x \in I$.*
2. *$\partial^- f(x) \leq \partial^+ f(x)$, for all $x \in I$.*
3. *$\partial^+ f(x) \leq \partial^- f(y)$ for all $x < y$ in I.*
4. *$\partial^+ f$ and $\partial^- f$ are nondecreasing.*
5. *$\partial^+ f$ is right-continuous, $\partial^- f$ is left-continuous.*
6. *$\{x : \partial^+ f(x) \neq \partial^- f(x)\}$ is at most countable.*
7. *Let $(g_n)_{n \geq 1}$ be a sequence of convex functions from I to $\mathbb{R}$ converging pointwise to a function g. If g is differentiable at x, then $\lim_{n\to\infty} \partial^+ g_n(x) = \lim_{n\to\infty} \partial^- g_n(x) = g'(x)$.*

Note that Item 6 shows that a convex function $f: I \to \mathbb{R}$ is differentiable everywhere outside an at most countable set.

Proof From (B.6), we see that

$$x \mapsto \frac{f(y) - f(x)}{y - x} \quad \text{and} \quad y \mapsto \frac{f(y) - f(x)}{y - x} \quad \text{are nondecreasing.} \tag{B.8}$$

1. In I, consider $x < y$ and a decreasing sequence $(z_k)_{k \geq 1}$ with $z_k > y$, for all k, and $z_k \downarrow y$. From (B.8), the sequence $\big((f(z_k) - f(y))/(z_k - y)\big)_{k \geq 1}$ is nonincreasing, and (B.6) implies that it is bounded below by $\big(f(y) - f(x)\big)/(y - x)$. It follows that the sequence converges, which establishes the existence of $\partial^+ f(y)$. A similar argument proves the existence of $\partial^- f(y)$.

2. Taking $x \uparrow y$ in the left-hand side, followed by $z \downarrow y$ in the right-hand side of (B.6) gives $\partial^- f(y) \le \partial^+ f(y)$.
3. Let $x < y$ in I. It follows from (B.8) that
$$\partial^+ f(x) \le \frac{f(y) - f(x)}{y - x} \le \partial^- f(y) . \tag{B.9}$$
4. This is a consequence of the second and third points.
5. We prove the claim for $\partial^+ f$; the other one is treated in the same way. On the one hand, it follows from the monotonicity of $\partial^+ f$ that $\lim_{y \downarrow x} \partial^+ f(y)$ exists and $\lim_{y \downarrow x} \partial^+ f(y) \ge \partial^+ f(x)$. On the other hand, we know from Proposition B.9 that f is continuous. It thus follows from (B.9) that, for each $z > x$,
$$\frac{f(z) - f(x)}{z - x} = \lim_{y \downarrow x} \frac{f(z) - f(y)}{z - y} \ge \lim_{y \downarrow x} \partial^+ f(y) .$$
Letting $z \downarrow x$, we obtain that $\partial^+ f(x) \ge \lim_{y \downarrow x} \partial^+ f(y)$ and the claim follows.
6. Since I can be written as the union of countably many closed intervals, and since a countable union of countable sets is countable, it is enough to prove the statement for an arbitrary closed interval $[a, b]$ contained in I. Let $\epsilon > 0$ such that $[a - \epsilon, b + \epsilon] \subset I$. Since f is continuous, $M \stackrel{\text{def}}{=} \sup_{x \in [a-\epsilon, b+\epsilon]} |f(x)| < \infty$. It thus follows from (B.9) that
$$\partial^+ f(b) \le \frac{f(b + \epsilon) - f(b)}{\epsilon} \le \frac{2M}{\epsilon}$$
and
$$\partial^- f(a) \ge \frac{f(a) - f(a - \epsilon)}{\epsilon} \ge -\frac{2M}{\epsilon} .$$
By what we saw above, $\partial^- f(a) \le \partial^\pm f(x) \le \partial^+ f(b)$ for all $x \in [a, b]$, we deduce that $\sup_{x \in [a,b]} |\partial^\pm f(x)| \le 2M/\epsilon$. For $r \in \mathbb{N}$, let
$$\mathscr{A}_r = \left\{ x \in [a, b] \, : \, \partial^+ f(x) - \partial^- f(x) \ge \tfrac{1}{r} \right\} .$$
Since
$$\left\{ x \in [a, b] \, : \, \partial^+ f(x) > \partial^- f(x) \right\} = \bigcup_{r \ge 1} \mathscr{A}_r ,$$
it suffices to prove that each $\mathscr{A}_r$ is finite. Consider n distinct points $x_1 < x_2 < \ldots < x_n$ from $\mathscr{A}_r$. Then,
$$\partial^+ f(x_n) - \partial^- f(x_1) = \sum_{k=1}^{n} \left(\partial^+ f(x_k) - \partial^- f(x_k) \right) \ge \frac{n}{r} ,$$
which implies $n \le r\left(\partial^+ f(x_n) - \partial^- f(x_1)\right) \le 4Mr/\epsilon$; $\mathscr{A}_r$ is therefore finite.
7. Using again (B.9), for any $h > 0$,
$$\limsup_{n \to \infty} \partial^+ g_n(x) \le \limsup_{n \to \infty} \frac{g_n(x + h) - g_n(x)}{h} = \frac{g(x + h) - g(x)}{h} .$$

Letting $h \downarrow 0$ gives $\partial^+ g(x) \geq \limsup_{n\to\infty} \partial^+ g_n(x)$. A similar argument yields $\partial^- g(x) \leq \liminf_{n\to\infty} \partial^- g_n(x)$. Therefore,

$$\partial^- g(x) \leq \liminf_{n\to\infty} \partial^- g_n(x) \leq \limsup_{n\to\infty} \partial^+ g_n(x) \leq \partial^+ g(x),$$

and the differentiability of g at x indeed implies that

$$g'(x) = \lim_{n\to\infty} \partial^- g_n(x) = \lim_{n\to\infty} \partial^+ g_n(x). \qquad \square$$

We say that $f: I \mapsto \mathbb{R}$ has a **supporting line of slope** m **at** x_0 if

$$f(x) \geq m(x - x_0) + f(x_0), \quad \forall x \in I. \tag{B.10}$$

Theorem B.13. *A function $f: I \to \mathbb{R}$ is convex if and only if f has a supporting line at each point $x \in I$. Moreover, in that case, there is a supporting line at x of slope m for all $m \in [\partial^- f(x), \partial^+ f(x)]$.*

Proof Suppose first that f has a supporting line at each point of I. Let $x < y$ be two points of I, $\alpha \in [0, 1]$ and $z = \alpha x + (1-\alpha)y$. By assumption, there exists m such that $f(u) \geq f(z) + m(u - z)$ for all $u \in I$. Applying this at x and y, we deduce that

$$\alpha f(x) + (1-\alpha)f(y) \geq f(z) + m\big(\underbrace{\alpha(x-z) + (1-\alpha)(y-z)}_{=0}\big),$$

which implies that $\alpha f(x) + (1-\alpha)f(y) \geq f(\alpha x + (1-\alpha)y)$ as desired.

Assume now that f is convex and let $x_0 \in I$. Let $m \in [\partial^- f(x_0), \partial^+ f(x_0)]$. By (B.9), $\frac{f(x)-f(x_0)}{x-x_0} \geq \partial^+ f(x_0) \geq m$ for all $x > x_0$, and $\frac{f(x)-f(x_0)}{x-x_0} \leq \partial^- f(x_0) \leq m$ for all $x < x_0$, which implies $f(x) \geq m(x - x_0) + f(x_0)$ for all x. $\square$

We also remind the reader of a well-known property that relates convexity to the positivity of the second derivative of a twice-differentiable function:

Exercise B.4. Let f be twice differentiable at each point of I. Show that f is convex if and only if $f''(x) \geq 0$ for all $x \in I$.

Note that a sequence of strictly convex functions $(f_n)_{n\geq 1}$ converging pointwise can have a limit that is not strictly convex; consider, for example, $f_n(x) = |x|^{1+1/n}$. A twice-differentiable function f for which one can find $c > 0$ such that $f''(x) > c$ for all x is said to be **strongly convex**. Note that a function can be strictly convex and fail to be strongly convex, for example $x \mapsto x^4$.

Exercise B.5. Let $(f_n)_{n\geq 1}$ be a sequence of twice-differentiable strongly convex functions such that $f = \lim_n f_n(x)$ exists and is finite everywhere. Show that f is strictly convex.

B.2.3 The Legendre Transform

Definition B.14. Let $f: I \to \mathbb{R} \cup \{+\infty\}$. The **Legendre–Fenchel transform** (or simply **Legendre transform**[1]) of f is defined by

$$f^*(y) \stackrel{\text{def}}{=} \sup_{x \in I}\{yx - f(x)\}, \quad y \in \mathbb{R}. \tag{B.11}$$

We will always suppose, from now on, that there exists at least one point at which f is finite, which guarantees that $f^*(y) > -\infty$ for all $y \in \mathbb{R}$.

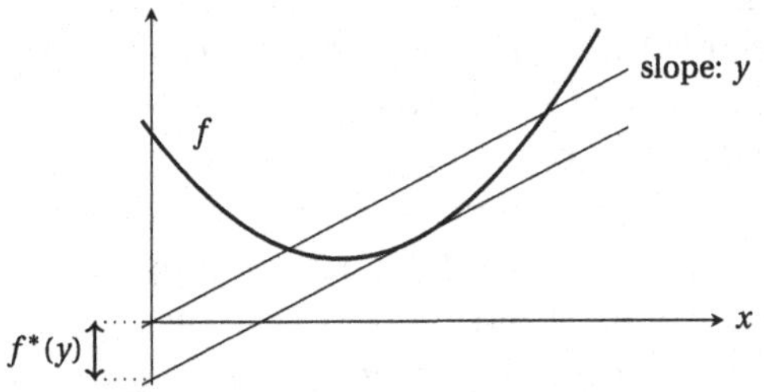

Figure B.2. Visualizing the Legendre transform: for a given $y \in \mathbb{R}$, $f^*(y)$ is the largest difference between the straight line $x \mapsto yx$ and the graph of f.

Exercise B.6. Show that any Legendre transform is convex.

Exercise B.7. Compute the Legendre transform f_i^* of each of the following functions:

$$f_1(x) = \tfrac{1}{2}x^2, \qquad f_2(x) = x^4, \qquad f_3(x) = \begin{cases} 0 & \text{if } x \in (-1,1), \\ +\infty & \text{if } x \notin (-1,1). \end{cases}$$

Compute also $f_i^{**} \stackrel{\text{def}}{=} (f_i^*)^*$ in each case. What do you observe?

As can be seen by solving the previous exercise, f^{**} is not always equal to f. Nevertheless,

---

Exercise B.8. Show that, for all $f: \mathbb{R} \to \mathbb{R} \cup \{+\infty\}$, $f^{**} \leq f$.

Let us see two more examples in which the geometrical effect of applying two successive Legendre transforms is made transparent:

[1] Actually, the latter form is usually reserved for a particular case; nevertheless, we use the term *Legendre transform* everywhere in this book.

Exercise B.9. If $f(x) = \big||x| - 1\big|$, show that

$$f^{**}(x) = \begin{cases} -x - 1 & \text{if } x < -1\,, \\ 0 & \text{if } |x| \leq 1\,, \\ +x - 1 & \text{if } x > +1\,. \end{cases}$$

Exercise B.10. Let $f(x) = x^4 - x^2$. Using the geometrical picture of Figure B.2, study qualitatively f^* and f^{**}.

With the above examples in mind, we now impose restrictions on f to guarantee that $f^{**} = f$.

We call f **lower semi-continuous** at x if, for any sequence $x_n \to x$,

$$\liminf_{n\to\infty} f(x_n) \geq f(x)\,.$$

Exercise B.11. Show that any Legendre transform is lower semi-continuous.

Lemma B.15. *Let $f\colon \mathbb{R} \to \mathbb{R} \cup \{+\infty\}$ be convex and lower semi-continuous. For all x_0, if $\alpha \in \mathbb{R}$ is such that $\alpha < f(x_0)$, then there exists an* ***affine function*** *$h(x) = ax + b$ such that $h \leq f$ and $h(x_0) \geq \alpha$.*

Proof For simplicity, we assume that $f(x_0) < +\infty$ (the case $f(x_0) = +\infty$ is treated similarly). If either $\partial^+ f(x_0)$, or $\partial^- f(x_0)$, is finite, then Theorem B.13 implies the result. If $\partial^- f(x_0) = +\infty$, convexity implies that $f(x) < f(x_0)$ for all $x \in (x_0 - \delta, x_0)$ (with $\delta > 0$ sufficiently small), and $f(x) = +\infty$ for all $x > x_0$. Let

$$a \stackrel{\text{def}}{=} \inf\big\{m \geq 0 \,:\, m(x - x_0) + \alpha \leq f(x),\ \forall x \in I\big\}\,.$$

We claim that $a < \infty$. Indeed, assume that $a = \infty$. Then, there would exist a sequence $x_n < x_0$, $x_n \uparrow x_0$, with $f(x_n) \leq \alpha < f(x_0)$, giving $\liminf_n f(x_n) < f(x_0)$, which would contradict the lower semi-continuity of f. When $a < \infty$, the affine function $h(x) = a(x - x_0) + \alpha$ satisfies the requirements. The remaining cases are treated similarly. □

Exercise B.12. Show that if f has a supporting line of slope m at x_0, then f^* has a supporting line of slope x_0 at m.

The **epigraph** of an arbitrary function $f\colon \mathbb{R} \to \mathbb{R} \cup \{+\infty\}$ is defined by

$$\operatorname{epi}(f) \stackrel{\text{def}}{=} \big\{(x, y) \in \mathbb{R}^2 \,:\, y \geq f(x)\big\}\,.$$

Exercise B.13. Let $f: I \to \mathbb{R} \cup \{+\infty\}$.

1. Show that f is convex if and only if epi (f) is convex.[2]
2. Show that f is lower semi-continuous if and only if epi (f) is closed.

Definition B.16. The **convex envelope** (or **convex hull**) of f, denoted CEf, is defined as the unique convex function g whose epigraph is

$$C \stackrel{\text{def}}{=} \bigcap \{F \subset \mathbb{R}^2 : F \text{ closed, convex, } F \supset \text{epi}\,(f)\} . \tag{B.12}$$

That is,

$$\text{CE}f(x) \stackrel{\text{def}}{=} \inf\{y : (x, y) \in C\} .$$

Clearly, if f is convex and lower semi-continuous, then CE$f = f$.

Observe that, since C is closed, $(x, \text{CE}f(x)) \in C$ for all x. Moreover, C is convex, which implies that $x \mapsto \text{CE}f(x)$ is convex. In fact, epi (CEf) $= C$. Since C is closed, this implies (Exercise B.13) that CEf is lower semi-continuous.

In words, as will be seen in the next exercise, CEf is the largest convex function g such that $g \leq f$:

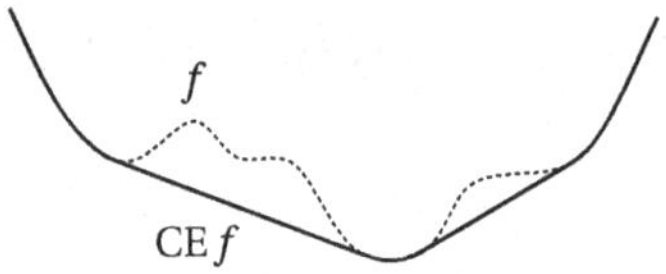

Exercise B.14. If g is convex, lower semi-continuous and $g \leq f$, then $g \leq \text{CE}f$.

Theorem B.17. *If $f: \mathbb{R} \to \mathbb{R}$ is lower semi-continuous,*

$$f^{**} = \text{CE}f .$$

Proof We have already seen that $f^{**} \leq f$. Since f^{**} is convex and lower semi-continuous, this implies $f^{**} \leq \text{CE}f$ (Exercise B.14). To establish the reverse inequality at a point x_0, $f^{**}(x_0) \geq \text{CE}f(x_0)$, we must show that, for all $\alpha \in \mathbb{R}$ satisfying $\alpha < \text{CE}f(x_0)$,

$$\text{there exists } y \in \mathbb{R} \text{ such that } x_0 y - f^*(y) \geq \alpha . \tag{B.13}$$

Since CEf is also lower semi-continuous, there exists, by Lemma B.15, an affine function h such that (i) $h \leq f$ and (ii) $\alpha \leq h(x_0) \leq f(x_0)$. If $h(x) = ax + b$, (i) means that $ax + b \leq f(x)$ for all x, which gives $f^*(a) \leq -b$. Then, (ii) implies that $\alpha \leq ax_0 + b$. Combining these bounds gives $ax_0 - f^*(a) \geq \alpha$, which implies (B.13). □

[2] $A \subset \mathbb{R}^2$ is **convex** if $z_1, z_2 \in A$, $\lambda \in [0, 1]$ implies $\lambda z_1 + (1 - \lambda)z_2 \in A$.

Corollary B.18. *If $f : \mathbb{R} \to \mathbb{R} \cup \{+\infty\}$ is lower semi-continuous, then*

$$(\mathrm{CE}f)^* = f^* .$$

Proof By Theorem B.17, $\mathrm{CE}f = f^{**}$, and so $(\mathrm{CE}f)^* = f^{***}$. Since f^* is lower semi-continuous, we have again by Theorem B.17 that $f^{***} = (f^*)^{**} = \mathrm{CE}f^*$. But f^* is convex, which implies that $\mathrm{CE}f^* = f^*$. □

In particular, we proved:

Theorem B.19. *If $f : \mathbb{R} \to \mathbb{R} \cup \{+\infty\}$ is lower semi-continuous and convex,*

$$f^{**} = f .$$

B.2.4 Legendre Transform of Nondifferentiable Functions

We have seen that the right and left derivatives of a convex function f at a point x_*, $\partial^+ f(x_*)$ and $\partial^- f(x_*)$, are well defined (Theorem B.12). If f is not differentiable at x_*, then $\partial^- f(x_*) < \partial^+ f(x_*)$, so f can have more than one supporting line at x_*, which has an important consequence on the qualitative behavior of the Legendre transform.

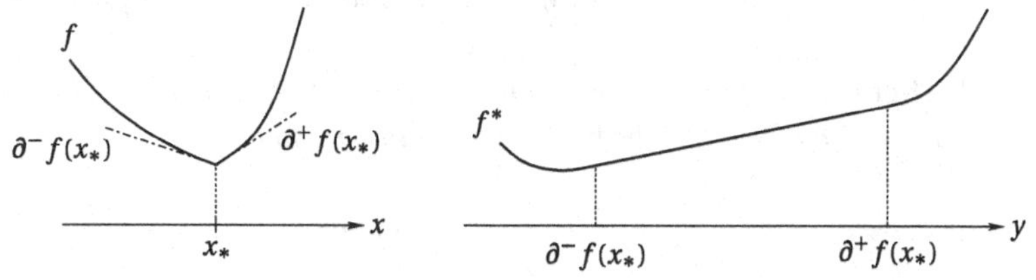

Theorem B.20. *Let f be a convex function. Then:*

1. *If f is not differentiable at x_*, then f^* is affine on the interval $[\partial^- f(x_*), \partial^+ f(x_*)]$.*
2. *If f is affine on some interval $[a, b]$, with a slope m, then f^* is nondifferentiable at m and $\partial^+ f^*(m) \geq b > a \geq \partial^- f^*(m)$.*

Note that if a continuous function f is not convex on an interval containing x, then $\mathrm{CE}f$ must be affine on that interval. In that case, the above theorem, combined with Corollary B.18, shows that f^* cannot be differentiable.

Proof By Theorem B.13, for each value $m \in [\partial^- f(x_*), \partial^+ f(x_*)]$, the line $x \mapsto m(x - x_*) + f(x_*)$ is a supporting line for f at x_*. By Exercise B.12, this implies that f^* admits, at each $m \in [\partial^- f(x_*), \partial^+ f(x_*)]$, a supporting line with the same slope x_*. Since f^* is convex, all these supporting lines actually coincide, which implies that f^* is affine on the interval.

If CEf is affine on $[a, b]$, with slope m, one has in particular that $f^*(m) = (\mathrm{CE}f)^*(m) = ma - f(a) = mb - f(b)$. Then, for all $\epsilon > 0$,

$$f^*(m+\epsilon) - f^*(m) \geq \{(m+\epsilon)b - f(b)\} - f^*(m) = \epsilon b,$$

and therefore $\partial^+ f^*(m) \geq b$. Similarly, $\partial^- f^*(m) \leq a$. □

B.3 Complex Analysis

Let $D \subset \mathbb{C}$ be a domain (that is, open and connected). Remember that a function $f : D \to \mathbb{C}$ is **holomorphic** if

$$f'(z) \stackrel{\text{def}}{=} \lim_{w \to z} \frac{f(w) - f(z)}{w - z}$$

exists and is finite at each $z \in D$. It is well known that holomorphic functions have derivatives of all orders, and that f is holomorphic if and only if it is **analytic**, that is, if and only if it can be represented at each point $z_0 \in D$ by a convergent Taylor series:

$$f(z) = \sum_{n \geq 0} a_n (z - z_0)^n,$$

where $a_n = \frac{1}{n!} f^{(n)}(z_0)$ and z belongs to a small disk around z_0. Therefore, holomorphic and analytic should be considered as synonyms in this section.

We start with the following fundamental result of complex analysis.

Theorem B.21 (Cauchy's Integral Theorem). *Let $D \subset \mathbb{C}$ be open and simply connected, and let f be holomorphic on D. Then*

$$\oint_\gamma f(\xi) \mathrm{d}\xi = 0,$$

for all closed paths $\gamma \subset D$.

Proof See, for example, [336, Theorem 4.14]. □

Corollary B.22. *Let $D \subset \mathbb{C}$ be open and simply connected, and let f be holomorphic on D. Then, there exists a function F, holomorphic on D, such that $F' = f$.*

Proof We fix some point $z_0 \in D$. Since D is open and connected, any other $z \in D$ can be joined from z_0 by a continuous path $\gamma_z \subset D$. Let

$$F(z) \stackrel{\text{def}}{=} \int_{\gamma_z} f(\xi) \mathrm{d}\xi.$$

By Theorem B.21, this definition does not depend on the choice of the path γ_z. Choose $r > 0$ small enough to ensure that the disk $B(z, r) \stackrel{\text{def}}{=} \{w \in \mathbb{C} : |w - z| \leq r\} \subset D$. Then,

$$F(w) = F(z) + \int_{[z,w]} f(\xi)\mathrm{d}\xi, \qquad \forall w \in B(z,r),$$

where $[z,w]$ is the straight line segment connecting z to w. But, f being holomorphic implies in particular that $f(\xi) = f(z) + O(|\xi - z|)$ for all $\xi \in B(z,r)$, and so

$$\lim_{w\to z} \frac{F(w) - F(z)}{w - z} = \lim_{w\to z} \frac{1}{w-z} \int_{[z,w]} f(\xi)\mathrm{d}\xi = f(z).$$

This implies that F is holomorphic and that $F' = f$. □

Theorem B.23. *Let f be a holomorphic function on a simply connected open set $D \subset \mathbb{C}$, which has no zeros on D. Then, there exists a function g analytic on D, called a* **branch of the logarithm of f on** *D, such that $f = e^g$.*

Proof Our assumptions imply that f'/f is holomorphic on D. Corollary B.22 thus implies the existence of a function F, holomorphic on D, such that $F' = f'/f$. In particular,

$$(fe^{-F})' = f'e^{-F} - fF'e^{-F} \equiv 0.$$

Therefore, there exists $c \in \mathbb{C}$ such that $fe^{-F} = e^c$, or equivalently $f = e^{F+c}$. □

Remark B.24. 1. Let g be a branch of the logarithm of f on D. Then $\mathfrak{Re}\, g = \log|f|$. Indeed,

$$|f| = |e^g| = |e^{\mathfrak{Re}\, g} e^{\mathrm{i}\, \mathfrak{Im}\, g}| = e^{\mathfrak{Re}\, g}.$$

2. Let g_1 and g_2 be two branches of the logarithm of f on D. Since, for each $z \in D$,

$$e^{g_2(z)-g_1(z)} = \frac{e^{g_2(z)}}{e^{g_1(z)}} = \frac{f(z)}{f(z)} = 1,$$

we conclude that $g_2(z) = g_1(z) + 2\mathrm{i}k(z)\pi$ for some $k(z) \in \mathbb{Z}$. However, $z \mapsto k(z) = (g_2(z) - g_1(z))/2\mathrm{i}\pi$ is continuous and integer-valued; it is therefore constant on D. This implies that $g_2 = g_1 + 2\mathrm{i}k\pi$ for some $k \in \mathbb{Z}$.
3. Assume that the domain D in Theorem B.23 is such that $D \cap \mathbb{R}$ is connected. Suppose also that $f(z) \in \mathbb{R}_{>0}$ for $z \in D \cap \mathbb{R}$. Then there is a branch g of the logarithm of f on D such that $g(z) \in \mathbb{R}$ for all $z \in D \cap \mathbb{R}$; in particular, g coincides with the usual logarithm of f (seen as a real function) on $D \cap \mathbb{R}$. Indeed, it suffices to observe that the function F in the proof can be constructed by starting from a point $z_0 \in D \cap \mathbb{R}$ at which one can fix $g(z_0) = \log|f(z_0)|$ and use the fact that $F(z) = \int_{z_0}^{z} f'(x)/f(x)\,\mathrm{d}x$ for $z \in D \cap \mathbb{R}$. ◇

It is well known that the limit of a sequence of analytic functions need not be analytic. Let us see how additional conditions can be imposed to guarantee that the limiting function is also analytic.

Remember that a family $\mathscr{A}$ of functions on $\mathbb{C}$ is **locally uniformly bounded** on a set $D \subset \mathbb{C}$ if, for each $z \in D$, there exists a real number M and a neighborhood $\mathscr{U}$ of z such that $|f(w)| \leq M$ for all $w \in \mathscr{U}$ and all $f \in \mathscr{A}$.

Theorem B.25 (Vitali Convergence Theorem). *Let D be an open, connected subset of $\mathbb{C}$ and $(f_n)_{n\geq 1}$ be a sequence of analytic functions on D, which are locally uniformly bounded and converge on a set having a cluster point in D. Then the sequence $(f_n)_{n\geq 1}$ converges locally uniformly on D to an analytic function.*

Proof See [71, p. 154]. □

Theorem B.26 (Hurwitz Theorem). *Let D be an open subset of $\mathbb{C}$ and $(f_n)_{n\geq 1}$ be a sequence of analytic functions, which converge, locally uniformly, on D to an analytic function f. If $f_n(z) \neq 0$, for all $z \in D$ and for all n, then either f vanishes identically, or f is never zero on D.*

Proof See [71, Corollary 2.6]. □

The following theorem is the complex counterpart to Theorem B.7. (Notice that, in the complex case, no control is needed on the series of the derivatives.)

Theorem B.27 (Weierstrass' Theorem on uniformly convergent series of analytic functions). *Let $D \subset \mathbb{C}$ be a domain. For each k, let $f_k \colon D \to \mathbb{C}$ be an analytic function. If the series*

$$f(z) \stackrel{\text{def}}{=} \sum_k f_k(z)$$

is uniformly convergent on every compact subset $K \subset D$, then it defines an analytic function on D. Moreover, for each $n \in \mathbb{N}$, $\sum_k f_k^{(n)}$ converges uniformly on every compact $K \subset D$ and

$$f^{(n)}(z) = \sum_k f_k^{(n)}(z), \quad \forall z \in D.$$

Proof See [228, Volume I, Theorem 15.6]. □

Let $U, V \subset \mathbb{C}$. A continuous function $F \colon U \times V \to \mathbb{C}$ is said to be **analytic** on $U \times V$ if $F(\cdot, z)$ is analytic on U for any fixed $z \in V$ and $F(z, \cdot)$ is analytic on V for any fixed $z \in U$.

Theorem B.28 (Implicit Function Theorem). *Let $(\omega, z) \mapsto F(\omega, z)$ be an analytic function on an open domain $U \times V \subset \mathbb{C}^2$. Let $(\omega_0, z_0) \in U \times V$ be such that $F(\omega_0, z_0) = 0$ and $\frac{\partial F}{\partial z}(\omega_0, z_0) \neq 0$. Then there exists an open subset $U_0 \subset U$ containing ω_0 and an analytic map $\varphi \colon U_0 \to V$ such that*

$$F(\omega, \varphi(\omega)) = 0 \quad \textit{for all } \omega \in U_0.$$

Proof See [228, Volume II, Theorem 3.11]. □

B.4 Metric Spaces

All topological notions used in the book (in particular those of Chapter 6) concern topologies induced by a *metric*. Let χ be an arbitrary set. A map $d\colon \chi \times \chi \to \mathbb{R}_{\geq 0}$ is a **metric (on χ)** (or **distance**) if it satisfies: (i) $d(x,y) = 0$ if and only if $x = y$, (ii) $d(x,y) = d(y,x)$ for all $x, y \in \chi$, (iii) $d(x,y) \leq d(x,z) + d(z,y)$ for all $x, y, z \in \chi$. The pair (χ, d) is then called a **metric space**.

The **open ball centered at $x \in \chi$ of radius $\epsilon > 0$** is $B_\epsilon(x) \stackrel{\text{def}}{=} \{y \in \chi : d(y,x) < \epsilon\}$. A set $A \subset \chi$ is **open** if, for each $x \in A$, there exists $\epsilon > 0$ such that $B_\epsilon(r) \subset A$. A set A is **closed** if $A^c \stackrel{\text{def}}{=} \chi \setminus A$ is open. Arbitrary unions and finite intersections of open sets are open. A sequence $(x_n)_{n\geq 1} \subset \chi$ **converges** to $x_* \in \chi$ (denoted $x_n \to x_*$) if, for all $\epsilon > 0$, there exists n_0 such that $x_n \in B_\epsilon(x_*)$ for all $n \geq n_0$. A set $F \subset \chi$ is closed if and only if $(x_n)_{n\geq 1} \subset F$, $x_n \to x_*$ implies $x_* \in F$. A set $D \subset \chi$ is **dense** if, for all $x \in \chi$ and all $\epsilon > 0$, $D \cap B_\epsilon(x) \neq \varnothing$. χ is **separable** if there exists a countable dense subset $D \subset \chi$.

On $\chi = \mathbb{R}^n$, one usually uses the *Euclidean metric* inherited from the Euclidean norm: $d(x,y) \stackrel{\text{def}}{=} \|x - y\|_2$; on $\chi = \mathbb{C}$, one uses the modulus: $d(w,z) \stackrel{\text{def}}{=} |w - z|$.

A function $f\colon \chi \to \chi'$ is **continuous** if $f(x_n) \to f(x_*)$ whenever $x_n \to x_*$. Equivalently, f is continuous if and only if $f^{-1}(A') \subset \chi$ is open for each open set $A' \subset \chi'$.

A metric space (χ, d) is **sequentially compact** (or simply **compact**) if there exists, for each sequence $(x_n)_{n\geq 1} \subset \chi$, a subsequence $(x_{n_k})_{k\geq 1}$ and some $x_* \in \chi$ such that $x_{n_k} \to x_*$ when $k \to \infty$. A compact metric space is always separable.

An introduction to metric spaces can be found in [284, Chapter 1].

B.5 Measure Theory

This section and the two following ones contain several definitions and results concerning measure theory and integration. Many detailed books exist on the subject, including the one by Bogachev [30].

B.5.1 Measures and Probability Measures

Throughout this section, Ω denotes an arbitrary set and $\mathscr{P}(\Omega)$ the collection of all subsets of Ω. The complement of a set $A \subset \Omega$ will be denoted $A^c \stackrel{\text{def}}{=} \Omega \setminus A$.

Definition B.29. A collection $\mathscr{A} \subset \mathscr{P}(\Omega)$ is an **algebra** if (i) $\varnothing \in \mathscr{A}$, (ii) $A \in \mathscr{A}$ implies $A^c \in \mathscr{A}$, and (iii) $A, B \in \mathscr{A}$ implies $A \cup B \in \mathscr{A}$.

Definition B.30. A collection $\mathscr{F} \subset \mathscr{P}(\Omega)$ is a **σ-algebra** if (i) $\varnothing \in \mathscr{A}$, (ii) $A \in \mathscr{A}$ implies $A^c \in \mathscr{F}$, and (iii) $(A_n)_{n\geq 1} \subset \mathscr{F}$ implies $\bigcup_{n\geq 1} A_n \in \mathscr{F}$.

Given an arbitrary collection $\mathscr{S} \subset \mathscr{P}(\Omega)$ of subsets of Ω, there exists a smallest σ-algebra containing $\mathscr{S}$, called the **σ-algebra generated by** $\mathscr{S}$, denoted $\sigma(\mathscr{S})$ and given by

$$\sigma(\mathscr{S}) \stackrel{\text{def}}{=} \bigcap \{\mathscr{F} : \mathscr{F} \text{ a } \sigma\text{-algebra containing } \mathscr{S}\} .$$

(Note that the intersection of an arbitrary collection of σ-algebras is a σ-algebra.)

Example B.31. If (χ, d) is a metric space whose collection of open sets is denoted by $\mathscr{O}$, then $\mathscr{B} \stackrel{\text{def}}{=} \sigma(\mathscr{O})$ is called the **σ-algebra of Borel sets on χ**. When (χ, d) is the Euclidean space $\mathbb{R}^n$ (equipped with the Euclidean metric), this σ-algebra is denoted $\mathscr{B}(\mathbb{R}^n)$. ◇

A pair $(\Omega, \mathscr{F})$, where $\mathscr{F}$ is a σ-algebra of subsets of Ω, is called a **measurable space** and the sets $A \in \mathscr{F}$ are called **measurable.**

Definition B.32. On a measurable space $(\Omega, \mathscr{F})$, a set function $\mu \colon \mathscr{F} \to [0, +\infty]$ is called a **measure** if the following holds:

1. $\mu(\varnothing) = 0$.
2. (**σ-additivity**) If $(A_n)_{n\geq 1} \subset \mathscr{F}$ is a sequence of pairwise disjoint sets, then $\mu(\bigcup_n A_n) = \sum_n \mu(A_n)$.

The measure μ is **finite** if $\mu(\Omega) < \infty$; μ is a **probability measure** if $\mu(\Omega) = 1$. If there exists a sequence $(A_n)_{n\geq 1} \subset \mathscr{F}$ such that $\bigcup_{n\geq 1} A_n = \Omega$ and $\mu(A_n) < \infty$ for each n, then μ is **σ-finite**.

Let us remind the reader of two straightforward consequences of the above definition. First, by the σ-additivity of item 2 above,

$$\mu\Big(\bigcup_n A_n\Big) \leq \sum_n \mu(A_n),$$

for any sequence $A_n \in \mathscr{F}$. In particular, if $\mu(A_n) = 0$ for all n, then

$$\mu\Big(\bigcup_n A_n\Big) = 0 .$$

A property A, defined for each element $\omega \in \Omega$ is said to occur **μ-almost everywhere** (or **for μ-almost all ω**) if there exists $B \in \mathscr{F}$ such that $\{\omega \in \Omega : A \text{ does not hold for } \omega\} \subset B$ and $\mu(B) = 0$. When μ is a probability measure, one usually says **μ-almost surely**.

Measures are usually constructed by defining a *finitely additive* set function on an algebra $\mathscr{A}$ and by extending it to the σ-algebra generated by $\mathscr{A}$.

Let $\mathscr{A}$ be an algebra. A set function $\mu_0 \colon \mathscr{A} \to [0, +\infty]$ is said to be **finitely additive** if $\mu_0(A \cup B) = \mu_0(A) + \mu_0(B)$ for all pairs of disjoint measurable sets; μ_0 is a **measure** if $\mu_0(\varnothing) = 0$ and if $\mu_0(\bigcup_{n\geq 1} A_n) = \sum_{n\geq 1} \mu_0(A_n)$ holds for all sequences $(A_n)_{n\geq 1} \subset \mathscr{A}$ of pairwise disjoint sets for which $\bigcup_{n\geq 1} A_n \in \mathscr{A}$.

Theorem B.33 (Carathéodory's Extension Theorem). *Let $\mu_0 \colon \mathscr{A} \to [0,+\infty]$ be a σ-finite measure on an algebra $\mathscr{A}$ and let $\mathscr{F} \stackrel{\text{def}}{=} \sigma(\mathscr{A})$. Then there exists a unique measure $\mu \colon \mathscr{F} \to [0,+\infty]$, called the* **extension** *of μ_0, which coincides with μ_0 on $\mathscr{A}$: $\mu(A) = \mu_0(A)$ for all $A \in \mathscr{A}$.*

The σ-algebra $\mathscr{F} = \sigma(\mathscr{A})$ is in general a much larger collection of sets than $\mathscr{A}$; nevertheless, each set $B \in \mathscr{F}$ can be approximated arbitrarily well by sets in $A \in \mathscr{A}$ in the sense of measure theory:

Lemma B.34. *Let μ be a probability measure on $(\Omega, \mathscr{F})$, where $\mathscr{F}$ is generated by an algebra $\mathscr{A}$: $\mathscr{F} = \sigma(\mathscr{A})$. Then, for all $B \in \mathscr{F}$ and all $\epsilon > 0$, there exists $A \in \mathscr{A}$ such that $\mu(B \triangle A) \leq \epsilon$.*

Proof Let $\mathscr{G} \stackrel{\text{def}}{=} \{B \in \mathscr{F} : \forall \epsilon > 0, \exists A \in \mathscr{A} \text{ s.t. } \mu(B\triangle A) \leq \epsilon\}$. Since, obviously, $\mathscr{G} \supset \mathscr{A}$, it suffices to show that $\mathscr{G}$ is a σ-algebra. Since $\mu(B\triangle A) = \mu(B^{\mathrm{c}} \triangle A^{\mathrm{c}})$, we see that $\mathscr{G}$ is stable under taking complements. Let $(B_n)_{n\geq 1} \subset \mathscr{G}$ and set $B = \bigcup_{n\geq 1} B_n$. Fix $\epsilon > 0$. For each n, let $A_n \in \mathscr{A}$ be such that $\mu(B_n \triangle A_n) \leq \epsilon/2^n$. Then, let $A = \bigcup_{n=1}^N A_n \in \mathscr{A}$. If N is large enough,

$$\mu(B\triangle A) \leq \sum_{n\geq 1} \mu(B_n \triangle A_n) \leq \epsilon\,.$$

Therefore, $B \in \mathscr{G}$. This shows that $\mathscr{G}$ is a σ-algebra. □

In measure theory, it is often useful to determine whether some property is verified by each measurable set of a σ-algebra $\mathscr{F}$. If $\mathscr{F}$ is generated by an algebra, then this can be done by checking conditions that are easier to verify than testing *each* $B \in \mathscr{F}$.

A collection $\mathscr{M} \subset \mathscr{P}(\Omega)$ is a **monotone class** if (i) $\Omega \in \mathscr{M}$, (ii) for any sequence $(A_n)_{n\geq 1} \subset \mathscr{M}$ such that $A_n \uparrow A$, one has $A \in \mathscr{M}$, and (iii) for any sequence $(A_n)_{n\geq 1} \subset \mathscr{M}$ such that $A_n \downarrow A$, one has $A \in \mathscr{M}$. As before, there always exists a smallest monotone class generated by a collection $\mathscr{S}$, denoted $\mathscr{M}(\mathscr{S})$.

Theorem B.35. *If $\mathscr{A}$ is an algebra, then $\mathscr{M}(\mathscr{A}) = \sigma(\mathscr{A})$.*

A similar result holds for a slightly different notion of class. A collection $\mathscr{D} \subset \mathscr{P}(\Omega)$ is a **Dynkin system** if (i) $\varnothing \in \mathscr{D}$, (ii) $A, B \in \mathscr{D}$ with $A \subset B$ implies $B \setminus A \in \mathscr{D}$, and (iii) for any sequence $(A_n)_{n\geq 1} \subset \mathscr{D}$ such that $A_n \uparrow A$, one has $A \in \mathscr{D}$. Again, there always exists a smallest Dynkin system generated by a collection $\mathscr{S}$, denoted $\delta(\mathscr{S})$. A collection $\mathscr{C} \subset \mathscr{P}(\Omega)$ is **$\cap$-stable** if $A, B \in \mathscr{C}$ implies $A \cap B \in \mathscr{C}$.

Theorem B.36. *If $\mathscr{C}$ is $\cap$-stable (in particular, if $\mathscr{C}$ is an algebra), then $\delta(\mathscr{C}) = \sigma(\mathscr{C})$.*

This result can be used to determine when two measures are identical.

Corollary B.37. *Let $(\Omega, \mathscr{F})$ be a measurable space. Let $\mathscr{C}$ be a collection of sets that is $\cap$-stable and generates $\mathscr{F}$: $\mathscr{F} = \sigma(\mathscr{C})$. If μ and ν are two probability measures on $(\Omega, \mathscr{F})$ that coincide on $\mathscr{C}$ ($\mu(C) = \nu(C)$ for all $C \in \mathscr{C}$), then $\mu = \nu$.*

B.5.2 Measurable Functions

Let $(\Omega, \mathscr{F})$ and $(\Omega', \mathscr{F}')$ be two measurable spaces. A map $f\colon \Omega \to \Omega'$ is $\mathscr{F}/\mathscr{F}'$-**measurable** if $f^{-1}(B') \in \mathscr{F}$ for each $B' \in \mathscr{F}'$.

Let $(\Omega', \mathscr{F}')$ be a measurable space. For any set Ω, given an arbitrary map $h\colon \Omega \to \Omega'$, we denote by $\sigma(h)$ the smallest σ-algebra on Ω with respect to which h is $\mathscr{F}/\mathscr{F}'$-measurable: $\sigma(h) \stackrel{\text{def}}{=} \{h^{-1}(B') : B' \in \mathscr{F}'\}$. $\sigma(h)$ is called the σ-algebra **generated by** h.

Lemma B.38 (Doob–Dynkin lemma). *Let $(\Omega, \mathscr{F})$, $(\Omega', \mathscr{F}')$ be measurable spaces, where $\mathscr{F} = \sigma(h)$ for some $h\colon \Omega \to \Omega'$. For any $\mathscr{F}/\mathscr{B}(\mathbb{R})$-measurable map $g\colon \Omega \to \mathbb{R}$, there exists an $\mathscr{F}'/\mathscr{B}(\mathbb{R})$-measurable map $\varphi\colon \Omega' \to \mathbb{R}$ such that $g = \varphi \circ h$.*

Proof See [186, Lemma 1.13]. □

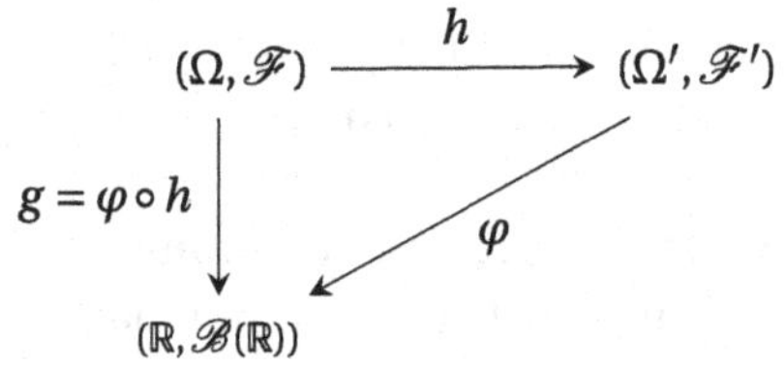

Figure B.3. The setting of Lemma B.38.

B.6 Integration

Let $\mathscr{F} \subset \mathscr{P}(\Omega)$ be a σ-algebra. When integrating real-valued functions, it is convenient to include the possibility of these taking the values $\pm\infty$. Let therefore $\overline{\mathbb{R}} \stackrel{\text{def}}{=} \mathbb{R} \cup \{\pm\infty\}$, together with the σ-algebra $\mathscr{B}(\overline{\mathbb{R}})$ containing sets of the form B, $B \cup \{+\infty\}$, $B \cup \{-\infty\}$ or $B \cup \{+\infty\} \cup \{-\infty\}$, where $B \in \mathscr{B}(\mathbb{R})$. An $\mathscr{F}/\mathscr{B}(\overline{\mathbb{R}})$-measurable function $f\colon \Omega \to \overline{\mathbb{R}}$ will simply be called **measurable**:

$f^{-1}(I) \in \mathscr{F}$ for all $I \in \mathscr{B}(\overline{\mathbb{R}})$. To be measurable, f need only satisfy $f^{-1}(\{\pm\infty\}) \in \mathscr{F}$ and $f^{-1}((-\infty, x]) \in \mathscr{F}$ for all $x \in \mathbb{R}$.

Integration is first defined for nonnegative functions. A measurable function $\varphi\colon \Omega \to \mathbb{R} \cup \{+\infty\}$ is **simple** if it takes a finite set of values; it can therefore be written as a finite linear combination

$$\varphi = \sum_{k=1}^{n} a_k \mathbf{1}_{E_k},$$

where $E_k = \{\omega \in \Omega : \varphi(\omega) = a_k\} \in \mathscr{F}$, where $\varphi(\mathbb{R}) = \{a_1, \dots, a_n\} \subset \mathbb{R} \cup \{+\infty\}$. A measurable map $f\colon \Omega \to [0, +\infty]$ can always be written as a limit of an increasing sequence of simple functions $\varphi_n \uparrow f$. The **integral of φ with respect to μ** is

$$\int \varphi \, \mathrm{d}\mu \stackrel{\text{def}}{=} \sum_{k=1}^{n} a_k \mu(E_k).$$

In this definition, we make the convention that $0 \cdot \infty = 0$. If $f\colon \Omega \to \mathbb{R} \cup \{+\infty\}$ is measurable and nonnegative, its **integral with respect to μ** is

$$\int f \, \mathrm{d}\mu \stackrel{\text{def}}{=} \sup\left\{\int \varphi \, \mathrm{d}\mu \, : \, \varphi \text{ simple}, 0 \le \varphi \le f\right\}.$$

For an arbitrary measurable function f, let $f^+ \stackrel{\text{def}}{=} f \mathbf{1}_{\{f \ge 0\}}$, $f^- \stackrel{\text{def}}{=} (-f)^+$. We say that f is **integrable** if $\int f^+ \, \mathrm{d}\mu < \infty$ and $\int f^- \, \mathrm{d}\mu < \infty$. The set of integrable functions is denoted by $L^1(\mu)$ (we sometimes omit the measure when it is clear from the context). The **integral** of $f \in L^1(\mu)$ is

$$\int f \, \mathrm{d}\mu \stackrel{\text{def}}{=} \int f^+ \, \mathrm{d}\mu - \int f^- \, \mathrm{d}\mu.$$

In this book, we also use alternative notations for $\int f \, \mathrm{d}\mu$, such as $\mu(f)$ or $\langle f \rangle_\mu$. We list below a few properties of the integral.

- If $f \ge 0$ and $\int f \, \mathrm{d}\mu < \infty$, then f is μ-almost everywhere finite.
- If $f \ge 0$ and $\int f \, \mathrm{d}\mu = 0$, then $f = 0$ μ-almost everywhere.
- If $f, g \in L^1(\mu)$, $\int (f + g) \, \mathrm{d}\mu = \int f \, \mathrm{d}\mu + \int g \, \mathrm{d}\mu$.
- $f \in L^1(\mu)$ if and only if $\int |f| \, \mathrm{d}\mu < \infty$, and when this occurs, $|\int f \, \mathrm{d}\mu| \le \int |f| \, \mathrm{d}\mu$.
- If $0 \le f \le g$ μ-almost everywhere, then $\int f \, \mathrm{d}\mu \le \int g \, \mathrm{d}\mu$.
- If $f, g \in L^1(\mu)$, $f = g$ μ-almost everywhere, then $\int f \, \mathrm{d}\mu = \int g \, \mathrm{d}\mu$.

Theorem B.39 (Monotone Convergence Theorem). *Let $(f_n)_{n \ge 1}$ be a sequence of nonnegative measurable functions such that $f_n \le f_{n+1}$ μ-almost everywhere. Then*

$$\int \lim_{n \to \infty} f_n \, \mathrm{d}\mu = \lim_{n \to \infty} \int f_n \, \mathrm{d}\mu.$$

Theorem B.40 (Dominated Convergence Theorem). *Let $(f_n)_{n\geq 1} \subset L^1(\mu)$. Assume there exists $g \in L^1(\mu)$ such that $|f_n| \leq g$ μ-almost everywhere, for all $n \geq 1$. If $f_n \to f$ μ-almost everywhere, then $f \in L^1(\mu)$ and*

$$\int f \, \mathrm{d}\mu = \lim_{n\to\infty} \int f_n \, \mathrm{d}\mu .$$

Exercise B.15. Let $(\xi_k)_{k\geq 1}$ be a sequence of real functions defined on some open interval $I \subset \mathbb{R}$ and $x_0 \in I$. Let that sequence be such that, for each k, $\lim_{x\to x_0} \xi_k(x)$ exists. Assuming there exists a summable sequence $(\epsilon_k)_{k\geq 1} \subset \mathbb{R}_{\geq 0}$ such that $\sup_{x\in I} |\xi_k(x)| \leq \epsilon_k$, show that

$$\lim_{x\to x_0} \sum_k \xi_k(x) = \sum_k \lim_{x\to x_0} \xi_k(x) . \tag{B.14}$$

Let μ, ν be two finite measures. ν is **absolutely continuous with respect to μ** if, for all $A \in \mathscr{F}$, $\mu(A) = 0$ implies $\nu(A) = 0$; we then write $\nu \ll \mu$. When both $\mu \ll \nu$ and $\nu \ll \mu$, the measures are said to be **equivalent**. If there exists $A \in \mathscr{F}$ such that $\mu(A) = 0$, $\nu(A^{\mathrm{c}}) = 0$, then μ and ν are **singular**.

Theorem B.41 (Radon-Nikodým Theorem). *Let μ and ν be two finite measures such that $\nu \ll \mu$. There exists a measurable function $f \geq 0$ such that*

$$\forall B \in \mathscr{F}, \quad \nu(B) = \int_B f \, \mathrm{d}\mu .$$

f is called the ***Radon-Nikodým derivative of ν with respect to μ*** *and is often denoted $\frac{\mathrm{d}\nu}{\mathrm{d}\mu}$.*

Any two versions of the Radon-Nikodým derivative coincide μ-almost everywhere; this is a consequence of the following lemma.

Lemma B.42. *Let $f, g \in L^1(\mu)$ be such that $\int_B f \, \mathrm{d}\mu = \int_B g \, \mathrm{d}\mu$ for all $B \in \mathscr{F}$. Then $f = g$ almost everywhere.*

The Radon-Nikodým derivative enjoys properties similar to that of the ordinary derivative. First, if $\nu_1, \nu_2 \ll \mu$, then $\nu_1 + \nu_2 \ll \mu$ and

$$\frac{\mathrm{d}(\nu_1 + \nu_2)}{\mathrm{d}\mu} = \frac{\mathrm{d}\nu_1}{\mathrm{d}\mu} + \frac{\mathrm{d}\nu_2}{\mathrm{d}\mu} . \tag{B.15}$$

Then, a property similar to the chain rule holds: if ν, μ, ρ satisfy $\nu \ll \mu \ll \rho$, then

$$\frac{\mathrm{d}\nu}{\mathrm{d}\rho} = \frac{\mathrm{d}\nu}{\mathrm{d}\mu} \frac{\mathrm{d}\mu}{\mathrm{d}\rho} . \tag{B.16}$$

B.6.1 Product Spaces

Given two measurable spaces $(\Omega, \mathscr{F})$, $(\Omega', \mathscr{F}')$, we can consider the product

$$\Omega \times \Omega' \stackrel{\text{def}}{=} \{(\omega, \omega') : \omega \in \Omega, \omega' \in \Omega'\},$$

equipped with the **product σ-algebra** $\mathscr{F} \otimes \mathscr{F}'$, generated by the algebra of finite unions of **rectangles**, that is, sets of the form $A \times A'$ with $A \in \mathscr{F}$ and $A \in \mathscr{F}'$. If μ is a measure on $(\Omega, \mathscr{F})$ and μ' is a measure on $(\Omega', \mathscr{F}')$, we can define, for a rectangle,

$$(\mu \otimes \mu')(A \times A') \stackrel{\text{def}}{=} \mu(A)\mu'(A').$$

Using Theorem B.33, it can be shown that, when μ and ν are σ-finite, $\mu \otimes \mu'$ has a unique extension to $\mathscr{F} \otimes \mathscr{F}'$; we call it the **product measure**.

> **Theorem B.43** (Fubini–Tonelli Theorem). *If μ and μ' are σ-finite and if $F: \Omega \times \Omega' \to \mathbb{R}_{\geq 0}$ is $\mathscr{F} \otimes \mathscr{F}'$-measurable, then the functions*
>
> $$\omega \mapsto \int_{\Omega'} F(\omega, \omega')\mu'(d\omega') \quad \text{and} \quad \omega' \mapsto \int_{\Omega} F(\omega, \omega')\mu(d\omega)$$
>
> *are $\mathscr{F}$- and $\mathscr{F}'$-measurable, respectively. Moreover,*
>
> $$\begin{aligned}\int_{\Omega\times\Omega'} F\, d(\mu \otimes \mu') &= \int_{\Omega}\left\{\int_{\Omega'} F(\omega, \omega')\, \mu'(d\omega')\right\}\mu(d\omega) \\ &= \int_{\Omega'}\left\{\int_{\Omega} F(\omega, \omega')\, \mu(d\omega)\right\}\mu'(d\omega').\end{aligned}$$

The above construction extends to the product of an arbitrary finite number of σ-finite measurable spaces: $(\Omega_1, \mathscr{F}_1), \dots, (\Omega_n, \mathscr{F}_n)$.

B.7 Lebesgue Measure

The Lebesgue measure is first constructed on the real line, by extending to all Borel sets the basic notion of *length* of bounded intervals:

$$\ell([a, b)) \stackrel{\text{def}}{=} b - a.$$

(Unbounded intervals are defined to have measure $+\infty$.) This allows us to define a natural measure on the algebra of finite unions of such intervals, which can be extended to all Borel sets $\mathscr{B}(\mathbb{R})$ using Theorem B.33. The resulting σ-finite measure ℓ on $(\mathbb{R}, \mathscr{B}(\mathbb{R}))$ is called the **Lebesgue measure.**

On $\mathbb{R}^n = \mathbb{R} \times \cdots \times \mathbb{R}$, equipped with the Borel σ-algebra $\mathscr{B}(\mathbb{R}^n)$, the Lebesgue measure is defined as the product measure, that is, it is first defined on parallelepipeds:

$$\ell^n([a_1, b_1) \times [a_2, b_2) \times \cdots \times [a_n, b_n)) \stackrel{\text{def}}{=} \prod_{i=1}^{n}(b_i - a_i),$$

and then extended. The Lebesgue measure is **translation invariant**, $\ell^n(B+\mathbf{x}) = \ell^n(B)$ for all $B \in \mathscr{B}(\mathbb{R}^n)$, $\mathbf{x} \in \mathbb{R}^n$, and enjoys the following **scaling property**: $\ell^n(\alpha B) = \alpha^n \ell^n(B)$ for all $B \in \mathscr{B}(\mathbb{R}^n)$ and all scaling factors $\alpha > 0$.

One usually writes $\mathrm{d}\mathbf{x}$ instead of $\ell^n(\mathrm{d}\mathbf{x})$. For instance, the integration of a function $f : \mathbb{R}^n \to \mathbb{R}$ with respect to ℓ^n is written $\int f(\mathbf{x})\,\mathrm{d}\mathbf{x}$.

B.8 Probability

We remind the reader of some basic elements from probability theory. The books by Kallenberg [186] or Grimmett and Stirzaker [152] provide good references.

In probability theory, a **probability space** is a triple $(\Omega, \mathscr{F}, P)$, where $(\Omega, \mathscr{F})$ is a measurable space and $P : \mathscr{F} \to [0,1]$ is a probability measure. Each $\omega \in \Omega$ is to be interpreted as the outcome of a random experiment and each measurable set $A \in \mathscr{F}$ is interpreted as an **event**, with $P(A)$ measuring the a-priori likelihood of the occurrence of A when sampling some $\omega \in \Omega$.

B.8.1 Random Variables and Vectors

A measurable map $X : \Omega \to \overline{\mathbb{R}}$ is called a **random variable**. The **distribution of** $X : \Omega \to \overline{\mathbb{R}}$ is the probability measure P_X on $\overline{\mathbb{R}}$ defined by $P_X(I) \overset{\text{def}}{=} P(X \in I)$, for all $I \in \mathscr{B}(\overline{\mathbb{R}})$. The **cumulative distribution function of** $X : \Omega \to \overline{\mathbb{R}}$ is $F_X(x) \overset{\text{def}}{=} P(X \leq x)$, $x \in \mathbb{R}$. X has a **density (with respect to the Lebesgue measure)** if there exists a measurable function $f_X : \mathbb{R} \to \mathbb{R}_{\geq 0}$ such that

$$P(X \in B) = \int_B f_X(x)\,\mathrm{d}x, \quad \forall B \in \mathscr{B}(\mathbb{R}).$$

The integral of a random variable $X \in L^1(P)$ is denoted

$$E[X] \overset{\text{def}}{=} \int X\,\mathrm{d}P,$$

and is called the **expectation** of X with respect to P. The **variance** of X is then defined by

$$\mathrm{Var}(X) \overset{\text{def}}{=} E\big[(X - E[X])^2\big].$$

We list here a few inequalities that are used frequently:

- **Jensen's Inequality**: If $X \in L^1(P)$ and if $\phi : \mathbb{R} \to \mathbb{R}$ is convex and such that $\phi(X) \in L^1(P)$, then $\phi(E[X]) \leq E[\phi(X)]$. When ϕ is strictly convex, equality holds if and only if X is almost surely constant.
- **Markov's Inequality**: for all nonnegative $X \in L^1(P)$ and all $\lambda > 0$,

$$P(X \geq \lambda) \leq \frac{E[X]}{\lambda}. \tag{B.17}$$

- **Chebyshev's Inequality**: for all X and all $\lambda > 0$,

$$P(|X - E[X]| \geq \lambda) \leq \frac{\mathrm{Var}(X)}{\lambda^2}. \tag{B.18}$$

- **Chernov's Inequality**: for all X and all $\lambda > 0$,

$$P(X \geq \lambda) \leq \inf_{t>0} \frac{E[e^{tX}]}{e^{t\lambda}}. \tag{B.19}$$

There are various ways by which a sequence of random variables $(X_n)_{n\geq 1}$ can *converge* to a limiting random variable X.

- $(X_n)_{n\geq 1}$ **converges to** X **almost surely** if there exists $C \in \mathscr{F}$, $P(C) = 1$, such that $X_n(\omega) \to X(\omega)$ for all $\omega \in C$.
- $(X_n)_{n\geq 1}$ **converges to** X **in probability** if, for all $\epsilon > 0$, $P(|X_n - X| \geq \epsilon) \to 0$ as $n \to \infty$.
- Let $p \geq 1$. $(X_n)_{n\geq 1}$ **converges to** X **in** L^p if $E[|X|^p] < \infty$, $E[|X_n^p|] < \infty$, for all n, and $E[|X_n - X|^p] \to 0$ when $n \to \infty$.
- $(X_n)_{n\geq 1}$ **converges to** X **in distribution** if $F_{X_n}(x) \to F_X(x)$ when $n \to \infty$, for all x at which F_X is continuous.

Almost-sure convergence and convergence in L^p both imply convergence in probability, which in turn implies convergence in distribution. The remaining implications do not hold in general.

B.8.2 Independence

A collection of events $(A_i)_{i\in I}$ is **independent** if $P(\cap_{j\in J} A_j) = \cap_{j\in J} P(A_j)$ for all $J \subset I$ finite. A collection of random variables, $(X_i)_i \in I$ is **independent** if the events $(\{X_i \leq \alpha_i\})_{i\in I}$ are independent for all $(\alpha_i)_{i\in I} \subset \mathbb{R}$. If, moreover, all the variables X_k have the same distribution, we say that $(X_k)_{k\in I}$ is **i.i.d. (independent, identically distributed)**.

We now state two central results of probability theory.

Theorem B.44 (Law of Large Numbers). *Let $(X_n)_{n\geq 1} \subset L^1(P)$ be an i.i.d. sequence. Then, as $n \to \infty$,*

$$\frac{X_1 + \cdots + X_n}{n} \to E[X_1] \quad P\text{-almost surely.}$$

Remember that X is a **standard normal** random variable, $X \sim \mathcal{N}(0, 1)$, if it has a density with respect to the Lebesgue measure dt, given by $\frac{1}{\sqrt{2\pi}} e^{-\frac{t^2}{2}}$; in particular, its cumulative distribution function is

$$F_X(x) = \frac{1}{\sqrt{2\pi}} \int_{-\infty}^{x} e^{-\frac{t^2}{2}}\, \mathrm{d}t, \quad \forall x \in \mathbb{R}.$$

Theorem B.45 (Central Limit Theorem). *Let $(X_n)_{n\geq 1}$ be an i.i.d. sequence with $m \overset{\text{def}}{=} E[X_1] < \infty$ and $\sigma^2 \overset{\text{def}}{=} \mathrm{Var}(X_1) < \infty$. Then, as $n \to \infty$,*

$$\frac{(X_1 - m) + \cdots + (X_n - m)}{\sigma\sqrt{n}} \to \mathcal{N}(0,1) \quad \textit{in distribution.}$$

In particular, for all $a < b$,

$$\mathrm{P}\Big(a \leq \frac{(X_1 - m) + \cdots + (X_n - m)}{\sigma\sqrt{n}} \leq b\Big) \to \frac{1}{\sqrt{2\pi}} \int_{a}^{b} e^{-\frac{t^2}{2}}\, \mathrm{d}t.$$

B.8.3 Moments and Cumulants of Random Variables

Let X be a random variable. If $r \in \mathbb{N}$, the **rth moment** of X is defined by

$$m_r(X) \overset{\text{def}}{=} E[X^r],$$

provided the expectation exists. The **moment generating function** associated with X is the function

$$t \mapsto M_X(t) \overset{\text{def}}{=} E[e^{tX}], \quad t \in \mathbb{R}.$$

If $M_X(t)$ possesses a convergent MacLaurin expansion, then all its moments exist and can be recovered from the formula

$$m_r(X) = \frac{\mathrm{d}^r}{\mathrm{d}t^r} M_X(t)|_{t=0}.$$

Under suitable conditions, the moments $(m_r(X))_{r\geq 1}$ completely characterize the distribution of X.

Theorem B.46. *Assume that all moments $m_r(X)$, $r \geq 1$, exist. If there exists some $\epsilon > 0$ such that $\sum_{r\geq 1} \frac{1}{r!} m_r(X) t^r$ converges for all $t \in (-\epsilon, \epsilon)$, then $M_X(t) = \sum_{r\geq 1} \frac{1}{r!} m_r(X) t^r$ on that interval, and any random variable Y with $m_r(Y) = m_r(X)$, for all $r \geq 1$, has the same distribution as X.*

Proof See [186, Exercise 10, Chapter 5]. □

Let us now consider the **cumulant generating function** (also known as the **log-moment generating function**) $C_X(t) = \log M_X(t)$. The coefficients of its MacLaurin expansion (if it has one) are called the cumulants of the random variable X: for $r \in \mathbb{N}$, the **rth cumulant** of X is defined by

$$c_r(X) \overset{\text{def}}{=} \frac{\mathrm{d}^r}{\mathrm{d}t^r} C_X(t)|_{t=0}.$$

Cumulants possess a variety of other names, depending on the context. When $r \geq 2$, they are also called **semi-invariants**, thanks to the following remarkable property: for any $a, b \in \mathbb{R}$,

$$c_r(aX + b) = a^r c_r(X). \tag{B.20}$$

(This of course doesn't hold for $r = 1$, since $c_1(aX + b) = ac_1(X) + b$.) In statistical mechanics, cumulants are often called **Ursell functions, truncated correlation functions** or **connected correlation functions.**

Exercise B.16. Show that cumulants can be expressed in terms of moments using the following recursion formula:

$$c_r = m_r - \sum_{m=1}^{r-1} \binom{r-1}{m-1} c_m m_{r-m}.$$

In particular,

$$c_1 = m_1, \quad c_2 = m_2 - m_1^2, \quad c_3 = m_3 - 3m_2 m_1 + 2m_1^3, \quad \ldots$$

Cumulants of a random variable X characterize the distribution of X whenever its moments do. The advantage of cumulants compared to moments, in addition to satisfying (B.20), is the way they act on sums of independent random variables: if X and Y are independent random variables, then

$$c_r(X + Y) = c_r(X) + c_r(Y).$$

This follows immediately from the identity $C_{X+Y} = C_X + C_Y$.

B.8.4 Characteristic Function

The **characteristic function** of a random variable X is defined by

$$\varphi_X(t) \stackrel{\text{def}}{=} E[e^{itX}] \stackrel{\text{def}}{=} E[\cos(tX)] + iE[\sin(tX)].$$

Note that, since

$$E[|e^{itX}|] = 1,$$

the characteristic function is well defined for all random variables. If there exists $\epsilon > 0$ such that the moment generating function $M_X(t)$ is finite for all $|t| < \epsilon$, then $\varphi_X(-it) = M_X(t)$.

Characteristic functions owe their name to the fact that they characterize the distribution of a random variable: $\varphi_X = \varphi_Y$ if and only if X and Y have the same distribution.

If $X_1, \ldots, X_n$ are independent random variables, then

$$\varphi_{X_1+\cdots+X_n}(s) = \varphi_{X_1}(s) \cdots \varphi_{X_n}(s).$$

Theorem B.47 (Lévy's Continuity Theorem). *Let X_n be a sequence of random variables. Assume that $\varphi(t) \stackrel{\text{def}}{=} \lim_{n\to\infty} \varphi_{X_n}(t)$ exists, for all $t \in \mathbb{R}$, and that φ is continuous at $t = 0$. Then there exists a random variable X such that $\varphi_X = \varphi$ and X_n converges to X in distribution.*

B.8.5 Conditional Expectation

Conditional expectation is a fundamental concept in probability theory and plays a central role in our study of infinite-volume Gibbs measures in Chapter 6. Before giving its formal definition, we motivate it starting from the simplest possible case.

In elementary probability, the conditional probability of an event A with respect to an event B with $P(B) > 0$ is defined by

$$P(A \mid B) \stackrel{\text{def}}{=} \frac{P(A \cap B)}{P(B)}.$$

This defines a new probability measure $P(\cdot \mid B)$ under which random variables can be integrated, yielding a **conditional expectation given** B: for $X \in L^1(P)$,

$$E[X \mid B] \stackrel{\text{def}}{=} \int X(\omega) P(\mathrm{d}\omega \mid B).$$

Often, one is more interested in considering the conditional expectation with respect to a *collection* of events, associated with some *partial information* in a random experiment.

Example B.48. Consider an experiment in which two dice are rolled, modeled by two independent random variables X_1, X_2 on a probability space Ω, taking values in $\{1, 2, \dots, 6\}$. Assume that some partial information is given about the outcome of the sum $S = X_1 + X_2$, namely whether $S > 5$ or $S \leq 5$. Given this partial information, the expectation of X_1 is $E[X_1 \mid S > 5]$ if $\{S > 5\}$ occurred and $E[X_1 \mid S \leq 5]$ if $\{S \leq 5\}$ occurred. It thus appears natural to encode this information in a *random variable*

$$\omega \mapsto E[X_1 \mid S > 5] \mathbf{1}_{\{S>5\}}(\omega) + E[X_1 \mid S \leq 5] \mathbf{1}_{\{S\leq 5\}}(\omega). \qquad \diamond$$

This example leads to a first generalization of conditional expectation, as follows. Let $(B_k)_k \subset \mathscr{F}$ be a countable partition of Ω: $B_k \cap B_{k'} = \varnothing$ if $k \neq k'$ and $\bigcup_k B_k = \Omega$. This means that, for each outcome ω of the experiment, exactly one event B_k occurs. For convenience, let $\mathscr{B} \subset \mathscr{F}$ denote the sub-σ-algebra containing the events that are unions of sets B_k. The occurrence of some $B \in \mathscr{B}$ provides some information on the occurrence of some events B_k.

Now, if we also assume that $P(B_k) > 0$ for all k, we can define, for $X \in L^1(P)$,

$$E[X \mid \mathscr{B}](\omega) \stackrel{\text{def}}{=} \sum_k E[X \mid B_k] \mathbf{1}_{B_k}(\omega).$$

Exercise B.17. Show that, as a random variable on Ω, $E[X \mid \mathscr{B}]$ satisfies the following properties:

$$\omega \mapsto E[X \mid \mathscr{B}](\omega) \qquad \text{is } \mathscr{B}\text{-measurable}, \tag{B.21}$$

$$E\big[E[X \mid \mathscr{B}]\mathbf{1}_B\big] = E[X\mathbf{1}_B] \qquad \text{for all } B \in \mathscr{B}. \tag{B.22}$$

In particular,

$$E\big[E[X \mid \mathscr{B}]\big] = E[X].$$

The above definition, although natural, is not yet suited to our needs, its main defect being the necessity to assume that $P(B_k) > 0$. Indeed, the theory of infinite-volume Gibbs measures, detailed in Chapter 6, requires conditioning on a fixed configuration outside a finite region, an event that always has zero probability. We therefore need a definition of conditional expectation that allows us to condition with respect to events of zero probability.

It turns out that (B.21) and (B.22) characterize $E[X \mid \mathscr{B}]$ in an essentially unique manner. This can be used to define conditional expectation in much greater generality:

Lemma B.49. *Let $(\Omega, \mathscr{F}, P)$ be a probability space. Consider $X \in L^1(P)$ and a sub-σ-algebra $\mathscr{G} \subset \mathscr{F}$. There exists a random variable $Y \in L^1(P)$ for which the following conditions hold:*

1. *Y is $\mathscr{G}$-measurable.*
2. *For all $G \in \mathscr{G}$, $E[Y\mathbf{1}_G] = E[X\mathbf{1}_G]$.*

If Y' is another variable satisfying these properties, then $P(Y \neq Y') = 0$. Any one of them is called a ***version of the conditional expectation of X with respect to $\mathscr{G}$*** *and is denoted by $E[X \mid \mathscr{G}]$.*

We list the main properties of conditional expectation. In view of the almost-sure uniqueness, all the properties are to be understood as holding almost surely. All the random variables below are assumed to be integrable.

1. $E[a_1X_1 + a_2X_2 \mid \mathscr{G}] = a_1E[X_1 \mid \mathscr{G}] + a_2E[X_2 \mid \mathscr{G}]$.
2. If $X \leq X'$, then $E[X \mid \mathscr{G}] \leq E[X' \mid \mathscr{G}]$.
3. $|E[X \mid \mathscr{G}]| \leq E[|X| \mid \mathscr{G}]$.
4. (**Tower property**) If $\mathscr{G} \subset \mathscr{H}$, then

$$E\big[E[X \mid \mathscr{G}] \,\big|\, \mathscr{H}\big] = E[X \mid \mathscr{G}] = E\big[E[X \mid \mathscr{H}] \,\big|\, \mathscr{G}\big].$$

5. If Z is $\mathscr{G}$-measurable, then $E[XZ \mid \mathscr{G}] = ZE[X \mid \mathscr{G}]$.

Conditional expectation can be characterized equivalently in the following way:

Lemma B.50. *Let $X \in L^1(P)$, $\mathscr{G} \subset \mathscr{F}$ a sub-σ-algebra. Then $E[X \mid \mathscr{G}]$ is the (almost sure) unique $\mathscr{G}$-measurable random variable with the property that*

$$E\big[(X - E[X \mid \mathscr{G}])Z\big] = 0 \quad \text{for all } \mathscr{G}\text{-measurable } Z \in L^1(P). \tag{B.23}$$

Remark B.51. The above definition provides a nice geometrical interpretation of the conditional expectation of a random variable with finite variance. Let us denote by $L^2(P)$ the (real) vector space of all random variables such that $E[X^2] < \infty$ (or, equivalently, with finite variance). The space $L^2(P)$ is a Hilbert space for the inner product $(X, Y) \mapsto E[XY]$. (B.23) can then be interpreted as stating that the vector $X - E[X \mid \mathscr{G}]$ is orthogonal to the linear subspace $\{Z \in L^2(P) : Z \text{ is } \mathscr{G}\text{-measurable}\}$. This implies that $E[X \mid \mathscr{G}]$ coincides with the orthogonal projection of X on this subspace; see Figure B.4. ◇

Finally, we will occasionally need the following classical result whose proof can be found in [351].

Theorem B.52 (Backward martingale convergence). *Let $X \in L^1$ and let $\mathscr{G}_n$ be a decreasing sequence of σ-algebras, $\mathscr{G}_n \supset \mathscr{G}_{n+1}$, and set $\mathscr{G}_\infty \stackrel{\text{def}}{=} \bigcap_n \mathscr{G}_n$. Then,*

$$E[X \mid \mathscr{G}_n] \to E[X \mid \mathscr{G}_\infty] \quad \text{in } L^1 \text{ and almost surely.}$$

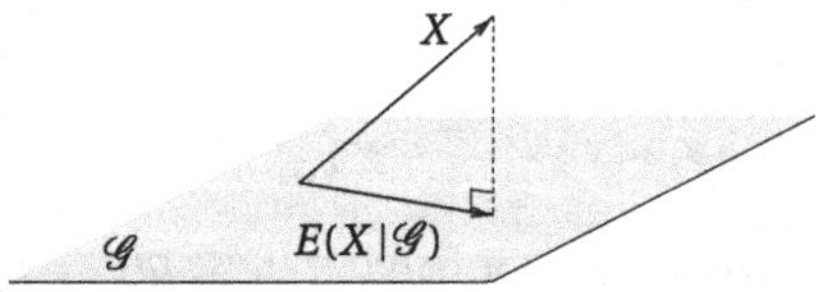

Figure B.4. Restricted to random variables with finite variance, the conditional expectation $E(X \mid \mathscr{G})$ corresponds to the orthogonal projection of X onto the linear subspace of all $\mathscr{G}$-measurable random variables.

B.8.6 Conditional Probability

Let $\mathscr{G} \subset \mathscr{F}$. The **conditional probability of** $A \in \mathscr{F}$ **with respect to** $\mathscr{G}$ is defined by the (almost surely unique) random variable

$$P(A \mid \mathscr{G})(\omega) \stackrel{\text{def}}{=} E[\mathbf{1}_A \mid \mathscr{G}](\omega).$$

By definition, $P(A \mid \mathscr{G})$ inherits many of the properties of the conditional expectation. In particular it is, up to almost-sure equivalence, the unique $\mathscr{G}$-measurable random variable for which

$$P(A \cap G) = \int_G P(A \mid \mathscr{G})\, \mathrm{d}P, \qquad \forall G \in \mathscr{G}.$$

Remember that, by linearity of the conditional expectation, one has, for disjoint events $A, B \in \mathscr{F}$, $P(A \cup B \mid \mathscr{G})(\omega) = P(A \mid \mathscr{G})(\omega) + P(B \mid \mathscr{G})(\omega)$. It is important to notice that even though this equality holds for P-almost all ω, the set of such ωs depends in general on A and B. Since there are usually uncountably many events in $\mathscr{F}$, one should therefore not expect, for a fixed ω, $P(\cdot \mid \mathscr{G})(\omega)$ to define a probability measure on $(\Omega, \mathscr{F})$. This leads to the following definition. A map $\widehat{P}(\cdot \mid \mathscr{G})(\cdot)\colon \mathscr{F} \times \Omega \to [0,1]$ is called a **regular conditional probability with respect to** $\mathscr{G}$ if (i) for each $\omega \in \Omega$, $\widehat{P}(\cdot \mid \mathscr{G})(\omega)$ is a probability distribution on $(\Omega, \mathscr{F})$, and (ii) for each $A \in \mathscr{F}$, $\widehat{P}(A \mid \mathscr{G})(\cdot)$ is a version of $P(A \mid \mathscr{G})$.

Regular conditional probabilities exist under fairly general assumptions, which can be found for example in [186].

The Gibbs measures constructed and studied in Chapter 6 are examples of regular conditional probabilities. Indeed, when $\mu \in \mathscr{G}(\pi)$ is conditioned with respect to the values taken by the spins outside a finite region Λ, the kernel $\pi_\Lambda(\cdot \mid \omega)$ is a version of $\mu(\cdot \mid \mathscr{F}_{\Lambda^c})(\omega)$. But, by definition, $\pi_\Lambda(\cdot \mid \omega)$ is a probability measure for each $\omega \in \Omega$. See the comments of Section 6.3.1. ◇

B.8.7 Random Vectors

Most of what was said for random variables can be adapted to the case of measurable functions taking values in a space of larger dimension: $\mathbf{X}\colon \Omega \to \mathbb{R}^n$ is a **random vector** if it is $\mathscr{F}/\mathscr{B}(\mathbb{R}^n)$-measurable. The **distribution** of $\mathbf{X}$ is the probability measure $P_{\mathbf{X}}$ on $(\mathbb{R}^n, \mathscr{B}(\mathbb{R}^n))$ defined by $P_{\mathbf{X}}(B) \stackrel{\text{def}}{=} P(\mathbf{X} \in B)$, $B \in \mathscr{B}(\mathbb{R}^n)$. The expectation $E[\mathbf{X}]$ is to be understood as coordinate-wise integration. A random vector has a **density (with respect to the Lebesgue measure)** if there exists a measurable $f_{\mathbf{X}}\colon \mathbb{R}^n \to \mathbb{R}_{\geq 0}$ such that

$$P(\mathbf{X} \in B) = \int_B f_{\mathbf{X}}(\mathbf{x})\,\mathrm{d}\mathbf{x}, \qquad \forall B \in \mathscr{B}(\mathbb{R}^n).$$

Random variables $X_1, \dots, X_n$ with density $f_{(X_1,\dots,X_n)}$ are independent if and only if

$$f_{(X_1,\dots,X_n)}(x_1,\dots,x_n) = f_{X_1}(x_1)\cdots f_{X_n}(x_n).$$

The different types of *convergence* defined earlier for random variables have direct analogues for random vectors. Moreover, an equivalent version of Theorem B.47 holds.

B.9 Gaussian Vectors and Fields

In this section, we recall some basic definitions and properties related to Gaussian fields. A good reference is the first chapter of Le Gall's book [212].

B.9.1 Basic Definitions and Properties

We defined a normal $\mathscr{N}(0,1)$ earlier. More generally, given $m \in \mathbb{R}$ and $\sigma^2 \in \mathbb{R}_{\geq 0}$, a random variable X is called a **Gaussian with mean m and variance σ^2**, $X \sim \mathscr{N}(m, \sigma^2)$, if it admits the density

$$f_X(x) = \frac{1}{\sqrt{2\pi\sigma^2}} e^{-(x-m)^2/2\sigma^2}.$$

As is easily verified, $E[X] = m$ and $\mathrm{Var}(X) = \sigma^2$.

Exercise B.18. Show that $X \sim \mathcal{N}(m, \sigma^2)$ if and only if its characteristic function is given by

$$E[e^{itX}] = \exp\left(-\tfrac{1}{2}\sigma^2 t^2 + imt\right).$$

Exercise B.19. Let $X_1, \dots, X_n$ be independent Gaussian random variables with $X_i \sim \mathcal{N}(m_i, \sigma_i^2)$ and let $t_1, \dots, t_n \in \mathbb{R}$. Show that

$$\sum_{i=1}^{n} t_i X_i \sim \mathcal{N}(t_1 m_1 + \cdots + t_n m_n, t_1^2\sigma_1^2 + \cdots + t_n^2\sigma_n^2).$$

Let us now introduce $\mathbb{R}^n$-valued *Gaussian vectors*. As before, elements of $\mathbb{R}^n$ will be denoted using bold letters: $\mathbf{x}, \mathbf{y}, \dots$ and the scalar product will be denoted $\mathbf{x} \cdot \mathbf{y}$.

Definition B.53. A random vector $\mathbf{X}\colon \Omega \to \mathbb{R}^n$ is **Gaussian** if the random variable $\mathbf{t} \cdot \mathbf{X}$ is Gaussian for each $\mathbf{t} \in \mathbb{R}^n$.

By Exercise B.18, this is equivalent to requiring that, for all $\mathbf{t} \in \mathbb{R}^n$,

$$E[e^{i\mathbf{t}\cdot\mathbf{X}}] = \exp\left(-\tfrac{1}{2}\mathrm{Var}(\mathbf{t}\cdot\mathbf{X}) + iE[\mathbf{t}\cdot\mathbf{X}]\right) = \exp\left(-\tfrac{1}{2}\mathbf{t}\cdot\Sigma\mathbf{t} + i\,\mathbf{m}\cdot\mathbf{t}\right),$$

where $\mathbf{m} = (m_1, \dots, m_n)$ with $m_i = E[X_i]$ and Σ is the $n \times n$ matrix with elements $\Sigma(i,j) = \mathrm{Cov}(X_i, X_j)$. $\mathbf{m}$ and Σ are called the **mean** and **covariance matrix** of $\mathbf{X}$. We write in this case $\mathbf{X} \sim \mathcal{N}(\mathbf{m}, \Sigma)$.

Since $1 \geq |E[e^{i\mathbf{t}\cdot\mathbf{X}}]| = \exp(-\frac{1}{2}\mathbf{t}\cdot\Sigma\mathbf{t})$, we have $\mathbf{t}\cdot\Sigma\mathbf{t} \geq 0$: the covariance matrix is nonnegative definite.

Lemma B.54. *Let Σ be an $n \times n$ nonnegative-definite symmetric matrix. Then there exists an $n \times n$ matrix A such that $\Sigma = AA^\top$ where $A^\top$ denotes the transpose of A. Moreover, if Σ is invertible, then so is A.*

Proof Let us denote by $\lambda_1, \dots, \lambda_n$ the eigenvalues of Σ; observe that $\lambda_i \geq 0$, $i = 1, \dots, n$, since Σ is nonnegative definite. Symmetry of Σ implies the existence of an orthogonal matrix O such that $\Sigma = O^\top D O$, where $D = \mathrm{diag}(\lambda_1, \dots, \lambda_n)$. Let $D^{1/2} \stackrel{\text{def}}{=} \mathrm{diag}(\sqrt{\lambda_1}, \dots, \sqrt{\lambda_n})$ and $A = O^\top D^{1/2}$. It then follows that $AA^\top = O^\top D^{1/2} D^{1/2} O = O^\top D O = \Sigma$, as required.

If Σ is invertible, then $\lambda_i > 0$, $i = 1, \dots, n$. This implies that $D^{1/2}$ is invertible. Since O is also invertible, it follows that so is A. □

Exercise B.20. Show that the components of a Gaussian vector $\mathbf{X} \sim \mathcal{N}(\mathbf{m}, \Sigma)$ are independent if and only if Σ is diagonal.

Exercise B.21. Show that $\mathbf{X} \sim \mathcal{N}(\mathbf{m}, \Sigma)$ if and only if $\mathbf{X} = A\mathbf{Y} + \mathbf{m}$, with $\mathbf{Y}$ a random vector with independent $\mathcal{N}(0, 1)$ components and A an $n \times n$ matrix such that $AA^\intercal = \Sigma$.

Proposition B.55. *Let Σ be a positive-definite symmetric $n \times n$ matrix and $\mathbf{m} \in \mathbb{R}^n$. Then $\mathbf{X} \sim \mathcal{N}(\mathbf{m}, \Sigma)$ if and only if it possesses the following density with respect to the Lebesgue measure* $\mathrm{d}\mathbf{x}$*:*

$$\mathbf{x} \mapsto \frac{1}{(2\pi)^{n/2}\sqrt{|\det \Sigma|}} \exp\left(-\tfrac{1}{2}(\mathbf{x}-\mathbf{m}) \cdot \Sigma^{-1}(\mathbf{x}-\mathbf{m})\right).$$

Proof Using Exercise B.21, $\mathbf{X} \sim \mathcal{N}(\mathbf{m}, \Sigma)$ if and only if $\mathbf{X} = A\mathbf{Y}+\mathbf{m}$, with $\Sigma = AA^\intercal$ and $\mathbf{Y}$ a Gaussian random vector with i.i.d. $\mathcal{N}(0, 1)$ components. The density of $\mathbf{Y}$ is given by

$$f_{\mathbf{Y}}(\mathbf{y}) = \frac{1}{(2\pi)^{n/2}} \exp\left(-\tfrac{1}{2}\|\mathbf{y}\|_2^2\right).$$

Note that, by Lemma B.54, A is invertible. The claim therefore follows from the change of variable formula, $f_{\mathbf{X}}(\mathbf{x}) = f_{\mathbf{Y}}(A^{-1}(\mathbf{x} - \mathbf{m}))\,|\det \Sigma|^{-1/2}$, where we have used the fact that the absolute value of the Jacobian of the transformation is equal to $|\det(A^{-1})| = |\det A|^{-1} = |\det \Sigma|^{-1/2}$. □

Exercise B.22. Use the method shown in the previous proof to prove (8.61).

B.9.2 Convergence of Gaussian Vectors

The following result shows that limits of convergent sequences of Gaussian random vectors are themselves Gaussian.

Proposition B.56. *Let $(\mathbf{X}^{(k)})_{k\geq 1}$ be a sequence of Gaussian random vectors, with mean $\mathbf{m}^{(k)}$ and covariance matrix $\Sigma^{(k)}$. Then $\mathbf{X}^{(k)}$ converges to a random vector $\mathbf{X}$ in distribution if and only if the limits $\mathbf{m} = \lim_{k\to\infty} \mathbf{m}^{(k)}$ and $\Sigma = \lim_{k\to\infty} \Sigma^{(k)}$ both exist. In that case, $\mathbf{X}$ is also a Gaussian vector, with mean $\mathbf{m}$ and covariance matrix Σ.*

Proof Assume that $\mathbf{m} = \lim_{k\to\infty} \mathbf{m}^{(k)}$ and $\Sigma = \lim_{k\to\infty} \Sigma^{(k)}$ exist. Then,

$$\lim_{k\to\infty} E[e^{\mathrm{i}\mathbf{t}\cdot\mathbf{X}^{(k)}}] = \lim_{k\to\infty} \exp\left(-\tfrac{1}{2}\mathbf{t}\cdot\Sigma^{(k)}\mathbf{t} + \mathrm{i}\mathbf{m}^{(k)}\cdot\mathbf{t}\right) = \exp\left(-\tfrac{1}{2}\mathbf{t}\cdot\Sigma\mathbf{t} + \mathrm{i}\mathbf{m}\cdot\mathbf{t}\right)$$

exists and is continuous at $\mathbf{t} = 0$. It thus follows from Levy's Continuity Theorem (n-dimensional version of Theorem B.47) that the sequence $(\mathbf{X}^{(k)})_{k\geq 1}$ converges in distribution and that the limit is a Gaussian random vector with mean $\mathbf{m}$ and covariance matrix Σ.

Assume now that $\mathbf{X}^{(k)} \to \mathbf{X}$ in distribution. The characteristic function of $\mathbf{X}$ satisfies, for any $\mathbf{t} \in \mathbb{R}^n$,

$$E[e^{\mathrm{i}\mathbf{t}\cdot\mathbf{X}}] = \lim_{k\to\infty} E[e^{\mathrm{i}\mathbf{t}\cdot\mathbf{X}^{(k)}}] = \lim_{k\to\infty} \exp\big(-\tfrac{1}{2}\mathbf{t}\cdot\Sigma^{(k)}\mathbf{t} + \mathrm{i}\mathbf{m}^{(k)}\cdot\mathbf{t}\big).$$

In particular, choosing $\mathbf{t} = t\mathbf{e}_i$, this yields

$$\lim_{k\to\infty} \exp\big(-\tfrac{1}{2}t^2\Sigma^{(k)}(i,i)\big) = |E[e^{\mathrm{i}t\mathbf{X}\cdot\mathbf{e}_i}]| \le 1.$$

This implies that the limits $\lim_{k\to\infty}\Sigma^{(k)}(i,i)$ $(i = 1,\dots,n)$ exist in $[0,\infty]$; moreover, the value $+\infty$ can be excluded, since it would contradict the continuity at $\mathbf{t} = 0$ of the characteristic function of $\mathbf{X}$.

Similarly, letting $\mathbf{t} = t(\mathbf{e}_i + \mathbf{e}_j)$, we obtain the existence of $\lim_{k\to\infty}\Sigma^{(k)}(i,j)$ for all $i \neq j$. This in turn implies the existence and continuity of

$$\lim_{k\to\infty} \exp\big(\mathrm{i}\mathbf{m}^{(k)}\cdot\mathbf{t}\big) = \exp\big(-\tfrac{1}{2}\mathbf{t}\cdot\Sigma\mathbf{t}\big)E[e^{\mathrm{i}\mathbf{t}\cdot\mathbf{X}}].$$

Consequently, $\lim_{k\to\infty}\mathbf{m}^{(k)}$ also exists. This proves the claim. □

Definition B.57. A collection of random variables $\varphi = (\varphi_i)_{i\in S}$ indexed by a countable set S is a **Gaussian random field** (or simply **Gaussian field**) if all its finite-dimensional distributions are Gaussian, that is, if

$$E[e^{\mathrm{i}\sum_{i\in S}t_i\varphi_i}] = \exp\Big(-\tfrac{1}{2}\sum_{i,j\in S} t_it_j\,\mathrm{Cov}(\varphi_i,\varphi_j) + \mathrm{i}\sum_{i\in S} t_iE[\varphi_i]\Big),$$

for all $(t_i)_{i\in S}$ taking only finitely many nonzero values.

As follows from the definition, Proposition B.56 and the Kolmogorov Extension Theorem, a sequence of Gaussian random fields $\varphi^{(k)}$ on S converges to a random field φ on S if and only if the limits

$$\lim_{k\to\infty} E[\varphi_i^{(k)}], \qquad \lim_{k\to\infty}\mathrm{Cov}(\varphi_i^{(k)},\varphi_j^{(k)})$$

exist for all $i,j \in S$. Moreover, in that case, $(\varphi_i)_{i\in S}$ is Gaussian with

$$E[\varphi_i] = \lim_{k\to\infty} E[\varphi_i^{(k)}], \qquad \mathrm{Cov}(\varphi_i,\varphi_j) = \lim_{k\to\infty}\mathrm{Cov}(\varphi_i^{(k)},\varphi_j^{(k)}),$$

for all $i,j \in S$.

B.9.3 Gaussian Fields and Independence

For $T \subset S$, let $\mathscr{F}_T \stackrel{\text{def}}{=} \sigma(\varphi_j, j \in T)$ (defined as the smallest σ-algebra on Ω such that each φ_j, $j \in T$, is measurable).

Proposition B.58. *Let $\varphi = (\varphi_i)_{i\in S}$ be a Gaussian field and $T \subset S$. Then $\mathscr{F}_T$ and $\mathscr{F}_{S\setminus T}$ are independent if and only if* $\mathrm{Cov}(\varphi_i, \varphi_j) = 0$ *for all $i \in T, j \in S \setminus T$.*

Proof See [212, Section 1.3]. □

B.10 The Total Variation Distance

There are various ways by which one can measure the similarity of two probability measures. The simplest is the total variation distance.

Definition B.59. The **total variation distance** between two probability measures μ and ν on $(\Omega, \mathscr{F})$ is defined by

$$\|\mu - \nu\|_{TV} \stackrel{\text{def}}{=} 2 \sup_{A\in\mathscr{F}} |\mu(A) - \nu(A)| .$$

We warn the reader that some authors define $\|\mu - \nu\|_{TV}$ without the factor 2.

Lemma B.60. *Let μ and ν be two probability measures on $(\Omega, \mathscr{F})$, with $\mu \ll \nu$. Then,*

$$\|\mu - \nu\|_{TV} = \left\langle \left|1 - \frac{\mathrm{d}\mu}{\mathrm{d}\nu}\right| \right\rangle_\nu = \sup_{f:\, \|f\|_\infty \le 1} \left| \langle f \rangle_\mu - \langle f \rangle_\nu \right| .$$

Proof If $\rho = \mathrm{d}\mu/\mathrm{d}\nu$, then

$$\mu(A) - \nu(A) = \int_A (\rho - 1)\mathrm{d}\nu \le \int (\rho - 1)_+ \mathrm{d}\nu .$$

Since the inequality is saturated for $A = \{\rho \ge 1\}$, we get

$$\sup_{A\in\mathscr{F}} \{\mu(A) - \nu(A)\} = \int (\rho - 1)_+ \mathrm{d}\nu .$$

In the same way,

$$\sup_{A\in\mathscr{F}} \{\nu(A) - \mu(A)\} = \int (\rho - 1)_- \mathrm{d}\nu .$$

But since $\int (\rho - 1)_+ \mathrm{d}\nu - \int (\rho - 1)_- \mathrm{d}\nu = \int (\rho - 1)\mathrm{d}\nu = 0$, this gives

$$\sup_{A\in\mathscr{F}} \{\mu(A) - \nu(A)\} = \sup_{A\in\mathscr{F}} \{\nu(A) - \mu(A)\} = \sup_{A\in\mathscr{F}} |\mu(A) - \nu(A)| .$$

We conclude that

$$\int |\rho - 1|\mathrm{d}\nu = \int (\rho - 1)_+ \mathrm{d}\nu + \int (\rho - 1)_- \mathrm{d}\nu = 2 \sup_{A\in\mathscr{F}} |\mu(A) - \nu(A)| ,$$

which proves the first identity. The second is a consequence of the first:

$$\sup_{f:\, \|f\|_\infty \le 1} \left| \int f \mathrm{d}\mu - \int f \mathrm{d}\nu \right| = \sup_{f:\, \|f\|_\infty \le 1} \left| \int f(\rho - 1) \mathrm{d}\nu \right| = \int |\rho - 1| \mathrm{d}\nu = \|\mu - \nu\|_{TV},$$

the supremum being achieved by the function $\mathbf{1}_{\{\rho \ge 1\}} - \mathbf{1}_{\{\rho < 1\}}$. □

B.11 Shannon Entropy

The Shannon entropy $\mathrm{S}_{\mathrm{Sh}}(\cdot)$ is the central object for the implementation of the maximum entropy principle, which was used in Chapter 1 to motivate the Gibbs distribution. In this section, we show that $\mathrm{S}_{\mathrm{Sh}}(\cdot)$ is unique, up to a multiplicative constant, among a class of functions $\mathrm{S}\colon \mathscr{M}_1(\Omega) \to \mathbb{R}$ satisfying a certain set of conditions, one of which is to be maximal for the uniform distribution. We follow the approach of Khinchin [191].

Consider a random experiment modeled by some probability space $(\Omega, \mathscr{F}, P)$. Consider a partition A of Ω into a finite number of events, called **atoms**. When A is a partition with k atoms, we will write $\mathsf{A} = \{A_1, \dots, A_k\}$. For convenience, we allow some atoms to be empty.

Such a partition can correspond to the outcome of a particular type of event in the experiment. For example, when throwing a dice, two possible partitions are $\mathsf{A} = \{A_1, \dots, A_6\}$, where $A_i = \{\text{the outcome is } i\}$ and $\mathsf{B} = \{B_1, B_2\}$, where $B_1 = \{\text{the outcome is even}\}$, $B_2 = \{\text{the outcome is odd}\}$. (In this example, we will say later that A is *finer* than B, see below.)

Our aim is to define the *unpredictability* of the outcome of the measurement corresponding to a partition A. Since the probability P is fixed, the unpredictability associated with the partition $\mathsf{A} = \{A_1, \dots, A_k\}$ will be defined through a function $\mathrm{S}(P(A_1), \dots, P(A_k))$, usually denoted simply $\mathrm{S}(\mathsf{A})$ or $\mathrm{S}(A_1, \dots, A_k)$ and called a function of the partition A. Notice that, since k is arbitrary, we are actually looking for a *collection* of functions.

Below, we define four conditions, most of which will be natural in terms of unpredictability, and then show that the only function that satisfies these conditions is, up to a positive multiplicative constant, the Shannon entropy.

The first three assumptions are natural. First, for all partitions $\mathsf{A} = \{A_1, \dots, A_k\}$,

$$\mathrm{S}(A_1, \dots, A_k) \quad \text{is continuous in } (P(A_1), \dots, P(A_k)). \tag{U1}$$

Second, we assume that unpredictability is not sensitive to the presence of atoms that have zero probability; for a partition $\mathsf{A} = \{A_1, \dots, A_k\}$,

$$P(A_k) = 0 \quad \text{implies} \quad \mathrm{S}(A_1, \dots, A_{k-1}, A_k) = \mathrm{S}(A_1, \dots, A_{k-2}, A_{k-1} \cup A_k). \tag{U2}$$

Third, as discussed in Section 1.2.2, we want the unpredictability to be maximal for partitions whose atoms have equal probabilities. Namely, call a partition $\mathsf{U} = \{U_1, \dots, U_m\}$ **uniform** if $P(U_i) = P(U_j) = \frac{1}{m}$ for all i, j.

$$\text{Among partitions with } m \text{ atoms, S is maximal for the uniform partitions.} \tag{U3}$$

To motivate the fourth assumption, we introduce some more terminology. A partition A is **finer** than a partition B if each atom of B is a union of atoms of A. When realizing the random experiment, if A is finer than B, information about the outcome $\omega \in \Omega$ can be revealed in two stages: first, by revealing the atom B_j such that $B_j \ni \omega$, and then, given B_j, one reveals the atom A_i such that $B_j \supset A_i \ni \omega$.

The unpredictability associated with the first stage is measured by S(B). After observing the result of the first stage, one should update our probability measure: assuming that the atom B_j occurred in the first stage, the relevant probability measure is $P(\cdot \mid B_j)$. The unpredictability of the second stage is thus measured by

$$\mathsf{S}(\mathsf{A} \mid B_j) \stackrel{\text{def}}{=} \mathsf{S}\big(P(A_1 \mid B_j), \dots, P(A_n \mid B_j)\big).$$

Averaging over the possible outcomes B_j of the first stage, we are led to define the entropy of the second stage by

$$\mathsf{S}(\mathsf{A} \mid \mathsf{B}) \stackrel{\text{def}}{=} \sum_j P(B_j)\mathsf{S}(\mathsf{A} \mid B_j).$$

Now, the unpredictability of the complete experience should not depend on the way the experiment was conducted (in one stage or in two stages). It is therefore natural to assume that

$$\mathsf{S}(\mathsf{A}) = \mathsf{S}(\mathsf{B}) + \mathsf{S}(\mathsf{A} \mid \mathsf{B}). \tag{U4}$$

Lemma B.61. *The* ***Shannon entropy****, defined by*

$$\mathsf{S}_{\text{Sh}}(\mathsf{B}) \stackrel{\text{def}}{=} -\sum_{j=1}^{k} P(B_j)\log P(B_j) \tag{B.24}$$

for all partitions $\mathsf{B} = \{B_1, \dots, B_k\}$*, satisfies* (U1)-(U4).

Proof (U1) and (U2) are clearly satisfied, (U3) has been shown in Lemma 1.9, and (U4) can be verified easily. □

Theorem B.62. *Let* $\mathsf{S}(\cdot)$ *be a function on finite partitions satisfying* (U1)-(U4). *Then there exists a constant* $\lambda > 0$ *such that*

$$\mathsf{S}(\cdot) = \lambda \mathsf{S}_{\text{Sh}}(\cdot).$$

We express (U4) in a slightly different way, better suited for the computations to come. For two arbitrary partitions A, B, consider the **composite** partition

$$\mathsf{A} \vee \mathsf{B} \stackrel{\text{def}}{=} \{A \cap B : A \in \mathsf{A}, B \in \mathsf{B}\}.$$

Then, (U2) implies that $\mathsf{S}(\mathsf{A} \vee \mathsf{B} \mid \mathsf{B}) = \mathsf{S}(\mathsf{A} \mid \mathsf{B})$, and (U4) can be used in the following form:

$$\mathsf{S}(\mathsf{A} \vee \mathsf{B}) = \mathsf{S}(\mathsf{B}) + \mathsf{S}(\mathsf{A} \mid \mathsf{B}). \tag{U4}$$

Notice that if $P(A \cap B) = P(A)P(B)$ for all $A \in \mathsf{A}$ and $B \in \mathsf{B}$ then $\mathsf{S}(\mathsf{A}\,|\,\mathsf{B}) = \mathsf{S}(\mathsf{A})$ and (U4) implies

$$\mathsf{S}(\mathsf{A} \vee \mathsf{B}) = \mathsf{S}(\mathsf{A}) + \mathsf{S}(\mathsf{B}). \tag{B.25}$$

We will start by proving a version of Theorem B.62 for uniform partitions U. Below, $|\mathsf{U}|$ denotes the number of atoms in U.

Proposition B.63. *Let* $\mathsf{S}(\cdot)$ *be a function defined on uniform partitions, which is monotone increasing in* $|\mathsf{U}|$ *and which is additive in the sense that if* $P(U \cap U') = P(U)P(U')$ *for all* $\mathrm{U} \in \mathsf{U}$ *and* $\mathrm{U}' \in \mathsf{U}'$. *Then*

$$\mathsf{S}(\mathsf{U} \vee \mathsf{U}') = \mathsf{S}(\mathsf{U}) + \mathsf{S}(\mathsf{U}'). \tag{B.26}$$

Then there exists $\lambda > 0$ *such that* $\mathsf{S}(\mathsf{U}) = \lambda \log |\mathsf{U}|$ *for all uniform partitions* U.

Proof Since $\mathsf{S}(\cdot)$ is constant on partitions with the same number of atoms, we define $L(k) \stackrel{\text{def}}{=} \mathsf{S}(\mathsf{U})$ if $|\mathsf{U}| = k$. By the assumption, $L(k)$ is increasing in k. Then let $\mathsf{U}^1, \dots, \mathsf{U}^n$ be independent partitions, each containing k atoms. On the one hand, $\mathsf{U}^1 \vee \cdots \vee \mathsf{U}^n$ is also a uniform partition which contains k^n atoms and, therefore, $\mathsf{S}(\mathsf{U}^1 \vee \cdots \vee \mathsf{U}^n) = L(k^n)$. On the other hand, by (B.26),

$$\mathsf{S}(\mathsf{U}^1 \vee \cdots \vee \mathsf{U}^n) = \sum_{j=1}^{n} \mathsf{S}(\mathsf{U}^j) = n\,L(k),$$

and so $L(k^n) = nL(k)$. We verify that $L(\cdot)$ is necessarily of the form $L(k) = \lambda \log k$, for some $\lambda > 0$. Namely, fix two arbitrary integers $k, \ell \geq 2$. Choose some large integer $m \geq 1$ and find some integer n so that $k^n \leq \ell^m < k^{n+1}$. On the one hand, $n \log k \leq m \log \ell < (n+1) \log k$. On the other hand, the monotonicity of $L(\cdot)$ implies that $nL(k) = L(k^n) \leq L(\ell^m) = mL(\ell)$ and, similarly, $mL(\ell) \leq (n+1)L(k)$, which with the previous set of inequalities gives

$$\left| \frac{L(\ell)}{L(k)} - \frac{\log \ell}{\log k} \right| \leq \frac{1}{m}.$$

Since m was arbitrary, this shows that $L(k)/\log k$ does not depend on k and must be equal to a constant. □

Proof of Theorem B.62. Assume $\mathsf{S}(\cdot)$ satisfies (U1)–(U4). Let us temporarily denote by $\mathsf{S}_k(\cdot)$ the function $\mathsf{S}(\cdot)$ when restricted to partitions with k atoms. Using (U2), followed by (U3),

$$\mathsf{S}_k\big(\tfrac{1}{k}, \dots, \tfrac{1}{k}\big) = \mathsf{S}_{k+1}\big(\tfrac{1}{k}, \dots, \tfrac{1}{k}, 0\big) \leq \mathsf{S}_{k+1}\big(\tfrac{1}{k+1}, \dots, \tfrac{1}{k+1}\big).$$

Note that (B.25) guarantees (B.26). This shows that, when restricted to uniform partitions, $\mathsf{S}(\cdot)$ satisfies the hypotheses of Proposition B.63, yielding the existence of a constant $\lambda > 0$ such that $\mathsf{S}(\mathsf{U}) = \lambda \log |\mathsf{U}|$ for all uniform partitions U.

Let us now consider an arbitrary partition $\mathsf{B} = \{B_1, \dots, B_k\}$. By (U1), we can safely assume that the probabilities $P(B_j) \in \mathbb{Q}$. If we consider a collection of integers

$w_1, \dots, w_k$ such that $P(B_j) = \frac{w_j}{Z}$, where $Z = w_1 + \cdots + w_k$, the partition B can be reinterpreted as follows. Consider a collection of Z labeled balls, each of a specific color, among k different colors. Assume that there are exactly w_j balls of color j, $j = 1, \dots, k$. A ball is sampled at random, uniformly. Then, clearly, the color of the ball sampled has color j with probability $\frac{w_j}{Z} = P(B_j)$. We therefore reinterpret B_j as the event "the sampled ball has color j" and use this to compute $\mathsf{S}(\mathsf{B})$.

In this same experiment, consider now the partition $\mathsf{A} = \{A_1, \dots, A_Z\}$ defined by $A_i = \{\text{the ball } i \text{ was sampled}\}$.

Since A is finer than B we have $\mathsf{A} \vee \mathsf{B} = \mathsf{A}$ and, since A is uniform, $\mathsf{S}(\mathsf{B} \vee \mathsf{A}) = \mathsf{S}(\mathsf{A}) = \lambda \log Z$.

Now, observe that

$$P(A_i \mid B_j) = \begin{cases} \frac{1}{w_j} & \text{if } i \in B_j\,, \\ 0 & \text{otherwise.} \end{cases}$$

Therefore, using (U2), $\mathsf{S}(\mathsf{A} \mid B_j) = \lambda \log w_j = \lambda \log P(B_j) + \lambda \log Z$, and so

$$\mathsf{S}(\mathsf{A} \mid \mathsf{B}) = \lambda \sum_{j=1}^{k} P(B_j) \log P(B_j) + \lambda \log Z\,.$$

This proves the claim, since assumption (U4) implies

$$\mathsf{S}(\mathsf{B}) = \mathsf{S}(\mathsf{A}) - \mathsf{S}(\mathsf{A} \mid \mathsf{B}) = -\lambda \sum_{j=1}^{k} P(B_j) \log P(B_j)\,.$$

□

B.12 Relative Entropy

B.12.1 Definition, Basic Properties

We have seen that, when μ, ν are two probability measures such that $\mu \ll \nu$, then there exists a nonnegative measurable function $\mathrm{d}\mu/\mathrm{d}\nu$, the Radon–Nikodým derivative of μ with respect to ν, such that $\mu(A) = \int_A \frac{\mathrm{d}\mu}{\mathrm{d}\nu}\mathrm{d}\nu$ for all $A \in \mathscr{F}$.

Definition B.64. The **relative entropy** $\mathsf{h}(\mu \mid \nu)$ of μ with respect to ν is defined as

$$\mathsf{h}(\mu \mid \nu) \stackrel{\text{def}}{=} \begin{cases} \left\langle \frac{\mathrm{d}\mu}{\mathrm{d}\nu} \log \frac{\mathrm{d}\mu}{\mathrm{d}\nu} \right\rangle_\nu & \text{if } \mu \ll \nu, \\ \infty & \text{otherwise.} \end{cases}$$

Since $x \log x \geq -e^{-1}$ on $\mathbb{R}_{>0}$, $\mathsf{h}(\mu \mid \nu)$ is always well defined (but can be equal to $+\infty$).

Lemma B.65. $\mathsf{h}(\mu \mid \nu) \geq 0$, *with equality if and only if* $\mu = \nu$.

Proof We can assume that $\mathsf{h}(\mu \mid \nu) < \infty$. Since $\Psi(x) = x \log x$ is strictly convex on $(0, \infty)$, Jensen's inequality implies that

$$\mathsf{h}(\mu \mid \nu) = \big\langle \Psi(\tfrac{\mathrm{d}\mu}{\mathrm{d}\nu}) \big\rangle_\nu \geq \Psi\big(\langle \tfrac{\mathrm{d}\mu}{\mathrm{d}\nu} \rangle_\nu\big) = \Psi(1) = 0 .$$

Moreover, Jensen's inequality is an equality if and only if $\frac{\mathrm{d}\mu}{\mathrm{d}\nu}$ is almost surely a constant, and the latter can only be 1. □

Proposition B.66. *1.* $(\mu, \nu) \mapsto \mathsf{h}(\mu \mid \nu)$ *is convex.*
2. $\mu \mapsto \mathsf{h}(\mu \mid \nu)$ *is strictly convex.*

To prove this proposition, we will need the following elementary inequality.

Exercise B.23. Let a_i, b_i, $i = 1, \dots, n$, be nonnegative real numbers. Set $A \stackrel{\mathrm{def}}{=} \sum_{i=1}^n a_i$ and $B \stackrel{\mathrm{def}}{=} \sum_{i=1}^n b_i$. Then,

$$\sum_{i=1}^n a_i \log \frac{a_i}{b_i} \geq A \log \frac{A}{B} ,$$

with equality if and only if there exists λ such that $a_i = \lambda b_i$ for all $1 \leq i \leq n$.
Hint: Use Lemma B.65.

Proof of Proposition B.66. 1. Let $\alpha \in (0, 1)$ and take four probability measures μ_1, μ_2, ν_1, ν_2. Set $\mu \stackrel{\mathrm{def}}{=} \alpha\mu_1 + (1-\alpha)\mu_2$ and $\nu \stackrel{\mathrm{def}}{=} \alpha\nu_1 + (1-\alpha)\nu_2$. We need to prove that

$$\mathsf{h}(\mu \mid \nu) \leq \alpha \mathsf{h}(\mu_1 \mid \nu_1) + (1-\alpha)\mathsf{h}(\mu_2 \mid \nu_2) . \tag{B.27}$$

We can assume that $\mu_i \ll \nu_i$, $i = 1, 2$, so that the right-hand side is finite and the following Radon–Nikodým derivatives are well defined: for $i = 1, 2$,

$$f_i \stackrel{\mathrm{def}}{=} \frac{\mathrm{d}\mu_i}{\mathrm{d}\nu} , \quad g_i \stackrel{\mathrm{def}}{=} \frac{\mathrm{d}\nu_i}{\mathrm{d}\nu} , \quad h_i \stackrel{\mathrm{def}}{=} \frac{\mathrm{d}\mu_i}{\mathrm{d}\nu_i} , \quad \phi \stackrel{\mathrm{def}}{=} \frac{\mathrm{d}\mu}{\mathrm{d}\nu} .$$

With this notation, (B.27) can be rewritten, thanks to (B.16), as

$$\begin{aligned} \langle \phi \log \phi \rangle_\nu &\leq \alpha \langle h_1 \log h_1 \rangle_{\nu_1} + (1-\alpha)\langle h_2 \log h_2 \rangle_{\nu_2} \\ &= \Big\langle \alpha f_1 \log \frac{f_1}{g_1} + (1-\alpha) f_2 \log \frac{f_2}{g_2} \Big\rangle_\nu . \end{aligned} \tag{B.28}$$

By (B.15), we have $\alpha f_1 + (1-\alpha) f_2 = \phi$ and $\alpha g_1 + (1-\alpha) g_2 = 1$, so Exercise B.23 implies that

$$\alpha f_1 \log \frac{f_1}{g_1} + (1-\alpha) f_2 \log \frac{f_2}{g_2} = \alpha f_1 \log \frac{\alpha f_1}{\alpha g_1} + (1-\alpha) f_2 \log \frac{(1-\alpha) f_2}{(1-\alpha) g_2} \geq \phi \log \phi ,$$

pointwise in Ω. Integrating this inequality with respect to ν yields (B.28).

2. This claim follows immediately from the corresponding properties of the function $x \mapsto x \log x$. □

B.12.2 Two Useful Inequalities

Pinsker's Inequality

The relative entropy is a measure of the similarity of two measures μ and ν. However, it is not a metric, as it is not even symmetric in its two arguments. Actually, even its symmetrized version $\mathsf{h}(\nu \mid \mu) + \mathsf{h}(\mu \mid \nu)$ fails to be a metric, as it violates the triangle inequality. Nevertheless, the smallness of the relative entropy between two measures allows one to control their total variation distance (see Section B.10).

Lemma B.67 (Pinsker's inequality). *Let μ and ν be two probability measures on the same measurable space, with $\mu \ll \nu$. Then*

$$\|\mu - \nu\|_{TV} \le \sqrt{2\mathsf{h}(\mu \mid \nu)}. \tag{B.29}$$

Proof Notice that, by applying Jensen's inequality,

$$\begin{aligned}(1+x)\log(1+x) - x &= x^2 \int_0^1 \mathrm{d}t \int_0^t \mathrm{d}s \, \frac{1}{1+xs} \\ &\ge \tfrac{1}{2}x^2 \frac{1}{1 + x\int_0^1 \mathrm{d}t \int_0^t \mathrm{d}s\, 2s} = \frac{x^2}{2(1+\frac{x}{3})}.\end{aligned} \tag{B.30}$$

Let $m \stackrel{\text{def}}{=} \frac{\mathrm{d}\mu}{\mathrm{d}\nu} - 1$. Then $\langle m \rangle_\nu = 0$ and, using (B.30),

$$\mathsf{h}(\mu \mid \nu) = \langle (1+m)\log(1+m)\rangle_\nu = \langle (1+m)\log(1+m) - m\rangle_\nu \ge \Big\langle \frac{m^2}{2(1+\frac{m}{3})}\Big\rangle_\nu .$$

But, using Lemma B.60 and the Cauchy–Schwartz inequality,

$$(\|\mu - \nu\|_{TV})^2 = \langle |m| \rangle_\nu^2 = \Big\langle \frac{|m|}{(1+\frac{m}{3})^{1/2}}\big(1+\tfrac{m}{3}\big)^{1/2}\Big\rangle_\nu^2 \le \Big\langle \frac{m^2}{1+\frac{m}{3}}\Big\rangle_\nu \big\langle 1 + \tfrac{m}{3}\big\rangle_\nu .$$

Since $\langle 1 + \frac{m}{3}\rangle_\nu = 1$, this proves (B.29). □

An Exponential Inequality

Pinsker's inequality (Lemma B.67) allows one to control the differences $|\mu(A) - \nu(A)|$ uniformly in $A \in \mathscr{F}$ in terms of the relative entropy between the two measures. Sometimes, however, we need to control the *ratio* of such probabilities. The following result can then be useful.

Lemma B.68. *Let μ and ν be two equivalent probability measures on some measurable space $(\Omega, \mathscr{F})$. If $\nu(A) > 0$, then*

$$\frac{\mu(A)}{\nu(A)} \ge \exp\Big(-\frac{\mathsf{h}(\nu \mid \mu) + e^{-1}}{\nu(A)}\Big).$$

Proof From Jensen's inequality and the inequality $x \log x \ge -e^{-1}$, which holds for all $x > 0$, we can write

$$\log\frac{\mu(A)}{\nu(A)} = \log\frac{\langle\frac{\mathrm{d}\mu}{\mathrm{d}\nu}\mathbf{1}_A\rangle_\nu}{\langle\mathbf{1}_A\rangle_\nu} = \log\langle\tfrac{\mathrm{d}\mu}{\mathrm{d}\nu}\,|\,A\rangle_\nu$$

$$\geq \langle\log\tfrac{\mathrm{d}\mu}{\mathrm{d}\nu}\,|\,A\rangle_\nu = -\frac{\langle\frac{\mathrm{d}\nu}{\mathrm{d}\mu}\log\frac{\mathrm{d}\nu}{\mathrm{d}\mu}\mathbf{1}_A\rangle_\mu}{\nu(A)}$$

$$\geq -\frac{\langle\frac{\mathrm{d}\nu}{\mathrm{d}\mu}\log\frac{\mathrm{d}\nu}{\mathrm{d}\mu}\rangle_\mu + e^{-1}}{\nu(A)}. \qquad \square$$

B.13 The Symmetric Simple Random Walk on $\mathbb{Z}^d$

Good references for these topics are the books by Spitzer [319], Lawler [209] and Lawler and Limic [211].

Let $(\xi_n)_{n\geq 1}$ be an i.i.d. sequence of random vectors uniformly distributed in the set $\{j \in \mathbb{Z}^d : j \sim 0\}$. The **simple random walk on $\mathbb{Z}^d$ started at $i \in \mathbb{Z}^d$** is the random process $(X_n)_{n\geq 0}$ with $X_0 = i$ and defined by

$$X_n \stackrel{\text{def}}{=} i + \sum_{k=1}^{n} \xi_k .$$

We denote the distribution of this process by $\mathbb{P}_i$.

B.13.1 Stopping Times and the Strong Markov Property

For each $n \geq 0$, we consider the σ-algebra $\mathscr{F}_n \stackrel{\text{def}}{=} \sigma(X_0, \dots, X_n)$. A random variable T with values in $\mathbb{Z}_{\geq 0} \cup \{+\infty\}$ is a **stopping time** if $\{T \leq n\} \in \mathscr{F}_n$ for all n, that is, if the occurrence of the event $\{T \leq n\}$ can be decided by considering only the first n steps of the walk. Given a stopping time T, let $\mathscr{F}_T$ denote the σ-algebra containing all events A such that $A \cap \{T \leq n\} \in \mathscr{F}_n$ for all n. That is, $\mathscr{F}_T$ contains all events that depend only on the part of the trajectory of the random walk up to time T.

We then have the following result.

Theorem B.69 (Strong Markov property)**.** *Let T be a stopping time. Then, on the event $\{T < \infty\}$, the random process $(X_{T+n} - X_T)_{n\geq 0}$ has the same distribution as a simple random walk started at 0 and is independent of $\mathscr{F}_T$.*

B.13.2 Local Limit Theorem

Theorem B.70. *There exists $\rho > 0$ such that, for any $i = (i_1, \dots, i_d) \in \mathbb{Z}^d$ such that $\sum_{k=1}^d i_k$ and n have the same parity and $\|i\|_2 < \rho n$,*

$$\mathbb{P}_0(X_n = i) = 2(2\pi n/d)^{-d/2} \exp\Big(-\frac{d\|i\|_2^2}{2n} + O(n^{-1}) + O(\|i\|_2^4 n^{-3})\Big). \qquad \text{(B.31)}$$

Proof See, for example, [211, Theorem 2.3.11], using the fact that the random walk $(X_{2n})_{n\geq 0}$ is aperiodic (see [211, Theorem 2.1.3] for a similar argument). □

B.13.3 Recurrence and Transience

Given $A \subset \mathbb{Z}^d$, we consider the first entrance times in A, $\tau_A \stackrel{\text{def}}{=} \inf\{n \geq 0 : X_n \in A\}$ and $\tau_A^+ \stackrel{\text{def}}{=} \inf\{n \geq 1 : X_n \in A\}$, with the usual convention that $\inf \varnothing = +\infty$. When $A = \{k\}$, we write simply τ_k , τ_k^+.

Definition B.71. The random walk is **recurrent** if $\mathbb{P}_0(\tau_0^+ < \infty) = 1$. Otherwise, it is **transient**.

Theorem B.72. *The simple random walk on $\mathbb{Z}^d$ is transient if and only if*

$$\int_{[-\pi,\pi]^d} \Big\{1 - \frac{1}{2d}\sum_{j\sim 0}\cos(p\cdot j)\Big\}^{-1} \mathrm{d}p < \infty\,. \tag{B.32}$$

Proof Let $(X_n)_{n\geq 0}$ be the walk starting at 0 and let $p \stackrel{\text{def}}{=} \mathbb{P}_0(\tau_0^+ < \infty)$. First observe that, by the strong Markov property, the number N_0 of returns of the walk to 0 satisfies, for all $k \geq 0$, $\mathbb{P}_0(N_0 = k) = p^k(1-p)$. In particular,

$$\sum_{n\geq 1}\mathbb{P}_0(X_n = 0) = \mathbb{E}_0\Big[\sum_{n\geq 1}\mathbf{1}_{\{X_n=0\}}\Big] = \mathbb{E}_0[N_0]$$

is finite if and only if $p < 1$; that is, if and only if X is transient. We show that the convergence of this series is equivalent to (B.32).

Using the identity $(2\pi)^{-d}\int_{[-\pi,\pi]^d} e^{\mathrm{i}p\cdot j}\,\mathrm{d}p = \mathbf{1}_{\{j=0\}}$, for all $j \in \mathbb{Z}^d$, we can rewrite $\mathbb{P}_0(X_n = 0) = (2\pi)^{-d}\int_{[-\pi,\pi]^d}\mathbb{E}_0\big[e^{\mathrm{i}p\cdot X_n}\big]\,\mathrm{d}p$. Now observe that

$$\mathbb{E}_0\big[e^{\mathrm{i}p\cdot X_n}\big] = \mathbb{E}\big[e^{\mathrm{i}p\cdot(\xi_1+\cdots+\xi_n)}\big] = \Big(\tfrac{1}{2d}\sum_{j\sim 0}\cos(p\cdot j)\Big)^n \stackrel{\text{def}}{=} (\phi_\xi(p))^n\,.$$

Therefore, for any $\lambda \in (0,1)$,

$$\sum_{n\geq 1}\lambda^n\mathbb{P}_0(X_n = 0) = \int_{[-\pi,\pi]^d}\sum_{n\geq 1}(\lambda\phi_\xi(p))^n\frac{\mathrm{d}p}{(2\pi)^d} = \int_{[-\pi,\pi]^d}\frac{\lambda\phi_\xi(p)}{1-\lambda\phi_\xi(p)}\frac{\mathrm{d}p}{(2\pi)^d}\,.$$

Clearly, $\lim_{\lambda\uparrow 1}\sum_{n\geq 1}\lambda^n\mathbb{P}_0(X_n = 0) = \sum_{n\geq 1}\mathbb{P}_0(X_n = 0)$. It thus only remains for us to show that the limit can be taken inside the integral in the right-hand side. To do that, first observe that $\phi_\xi(p)$ is positive for all $p \in [-\delta,\delta]^d$, as soon as $0 < \delta < \frac{\pi}{2}$. Therefore, by monotone convergence,

$$\lim_{\lambda\uparrow 1}\int_{[-\delta,\delta]^d}\frac{\lambda\phi_\xi(p)}{1-\lambda\phi_\xi(p)}\frac{\mathrm{d}p}{(2\pi)^d} = \int_{[-\delta,\delta]^d}\frac{\phi_\xi(p)}{1-\phi_\xi(p)}\frac{\mathrm{d}p}{(2\pi)^d}\,.$$

To deal with the integral over $[-\pi,\pi]^d \setminus [-\delta,\delta]^d$, observe that, on this domain, the sequence of functions $\big(\lambda\phi_\xi(p)/(1-\lambda\phi_\xi(p))\big)_{0<\lambda<1}$ converges pointwise as $\lambda \uparrow 1$ and is uniformly bounded. Thus, by dominated convergence,

$$\lim_{\lambda\uparrow 1}\int_{[-\pi,\pi]^d\setminus[-\delta,\delta]^d}\frac{\lambda\phi_\xi(p)}{1-\lambda\phi_\xi(p)}\frac{\mathrm{d}p}{(2\pi)^d}=\int_{[-\pi,\pi]^d\setminus[-\delta,\delta]^d}\frac{\phi_\xi(p)}{1-\phi_\xi(p)}\frac{\mathrm{d}p}{(2\pi)^d}$$

and we are done. □

The following corollary, a result originally due to Pólya, shows that the simple random walk behaves very differently in low dimensions ($d = 1, 2$) and in high dimensions ($d \geq 3$).

Corollary B.73. *The simple random walk X is transient if and only if* $d \geq 3$.

Proof A Taylor expansion yields $\cos(x) = 1 - \frac{1}{2}x^2 + \frac{1}{24}x_0^4$ for some $0 \leq x_0 \leq x$. It follows that, for any $x \in [-1, 1]$, $1 - \frac{1}{2}x^2 \leq \cos(x) \leq 1 - \frac{11}{24}x^2$. Therefore, changing variables to spherical coordinates, we see that the integral in (B.32) is convergent if and only if

$$\int_0^1 r^{-2}r^{d-1}\,\mathrm{d}r=\int_0^1 r^{d-3}\,\mathrm{d}r<\infty,$$

which is true if and only if $d > 2$. □

By definition, a recurrent random walk returns to its starting point with probability 1. The next result quantifies the probability that it manages to travel far away before the first return.

Theorem B.74. *For all* $n \geq 1$,

$$\mathbb{P}_0\big(\tau_{\mathrm{B}(n)^{\mathrm{c}}}<\tau_0^+\big)=\begin{cases}\frac{1}{n+1} & \text{in } d=1,\\ O\big(\frac{1}{\log n}\big) & \text{in } d=2.\end{cases}$$

Proof The first statement is a particular instance of the gambler's ruin estimate; it is discussed, for example, in [209, equation (1.20)]. The second estimate can be found in [209, Proposition 1.6.7]. □

The next result shows that, while a recurrent random walk visits a.s. all vertices, a transient one will a.s. miss arbitrarily large regions on its way to infinity.

Theorem B.75. *For any* $r \geq 0$ *and any* $i \in \mathbb{Z}^d \setminus \mathrm{B}(r-1)$,

$$\lim_{n\to\infty}\mathbb{P}_i\big(\tau_{\mathrm{B}(n)^{\mathrm{c}}}>\tau_{\mathrm{B}(r)}^+\big)=1,$$

if and only if X is recurrent.

Proof See, for example, [209, Chapter 2]. □

B.13.4 Discrete Potential Theory

The n-**step Green function** is defined by

$$G_n(i,j) \stackrel{\text{def}}{=} \mathbb{E}_i\Big[\sum_{k=0}^{n} \mathbf{1}_{\{X_k=j\}}\Big], \qquad i,j \in \mathbb{Z}^d.$$

Let A be a nonempty, proper subset of $\mathbb{Z}^d$. The **Green function in** A is defined by

$$G_A(i,j) \stackrel{\text{def}}{=} \mathbb{E}_i\Big[\sum_{k=0}^{\tau_{A^c}-1} \mathbf{1}_{\{X_k=j\}}\Big], \qquad i,j \in \mathbb{Z}^d.$$

In the transient case, $d \geq 3$, the **Green function** is defined by

$$G(i,j) \stackrel{\text{def}}{=} \mathbb{E}_i\Big[\sum_{n=0}^{\infty} \mathbf{1}_{\{X_n=j\}}\Big], \qquad i,j \in \mathbb{Z}^d.$$

In the recurrent case, $d \leq 2$, the **potential kernel** is defined by

$$a(i,j) \stackrel{\text{def}}{=} \lim_{n\to\infty} \big\{G_n(i,j) - G_n(i,i)\big\}, \qquad i,j \in \mathbb{Z}^d.$$

We will also use the shorter notations $G_n(i) \equiv G_n(0,i)$, $G_A(i) \equiv G_A(0,i)$, $G(i) \equiv G(0,i)$, $a(i) \equiv a(0,i)$.

Theorem B.76. *1. In $d = 1$,*

$$G_{B(n)}(0) = a(n+1) = n+1, \qquad \forall n \geq 1.$$

2. In $d = 2$,

$$G_{B(n)}(0) = \frac{2}{\pi}\log n + O(1), \qquad a(i) = \frac{2}{\pi}\log \|i\|_2 + O(1).$$

3. In $d \geq 3$,

$$G_{B(n)}(0) = G(0) + O(n^{2-d}), \qquad G(i) = a_d\|i\|_2^{2-d} + O(\|i\|_2^{-d}),$$

where $a_d \stackrel{\text{def}}{=} \frac{d}{2}\Gamma\big(\frac{d}{2}-1\big)\pi^{-d/2}$ (Γ denotes here the gamma function).

Proof The claim in $d = 1$ is proved in [209, Theorem 1.6.4]. Those in $d = 2$ can be found in [209, Theorems 1.6.2 and 1.6.6], and those in higher dimensions are established in [209, Theorem 1.5.4 and Proposition 1.5.8]. □

Exercise B.24. Show that, in $d = 1, 2$,

$$\lim_{n\to\infty} \big(G_{B(n)}(i) - G_{B(n)}(0)\big) = a(i).$$

Finally, we will need the following estimate on the spatial variation of the Green function.

Theorem B.77. *There exists $C < \infty$ such that, for any $A \Subset \mathbb{Z}^2$ and any neighbors $i,j \in \mathbb{Z}^d$,*

$$G_A(i) - G_A(j) \leq C\,.$$

Proof This follows from [209, Proposition 1.6.3] and the asymptotic behavior of the potential kernel. □

B.14 The Isoperimetric Inequality on $\mathbb{Z}^d$

In this section, we provide a version of the isoperimetric inequality in $\mathbb{Z}^d$. Given $S \subset \mathbb{Z}^d$, we denote by $\partial_e S \stackrel{\text{def}}{=} \{\{i,j\} \in \mathscr{E}_{\mathbb{Z}^d} : i \in S, j \notin S\}$ the **edge boundary of** S.

Theorem B.78. *For any $S \Subset \mathbb{Z}^d$,*

$$|\partial_e S| \geq 2d|S|^{(d-1)/d}\,. \tag{B.33}$$

Notice that (B.33) is saturated for cubes, for example when $S = \mathrm{B}(n)$.

For simplicity, let $|D| \stackrel{\text{def}}{=} \ell^d(D)$ denote the Lebesgue measure of $D \subset \mathbb{R}^d$. The scaling property of the Lebesgue measure then reads $|\lambda D| = \lambda^d|D|$ for all $\lambda > 0$. If $A, B \subset \mathbb{R}^d$, let $A + B \stackrel{\text{def}}{=} \{x + y : x \in A, y \in B\}$.

The following is a weak version of the **Brunn–Minkowski inequality**, adapted from [131, Theorem 4.1]. Let $\mathscr{P}$ denote the collection of all parallelpipeds of $\mathbb{R}^d$ whose faces are perpendicular to the coordinate axes.

Proposition B.79. *If $A, B \subset \mathbb{R}^d$ are finite unions of elements of $\mathscr{P}$, then*

$$|A + B|^{1/d} \geq |A|^{1/d} + |B|^{1/d}\,. \tag{B.34}$$

Proof First observe that, for all $x \in \mathbb{R}^d$, $|A + B| = |A + B + x| = |A + (B + x)|$. Therefore, one can always translate A or B in an arbitrary way. In particular, one can always assume A and B to be disjoint.

Now if A and B are arbitrary unions of parallelepipeds in $\mathscr{P}$, we can express $A \cup B$ as a union $\bigcup_{k=1}^n C_k$, where $C_K \in \mathscr{P}$, and the interior of the C_ks are nonoverlapping (they can, however, share points on their boundaries). We will prove the statement by induction on n.

To prove the claim for $n = 2$, assume that $A \in \mathscr{P}$ has volume $\prod_{i=1}^d a_i$ and that $\mathsf{B} \in \mathscr{P}$, disjoint from A, and of volume $\prod_{i=1}^d b_i$. Then, $|A + B| = \prod_{i=1}^d (a_i + b_i)$ and, since $(\prod_{i=1}^d x_i)^{1/d} \leq \frac{1}{d}\sum_{i=1}^d x_i$, see (B.1), we have

$$\Bigl(\prod_{i=1}^d \frac{a_i}{a_i + b_i}\Bigr)^{1/d} + \Bigl(\prod_{i=1}^d \frac{b_i}{a_i + b_i}\Bigr)^{1/d} \leq \frac{1}{d}\sum_{i=1}^d \frac{a_i}{a_i + b_i} + \frac{1}{d}\sum_{i=1}^d \frac{b_i}{a_i + b_i} = 1\,,$$

which proves (B.34) for those particular sets.

Let us then suppose that the claim has been proved up to n and assume that A and B are such that their union can be expressed as a union of $n+1$ nonoverlapping parallelepipeds: $A \cup B = \bigcup_{i=1}^{n+1} C_i$. Since A and B can be assumed to be far apart, A and B can each be expressed using a subset of $\{C_1, \ldots, C_{n+1}\}$. For simplicity, assume that $A = \bigcup_{i=1}^{l} C_i$, $l \geq 2$ and $B = \bigcup_{i=l+1}^{n+1} C_i$.

Observe that C_1 and C_2 can always be separated by some plane π, perpendicular to one of the coordinate axes. Denoting by Π^+ and Π^- the two closed half-spaces delimited by π, let $A^\pm \stackrel{\text{def}}{=} A \cap \Pi^\pm$ and $B^\pm \stackrel{\text{def}}{=} B \cap \Pi^\pm$. Again using the fact that B can be translated in an arbitrary manner, we can assume that

$$\frac{|B^\pm|}{|B|} = \frac{|A^\pm|}{|A|}.$$

Now, observe that $A^+ \cup B^+$ and $A^- \cup B^-$ can each be expressed as unions of *at most* n parallelepipeds. We can therefore use the induction hypothesis as follows:

$$\begin{aligned}
|A \cup B| &= |A^+ \cup B^+| + |A^- \cup B^-| \\
&\geq \left(|A^+|^{1/d} + |B^+|^{1/d}\right)^d + \left(|A^-|^{1/d} + |B^-|^{1/d}\right)^d \\
&= |A^+|\left\{1 + \left(\frac{|B^+|}{|A^+|}\right)^{1/d}\right\}^d + |A^-|\left\{1 + \left(\frac{|B^-|}{|A^-|}\right)^{1/d}\right\}^d \\
&= |A|\left\{1 + \left(\frac{|B|}{|A|}\right)^{1/d}\right\}^d \\
&= \left(|A|^{1/d} + |B|^{1/d}\right)^d.
\end{aligned}$$

□

Proof of Theorem B.78 Remember the notation $\mathscr{S}_0 = [-\frac{1}{2}, \frac{1}{2}]^d$ used for the closed unit cube of $\mathbb{R}^d$. We can always identify $S \Subset \mathbb{Z}^d$ with $A_S \stackrel{\text{def}}{=} \bigcup_{i \in S}\{i + \mathscr{S}_0\} \subset \mathbb{R}^d$. Note that the (Euclidean) boundary of A_S is made of $(d-1)$-dimensional unit cubes, which are crossed in their middle by the edges of $\partial_e S$. Notice also that, for all small $\epsilon > 0$, $(A_S + \epsilon\mathscr{S}_0) \setminus A_S$ is a thin layer wrapping A_S, of thickness $\epsilon/2$. We can therefore count the number of edges in $\partial_e S$ by computing the following limit:

$$|\partial_e S| = \lim_{\epsilon \downarrow 0} \frac{|A_S + \epsilon\mathscr{S}_0| - |A_S|}{\epsilon/2}. \tag{B.35}$$

For a fixed $\epsilon > 0$, we use (B.34) as follows:

$$|A_S + \epsilon\mathscr{S}_0| = \left(|A_S + \epsilon\mathscr{S}_0|^{1/d}\right)^d \geq |A_S| + \epsilon d |A_S|^{(d-1)/d}.$$

In the last step, we used $(a+b)^n \geq a^n + na^{n-1}b$ for $a, b \geq 0$, the scaling property of the Lebesgue measure and $|\mathscr{S}_0| = 1$. Using this in (B.35), we get (B.33) since $|A_S| = |S|$. □

Let us finally state an immediate consequence, which is used in Chapter 7.

Corollary B.80. *Let $S \Subset \mathbb{Z}^d$ and write $\partial^{\mathrm{ex}} S \stackrel{\mathrm{def}}{=} \{i \in S^c : d_\infty(i, S) \le 1\}$. Then,*

$$|\partial^{\mathrm{ex}} S| \ge |S|^{\frac{d-1}{d}}.$$

Proof Since there can be at most $2d$ edges of $\partial_e S$ incident at a given vertex of $\partial^{\mathrm{ex}} S$, we have $|\partial_e S| \le 2d |\partial^{\mathrm{ex}} S|$. The conclusion thus follows from (B.33). □

B.15 A Result on the Boundary of Subsets of $\mathbb{Z}^d$

In this section, we provide the tools needed to prove Lemma 7.19.

Consider the set of **★-edges** of $\mathbb{Z}^d$, defined by

$$\mathscr{E}^{\star}_{\mathbb{Z}^d} \stackrel{\mathrm{def}}{=} \{\{i, j\} \in \mathbb{Z}^d \times \mathbb{Z}^d : \|j - i\|_\infty = 1\}.$$

That is, $\mathscr{E}^{\star}_{\mathbb{Z}^d}$ contains all edges between pairs of vertices that are corners of the same unit cube in $\mathbb{Z}^d$.

Given $E \subset \mathscr{E}^{\star}_{\mathbb{Z}^d}$ and a vertex $i \in \mathbb{Z}^d$, we denote by $I(i; E)$ the number of ★-edges of E having i as an endpoint. The **boundary** of E is then defined as the set $\partial E \stackrel{\mathrm{def}}{=} \{i \in \mathbb{Z}^d : I(i; E) \text{ is odd}\}$.

A **★-path** between two vertices $i, j \in \mathbb{Z}^d$ is a set $E \subset \mathscr{E}^{\star}_{\mathbb{Z}^d}$ with $\partial E = \{i, j\}$. A **★-cycle** is a nonempty set $E \subset \mathscr{E}^{\star}_{\mathbb{Z}^d}$ with $\partial E = \varnothing$.

Two vertices $i, j \in \mathbb{Z}^d$ are **★-connected in** $A \subset \mathbb{Z}^d$ if there exists a ★-path between i and j, all of whose ★-edges are made of two vertices of A. (A vertex i is always considered to be ★-connected to itself.) A set $A \subset \mathbb{Z}^d$ is **★-connected** if all pairs of vertices $i, j \in A$ are ★-connected in A. A set $A \subset \mathbb{Z}^d$ is **c-connected** if $A^c \stackrel{\mathrm{def}}{=} \mathbb{Z}^d \setminus A$ is ★-connected.

The **★-interior-boundary** of $A \subset \mathbb{Z}^d$ is defined by $\partial^{\mathrm{in}}_{\star} A \stackrel{\mathrm{def}}{=} \{i \in A : \exists j \notin A, \{i, j\} \in \mathscr{E}^{\star}_{\mathbb{Z}^d}\}$. The **★-exterior-boundary** of $A \subset \mathbb{Z}^d$ is defined by $\partial^{\mathrm{ex}}_{\star} A \stackrel{\mathrm{def}}{=} \{i \notin A : \exists j \in A, \{i, j\} \in \mathscr{E}^{\star}_{\mathbb{Z}^d}\}$. The **★-edge-boundary** of $A \subset \mathbb{Z}^d$ is defined by $\partial_{\star} A \stackrel{\mathrm{def}}{=} \{\{i, j\} \in \mathscr{E}^{\star}_{\mathbb{Z}^d} : i \in A, j \notin A\}$.

Let the set of **★-triangles** be defined by

$$\mathscr{T} \stackrel{\mathrm{def}}{=} \{[i, j, k] \stackrel{\mathrm{def}}{=} \{\{i, j\}, \{j, k\}, \{k, i\}\} \subset \mathscr{E}^{\star}_{\mathbb{Z}^d}\}.$$

That is, a ★-triangle is a cycle built out of three distinct ★-edges whose endpoints all belong to the vertices of a common unit cube in $\mathbb{Z}^d$.

In the following, it will be convenient to identify a subset $E \subset \mathscr{E}^{\star}_{\mathbb{Z}^d}$ with the element of $\{0, 1\}^{\mathscr{E}^{\star}_{\mathbb{Z}^d}}$ equal to 1 at each ★-edge $e \in E$ and 0 everywhere else. The set $\{0, 1\}^{\mathscr{E}^{\star}_{\mathbb{Z}^d}}$ can be seen as a group for the coordinate-wise addition modulo 2, which we denote by $\oplus$. With this identification, the symmetric difference between two sets E and F can be expressed as $E \triangle F = E \oplus F$. In particular, $E \oplus E = \varnothing$.

The following is a discrete version of (a special case of) the *Poincaré lemma* of differential topology. Informally, it states that any ★-cycle can be realized as the boundary of a surface built out of ★-triangles.

Lemma B.81. *Let C be a bounded $\star$-cycle. There exists a finite collection of $\star$-triangles $\mathscr{T}' \subset \mathscr{T}$ such that*

$$C = \bigoplus_{T \in \mathscr{T}'} T.$$

The constructive proof given below uses the following elementary property: if C is a cycle and T is a triangle, then $C \oplus T$ is again a cycle or is empty.

Proof of Lemma B.81. We construct $\mathscr{T}'$ using the following algorithm.

Step 0. Set $\mathscr{T}' = \varnothing$.
Step 1. If C is empty, then stop. Otherwise, go to Step 2.
Step 2. If there exist two $\star$-edges $e = \{i,j\}, e' = \{j,k\}$ in C with $\|k - i\|_\infty = 1$, then:

- $T = [i,j,k]$ is a $\star$-triangle;
- replace $\mathscr{T}'$ by $\mathscr{T}' \cup \{T\}$;
- replace C by $C \oplus T$. Note that the number of $\star$-edges in C decreases at least by 1 in this operation.
- Go to Step 1.

Otherwise go to Step 3.

Step 3. Let us denote by $[C]$ the smallest (with respect to inclusion) parallelepiped $\{a_1,\ldots,b_1\} \times \cdots \times \{a_d,\ldots,b_d\} \subset \mathbb{Z}^d$, $a_m \le b_m$, such that C is a $\star$-cycle in $[C]$. Let $\ell = \min\{1 \le m \le d : a_m < b_m\}$. Let $e = \{i,j\}, e' = \{j,k\}$ be two $\star$-edges in C such that $j \in \partial_\star^{\text{in}}[C]$ and the ℓth component of j is equal to b_ℓ. Note that, necessarily, $\|k-i\|_\infty = 2$. Let $j' \in [C] \setminus \partial_\star^{\text{in}}[C]$ such that $\|j'-i\|_\infty = \|j' - k\|_\infty = 1$. We add to $\mathscr{T}'$ the two triangles $T_1 = [i,j,j']$ and $T_2 = [j,k,j']$ and replace C by $C \oplus T_1 \oplus T_2$. Note that, during this operation, the number of $\star$-edges in C does not increase and either (i) $[C]$ decreases (with respect to inclusion), or (ii) the number of vertices in $C \cap \partial_\star^{\text{in}}[C]$ decreases. Go to Step 1.

The algorithm terminates after finitely many steps, yielding a finite set of triangles $\mathscr{T}'$ such that $C = \bigoplus_{T \in \mathscr{T}'} T$. □

Proposition B.82. *Let $A \subset \mathbb{Z}^d$ be $\star$-connected and* c*-connected. Then $\partial_\star^{\text{in}} A$ and $\partial_\star^{\text{ex}} A$ are $\star$-connected.*

The idea used in the proof is due to [333]. It is based on

Lemma B.83. *Let $\partial_\star A = E_1 \cup E_2$ be an arbitrary partition of $\partial_\star A$. Then there exists a $\star$-triangle containing at least one $\star$-edge from both E_1 and E_2.*

Proof of Proposition B.82. To prove that $\partial_\star^{\text{in}} A$ is $\star$-connected, consider an arbitrary partition $\partial_\star^{\text{in}} A = B_1 \cup B_2$. This partition induces a natural partition of $\partial_\star A$: E_k, $k =$

1, 2, is the set of all $\star$-edges of $\partial_\star A$ with one endpoint in B_k. By Lemma B.83, there exists a $\star$-triangle containing at least one $\star$-edge of both E_1 and E_2. This implies that there exist $u \in B_1$ and $v \in B_2$ with $\{u, v\} \in \mathscr{E}^\star_{\mathbb{Z}^d}$. Since the partition was arbitrary, the conclusion follows. The same argument can be made for $\partial^{\text{ex}}_\star A$. □

Proof of Lemma B.83. Consider two arbitrary vertices $i \in A, j \notin A$. Let π_1 be a $\star$-path between i and j which does not cross E_2 and π_2 a $\star$-path between i and j which does not cross E_1. The existence of such $\star$-paths follows from our assumptions: given any $\star$-edge $\{u, v\} \in \partial_\star A$ with $u \in A$ and $v \notin A$, i is $\star$-connected to u in A (since A is $\star$-connected), while v is $\star$-connected to j in A^c (since A^c is $\star$-connected).

Since every vertex has an even number of incident $\star$-edges in $\pi_1 \oplus \pi_2$, the latter set is a $\star$-cycle. Therefore, by Lemma B.81, there exists $\mathscr{T}_{\pi_1,\pi_2} \subset \mathscr{T}$ such that

$$\pi_1 \oplus \pi_2 = \bigoplus_{T \in \mathscr{T}_{\pi_1,\pi_2}} T. \tag{B.36}$$

Let us denote by $\mathscr{T}'$ the subset of $\mathscr{T}_{\pi_1,\pi_2}$ composed of all $\star$-triangles containing at least one $\star$-edge of E_1 and set $\mathscr{T}'' \stackrel{\text{def}}{=} \mathscr{T}_{\pi_1,\pi_2} \setminus \mathscr{T}'$. Identity (B.36) can then be rewritten as

$$\pi_1 \oplus \bigoplus_{T \in \mathscr{T}'} T = \pi_2 \oplus \bigoplus_{T \in \mathscr{T}''} T \stackrel{\text{def}}{=} F. \tag{B.37}$$

Since i and j are the only vertices with an odd number of incident $\star$-edges, F must contain a path $\tilde{\pi}$ between i and j. Removing the latter's $\star$-edges from F, one is left with a cycle $\tilde{C}$, which can be decomposed as $\tilde{C} = \bigoplus_{T \in \tilde{\mathscr{T}}} T$.

By construction, neither π_2 nor any $\star$-triangle in $\mathscr{T}''$ contains a $\star$-edge of E_1. This implies that $\tilde{\pi}$ must contain an odd number of $\star$-edges of E_2 (since each such $\star$-edge connects a vertex of A and a vertex of A^c), while each of the $\star$-triangles in $\tilde{\mathscr{T}}$ must contain either 0 or 2. We conclude that F contains an odd number of $\star$-edges of E_2 and therefore $F \cap E_2 \neq \varnothing$.

Returning to (B.37), this implies that at least one of the $\star$-triangles in $\mathscr{T}'$ contains a $\star$-edge of E_2, since π_1 does not contain any $\star$-edge of E_2. However, by definition, every triangle of $\mathscr{T}'$ contains at least one $\star$-edge of E_1. This proves the claim. □

Appendix C Solutions to Exercises

In this appendix are regrouped the solutions to many of the exercises stated in the main body of the book. Some solutions are given with full details, while others are only sketched. In all cases, we recommend that the reader at least spends some time thinking about these problems before reading the solutions.

Solutions for Chapter 1

Exercise 1.1: Fix $n \in \mathbb{Z}_{>0}$ and observe first that our system Σ, with parameters U, V, N, can be seen as a system Σ' composed of two subsystems Σ_1, Σ_2 with parameters $\frac{1}{n}U, \frac{1}{n}V, \frac{1}{n}N$ and $\frac{n-1}{n}U, \frac{n-1}{n}V, \frac{n-1}{n}N$. Then, by additivity,

$$\begin{aligned} S^{\Sigma}(U,V,N) &= S^{\Sigma'}\left(\tfrac{1}{n}U, \tfrac{1}{n}V, \tfrac{1}{n}N, \tfrac{n-1}{n}U, \tfrac{n-1}{n}V, \tfrac{n-1}{n}N\right) \\ &= S^{\Sigma}\left(\tfrac{1}{n}U, \tfrac{1}{n}V, \tfrac{1}{n}N\right) + S^{\Sigma}\left(\tfrac{n-1}{n}U, \tfrac{n-1}{n}V, \tfrac{n-1}{n}N\right), \end{aligned}$$

where we used the fact that each of the two subsystems is of the same type as the original system and is therefore associated with the same entropy function. Iterating this, we get

$$S^{\Sigma}(U,V,N) = nS^{\Sigma}(\tfrac{1}{n}U, \tfrac{1}{n}V, \tfrac{1}{n}N).$$

Using this relation twice, we conclude that, for any $m, n \in \mathbb{Z}_{>0}$,

$$S^{\Sigma}\left(\tfrac{m}{n}U, \tfrac{m}{n}V, \tfrac{m}{n}N\right) = mS^{\Sigma}\left(\tfrac{1}{n}U, \tfrac{1}{n}V, \tfrac{1}{n}N\right) = \tfrac{m}{n}S^{\Sigma}(U,V,N).$$

This proves (1.7) for $\lambda \in \mathbb{Q}$. Since S^{Σ} is assumed to be differentiable, it is also continuous. We can therefore approximate any real $\lambda > 0$ by a sequence $(\lambda_n)_{n\geq 1} \subset \mathbb{Q}$, $\lambda_n \to \lambda$, and get

$$S^{\Sigma}(\lambda U, \lambda V, \lambda N) = \lim_{n\to\infty} S^{\Sigma}(\lambda_n U, \lambda_n V, \lambda_n N) = \lim_{n\to\infty} \lambda_n S^{\Sigma}(U,V,N) = \lambda S^{\Sigma}(U,V,N).$$

Exercise 1.2: Decompose the system into two subsystems Σ_1, Σ_2. By the postulate, $S^{\Sigma}(U,V,N)$ maximizes $S^{\Sigma}(\widetilde{U}_1, \widetilde{V}_1, \widetilde{N}_1) + S^{\Sigma}(\widetilde{U}_2, \widetilde{V}_2, \widetilde{N}_2)$ over all possible ways of partitioning U, V, N into $\widetilde{U}_1 + \widetilde{U}_2$, $\widetilde{V}_1 + \widetilde{V}_2$ and $\widetilde{N}_1 + \widetilde{N}_2$. This implies in particular that

$$S^{\Sigma}(U,V,N) \geq S^{\Sigma}(\alpha U_1,\alpha V_1,\alpha N_1) + S^{\Sigma}((1-\alpha)U_2,(1-\alpha)V_2,(1-\alpha)N_2)$$
$$= \alpha S^{\Sigma}(U_1,V_1,N_1) + (1-\alpha)S^{\Sigma}(U_2,V_2,N_2),$$

where the equality is a consequence of (1.7).

Exercise 1.3: Fix V, N, β_1, β_2 and $\alpha \in [0,1]$. For all U,

$$\{\alpha\beta_1+(1-\alpha)\beta_2\}U - S(U,V,N) = \alpha\underbrace{\{\beta_1 U - S(U,V,N)\}}_{\geq \hat{F}(\beta_1,V,N)} + (1-\alpha)\underbrace{\{\beta_2 U - S(U,V,N)\}}_{\geq \hat{F}(\beta_2,V,N)}.$$

Taking the infimum over U on the left-hand side,

$$\hat{F}(\alpha\beta_1 + (1-\alpha)\beta_2, V, N) \geq \alpha\hat{F}(\beta_1,V,N) + (1-\alpha)\hat{F}(\beta_2,V,N),$$

so $\hat{F}$ is concave in β. A similar argument, exploiting the concavity of S, shows that $\hat{F}$ is convex in V, N.

Exercise 1.4: The extremum principle follows from the one postulated for S. Indeed, suppose that we keep our system isolated, with a total energy U (and the subsystems can exchange energy, which can always be assumed in the present setting, since they can do that through the reservoir). Then, the equilibrium values are those maximizing

$$S(U^1,V^1,N^1) + S(U^2,V^2,N^2),$$

among all values satisfying the constraints on V^1, N^1, V^2, N^2 as well as $U^1 + U^2 = U$. Therefore, the same values minimize

$$\beta U - \big(S(U^1,V^1,N^1) + S(U^2,V^2,N^2)\big) = \big(\beta U^1 - S(U^1,V^1,N^1)\big) + \big(\beta U^2 - S(U^2,V^2,N^2)\big),$$

under the same conditions. Taking now the infimum over U yields the desired result, since this removes the constraint $U^1 + U^2 = U$.

Exercise 1.5: The critical points of the function $v \mapsto p(v)$ are given by the solutions of the equation $RTv^3 = 2a(v-b)^2$, which is of the form $f(v) = g(v)$. When $v > b$, this equation has zero, one or two solutions depending on the value of T. The critical case corresponds to when there is exactly one solution (at which $f(v) = g(v)$ and $f'(v) = g'(v)$). This happens when $T = \frac{8a}{27Rb}$.

Exercise 1.6: Writing $S_{Sh}(\mu) = \sum_{\omega\in\Omega}\psi(\mu(\omega))$, where $\psi(x) \stackrel{\text{def}}{=} -x\log x$, we see that S_{Sh} is concave.

Exercise 1.8: The desired probabilities are given by $\mu(i) = e^{-\beta i}/\mathbf{Z}_\beta$, where $\mathbf{Z}_\beta = \sum_{i=1}^{6} e^{-\beta i}$ and β must be chosen such that $\sum_i i\mu(i) = 4$. Numerically, one finds that

$$\mu(1) \cong 0.10\,,\ \mu(2) \cong 0.12\,,\ \mu(3) \cong 0.15\,,\ \mu(4) \cong 0.17\,,\ \mu(5) \cong 0.21\,,\ \mu(6) \cong 0.25\,.$$

Exercise 1.9: Letting $V' = V - \frac{N}{2}$ and writing $N_1 = \frac{N}{2} + m$, $N_2 = \frac{N}{2} - m$, we need to show that

$$m \mapsto (\tfrac{N}{2} + m)!\,(\tfrac{N}{2} - m)!(V' + m)!(V' - m)!$$

is minimal when $m = 0$. But this follows by simple termwise comparison. For the second part, expressing the desired probability using Stirling's formula (Lemma B.3) shows that there exist constants $c_- < c_+$ such that if V and N are both large, with $\frac{N}{2V}$ bounded away from 0 and 1, then

$$\frac{c_-}{\sqrt{N}} \le \frac{\binom{V}{\frac{N}{2}}\binom{V}{\frac{N}{2}}}{\binom{2V}{N}} \le \frac{c_+}{\sqrt{N}}\,.$$

Exercise 1.10: Note that the second derivative of $\log \mathbf{Q}_{\Lambda;\beta,N}$ with respect to β yields the variance of $\mathscr{H}$ under the canonical distribution and is thus nonnegative. We conclude that $\beta \mapsto -\log \mathbf{Q}_{\Lambda;\beta,N}$ is concave. Moreover, since the limit of a sequence of concave functions is concave (see Exercise B.3), this implies that $\hat{f}$ is concave in β.

Exercise 1.12: Plugging $\mu_{\Lambda;\beta_U,N}$ in the definition of $\mathrm{S}_{\mathrm{Sh}}(\cdot)$ gives

$$\mathrm{S}_{\mathrm{Sh}}(\mu_{\Lambda;\beta_U,N}) = \beta_U \langle \mathscr{H} \rangle_{\mu_{\Lambda;\beta_U,N}} + \log \mathbf{Z}_{\Lambda;\beta_U,N} = \beta_U U + \log \mathbf{Z}_{\Lambda;\beta_U,N}\,. \tag{C.1}$$

By the Implicit Function Theorem, $U \mapsto \beta_U$ is differentiable. So, differentiating with respect to U,

$$\frac{\partial \mathrm{S}_{\mathrm{Sh}}(\mu_{\Lambda;\beta_U,N})}{\partial U} = \frac{\partial \beta_U}{\partial U} U + \beta_U + \underbrace{\frac{\partial}{\partial \beta} \log \mathbf{Z}_{\Lambda;\beta,N}\Big|_{\beta=\beta_U}}_{=-U} \frac{\partial \beta_U}{\partial U} = \beta_U\,,$$

as one expects from the definition of the inverse temperature in (1.3). Then,

$$U - T_U \mathrm{S}_{\mathrm{Sh}}(\mu_{\Lambda;\beta_U,N}) = U - T_U\{\beta_U U + \log \mathbf{Z}_{\Lambda;\beta_U,N}\} = -\tfrac{1}{\beta_U} \log \mathbf{Z}_{\Lambda;\beta_U,N}\,,$$

in accordance with the definition of free energy given earlier.

Exercise 1.13: Since $M_\Lambda(-\omega) = -M_\Lambda(\omega)$,

$$\langle M_\Lambda \rangle_{\Lambda;\beta,0} = \sum_{\omega \in \Omega_\Lambda} M_\Lambda(\omega) \mu_{\Lambda;\beta,0}(\omega) = \tfrac{1}{2} \sum_{\omega \in \Omega_\Lambda} M_\Lambda(\omega) \underbrace{\{\mu_{\Lambda;\beta,0}(\omega) - \mu_{\Lambda;\beta,0}(-\omega)\}}_{=0} = 0\,.$$

Solutions for Chapter 2

Exercise 2.2: If one writes $\widetilde{\mathscr{H}}_{N;\beta,0} = \mathscr{H}^{\mathrm{CW}}_{N;\beta(N),0}$, where $\beta(N) \overset{\mathrm{def}}{=} N\beta/\zeta(N)$, then either $\beta(N) \uparrow +\infty$ or $\beta(N) \downarrow 0$. The conclusion now follows from our previous analysis.

Exercise 2.3: The analyticity of $h \mapsto m^{\mathrm{CW}}_\beta(h)$ follows from the Implicit Function Theorem (Theorem B.28).

Exercise 2.6: Let us write $\varphi(y) \equiv \varphi_{\beta,h}(y)$. Notice that, since $\beta > 0$, $\varphi(y) \uparrow +\infty$ as $y \to \pm\infty$ sufficiently fast to ensure that $\int_{-\infty}^{+\infty} e^{-c\varphi(y)}\,dy < \infty$ for all $c > 0$. Depending on β, φ has either one or two global minima. For simplicity, consider the case in which there is a unique global minimum y_*. Let $\widetilde{\varphi}(y) \stackrel{\text{def}}{=} \varphi(y) - \varphi(y_*)$, $B_\epsilon(y_*) \stackrel{\text{def}}{=} [y_* - \epsilon, y_* + \epsilon]$ and write

$$\int_{-\infty}^{\infty} e^{-N(\varphi_{\beta,h}(y) - \min_y \varphi_{\beta,h}(y))}\,dy \geq \int_{B_\epsilon(y_*)} e^{-N\widetilde{\varphi}(y)}\,dy\,.$$

Let $c > 0$ be such that $\widetilde{\varphi}(y) \leq c(y - y_*)^2$ for all $y \in B_\epsilon(y_*)$. Then

$$\sqrt{N}\int_{B_\epsilon(y_*)} e^{-N\widetilde{\varphi}(y)}\,dy \geq \sqrt{N}\int_{B_\epsilon(y_*)} e^{-cN(y-y_*)^2}\,dy = \frac{1}{\sqrt{2c}}\int_{-\epsilon\sqrt{2cN}}^{+\epsilon\sqrt{2cN}} e^{-x^2/2}\,dx\,,$$

and this last expression converges to $\sqrt{\pi/c}$ when $N \to \infty$.

Solutions for Chapter 3

Exercise 3.1: Notice that $|\mathrm{B}(n)| = (2n+1)^d$ and that

$$|\partial^{\text{in}}\mathrm{B}(n)| = |\mathrm{B}(n) \setminus \mathrm{B}(n-1)| = (2n+1)^d - (2n-1)^d \leq d(2n+1)^{d-1}\,,$$

which shows that $\frac{|\partial^{\text{in}}\mathrm{B}(n)|}{|\mathrm{B}(n)|} \to 0$. Any sequence $\Lambda_n \uparrow \mathbb{Z}^d$ whose boundary grows as fast as its volume, such as $\Lambda_n = \mathrm{B}(n) \cup \{(i,0,\dots,0) \in \mathbb{Z}^d : 0 \leq i \leq e^n\}$, will not converge in the sense of van Hove.

Exercise 3.5: By a straightforward computation, $m_\beta(h) = \sinh(h)/\sqrt{\sinh^2(h) + e^{-4\beta}}$.

Exercise 3.6: 1. The partition function with free boundary condition can be expressed as

$$\mathbf{Z}^{\varnothing}_{\mathrm{B}(n);\beta,h} = \sum_{\substack{\omega_i=\pm1\\ i\in\mathrm{B}(n)}} \prod_{i=-n}^{n-1} e^{\beta\omega_i\omega_{i+1}} = e^{2\beta n} \sum_{\substack{\omega_i=\pm1\\ i\in\mathrm{B}(n)}} \prod_{i=-n}^{n-1} e^{\beta(\omega_i\omega_{i+1}-1)}\,.$$

Each factor in the last product is either equal to 1 (if $\omega_i = \omega_{i+1}$) or to $e^{-2\beta}$. Therefore,

$$\mathbf{Z}^{\varnothing}_{\mathrm{B}(n);\beta,h} = 2\,e^{2\beta n}\sum_{k=0}^{2n}\binom{2n}{k}(e^{-2\beta})^k = 2\,e^{2\beta n}\,(1+e^{-2\beta})^{2n}\,.$$

This yields $\psi(\beta) = \log\cosh(\beta) + \log 2$, which of course coincides with (3.10).

2. In terms of the variables τ_i,

$$\mathbf{Z}^{\varnothing}_{\mathrm{B}(n);\beta,h} = \sum_{\omega_{-n}=\pm1} \sum_{\substack{\tau_i=\pm1\\ i=-n+1,\dots,n}} \prod_{i=-n+1}^{n} e^{\beta\tau_i} = 2(e^\beta + e^{-\beta})^{2n}\,.$$

Exercise 3.8: Notice that any local function can be expressed as a finite linear combination of **cylinder functions**, which are of the following form: $f(\omega) = 1$ if ω coincides, on a finite region Λ, with some configuration τ, and zero otherwise. Since each spin ω_i takes only two values, there are countably many cylinder functions; we denote them by $f_1, f_2, \ldots$ Since, for each j, the sequence $(\langle f_j\rangle^{\eta_n}_{\Lambda_n;\beta,h})_{n\geq 1}$ is bounded, a standard diagonalization argument (this type of argument will be explained in more detail later, for instance in the proof of Proposition 6.20) allows one to extract a subsequence $(n_k)_{k\geq 1}$ such that $\lim_{k\to\infty}\langle f_j\rangle^{\eta_{n_k}}_{\Lambda_{n_k};\beta,h}$ exists for all j. The existence of $\langle f\rangle \stackrel{\text{def}}{=} \lim_{k\to\infty}\langle f\rangle^{\eta_{n_k}}_{\Lambda_{n_k};\beta,h}$ for all local functions f follows by linearity and defines a Gibbs state.

Exercise 3.9: Simply differentiate $\langle\sigma_A\rangle^+_{\Lambda;\mathbf{J},\mathbf{h}}$ with respect to J_{ij} or h_i and use (3.22).

Exercise 3.10: Observe that, to show that f is nondecreasing, it suffices to show that $f(\omega) \leq f(\omega')$ whenever there exists $i \in \mathbb{Z}^d$ such that $\omega_i = -1$, $\omega'_i = 1$ and $\omega_j = \omega'_j$ for all $j \neq i$. The exercise is then straightforward.

Exercise 3.11: We will come back to this important property in Chapter 6 and prove it in a more general setting (see Lemma 6.7). For simplicity, assume $h = 0$. The numerator appearing in $\mu^\eta_{\Lambda;\beta,h}\big(\omega \mid \sigma_i = \omega'_i, \forall i \in \Lambda\setminus\Delta\big)$ contains the term

$$\exp\Big(\beta \sum_{\{i,j\}\in\mathscr{E}^{\mathrm{b}}_\Lambda} \omega_i\omega_j\Big) = \exp\Big(\beta \sum_{\{i,j\}\in\mathscr{E}^{\mathrm{b}}_\Delta} \omega_i\omega_j\Big)\exp\Big(\beta \sum_{\substack{\{i,j\}\in\mathscr{E}^{\mathrm{b}}_\Lambda\\ \{i,j\}\cap\Delta=\varnothing}} \omega_i\omega_j\Big).$$

The first term, containing the sum over $\{i,j\} \in \mathscr{E}^{\mathrm{b}}_\Delta$, is used to form $\mu^{\omega'}_{\Delta;\beta,h}(\omega)$. The same decomposition can be used for the partition functions; the second factor then cancels out.

Exercise 3.12: Let $D \subset \mathscr{E}_{\Lambda_2}$ be the set of edges $\{i,j\}$ with $i \in \Lambda_2\setminus\Lambda_1$, $j \in \Lambda_1$. Consider, for $s \in [0,1]$, the Hamiltonian

$$\mathscr{H}^s_{\Lambda_2;\beta,h} \stackrel{\text{def}}{=} -\beta \sum_{\substack{\{i,j\}\in\mathscr{E}_{\Lambda_2}\\ \{i,j\}\notin D}} \sigma_i\sigma_j - s\beta\sum_{\{i,j\}\in D}\sigma_i\sigma_j - h\sum_{i\in\Lambda_2}\sigma_i\,.$$

Let $\langle\cdot\rangle^s_{\Lambda_2;\beta,h}$ denote the corresponding Gibbs distribution. Observe that, when $A \subset \Lambda_1$, $\langle\sigma_A\rangle^\varnothing_{\Lambda_2;\beta,h} = \langle\sigma_A\rangle^{s=1}_{\Lambda_2;\beta,h}$ and $\langle\sigma_A\rangle^\varnothing_{\Lambda_1;\beta,h} = \langle\sigma_A\rangle^{s=0}_{\Lambda_2;\beta,h}$. The conclusion follows since, by Exercise 3.9, $\langle\sigma_A\rangle^{s=0}_{\Lambda_2;\beta,h} \leq \langle\sigma_A\rangle^{s=1}_{\Lambda_2;\beta,h}$. For the other claim, add a magnetic field h' acting on spins in $\Lambda_2\setminus\Lambda_1$, and let $h' \longrightarrow \infty$.

Exercise 3.15: First, the FKG inequality and translation invariance yield, for any i,

$$\langle n_A n_{B+i}\rangle^+_{\beta,h} \geq \langle n_A\rangle^+_{\beta,h}\langle n_B\rangle^+_{\beta,h}\,.$$

Fix L large enough to ensure that $A, B \subset \mathrm{B}(L)$. Taking $\|i\|_1$ sufficiently large, we can guarantee that $\mathrm{B}(L+1)\cap(i+\mathrm{B}(L)) = \varnothing$. Fixing all the spins on $\partial^{\mathrm{ex}}\mathrm{B}(L)\cup\partial^{\mathrm{ex}}(i+\mathrm{B}(L))$ to $+1$, it follows from the FKG inequality that

$$\langle n_A n_{B+i}\rangle^+_{\beta,h} \leq \langle n_A\rangle^+_{\mathrm{B}(L);\beta,h}\langle n_{B+i}\rangle^+_{i+\mathrm{B}(L);\beta,h} = \langle n_A\rangle^+_{\mathrm{B}(L);\beta,h}\langle n_B\rangle^+_{\mathrm{B}(L);\beta,h}\,.$$

We conclude that

$$\begin{aligned}\langle n_A\rangle^+_{\beta,h}\langle n_B\rangle^+_{\beta,h} &\leq \liminf_{\|i\|_1\to\infty}\langle n_A n_{B+i}\rangle^+_{\beta,h}\\ &\leq \limsup_{\|i\|_1\to\infty}\langle n_A n_{B+i}\rangle^+_{\beta,h} \leq \langle n_A\rangle^+_{\mathrm{B}(L);\beta,h}\langle n_B\rangle^+_{\mathrm{B}(L);\beta,h}\,.\end{aligned}$$

The desired conclusion follows by letting $L \to \infty$ in the right-hand side. The case of general local functions f and g follows from Lemma 3.19.

Exercise 3.16: Follow the steps of the proof of Theorem 3.17, using Exercise 3.12 for the existence of the thermodynamic limit (use $\langle\sigma_A\rangle^{\varnothing}_{\Lambda_n;\beta,h} = (-1)^{|A|}\langle\sigma_A\rangle^{\varnothing}_{\Lambda_n;\beta,-h}$ when dealing with $h < 0$).

Exercise 3.17: Proceed as in the proof of Lemma 3.31, using the monotonicity results established in Exercises 3.9 and 3.12.

Exercise 3.18: 1. This is a consequence of (3.34). Indeed, let us denote by $\mathscr{A}_\ell$ the set of all contours γ (in $\mathrm{B}(n)$) with length ℓ. Then,

$$\mu^+_{\mathrm{B}(n);\beta,0}(\exists\gamma\in\Gamma \text{ with } |\gamma|\geq K\log n) \leq \sum_{\ell\geq K\log n}|\mathscr{A}_\ell|e^{-2\beta\ell}\,.$$

Now, the number of contours of length ℓ passing through a given point is bounded above by 4^ℓ, and the number of translates of such a contour entirely contained inside $\mathrm{B}(n)$ is bounded above by $4n^2$. Therefore, the probability we are interested in is bounded above by

$$4n^2\sum_{\ell\geq K\log n}(4e^{-2\beta})^\ell \leq 8n^{2-K(2\beta-\log 4)}\,,$$

for all $\beta\geq\log 3$, say. This bound can be made smaller than n^{-c}, for any fixed $c > 0$, by taking K sufficiently large (uniformly in $\beta\geq\log 3$).

2. Partition each row of $\mathrm{B}(n)$ into intervals of length $K\log n$ (and, possibly, a remaining shorter interval that we ignore). We denote these intervals by I_k, $k = 1,\dots,N$, and consider the event

$$\mathscr{I}_k = \{\sigma_i = -1\ \forall i\in I_k\}\,.$$

Of course, there exists $C = C(\beta)$ such that $\mu^+_{I_k;\beta,0}(\mathscr{I}_k)\geq e^{-CK\log n} = n^{-CK}$. Now,

$$\mu^+_{\mathrm{B}(n);\beta,0}(\exists\gamma\in\Gamma \text{ with } |\gamma|\geq K\log n) \geq \mu^+_{\mathrm{B}(n);\beta,0}\Big(\bigcup_{k=1}^N\mathscr{I}_k\Big) = 1-\mu^+_{\mathrm{B}(n);\beta,0}\Big(\bigcap_{k=1}^N\mathscr{I}_k^{\mathrm{c}}\Big)\,.$$

Notice that

$$\mu^+_{\mathrm{B}(n);\beta,0}\Big(\bigcap_{k=1}^N\mathscr{I}_k^{\mathrm{c}}\Big) = \prod_{m=1}^N\mu^+_{\mathrm{B}(n);\beta,0}\Big(\mathscr{I}_m^{\mathrm{c}}\,\Big|\,\bigcap_{k=1}^{m-1}\mathscr{I}_k^{\mathrm{c}}\Big)$$

$$= \prod_{m=1}^{N} \Big\{ 1 - \mu^{+}_{B(n);\beta,0}\Big(\mathscr{I}_m \;\Big|\; \bigcap_{k=1}^{m-1} \mathscr{I}^{c}_{k} \Big) \Big\} .$$

By the FKG inequality,

$$\mu^{+}_{B(n);\beta,0}\Big(\mathscr{I}_m \;\Big|\; \bigcap_{k=1}^{m-1} \mathscr{I}^{c}_{k} \Big) \geq \mu^{+}_{I_m;\beta,0}(\mathscr{I}_m) \geq n^{-CK} ,$$

so that

$$\mu^{+}_{B(n);\beta,0}\Big(\exists \gamma \in \Gamma \text{ with } |\gamma| \geq K \log n \Big) \geq 1 - (1 - n^{-CK})^{N} \geq 1 - e^{-n^{-CK} N} .$$

The conclusion follows since $N = (2n+1)\lfloor (2n+1)/K \log n \rfloor \geq n^{2-c/2}/K$ for $n > n_0(c)$ and $n^{-CK}/K \geq n^{-c/2}$ if $K \leq K_1(\beta, c)$.

Exercise 3.20: In higher dimensions, the deformation operation leading to contours is less convenient, so we will avoid it. For the sake of concreteness, we consider the case $d = 3$. The bounds we give below are very rough and can be improved. The three-dimensional analogues of the contours described above are sets of **plaquettes**, which are the squares that form the boundary of the cubic cells of $\mathbb{Z}^3$. For a given configuration ω, the set $\partial \mathscr{M}(\omega)$ can be defined as before and decomposed into maximal connected sets of plaquettes: $\partial \mathscr{M}(\omega) = \hat{\gamma}_1 \cup \cdots \cup \hat{\gamma}_n$. The analogue of (3.38) then becomes

$$\mu^{+}_{B(n);\beta,0}(\sigma_0 = -1) \leq \sum_{k \geq 6} e^{-2\beta k} \, \# \left\{ \hat{\gamma}^* \,:\, \text{dist}(\hat{\gamma}^*, 0) \leq k, |\hat{\gamma}^*| = k \right\} .$$

With each $\hat{\gamma}^*$ in the latter set, we associate a connected graph G^* whose set of vertices V^* is formed by all the centers of the plaquettes of $\hat{\gamma}^*$ and in which two vertices $u, v \in V^*$ are connected by an edge if the corresponding plaquettes share a common edge. The above sum is then bounded by (observe that a vertex of V^* has at most 12 neighbors and that each edge is shared by two vertices, so that $|E'| \leq 6k$)

$$\sum_{k \geq 6} e^{-2\beta k} \, \# \left\{ G^* \,:\, |V^*| = k \right\} \leq \sum_{k \geq 6} e^{-2\beta k} \cdot k^3 \cdot 12^{6k} .$$

This last inequality was obtained using Lemma 3.38. As in the two-dimensional case, the series is smaller than $\frac{1}{2}$ once β is large enough.

Exercise 3.21: Define τ by $e^{-\tau} \overset{\text{def}}{=} 3e^{-2\beta}$. Notice that (3.40) can be written $(4e^{-4\tau} - 3e^{-5\tau})/(1 - e^{-\tau})^2 < \frac{3}{4}$. The first point follows by verifying that this holds once $\beta > 0.88$.

Let us turn to the second point. Write $\mu^{\pm}_{B(n);\beta,0}(A) = \frac{\mathbf{Z}^{\pm}_{B(n);\beta,0}[A]}{\mathbf{Z}^{\pm}_{B(n);\beta,0}[\Omega^{\pm}_{B(n)}]}$, where

$$\mathbf{Z}^{\pm}_{B(n);\beta,0}[A] \overset{\text{def}}{=} \sum_{\omega \in \Omega^{\pm}_{B(n)}} \mathbf{1}_A(\omega) \prod_{\gamma \in \Gamma(\omega)} e^{-2\beta|\gamma|} .$$

Let $A^{\pm} \overset{\text{def}}{=} \{\sigma_i = \pm 1 \, \forall i \in B(R)\}$. Under $\mu^{+}_{B(n);\beta,0}$, the occurrence of A^- forces the presence of at least one self-avoiding closed path $\pi^* \subset \partial \mathscr{M}$ surrounding $B(R)$.

Therefore, by flipping all the spins located inside the region delimited by π^*, one gets

$$\mathbf{Z}^+_{\mathrm{B}(n);\beta,0}[A^-] \le \sum_{\pi^*} \mathbf{Z}^+_{\mathrm{B}(n);\beta,0}[\pi^* \subset \partial\mathscr{M}, A^-] \le \sum_{\pi^*} e^{-2\beta|\pi^*|}\,\mathbf{Z}^+_{\mathrm{B}(n);\beta,0}[A^+].$$

But

$$\sum_{\pi^*} e^{-2\beta|\pi^*|} \le \sum_{k\ge 8R} k e^{-2\beta k} C_k .$$

Now, for all $\epsilon > 0$, $C_k \le (\mu+\epsilon)^k$ for all large enough k. Therefore,

$$\frac{\mu^+_{\mathrm{B}(n);\beta,0}(\sigma_i = -1\,\forall i \in \mathrm{B}(R))}{\mu^-_{\mathrm{B}(n);\beta,0}(\sigma_i = -1\,\forall i \in \mathrm{B}(R))} = \frac{\mathbf{Z}^+_{\mathrm{B}(n);\beta,0}[A^-]}{\mathbf{Z}^-_{\mathrm{B}(n);\beta,0}[A^-]} = \frac{\mathbf{Z}^+_{\mathrm{B}(n);\beta,0}[A^-]}{\mathbf{Z}^+_{\mathrm{B}(n);\beta,0}[A^+]} \le \sum_{k\ge 8R} k e^{-2\beta k}(\mu+\epsilon)^k .$$

If $e^{-2\beta}\mu < 1$, ϵ can be chosen such that the last series converges. Taking R sufficiently large allows us to make the whole sum < 1.

Exercise 3.23: We only provide the answer for the free boundary condition.
Let $\mathfrak{E}^{\mathrm{even}}_\Lambda \overset{\mathrm{def}}{=} \{E \subset \mathscr{E}_\Lambda : I(i,E) \text{ is even for all } i \in \Lambda\}$. Then,

$$\mathbf{Z}^{\varnothing}_{\Lambda;\beta,0} = 2^{|\Lambda|}\cosh(\beta)^{|\mathscr{E}_\Lambda|} \sum_{E\in\mathfrak{E}^{\mathrm{even}}_\Lambda} \tanh(\beta)^{|E|} .$$

Moreover,

$$\langle\sigma_i\sigma_j\rangle^{\varnothing}_{\Lambda;\beta,0} = \sum_{\substack{E_0\in\mathfrak{E}^{i,j}_\Lambda\\ \text{connected}, E_0\ni i}} \tanh(\beta)^{|E_0|} \frac{\sum_{E'\in\mathfrak{E}^{\mathrm{even}}_\Lambda : E'\subset\Delta(E_0)} \tanh(\beta)^{|E'|}}{\sum_{E\in\mathfrak{E}^{\mathrm{even}}_\Lambda} \tanh(\beta)^{|E|}},$$

where

$$\mathfrak{E}^{i,j}_\Lambda \overset{\mathrm{def}}{=} \{E \subset \mathscr{E}_\Lambda : I(k,E) \text{ is even for all } k \in \mathrm{B}(n)\setminus\{i,j\}, \text{ but } I(i,E) \text{ and } I(j,E) \text{ are odd}\} .$$

Exercise 3.24: Since the Gibbs state is unique, we can consider the free boundary condition. Proceeding as we did for the representation of $\langle\sigma_0\rangle^+_{\Lambda;\beta,h}$ in terms of a sum over graphs in (3.47), we get, for $i,j \in \mathrm{B}(n)$,

$$\langle\sigma_i\sigma_j\rangle^{\varnothing}_{\mathrm{B}(n);\beta,0} \le \sum_{\substack{E\in\mathfrak{E}^{i,j}_{\mathrm{B}(n)}\\ \text{connected}, E_0\ni i}} \tanh(\beta)^{|E|} .$$

All graphs $E \in \mathfrak{E}^{i,j}_{\mathrm{B}(n)}$ have at least $\|i-j\|_1$ edges. Proceeding as in (3.49), we derive the exponential decay once β is sufficiently small.

Exercise 3.25: Fix a shortest path $\pi = (i = i_1, i_2, \dots, i_m = j)$ from i to j and introduce $\mathscr{E}_\pi \overset{\mathrm{def}}{=} \{\{i_k, i_{k+1}\} : 1 \le k < m\}$. For $s \in [0,1]$, set

$$J_{uv} = \begin{cases} \beta & \text{if } \{u,v\} \in \mathscr{E}_\pi, \\ s\beta & \text{otherwise.} \end{cases}$$

Denote by $\mu^{\varnothing,s}_{B(n);\beta,h}$ the distribution of the Ising model in $B(n) \subset \mathbb{Z}^d$ with these coupling constants and free boundary condition. Check that

$$\langle\sigma_i\sigma_j\rangle^{\varnothing,s=1}_{B(n);\beta,0} = \langle\sigma_i\sigma_j\rangle^{\varnothing}_{B(n);\beta,0} \quad \text{and} \quad \langle\sigma_i\sigma_j\rangle^{\varnothing,s=0}_{B(n);\beta,0} = \langle\sigma_0\sigma_{\|j-i\|_1}\rangle^{d=1}_{\Lambda_{ij};\beta,0}.$$

Conclude, using the fact that, by GKS inequalities, $\langle\sigma_i\sigma_j\rangle^{\varnothing,s=1}_{B(n);\beta,0} \geq \langle\sigma_i\sigma_j\rangle^{\varnothing,s=0}_{B(n);\beta,0}$.

Exercise 3.26:

$$\begin{aligned}\mathbf{Z}^{+}_{B(n);\beta,0} &= 2^{2n+1}(\cosh\beta)^{2n+2}\big(1+(\tanh\beta)^{2n+2}\big),\\ \mathbf{Z}^{\varnothing}_{B(n);\beta,0} &= 2^{2n+1}(\cosh\beta)^{2n},\\ \mathbf{Z}^{\text{per}}_{B(n);\beta,0} &= 2^{2n+1}(\cosh\beta)^{2n+1}\big(1+(\tanh\beta)^{2n+1}\big).\end{aligned}$$

Exercise 3.27: Notice that, by a straightforward computation,

$$|\alpha z+1|^2 - |\alpha+z|^2 = (1-|z|^2)(1-\alpha^2).$$

Since $1-\alpha^2 > 0$, all the claims can be deduced from this identity. For example, $|z| < 1$ implies $1-|z|^2 > 0$ and, therefore, $|\alpha z+1|^2 - |\alpha+z|^2 > 0$, that is, $|\varphi(z)| = |(\alpha z+1)/(\alpha+z)| > 1$.

Exercise 3.28: Since the argument of the logarithm in (3.10) is always larger than 1, only the square root can be responsible for the singularities of the pressure. But the square root vanishes at the values $h \in \mathbb{C}$ at which $e^{\beta}\cosh(h) = 2\sinh(2\beta)$. Since we know that all singularities lie on the imaginary axis, they can be expressed as $h = \mathrm{i}(\pm t + k2\pi)$, where $t = \arccos\sqrt{1-e^{-4\beta}}$, $k \in \mathbb{Z}$. Observe that, as $\beta \to \infty$, the two singularities at $\pm \mathrm{i}t$ converge from above and from below to $h = 0$. This is compatible with the fact that, in that limit, a singularity appears at $h = 0$. Namely, using (3.10),

$$\lim_{\beta\to\infty} \frac{\psi_\beta(\beta h)}{\beta} = |h| + 1,$$

which is nonanalytic at $h = 0$.

Exercise 3.29: Duplicating the system, we can write

$$|\mathbf{Z}^{\varnothing}_{\Lambda;\beta,h}|^2 = \sum_{\omega,\omega'} e^{\beta\sum_{\{i,j\}\in\mathscr{E}_\Lambda}(\omega_i\omega_j+\omega'_i\omega'_j)+\sum_{i\in\Lambda}(h\omega_i+\bar{h}\omega'_i)}.$$

Define the variables $\theta_i \in \{0,\pi/2,\pi,3\pi/2\}$, $i \in \Lambda$, by $\cos\theta_i = \frac{1}{2}(\omega_i+\omega'_i)$ and $\sin\theta_i = \frac{1}{2}(\omega_i-\omega'_i)$. It is easy to check that

$$\begin{aligned}\omega_i\omega_j+\omega'_i\omega'_j &= 2\cos(\theta_i-\theta_j) = e^{\mathrm{i}(\theta_i-\theta_j)} + e^{-\mathrm{i}(\theta_i-\theta_j)},\\ h\omega_i+\bar{h}\omega'_i &= 2\,\mathfrak{Re}\,h\cos(\theta_i) + 2\mathrm{i}\,\mathfrak{Im}\,h\sin(\theta_i)\\ &= (\mathfrak{Re}\,h+\mathfrak{Im}\,h)e^{\mathrm{i}\theta_i} + (\mathfrak{Re}\,h-\mathfrak{Im}\,h)e^{-\mathrm{i}\theta_i}.\end{aligned}$$

Substituting these expressions yields

$$|\mathbf{Z}^{\varnothing}_{\Lambda;\beta,h}|^2 = \sum_{(\theta_i)_{i\in\Lambda}} \exp\Big\{ \sum_{\substack{\mathbf{m}=(m_i)_{i\in\Lambda}\\ m_i\in\{0,1,2,3\}}} \alpha_{\mathbf{m}} e^{\mathrm{i}\sum_{i\in\Lambda} m_i\theta_i}\Big\},$$

for some nonnegative coefficients $\alpha_{\mathbf{m}}$ which are nondecreasing both in $\mathfrak{Re}\, h+\mathfrak{Im}\, h$ and $\mathfrak{Re}\, h-\mathfrak{Im}\, h$. Consequently, expanding the exponential gives

$$|\mathbf{Z}^{\varnothing}_{\Lambda;\beta,h}|^2 = \sum_{(\theta_i)_{i\in\Lambda}} \sum_{\substack{\mathbf{m}=(m_i)_{i\in\Lambda}\\ m_i\in\{0,1,2,3\}}} \widehat{\alpha}_{\mathbf{m}} e^{\mathrm{i}\sum_{i\in\Lambda} m_i\theta_i},$$

where the coefficients $\widehat{\alpha}_{\mathbf{m}}$ are still nonnegative and nondecreasing in both $\mathfrak{Re}\, h+\mathfrak{Im}\, h$ and $\mathfrak{Re}\, h-\mathfrak{Im}\, h$. Now, observe that

$$\sum_{(\theta_i)_{i\in\Lambda}} e^{\mathrm{i}\sum_{i\in\Lambda} m_i\theta_i} = \prod_{i\in\Lambda}\sum_{\theta_i} e^{\mathrm{i}m_i\theta_i} = \begin{cases} 4^{|\Lambda|} & \text{if } m_i=0,\ \forall i\in\Lambda,\\ 0 & \text{otherwise.}\end{cases}$$

We deduce that $|\mathbf{Z}^{\varnothing}_{\Lambda;\beta,h}|^2 = 4^{|\Lambda|}\widehat{\alpha}_{(0,0,\dots,0)}$ and, thus, that $|\mathbf{Z}^{\varnothing}_{\Lambda;\beta,h}|^2$ is nondecreasing in both $\mathfrak{Re}\, h+\mathfrak{Im}\, h$ and $\mathfrak{Re}\, h-\mathfrak{Im}\, h$. Since $\mathfrak{Re}\, h - |\mathfrak{Im}\, h| = \min(\mathfrak{Re}\, h+\mathfrak{Im}\, h, \mathfrak{Re}\, h-\mathfrak{Im}\, h)$, this proves that

$$|\mathbf{Z}^{\varnothing}_{\Lambda;\beta,h}| \geq \mathbf{Z}^{\varnothing}_{\Lambda;\beta,\mathfrak{Re}\, h-|\mathfrak{Im}\, h|} > 0\,.$$

Exercise 3.31: We write

$$\begin{aligned}(\mathbf{Z}_{\Lambda;\mathbf{K}}\mathbf{Z}_{\Lambda;\mathbf{K}'})\langle\sigma_A-\sigma'_A\rangle_{\nu_{\Lambda;\mathbf{K}}\otimes\nu_{\Lambda;\mathbf{K}'}} &= \sum_{\omega,\omega'}(\omega_A-\omega'_A)\prod_{C\subset\Lambda} e^{K_C\omega_C+K'_C\omega'_C}\\ &= \sum_{\omega''}(1-\omega''_A)\sum_{\omega}\omega_A\prod_{C\subset\Lambda} e^{(K_C+K'_C\omega''_C)\omega_C},\end{aligned}$$

and we can conclude as in the proof of (3.55), since $K_C+K'_C\omega''_C\geq 0$ by assumption.

Exercise 3.35: The only delicate part is showing that, for all $E,E'\subset\mathscr{E}^{\mathrm{b}}_{\Lambda}$,

$$N^{\mathrm{w}}_{\Lambda}(E)+N^{\mathrm{w}}_{\Lambda}(E')\leq N^{\mathrm{w}}_{\Lambda}(E\cup E')+N^{\mathrm{w}}_{\Lambda}(E\cap E').\tag{C.2}$$

In order to establish (C.2), it is sufficient to prove that

$$E'\mapsto N^{\mathrm{w}}_{\Lambda}(E\cup E')-N^{\mathrm{w}}_{\Lambda}(E') \text{ is nondecreasing.}\tag{C.3}$$

Indeed, (C.3) implies that

$$N^{\mathrm{w}}_{\Lambda}(E\cup E')-N^{\mathrm{w}}_{\Lambda}(E')\geq N^{\mathrm{w}}_{\Lambda}(E\cup(E'\cap E))-N^{\mathrm{w}}_{\Lambda}(E'\cap E)=N^{\mathrm{w}}_{\Lambda}(E)-N^{\mathrm{w}}_{\Lambda}(E'\cap E),$$

which is equivalent to (C.2). Let $E=\{e_1,\dots,e_n\}\subset\mathscr{E}^{\mathrm{b}}_{\Lambda}$. Since

$$N^{\mathrm{w}}_{\Lambda}(E\cup E')-N^{\mathrm{w}}_{\Lambda}(E')=\sum_{k=1}^{n}\{N^{\mathrm{w}}_{\Lambda}(\{e_1,\dots,e_k\}\cup E')-N^{\mathrm{w}}_{\Lambda}(\{e_1,\dots,e_{k-1}\}\cup E')\},$$

it is sufficient to show that each summand in the right-hand side verifies (C.3). But this is immediate, since, if $e_k=\{i,j\}$,

$$N^{\mathrm{w}}_{\Lambda}(\{e_1,\dots,e_k\}\cup E')-N^{\mathrm{w}}_{\Lambda}(\{e_1,\dots,e_{k-1}\}\cup E')=\begin{cases}0 & \text{if } i\leftrightarrow j \text{ in } \{e_1,\dots,e_{k-1}\}\cup E',\\ -1 & \text{otherwise.}\end{cases}$$

Exercise 3.37: Since $\lim_{\Lambda\uparrow\mathbb{Z}^d}\langle\sigma_0\rangle^+_{\Lambda;\beta,0} = \langle\sigma_0\rangle^+_{\beta,0}$, it follows from Exercise 3.34 that

$$\lim_{\Lambda\uparrow\mathbb{Z}^d} \nu^{\mathrm{FK,w}}_{\Lambda;p_\beta,2}(0 \leftrightarrow \partial^{\mathrm{ex}}\Lambda) = \lim_{\Lambda\uparrow\mathbb{Z}^d}\langle\sigma_0\rangle^+_{\Lambda;\beta,0} = \langle\sigma_0\rangle^+_{\beta,0}\,.$$

Therefore, we only have to check that

$$\lim_{\Lambda\uparrow\mathbb{Z}^d} \nu^{\mathrm{FK,w}}_{\Lambda;p_\beta,2}(0 \leftrightarrow \partial^{\mathrm{ex}}\Lambda) = \nu^{\mathrm{FK,w}}_{p_\beta,2}(0 \leftrightarrow \infty)\,.$$

Observe that, for all $0 \in \Delta \subset \Lambda \Subset \mathbb{Z}^d$,

$$\nu^{\mathrm{FK,w}}_{p_\beta,2}(0 \leftrightarrow \partial^{\mathrm{ex}}\Lambda) \le \nu^{\mathrm{FK,w}}_{\Lambda;p_\beta,2}(0 \leftrightarrow \partial^{\mathrm{ex}}\Lambda) \le \nu^{\mathrm{FK,w}}_{\Lambda;p_\beta,2}(0 \leftrightarrow \partial^{\mathrm{ex}}\Delta)\,,$$

the first inequality resulting from the FKG inequality (as can be checked by the reader) and the second one from the inclusion $\{0 \leftrightarrow \partial^{\mathrm{ex}}\Lambda\} \subset \{0 \leftrightarrow \partial^{\mathrm{ex}}\Delta\}$. The desired result follows by taking the limit $\Lambda \uparrow \mathbb{Z}^d$ and then the limit $\Delta \uparrow \mathbb{Z}^d$.

Solutions for Chapter 4

Exercise 4.2: Let $\epsilon > 0$ and let ℓ be such that $\sum_{j\notin \mathrm{B}(\ell)} K(0,j) \le \epsilon$. Let $\Lambda_* \subset \Lambda$ be a parallelepiped, large enough to contain $\lceil\rho|\Lambda|\rceil$ particles, but such that if either of its sides is reduced by 1, then it becomes too small to contain those $\lceil\rho|\Lambda|\rceil$ particles. Then $|\Lambda_*| = \rho|\Lambda| + O(|\partial^{\mathrm{in}}\Lambda|)$. If η_* denotes the configuration obtained by densely filling Λ_* with particles (except possibly along its boundary), we get

$$-\mathscr{H}_{\Lambda;K}(\eta_*) = \tfrac{1}{2}\sum_{i\in\Lambda_*}\sum_{\substack{j\in\Lambda_*\\ j\neq i}} K(i,j) + O(|\partial^{\mathrm{in}}\Lambda|)\,.$$

Then let Λ_*^- denote the set of vertices $i \in \Lambda_*$ for which $\mathrm{B}(\ell) + i \subset \Lambda_*$. Note that, whenever $i \in \Lambda_*^-$, we have $\big|\sum_{\substack{j\in\Lambda_*\\ j\neq i}} K(i,j) - \kappa\big| \le \epsilon$ and thus, since $|\Lambda_* \setminus \Lambda_*^-| \le \ell|\partial^{\mathrm{in}}\Lambda|$,

$$\big|\mathscr{H}_{\Lambda;K}(\eta_*) - (-\tfrac{1}{2}\kappa\rho|\Lambda|)\big| \le \epsilon|\Lambda| + O(|\partial^{\mathrm{in}}\Lambda|)\,.$$

We conclude that $\lim_{\Lambda\uparrow\mathbb{Z}^d}\big|\mathscr{H}_{\Lambda;K}(\eta_*) - (-\tfrac{1}{2}\kappa\rho|\Lambda|)\big| \,/\, |\Lambda| \le \epsilon$. Since ϵ is arbitrary, the claim follows.

Exercise 4.4: The proof is similar to the one for the free energy: if Λ_1 and Λ_2 are two adjacent parallelepipeds, ignoring the interactions between pairs composed of one particle in Λ_1 and one in Λ_2 gives

$$\Theta_{\Lambda_1\cup\Lambda_2;\beta,\mu} \ge \Theta_{\Lambda_1;\beta,\mu}\Theta_{\Lambda_2;\beta,\mu}\,.$$

We conclude, as before, that the thermodynamic limit exists along any increasing sequence of parallelepipeds.

Exercise 4.5: Let us denote by η^1 (resp. η^0) the configuration in which $\eta_j = m_j$ for each $j \neq i$ and $\eta_i = 1$ (resp. $\eta_i = 0$). The difference

$$\{\mathscr{H}_{\Lambda;K}(\eta^1) - \mu N_\Lambda(\eta^1)\} - \{\mathscr{H}_{\Lambda;K}(\eta^0) - \mu N_\Lambda(\eta^0)\} = -\sum_{j\in\Lambda, j\neq i} K(i,j)m_j - \mu$$

belongs to the interval $(-\kappa - \mu, -\mu)$. Therefore, $\nu_{\Lambda;\beta,\mu}(\eta_i = 1 \mid \eta_j = m_j, \forall j \in \Lambda\setminus\{i\})$ belongs to the interval $\big(1/(1+e^{-\beta\mu}), 1/(1+e^{-\beta(\kappa+\mu)})\big)$.

Exercise 4.6: Since

$$1 \leq \Theta_{\Lambda;\beta,\mu} \leq \sum_{N=0}^{|\Lambda|} \binom{|\Lambda|}{N} e^{\beta(\frac{\kappa}{2}+\mu)N} = \big(1+e^{\beta(\frac{\kappa}{2}+\mu)}\big)^{|\Lambda|},$$

we have $0 \leq p_\beta(\mu) \leq \beta^{-1}\log(1+e^{\beta(\frac{\kappa}{2}+\mu)})$. To bound $\Theta_{\Lambda;\beta,\mu}$ from below, we keep only the configuration in which $\eta_i = 1$ for each $i \in \Lambda$. This leads to $p_\beta(\mu) \geq \frac{\kappa}{2} + \mu$. The first two claims follow. The last two claims about ρ_β follow from the convexity of p_β.

Exercise 4.7: As we did earlier, let $\epsilon > 0$ and take ℓ such that $\sum_{j\notin \mathrm{B}(\ell)} K(i,j) \leq \epsilon$. Then

$$\sum_{\substack{i\in\Lambda':\\ i+\mathrm{B}(\ell)\subset\Lambda'}} \sum_{j\in\Lambda''} K(i,j) \leq \epsilon|\Lambda'|$$

and, since

$$\sum_{\substack{i\in\Lambda':\\ i+\mathrm{B}(\ell)\not\subset\Lambda'}} \sum_{j\in\Lambda''} K(i,j) \leq \kappa\ell|\partial^{\mathrm{in}}\Lambda'|,$$

the conclusion follows easily.

Exercise 4.9: Consider the gas branch: $\rho < \rho_g$. By the strict convexity of the pressure and the equivalence of ensembles, there exists a unique $\mu(\rho)$ such that

$$f_\beta(\rho) = \mu(\rho)\rho - p_\beta(\mu(\rho)).$$

Since $\rho < \rho_g$, we have $\mu(\rho) < \mu_*$ and $\mu(\rho)$ is the solution of $\rho = \frac{\partial p_\beta}{\partial \mu}$. Then, we use (i) the analyticity of the pressure, which implies in particular that its first and second derivatives exist, outside μ_*, (ii) the fact, proved in Theorem 4.12, that $\frac{\partial^2 p_{\Lambda;\beta}}{\partial\mu^2} \geq \beta c(1-c) > 0$, which implies that $\frac{\partial^2 p_\beta}{\partial \mu^2} > 0$ whenever it exists, and (iii) the Implicit Function Theorem (Section B.28), to conclude that $\mu(\cdot)$ is also analytic in a neighborhood of ρ. Since the composition of analytic maps is also analytic, this shows that $f_\beta(\cdot)$ is analytic in the neighborhood of ρ.

Exercise 4.13: We only consider the case $d = 1$; the general case can be treated in the same way. Let us identify each $\Lambda^{(\alpha)} \subset \mathbb{Z}$ with the interval $J^{(\alpha)} = \{x \in \mathbb{R} :$

$\mathrm{dist}(x, \Lambda^{(\alpha)}) \leq \frac{1}{2}\}$, whose length equals $|J^{(\alpha)}| = \ell$, and let $J_\gamma^{(\alpha)} \stackrel{\text{def}}{=} \{\gamma x : x \in J^{(\alpha)}\}$. We have (up to terms that vanish in the van der Waals limit)

$$\sum_{\alpha'>1} |J_\gamma^{(\alpha')}| \inf_{x \in J_\gamma^{(\alpha')}} \varphi(x) \leq |\Lambda^{(1)}| \sum_{\alpha'>1} \underline{K}_\gamma(1,\alpha') \\ \leq |\Lambda^{(1)}| \sum_{\alpha'>1} \overline{K}_\gamma(1,\alpha') \leq \sum_{\alpha'>1} |J_\gamma^{(\alpha')}| \sup_{x \in J_\gamma^{(\alpha')}} \varphi(x) . \tag{C.4}$$

The conclusion follows, since the first and last sums of this last display are Darboux sums that converge to $\int \varphi(x)\,\mathrm{d}x$ as $|J_\gamma^{(\alpha')}| = \gamma\ell \downarrow 0$.

Exercise 4.14: Let $N = \lceil \rho|\Lambda| \rceil$. Since $\mathscr{N}(N; M)$ counts the number of ways N identical balls can be distributed in M boxes, with at most $|\Lambda^{(1)}|$ balls per box, this number is obviously smaller than the number of ways of putting N identical balls in M boxes, without restrictions on the number of balls per box. The latter equals

$$\binom{N+M-1}{M-1}.$$

Since

$$M = \frac{|\Lambda|}{|\Lambda^{(1)}|} \stackrel{\text{def}}{=} \delta_{\Lambda,\ell} N ,$$

and $\lim_{\Lambda \Uparrow \mathbb{Z}^d} \delta_{\Lambda,\ell} = \frac{1}{\rho|\Lambda^{(1)}|} \stackrel{\text{def}}{=} \delta_\ell$, Stirling's formula gives

$$\lim_{\ell\to\infty} \lim_{N\to\infty} \frac{1}{N} \log \binom{N+M-1}{M-1} = \lim_{\ell\to\infty} \{(1+\delta_\ell)\log(1+\delta_\ell) + \delta_\ell \log \delta_\ell\} = 0 .$$

Exercise 4.15: Let $\epsilon > 0$ and n be large enough to ensure that $f(x) - \epsilon \leq f_n(x) \leq f(x) + \epsilon$ for all $x \in [a,b]$. Since $\mathrm{CE}\, g \leq \mathrm{CE}\, h$ whenever $g \leq h$, this implies $\mathrm{CE} f(x) - \epsilon \leq \mathrm{CE} f_n(x) \leq \mathrm{CE} f(x) + \epsilon$ for all $x \in [a,b]$, which gives the result.

Exercise 4.16: Let $a_n \stackrel{\text{def}}{=} e^{-\beta 2dn^{(d-1)/d}}$. For all compact $K \subset H^+$,

$$\sup_{h\in K} \Big| \sum_{n\geq 1} a_n e^{-hn} - \sum_{n=1}^{N} a_n e^{-hn} \Big| \leq \sup_{h\in K} \sum_{n>N} a_n e^{-\Re\mathfrak{e}\, h n} \leq \sum_{n>N} a_n e^{-x_0 n} ,$$

where $x_0 \stackrel{\text{def}}{=} \inf\{\Re\mathfrak{e}\, h : h \in K\} > 0$, and this last series goes to zero when $N \to \infty$. This implies that the series defining ψ_β converges uniformly on compacts. Since $h \mapsto e^{-hn}$ is analytic on H^+, Theorem B.27 implies that ψ_β is analytic on H^+. Moreover, it can be differentiated term by term an arbitrary number of times, yielding, when $h \in \mathbb{R}_{>0}$,

$$\Big| \lim_{h\downarrow 0} \frac{\mathrm{d}^k \psi_\beta}{\mathrm{d}h^k} \Big| = \Big| (-1)^k \lim_{h\downarrow 0} \sum_{n\geq 1} n^k a_n e^{-hn} \Big| = \sum_{n\geq 1} n^k a_n .$$

A lower bound on the sum is obtained by keeping only its largest term. Notice that $x \mapsto x^k e^{-2d\beta x^{(d-1)/d}}$ is maximal at

$$x_* = x_*(k,\beta,d) \stackrel{\text{def}}{=} \Big(\frac{k}{2(d-1)\beta}\Big)^{d/(d-1)} .$$

Keeping the term $n_* \stackrel{\text{def}}{=} \lceil x_* \rceil$, reorganizing the terms and using Stirling's formula, we get

$$\sum_{n\geq 1} n^k a_n \geq n_*^k a_{n_*} \geq C_-^k k!^{d/(d-1)},$$

for some $C_- = C_-(\beta, d) > 0$. The reader may check that an upper bound of the same kind holds, with a constant $C_+ < \infty$.

Solutions for Chapter 5

Exercise 5.1: In (5.9), just distinguish the case $k = 1$ from $k \geq 2$.

Exercise 5.2: We proceed by induction. The case $n = 1$ is trivial. Now, if the claim holds for n, it can be shown to hold for $n + 1$ too, by writing

$$\Bigl(\prod_{k=1}^{n+1}(1 + \alpha_k)\Bigr) - 1 = (1 + \alpha_{n+1})\Bigl(\prod_{k=1}^{n}(1 + \alpha_k) - 1\Bigr) + \alpha_{n+1}.$$

Exercise 5.5: When using more general boundary conditions, the same sets S_i can be used, but the surface term $e^{-2\beta|\partial_e S_i|}$ in their weights might have to be modified if $S_i \cap \partial^{\text{in}}\Lambda \neq \varnothing$. The condition (5.26) can nevertheless be seen to hold since the surface term was ignored in our analysis. Then, the contributions to $\log \Xi^{\text{LF}}_{\Lambda;\beta,h}$ coming from clusters containing sets S_i that intersect $\partial^{\text{in}}\Lambda$ is a surface contribution that vanishes in the thermodynamic limit, yielding the same expression for the pressure.

Exercise 5.6: First,

$$\begin{aligned}
\phi(\phi^{-1}(z)) &= \sum_{n\geq 1}\tilde{a}_n\Bigl(\sum_{k\geq 1}c_k z^k\Bigr)^n = \sum_{n\geq 1}\tilde{a}_n \sum_{k_1,\dots,k_n\geq 1}\prod_{i=1}^{n} c_{k_i} z^{k_i} \\
&= \sum_{n\geq 1}\tilde{a}_n \sum_{m\geq n} z^m \sum_{\substack{k_1,\dots,k_n\geq 1\\ k_1+\dots+k_n=m}}\prod_{i=1}^{n} c_{k_i} = \sum_{m\geq 1}\Bigl\{\sum_{n=1}^{m}\tilde{a}_n \sum_{\substack{k_1,\dots,k_n\geq 1\\ k_1+\dots+k_n=m}}\prod_{i=1}^{n} c_{k_i}\Bigr\} z^m .
\end{aligned}$$

However, since $\phi(\phi^{-1}(z)) = z$ by definition, we conclude that the coefficient of z in the last sum, which is $\tilde{a}_1 c_1$, must be equal to 1, while the coefficient of z^m, $m \geq 2$, must vanish. The claim follows.

Exercise 5.7: The procedure is identical to that used in the proof of Lemma 5.10. The expansion, up to the second nontrivial order, is given by

$$\begin{aligned}
\psi_\beta(0) - d\log(\cosh\beta) - \log 2 &= \tfrac{1}{2}d(d-1)(\tanh\beta)^4 \\
&\quad + \tfrac{1}{3}d(d-1)(8d-13)(\tanh\beta)^6 + O(\tanh\beta)^8 .
\end{aligned} \tag{C.5}$$

These two terms correspond, respectively, to sets of 4 and 6 edges. In the terminology used in the proof of Lemma 5.10, one has: $A = 4, B = 1, C = 1$ for the first term and $A = 6, B = 1, C = 1$ for the second. Therefore, the only thing left to do in order to derive (C.5) is to determine the number of such sets containing the origin, which is a purely combinatorial task left to the reader.

Exercise 5.8: First, the high-temperature representation (5.38) needs to be adapted to the presence of a magnetic field. Indeed, (3.44) must be replaced by

$$\sum_{\omega_i=\pm 1} \omega_i^{I(i,E)} e^{h\omega_i} = \begin{cases} 2\cosh(h) & \text{if } I(i,E) \text{ is even,} \\ 2\sinh(h) & \text{if } I(i,E) \text{ is odd.} \end{cases}$$

Then, the class of sets E that contribute to the partition function is larger (the incidence numbers $I(i,E)$ are allowed to be odd), giving

$$\mathbf{Z}^{\varnothing}_{\Lambda;\beta,0} = (2\cosh h)^{|\Lambda|}(\cosh\beta)^{|\mathscr{E}_\Lambda|} \sum_{E\subset\mathscr{E}_\Lambda} (\tanh\beta)^{|E|}(\tanh h)^{|\partial E|},$$

where $\partial E \stackrel{\text{def}}{=} \{ i \in \mathbb{Z}^d : I(i,E) \text{ is odd} \}$. Notice that $|\tanh h| \leq 1$ when $|h|$ is small enough. Then, the weights of the components are bounded by the same weight as the one used above, $(\tanh\beta)^{|E|}$, and the rest of the analysis is essentially the same (keeping in mind that the class of objects is larger).

Exercise 5.9: It is convenient to use the notion of interior of a contour depicted in Figure 5.2. Then, given a collection $\Gamma' = \{\gamma_1, \dots, \gamma_n\} \subset \Gamma_\Lambda$ of pairwise disjoint contours in Λ, consider the configuration

$$\omega_i = (-1)^{\#\{\gamma\in\Gamma' : i\in \operatorname{Int}\gamma\}}.$$

Since $\Lambda^c \cap \bigcup_{\gamma\in\Gamma'} \operatorname{Int}\gamma = \varnothing$ when Λ is c-connected, it follows that $\omega \in \Omega^+_\Lambda$. It is also easy to verify that $\Gamma'(\omega) = \Gamma'$. This shows that the collection is admissible.

When Λ is not c-connected, this implication is no longer true. For example, consider the set $\Lambda = \mathrm{B}(2n) \setminus \mathrm{B}(n)$. Because of the + boundary condition outside $\mathrm{B}(2n)$ *and inside* $\mathrm{B}(n)$, in any configuration $\omega \in \Omega^+_\Lambda$, the number of contours $\gamma \in \Gamma'(\omega)$ such that $\mathrm{B}(n) \subset \operatorname{Int}\gamma$ *has to be even.* Observe that the latter is a *global* constraint on the family of contours.

Exercise 5.10: See Exercise 3.20.

Exercise 5.11: As was done earlier, one can write for example

$$\sum_{\substack{X\sim A:\\ \overline{X}\subset\Lambda}} \Psi^A_\beta(X) = \sum_{X\sim A} \Psi^A_\beta(X) - \sum_{\substack{X\sim A:\\ \overline{X}\not\subset\Lambda}} \Psi^A_\beta(X).$$

The clusters that satisfy at the same time $X \sim A$ and $\overline{X} \not\subset \Lambda$ have a support of size at least $d(A, \Lambda^c)$. As before, one can show that their contribution vanishes when $\Lambda \uparrow \mathbb{Z}^d$.

Solutions for Chapter 6

Exercise 6.2: Clearly, the family of subsets $\Lambda \Subset \mathbb{Z}^d$ is at most countable. Since each $\mathscr{C}(\Lambda)$ is finite, and since a countable union of finite sets is countable, $\mathscr{C}_S$ is countable. To show that $\mathscr{C}_S$ is an algebra, observe that, whenever $A \in \mathscr{C}_S$, there exists some $\Lambda \Subset S$ and some $B \in \Omega_\Lambda$ such that $A = \Pi_\Lambda^{-1}(B)$. But, since $A^c = \Pi_\Lambda^{-1}(B^c)$, we also have $A^c \in \mathscr{C}_S$. Moreover, if $A, A' \in \mathscr{C}_S$, of the form $A = \Pi_\Lambda^{-1}(B)$, $A' = \Pi_{\Lambda'}^{-1}(B')$, then one can find some $\Lambda'' \Subset S$ containing Λ and Λ' (for example $\Lambda'' = \Lambda \cup \Lambda'$), use the hint to express $A = \Pi_{\Lambda''}^{-1}(B_1)$, $A' = \Pi_{\Lambda''}^{-1}(B_2)$, and write $A \cup A' = \Pi_{\Lambda''}^{-1}(B_1 \cup B_2)$. This implies $A \cup A' \in \mathscr{C}_S$.

Exercise 6.4: For example, consider $\Delta = \{0,1\} \times \{0\}$ and $\Lambda = \{0,1\}^2$. It then immediately follows from the high-temperature representation that

$$\langle \sigma_{(0,0)}\sigma_{(1,0)} \rangle_\Delta^\varnothing = \tanh\beta\,,$$

while

$$\langle \sigma_{(0,0)}\sigma_{(1,0)} \rangle_\Lambda^\varnothing = (\tanh\beta + \tanh^3\beta)/(1 + \tanh^4\beta)\,.$$

Since these two expressions do not coincide when $\beta > 0$, it follows that $\mu_\Lambda^\varnothing \circ (\Pi_\Delta^\Lambda)^{-1} \neq \mu_\Delta^\varnothing$.

Exercise 6.5: By definition,

$$\pi_\Lambda \pi_\Delta(A \,|\, \eta) = \sum_{\omega_\Lambda \in \Omega_\Lambda} \pi_\Lambda(\omega_\Lambda \,|\, \eta)\, \pi_\Delta(A \,|\, \omega_\Lambda \eta_{\Lambda^c})\,,$$

which only depends on η_{Λ^c}. This also immediately implies that $\pi_\Lambda\pi_\Delta$ is proper.

Exercise 6.6: If $f = \mathbf{1}_A$,

$$\mu\pi_\Lambda(\mathbf{1}_A) = \int \pi_\Lambda(A \,|\, \omega)\mu(\mathrm{d}\omega) = \int \pi_\Lambda \mathbf{1}_A(\omega)\mu(\mathrm{d}\omega) = \mu(\pi_\Lambda \mathbf{1}_A)\,.$$

For the general case, just approximate f by a sequence of simple functions of the form $\sum_i a_i \mathbf{1}_{A_i}$.

Exercise 6.7: The proof of the first claim is left to the reader. For the second, observe that, for any $A \in \mathscr{F}$,

$$\rho\pi_\Lambda(A) = \int \pi_\Lambda(A \,|\, \omega)\rho(\mathrm{d}\omega) = \int \mathbf{1}_A(\tau_\Lambda \omega_{\Lambda^c})\rho^\Lambda(\mathrm{d}\tau_\Lambda)\rho(\mathrm{d}\omega_{\Lambda^c}) = \rho(A)\,,$$

so that $\rho \in \mathscr{G}(\pi)$. To prove uniqueness, let $\mu \in \mathscr{G}(\pi)$ and consider an arbitrary cylinder $C = \Pi_\Lambda^{-1}(E)$ with base Λ. Then, one must have

$$\mu(C) = \mu\pi_\Lambda(C) = \int \pi_\Lambda(\Pi_\Lambda^{-1}(E) \,|\, \omega)\mu(\mathrm{d}\omega) = \int \rho^\Lambda(E)\mu(\mathrm{d}\omega) = \rho^\Lambda(E) = \rho(C)\,,$$

and therefore μ must coincide with ρ on all cylinders, which implies that $\mu = \rho$.

Exercise 6.8: To show absolute summability, it suffices to prove that

$$\sum_{i\in\mathbb{Z}^d\setminus\{0\}} J_{0i} = \sum_{r\geq 1} |\partial^{\text{in}}\mathrm{B}(r)| r^{-\alpha}$$

is bounded. Since $|\partial^{\text{in}}\mathrm{B}(r)|$ is of order r^{d-1}, the potential is absolutely summable if and only if $\alpha > d$.

Exercise 6.9: Clearly, $d(\omega,\eta) \geq 0$ with equality if and only if $\omega = \eta$. Since $\mathbf{1}_{\{\omega_i\neq\eta_i\}} \leq \mathbf{1}_{\{\omega_i\neq\tau_i\}} + \mathbf{1}_{\{\tau_i\neq\eta_i\}}$ for all $i \in \mathbb{Z}^d$, we have $d(\omega,\eta) \leq d(\omega,\tau) + d(\tau,\eta)$, so $d(\cdot,\cdot)$ is a distance.

Notice that if $\omega_{\mathrm{B}(r)} = \eta_{\mathrm{B}(r)}$, then $d(\omega,\eta) \leq 2d\sum_{k\geq r} k^{d-1}2^{-k} \stackrel{\text{def}}{=} \epsilon(r)$, with $\epsilon(r) \to 0$ as $r \to \infty$.

Suppose that $\omega^{(n)} \to \omega^*$. In this case, for any $r \geq 1$, there exists n_0 such that $\omega^{(n)}_{\mathrm{B}(r)} = \omega^*_{\mathrm{B}(r)}$ for all $n \geq n_0$. This implies that $d(\omega^{(n)},\omega^*) \to 0$ as $n \to \infty$.

Assume now that $d(\omega^{(n)},\omega^*) \to 0$. In that case, for any $k \geq 1$, one can find n_1 such that $d(\omega^{(n)},\omega^*) < 2^{-k}$ for all $n \geq n_1$. But this implies that $\mathbf{1}_{\{\omega_i^{(n)}\neq\omega_i^*\}} = 0$ each time $\|i\|_\infty \leq k$. This implies that $\omega^{(n)}_{\mathrm{B}(k)} = \omega^*_{\mathrm{B}(k)}$ for all $n \geq n_1$. Therefore, $\omega^{(n)} \to \omega^*$.

Exercise 6.10: Let $C = \Pi_\Lambda^{-1}(A)$ be a cylinder. If $\omega \in C$, then any configuration ω' that coincides with ω on Λ is also in C, which implies that C is open. Now let $G \subset \Omega$ be open. For each $\omega \in G$, one can find a cylinder C_ω such that $G \supset C_\omega \ni \omega$. Therefore, $G = \bigcup_{\omega\in G} C_\omega$. But, since $\mathscr{C}$ is countable (Exercise 6.2), that union is countable. This shows that $G \in \mathscr{F}$.

Exercise 6.11: Assume $f\colon \Omega \to \mathbb{R}$ is continuous but not uniformly continuous. There exists some $\epsilon > 0$, a sequence $(\delta_n)_{n\geq 1}$ decreasing to 0 and two sequences $(\omega^{(n)})_{n\geq 1}, (\eta^{(n)})_{n\geq 1} \subset \Omega$ such that $d(\omega^{(n)},\eta^{(n)}) \to 0$ and $|f(\omega^{(n)}) - f(\eta^{(n)})| \geq \epsilon$ for all n. By Proposition 6.20, there exists a subsequence $(\omega^{(n_k)})_{k\geq 1}$ and some ω_* such that $\omega^{(n_k)} \to \omega_*$. This implies also $d(\eta^{(n_k)},\omega_*) \leq d(\eta^{(n_k)},\omega^{(n_k)}) + d(\omega^{(n_k)},\omega_*) \to 0$. But, since $\epsilon \leq |f(\omega^{(n_k)}) - f(\omega_*)| + |f(\eta^{(n_k)}) - f(\omega_*)|$, at least one of the sequences $(|f(\omega^{(n_k)}) - f(\omega_*)|)_{k\geq 1}, (|f(\eta^{(n_k)}) - f(\omega_*)|)_{k\geq 1}$ cannot converge to zero. This implies that f is not continuous at ω_*, a contradiction. The other two facts are proved in a similar way.

Exercise 6.12: 1⇒2 is immediate since local functions can be expressed as finite linear combinations of indicators of cylinders.
2⇒3: Let $f \in C(\Omega)$. Fix $\epsilon > 0$, and let g be a local function such that $\|g - f\|_\infty \leq \epsilon$. Then $|\mu_n(f) - \mu(f)| \leq |\mu_n(g) - \mu(g)| + 2\epsilon$, and thus $\limsup_n |\mu_n(f) - \mu(f)| \leq 2\epsilon$. This implies that $\mu_n(f) \to \mu(f)$.
3⇒1 is immediate, since, for each $C \in \mathscr{C}, f = \mathbf{1}_C$ is continuous.
1⇒4: Let $m_n(k) \stackrel{\text{def}}{=} \max_{C\in\mathscr{C}(\mathrm{B}(k))} |\mu_n(C) - \mu(C)|$. Notice that $m_n(k) \leq 1$. Fix $\epsilon > 0$. Let $k_0 \stackrel{\text{def}}{=} \epsilon^{-1}$, Clearly, as $n \to \infty$,

$$\max_{1\leq k\leq k_0} m_n(k) \to 0\,.$$

On the other hand, if $k > k_0$, then $\frac{m_n(k)}{k} \leq \frac{1}{k} < \epsilon$. Therefore,

$$\limsup_{n\to\infty} \rho(\mu_n, \mu) = \limsup_{n\to\infty} \sup_{k\geq 1} \frac{m_n(k)}{k} \leq \epsilon .$$

4⇒1: Let $C \in \mathscr{C}$ and fix some $\epsilon > 0$. Let k be large enough so that $C \in \mathscr{C}(\mathrm{B}(k))$, and let n_0 be such that $\rho(\mu_n, \mu) \leq \frac{\epsilon}{k}$ for all $n \geq n_0$. For those values of n, we also have

$$|\mu_n(C) - \mu(C)| \leq \max_{C' \in \mathscr{C}(\mathrm{B}(k))} |\mu_n(C') - \mu(C')| \leq \epsilon .$$

Exercise 6.13: Writing $\pi_\Lambda f(\omega) = \sum_{\tau_\Lambda} f(\tau_\Lambda \omega_{\Lambda^c}) \pi_\Lambda(\tau_\Lambda \mid \omega)$ makes the statement obvious.

Exercise 6.14: The construction of $\mu^{\varnothing}_{\beta,h}$, using Exercise 3.16 and Theorem 6.5, is straightforward. We check that $\mu^{\varnothing}_{\beta,h} \in \mathscr{G}(\beta, h)$. Let f be some local function and take $\Delta \Subset \mathbb{Z}^d$ sufficiently large to contain the support of f. Lemma 6.7 (whose proof extends verbatim to the case of free boundary conditions) implies that, for any $\Lambda \Subset \mathbb{Z}^d$ containing Δ, $\langle f \rangle^{\varnothing}_{\Lambda;\beta,h} = \langle \langle f \rangle^{\cdot}_{\Delta;\beta,h} \rangle^{\varnothing}_{\Lambda;\beta,h}$. Again, since $\omega \mapsto \langle f \rangle^{\omega}_{\Delta;\beta,h}$ is local, one can let $\Lambda \uparrow \mathbb{Z}^d$, and obtain $\langle f \rangle^{\varnothing}_{\beta,h} = \langle \langle f \rangle^{\cdot}_{\Delta;\beta,h} \rangle^{\varnothing}_{\beta,h}$, from which the claim follows.

Exercise 6.15: Assume there exists $\mu \in \mathscr{G}(\pi)$. Notice that

$$\mu(N^+ = 0) = \mu(\{\eta^-\}) = \mu\pi_\Lambda(\{\eta^-\}) = \int \pi_\Lambda(\{\eta^-\} \mid \omega)\mu(\mathrm{d}\omega) = 0 .$$

Then, $\mu(N^+ = 1) = \sum_{i \in \mathbb{Z}^d} \mu(\{\eta^{-,i}\})$. However, for all $\Lambda \Subset \mathbb{Z}^d$ containing i,

$$\mu(\{\eta^{-,i}\}) = \mu\pi_\Lambda(\{\eta^{-,i}\}) \leq \frac{1}{|\Lambda|} ,$$

so that $\mu(\{\eta^{-,i}\}) = 0$. We conclude that $\mu(N^+ = 1) = 0$. Finally, $\mu(N^+ \geq 2) \leq \sum_{i \neq j} \mu(\{\omega_i = \omega_j = +1\}) = 0$, since $\mu(\{\omega_i = \omega_j = +1\}) = \mu\pi_{\{i,j\}}(\{\omega_i = \omega_j = +1\}) = 0$ for all $i \neq j$. All this implies that $\mu(\Omega) = 0$, which contradicts the assumption that μ is a probability measure.

Exercise 6.16: For example,

$$f(\omega) = \limsup_{n\to\infty} \frac{1}{|\mathrm{B}(n)|} \sum_{i \in \mathrm{B}(n)} \omega_i ,$$

has $\Delta(f) = 0$, but it is not continuous (see Exercise 6.65). In dimension $d = 2$, take $\epsilon > 0$ and consider, for example,

$$g(\omega) = \sum_{k \geq 1} \frac{1}{k^{1+\epsilon}} \Bigl(\max_{j \in \mathrm{B}(k)} \omega_j - \min_{j \in \mathrm{B}(k)} \omega_j \Bigr) .$$

Then $g \in C(\Omega)$, but $\Delta(g) = \infty$.

Exercise 6.17: By the FKG inequality, for any $\omega \in \Omega$,

$$1 \geq \mu^{\omega}_{\{i\};\beta,h}(\sigma_i = 1) \geq \mu^{-}_{\{i\};\beta,h}(\sigma_i = 1) = \{1 + e^{-2h+4d\beta}\}^{-1}.$$

Therefore,

$$\sum_{\omega_i=\pm 1} \left|\pi_i(\omega_i \,|\, \omega) - \pi_i(\omega_i \,|\, \omega')\right| \leq \frac{2}{1 + e^{2h-4d\beta}}.$$

Since the expression on the left-hand side is actually equal to 0 when $\omega_j = \omega'_j$ for all $j \sim i$, we obtain

$$c(\pi) \leq \frac{4d}{1 + e^{2h-4d\beta}},$$

which is indeed smaller than 1 as soon as $h > 2d\beta + \frac{1}{2}\log(4d - 1)$.

Exercise 6.19: Clearly, $c_{ij}(\pi) = 0$ whenever $j \not\sim i$. Let $j \sim i$ and consider two configurations ω, η such that $\omega_k = \eta_k$ for all $k \neq j$. When $s \in \{0, \dots, q-1\} \setminus \{\omega_j, \eta_j\}$, $\pi_i(\sigma_i = s \,|\, \eta) = \pi_i(\sigma_i = s \,|\, \omega)$. Let us therefore assume that $\omega_j = s \neq \eta_j$. In this case,

$$\pi_i(\sigma_i = s \,|\, \eta) - \pi_i(\sigma_i = s \,|\, \omega) = \frac{e^{-\beta\#\{k\sim i:\,\eta_k=s\}}}{\mathbf{Z}^{\eta}_{\{i\}}}\left\{1 - \frac{\mathbf{Z}^{\eta}_{\{i\}}}{\mathbf{Z}^{\omega}_{\{i\}}}e^{-\beta}\right\}.$$

Now, observe that

$$\frac{\mathbf{Z}^{\eta}_{\{i\}}}{\mathbf{Z}^{\omega}_{\{i\}}}e^{-\beta} = \left\langle e^{-\beta(\delta_{\sigma_i,\eta_j} - \delta_{\sigma_i,\omega_j} + 1)}\right\rangle^{\omega}_{\{i\}} \in [e^{-2\beta}, 1].$$

Therefore, $\left|\pi_i(\sigma_i = s \,|\, \eta) - \pi_i(\sigma_i = s \,|\, \omega)\right| \leq 1/\mathbf{Z}^{\eta}_{\{i\}} \leq 1/(q - 2d)$. This yields $c_{ij}(\pi) \leq 2/(q-2d)$ and thus $c(\pi) \leq 4d/(q-2d)$, which is indeed smaller than 1 as soon as $q > 6d$.

Exercise 6.20: We have seen in Exercise 6.8 that $\alpha > d$ is necessary for the potential to be absolutely summable. Then,

$$\begin{aligned} b = \sup_{i\in\mathbb{Z}^d} \sum_{B\ni i}(|B| - 1)\|\Phi_B\|_\infty &= \sum_{k\geq 1} \frac{1}{k^\alpha}\#\{j \in \mathbb{Z}^d \,:\, j \neq 0,\ \|j\|_\infty = k\} \\ &\leq 2d\sum_{k\geq 1}\frac{1}{k^{1+(\alpha-d)}} \stackrel{\text{def}}{=} b_0(\alpha, d). \end{aligned}$$

For all $\alpha > d$, we have uniqueness as soon as $\beta < \beta_0 \stackrel{\text{def}}{=} \frac{1}{2b_0}$. Observe that $\beta_0 \downarrow 0$ when $\alpha \downarrow d$.

Exercise 6.21: Using the invariance of π_Λ in the second equality,

$$\begin{aligned} (\theta_j\mu)\pi_\Lambda(A) &= \int \pi_\Lambda(A \,|\, \theta_j\omega)\mu(\mathrm{d}\omega) = \int \pi_{\theta_j^{-1}\Lambda}(\theta_j^{-1}A \,|\, \omega)\mu(\mathrm{d}\omega) \\ &= \mu\pi_{\theta_j^{-1}\Lambda}(\theta_j^{-1}A) = \mu(\theta_j^{-1}A) = \theta_j\mu(A). \end{aligned}$$

Exercise 6.22: By Theorem 6.24, we can consider a subsequence along which μ_n converges: $\mu_{n_k} \Rightarrow \mu_*$. To see that μ_* is translation invariant, $\theta_i\mu_* = \mu_*$ for all $i \in \mathbb{Z}^d$, it suffices to observe that, for any local function f,

$$\Big|\sum_{j\in \mathrm{B}(n_k)} \theta_{j+i}\mu(f) - \sum_{j\in \mathrm{B}(n_k)} \theta_j\mu(f)\Big| \le C\|f\|_\infty \|i\|_\infty^d \, |\partial^{\mathrm{ex}}\mathrm{B}(n_k)| \,.$$

Then, $\mu \in \mathscr{G}(\pi)$ and Exercise 6.21 imply that $\mu_{n_k} \in \mathscr{G}(\pi)$ for all k. Since $\mathscr{G}(\pi)$ is closed (Theorem 6.27), this implies that $\mu_* \in \mathscr{G}(\pi)$.

Exercise 6.23: Let $\omega \in \Omega$. Since f is nonconstant, there exists ω' such that $f(\omega) \neq f(\omega')$. Let $\omega^{(n)} = \omega_{\mathrm{B}(n)}\omega'_{\mathrm{B}(n)^c}$. Then $\omega^{(n)} \to \omega$. However, $f(\omega^{(n)}) = f(\omega')$ for all n and therefore $f(\omega^{(n)}) \not\to f(\omega)$.

Exercise 6.24: Let g be $\mathscr{F}_{\Lambda^c}$-measurable. We first assume that g is a finite linear combination $\sum_j \alpha_j \mathbf{1}_{A_j}$, with $A_j \in \mathscr{F}_{\Lambda^c}$. On the one hand,

$$(g\nu)\pi_\Lambda(A) = \int \pi_\Lambda(A\,|\,\omega) g(\omega)\nu(d\omega) = \sum_j \alpha_j \int_{A_j} \pi_\Lambda(A\,|\,\omega)\nu(d\omega)\,.$$

On the other hand,

$$g(\nu\pi_\Lambda)(A) = \int_A g(\omega')\nu\pi_\Lambda(d\omega') = \sum_j \alpha_j \nu\pi_\Lambda(A\cap A_j) = \sum_j \alpha_j \int \pi_\Lambda(A\cap A_j\,|\,\omega)\nu(d\omega)\,.$$

By Lemma 6.13, we have $\pi_\Lambda(A \cap A_j\,|\,\omega) = \pi_\Lambda(A\,|\,\omega)\mathbf{1}_{A_j}(\omega)$. This implies that $(g\nu)\pi_\Lambda = g(\nu\pi_\Lambda)$. In the general case, it suffices to consider a sequence of approximations g_n (each being a finite linear combination of the above type) with $\|g_n - g\|_\infty \to 0$, and use twice dominated convergence to compute

$$(g\nu)\pi_\Lambda(A) = \lim_{n\to\infty} (g_n\nu)\pi_\Lambda(A) = \lim_{n\to\infty} g_n(\nu\pi_\Lambda)(A) = g(\nu\pi_\Lambda)(A)\,.$$

The reader can find counter-examples that show that (6.63) does not hold in general when g is not $\mathscr{F}_{\Lambda^c}$-measurable.

Exercise 6.25: Since $\mathbf{1}_A = (1+\sigma_0)/2$ and $\mathbf{1}_{B_i} = (1+\sigma_i)/2$, $\mu(A\cap B_i) - \mu(A)\mu(B_i) = \frac{1}{4}\big(\mu(\sigma_0\sigma_i) - \mu(\sigma_0)\mu(\sigma_i)\big)$. By symmetry, $\mu^+_{\beta,0}(\sigma_0\sigma_i) = \mu^-_{\beta,0}(\sigma_0\sigma_i)$ and $\mu(\sigma_0) = (2\lambda-1)\mu^+_{\beta,0}(\sigma_0)$, $\mu(\sigma_i) = (2\lambda-1)\mu^+_{\beta,0}(\sigma_i)$. By the FKG inequality, $\mu^+_{\beta,0}(\sigma_0\sigma_i) \ge \mu^+_{\beta,0}(\sigma_0)\mu^+_{\beta,0}(\sigma_i)$. We therefore conclude that $\mu(\sigma_0\sigma_i) - \mu(\sigma_0)\mu(\sigma_i) \ge \big(1-(2\lambda-1)^2\big)\big(\mu^+_{\beta,0}(\sigma_0)\big)^2$, which is positive for all $\beta > \beta_c(2)$ and all $\lambda \in (0,1)$.

Exercise 6.26: Extremality of $\mu^+_{\beta,h}$ implies that, for any $\epsilon > 0$, there exists r such that $0 \le \langle\sigma_i;\sigma_j\rangle^+_{\beta,h} \le \epsilon$ for all $j \notin i + \mathrm{B}(r)$. Therefore,

$$\mathrm{Var}_{\mu^+_{\beta,h}}(m_{\mathrm{B}(n)}) = |\mathrm{B}(n)|^{-2} \sum_{i,j\in\mathrm{B}(n)} \langle\sigma_i;\sigma_j\rangle^+_{\beta,h}$$
$$\leq |\mathrm{B}(n)|^{-2} \sum_{i\in\mathrm{B}(n)} \Big\{ \underbrace{\sum_{j\in i+\mathrm{B}(r)} \langle\sigma_i;\sigma_j\rangle^+_{\beta,h}}_{\leq 1} + \underbrace{\sum_{\substack{j\in\mathrm{B}(n)\\ j\notin i+\mathrm{B}(r)}} \langle\sigma_i;\sigma_j\rangle^+_{\beta,h}}_{\leq\epsilon} \Big\} \leq \frac{|\mathrm{B}(r)|}{|\mathrm{B}(n)|} + \epsilon\,.$$

Letting $n \to \infty$ and then $\epsilon \to 0$ shows that $\lim_{n\to\infty} \mathrm{Var}_{\mu^+_{\beta,h}}(m_{\mathrm{B}(n)}) = 0$. The conclusion follows from Chebyshev's inequality (B.18).

Exercise 6.27: On the one hand, if ν is trivial on $\mathscr{T}_\infty$, then

$$\int_A \nu(B)\nu(\mathrm{d}\omega) = \nu(B)\nu(A) = \nu(A\cap B)\,,$$

for all $B \in \mathscr{F}$ and all $A \in \mathscr{T}_\infty$, since $\nu(A)$ is either 1 or 0. This shows that $\nu(B) = \nu(B\,|\,\mathscr{T}_\infty)$ ν-almost surely. On the other hand, if the latter condition holds, then, for any $A \in \mathscr{T}_\infty$,

$$\nu(A) = \int_A \mathbf{1}_A \mathrm{d}\nu = \int_A \nu(A\,|\,\mathscr{T}_\infty)\mathrm{d}\nu = \int_A \nu(A)\mathrm{d}\nu = \nu(A)^2\,,$$

which implies that $\nu(A) \in \{0,1\}$.

Exercise 6.28: A simple computation yields

$$\mathscr{W}_{V_n}(\mu_{V_n}) = |V_n|\Big\{d\beta m^2 + hm - \frac{1+m}{2}\log\frac{1+m}{2} - \frac{1-m}{2}\log\frac{1-m}{2}\Big\}\,,$$

where we have introduced $m \stackrel{\text{def}}{=} \langle\sigma_i\rangle_{\rho_i}$. It is now a matter of straightforward calculus to show that the unique maximum is attained when m satisfies $m = \tanh(2d\beta m + h)$.

Exercise 6.29: By (6.93), we have, for any $n \geq 0$,

$$\limsup_{k\to\infty} s(\mu_k) = \limsup_{k\to\infty} \inf_{\Lambda\in\mathscr{R}} \frac{S_\Lambda(\mu_k)}{|\Lambda|} \leq \limsup_{k\to\infty} \frac{S_{\mathrm{B}(n)}(\mu_k)}{|\mathrm{B}(n)|} = \frac{S_{\mathrm{B}(n)}(\mu)}{|\mathrm{B}(n)|}\,.$$

Letting $n \to \infty$ yields the desired result.

Exercise 6.32: We show that $\mu\pi^\Phi_\Lambda = \mu$ for all $\Lambda \Subset \mathbb{Z}^d$. For each local function f, we write

$$\mu\pi^\Phi_\Lambda(f) = \{\mu\pi^\Phi_\Lambda(f) - \mu^k\pi^\Phi_\Lambda(f)\} + \{\mu^k\pi^\Phi_\Lambda(f) - \mu^k\pi^{\Phi^k}_\Lambda(f)\} + \mu^k\pi^{\Phi^k}_\Lambda(f)\,.$$

Since Φ has finite range, $\omega \mapsto \pi^\Phi_\Lambda(f|\omega)$ is local. Therefore, $\mu^k \Rightarrow \mu$ implies that $\mu^k\pi^\Phi_\Lambda(f) \to \mu\pi^\Phi_\Lambda(f)$ as $k \to \infty$. For the second term, proceeding as in (6.35) gives

$$\big|\mu^k\pi^\Phi_\Lambda(f) - \mu^k\pi^{\Phi^k}_\Lambda(f)\big| \leq \int \big|\pi^\Phi_\Lambda(f\,|\,\omega) - \pi^{\Phi^k}_\Lambda(f\,|\,\omega)\big|\mu^k(\mathrm{d}\omega)$$
$$\leq 2|\Lambda|\|f\|_\infty \sum_{B\ni 0} \|\Phi_B - \Phi^k_B\|_\infty\,,$$

which tends to zero when $k \to \infty$. Finally, since $\mu^k \in \mathscr{G}(\Phi^k)$, $\mu^k \pi_\Lambda^{\Phi^k}(f) = \mu^k(f)$, and $\mu^k(f) \to \mu(f)$.

Exercise 6.33: Assume that there is a unique Gibbs measure at (β_0, h_0). Observe that, setting $g = \frac{1}{2d}\sum_{i\sim 0}\sigma_0\sigma_i$ and $\lambda = \beta - \beta_0$, we have

$$\psi(\lambda) \stackrel{\text{def}}{=} \lim_{\Lambda \Uparrow \mathbb{Z}^d} \frac{1}{|\Lambda(g)|} \log\Bigl\langle \exp\Bigl\{\lambda \sum_{j\in\Lambda(g)} g\circ\theta_j\Bigr\}\Bigr\rangle^{+}_{\Lambda;\beta_0,h_0} = \psi^{\text{Ising}}(\beta, h_0) - \psi^{\text{Ising}}(\beta_0, h_0).$$

We deduce that

$$\frac{\partial\psi}{\partial\lambda^-}\Big|_{\lambda=0} = \frac{\partial\psi^{\text{Ising}}(\beta,h_0)}{\partial\beta^-}\Big|_{\beta=\beta_0}, \qquad \frac{\partial\psi}{\partial\lambda^+}\Big|_{\lambda=0} = \frac{\partial\psi^{\text{Ising}}(\beta,h_0)}{\partial\beta^+}\Big|_{\beta=\beta_0}.$$

Therefore, if $\psi^{\text{Ising}}(\beta, h_0)$ was not differentiable at β_0, then the same would be true of ψ and Proposition 6.91 would imply the existence of multiple Gibbs measures at (β_0, h_0), which would contradict our assumption.

Solutions for Chapter 7

Exercise 7.2: It suffices to show that η enjoys the following property. For each $k \geq 1$, η is a minimizer (possibly not unique) of $\mathscr{H}_{\mathrm{B}(k);\Phi^0}$ among all configurations of $\Omega^\eta_{\mathrm{B}(k)}$. To prove this, observe that the configuration η possesses a unique Peierls contour γ and check that the length of $\gamma \cap \{x \in \mathbb{R}^2 : \|x\|_\infty \leq k\}$ cannot be decreased by flipping spins in $\mathrm{B}(k-1)$.

Exercise 7.3: The following construction relies on a diagonalization argument, as already done earlier in the book. Fix some arbitrary configuration $\eta \in \Omega$. For each $n \geq 0$, let $\omega^{(n)}$ be a configuration coinciding with η outside $\mathrm{B}(n)$ and minimizing $\mathscr{H}_{\mathrm{B}(n);\Phi}$. Order the vertices of $\mathbb{Z}^d$: $i_1, i_2, \dots$ Let $(n_{1,k})_{k\geq 1}$ be a sequence such that $\omega_{i_1}^{(n_{1,k})}$ converges as $k \to \infty$. Then let $(n_{2,k})_{k\geq 2}$ be a subsequence of $(n_{1,k})_{k\geq 1}$ such that $\omega_{i_2}^{(n_{2,k})}$ converges. We proceed in the same way for all vertices of $\mathbb{Z}^d$: for each $m \geq 1$, the sequences $(\omega_{i_m}^{(n_{m,k})})_{k\geq 1}$ converge as $k \to \infty$. We claim that the configuration ω defined by

$$\omega_i \stackrel{\text{def}}{=} \lim_{m\to\infty} \omega_i^{(n_{m,m})}, \qquad \forall i \in \mathbb{Z}^d,$$

is a ground state. Indeed, let $\omega' \stackrel{\infty}{=} \omega$ and choose n so large that ω and ω' coincide outside $\mathrm{B}(n)$. Let N be so large that ω coincides with $\omega^{(N)}$ on $\mathrm{B}(n + r(\Phi))$. Then, by our choice of $\omega^{(N)}$,

$$\begin{aligned}\mathscr{H}_\Phi(\omega' \mid \omega) &= \sum_{B\cap \mathrm{B}(n)\neq\varnothing} \{\Phi_B(\omega') - \Phi_B(\omega)\} \\ &= \sum_{B\cap \mathrm{B}(n)\neq\varnothing} \{\Phi_B(\omega'_{\mathrm{B}(N)}\eta_{\mathrm{B}(N)^c}) - \Phi_B(\omega^{(N)})\} = \mathscr{H}_\Phi(\omega'_{\mathrm{B}(N)}\eta_{\mathrm{B}(N)^c} \mid \omega^{(N)}) \geq 0.\end{aligned}$$

Exercise 7.4: 1. Consider the pressure constructed using a boundary condition $\eta \in \mathrm{g}^{\mathrm{per}}(\Phi)$. On the one hand, $\mathbf{Z}^{\eta}_{\Phi}(\Lambda) \geq e^{-\beta\mathscr{H}_{\Lambda;\Phi}(\eta)}$, which gives $\psi(\Phi) \geq -\mathrm{e}_{\Phi}(\eta)$. On the other hand, for any $\omega \in \Omega^{\eta}_{\Lambda}$,

$$\mathscr{H}_{\Lambda;\Phi}(\omega) = \mathscr{H}_{\Lambda;\Phi}(\eta) + \{\mathscr{H}_{\Lambda;\Phi}(\omega) - \mathscr{H}_{\Lambda;\Phi}(\eta)\} = \mathscr{H}_{\Lambda;\Phi}(\eta) + \mathscr{H}_{\Phi}(\omega \,|\, \eta) \geq \mathscr{H}_{\Lambda;\Phi}(\eta).$$

This gives

$$\mathbf{Z}^{\eta}_{\Phi}(\Lambda) \leq e^{-\beta\mathscr{H}_{\Lambda;\Phi}(\eta)}|\Omega^{\eta}_{\Lambda}|.$$

Since $|\Omega^{\eta}_{\Lambda}| = |\Omega_0|^{|\Lambda|}$, this yields $\psi(\Phi) \leq -\mathrm{e}_{\Phi}(\eta) + \beta^{-1}\log|\Omega_0|$.

2. Observe that a configuration $\omega \in \Omega^{\eta}_{\Lambda}$ is completely characterized by its restriction to $\mathscr{B}(\omega)$. Therefore,

$$\begin{aligned}
\mathbf{Z}^{\eta}_{\Phi}(\Lambda) &= \sum_{\omega\in\Omega^{\eta}_{\Lambda}} e^{-\beta\mathscr{H}_{\Lambda;\Phi}(\omega)} \leq e^{-\beta\mathscr{H}_{\Lambda;\Phi}(\eta)} \sum_{\omega\in\Omega^{\eta}_{\Lambda}} e^{-\beta\rho|\mathscr{B}(\omega)|} \\
&= e^{-\beta\mathscr{H}_{\Lambda;\Phi}(\eta)} \sum_{B\subset\Lambda} \sum_{\substack{\omega\in\Omega^{\eta}_{\Lambda}:\\ \mathscr{B}(\omega)\cap\Lambda=B}} e^{-\beta\rho|B|} \\
&\leq e^{-\beta\mathscr{H}_{\Lambda;\Phi}(\eta)} \sum_{n=0}^{|\Lambda|} \binom{|\Lambda|}{n} \left(|\Omega_0|e^{-\beta\rho}\right)^n = e^{-\beta\mathscr{H}_{\Lambda;\Phi}(\eta)}\left(1 + |\Omega_0|e^{-\beta\rho}\right)^{|\Lambda|}.
\end{aligned}$$

This gives $\psi(\Phi) \leq -\underline{\mathrm{e}}_{\Phi} + \beta^{-1}\log\left(1 + |\Omega_0|e^{-\beta\rho}\right) \leq -\underline{\mathrm{e}}_{\Phi} + \beta^{-1}|\Omega_0|e^{-\beta\rho}$.

Exercise 7.5: It is convenient to work with the following equivalent potential:

$$\tilde{\Phi}_B(\omega) \stackrel{\text{def}}{=} \begin{cases} \omega_i\omega_j - \frac{h}{2d}(\omega_i + \omega_j) & \text{if } B = \{i,j\},\ i \sim j, \\ 0 & \text{otherwise.} \end{cases}$$

We are going to determine $g_m(\Phi)$. For any pair $i \sim j$,

$$\phi_{\{i,j\}} = \min_{\omega} \tilde{\Phi}_{\{i,j\}}(\omega) = \begin{cases} 1 - (h/d) & \text{if } h \geq +2d, \\ -1 & \text{if } |h| \leq 2d, \\ 1 + (h/d) & \text{if } h \leq -2d. \end{cases}$$

(The three cases correspond to $\omega_i = \omega_j = 1$, $\omega_i = \omega_j = -1$ and $\omega_i \neq \omega_j$ respectively.) The cases $h = \pm 2d$ are discussed below; for all other cases:

$$g_m(\Phi) = \begin{cases} \{\eta^+\} & \text{if } h > +2d, \\ \{\eta^{\pm}, \eta^{\mp}\} & \text{if } |h| < 2d, \\ \{\eta^-\} & \text{if } h < -2d, \end{cases}$$

where $\eta^{\pm}, \eta^{\mp}$ are the two chessboard configurations defined by $\eta^{\pm}_i \stackrel{\text{def}}{=} (-1)^{\sum_{k=1}^{d} i_k}$ and $\eta^{\mp} \stackrel{\text{def}}{=} -\eta^{\pm}$. When $h = \pm 2d$, $g_m(\Phi)$ contains infinitely many ground states. For example, if $h = +2d$,

$$g_m(\Phi) = \left\{\omega \in \Omega : \nexists i, j \in \mathbb{Z}^d, i \sim j, \text{ such that } \omega_i = \omega_j = -1\right\}.$$

Exercise 7.7: For all $\{i,j\} \in \mathscr{T}$, $\omega_i\omega_j$ is minimal if and only if $\omega_i \neq \omega_j$, and this cannot be realized simultaneously for all three pairs of spins living at the vertices of any given triangle. This implies that Φ is not an m-potential.

For the triangle $T = \{(0,0),(0,1),(1,1)\}$, let

$$\widetilde{\Phi}_T(\omega) = \omega_{(0,0)}\omega_{(0,1)} + \omega_{(0,1)}\omega_{(1,1)} + \omega_{(0,0)}\omega_{(1,1)}\,.$$

Define $\widetilde{\Phi}$ similarly on all translates of T. Then $\widetilde{\Phi}$ is clearly equivalent to Φ, and can be easily seen to be an m-potential; each $\widetilde{\Phi}_T$ being minimized if the configuration on T contains at least one spin of each sign. This allows us to construct infinitely many periodic ground states for $\widetilde{\Phi}$. For example, any configuration obtained by alternating the spin values along every column necessarily belongs to $g_m(\widetilde{\Phi})$. Since this yields two possible configurations for each column, one can alternate them in order to construct configurations on $\mathbb{Z}^2$ of arbitrarily large period.

Exercise 7.8: Clearly, the constant configurations η^+ and η^- are periodic ground states, and their energy density equals $\underline{e}_\Phi = \alpha$. Then, any other periodic configuration will necessarily contain (infinitely many) plaquettes whose energy is $\delta > \alpha$. By Lemma 7.13, this implies that $g^{\mathrm{per}}(\Phi) = \{\eta^+, \eta^-\}$. Examples of nonperiodic ground states are easily obtained by patching plaquettes with minimal energy.

To see that Peierls' condition is not satisfied, consider a configuration $\omega \stackrel{\infty}{=} \eta^-$, which coincides everywhere with η^- except on a triangular region of the following type, with L large:

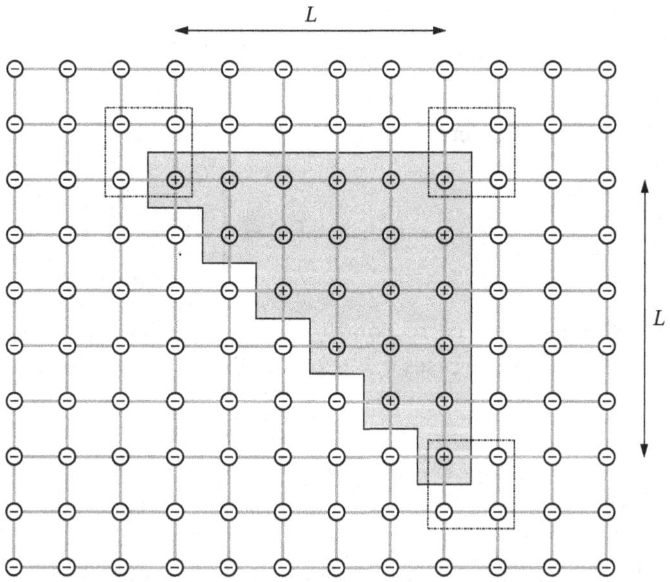

Notice that all points along the boundary of the triangle are incorrect, which implies that $|\mathscr{B}(\omega)|$ (and therefore $|\Gamma(\omega)|$) grows linearly with L. Nevertheless, for each L, there are exactly three plaquettes with a nonzero contribution to $\mathscr{H}_\Phi(\omega \mid \eta)$, indicated at the three corners of the triangle. This means that $\mathscr{H}_\Phi(\omega \mid \eta)$ is bounded, uniformly in L: Peierls' condition is not satisfied.

Exercise 7.9: Let

$$W_B^i(\omega) \stackrel{\text{def}}{=} \begin{cases} |\mathrm{B}(r_*)|^{-1}\mathbf{1}_{\{\omega_B=\eta_B^i\}} & \text{if } B \text{ is a translate of } \mathrm{B}(r_*),\\ 0 & \text{otherwise.}\end{cases}$$

Suppose first that $m \in I$. In that case, setting $\lambda^i = 0$ for all $i \in I$ and $\lambda^j = \lambda > 0$ for all $j \in \{1,\dots,m\} \setminus I$, we obtain, for each $k \in \{1,2,\dots,m\}$,

$$\mathrm{e}_{\Phi^0+\sum_{i=1}^{m-1}\lambda^i W^i}(\eta^k) = \begin{cases} \underline{\mathrm{e}}_{\Phi^0} + \lambda & \text{if } k \in \{1,\dots,m\} \setminus I,\\ \underline{\mathrm{e}}_{\Phi^0} & \text{if } k \in I.\end{cases}$$

When $m \notin I$, we proceed similarly by setting $\lambda^i = 0$ for all $i \in \{1,\dots,m\} \setminus I$ and $\lambda^j = \lambda < 0$ for all $j \in I$.

The reader can check that these potentials do not create new ground states.

Exercise 7.10: By construction,

$$\mathscr{H}_{\hat\Phi}(\hat\omega \mid \hat\eta) = \mathscr{H}_\Phi(\omega \mid \eta) \geq \rho|\Gamma(\omega)| \geq \rho|\mathscr{B}(\omega)|,$$

where we used Peierls' condition for Φ. Now, observe that if a vertex $\hat{i}$ of the renormalized lattice is not $\hat{\#}$-correct, there must exist a vertex of the original lattice such that $j \in \hat{i}r_* + \mathrm{B}(3r_*)$ and j is not #-correct. Therefore, $|\mathscr{B}(\hat\omega)| \geq |\mathrm{B}(3r_*)|^{-1}|\mathscr{B}(\omega)| \geq 3^{-d}|\mathrm{B}(r_*)|^{-1}|\mathscr{B}(\omega)|$. Thus, since $|\Gamma(\hat\omega)| \leq 3^d|\mathscr{B}(\hat\omega)|$,

$$\rho|\mathscr{B}(\omega)| \geq \rho 3^{-d}|\mathrm{B}(r_*)|^{-1}|\mathscr{B}(\hat\omega)| \geq \rho 3^{-2d}|\mathrm{B}(r_*)|^{-1}|\Gamma(\hat\omega)|.$$

We conclude that Peierls' condition holds for $\hat\Phi$ with a constant $\rho 3^{-2d}|\mathrm{B}(r_*)|^{-1}$.

Exercise 7.11: Let $i,j \in A_\ell^{\mathrm{c}}$ and consider a path $\pi = (i_1 = i, i_2, \dots, i_{n-1}, i_n = j)$, with $d_\infty(i_k, i_{k+1}) = 1$. If π exits A_ℓ^{c}, let $s_1 \stackrel{\text{def}}{=} \min\{k : i_k \in A_\ell\} - 1$ and $s_2 \stackrel{\text{def}}{=} \max\{k : i_k \in A_\ell\} + 1$. By construction, $i_{s_1}, i_{s_2} \in \overline{\gamma}$. Since $\overline{\gamma}$ is connected, there exists a path from i_{s_1} to i_{s_2} entirely contained inside $\overline{\gamma}$. But this allows us to deform π so that it is entirely contained in A_ℓ^{c}.

Exercise 7.12: For ease of notation, we treat the case of a single contour; the same argument applies in the general situation. Proceeding exactly as we did to arrive at (7.34), treating separately the numerator and the denominator, we arrive at the following representation:

$$\mu^{\#}_{\Lambda;\Phi}(\Gamma' \ni \gamma') = \frac{\sum_{\Gamma\in\mathscr{A}_1}\prod_{\gamma\in\Gamma}\mathrm{w}^{\#}(\gamma)}{\sum_{\Gamma\in\mathscr{A}_0}\prod_{\gamma\in\Gamma}\mathrm{w}^{\#}(\gamma)} = \mathrm{w}^{\#}(\gamma')\frac{\sum_{\Gamma\in\mathscr{A}_2}\prod_{\gamma\in\Gamma}\mathrm{w}^{\#}(\gamma)}{\sum_{\Gamma\in\mathscr{A}_0}\prod_{\gamma\in\Gamma}\mathrm{w}^{\#}(\gamma)} \leq \mathrm{w}^{\#}(\gamma'),$$

where we have introduced the following families of contours:

$$\mathscr{A}_0 \stackrel{\text{def}}{=} \{\Gamma \text{ compatible}\}, \qquad \mathscr{A}_1 \stackrel{\text{def}}{=} \{\Gamma \in \mathscr{A}_0 : \gamma' \text{ is an external contour of } \Gamma\},$$

while $\mathscr{A}_2$ is the set of all $\Gamma \in \mathscr{A}_0$ such that each $\gamma \in \Gamma$ is compatible with γ' and there does not exist $\gamma \in \Gamma$ such that $\overline{\gamma}' \subset \mathrm{int}\gamma$.

Exercise 7.13: Notice that if R_1, R_2 are two parallelepipeds such that $R_1 \cup R_2$ is also a parallelepiped, then $\Xi(R_1 \cup R_2) \geq \Xi(R_1)\Xi(R_2)$. Namely, the union of two compatible families contributing to $\Xi(R_1)$ and $\Xi(R_2)$ is always a family contributing to $\Xi(R_1 \cup R_2)$. One can then use Lemma B.6.

Solutions for Chapter 8

Exercise 8.1: We can suppose that Λ is connected. One could use a Gaussian integration formula, but we prefer to provide an argument that also works for more general gradient models. Consider a spanning tree[1] T of the graph $(\Lambda, \mathscr{E}_\Lambda)$ and denote by T_0 the tree obtained by adding to T one vertex $i_0 \in \partial^{\rm ex}\Lambda$ and an edge of $\mathscr{E}^{\rm b}_\Lambda$ between i_0 and one of the vertices of T; we consider i_0 to be the root of the tree $T_0 = (V_0, E_0)$. Clearly,

$$\mathscr{H}_{\Lambda;\beta,m}(\omega) \geq \frac{\beta}{4d} \sum_{\{i,j\} \in E_0} (\omega_i - \omega_j)^2 \stackrel{\text{def}}{=} \tilde{\mathscr{H}}_{T_0;\beta}(\omega)\,.$$

Of course,

$$\mathbf{Z}^\eta_{\Lambda;\beta,m} \leq \int e^{-\tilde{\mathscr{H}}_{T_0;\beta}(\omega_{V_0}\eta_{i_0})} \prod_{i \in \Lambda} \mathrm{d}\omega_i\,.$$

Let $i \in \Lambda$ be a leaf[2] of the tree T_0. Then, denoting by j the unique neighbor of i in T_0,

$$\int_{-\infty}^{\infty} e^{-\frac{\beta}{4d}(\omega_i - \omega_j)^2} \mathrm{d}\omega_i = \int_{-\infty}^{\infty} e^{-\frac{\beta}{4d}x^2} \mathrm{d}x = \sqrt{4\pi d/\beta}\,.$$

We can thus integrate over each variable in Λ, removing one leaf at a time. The end result is the upper bound

$$\mathbf{Z}^\eta_{\Lambda;\beta,m} \leq \{4\pi d/\beta\}^{|\Lambda|/2}\,.$$

Exercise 8.2: The problem arises from the fact that, when no spins are fixed on the boundary, all spins inside Λ can be shifted by the same amount without changing the energy. This can already be seen in the simple case where $\Lambda = \{0, 1\} \subset \mathbb{Z}$, with free boundary condition:

$$\mathbf{Z}^\varnothing_{\Lambda;\beta,0} = \int \Big\{ \int e^{-\frac{\beta}{4d}(\omega_0 - \omega_1)^2} \mathrm{d}\omega_1 \Big\} \mathrm{d}\omega_0 = \sqrt{\frac{4d\pi}{\beta}} \int \mathrm{d}\omega_0 = +\infty\,.$$

Exercise 8.3: The first claim follows from the fact $\varphi_{i_1} \cdots \varphi_{i_{2n+1}}$ is an odd function, so its integral with respect to the density (8.10) (with $a_\Lambda = 0$) vanishes.

Let us turn to the second claim. First, observe that one can assume that all vertices $i_1, \ldots, i_{2n}$ are distinct; otherwise for each vertex j appearing $r_j > 1$ times, introduce $r_j - 1$ new random variables, perfectly correlated with φ_j.

[1] A **spanning tree** of a graph $G = (V, E)$ is a connected subgraph of G which is a tree and contains all vertices of G.

[2] A **leaf** of a tree is a vertex of degree 1 distinct from the root.

The desired expectation can be obtained from the moment generating function by differentiation:

$$E_\Lambda[\varphi_{i_1}\cdots\varphi_{i_{2n}}] = \frac{\partial^{2n}}{\partial t_{i_1}\cdots\partial t_{i_{2n}}} E_\Lambda\left[e^{t_\Lambda\cdot\varphi_\Lambda}\right]\Big|_{t_\Lambda\equiv 0}.$$

The identity (8.9) allows one to perform this computation in another way. First,

$$\exp\left\{\tfrac{1}{2}t_\Lambda\cdot\Sigma_\Lambda t_\Lambda\right\} = \sum_{n\geq 0}\tfrac{1}{n!}2^{-n}\Big\{\sum_{j,k\in\Lambda}\Sigma_\Lambda(j,k)t_j t_k\Big\}^n.$$

Then,

$$\frac{\partial^{2n}}{\partial t_{i_1}\cdots\partial t_{i_{2n}}}\exp\left\{\tfrac{1}{2}t_\Lambda\cdot\Sigma_\Lambda t_\Lambda\right\}\Big|_{t_\Lambda\equiv 0} = \tfrac{1}{n!}\sum_{\substack{\{j_1,k_1\},\dots,\{j_n,k_n\}\subset\Lambda\\ \bigcup_{m=1}^n\{j_m,k_m\}=\{i_1,\dots,i_{2n}\}}}\prod_{m=1}^n\Sigma_\Lambda(j_m,k_m),$$

where the factor 2^{-n} was canceled by the factor 2^n accounting for the possible interchange of j_m and k_m for each $m = 1,\dots,n$. Now, we can rewrite the latter sum in terms of pairings as in the claim. Note that, doing so, we lose the ordering of the n pairs $\{j_m, k_m\}$, so that we have to introduce an additional factor of $n!$, canceling the factor $1/n!$. The claim follows.

Exercise 8.4: The procedure is very similar to the one used in the previous exercise. We write $C_{ij} = E(\varphi_i\varphi_j)$. Then, using (8.11) and (8.12),

$$\begin{aligned}
E\left[e^{t_\Lambda\cdot\varphi_\Lambda}\right] &= \sum_{n\geq 0}\sum_{\substack{(r_i)_{i\in\Lambda}\subset\mathbb{Z}_{\geq 0}:\\ \sum_i r_i=2n}} E\Big[\prod_i\varphi_i^{r_i}\Big]\prod_i t_i^{r_i}\\
&= \sum_{n\geq 0}\sum_{\substack{(r_i)_{i\in\Lambda}\subset\mathbb{Z}_{\geq 0}:\\ \sum_i r_i=2n}}\Big\{\sum_{\mathscr{P}}\prod_{\{\ell,\ell'\}\in\mathscr{P}} C_{\ell\ell'}\Big\}\prod_i t_i^{r_i}\\
&= \sum_{n\geq 0}\sum_{\substack{(r_i)_{i\in\Lambda}\subset\mathbb{Z}_{\geq 0}:\\ \sum_i r_i=2n}}\tfrac{1}{n!}2^{-n}\Big\{\sum_{\substack{\{j_1,k_1\},\dots,\{j_n,k_n\}\subset\Lambda\\ \sum_{m=1}^n(\mathbf{1}_{\{j_m=i\}}+\mathbf{1}_{\{k_m=i\}})=r_i,\,\forall i\in\Lambda}}\prod_{m=1}^n C_{i_m j_m}\Big\}\prod_i t_i^{r_i}\\
&= \sum_{n\geq 0}\tfrac{1}{n!}2^{-n}\Big\{\sum_{\{j_1,k_1\},\dots,\{j_n,k_n\}\subset\Lambda}\prod_{m=1}^n C_{i_m j_m}t_{i_m}t_{j_m}\Big\}\\
&= \sum_{n\geq 0}\tfrac{1}{n!}2^{-n}\Big\{\sum_{i,j\in\Lambda}C_{ij}t_i t_j\Big\}^n = \exp\Big\{\sum_{i,j\in\Lambda}\tfrac{1}{2}C_{ij}t_i t_j\Big\}.
\end{aligned}$$

Exercise 8.6: Let $(\xi_i)_{i=-n-1,\dots,n}$ be i.i.d. random variables with distribution $\xi_i \sim \mathscr{N}(0,2)$. Let $L_n \stackrel{\text{def}}{=} \xi_{-n-1}+\cdots+\xi_{-1}$ and $R_n \stackrel{\text{def}}{=} \xi_0+\cdots+\xi_n$. The density of φ_0 coincides with the conditional probability density of L_n given that $L_n + R_n = 0$, which is equal to

$$\frac{f_{L_n}(x)f_{R_n}(-x)}{f_{L_n+R_n}(0)} = \frac{\left\{\frac{1}{\sqrt{4\pi(n+1)}}e^{-\frac{1}{2}\frac{x^2}{2(n+1)}}\right\}^2}{\frac{1}{\sqrt{8\pi(n+1)}}} = \frac{1}{\sqrt{2\pi(n+1)}}e^{-\frac{1}{2}\frac{x^2}{(n+1)}},$$

so that $\varphi_0 \sim \mathscr{N}(0, n+1)$.

Exercise 8.7: Since φ is centered, $\widetilde{\varphi}$ also is. Then, observe that

$$E^0_{\Lambda;0}\big[e^{\mathrm{i}t_\Lambda \cdot \tilde{\varphi}_\Lambda}\big] = E^0_{\Lambda;0}\big[e^{\mathrm{i}\tilde{t}_\Lambda \cdot \varphi_\Lambda}\big],$$

where $\tilde{t}_i \stackrel{\text{def}}{=} t_i$ for all $i \neq 0$ and $\tilde{t}_0 = -\sum_{j\in\Lambda\setminus\{0\}} t_j$. From (8.8), we get

$$E^0_{\Lambda;0}\big[e^{\mathrm{i}\tilde{t}_\Lambda \cdot \varphi_\Lambda}\big] = \exp\Big\{-\tfrac{1}{2}\sum_{i,j\in\Lambda} G_\Lambda(i,j)\tilde{t}_i\tilde{t}_j\Big\} = \exp\Big\{-\tfrac{1}{2}\sum_{i,j\in\Lambda\setminus\{0\}} \tilde{G}_\Lambda(i,j)t_it_j\Big\},$$

with $\tilde{G}_\Lambda(i,j)$ given in (8.34).

Exercise 8.10: When $d = 1$, $(-\frac{1}{2d}\Delta + m^2)u = 0$ becomes

$$u_{k+1} = 2(1+m^2)u_k - u_{k-1}.$$

For any pair of initial values $u_0, u_1 \in \mathbb{R}$, we can then easily verify that u_k, $k \geq 2$, is of the form

$$u_k = Az_+^k + Bz_-^k,$$

where $z_\pm = 1 + m^2 \pm \sqrt{2m^2 + m^4}$ and A, B are functions of u_0, u_1. The conclusion follows, since $z_- = 1/z_+$.

Solutions for Chapter 9

Exercise 9.1: Since $\langle \mathbf{S}_i \cdot \mathbf{S}_j\rangle_\mu \to 0$ uniformly as $\|j-i\|_2 \to \infty$, we can find, for any $\epsilon > 0$, a number R such that $\langle \mathbf{S}_i \cdot \mathbf{S}_j\rangle_\mu \leq \epsilon$ for all i,j such that $\|j-i\|_2 > R$. Consequently, for any $i \in \mathrm{B}(n)$,

$$\sum_{j\in\mathrm{B}(n)} \langle \mathbf{S}_i \cdot \mathbf{S}_j\rangle_\mu \leq |\mathrm{B}(R)| + \epsilon|\mathrm{B}(n)|.$$

It follows that

$$\begin{aligned}\limsup_{n\to\infty}\langle \|m_{\mathrm{B}(n)}\|_2^2\rangle_\mu &= \limsup_{n\to\infty} |\mathrm{B}(n)|^{-2} \sum_{i,j\in\mathrm{B}(n)} \langle \mathbf{S}_i \cdot \mathbf{S}_j\rangle_\mu \\ &\leq \limsup_{n\to\infty} |\mathrm{B}(n)|^{-1} \max_{i\in\mathrm{B}(n)} \sum_{j\in\mathrm{B}(n)} \langle \mathbf{S}_i \cdot \mathbf{S}_j\rangle_\mu \leq \epsilon.\end{aligned}$$

Since ϵ was arbitrary, the conclusion follows.

Exercise 9.3: Simply take $\theta_i^{\mathrm{SW}} = \big(1 - (\|i\|_\infty/n)\big)\pi$.

Exercise 9.4: First, observe that

$$\sum_{\{i,j\}\in\mathscr{E}^{\mathrm{b}}_{\mathrm{B}(n)}} \big(f(\|i\|_\infty)-f(\|j\|_\infty)\big)^2 \geq \sum_{k=\ell}^{n-1} k^{d-1}\big(f(k)-f(k+1)\big)^2 .$$

By the Cauchy–Schwarz inequality,

$$\Big\{\sum_{k=\ell}^{n-1}\big(f(k)-f(k+1)\big)\Big\}^2 \leq \Big\{\sum_{k=\ell}^{n-1} k^{d-1}\big(f(k)-f(k+1)\big)^2\Big\}\Big\{\sum_{k=\ell}^{n-1} k^{-(d-1)}\Big\}.$$

Therefore,

$$\begin{aligned}\sum_{\{i,j\}\in\mathscr{E}^{\mathrm{b}}_{\mathrm{B}(n)}} \big(f(\|i\|_\infty)-f(\|j\|_\infty)\big)^2 &\geq \Big\{\sum_{k=\ell}^{n-1}\big(f(k)-f(k+1)\big)\Big\}^2\Big\{\sum_{k=\ell}^{n-1} k^{-(d-1)}\Big\}^{-1}\\ &\geq \big(f(\ell)-f(n)\big)^2\Big\{\sum_{k\geq\ell} k^{-(d-1)}\Big\}^{-1} = \Big\{\sum_{k\geq\ell} k^{-(d-1)}\Big\}^{-1}.\end{aligned}$$

Exercise 9.5: Fix $M\geq 2$ and partition $(-\pi,\pi]$ into intervals $I_1,\dots,I_M$ of length $2\pi/M$. Write $\nu_r \stackrel{\text{def}}{=} \mu^{\eta}_{\mathrm{B}(n);\beta}(\vartheta_0\in I_r)$. Then, (9.9) implies that $|\nu_r-\nu_s|\leq c/T_n(d)$ for all $1\leq r,s\leq M$ and thus $|\nu_r-\frac{1}{M}| = \frac{1}{M}\big|\sum_{s=1}^{M}(\nu_r-\nu_s)\big|\leq c/T_n(d)$. The first claim follows. The second claim is an immediate consequence of the first one.

Exercise 9.7: Writing $\mathbf{S}_i=(\cos\vartheta_i,\sin\vartheta_i)$ gives $\mathbf{S}_i\cdot\mathbf{S}_j=\cos(\vartheta_i-\vartheta_j)$, so the partition function with free boundary condition can be written as

$$\mathbf{Z}^{\varnothing}_{\mathrm{B}(n);\beta} = \int_{-\pi}^{\pi}\mathrm{d}\theta_{-n}\cdots\int_{-\pi}^{\pi}\mathrm{d}\theta_n \prod_{i=-n+1}^{n} e^{\beta\cos(\theta_i-\theta_{i-1})}.$$

Now, observe that

$$\int_{-\pi}^{\pi}\mathrm{d}\theta_n e^{\beta\cos(\theta_n-\theta_{n-1})} = \int_{-\pi}^{\pi}\mathrm{d}\tau\, e^{\beta\cos\tau} = 2\pi I_0(\beta).$$

One can then continue integrating successively over $\theta_{n-1},\dots,\theta_{-n+1}$, each time getting a factor $2\pi I_0(\beta)$, with a final factor 2π for the last integration (over θ_{-n}). Therefore,

$$\mathbf{Z}^{\varnothing}_{\mathrm{B}(n);\beta} = (2\pi)^{|\mathrm{B}(n)|} I_0(\beta)^{|\mathrm{B}(n)|-1},$$

and, thus, $\psi(\beta)=\lim_{n\to\infty}|\mathrm{B}(n)|^{-1}\log\mathbf{Z}^{\varnothing}_{\mathrm{B}(n);\beta}=\log(2\pi)+\log I_0(\beta)$. The computation of the numerator of the correlation function $\langle\mathbf{S}_0\cdot\mathbf{S}_i\rangle^{\varnothing}_{\mathrm{B}(n);\beta}$ is similar. We assume that $i>0$. Integration over $\theta_n,\dots,\theta_{i+1}$ is carried out as before and yields $I_0(\beta)^{n-i}$. The integration over θ_i yields, using the identity $\cos(s+t)=\cos s\cos t+\sin s\sin t$,

$$\int_{-\pi}^{\pi} d\theta_i e^{\beta \cos(\theta_i - \theta_{i-1})} \cos(\theta_i - \theta_0) = \int_{-\pi}^{\pi} d\tau e^{\beta \cos \tau} \cos(\tau + \theta_{i-1} - \theta_0)$$
$$= \cos(\theta_{i-1} - \theta_0) \int_{-\pi}^{\pi} d\tau e^{\beta \cos \tau} \cos(\tau) = 2\pi I_1(\beta) \cos(\theta_{i-1} - \theta_0).$$

The integration over $\theta_{i-1}, \dots, \theta_1$ is performed identically. Then, the integration over $\theta_0, \dots, \theta_{-n}$ is done as for the partition function. We thus get, after simplification,

$$\langle \mathbf{S}_0 \cdot \mathbf{S}_i \rangle^{\varnothing}_{\mathrm{B}(n);\beta} = \frac{2\pi \left(2\pi I_0(\beta)\right)^n \left(2\pi I_1(\beta)\right)^i \left(2\pi I_0(\beta)\right)^{n-i}}{2\pi \left(2\pi I_0(\beta)\right)^{2n}} = \left(\frac{I_1(\beta)}{I_0(\beta)}\right)^{|i|}.$$

Letting $n \to \infty$ yields $\langle \mathbf{S}_0 \cdot \mathbf{S}_i \rangle_\mu = (I_1(\beta)/I_0(\beta))^i$.

Solutions for Chapter 10

Exercise 10.2: Suppose first that $\Theta(\mu) = \mu$. In this case, for any $f, g \in \mathfrak{A}_+(\Theta)$,

$$\langle f\Theta(g)\rangle_\mu = \langle \Theta(f\Theta(g))\rangle_\mu = \langle \Theta(f)g\rangle_\mu .$$

Conversely, suppose that $\langle f\,\Theta(g)\rangle_\mu = \langle g\,\Theta(f)\rangle_\mu$, for all $f, g \in \mathfrak{A}_+(\Theta)$. Let $A \subset \Omega_0^{\mathbb{T}_{L,+}(\Theta)}$, $B \subset \Omega_0^{\mathbb{T}_{L,-}(\Theta)}$ be arbitrary measurable sets and set $\bar{A} \stackrel{\text{def}}{=} A \times \Omega_0^{\mathbb{T}_{L,-}(\Theta)}$, $\bar{B} \stackrel{\text{def}}{=} \Omega_0^{\mathbb{T}_{L,+}(\Theta)} \times B$, $f = \mathbf{1}_{\bar{A}}$ and $g = \mathbf{1}_{\Theta(\bar{B})}$. We then have

$$\mu(\bar{A} \cap \bar{B}) = \langle f\Theta(g)\rangle_\mu = \langle \Theta(f)g\rangle_\mu = \mu(\Theta(\bar{A} \cap \bar{B})).$$

Since events of the form $\bar{A} \cap \bar{B}$ generate the product σ-algebra, we have shown that $\mu = \Theta(\mu)$.

Exercise 10.3: Consider $\Omega_0 = \{\pm 1\}$. Let $\omega', \omega'' \in \Omega_L$ be defined as follows: $\omega'_i = (-1)^{\sum_{k=1}^d i_k}$, $\omega''_i = -\omega'_i$. Let $\mu \stackrel{\text{def}}{=} \frac{1}{2}(\delta_{\omega'} + \delta_{\omega''})$. Then μ is translation invariant but not reflection positive. Namely, let Θ be any reflection through the edges and let $e = \{i, j\} \in \mathscr{E}_L$ be such that $j = \Theta(i)$. Let $f(\omega) \stackrel{\text{def}}{=} \omega_i$. Then $\langle f\Theta(f)\rangle_\mu = -1 < 0$.

Exercise 10.4: As a counter-example to such an identity, one can consider, for example, the Ising model on $\mathbb{T}_8$ with blocks of length $B = 2$ and the four Λ_2-local functions given by $f_0 = \mathbf{1}_{++}$, $f_1 = \mathbf{1}_{+-}$, $f_2 = \mathbf{1}_{--}$, $f_3 = \mathbf{1}_{-+}$, where $\mathbf{1}_{ss'} \stackrel{\text{def}}{=} \mathbf{1}_{\{\sigma_0 = s, \sigma_1 = s'\}}$.

Exercise 10.5: Write $f(\mathbf{S}_i, \mathbf{S}_j) = -\alpha \mathbf{S}_i \cdot \mathbf{S}_j - (1-\alpha) S_i^1 S_j^1$. The first term is minimal if and only if $\mathbf{S}_i = \mathbf{S}_j$. The second term is minimal if and only if $S_i^1 = S_j^1 = \pm 1$. The claim follows.

Exercise 10.6: This is immediate using the following elementary identities:

$$|\mathbb{T}_L^\star|^{-1} \sum_{p \in \mathbb{T}_L^\star} e^{\mathrm{i}(j-k)\cdot p} = \delta_{j,k} \quad \text{for all } j, k \in \mathbb{T}_L, \quad \text{and}$$
$$|\mathbb{T}_L|^{-1} \sum_{j \in \mathbb{T}_L} e^{\mathrm{i}(p-p')\cdot j} = \delta_{p,p'} \quad \text{for all } p, p' \in \mathbb{T}_L^\star.$$

Exercise 10.7: We will use twice an adaptation of the discrete Green identity (8.14) on the torus. First, since (using $\sum_{i\in\mathbb{T}_L}(\Delta h)_i = 0$ for the last identity)

$$\begin{aligned}\frac{\mathbf{Z}_{L;\beta}(h)}{\mathbf{Z}_{L;\beta}(0)} &= \Big\langle \exp\Big\{-2\beta \sum_{\{i,j\}\in\mathscr{E}_L} (\mathbf{S}_i-\mathbf{S}_j)\cdot(h_i-h_j) - \beta\sum_{\{i,j\}\in\mathscr{E}_L}\|h_i-h_j\|_2^2\Big\}\Big\rangle_{L;\beta} \\ &= \Big\langle \exp\Big\{2\beta \sum_{i\in\mathbb{T}_L} \mathbf{S}_i\cdot(\Delta h)_i - \beta\sum_{\{i,j\}\in\mathscr{E}_L}\|h_i-h_j\|_2^2\Big\}\Big\rangle_{L;\beta} \\ &= \Big\langle \exp\Big\{2\beta \sum_{i\in\mathbb{T}_L} (\mathbf{S}_i-\mathbf{S}_0)\cdot(\Delta h)_i - \beta\sum_{\{i,j\}\in\mathscr{E}_L}\|h_i-h_j\|_2^2\Big\}\Big\rangle_{L;\beta},\end{aligned}$$

the equivalence of (10.43) and (10.45) follows.

Now, observe that the Boltzmann weight appearing in $\nu_{L;\beta}$ can be written as a product:

$$\exp\Big(-\beta\sum_{\{i,j\}\in\mathscr{E}_L}\|\mathbf{S}_i-\mathbf{S}_j\|_2^2\Big) = \prod_{k=1}^{\nu}\exp\Big(-\tfrac{1}{4d}\sum_{\{i,j\}\in\mathscr{E}_L}(\varphi_i^k-\varphi_j^k)^2\Big),$$

where we defined the collections $\varphi_i^k \stackrel{\text{def}}{=} \sqrt{4d\beta}\,S_i^k$. Therefore, the families $(\varphi_i^k)_{i\in\mathbb{T}_L}$ and $(\varphi_i^\ell)_{i\in\mathbb{T}_L}$ are independent of each other if $k\neq\ell$ and each is distributed as a massless Gaussian Free Field on $\mathbb{T}_L$. Of course, the latter is ill defined on the torus. However, notice that the expectation we are interested in only involves the field $\tilde\varphi_i^k \stackrel{\text{def}}{=} \varphi_i^k-\varphi_0^k$. Adapting the arguments of Chapter 8 (working on $\mathbb{T}_L$ with 0 boundary condition at the vertex $\{0\}$), the reader can check that the latter is a well-defined centered Gaussian field with covariance matrix

$$G_{\mathbb{T}_L\setminus\{0\}}(i,j) \stackrel{\text{def}}{=} \sum_{n\geq 0}\mathbb{P}_i(X_n=j,\tau_0>n),$$

that is, the Green function of the simple random walk on $\mathbb{T}_L$, killed at its first visit at 0. Moreover, this Green function is the inverse of the discrete Laplacian $-\frac{1}{2d}\Delta$ on $\mathbb{T}_L$. We can thus fix $k\in\{1,2,\dots,\nu\}$, define

$$h^k=(h_i^k)_{i\in\mathbb{T}_L}, \quad t_L^k \stackrel{\text{def}}{=} \big(\sqrt{\beta/d}(\Delta h^k)_i\big)_{i\in\mathbb{T}_L}, \quad \varphi_L^k \stackrel{\text{def}}{=} (\varphi_i^k)_{i\in\mathbb{T}_L},$$

and use (8.9):

$$\Big\langle \exp\Big\{2\beta\sum_{i\in\mathbb{T}_L}(\Delta h^k)_i(S_i^k-S_0^k)\Big\}\Big\rangle_{\nu_{L;\beta}} = \big\langle e^{t_L^k\cdot\tilde\varphi_L^k}\big\rangle_{\nu_{L;\beta}} = \exp\big(\tfrac12 t_L^k\cdot G_{\mathbb{T}_L\setminus\{0\}}t_L^k\big).$$

Changing back to the original variables and using the fact that the Green function is the inverse of the discrete Laplacian, the conclusion follows:

$$\tfrac12 t_L^k\cdot G_{\mathbb{T}_L\setminus\{0\}}t_L^k = -\beta\Delta h^k\cdot\big\{G_{\mathbb{T}_L\setminus\{0\}}\big(-\tfrac{1}{2d}\Delta\big)h^k\big\} = -\beta\Delta h^k\cdot h^k.$$

Solutions for Appendix B

Exercise B.1: For the first inequality, it suffices to write y as $y=\alpha x+(1-\alpha)z$, with $\alpha=(z-y)/(z-x)$. The second follows by subtracting $f(x)$ on both sides of (B.5).

Exercise B.4: Assume first that $f''(x) \geq 0$ for all $x \in I$. Then, for all $x, y \in I$,

$$f(y) = f(x) + \int_x^y f'(u)\mathrm{d}u = f(x) + \int_x^y \Big\{f'(x) + \int_x^u f''(v)\mathrm{d}v\Big\}\mathrm{d}u \geq f(x) + f'(x)(y - x).$$

This implies that f has a supporting line at each point of I and is thus convex by Theorem B.13. Now if f is convex, then, for all $x \in I$ and all $h \neq 0$ (small enough), $(f'(x + h) - f'(x))/h \geq 0$, since f' is increasing. By letting $h \to 0$, it follows that $f''(x) \geq 0$.

Exercise B.5: Assume f is affine on some interval $I = [a, b]$, and consider $a < a_0 < b_0 < b$. On the one hand, by Theorem B.12 and since each f_n is differentiable, $0 = f'(b_0) - f'(a_0) = \lim_n(f_n'(b_0) - f_n'(a_0))$. On the other hand, $f_n'' \geq c$ and the Mean Value Theorem implies that, uniformly in n, $f_n'(b_0) - f_n'(a_0) \geq c(b_0 - a_0) > 0$, a contradiction.

Exercise B.6: It suffices to write, for all x, y_1, y_2 and $\alpha \in [0, 1]$,

$$x(\alpha y_1 + (1-\alpha)y_2) - f(x) = \alpha\{xy_1 - f(x)\} + (1-\alpha)\{xy_2 - f(x)\} \leq \alpha f^*(y_1) + (1-\alpha)f^*(y_2).$$

Exercise B.7: By explicit computation: $f_1^*(y) = \frac{1}{2}y^2$, $f_2^*(y) = \frac{3}{4^{4/3}}y^{4/3}$, $f_3^*(y) = |y|$, which are all convex. Furthermore, $f_1^{**} = f_1$, $f_2^{**} = f_2$ but $f_3^{**} \neq f_3$ since $f_3^{**}(x) = 0$ if $|x| \leq 1$, $+\infty$ otherwise.

Exercise B.8:

$$f^{**}(x) = \sup_y\{xy - \underbrace{\sup_z(yz - f(z))}_{\geq yx - f(x)}\} \leq f(x).$$

Exercise B.10:

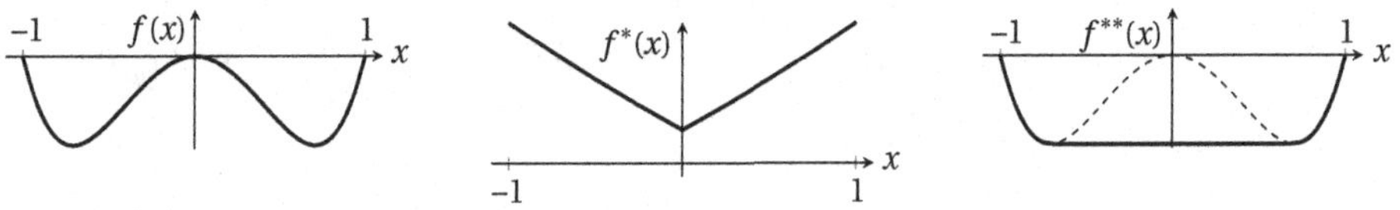

Exercise B.11: Let $x_n \to x$. Then, for any $z \in I$,

$$\liminf_{n\to\infty} f^*(x_n) = \liminf_{n\to\infty} \sup_{y\in I}\{x_n y - f(y)\} \geq \liminf_{n\to\infty} \{x_n z - f(z)\} = xz - f(z).$$

Therefore, $\liminf_{n\to\infty} f^*(x_n) \geq \sup_{z\in I}\{xz - f(z)\} = f^*(x)$.

Exercise B.12: Since $f(x) \geq f(x_0) + m(x - x_0)$ for all x, we have $f^*(m) = x_0 m - f(x_0)$. By definition, $f^*(y) = \sup_x\{yx - f(x)\}$, and so

$$f^*(y) \geq x_0 y - f(x_0) = x_0(y - m) + (x_0 m - f(x_0)) = x_0(y - m) + f^*(m).$$

Exercise B.14: Since epi (g) is convex, closed and contains epi (f) (since $g \leq f$), we have

$$\mathrm{CE}f(x) = \inf\{y : (x,y) \in C\} \geq \inf\{y : (x,y) \in \mathrm{epi}\,(g)\} = g(x)\,.$$

Exercise B.15: Let μ be the counting measure on $(\mathbb{N}, \mathscr{P}(\mathbb{N}))$: $\mu(n) \stackrel{\text{def}}{=} 1$ for all $n \in \mathbb{N}$. Let $(x_n)_{n\geq 1} \subset I$ be any sequence converging to x_0 and consider the sequence $(f_n)_{n\geq 1}$ of functions $f_n\colon \mathbb{N} \to \mathbb{R}$ defined by $f_n(k) \stackrel{\text{def}}{=} \xi_k(x_n)$. Then $\sum_k \xi_k(x_n) = \int f_n \,\mathrm{d}\mu$, so the result follows from Theorem B.40.

References

[1] A. A. Abrikosov. *Fundamentals of the theory of metals*. North-Holland, Amsterdam, New York, 1988.

[2] R. Ahlswede and D. E. Daykin. An inequality for the weights of two families of sets, their unions and intersections. *Z. Wahrsch. Verw. Gebiete*, 43(3):183–185, 1978.

[3] M. Aizenman. Translation invariance and instability of phase coexistence in the two-dimensional Ising system. *Comm. Math. Phys.*, 73(1):83–94, 1980.

[4] M. Aizenman. Geometric analysis of φ^4 fields and Ising models. I, II. *Comm. Math. Phys.*, 86(1):1–48, 1982.

[5] M. Aizenman, D. J. Barsky, and R. Fernández. The phase transition in a general class of Ising-type models is sharp. *J. Stat. Phys.*, 47(3–4):343–374, 1987.

[6] M. Aizenman, J. Bricmont, and J. L. Lebowitz. Percolation of the minority spins in high-dimensional Ising models. *J. Stat. Phys.*, 49(3–4):859–865, 1987.

[7] M. Aizenman and R. Fernández. On the critical behavior of the magnetization in high-dimensional Ising models. *J. Stat. Phys.*, 44(3–4):393–454, 1986.

[8] M. Aizenman, H. Duminil-Copin, and V. Sidoravicius. Random currents and continuity of Ising model's spontaneous magnetization. *Comm. Math. Phys.*, 334(2):719–742, 2015.

[9] M. Aizenman and R. Graham. On the renormalized coupling constant and the susceptibility in φ_4^4 field theory and the Ising model in four dimensions. *Nucl. Phys. B*, 225(2, FS 9):261–288, 1983.

[10] N. Alon and J. H. Spencer. *The probabilistic method*. Wiley-Interscience Series in Discrete Mathematics and Optimization. John Wiley & Sons, Inc., Hoboken, NJ, third edition, 2008.

[11] A. F. Andreev. Singularity of thermodynamic quantities at a first order phase transition point. *Sov. Phys. JETP*, 18(5):1415–1416, 1964.

[12] T. Andrews. The Bakerian Lecture: On the continuity of the gaseous and liquid states of matter. *Phil. Trans. R. Soc. London*, 159:575-590, 1869.

[13] T. Asano. Theorems on the partition functions of the Heisenberg ferromagnets. *J. Phys. Soc. Japan*, 29:350-359, 1970.

[14] N. W. Ashcroft and N. D. Mermin. *Solid state physics*. Saunders College, Philadelphia, PA, 1976.

[15] R. J. Baxter. Onsager and Kaufman's calculation of the spontaneous magnetization of the Ising model. *J. Stat. Phys.*, 145(3):518–548, 2011.

[16] R. J. Baxter. Onsager and Kaufman's calculation of the spontaneous magnetization of the Ising model: II. *J. Stat. Phys.*, 149(6):1164–1167, 2012.

[17] R. J. Baxter. *Exactly solved models in statistical mechanics*. Academic Press, Inc. [Harcourt Brace Jovanovich], London, 1989. Reprint of the 1982 original.

[18] O. Benois, T. Bodineau, and E. Presutti. Large deviations in the van der Waals limit. *Stoch. Proc. Appl.*, 75(1):89–104, 1998.

[19] V. Berezinskiĭ. Destruction of long-range order in one-dimensional and two-dimensional systems having a continuous symmetry group I. Classical systems. *Sov. Phys. JETP*, 32, 1971.

[20] M. Biskup, C. Borgs, J. T. Chayes, L. J. Kleinwaks, and R. Kotecký. Partition function zeros at first-order phase transitions: a general analysis. *Comm. Math. Phys.*, 251(1):79–131, 2004.

[21] M. Biskup, C. Borgs, J. T. Chayes, and R. Kotecký. Partition function zeros at first-order phase transitions: Pirogov–Sinai theory. *J. Stat. Phys.*, 116(1–4):97–155, 2004.

[22] M. Biskup. Reflection positivity and phase transitions in lattice spin models. In *Methods of contemporary mathematical statistical physics*, volume 1970 of Lecture Notes in Mathematics, pages 1–86. Springer, Berlin, 2009.

[23] M. Biskup and L. Chayes. Rigorous analysis of discontinuous phase transitions via mean-field bounds. *Comm. Math. Phys.*, 238(1–2):53–93, 2003.

[24] M. Biskup, L. Chayes, and N. Crawford. Mean-field driven first-order phase transitions in systems with long-range interactions. *J. Stat. Phys.*, 122(6):1139–1193, 2006.

[25] M. Blume. Theory of the first-order magnetic phase change in UO_2. *Phys. Rev.*, 141:517–524, 1966.

[26] T. Bodineau. The Wulff construction in three and more dimensions. *Comm. Math. Phys.*, 207(1):197–229, 1999.

[27] T. Bodineau. Translation invariant Gibbs states for the Ising model. *Probab. Theory Related Fields*, 135(2):153–168, 2006.

[28] T. Bodineau, D. Ioffe, and Y. Velenik. Rigorous probabilistic analysis of equilibrium crystal shapes. *J. Math. Phys.*, 41(3):1033–1098, 2000.

[29] T. Bodineau, D. Ioffe, and Y. Velenik. Winterbottom construction for finite range ferromagnetic models: an $\mathbb{L}_1$-approach. *J. Stat. Phys.*, 105(1–2):93–131, 2001.

[30] V. I. Bogachev. *Measure theory. Vols. I, II.* Springer-Verlag, Berlin, 2007.

[31] B. Bollobás and O. Riordan. *Percolation.* Cambridge University Press, New York, 2006.

[32] C. A. Bonato, J. Fernando Perez, and A. Klein. The Mermin-Wagner phenomenon and cluster properties of one- and two-dimensional systems. *J. Stat. Phys.*, 29(2):159–175, 1982.

[33] J. Borcea and P. Brändén. The Lee-Yang and Pólya-Schur programs. I. Linear operators preserving stability. *Invent. Math.*, 177(3):541–569, 2009.

[34] C. Borgs and J. Z. Imbrie. A unified approach to phase diagrams in field theory and statistical mechanics. *Comm. Math. Phys.*, 123(2):305–328, 1989.

[35] C. Borgs and R. Kotecký. A rigorous theory of finite-size scaling at first-order phase transitions. *J. Stat. Phys.*, 61(1–2):79–119, 1990.

[36] M. Born and K. Fuchs. The statistical mechanics of condensing systems. *Proc. R. Soc. London A*, 166(926):391–414, 1938.

[37] A. Bovier. *Statistical mechanics of disordered systems: A mathematical perspective.* Cambridge Series in Statistical and Probabilistic Mathematics. Cambridge University Press, Cambridge, 2006.

[38] A. Bovier and F. den Hollander. *Metastability: A potential-theoretic approach*, volume 351 of Grundlehren der Mathematischen Wissenschaften [Fundamental Principles of Mathematical Sciences]. Springer, Cham, 2015.

[39] A. Bovier and M. Zahradník. The low-temperature phase of Kac–Ising models. *J. Stat. Phys.*, 87(1-2):311–332, 1997.

[40] R. Bowen. *Equilibrium states and the ergodic theory of Anosov diffeomorphisms*, volume 470 of Lecture Notes in Mathematics. Springer-Verlag, Berlin, revised edition, 2008.

[41] H. J. Brascamp, E. H. Lieb, and J. L. Lebowitz. The statistical mechanics of anharmonic lattices. In *Proceedings of the 40th Session of the International Statistical Institute (Warsaw, 1975), Vol. 1. Invited papers*, volume 46, pages 393–404, 1975.

[42] H. J. Brascamp and E. H. Lieb. On extensions of the Brunn–Minkowski and Prékopa–Leindler theorems, including inequalities for log concave functions, and with an application to the diffusion equation. *J. Functional Anal.*, 22(4):366–389, 1976.

[43] O. Bratteli and D. W. Robinson. *Operator algebras and quantum statistical mechanics. 1*. Texts and Monographs in Physics. Springer-Verlag, New York, second edition, 1987.

[44] O. Bratteli and D. W. Robinson. *Operator algebras and quantum statistical mechanics. 2*. Texts and Monographs in Physics. Springer-Verlag, Berlin, second edition, 1997.

[45] P. Brémaud. *Markov chains: Gibbs fields, Monte Carlo simulation, and queues*, volume 31 of Texts in Applied Mathematics. Springer-Verlag, New York, 1999.

[46] J. Bricmont, A. El Mellouki, and J. Fröhlich. Random surfaces in statistical mechanics: roughening, rounding, wetting,.... *J. Stat. Phys.*, 42(5-6):743–798, 1986.

[47] J. Bricmont, J. R. Fontaine, and L. J. Landau. On the uniqueness of the equilibrium state for plane rotators. *Comm. Math. Phys.*, 56(3):281–296, 1977.

[48] J. Bricmont, K. Kuroda, and J. L. Lebowitz. First order phase transitions in lattice and continuous systems: extension of Pirogov-Sinaĭ theory. *Comm. Math. Phys.*, 101(4):501–538, 1985.

[49] J. Bricmont, J. L. Lebowitz, and C. E. Pfister. On the local structure of the phase separation line in the two-dimensional Ising system. *J. Stat. Phys.*, 26(2):313–332, 1981.

[50] J. Bricmont and J. Slawny. First order phase transitions and perturbation theory. In *Statistical mechanics and field theory: mathematical aspects (Groningen, 1985)*, volume 257 of Lecture Notes in Physics, pages 10–51. Springer, Berlin, 1986.

[51] J. Bricmont and J. Slawny. Phase transitions in systems with a finite number of dominant ground states. *J. Stat. Phys.*, 54(1-2):89–161, 1989.

[52] J. Bricmont, J.-R. Fontaine, and J. L. Lebowitz. Surface tension, percolation, and roughening. *J. Stat. Phys.*, 29(2):193–203, 1982.

[53] J. Bricmont, J. L. Lebowitz, and C. E. Pfister. On the surface tension of lattice systems. In *Third International Conference on Collective Phenomena (Moscow, 1978)*, volume 337 of Annals of the New York Academy of Sciences, pages 214–223. New York Academy of Sciences, New York, 1980.

[54] S. G. Brush. *The kind of motion we call heat: a history of the kinetic theory of gases in the 19th century*, volume 6. North-Holland, New York, 1976.

[55] S. G. Brush. History of the Lenz-Ising model. *Rev. Mod. Phys.*, 39:883–893, 1967.

[56] D. Brydges, J. Fröhlich, and T. Spencer. The random walk representation of classical spin systems and correlation inequalities. *Comm. Math. Phys.*, 83(1):123–150, 1982.

[57] D. C. Brydges. Lectures on the renormalisation group. In *Statistical mechanics*, volume 16 of IAS/Park City Mathematics Series, pages 7-93. American Mathematical Society, Providence, RI, 2009.

[58] H. B. Callen. *Thermodynamics*. John Wiley & Sons, Inc., New York, 1960.

[59] F. Camia, C. Garban, and C. M. Newman. The Ising magnetization exponent on $\mathbb{Z}^2$ is 1/15. *Probab. Theory Related Fields*, 160(1–2):175–187, 2014.

[60] M. Campanino, D. Ioffe, and Y. Velenik. Ornstein–Zernike theory for finite-range Ising models above T_c. *Probab. Theory Related Fields*, 125(3):305–349, 2003.

[61] H. W. Capel. On the possibility of first-order phase transitions in Ising systems of triplet ions with zero-field splitting. *Physica*, 32(5):966–988, 1966.

[62] M. Cassandro and E. Presutti. Phase transitions in Ising systems with long but finite range interactions. *Markov Process. Related Fields*, 2(2):241–262, 1996.

[63] C. Cercignani. *Ludwig Boltzmann: The man who trusted atoms*. Oxford University Press, Oxford, 1998.

[64] R. Cerf and Á. Pisztora. On the Wulff crystal in the Ising model. *Ann. Probab.*, 28(3):947–1017, 2000.

[65] P. M. Chaikin and T. C. Lubensky. *Principles of condensed matter physics*. Cambridge University Press, Cambridge, 1995.

[66] S. Chatterjee. Stein's method for concentration inequalities. *Probab. Theory Related Fields*, 138(1–2):305–321, 2007.

[67] J. T. Chayes, L. Chayes, and R. H. Schonmann. Exponential decay of connectivities in the two-dimensional Ising model. *J. Stat. Phys.*, 49(3–4):433–445, 1987.

[68] L. Chayes. Mean field analysis of low-dimensional systems. *Comm. Math. Phys.*, 292(2):303–341, 2009.

[69] L. Chayes and J. Machta. Graphical representations and cluster algorithms II. *Physica A: Stat. Mech. Appl.*, 254(3–4):477–516, 1998.

[70] D. Cimasoni and H. Duminil-Copin. The critical temperature for the Ising model on planar doubly periodic graphs. *Electron. J. Probab.*, 18(44), 2013.

[71] J. B. Conway. *Functions of one complex variable*, volume 11 of Graduate Texts in Mathematics. Springer-Verlag, New York, second edition, 1978.

[72] L. Coquille. Examples of DLR states which are not weak limits of finite volume Gibbs measures with deterministic boundary conditions. *J. Stat. Phys.*, 159(4):958–971, 2015.

[73] L. Coquille and Y. Velenik. A finite-volume version of Aizenman–Higuchi theorem for the 2d Ising model. *Probab. Theory Related Fields*, 153(1–2):25–44, 2012.

[74] A. Dembo and O. Zeitouni. *Large deviations techniques and applications*, volume 38 of Stochastic Modelling and Applied Probability. Springer-Verlag, Berlin, 2010. Corrected reprint of the second (1998) edition.

[75] F. den Hollander. *Large deviations*, volume 14 of Fields Institute Monographs. American Mathematical Society, Providence, RI, 2000.

[76] J.-D. Deuschel, G. Giacomin, and D. Ioffe. Large deviations and concentration properties for $\nabla\phi$ interface models. *Probab. Theory Related Fields*, 117(1):49–111, 2000.

[77] J.-D. Deuschel and D. W. Stroock. *Large deviations*, volume 137 of Pure and Applied Mathematics. Academic Press, Boston, MA, 1989.

[78] J.-D. Deuschel, D. W. Stroock, and H. Zessin. Microcanonical distributions for lattice gases. *Comm. Math. Phys.*, 139(1):83–101, 1991.

[79] R. Dobrushin, R. Kotecký, and S. Shlosman. *Wulff construction: A global shape from local interaction*, volume 104 of Translations of Mathematical Monographs. American Mathematical Society, Providence, RI, 1992. Translated from the Russian by the authors.

[80] R. L. Dobrushin. Existence of a phase transition in the two-dimensional and three-dimensional Ising models. *Sov. Phys. Dokl.*, 10:111–113, 1965.

[81] R. L. Dobrushin. Gibbs states describing a coexistence of phases for the three-dimensional ising model. *Theory Probab. Appl.*, 17(3):582–600, 1972.

[82] R. L. Dobrushin. Estimates of semi-invariants for the Ising model at low temperatures. In *Topics in statistical and theoretical physics*, volume 177 of American Mathematical Society Translations Series 2, pages 59–81. American Mathematical Society, Providence, RI, 1996.

[83] R. L. Dobrushin and S. B. Shlosman. Absence of breakdown of continuous symmetry in two-dimensional models of statistical physics. *Comm. Math. Phys.*, 42:31–40, 1975.

[84] R. L. Dobrushin and S. B. Shlosman. Nonexistence of one- and two-dimensional Gibbs fields with noncompact group of continuous symmetries. In *Multicomponent random systems*, volume 6 of Advances in Probability and Related Topics, pages 199–210. Dekker, New York, 1980.

[85] R. L. Dobrushin and S. B. Shlosman. The problem of translation invariance of Gibbs states at low temperatures. In *Mathematical physics reviews*, volume 5 of Soviet Sci. Rev. Sect. C Math. Phys. Rev., pages 53–195. Harwood Academic, Chur, 1985.

[86] R. L. Dobrushin and S. B. Shlosman. Completely analytical interactions: constructive description. *J. Stat. Phys.*, 46(5–6):983–1014, 1987.

[87] R. L. Dobrushin and M. Zahradník. Phase diagrams for continuous-spin models: an extension of the Pirogov-Sinaĭ theory. In *Mathematical problems of statistical mechanics and dynamics*, volume 6 of Mathematics and its Applications (Soviet Ser.), pages 1–123. Reidel, Dordrecht, 1986.

[88] R. L. Dobrushin. Description of a random field by means of conditional probabilities and conditions for its regularity. *Teor. Verojatnost. i Primenen*, 13:201–229, 1968.

[89] C. Domb. *The critical point: A historical introduction to the modern theory of critical phenomena*. CRC Press, Boca Raton, FL, 1996.

[90] T. C. Dorlas. *Statistical mechanics: Fundamentals and model solutions*. IOP Publishing, Bristol, 1999.

[91] H. Duminil-Copin. *Geometric representations of lattice spin models*. Cours Peccot du Collège de France. Éditions Spartacus, 2015.

[92] H. Duminil-Copin. *Parafermionic observables and their applications to planar statistical physics models*, volume 25 of Ensaios Matemáticos [Mathematical Surveys]. Sociedade Brasileira de Matemática, Rio de Janeiro, 2013.

[93] H. Duminil-Copin. A proof of first order phase transition for the planar random-cluster and Potts models with $q \gg 1$. In *Proceedings of Stochastic Analysis on Large Scale Interacting Systems in RIMS Kokyuroku Besssatsu*, 2016.

[94] H. Duminil-Copin and S. Smirnov. Conformal invariance of lattice models. In *Probability and statistical physics in two and more dimensions*, volume 15 of Clay Mathematics Proceedings, pages 213–276. American Mathematical Society, Providence, RI, 2012.

[95] M. Duneau, D. Iagolnitzer, and B. Souillard. Strong cluster properties for classical systems with finite range interaction. *Comm. Math. Phys.*, 35:307–320, 1974.

[96] F. Dunlop. Zeros of partition functions via correlation inequalities. *J. Stat. Phys.*, 17(4):215–228, 1977.

[97] E. B. Dynkin. Sufficient statistics and extreme points. *Ann. Probab.*, 6(5):705–730, 1978.

[98] F. J. Dyson, E. H. Lieb, and B. Simon. Phase transitions in the quantum Heisenberg model. *Phys. Rev. Lett.*, 37(3):120–123, 1976.

[99] F. J. Dyson. Existence of a phase-transition in a one-dimensional Ising ferromagnet. *Comm. Math. Phys.*, 12(2):91–107, 1969.

[100] R. S. Ellis. *Entropy, large deviations, and statistical mechanics*. Classics in Mathematics. Springer-Verlag, Berlin, 2006. Reprint of the 1985 original.

[101] R. Fernández. Gibbsianness and non-Gibbsianness in lattice random fields. In *Mathematical statistical physics*, pages 731–799. Elsevier B. V., Amsterdam, 2006.

[102] R. Fernández, J. Fröhlich, and A. D. Sokal. *Random walks, critical phenomena, and triviality in quantum field theory*. Texts and Monographs in Physics. Springer-Verlag, Berlin, 1992.

[103] R. Fernández and A. Procacci. Cluster expansion for abstract polymer models. New bounds from an old approach. *Comm. Math. Phys.*, 274(1):123–140, 2007.

[104] R. Fernández. *Contour ensembles and the description of Gibbsian probability distributions at low temperature*. 21° Colóquio Brasileiro de Matemática. [21st Brazilian Mathematics Colloquium]. Instituto de Matemática Pura e Aplicada (IMPA), Rio de Janeiro, 1997.

[105] M. E. Fisher. The theory of equilibrium critical phenomena. *Rep. Prog. Phys.*, 30(2):615, 1967.

[106] M. E. Fisher. The theory of condensation and the critical point. *Physics*, 3(5):255+, 1967.

[107] M. E. Fisher. Walks, walls, wetting, and melting. *J. Stat. Phys.*, 34(5–6):667–729, 1984.

[108] H. Föllmer. Random fields and diffusion processes. In *École d'Été de Probabilités de Saint-Flour XV–XVII, 1985–87*, volume 1362 of Lecture Notes in Mathematics, pages 101–203. Springer, Berlin, 1988.

[109] C. M. Fortuin and P. W. Kasteleyn. On the random-cluster model. I. Introduction and relation to other models. *Physica*, 57:536–564, 1972.

[110] C. M. Fortuin, P. W. Kasteleyn, and J. Ginibre. Correlation inequalities on some partially ordered sets. *Comm. Math. Phys.*, 22:89–103, 1971.

[111] S. H. Friedberg, A. J. Insel, and L. E. Spence. *Linear algebra*. Prentice Hall, Upper Saddle River, NJ, third edition, 1997.

[112] S. Friedli. *On the non-analytic behavior of thermodynamic potentials at first-order phase transitions*. PhD thesis, École Polytechnique Fédérale de Lausanne, 2003.

[113] S. Friedli and C.-E. Pfister. Non-analyticity and the van der Waals limit. *J. Stat. Phys.*, 114(3–4):665–734, 2004.

[114] S. Friedli and C.-É. Pfister. On the singularity of the free energy at a first order phase transition. *Comm. Math. Phys.*, 245(1):69–103, 2004.

[115] J. Fröhlich, R. Israel, E. H. Lieb, and B. Simon. Phase transitions and reflection positivity. I. General theory and long range lattice models. *Comm. Math. Phys.*, 62(1):1–34, 1978.

[116] J. Fröhlich, R. Israel, E. H. Lieb, and B. Simon. Phase transitions and reflection positivity. II. Lattice systems with short-range and Coulomb interactions. *J. Stat. Phys.*, 22(3):297–347, 1980.

[117] J. Fröhlich and E. H. Lieb. Phase transitions in anisotropic lattice spin systems. *Comm. Math. Phys.*, 60(3):233–267, 1978.

[118] J. Fröhlich, B. Simon, and T. Spencer. Infrared bounds, phase transitions and continuous symmetry breaking. *Comm. Math. Phys.*, 50(1):79–95, 1976.

[119] J. Fröhlich and C. Pfister. On the absence of spontaneous symmetry breaking and of crystalline ordering in two-dimensional systems. *Comm. Math. Phys.*, 81(2):277–298, 1981.

[120] J. Fröhlich and P.-F. Rodriguez. Some applications of the Lee-Yang theorem. *J. Math. Phys.*, 53(9):095218, 15, 2012.

[121] J. Fröhlich and P.-F. Rodríguez. On cluster properties of classical ferromagnets in an external magnetic field. *J. Stat. Phys.*, 166(3):828–840, 2017.

[122] J. Fröhlich and T. Spencer. The Kosterlitz–Thouless transition in two-dimensional abelian spin systems and the Coulomb gas. *Comm. Math. Phys.*, 81(4):527–602, 1981.

[123] J. Fröhlich and T. Spencer. The phase transition in the one-dimensional Ising model with $1/r^2$ interaction energy. *Comm. Math. Phys.*, 84(1):87–101, 1982.

[124] T. Funaki and H. Spohn. Motion by mean curvature from the Ginzburg–Landau $\nabla\phi$ interface model. *Comm. Math. Phys.*, 185(1):1–36, 1997.

[125] T. Funaki. Stochastic interface models. In *Lectures on probability theory and statistics*, volume 1869 of Lecture Notes in Mathematics, pages 103–274. Springer, Berlin, 2005.

[126] M. Gagnebin and Y. Velenik. Upper bound on the decay of correlations in a general class of $O(N)$-symmetric models. *Comm. Math. Phys.*, 332(3):1235–1255, 2014.

[127] G. Gallavotti. Instabilities and phase transitions in the Ising model: a review. *Riv. Nuovo Cimento*, 2(2):133–169, 1972.

[128] G. Gallavotti and S. Miracle-Solé. Statistical mechanics of lattice systems. *Comm. Math. Phys.*, 5:317–323, 1967.

[129] G. Gallavotti and S. Miracle-Solé. Equilibrium states of the Ising model in the two-phase region. *Phys. Rev. B*, 5:2555–2559, 1972.

[130] G. Gallavotti. *Statistical mechanics: A short treatise.* Texts and Monographs in Physics. Springer-Verlag, Berlin, 1999.

[131] R. J. Gardner. The Brunn–Minkowski inequality. *Bull. Amer. Math. Soc. (N.S.)*, 39(3):355–405, 2002.

[132] H.-O. Georgii, O. Häggström, and C. Maes. The random geometry of equilibrium phases. In *Phase transitions and critical phenomena*, volume 18 of *Phase Transitions and Critical Phenomena*, pages 1–142. Academic Press, San Diego, CA, 2001.

[133] H.-O. Georgii. The equivalence of ensembles for classical systems of particles. *J. Stat. Phys.*, 80(5–6):1341–1378, 1995.

[134] H.-O. Georgii. *Gibbs measures and phase transitions*, volume 9 of *de Gruyter Studies in Mathematics*. Walter de Gruyter & Co., Berlin, second edition, 2011.

[135] H.-O. Georgii and Y. Higuchi. Percolation and number of phases in the two-dimensional Ising model. *J. Math. Phys.*, 41(3):1153–1169, 2000.

[136] G. Giacomin. Aspects of statistical mechanics of random surfaces. Notes of lectures given in Fall 2001 at IHP, 2002.

[137] J. W. Gibbs. *Elementary principles in statistical mechanics: developed with especial reference to the rational foundation of thermodynamics*. Dover, New York, 1960. Original publication year: 1902.

[138] J. Ginibre. Simple proof and generalization of Griffiths' second inequality. *Phys. Rev. Lett.*, 23:828–830, 1969.

[139] J. Ginibre. General formulation of Griffiths' inequalities. *Comm. Math. Phys.*, 16:310–328, 1970.

[140] J. Glimm and A. Jaffe. *Quantum physics: A functional integral point of view*. Springer-Verlag, New York, second edition, 1987.

[141] L. Greenberg and D. Ioffe. On an invariance principle for phase separation lines. *Ann. Inst. H. Poincaré Probab. Statist.*, 41(5):871–885, 2005.

[142] R. B. Griffiths. Correlation in Ising ferromagnets I, II. *J. Math. Phys.*, 8:478–489, 1967.

[143] R. B. Griffiths, C. A. Hurst, and S. Sherman. Concavity of magnetization of an Ising ferromagnet in a positive external field. *J. Math. Phys.*, 11:790–795, 1970.

[144] R. B. Griffiths. Peierls proof of spontaneous magnetization in a two-dimensional Ising ferromagnet. *Phys. Rev. (2)*, 136:A437–A439, 1964.

[145] R. B. Griffiths. A proof that the free energy of a spin system is extensive. *J. Math. Phys.*, 5:1215–1222, 1964.

[146] R. B. Griffiths. Rigorous results and theorems. In *Phase transitions and critical phenomena*, volume 1, pages 7–109. Academic Press, New York, 1972.

[147] R. B. Griffiths and P. A. Pearce. Mathematical properties of position-space renormalization-group transformations. *J. Stat. Phys.*, 20(5):499–545, 1979.

[148] C. Grillenberger and U. Krengel. On the spatial constant of superadditive set functions in $\mathbf{R}^d$. In *Ergodic theory and related topics (Vitte, 1981)*, volume 12 of Mathematical Research, pages 53–57. Akademie-Verlag, Berlin, 1982.

[149] G. Grimmett. *Percolation*, volume 321 of Grundlehren der Mathematischen Wissenschaften [Fundamental Principles of Mathematical Sciences]. Springer-Verlag, Berlin, second edition, 1999.

[150] G. Grimmett. The random-cluster model, volume 333 of Grundlehren der Mathematischen Wissenschaften [Fundamental Principles of Mathematical Sciences]. Springer-Verlag, Berlin, 2006.

[151] G. Grimmett. *Probability on graphs: Random processes on graphs and lattices*, volume 1 of Institute of Mathematical Statistics Textbooks. Cambridge University Press, Cambridge, 2010.

[152] G. R. Grimmett and D. R. Stirzaker. *Probability and random processes*. Oxford University Press, New York, third edition, 2001.

[153] J. Groeneveld. Two theorems on classical many-particle systems. *Phys. Lett.*, 3:50–51, 1962.

[154] C. Gruber, A. Hintermann, and D. Merlini. *Group analysis of classical lattice systems*, volume 60 of Lecture Notes in Physics. Springer-Verlag, Berlin, New York, 1977.

[155] C. Gruber and H. Kunz. General properties of polymer systems. *Comm. Math. Phys.*, 22:133–161, 1971.

[156] O. Häggström. *Finite Markov chains and algorithmic applications*, volume 52 of London Mathematical Society Student Texts. Cambridge University Press, Cambridge, 2002.

[157] G. C. Hegerfeldt and C. R. Nappi. Mixing properties in lattice systems. *Comm. Math. Phys.*, 53(1):1–7, 1977.

[158] B. Helffer and J. Sjöstrand. On the correlation for Kac-like models in the convex case. *J. Stat. Phys.*, 74(1-2):349–409, 1994.

[159] P. C. Hemmer, H. Holden, and S. Kjelstrup Ratkje. Autobiographical commentary of Lars Onsager. In *The collected works of Lars Onsager (with commentary)*, volume 17 of World Scientific Series in 20th Century Physics. World Scientific, Singapore, 1996.

[160] Y. Higuchi. On the absence of non-translation invariant Gibbs states for the two-dimensional Ising model. In *Random fields, Vol. I, II (Esztergom, 1979)*, volume 27 of Colloquia Mathematica Societatis János Bolyai, pages 517–534. North-Holland, Amsterdam, 1981.

[161] Y. Higuchi. On some limit theorems related to the phase separation line in the two-dimensional Ising model. *Z. Wahrsch. Verw. Gebiete*, 50(3):287–315, 1979.

[162] P. Holický, R. Kotecký, and M. Zahradník. Phase diagram of horizontally invariant Gibbs states for lattice models. *Ann. Inst. H. Poincaré*, 3(2):203–267, 2002.

[163] R. Holley. Remarks on the FKG inequalities. *Comm. Math. Phys.*, 36:227–231, 1974.

[164] W. Holsztynski and J. Slawny. Peierls condition and number of ground states. *Comm. Math. Phys*, 61:177–190, 1978.

[165] K. Huang. *Statistical mechanics*. John Wiley & Sons, New York, second edition, 1987.

[166] J. Hubbard. Calculation of partition functions. *Phys. Rev. Lett.*, 3:77–78, 1959.

[167] K. Husimi. Statistical mechanics of condensation. In H. Yukawa, editor, *Proceedings of the International Conference of Theoretical Physics*. Science Council of Japan, 1954.

[168] J. Z. Imbrie. Phase diagrams and cluster expansions for low temperature $\mathscr{P}(\varphi)_2$ models. I. The phase diagram. *Comm. Math. Phys.*, 82(2):261–304, 1981/82.

[169] D. Ioffe, S. Shlosman, and Y. Velenik. 2D models of statistical physics with continuous symmetry: the case of singular interactions. *Comm. Math. Phys.*, 226(2):433–454, 2002.

[170] D. Ioffe. Large deviations for the 2D Ising model: a lower bound without cluster expansions. *J. Stat. Phys.*, 74(1-2):411–432, 1994.

[171] D. Ioffe. Exact large deviation bounds up to T_c for the Ising model in two dimensions. *Probab. Theory Related Fields*, 102(3):313–330, 1995.

[172] D. Ioffe. Stochastic geometry of classical and quantum Ising models. In *Methods of contemporary mathematical statistical physics*, volume 1970 of Lecture Notes in Mathematics, pages 87–127. Springer, Berlin, 2009.

[173] D. Ioffe and R. H. Schonmann. Dobrushin–Kotecký–Shlosman theorem up to the critical temperature. *Comm. Math. Phys.*, 199(1):117–167, 1998.

[174] S. N. Isakov. Nonanalytic features of the first order phase transition in the Ising model. *Comm. Math. Phys.*, 95(4):427–443, 1984.

[175] E. Ising. Beitrag zur Theorie des Ferromagnetismus. *Zeitschrift für Physik*, 31(1):253–258, 1925.

[176] R. B. Israel. *Convexity in the theory of lattice gases*. Princeton University Press, Princeton, NJ, 1979.

[177] R. B. Israel. Banach algebras and Kadanoff transformations. In *Random Fields*, pages 593–608. North Holland, Amsterdam, 1981.

[178] R. B. Israel. Existence of phase transitions for long-range interactions. *Comm. Math. Phys.*, 43:59–68, 1975.

[179] K. R. Ito. Clustering in low-dimensional SO(N)-invariant statistical models with long-range interactions. *J. Stat. Phys.*, 29(4):747–760, 1982.

[180] J. L. Jacobsen, C. R. Scullard, and A. J. Guttmann. On the growth constant for square-lattice self-avoiding walks. *J. Phys. A: Math. Theor.*, 49, 494004, 2016.

[181] E. T. Jaynes. Information theory and statistical mechanics. *Phys. Rev. (2)*, 106:620–630, 1957.

[182] I. Jensen. Improved lower bounds on the connective constants for two-dimensional self-avoiding walks. *J. Phys. A*, 37(48):11521–11529, 2004.

[183] M. Kać. Mathematical mechanisms of phase transitions. In M. Chretien, E. P. Gross, and S. Deser, editors, *Statistical physics: phase transitions and superfluidity*, volume 1. Gordon and Breach, New York, 1968.

[184] M. Kać, G. E. Uhlenbeck, and P. C. Hemmer. On the van der Waals theory of the vapor-liquid equilibrium. I. Discussion of a one-dimensional model. *J. Math. Phys.*, 4:216–228, 1963.

[185] B. Kahn and G.E. Uhlenbeck. On the theory of condensation. *Physica*, 5(5):399–416, 1938.

[186] O. Kallenberg. *Foundations of modern probability*. Probability and its Applications. Springer-Verlag, New York, second edition, 2002.

[187] G. Keller. *Equilibrium states in ergodic theory*, volume 42 of London Mathematical Society Student Texts. Cambridge University Press, Cambridge, 1998.

[188] D. G. Kelly and S. Sherman. General Griffiths's inequality on correlation in Ising ferromagnets. *J. Math. Phys.*, 9:466–484, 1968.

[189] H. Kesten. *Percolation theory for mathematicians*, volume 2 of Progress in Probability and Statistics. Birkhäuser, Boston, MA, 1982.

[190] A. I. Khinchin. *Mathematical foundations of statistical mechanics*. Dover, New York, 1949. Translated by G. Gamow.

[191] A. I. Khinchin. *Mathematical foundations of information theory*. Dover New York, 1957. Translated by R. A. Silverman and M. D. Friedman.

[192] R. Kindermann and J. L. Snell. *Markov random fields and their applications*, volume 1 of Contemporary Mathematics. American Mathematical Society, Providence, RI, 1980.

[193] A. Klein, L. J. Landau, and D. S. Shucker. On the absence of spontaneous breakdown of continuous symmetry for equilibrium states in two dimensions. *J. Stat. Phys.*, 26(3):505–512, 1981.

[194] F. Kos, D. Poland, D. Simmons-Duffin, and A. Vichi. Precision islands in the Ising and $O(n)$ models. *J. High Energy Phys.*, 2016.

[195] J. M. Kosterlitz and D. J. Thouless. Ordering, metastability and phase transitions in two-dimensional systems. *J. Phys. C: Solid State Phys.*, 6(7):1181, 1973.

[196] R. Kotecký and D. Preiss. Cluster expansion for abstract polymer models. *Comm. Math. Phys.*, 103(3):491–498, 1986.

[197] R. Kotecký and S. B. Shlosman. First-order phase transitions in large entropy lattice models. *Comm. Math. Phys.*, 83(4):493–515, 1982.

[198] R. Kotecký, A. D. Sokal, and J. M. Swart. Entropy-driven phase transition in low-temperature antiferromagnetic Potts models. *Comm. Math. Phys.*, 330(3):1339–1394, 2014.

[199] F. Koukiou, D. Petritis, and M. Zahradník. Extension of the Pirogov-Sinaĭ theory to a class of quasiperiodic interactions. *Comm. Math. Phys.*, 118(3):365–383, 1988.

[200] H. A. Kramers and G. H. Wannier. Statistics of the two-dimensional ferromagnet. I. *Phys. Rev. (2)*, 60:252–262, 1941.

[201] H. Kunz and C.-E. Pfister. First order phase transition in the plane rotator ferromagnetic model in two dimensions. *Comm. Math. Phys.*, 46(3):245–251, 1976.

[202] H. Kunz, C. E. Pfister, and P.-A. Vuillermot. Inequalities for some classical spin vector models. *J. Phys. A*, 9(10):1673–1683, 1976.

[203] L. Laanait, A. Messager, S. Miracle-Solé, J. Ruiz, and S. Shlosman. Interfaces in the Potts model. I. Pirogov–Sinai theory of the Fortuin–Kasteleyn representation. *Comm. Math. Phys.*, 140(1):81–91, 1991.

[204] O. E. Lanford, III and D. Ruelle. Observables at infinity and states with short range correlations in statistical mechanics. *Comm. Math. Phys.*, 13:194–215, 1969.

[205] O. E. Lanford. Entropy and equilibrium states in classical statistical mechanics. In *Statistical mechanics and mathematical problems*, pages 1–113. Springer, Berlin, Heidelberg, 1973.

[206] J. S. Langer. Statistical theory of the decay of metastable states. *Ann. Phys.*, 54(2):258–275, 1969.

[207] D. A. Lavis. *Equilibrium statistical mechanics of lattice models*. Theoretical and Mathematical Physics. Springer, Dordrecht, 2015.

[208] G. Lawler. Schramm–Loewner evolution (SLE). In *Statistical mechanics*, volume 16 of IAS/Park City Mathematics Series, pages 231–295. American Mathematical Society, Providence, RI, 2009.

[209] G. F. Lawler. *Intersections of random walks: Probability and its applications.* Birkhäuser, Boston, MA, 1991.

[210] G. F. Lawler. *Conformally invariant processes in the plane*, volume 114 of Mathematical Surveys and Monographs. American Mathematical Society, Providence, RI, 2005.

[211] G. F. Lawler and V. Limic. *Random walk: a modern introduction*, volume 123 of Cambridge Studies in Advanced Mathematics. Cambridge University Press, Cambridge, 2010.

[212] J.-F. Le Gall. *Mouvement brownien, martingales et calcul stochastique*, volume 71 of Mathématiques & Applications (Berlin) [Mathematics & Applications]. Springer, Heidelberg, 2013.

[213] A. Le Ny. *Introduction to (generalized) Gibbs measures*, volume 15 of Ensaios Matemáticos [Mathematical Surveys]. Sociedade Brasileira de Matemática, Rio de Janeiro, 2008.

[214] J. L. Lebowitz, A. Mazel, and E. Presutti. Liquid-vapor phase transitions for systems with finite-range interactions. *J. Stat. Phys.*, 94(5–6):955–1025, 1999.

[215] J. L. Lebowitz and O. Penrose. Rigorous treatment of the van der Waals-Maxwell theory of the liquid-vapor transition. *J. Math. Phys.*, 7:98–113, 1966.

[216] J. L. Lebowitz and O. Penrose. Analytic and clustering properties of thermodynamic functions and distribution functions for classical lattice and continuum systems. *Comm. Math. Phys.*, 11:99–124, 1968/1969.

[217] J. L. Lebowitz and C. E. Pfister. Surface tension and phase coexistence. *Phys. Rev. Lett.*, 46(15):1031–1033, 1981.

[218] J. L. Lebowitz. Coexistence of phases in Ising ferromagnets. *J. Stat. Phys.*, 16(6):463–476, 1977.

[219] J. L. Lebowitz and A. Martin-Löf. On the uniqueness of the equilibrium state for Ising spin systems. *Comm. Math. Phys.*, 25:276–282, 1972.

[220] T. D. Lee and C. N. Yang. Statistical theory of equations of state and phase transitions. II. Lattice gas and Ising model. *Phys. Rev.*, 87(3):410–419, 1952.

[221] W. Lenz. Beiträg zum Verständnis der magnetischen Eigenschaften in festen Körpern, *Phys. Z.*, 21: 613–615, 1920.

[222] J. T. Lewis, C.-E. Pfister, and W. G. Sullivan. The equivalence of ensembles for lattice systems: some examples and a counterexample. *J. Stat. Phys.*, 77(1–2):397–419, 1994.

[223] J. T. Lewis, C.-E. Pfister, and W. G. Sullivan. Entropy, concentration of probability and conditional limit theorems. *Markov Proc. Related Fields*, 1(3):319-386, 1995.

[224] E. H. Lieb and J. Yngvason. The physics and mathematics of the second law of thermodynamics. *Phys. Rep.*, 310(1):96, 1999.

[225] T. M. Liggett. *Interacting particle systems.* Classics in Mathematics. Springer-Verlag, Berlin, 2005. Reprint of the 1985 original.

[226] V. A. Malyshev. Phase transitions in classical Heisenberg ferromagnets with arbitrary parameter of anisotropy. Comm. Math. Phys., 40:75–82, 1975.

[227] V. A. Malyshev and R. A. Minlos. *Gibbs random fields: Cluster expansions*, volume 44 of Mathematics and its Applications (Soviet Series). Kluwer Academic Publishers Group, Dordrecht, 1991. Translated from the Russian by R. Kotecký and P. Holický.

[228] A. I. Markushevich. *Theory of functions of a complex variable. Volumes I, II, III.* Chelsea Publishing, New York, English edition, 1977. Translated and edited by Richard A. Silverman.

[229] A. Martin-Löf. Mixing properties, differentiability of the free energy and the central limit theorem for a pure phase in the Ising model at low temperature. *Comm. Math. Phys.*, 32:75–92, 1973.

[230] A. Martin-Löf. *Statistical mechanics and the foundations of thermodynamics*, volume 101 of Lecture Notes in Physics. Springer-Verlag, Berlin-New York, 1979.

[231] F. Martinelli, E. Olivieri, and R. H. Schonmann. For 2-D lattice spin systems weak mixing implies strong mixing. *Comm. Math. Phys.*, 165(1):33–47, 1994.

[232] F. Martinelli. Lectures on Glauber dynamics for discrete spin models. In *Lectures on probability theory and statistics (Saint-Flour, 1997)*, volume 1717 of Lecture Notes in Mathematics, pages 93–191. Springer, Berlin, 1999.

[233] D. H. Martirosian. Translation invariant Gibbs states in the q-state Potts model. *Comm. Math. Phys.*, 105(2):281–290, 1986.

[234] V. Mastropietro. *Non-perturbative renormalization.* World Scientific, Hackensack, NJ, 2008.

[235] J. C. Maxwell. On the dynamical evidence of the molecular constitution of bodies. *Nature*, 11:357–359, 1875.

[236] J. E. Mayer. The statistical mechanics of condensing systems. I. *J. Chem. Phys.*, 5(1):67–73, 1937.

[237] O. A. McBryan and T. Spencer. On the decay of correlations in SO(n)-symmetric ferromagnets. *Comm. Math. Phys.*, 53(3):299–302, 1977.

[238] O. A. McBryan and J. Rosen. Existence of the critical point in ϕ^4 field theory. *Comm. Math. Phys.*, 51(2):97–105, 1976.

[239] B. McCoy and T. T. Wu. *The two-dimensional Ising model.* Harvard University Press, Harvard, MA, 1973.

[240] F. Merkl and H. Wagner. Recurrent random walks and the absence of continuous symmetry breaking on graphs. *J. Stat. Phys.*, 75(1–2):153–165, 1994.

[241] N. D. Mermin. Absence of ordering in certain classical systems. *J. Math. Phys.*, 8(5):1061–1064, 1967.

[242] N. D. Mermin and H. Wagner. Absence of ferromagnetism or antiferromagnetism in one- or two-dimensional isotropic Heisenberg models. *Phys. Rev. Lett.*, 17:1133–1136, 1966.

[243] A. Messager, S. Miracle-Solé, and J. Ruiz. Upper bounds on the decay of correlations in SO(N)-symmetric spin systems with long range interactions. *Ann. Inst. H. Poincaré Sect. A (N.S.)*, 40(1):85–96, 1984.

[244] A. Messager, S. Miracle-Solé, and J. Ruiz. Convexity properties of the surface tension and equilibrium crystals. *J. Stat. Phys.*, 67(3–4):449–470, 1992.

[245] M. Mézard and A. Montanari. *Information, physics, and computation*. Oxford Graduate Texts. Oxford University Press, Oxford, 2009.

[246] P. Miłoś and R. Peled. Delocalization of two-dimensional random surfaces with hard-core constraints. *Comm. Math. Phys.*, 340(1):1–46, 2015.

[247] R. A. Minlos. *Introduction to mathematical statistical physics*, volume 19 of University Lecture Series. American Mathematical Society, Providence, RI, 2000.

[248] R. A. Minlos and Ja. G. Sinaĭ. The phenomenon of "separation of phases" at low temperatures in certain lattice models of a gas. I. *Mat. Sb. (N.S.)*, 73 (115):375–448, 1967.

[249] M. Miyamoto. *Statistical mechanics: A mathematical approach*. Nippon-Hyoron-Sha Co., Ltd., 2004 (in Japanese).

[250] J. L. Monroe and P. A. Pearce. Correlation inequalities for vector spin models. *J. Stat. Phys.*, 21(6):615–633, 1979.

[251] P. Mörters and Y. Peres. *Brownian motion*. Cambridge Series in Statistical and Probabilistic Mathematics. Cambridge University Press, Cambridge, 2010. With an appendix by Oded Schramm and Wendelin Werner.

[252] A. Naddaf. On the decay of correlations in non-analytic SO(n)-symmetric models. *Comm. Math. Phys.*, 184(2):387–395, 1997.

[253] C. M. Newman and L. S. Schulman. Complex free energies and metastable lifetimes. *J. Stat. Phys.*, 23(2):131–148, 1980.

[254] C. M. Newman. *Topics in disordered systems*. Lectures in Mathematics ETH Zürich. Birkhäuser Verlag, Basel, 1997.

[255] M. Niss. History of the Lenz–Ising model 1920–1950: from ferromagnetic to cooperative phenomena. *Arch. Hist. Exact Sci.*, 59(3):267–318, 2005.

[256] M. Niss. History of the Lenz–Ising model 1950–1965: from irrelevance to relevance. *Arch. Hist. Exact Sci.*, 63(3):243–287, 2009.

[257] M. Niss. History of the Lenz–Ising model 1965–1971: the role of a simple model in understanding critical phenomena. *Arch. Hist. Exact Sci.*, 65(6):625–658, 2011.

[258] E. Olivieri and M. E. Vares. *Large deviations and metastability*, volume 100 of Encyclopedia of Mathematics and its Applications. Cambridge University Press, Cambridge, 2005.

[259] L. Onsager. Crystal statistics. I. A two-dimensional model with an order-disorder transition. *Phys. Rev. (2)*, 65:117–149, 1944.

[260] A. Pais. Einstein and the quantum theory. *Rev. Mod. Phys.*, 51:863–914, 1979.

[261] J. Palmer. *Planar Ising correlations*, volume 49 of Progress in Mathematical Physics. Birkhäuser Boston, Inc., Boston, MA, 2007.

[262] Y. M. Park. Extension of Pirogov-Sinai theory of phase transitions to infinite range interactions. I. Cluster expansion. *Comm. Math. Phys.*, 114(2):187–218, 1988.

[263] Y. M. Park. Extension of Pirogov-Sinai theory of phase transitions to infinite range interactions. II. Phase diagram. *Comm. Math. Phys.*, 114(2):219–241, 1988.

[264] R. K. Pathria. *Statistical mechanics*. Elsevier Science, 2011.

[265] E. A. Pechersky. The Peierls condition (or GPS condition) is not always satisfied. *Selecta Math. Soviet.*, 3(1):87–91, 1983/84.

[266] R. E. Peierls. On Ising's ferromagnet model. *Proc. Cambridge Phil. Soc.*, 32:477–481, 1936.

[267] A. Pelissetto and E. Vicari. Critical phenomena and renormalization-group theory. *Phys. Rep.*, 368(6):549–727, 2002.

[268] X. Peng, H. Zhou, B.-B. Wei, J. Cui, J. Du, and R.-B. Liu. Experimental observation of Lee-Yang zeros. *Phys. Rev. Lett.*, 114:010601, 2015.

[269] O. Penrose. Convergence of fugacity expansions for fluids and lattice gases. *J. Math. Phys.*, 4:1312–1320, 1963.

[270] C.-E. Pfister. Large deviations and phase separation in the two-dimensional Ising model. *Helv. Phys. Acta*, 64(7):953–1054, 1991.

[271] C.-E. Pfister. Interface free energy or surface tension: definition and basic properties, 2009. arXiv:0911.5232.

[272] C.-E. Pfister and Y. Velenik. Large deviations and continuum limit in the 2D Ising model. *Probab. Theory Related Fields*, 109(4):435–506, 1997.

[273] C.-E. Pfister. On the symmetry of the Gibbs states in two-dimensional lattice systems. *Comm. Math. Phys.*, 79(2):181–188, 1981.

[274] C.-E. Pfister. Thermodynamical aspects of classical lattice systems. In *In and out of equilibrium (Mambucaba, 2000)*, volume 51 of Progress in Probability, pages 393–472. Birkhäuser, Boston, MA, 2002.

[275] C.-E. Pfister. On the nature of isotherms at first order phase transitions for classical lattice models. *Ensaios Mat.*, 9:1–90, 2005.

[276] A. Pisztora. Surface order large deviations for Ising, Potts and percolation models. *Probab. Theory Related Fields*, 104(4):427–466, 1996.

[277] A. Pönitz and P. Tittmann. Improved upper bounds for self-avoiding walks in $\mathbb{Z}^d$. *Electron. J. Combin.*, 7:Research Paper 21, 10 pp. (electronic), 2000.

[278] C. Preston. *Random fields*, volume 534 of Lecture Notes in Mathematics. Springer-Verlag, Berlin, New York, 1976.

[279] E. Presutti. *Scaling limits in statistical mechanics and microstructures in continuum mechanics*. Theoretical and Mathematical Physics. Springer, Berlin, 2009.

[280] V. Privman and L. S. Schulman. Analytic continuation at first-order phase transitions. *J. Stat. Phys.*, 29(2):205–229, 1982.

[281] V. Privman and L. S. Schulman. Analytic properties of thermodynamic functions at first-order phase transitions. *J. Phys. A*, 15(5):L231–L238, 1982.

[282] B. Prum. *Processus sur un réseau et mesures de Gibbs: Applications*. Techniques Stochastiques [Stochastic Techniques]. Masson, Paris, 1986.

[283] F. Rassoul-Agha and T. Seppäläinen. *A course on large deviations with an introduction to Gibbs measures*, volume 162 of Graduate Studies in Mathematics. American Mathematical Society, Providence, RI, 2015.

[284] M. Reed and B. Simon. *Methods of modern mathematical physics. I. Functional analysis*. Academic Press, New York, second edition, 1980.

[285] H. Robbins. A remark on Stirling's formula. *Amer. Math. Monthly*, 62:26–29, 1955.

[286] A. W. Roberts and D. E. Varberg. *Convex functions*, volume 57 of Pure and Applied Mathematics. Academic Press, New York, London, 1973.

[287] R. T. Rockafellar. *Convex analysis*. Princeton Mathematical Series, No. 28. Princeton University Press, Princeton, NJ, 1970.

[288] U. A. Rozikov. *Gibbs measures on Cayley trees*. World Scientific, Hackensack, NJ, 2013.

[289] D. Ruelle. *Statistical mechanics: Rigorous results*. World Scientific, River Edge, NJ, 1999. Reprint of the 1989 edition.

[290] D. Ruelle. Extension of the Lee-Yang circle theorem. *Phys. Rev. Lett.*, 26:303–304, 1971.

[291] D. Ruelle. *Thermodynamic formalism: The mathematical structures of equilibrium statistical mechanics*. Cambridge Mathematical Library. Cambridge University Press, Cambridge, second edition, 2004.

[292] A. Sakai. Lace expansion for the Ising model. *Comm. Math. Phys.*, 272(2):283–344, 2007.

[293] O. M. Sarig. Thermodynamic formalism for countable Markov shifts. *Ergodic Theory Dynam. Syst.*, 19(6):1565–1593, 1999.

[294] O. M. Sarig. Lecture notes on thermodynamic formalism for topological Markov shifts. Penn State, PA, 2009.

[295] R. H. Schonmann. Projections of Gibbs measures may be non-Gibbsian. *Comm. Math. Phys.*, 124(1):1–7, 1989.

[296] R. H. Schonmann and S. B. Shlosman. Constrained variational problem with applications to the Ising model. *J. Stat. Phys.*, 83(5-6):867–905, 1996.

[297] R. H. Schonmann and S. B. Shlosman. Wulff droplets and the metastable relaxation of kinetic Ising models. *Comm. Math. Phys.*, 194(2):389–462, 1998.

[298] F. Schwabl. *Statistical mechanics*. Springer-Verlag, Berlin, second edition, 2006.

[299] J. P. Sethna. *Statistical mechanics: Entropy, order parameters, and complexity*, volume 14 of Oxford Master Series in Physics. Oxford University Press, Oxford, 2006.

[300] G. L. Sewell. *Quantum theory of collective phenomena*. Monographs on the Physics and Chemistry of Materials. The Clarendon Press, New York, 1986.

[301] C. E. Shannon. A mathematical theory of communication. *Bell Syst. Tech. J.*, 27:379–423, 623–656, 1948.

[302] S. Sheffield. Random surfaces. *Astérisque*, 304:vi+175, 2005.

[303] S. Sheffield. Gaussian free fields for mathematicians. *Probab. Theory Related Fields*, 139(3-4):521–541, 2007.

[304] S. B. Shlosman. Phase transitions for two-dimensional models with isotropic short-range interactions and continuous symmetries. *Comm. Math. Phys.*, 71(2):207–212, 1980.

[305] S. B. Shlosman. The method of reflection positivity in the mathematical theory of first-order phase transitions. *Russ. Math. Surv.*, 41(3):83–134, 1986.

[306] S. B. Shlosman. Signs of the Ising model Ursell functions. *Comm. Math. Phys.*, 102(4):679–686, 1986.

[307] B. Simon. Fifteen problems in mathematical physics. In *Perspectives in mathematics*, pages 423–454. Birkhäuser, Basel, 1984.

[308] B. Simon. *The statistical mechanics of lattice gases. Vol. I*. Princeton Series in Physics. Princeton University Press, Princeton, NJ, 1993.

[309] B. Simon. A remark on Dobrushin's uniqueness theorem. *Comm. Math. Phys.*, 68(2):183–185, 1979.

[310] B. Simon and R. B. Griffiths. The $(\phi^4)_2$ field theory as a classical Ising model. *Comm. Math. Phys.*, 33:145–164, 1973.

[311] Ja. G. Sinaĭ. Gibbs measures in ergodic theory. *Uspehi Mat. Nauk*, 27(4(166)):21–64, 1972.

[312] Y. G. Sinaĭ. *Theory of phase transitions: rigorous results*, volume 108 of International Series in Natural Philosophy. Pergamon Press, Oxford, 1982. Translated from the Russian by J. Fritz, A. Krámli, P. Major and D. Szász.

[313] Ya. G. Sinaĭ, editor. *Mathematical problems of statistical mechanics*, volume 2 of Advanced Series in Nonlinear Dynamics. World Scientific, Teaneck, NJ, 1991. Collection of papers.

[314] L. Sklar. *Physics and chance: Philosophical issues in the foundations of statistical mechanics*. Cambridge University Press, Cambridge, 1993.

[315] G. Slade. *The lace expansion and its applications*, volume 1879 of Lecture Notes in Mathematics. Springer-Verlag, Berlin, 2006. Lectures from the 34th Summer School on Probability Theory held in Saint-Flour, July 6–24, 2004.

[316] J. Slawny. Low-temperature properties of classical lattice systems: phase transitions and phase diagrams. In *Phase transitions and critical phenomena, Volume 11*, pages 127–205. Academic Press, London, 1987.

[317] S. B. Šlosman. Decrease of correlations in two-dimensional models with continuous group symmetry. *Teoret. Mat. Fiz.*, 37(3):427–430, 1978.

[318] A. D. Sokal. A rigorous inequality for the specific heat of an Ising or φ^4 ferromagnet. *Phys. Lett. A*, 71(5–6):451–453, 1979.

[319] F. Spitzer. *Principles of random walks*, volume 34 of Graduate Texts in Mathematics. Springer-Verlag, New York, second edition, 1976.

[320] F. L. Spitzer. Introduction aux processus de Markov à paramètre dans Z_ν. In *École d'Été de Probabilités de Saint-Flour, III-1973*, pages 114–189. Lecture Notes in Mathematics, Volume 390. Springer, Berlin, 1974.

[321] D. L. Stein and C. M. Newman. *Spin glasses and complexity*. Primers in Complex Systems. Princeton University Press, Princeton, NJ, 2013.

[322] R. L. Stratonovich. On a method of calculating quantum distribution functions. *Doklady Akad. Nauk S.S.S.R.*, 2:416, 1957.

[323] R. H. Swendsen and J.-S. Wang. Nonuniversal critical dynamics in Monte Carlo simulations. *Phys. Rev. Lett.*, 58:86–88, 1987.

[324] K. Symanzik. Euclidean quantum field theory. In R. Jost, editor, *Local quantum theory*. Academic Press, New York, 1969.

[325] M. Talagrand. *Spin glasses: a challenge for mathematicians. Cavity and mean field models*, volume 46 of Ergebnisse der Mathematik und ihrer Grenzgebiete. 3. Folge. A Series of Modern Surveys in Mathematics. Springer-Verlag, Berlin, 2003.

[326] M. Talagrand. *Mean field models for spin glasses. Volume I, Basic examples*, volume 54 of Ergebnisse der Mathematik und ihrer Grenzgebiete. 3. Folge. A Series of Modern Surveys in Mathematics. Springer-Verlag, Berlin, 2011.

[327] M. Talagrand. *Mean field models for spin glasses. Volume II. Advanced replica-symmetry and low temperature*, volume 55 of Ergebnisse der Mathematik und ihrer Grenzgebiete. 3. Folge. A Series of Modern Surveys in Mathematics. Springer, Heidelberg, 2011.

[328] H. N. V. Temperley. The Mayer theory of condensation tested against a simple model of the imperfect gas. *Proc. Phys. Soc., London, A*, 67(3):233, 1954.

[329] A. Thess. *The entropy principle: Thermodynamics for the unsatisfied*. Springer-Verlag, Berlin, 2011.

[330] C. J. Thompson. Upper bounds for Ising model correlation functions. *Comm. Math. Phys.*, 24:61–66, 1971/1972.

[331] C. J. Thompson. *Mathematical statistical mechanics.* Macmillan, New York; Collier-Macmillan, London, 1972.

[332] C. J. Thompson. Validity of mean-field theories in critical phenomena. *Prog. Theor. Phys.*, 87(3):535–559, 1992.

[333] Á. Timár. Boundary-connectivity via graph theory. *Proc. Amer. Math. Soc.*, 141(2):475–480, 2013.

[334] H. Touchette. The large deviation approach to statistical mechanics. *Phys. Rep.*, 478(1-3):1–69, 2009.

[335] D. Ueltschi. Cluster expansions and correlation functions. *Mosc. Math. J.*, 4(2):511–522, 536, 2004.

[336] D. C. Ullrich. *Complex made simple*, volume 97 of Graduate Studies in Mathematics. American Mathematical Society, Providence, RI, 2008.

[337] H. D. Ursell. The evaluation of Gibbs' phase-integral for imperfect gases. *Math. Proc. Cambridge Phil. Soc.*, 23:685–697, 1927.

[338] H. van Beijeren. Interface sharpness in the Ising system. *Comm. Math. Phys.*, 40(1):1–6, 1975.

[339] J. D. van der Waals. *Over de Continuiteit van de Gas- en Vloeistoftoestand.* PhD thesis, Hoogeschool te Leiden, 1873.

[340] B. L. van der Waerden. Die lange Reichweite der regelmäßigen Atomanordnung in Mischkristallen. *Z. Phys.*, 118(7):473–488, 1941.

[341] A. C. D. van Enter, R. Fernández, F. den Hollander, and F. Redig. Possible loss and recovery of Gibbsianness during the stochastic evolution of Gibbs measures. *Comm. Math. Phys.*, 226(1):101–130, 2002.

[342] A. van Enter, C. Maes, and S. Shlosman. Dobrushin's program on Gibbsianity restoration: weakly Gibbs and almost Gibbs random fields. In *On Dobrushin's way: From probability theory to statistical physics*, volume 198 of American Mathematical Society Translation Series 2, pages 59–70. American Mathematical Society, Providence, RI, 2000.

[343] A. C. D. van Enter, R. Fernández, and A. D. Sokal. Regularity properties and pathologies of position-space renormalization-group transformations: scope and limitations of Gibbsian theory. *J. Stat. Phys.*, 72(5-6):879–1167, 1993.

[344] A. C. D. van Enter and S. B. Shlosman. Provable first-order transitions for nonlinear vector and gauge models with continuous symmetries. *Comm. Math. Phys.*, 255(1):21–32, 2005.

[345] L. van Hove. Quelques propriétés générales de l'intégrale de configuration d'un système de particules avec interaction. *Physica*, 15:951–961, 1949.

[346] Y. Velenik. Entropic repulsion of an interface in an external field. *Probab. Theory Related Fields*, 129(1):83–112, 2004.

[347] Y. Velenik. Localization and delocalization of random interfaces. *Probab. Surv.*, 3:112–169, 2006.

[348] W. Werner. Topics on the two-dimensional Gaussian free field. Lecture notes, ETH Zurich, 2014.

[349] W. Werner. Random planar curves and Schramm–Loewner evolutions. In *Lectures on probability theory and statistics*, volume 1840 of Lecture Notes in Mathematics, pages 107–195. Springer, Berlin, 2004.

[350] W. Werner. *Percolation et modèle d'Ising*, volume 16 of Cours Spécialisés [Specialized Courses]. Société Mathématique de France, Paris, 2009.

[351] D. Williams. *Probability with martingales*. Cambridge Mathematical Textbooks. Cambridge University Press, Cambridge, 1991.

[352] C. N. Yang. The spontaneous magnetization of a two-dimensional Ising model. *Phys. Rev. (2)*, 85:808–816, 1952.

[353] C. N. Yang and T. D. Lee. Statistical theory of equations of state and phase transitions. I. Theory of condensation. *Phys. Rev. (2)*, 87:404–409, 1952.

[354] M. Zahradník. An alternate version of Pirogov-Sinaĭ theory. *Comm. Math. Phys.*, 93(4):559–581, 1984.

[355] M. Zahradník. Contour methods and Pirogov-Sinai theory for continuous spin lattice models. In *On Dobrushin's way: From probability theory to statistical physics*, volume 198 of American Mathematical Society Translation Series 2, pages 197–220. American Mathematical Society, Providence, RI, 2000.

[356] J. M. Ziman. *Principles of the theory of solids*. Cambridge University Press, Cambridge, second edition, 1972.

Index